CW01545838

Concrete Durability and Repair Technology

Proceedings of the International Conference held at the University of Dundee, Scotland, UK on 8-10 September 1999

Edited by

Ravindra K. Dhir
Director, Concrete Technology Unit
University of Dundee

and

Michael J. McCarthy
Lecturer, Concrete Technology Unit
University of Dundee

Published by Thomas Telford Publishing, Thomas Telford Limited, 1 Heron Quay, London E14 4JD.

URL: http://www.t-telford.co.uk

Distributors for Thomas Telford books are
USA: ASCE Press, 1801 Alexander Bell Drive, Reston, VA 20191-4400, USA
Japan: Maruzen Co. Ltd, Book Department, 3–10 Nihonbashi 2-chome, Chuo-ku, Tokyo 103
Australia: DA Books and Journals, 648 Whitehorse Road, Mitcham 3132, Victoria

First published 1999

The full list of titles from the 1999 International Congress 'Creating with Concrete' and available from Thomas Telford is as follows

- *Creating with concrete*
- *Radical design and concrete practices*
- *Role of interfaces in concrete*
- *Controlling concrete degradation*
- *Extending performance of concrete structures*
- *Exploiting wastes in concrete*
- *Modern concrete materials: binders, additions and admixtures*
- *Utilizing ready-mixed concrete and mortar*
- *Innovation in concrete structures: design and construction*
- *Specialist techniques and materials in concrete construction*
- *Concrete durability and repair technology*

A catalogue record for this book is available from the British Library

ISBN: 0 7277 2826 1

© The authors, except where otherwise stated

All rights, including translation, reserved. Except for fair copying, no part of this publication may be reproduced, stored in a retrieval system or transmitted in any form or by any means, electronic, mechanical, photocopying or otherwise, without the prior written permission of the Books Publisher, Thomas Telford Publishing, Thomas Telford Ltd, 1 Heron Quay, London E14 4JD.

This book is published on the understanding that the authors are solely responsible for the statements made and opinions expressed in it and that its publication does not necessarily imply that such statements and/or opinions are or reflect the views or opinions of the publishers or of the conference organizers.

Printed and bound in Great Britain by MPG Books, Bodmin, Cornwall

PREFACE

Concrete is the key material for Mankind to create the built environment, the requirements for which are both demanding in terms of technical performance and economy and yet greatly varied from architectural masterpieces to the simplest of utilities. This presents the greatest challenge and the question is how best to advance concrete and create imaginatively.

In response, the Concrete Technology Unit (CTU) of the University of Dundee organised this Congress following on from its established series of events, namely, Concrete in the Service of Mankind in 1996, Concrete 2000: Economic and Durable Concrete Construction Through Excellence in 1993 and Protection of Concrete in 1990.

Under the theme of Creating with Concrete, the Congress consisted of five Seminars: (i) Radical Design and Concrete Practices, (ii) Role of Interfaces in Concrete, (iii) Controlling Concrete Degradation, (iv) Extending Performance of Concrete Structures and (v) Exploiting Wastes in Concrete, and five Conferences: (i) Modern Concrete Materials: Binders, Additions and Admixtures, (ii) Utilising Ready-Mixed Concrete and Mortar, (iii) Innovation in Concrete Structures: Design and Construction, (iv) Specialist Techniques and Materials for Concrete and Construction and (v) Concrete Durability and Repair Technology. In all, a total of 421 papers were presented from 67 countries.

The Opening Addresses were given by Mr Henry McLeish, MP, MSP, Minister for Enterprise and Lifelong Learning, Scotland, Dr Ian Graham-Bryce, Principal and Vice Chancellor of Dundee University, Mrs Helen Wright, Lord Provost, City of Dundee and Professor Peter Hewlett, President of the Concrete Society. This was followed by four Opening Papers by leading international experts; Dr Bryant Mather, US Army Corps of Engineers, Professor Charles F Hendriks, Delft University of Technology, Netherlands, Dr Bjørn Jensen, Danish Technological Institute, Dr Oliver Kornadt, Philipp Holzmann AG, Germany, Professor Jurek Tolloczko, Concrete Society, UK, Mr Michael Téménidès, CIMBÉTON, France and Professor Yves Malier, Ecole Normale Superieure de Cachan, France. The Closing Address was given by Professor John Morris, University of the Witwatersrand, South Africa.

The support of 20 International Professional Institutions and 31 sponsors was a major contribution to the success of the Congress. An extensive Trade Fair, participated in by 50 organisations, formed an integral part of the Congress. The work of the Congress was an immense undertaking and all of those involved are gratefully acknowledged, in particular, the members of the Organising Committee for managing the event from start to finish; members of the International Advisory and National Technical Committees for advising on the selection and reviewing of papers; the Authors and the Chairmen of Technical Sessions for their invaluable contributions to the proceedings.

All of the proceedings have been prepared directly from the camera-ready manuscripts submitted by the authors and editing has been restricted to minor changes where it was considered absolutely necessary.

Dundee
September 1999

Ravindra K Dhir
Chairman, Congress Organising Committee

INTRODUCTION

Although durability has always been considered by engineers, the practical outcomes have not necessarily been successful. In turn, this has lead to a much greater awareness of the need to design in durability

Recent developments have seen an increasing range of available materials for use in concrete. Similarly refined mix design techniques aimed at maximising the packing of solids in concrete have also been devised. The effective combination of these approaches should, both by physical and chemical means, better equip concrete for specific environmental exposure conditions. In addition to these, attention is increasingly being given to the roles of water and fillers in relation to durability provisions.

Moves away from durability specification by prescription to performance-based methods, where concrete properties relating to serviceability requirements are considered, should gradually introduce quantification to the process. This represents an important element if whole life costing of structures is to be reliably achieved.

Similarly, it is generally recognised that maintenance and repair programmes are essential for many structures to keep them serviceable. One of the key issues of repair research is the focus on the durability of the repair itself. There are a number of areas which must be addressed, including, retention of the initial protection and how this can be assessed in-situ. It is also important that recognised and agreed standards are established to allow the determination of satisfactory repair performance.

The diagnosis of concrete deterioration is becoming ever more sophisticated, with new methodologies being applied. These should improve the estimation of residual life and assist with the determination of the correct repair or strengthening method.

The Proceedings of this Conference: *Concrete Durability and Repair Technologies* dealt with these issues and the subjects raised, under six clearly identified themes (i) Selection of Materials and Mix Composition, (ii) Performance Specification and Durability By Intent, (iii) Whole Life Cost and Durability Audit, (iv) Repair Materials and Methods, (v) Development and Diagnostic Technology and (vi) Contract Conditions/Management. The conference was opened by a Leader Paper, followed by a Keynote Paper in each theme, presented by the foremost exponents in their respective fields. A total of 76 papers were presented during the International Conference which have been compiled into these proceedings.

Dundee
September 1999

Ravindra K Dhir
Michael J McCarthy

ORGANISING COMMITTEE
Concrete Technology Unit

Professor R K Dhir, OBE (Chairman)

Dr M R Jones (Secretary)

Mr M D Newlands (Joint Secretary)

Professor P C Hewlett
British Board of Agrément

Dr N A Henderson
Mott MacDonald Ltd

Professor V K Rigopoulou
National Technical University of Athens, Greece

Dr S Y N Chan
Hong Kong Polytechnic University

Dr N Y Ho
L & M Structural Systems, Singapore

Dr M J McCarthy

Dr M C Limbachiya

Dr T D Dyer

Dr K A Paine

Dr T G Jappy

Mr P A J Tittle

Mr J C Knights

Mr S R Scott (Unit Assistant)

Miss A M Duncan (Unit Secretary)

INTERNATIONAL ADVISORY COMMITTEE

Dr H M Z-Al-Abideen
Deputy Minister
Ministry of Public Works and Housing, Saudi Arabia

Professor M S Akman
Emeritus Professor of Civil Engineering
Istanbul Technical University, Turkey

Dr R Amtsbüchler
Manager-Technical Services
Blue Circle Ltd (South Africa), South Africa

Professor C Andradé
Director
Institute of Construction Sciences, Spain

Professor J M J M Bijen
Director
INTRON B.V., The Netherlands

Professor A M Brandt
Head of Section
Polish Academy of Sciences, Poland

Dr J-M Chandelle
Managing Director
CEMBUREAU, Belgium

Professor P Helene
Head of Civil Construction Engineering Department
University of Sao Paulo, Brazil

Dr G C Hoff
Senior Engineering Consultant
Mobil Technology Company, USA

Professor I Holand
Senior Research Engineer
SINTEF, Norway

Professor B C Jensen
Director
Carl Bro, Denmark

Professor S Mirza
Professor of Civil Engineering and Applied Mechanics
McGill University, Canada

Professor S Nagataki
Professor of Civil Engineering and Architecture
Niigata University, Japan

Professor H Okamura
Vice President
Kochi University of Technology, Japan

Professor E A e Oliveira
Director
Laboratório Nacional de Eng Civil, Portugal

Professor J-P Ollivier
Director of LMDC-INSA
LMDC, France

INTERNATIONAL ADVISORY COMMITTEE (CONTINUED)

Professor R Park
Professor of Civil Engineering
University of Canterbury, New Zealand

Mr S A Reddi
Managing Director
Gammon India Limited, India

Professor H-W Reinhardt
Head of Construction Materials Institute
University of Stuttgart, Germany

Professor R Rivera-Villarreal
Chief of Concrete Technology Department
Ciudad Universitaria, Mexico

Professor A Samarin
Consultant
Sustainable Development Technological Sciences and Engineering, Australia

Professor A E Sarja
Research Professor
Technical Research Centre of Finland, Finland

Professor S P Shah
Walter P Murphy Professor or Civil Engineering
Northwestern University, USA

Professor H Sommer
Head of Research Institute
VÖZ, Austria

Professor I Soroka
Professor of Civil Engineering
National Building Research Institute, Israel

Professor M Tang
Research Professor
Nanjing University of Chemical Technology, China

Professor T Tassios
Professor
National Technical University of Athens, Greece

Professor K Tuutti
Vice President
Skanska Technik AB, Sweden

Professor T Vogel
Professor of Structural Engineering
Swiss Federal Institute of Technology ETH , Switzerland

Professor F H Wittmann
Head of Building Materials Laboratory
Swiss Federal Institute of Technology ETH , Switzerland

Professor A V Zabegayev
Head of Department RC Structures
Moscow State University of Civil Engineering, Russia

NATIONAL TECHNICAL COMMITTEE

Mr P Barber
Manager of the Scheme, The Quality Scheme for Ready Mixed Concrete

Professor A W Beeby
Professor of Structural Design, University of Leeds

Mr B V Brown
Divisional Technical Executive, Readymix (UK) Ltd.

Dr T W Broyd
Technology Development Director, W S Atkins Ltd.

Professor J H Bungey
Professor of Civil Engineering, University of Liverpool

Dr P S Chana
Director, CRIC, Imperial College of Science, Technology & Medicine

Professor J L Clarke
Principal Engineer, The Concrete Society

Dr P C Das
Group Manager, Structures Management, Highways Agency

Dr S B Desai, OBE
Principal Civil Engineer, Department of the Environment, Transport and the Regions

Professor R K Dhir, OBE (Chairman)
Director, Concrete Technology Unit, University of Dundee

Mr C R Ecob
Director Special Services Division, Mott MacDonald Ltd.

Professor F P Glasser
University of Aberdeen

Professor T A Harrison
Technical Consultant, Quarry Products Association

Professor P C Hewlett
Director, British Board of Agrément

Professor J Innes
Director of Roads, Scottish Office

NATIONAL TECHNICAL COMMITTEE (CONTINUED)

Mr K A L Johnson
Director, AMEC Civil Engineering Ltd.

Dr M R Jones
Senior Lecturer, Concrete Technology Unit, University of Dundee

Mr P Livesey
National Technical Services Manager, Castle Cement Ltd.

Professor A E Long
Director of School, Queens University of Belfast

Professor P S Mangat
Head of Research, Sheffield Hallam University

Mr G Masterton
Director, Babtie Group Ltd.

Professor G C Mays
Director of Civil Engineering, Cranfield University

Mr L H McCurrich
Technology Development Consultant, Fosroc Construction

Professor R S Narayanan
Partner, SB Tietz & Partners Consulting Engineers

Dr P J Nixon
Head, Centre for Concrete Construction, Building Research Establishment Ltd.

Dr W F Price
Senior Associate, Messrs Sandberg

Professor G Somerville, OBE
Director of Engineering, British Cement Association

Professor D C Spooner
Director, Materials and Standards, British Cement Association

Dr H P J Taylor
Director, Tarmac Precast Concrete Ltd.

Mr M Walker
Technical Manager, The Concrete Society

Dr R J Woodward
Senior Project Manager, Transport Research Laboratory

SUPPORTING INSTITUTIONS

American Concrete Institute, USA

American Society of Civil Engineers, USA

Australian Concrete Institute

Concrete Association of Finland

Concrete Society of Southern Africa

Concrete Society, UK

Danish Concrete Society, Denmark

Fédération de l'Industrie du Beton, France

German Concrete Association (DBV)

Hong Kong Institution of Engineers

Indian Concrete Institute

Institute of Concrete Technology, UK

Institution of Civil Engineers, UK

Instituto Brasileiro Do Concreto, Brazil

Japan Concrete Institute

Netherlands Concrete Society

New Zealand Concrete Society

Norwegian Concrete Association

Singapore Concrete Institute

Spanish Association for Structural Concrete

Swedish Concrete Association

SPONSORING ORGANISATIONS WITH EXHIBITION

AMEC Civil Engineering Ltd.

Babtie Group Ltd.

Bardon Aggregates

Blue Circle Cement

Blyth & Blyth

British Board of Agrément

British Cement Association

Building Research Establishment

Castle Cement Ltd.

Cementitious Slag Makers Association

CIMBÉTON, France

Du Pont de Nemours International S.A., Switzerland

ECC International Ltd.

Elkem Ltd. (Materials)

Fosroc International Ltd.

Grace Construction Products

HERACLES General Cement Co., Greece

John Doyle Group

Lafarge Aluminates

L M Scofield Europe Ltd.

Minelco Ltd.

Mott MacDonald Ltd.

O'Rourke Group

Ove Arup and Partners

SPONSORING ORGANISATIONS WITH EXHIBITION (CONTINUED)

Readymix (UK) Ltd.

Rugby Cement

Scottish Enterprise Tayside

Sika Ltd.

SKW - MBT Construction Chemicals

Thomas Telford Publishing Ltd.

United Kingdom Quality Ash Association

W A Fairhurst & Partners

ADDITIONAL EXHIBITORS

Christison Scientific Equipment Ltd.

CMS Pozament Limited

The Concrete Society

David Ball Group plc.

E & FN Spon

Flexcrete Ltd.

Germann Instruments A/S, Denmark

Natural Cement Distribution Limited

Palladian Publications Ltd.

Quality Scheme for Ready Mixed Concrete

UK Certification Authority for Reinforcing Steel

Wacker-Chemie GmbH, Germany

Wexham Developments

CONTENTS

THEME 4 REPAIR MATERIALS AND METHODS

Keynote Paper

LEADER PAPER

WHOLE LIFE DESIGN FOR DURABILITY AND SUSTAINABILITY. WHERE ARE WE GOING, AND HOW DO WE GET THERE?

G Somerville

British Cement Association

United Kingdom

ABSTRACT. It is argued that durability design is an integral part of a move towards greater sustainability. The definition of performance requirements is an essential first step, while relating technical performance more to the future operational requirements for the structure. There is also a need to radically review design methods - a cultural change, requiring less dependency on numerical analysis and more on providing satisfactory whole life performance at minimum total cost. This paper reviews the options for change, summarizes the major difficulties, and presents a strategy for future development - with the emphasis on a design-led holistic approach.

Keywords: Sustainability, Durability, Integrated design, Performance requirements.

Professor George Somerville is Director of Engineering at the British Cement Association, and a Visiting Professor, both at Imperial College, London, and at Kingston University.

INTRODUCTION

In 1986, the author [1] made a distinction between the production and placing of durable concrete, and the design and construction of structures which would be durable. The essence of the argument was that we tried to solve our durability problems via a prescriptive materials approach, and yet standards of design, detailing and construction were at least as important in practice.

Except for checks against Code requirements for concrete cover and grade, durability was not an integral part of the design process, which was perceived as the provision of adequate strength, stiffness, stability and serviceability - largely a numerate process, following well-trodden paths laid down in Codes of increasing complexity.

In the interim, that scenario has changed little - at least for routine everyday practice. However, there have been significant trends and developments, which, in time, could and should radically change the whole basis of design - both in terms of performance requirements and in our approach to satisfying these requirements. These developments include the following:-

- A growing demand, on the part of owners, for structures which better meet their operational needs, while recognising the importance of life cycle costing and the reality of planned maintenance, management and replacement;

- significant improvements in our ability to specify durable concretes for specific aggressive actions;

- the introduction of a wide range of relatively new protective measures for corrosion resistance (coatings, sealers, layers, inhibitors etc.);

- the emergence of a better perspective of the relative importance of the different deterioration mechanisms, and of predictive models to quantify their effects - mainly via more definitive state-of-the-art reports (eg.[2-5]);

- major attempts to develop numerate design concepts for durability, which differ only in how they deal with reliability, safety and margins (eg. [6-11]). Here, the approaches have a strong emphasis on design, while recognising to varying degrees the importance of a holistic approach, which also takes on board material, construction (quality) and maintenance issues;

- a relatively recent recognition of the need for greater sustainability, and the significance of durability design, in meeting that need.

There is little doubt that the groundswell created by these developments will accelerate a move towards rational durability design. What is less certain is the form that that should take.

The purpose of this paper is to examine the options, while identifying the areas of difficulty in taking these forward, before suggesting a strategy for future development.

THE BIG PICTURE - SUSTAINABILITY, DURABILITY AND PERFORMANCE REQUIREMENTS

In a broad, even global sense, sustainability is sometimes simply defined as:-

"Development that meets the need of the present, without compromising the ability of future generations to meet their own needs"

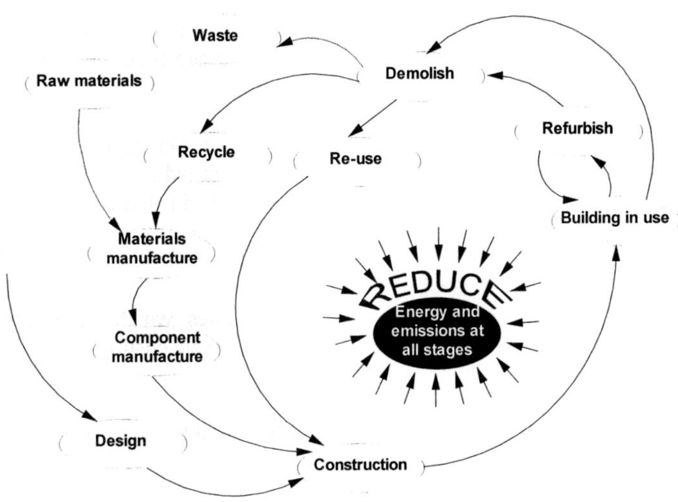

Figure 1 Material flows; cradle to the grave

This definition embraces climatic, environmental, social and economic issues. For construction, these translate into factors which include:- energy conservation, waste reduction, improved quality of life and environment, more efficient use of scarce or non-renewable resources, recycling, better operational performance, etc.

A simplistic, but commonly used, illustration of sustainability in construction is the cradle to the grave material flow shown in Figure 1. In particular, this emphasises the current focal points for sustainability - energy and emissions, with some interest in recycling; better use of waste materials is also an integral feature.

However, in the present context of durability and design life, a more relevant sustainability diagram is that shown in Figure 2, which plots 'sustainability' against 'quality', where quality is representative of a desired achievement in design, compared with required performance in service.

In simple terms, this suggests that we need to move from curve 1 (representative of present practice, where sustainability is not yet a driver) towards curve 2, thus creating a zone of sustainable construction with a definite plateau.

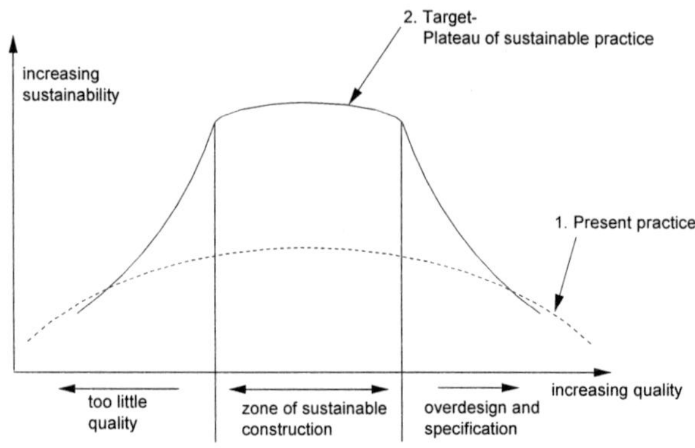

Figure 2 The plateau of sustainable practice

If, on the one hand, we put too little quality into our structure, and it fails to meet the owner's in-service performance requirements, then this constitutes 'failure', in sustainability terms. On the other hand, it is equally bad to over-design or over-specify, since this wastes scarce resources. The only solution to this dilemma is to get as close as we can in design, to an exact fit with clearly defined performance requirements.

Figure 1 presents an outline of the sustainable construction arena. Figure 2 gives the elements of a strategy, which can lead to a broad definition of performance requirements. In fact, that given in the CEB - FIP Model Code [12] fits within the sustainability concept; this is as follows:-

> *"Concrete structures shall be designed, constructed, and operated in such a way that, under the expected environmental influences, they maintain their safety, serviceability and appearance during an explicit or implicit period of time, without requiring unforeseen high costs for maintenance or repair"*

This represents a move towards a performance-based design approach, consistent with greater sustainability. Essential to this is the establishment of limiting performance criteria; this has to be at 2 levels:-

1. A requirement, which deals with the in-service time factor - possibly a target, somehow expressed in terms of 'life';

2. A requirement which directly relates to a minimum acceptable technical performance, over that life.

The 'pure' approach to 1.[6][7] is to use a reference period of, say, 50 years for buildings and relate it to a reliability index of 3.8 for the ultimate limit state (as given in ENV 1991-1) and of 1.5 for serviceability. This sets a mean service life, over which the normal imposed loads and load combinations are to be carried. This can lead to service lives which are well beyond our experience. It also begs the question of what constitutes minimum technical performance, due to effects caused by deterioration mechanisms, and how these might affect normal action effects such as bending and shear.

In considering how to move forward here, a broader view is necessary. Most cities, in most countries, have buildings dating from the last century; few will be exactly the same, due to changes in use, upgrading, new cladding, even some replacements in whole or in part. Obsolescence, and change in use, is therefore an issue.

To some extent, the situation is similar, for civil engineering structures, especially for highway structures on principal routes.

In addition, based on experience with modern infrastructure, owners are now more conscious of the need for regular inspection and good maintenance. Further, if structural changes have to be made for whatever reason, they wish the disruption to the operation of the structure to be kept to a minimum.

In developing this type of management and maintenance strategy, it is important to remember that a structure is not a single unit; it is made up of numerous elements, services, furniture and fittings - all of which have different lives.

Table 1 Basics of a performance profile plan for building

SYSTEM	CRITIC-ALITY	TARGET LIFE BEFORE REPLACEMENT, YEARS						CAPITAL COSTS*	COST IN USE TARGET: X YEARS AT Y PRICE**	MAINT-ENANCER EQUIR-EMENTS
		>5	5-10	10-20	20-40	40-100	>100			
Foundations	A	••••••	••••••	•••••••	•••••••	••••••	•••	—	—	Clear specif-ication for each system
Structure	A	••••••	••••••	••••••	••••••	•••		—	—	
External Cladding	B	••••••	••••••	••••••	••••••			—	—	
Glazing	B	••••••	••••••	••••••	•••			—	—	
Partitions	B	••••••	••••••	••••••	•••			—	—	
Heating & Ventilation	B	••••••	••••••	••••				—	—	
Water, Public Health Service	A/B	••••••	••••••	••••••	••••••			—	—	
Electrical	A	••••••	••••••	••••••	••••••			—	—	
Decorating	C	••••••	••••••					—	—	

* e.g., 100 units
** e.g., 1500 units over x years

The trend here is to plan for this in a systematic and economic way, using performance plans, shown simplistically in Table 1 for a building. Doing this properly involves whole life costing methods, with the objective of minimising total cost and maximising the operational use of the structure.

This type of thinking (obsolescence, systematic performance plans) promotes the idea that design life is not a single absolute value, but will vary, depending on the type of structure, on its use and on the maintenance/replacement plan (Table 1). Actual values will be established, in individual cases, at the conceptual design stage, between the client and the designer. The designer must then have available to him:-

a) A simple classification system for distinguishing between the life time requirements for different elements in the structure as a whole.

b) A design procedure which ensures that the performance requirements are met for each classification.

One possible classificaiton is:-

A. *Life-long* The element is so important that failure would cause cessation of function, and/or major disruption or cost, during remedial work. Highly critical.

B. *Repairable* Efficiency of operation would be reduced, but replacement/remedial work could be done economically. Medium criticality.

C. *Replaceable* Routine maintenance or replacement could be done at minimum cost and little inconvenience. Low criticality.

A British Standard exists [13], based on this principle, which simultaneously encourages the development of materials and components on a targeted service life basis; similar approaches are under development at a European level via CIB W60 and RILEM PSL140.

The key point being made here is that required life and performance should relate to the owner's operational plans and maintenance plans. It is often assumed that 'better' durability means longer life. This is not so. The sustainability argument implicit in Figure 2 makes it clear that the design objective is to match, as best as we can, the performance profile in Table 1.

A QUANTITATIVE APPROACH - MAJOR HURDLES TO BE OVERCOME

General

The first major hurdle - the definition of performance-based criteria - has been discussed in the previous section; real progress is being made here. What other difficulties exist? To appreciate this, it is necessary to resort to feedback from in-service performance. Figure 3, from reference 10, presents typical data.

Easily the most dominant mechanism is corrosion, from various causes. However, there are variations between the different types of structures. External chlorides dominate for both bridges and marine structures, and are the most significant contributor for car parks; for buildings, corrosion is mainly initiated by carbonation of the concrete.

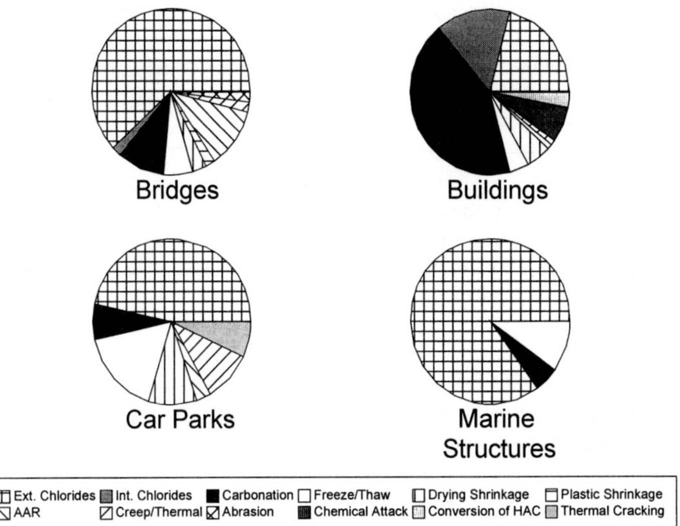

Figure 3 Reported Deterioration Mechanisms for Bridges, Buildings, Car Parks and Marine Structures

Table 2 Factors contributing to the failure of all structures reviewed in reference [10]

FACTORS CONTRIBUTING TO DETERIORATION	NO. OF CASES	NO. OF CASES %
Low Cover	47	11.6
Environment	156	38.5
Poor Quality Concrete	64	15.8
Poor Design Detailing	29	7.2
Poor Workmanship	17	4.2
Wrong specification	6	1.5
Failure of joint/waterproofing	31	7.7
Inadequate conceptual design	2	0.5
Wrong material selection	53	13
Total Number. of Cases	405	100

Of particular interest, in reference 10, is the analysis of the factors contributing to the deterioration. This is shown in Table 2. 'Environment' (natural or man-made) is the biggest single factor. However, if Low Cover, Poor Concrete Quality, Poor Design Detailing, and Poor Workmanship are added together then, collectively, they are of the same magnitude as for Environment. This feedback strongly suggests that, if durability design is to be significantly improved, two essential elements must be dealt with:-

1. Identification and quantification of critical aggressive actions.

2. Improving, or otherwise dealing with, quality of construction.

Quantifying Aggressive Actions

It is possible to do this for individual aggressive actions which directly attack the concrete. Table 3 shows the general approach which has been developed in Codes and Standards. This is a specification approach, which involves defining ranges of intensity for each action, and providing a corresponding prescriptive solution.

The major difficulty arises in dealing with corrosion. It is not difficult to identify the major causes of corrosion - chlorides and carbonation - but the effects of these, in penetrating the protective concrete cover, are much influenced by general and local climatic conditions. This involves an almost infinite combination of moisture, wind and temperature, creating a wide range of transport mechanisms of varying intensity.

As part of that, there is a strong interaction between the local climatic conditions and the outer fabric of the structure itself. Any planes of weakness (joints, cracks, etc) will quickly be found out - and therefore there is a link between durability and structural design and detailing.

Table 3 Types of aggressive action for which material specifications have been developed

AGGRESSIVE ACTION	GENERAL APPROACH	COMMENTS
Sulfate attack	Quantify the action	Specific material and mix proportions are recommended in most codes for defined ranges of sulfate concentration.
Alkali-silica reaction	Define ranges of intensity for it	The basic reaction and its possible effects are now well understood. Recommendations to minimize the risk of damage are published.
Freezing and thawing	Produce a specification for each range	Dealt with by choice of materials, mix proportions, and concrete grade. Air entrainment for lower grades. Detail to minimise exposure to moisture.
Abrasion		Specifications to cover aggregate properties, concrete grade and mix proportions, compaction and curing, methods of finishing, etc.

Table 4 Factors, and interactive effects, which influence 'loads' due to external chlorides

C2 Environmental loads due to chlorides

Actions
1) Chloride concentration. Type of cation
2) Humidity - Temperature
3) Oxygen concentration at rebar surface
4) Carbonation [hydroxide content]

Influencing concrete properties
a) Permeability [Diffusion coefficient]
b) Binding capacity [C_3A content]
c) Concrete humidity content
c) Alkalinity [Cl^-/OH^- threshold]

Climate classification

Micro Actions

A.	Wind direction
B.	Wash out from rain
C.	Condensation:- Sun orientation, Temperature, Closed spaces
D.	Leaks - Joints
E.	Irregular distribution of chlorides

Macro environment

I.	Marine	- Air - sheltered, exposed
		- Submerged
		- Tidal
		- Splash
II.	Deicing salts	- low cycling [Kg NaCl/year], frequent cycling
III.	Industrial	

How can we deal with this? Any attempt at an environmental classification, sufficiently simple for general use in Codes, is unlikely to represent real conditions. The real world, in terms of micro-climate and interactive effects, is typified by Table 4. Translating that into meaningful general Code guidance is difficult, as has been demonstrated by the CEB, in their most recent Bulletin on durability design [6].

Following on from that, we have the difficulty of dealing with the interactive effects between the aggressive actions and the outer fabric of the structure. This is implicit in Table 4, and shown graphically in Figure 4.

It is difficult to see a unique universal solution to this problem, and it may have to be addressed in different ways, as we shall see later. Nevertheless, one general principle can be stated here:- the need to consider water (moisture) alongside wind and temperature, in general design procedures - and especially for the detailing of facades in buildings and the outer exposed surfaces in other structures.

All structural engineers now consider wind and temperature in their designs (although neither is fully treated in Codes).

They would be concerned with safety and with stress limits, and, to some extent, with movement. Movement can lead to deformations and cracking, thus creating openings for wind-driven moisture. Removing, or keeping out water is fundamental to improved durability. It is a transport mechanism for aggressive actions, and the environmental conditions in Figure 4 can be translated into a range of penetration mechanisms, which involve pressure, capillary action, surface tension or gravity.

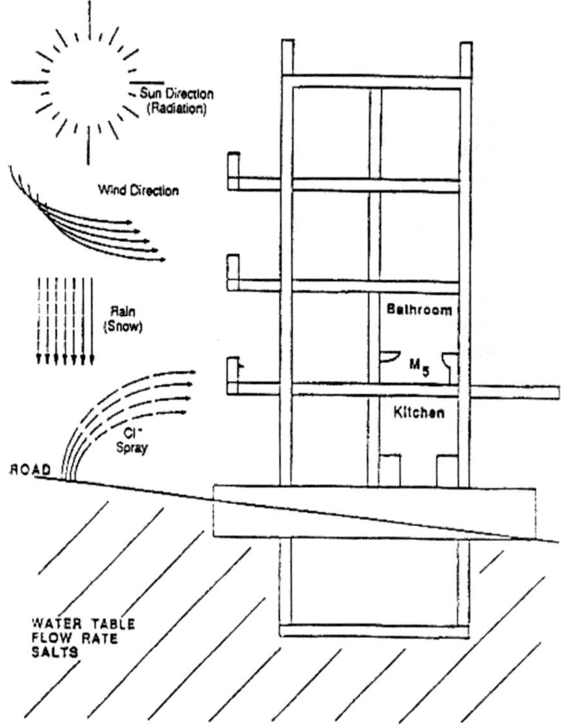

Figure 4 Possible influences on the outer micro-climate for a building

What is being described here is a fundamental difficulty in truly representing the conditions in Table 4 in Code terms, followed by finding detailing solutions to the outer fabric or surfaces of structures which either eliminate the effects of the environmental loads in Figure 4, or control them in a way which can be reliably taken into account in design. A durability design approach has to take these issues on board.

The Influence of Construction Quality

Figure 3 and Table 2 clearly illustrate the importance of construction quality and variability, particularly in providing resistance to corrosion. Space does not permit a thorough review of this subject, although it has been covered in a recent report [14].

In this context, variability relates both to the achievement of concrete cover, and to the quality of concrete itself. It should be recognised that there are only two basic ways of addressing the issue, in developing a quantitative generic solution. These are summarised in Table 5. Option 1 involves improvement, via the development of a quality package, within a systems approach, which raises standards to an acceptable (and repeatable) level. This should be an industry objective in any case. However it should be stressed that improvements must take place in design and detailing, as well as in site practice. Some design solutions are extremely difficult to build; others are deficient in providing adequate planned resistance to aggressive actions [14].

Table 5 Options for dealing with construction quality in durability design

OPTION	DESCRIPTION
1.	Raise standards - encourage better design and detailing (buildability) - improve communications from designer to site operative - education and training - bench marking best practice - method statements and certification
2.	Assume current variability - develop better defence strategies - introduce multi-layer protection systems - make more use of a 'design-out' approach

THE DESIGN APPROACH IN OUTLINE

Traditional structural design can be conveniently considered in two phases:-

1. Initial conceptual design, associated with preliminary sizing of elements and costing.

2. Final detailed design, essentially a numerical process, involving structural analysis, section design, and satisfying prescribed limit states.

Conscious design for durability has somehow to be fitted into this scheme of things. How can that be done? To begin with, it is probably necessary to subdivide phase 2, as follows:-

(2a) Detailing, either in terms of reinforcement arrangements, or the final shaping of cross-sections (for both buildability and durability), or integrating the basic structure into the artefact as a whole (e.g. cladding, services, finishes, etc., for buildings).

(2b) The numerate processes associated with limit states.

The prime reason for doing this is that phase (2a) is likely to be much more significant in durability terms, and it is not yet clear how valid detailed calculations might be, in satisfying defined durability limit states for phase (2b). In addition, conceptual design will be much more important for durability.

PHASE 1 - CONCEPTUAL DESIGN

The basic issues requiring consideration at this stage are listed in Table 6. It is intended to focus on section D in particular, in developing a defence strategy, while noting that section E will also influence this, both in terms of costs and in making judgements of possible levels of construction quality and of the known performance of the materials and components to be used.

Table 6 Conceptual design - issues and approach

A	Function and type of structure	(i)	Client's basic needs
		(ii)	Overall requirements (dimensions, etc.)
B	Performance requirements	(i)	Safety and serviceability criteria; minimum acceptable technical performance
		(ii)	Importance of continuity of function
		(iii)	Nominal design life (structure and/or elements)
		(iv)	Management and maintenance strategy
		(v)	Likelihood of future change of use, or upgrading
C	Loads	(i)	Dead and imposed loading
		(ii)	Wind, water and temperature effects
		(iii)	Aggressive actions
D	Overall design approach	(i)	Basic defence strategy
		(ii)	'Design out' versus 'provide resistance'
		(iii)	Articulation system: movement & joints
E	Preliminary evaluation of alternatives	(i)	Costs, including whole life costs
		(ii)	Construction method and quality issues
		(iii)	Known performance of materials and components

The essence of section D is to formulate a basic defence strategy. If we assume that we can deal with most aggressive actions that can directly affect the concrete, via a prescriptive approach, then the prime concern is corrosion, possibly affected by synergetic effects from other actions. The options for a defence strategy are then those shown in the flow diagram in Figure 5. None of these options are completely separate; the boxed feature for each of the three primary routes merely indicate the main component.

How might this work in practice, and which approach might be preferred in any given situation? To some extent, this will depend on the confidence that the designer has in:

1. His knowledge of possible aggressive actions, and the general and local environmental conditions.

2. The reliability of the various resistance options at his disposal.

An outline of some possible scenarios may illustrate the type of thinking required.

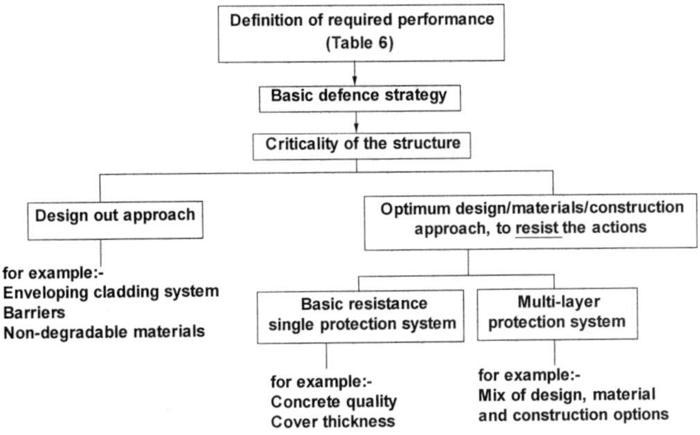

Definition of required performance (Table 6)

Basic defence strategy

Criticality of the structure

Design out approach

for example:-
Enveloping cladding system
Barriers
Non-degradable materials

Optimum design/materials/construction approach, to resist the actions

Basic resistance single protection system

for example:-
Concrete quality
Cover thickness

Multi-layer protection system

for example:-
Mix of design, material and construction options

Figure 5 Options for a defence strategy, in durability design

Buildings in General

A design out approach may be attractive here, since it is comparatively easy to provide an enveloping cladding system, and the structural concrete would then only require a basic resistance appropriate to indoor conditions. Alternatively, special attention could be given at the phase 2a design stage to the detailing of the facade, as part of a multi-layer design approach. In this instance, movement and joints are important considerations.

Civil Engineering Structures in General

The design out approach is less practical here as a single protective option, although both barriers and non-degradable materials can be part of a multi-layer protective system especially for critical or vulnerable zones. The most common approach will therefore be a resistance option of some kind, whose nature may depend on the type of structure, the scale of the project, and what is known about the nature of the environment (which is commonly severe). Some examples are given below.

Maritime Structures

The performance of concrete in maritime situations has been well researched, and specialist guidance is available, in terms of design and specifications for the critical tidal and splash zones. Based on this, general approaches can be developed, most probably via the multi-layer protection route in Figure 5.

Major Structures

If the scale (and cost) of structures is very large, then special development work can normally be undertaken to produce solutions unique to the conditions. Two recent examples illustrate this. These are the Oresund crossing between Sweden and Denmark [15], and the Confederation Bridge in Canada [16].

Other Structures

For all other structures, but especially civil engineering structures in severe environments, where the local micro-climate may be difficult to define with sufficient precision to permit in-depth modelling and analysis, a multi-layer protection approach will usually be necessary.

PHASE 2A - DESIGN DETAILING

There is little to add about this phase, at this time. It is essentially concerned with the issues raised in Table 4 and Figure 4, associated with the need for designers to consider water alongside wind and temperature in evolving details, which create effective barriers to transport mechanisms. It does not involve intense numerical calculations or modelling. It does involve building up (and using) a portfolio of good practical details, based on feedback.

The next step, in developing an overall methodology is to consider how this phase might relate to the different approaches in Figure 5, in moving from the initial conceptual phase to complete the detailed design.

PHASE 2B - POSSIBLE ROLE OF NUMERICAL CALCULATIONS & LIMIT STATES

The extensive literature on durability and deterioration is mostly concerned with the development of analytical models to predict the processes of deterioration. A key question is: to what extent can these techniques be used, in a design format?

To understand the potential for this, corrosion protection is taken as an example, and a list of possible resistance options produced, which might be considered as part of a multi-layer approach as suggested in Figure 5. A possible list is given in Table 7, under the sub-headings: Materials; Design: Construction. The basic defence strategy would involve a choice of options from each group, and the design skill comes, in assembling the best options in individual cases.

Most of the Design and Construction options in Table 7, are considered at the conceptual or phase 2a stages of design, in deciding what basic strategy to use, and in making judgements about construction quality. Largely, these are not numerate. However, the materials-based options would also be part of a multi-barrier approach. Here, the designer might wish to evaluate different combinations of, say, concrete cover and quality; in that context, numerical modelling could have a role to play.

Table 7 Shopping list of resistance options for corrosion

A Materials	Concrete quality	-	mix proportions
		-	mix ingredients
	Cover	-	minimum
		-	tolerances
	Permeable formwork		
	Concrete protection	-	sealers
		-	coatings
		-	penetrants
		-	layers
	Rebar protection	-	epoxy coating
	Special steels		
	Non-corrodible reinforcement		
	Cathodic protection		
B Design	Design concept		
	Structural detailing		
	Cladding, services, fittings, finishes		
	Articulation, joints, movement		
	Treatment of water, drainage		
	Control of the environment, barriers		
	Provision for inspection, maintenance, replacement		
	Accurate assessment of effects of deterioration mechanisms		
C Construction	Construction methods		
	Quality control		
	Certification		
	Testing		
	Rationalisation, standardisation, simplification		

The question then is:- should these predictive tools be developed so that they can be used with confidence, in individual designs - or should they be used to develop simplified generic guidance, on the relative merits of alternative options.

If we consider the first of these possibilities, then we would have to satisfy the usual design condition, expressed below in its simplest form:-

$$S_K.g_S \leq \frac{R_K}{g_R}$$

where

S_K is the characteristics value of the aggressive action
γ_S is a partial safety factor for this loading
R_K is the characteristic resistance
γ_R is a partial safety factor for that resistance

To apply this, we need to have:-

1. Detailed knowledge of the loads, pertaining to the particular environment.

2. A model which predicts the penetration mechanism of the aggressive action.

3. Agreed limiting performance criteria, eg. the reinforcement should not be depassivated for x years.

4. Agreed values for the partial safety factors.

Of these four items, serious work has only been done on item 2. Item 1. will always be a major difficulty (see Table 4), except for major structures where detailed investigation can be justified; at the other extreme, detailed analysis is not necessary for indoor concrete.

Where modelling can be justified is in the generic development and evaluation of the relative merits of options in a multi-layer protection system. This is where the research community should now be concentrating its efforts. What sort of guidance are designers looking for here? If we consider the options in Table 7, then, for a given situation, questions such as the following require quantitative answers:

1. How does the use of permeable formwork compare with protective coatings or with adding an extra 20mm of cover? Or

2. Can we add the benefit of a concrete protection system to a Grade 60 concrete, as well as to a Grade 30? Or

3. How long do these additional protective layers last, without having to be renewed - and what is the cost/benefit ratio for each?

It is in this context, in relation to Table 7, that numerical calculations and modelling should first be deployed. Because of uncertainties or variability in environmental loading and quality of construction, designers will frequently be attracted to the multi-layer protection option in Figure 5. If they then look at Table 7, they need some rational basis for making a choice:- how many options to use for different aggressivities; which ones are genuinely compatible and additive; which combination is then the best, in cost/benefit terms?

SUMMARY AND CONCLUDING REMARKS

The design approach outlined earlier will require significant development prior to routine use. What are the essentials, and how might it be introduced progressively in the future? Table 8 presents a summary.

Sustainable construction requires the best use of resources to accurately meet performance requirements for our infrastructure. This is especially true for performance-in-service-with time, where we need to put some effort into defining these requirements better, as well as finding good solutions to satisfy them.

Table 8 Essentials in developing durability design

1	A compatible approach, involving design, materials and construction.
2	Use a specification approach for aggressive actions, which affect the concrete (Table 3).
3	For corrosion, the progression, in time, is (a) Recipes now (Code values for cover and grade). (b) Multi-layer protection in the short-term (Figure 5; Table 7). (c) Complete design framework (longer term).
4	Design must lead in:- (a) Phases 1, 2a and 2b and in Table 6 and Figure 5. (b) Introduce new techniques for dealing with long-term performance, Whole Life Costing, Life Cycle Analysis; Risk Analysis; Value Engineering.

Durability design is fundamental in taking this forward. However, it is argued in this paper that this need not involve numerical calculations comparable with those for structural design and safety. In this context, design does not equate to analysis, and much more emphasis is required on conceptual design (in developing a coherent defensive strategy) and on design details, which both make construction easier (and hence better) and are known to work in terms of providing proper resistance to transport mechanisms.

Prime factors in generating this shift in emphasis are:-

1. Recognition that, for individual cases involving routine structures, it is virtually impossible to obtain reliable and accurate information on the nature and intensity of environmental loads, to justify the apparent precision of existing analytical models. Such models can only be used, within an effective design mode, if matched by comparable precision in the input.

2. The dominant influence of quality of construction, which, if ignored, could invalidate rigorous modelling in durability design.

It is questionable if the durability design solutions outlined here are suitable for inclusion in traditional Codes and Standards, with their heavy dependance on specifications and numerical solutions; design has to be interpreted more generally. What would help future developments considerably are the following:-

3. Codes/specifications which set out performance requirements for different types of structure in some detail, and in sustainability terms.

4. The development work outlined in relation to Table 7, in evaluating, to a common base, the relative merits of alternative resistance options - ideally leading to a register of components and materials with known properties and behaviour, under a range of known environments.

5. The development of a systems approach to construction quality, with defined standards (targets) and clear commitments and responsibilities (as in Table 5).

REFERENCES

1. SOMERVILLE, G. The design life of concrete structures. The Structural Engineer, Volume 64A, N°2. February 1986. pp. 60-71. The Institution of Structural Engineers, London.

2. KROPP, J AND HILSDORF, H K. Performance criteria for concrete durability. RILEM Report 12, 1995. E&FN Spon, London. p327.

3. CEMENTA AB. Durability of concrete in saline environment. Cementa AB, Danderyd, Sweden. May 1996. p 206.

4. BROOMFIELD, J P. Corrosion of steel in concrete: understanding, investigation and repair. E&FN Spon, London, 1997. p240.

5. REINHARDT, H W. (1997). Penetration and permeability of concrete. RILEM Report 16. E&FN Spon, London, 1997. p331.

6. COMITÉ EURO-INTERNATIONAL DU BETON (CEB). New approach to durability design. An example for carbonation induced corrosion. Bulletin 238. CEB, Lausanne. May 1997. p138.

7. SARJA, A. AND VESIKARI, E. Durability design of concrete structures. RILEM Report 14. E&FN Spon, London, 1996. p165.

8. SAKAI, K. Integrated design and environmental issues in concrete technology. E&FN Spon, London, 1996. p308.

9. JAPAN SOCIETY OF CIVIL ENGINEERS (1989). Proposed recommendation on durability design of concrete structures. Concrete Library International No. 14. JSCE, Tokyo, 1989.

10. BRITISH CEMENT ASSOCIATION. Development of an holistic approach to ensure the durability of new concrete construction. Final report to the Department of the Environment on Project 38/13/21 (cc 1031). BCA, Crowthorne. October 1997. p81.

11. SOMERVILLE, G. Engineering design and service life - a framework for the future. Prediction of concrete durability. Proceedings of STATS 21st Anniversary Conference. E&FN Spon, London, 1997. pp58-76.

12. COMITÉ EURO-INTERNATIONAL DE BETON (CEB). CEB-FIP Model Code 1990, Design Code. Thomas Telford, London 1993.

13. BRITISH STANDARDS INSTITUTION. Buildings-Service life planning. Part 1: General principles. Ref. No. HB10141. BSI, London. October 1997. p60.

14. BRITISH CEMENT ASSOCIATION. Durability by Design. Final report to the Department of the Environment (now the DETR) on Project CI.38/13/12 (cc 716). BCA, Crowthorne, UK. April 1998. p46.

15. MUNCH-PETERSEN, C. ET AL. Concrete strategy for the Oresund Crossing. Betonwerk & Fertigteil - Technik. Bft 11. November 1997. pp44-54.

16. DUNASZEGI, L. High performance concrete a key component in the design of the Confederation Bridge. Newsletter, Beton Concrete Canada. Vol 1, No. 3. December 1996. pp1-2.

THEME ONE:

SELECTION OF MATERIALS AND MIX COMPOSITION

Keynote Paper

ADVANCES IN CONCRETE MIXTURE OPTIMISATION

M Simon

Federal Highway Administration

K Snyder

G Frohnsdorff

National Institute of Standards and Technology

United States of America

ABSTRACT. The complexity of high performance concrete (HPC) mixture designs continues to increase, along with the number of criteria that a particular mixture must satisfy. Robust statistical methods exist that can be used to not only determine the mixture for a concrete that meets specifications, but does so while satisfying a number of additional user-specified constraints. Two such techniques are the mixture method and the response surface method of experimental design. These techniques are discussed and compared. Future developments in concrete mixture optimization based upon materials science-based models are discussed.

Keywords: High performance concrete (HPC), Materials science, Mixture design, Mixture optimization, Response surface methodology (RSM)

Marcia Simon is a Research Materials Engineer at the Federal Highway Administration (FHWA) Turner-Fairbank Highway Research Center.

Kenneth Snyder is a Physicist in the Building Materials Division of the National Institute of Standards and Technology (NIST).

Geoffrey Frohnsdorff is Chief of the Building Materials Division (BMD) of the National Institute of Standards and Technology

INTRODUCTION

As higher and higher performance is sought from concrete, obtaining the proper mixture proportion to achieve specific objectives is becoming more difficult and useful tools are needed to aid the process. The Partnership for High Performance Concrete Technology (PHPCT) of the National Institute of Standards and Technology (NIST) and the Federal Highway Administration (FHWA) seek to facilitate the use of high performance concrete (HPC) in both public and private construction, and are currently developing tools for optimizing HPC mixture proportions to meet a number of performance criteria (user-specified constraints) simultaneously. These performance criteria could include fresh concrete properties such as viscosity, yield stress, setting time, and temperature; mechanical properties such as strength, modulus of elasticity, creep and shrinkage; and durability-related properties such as resistance to freezing and thawing, abrasion, or chloride penetration.

The American Concrete Institute (ACI) defines HPC as "concrete meeting special combinations of performance and uniformity requirements that cannot be achieved routinely using conventional constituents and normal mixing, placing, and curing practices." [1]. Since fewer, or possibly no, prescriptive constraints, such as minimum cement contents or maximum water-cement ratios, will be included in performance specifications, a concrete producer or materials engineer will have greater than usual latitude in selecting constituent materials and defining their proportions. At the same time, the task of achieving the design specifications has become more complex. HPC mixtures are usually more expensive than conventional concrete mixtures, since they usually contain one or more of the following: (1) more cement, (2) higher dosages of chemical admixtures, and (3) mineral admixtures. The desire to optimize concrete mixture proportions, by meeting the performance criteria at the lowest cost, increases as the cost of materials increases. Furthermore, as the number of constituent materials increases, the problem of identifying optimal mixtures becomes increasingly complex. Not only are there more materials to consider, but more potential interactions among materials. Combined with several performance criteria to meet, the number of trial batches required to find optimal proportions using traditional methods could become prohibitive. Statistical and computational tools are required to provide cost-effective means of formulating optimized concrete mixtures.

The process of concrete mixture proportioning typically involves the following steps:

1. Identifying a starting set of mixture proportions

2. Performing a series of trial batches, starting with the mixture identified in Step 1, and adjusting the proportions in subsequent trial batches until all criteria are satisfied. Typically, this is performed by changing one component at a time.

The remainder of this paper discusses two types of tools that could be used to improve the process of mixture proportioning. The first of these is the application of statistical methods for product optimization. The second, a not-yet-fully-realized goal of the NIST program, is the application of materials science-based models [2] to predict the performance of concrete mixtures from knowledge of the properties and proportions of the ingredients and the processing of the mixtures [3].

APPLICATION OF STATISTICAL METHODS TO CONCRETE MIXTURE PROPORTIONING

Current practice in the United States for developing new concrete mixtures usually relies upon historical information (i.e., what has worked for the producer in the past) or the guidelines for mixture proportioning outlined in ACI 211.1 [4]. While both methods can yield a starting point for trial batches, neither method is a comprehensive procedure for optimizing mixtures. Not only does ACI 211 not account for interacting effects among the concrete constituents, there is no means by which one can efficiently achieve an optimized mixture for a given criterion. In contrast, statistical experimental design methods are rigorous techniques for both achieving desired properties and determining an optimized mixture for a given constraint, while minimizing the number of trials. They are used widely in industry to optimize products and processes [5].

Following the ACI 211 guidelines, an engineer would select and run a first trial batch (using proportions selected using ACI 211 or historical data), evaluate the results, adjust the proportions of various components and run further trial batches until all specified criteria were met. Employing statistical methods in the trial batch process would not change the overall approach, but would change how trial batching is done and speed the process. Rather than selecting one starting point, a set of trial batches covering a chosen range of proportions for each component is set up according to established statistical procedures [5]. Trial batches are performed and results are analyzed using standard statistical methods that yield reliable estimates of parameters from empirical models for each performance criterion. Each response, such as strength, slump, or cost, is expressed as an algebraic function of factors such as water-cement ratio (w/c), cement content, chemical admixture dosage, pozzolanic replacement, etc. Once a response can be characterized by an equation, any number of analyses are possible. For instance, the user could determine which mixture proportions would yield a desired response. Similarly, the user could optimize any response function subject to constraints on the others. For example, one could determine the lowest cost mixture with strength greater than the specified strength. A method for optimizing several responses simultaneously is described later in the paper.

Efforts have also been made to develop mixture proportioning methods based on mechanistic (or semi-mechanistic) models [6]. In this approach, models are developed from results of fundamental and applied materials research, and the user does not perform a series of trial batches from which empirical models are estimated. This approach has the advantage of eliminating the need for trial batches to obtain the models; however, some trial batches would most likely be needed to adjust proportions because of differences in a user's specific materials. It is unlikely that a mechanistic model would be able to account for all possible differences among materials from various localities. The advantage of the trial batch approach is that the project-specific materials are used and accounted for in the model.

The statistical approach has an additional advantage that is often overlooked in mixture design procedures: the expected responses can be characterized by an uncertainty. This has important implications for both mixture specification and for production. One could use the empirical model equations to determine a mixture design that yielded a desired strength. However, the model equation would only give the expected mean strength; if replicate mixtures were made, the model equation would predict the *mean* value. For producers to be relatively sure that most of the on-site tests would comply with the specifications, they would select target values for the mean strength to account for the variability and to ensure that, say, 95 percent of the time the concrete performance would be in compliance.

Background on Statistical Approaches to Optimization

There exist two primary approaches to the general problem of optimizing a mixture whose properties depend on the proportions of the component materials: the *mixture approach* and the *response surface approach* [5]. Each technique has advantages and disadvantages.

Mixture approach

Using a mixture approach, the total amount (mass or volume) of the product is fixed, and the settings of each of the q components are proportions. Because the total amount is constrained to sum to one, only q-1 of the component variables are independent.

As a simple example of a mixture experiment, consider concrete as a mixture of three components: water (x_1), cement (x_2), and aggregate (x_3), where each x_i represents the volume fraction of a component. Assume the coarse-to-fine aggregate ratio is held constant. The volume fractions of these components sum to one,

$$x_1 + x_2 + x_3 = 1 \tag{1}$$

and the region defined by this constraint is the triangle, or three-component simplex, shown in Figure 1a. The axis for each component x_i extends from the vertex it labels ($x_i = 1$) to the midpoint of the opposite side of the triangle ($x_i = 0$). The vertex represents the pure component. For example, the vertex labeled x_1 is the pure water mixture with $x_1 = 1$, $x_2 = 0$, and $x_3 = 0$, or (1,0,0). The coordinate where the three axes intersect is (1/3,1/3,1/3) and is called the *centroid.* A good experiment design for studying properties over the entire region of a three-component mixture would be the simplex-centroid design shown in Figure 1b (this example is included as an illustration only, since much of this region would not represent either feasible or workable concrete mixtures). The points shown in Figure 1b represent mixtures included in the experiment and include all vertices, midpoints of edges, and the overall centroid. For each mixture, all properties of interest would be measured, and empirical models for each property as a function of the components would be determined from regression analysis.

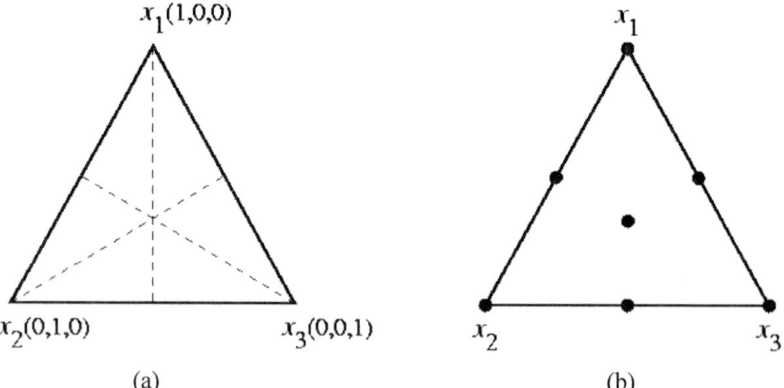

Figure 1 Three-component mixture plans: a) experimental region, and b) layout of simplex-centroid experiment design

Since feasible concrete mixtures do not exist over the entire region shown in Figures 1a and 1b, a subregion of the full simplex that contains the range of feasible mixtures must be defined by constraining the component proportions. An example of a possible subregion for the three-component example is shown in Figure 2. It is defined by the following volume fraction constraints (x_1 = water, x_2 = cement, x_3 = aggregate):

$$0.15 \leq x_1 \leq 0.25$$
$$0.10 \leq x_2 \leq 0.20$$
$$0.60 \leq x_3 \leq 0.70$$

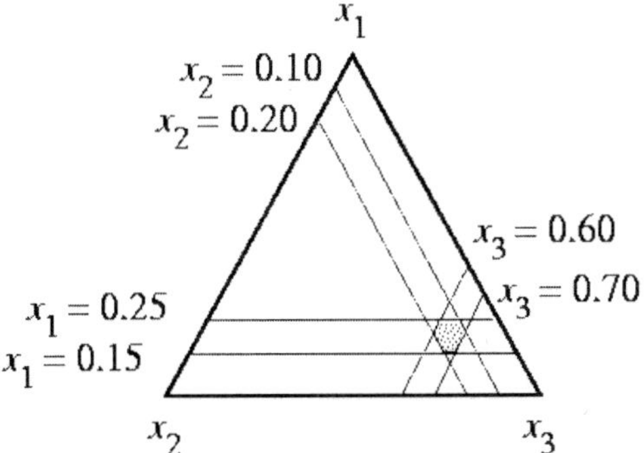

Figure 2 Example of constrained experimental region for a mixture with three components

In this case simplex designs are no longer appropriate and other designs are used [5].

The advantage of the mixture approach is that the experimental region of interest is more naturally defined; however, analysis of the results is more complicated, especially if the number of components is greater than three, as it usually will be.

Response surface approach

In the response surface approach, the q components of a mixture are reduced to q-1 independent variables by using the ratio of two components as an independent variable. In the case of concrete, the w/c ratio is a natural choice for this ratio variable. As a simple example, consider a concrete mixture composed of four components: water, cement, fine and coarse aggregate. Three factors, or independent variables, x_k, that can be selected to describe this system are x_1 = w/c ratio (by mass), x_2 = fine aggregate volume fraction, and x_3 = coarse aggregate volume fraction. Reasonable ranges for each variable might be:

$$0.40 \leq x_1 \leq 0.50$$
$$0.25 \leq x_2 \leq 0.30$$
$$0.40 \leq x_3 \leq 0.45$$

To simplify calculations and analysis, the actual variable ranges are usually transformed to dimensionless coded variables with a range of ±1. In this example, the actual range of $0.40 \leq x_1 \leq 0.50$ would translate to a coded range of $-1 \leq x_1 \leq 1$. Intermediate values of x_1 would translate similarly (e.g., the actual value of 0.45 would translate to a coded value of zero).

Suppose also that the specifications for this mixture require a slump of 75 mm to 150 mm and a 28-day strength of at least 30 MPa. These specified properties are the responses, or dependent variables, y_i, which are the performance criteria for optimizing the mixture.

A common response surface experimental plan that could be used in this scenario is a *central composite design* (CCD), illustrated schematically in Figure 3. The CCD for $k = 3$ independent variables consists of eight (2^k) factorial points (filled circles in Figure 3) representing all combinations of coded values $x_k = \pm 1$, six ($2k$) axial points (hollow circles in Figure 3) at a distance $\pm\alpha$ from the origin, and at least 3 center points (hatched circle in Figure 3) with coded values of zero for each x_k. The value of α is usually chosen to make the design rotatable (implies that at locations equidistant from the origin, predicted values should have equal variance) but there could be valid reasons for selecting other values for α [5].

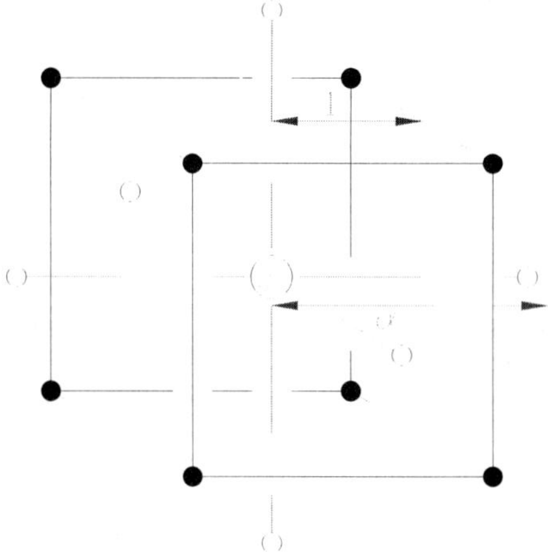

Figure 3 Central composite design for three independent variables

Empirical models

Use of an appropriate mixture experiment design or a CCD allows estimation of a full quadratic model for each response. Equation 2 shows a full quadratic model for $k = 3$ independent variables:

$$y = b_0 + b_1 x_1 + b_2 x_2 + b_3 x_3 + b_{12} x_1 x_2 + b_{13} x_1 x_3 + b_{23} x_2 x_3 + b_{11} x_1^2 + b_{22} x_2^2 + b_{33} x_3^2 + e \quad (2)$$

In Equation 2, the ten coefficients are represented by the b_k, and e is a random error term representing the combined effects of variables not included in the model. The interaction terms $x_i x_j$ and the quadratic terms x_i^2 in Equation 2 account for curvature in the response surface, which is often present when a response is at or near a maximum or minimum in the region of interest. A model with only linear terms would be sufficient if curvature was not present, and the factorial portion of the CCD is a valid design by itself in that case. Often, the presence or absence of significant curvature is not known with certainty at the start. An advantage of the CCD over the mixture approach is that the CCD can be run sequentially in two blocks. The first block would consist of the factorial points (all combinations of $x_i = \pm 1$) and some center points (at the origin), and the second block would consist of the axial points (points along each axis at distance α from the origin) and additional center points. This approach allows analysis of the factorial portion before the axial portion is run, providing an indication of whether the axial portion is necessary.

The number of coefficients in the quadratic model increases with k, and the number of trial batches required using a CCD begins to increase significantly for $k > 5$. Therefore, the use of a CCD to optimize a concrete mixture with six or more components may be uneconomical. In such cases, one could identify the most important factors and limit them to five or fewer. For example, if the cementitious materials and chemical admixtures were the most important components, they would be varied, while the amounts of coarse and fine aggregate would be held constant.

Application of the Response Surface Approach to Concrete Mixtures

As part of a current research project, NIST and FHWA are developing an interactive website that can be used to optimize concrete mixture proportions using the response surface approach. As part of this project, laboratory experiments were conducted using both the mixture approach and the response surface approach. Although both give comparable results, it was concluded that the response surface method is easier to use, and the interpretation is more straightforward. The following section describes the major steps in a response surface approach to mixture proportioning. These steps include planning (experiment design), executing trial batches, fitting and validating models, and determining optimal mixture proportions.

Experiment design

The first step in the planning process is to define the performance criteria to be met. These are usually defined in project specifications and might include (for example) slump range, fresh air content range, and minimum strength. Once the criteria are established, the next step is to select the materials to be used. Knowledge and experience are necessary here. If possible, the producer will want to use materials that he normally uses and stocks, but it is necessary to be confident that the performance criteria can be met using those materials. Otherwise, other locally available materials, or possibly special materials, will have to be procured. Comparisons of different materials (e.g., several possible choices of cement) would not usually be included in the response surface approach; therefore, if such a comparison is needed, it must be done ahead of time.

Once materials are selected, the independent variables must be chosen. One of these variables will be w/c ratio, and the rest will be volume or mass fractions of materials whose proportions will be varied. Which materials to vary will depend on the overall goal of the project and the budget allocated for mixture proportioning. Since the number of trial batches required increases exponentially ($2^k + 2k$) with k, making trial batches usually becomes prohibitive if more than six components are varied. In this case the most important components (those thought to influence properties significantly) should be identified.

Once materials are selected, the ranges for their proportions must be determined. ACI 211 or other guidelines and historical mixture information, combined with experience and knowledge, can be used to identify a starting point. The selection of ranges is important: too narrow a range may result in inability to meet all performance criteria simultaneously, too wide a range may fail to identify the best mixture.

The ranges can be defined in terms of volumes or volume fractions, mass or mass fractions. The volume fractions of all components will be constrained to sum to 1, so at some point calculation of corresponding volume fractions will be necessary. For batching, masses of materials will have to be calculated.

When the above steps are completed, the trial batching plan is constructed according to established procedures for constructing central composite experiment [5,7]. Several commercially available statistical software packages are available for this purpose, or it can be done by hand.

Execution of trial batches

Once the trial batch plan is constructed, the trial batches can be executed. As mentioned above, the CCD is usually set up in two blocks. The order of the batches within each block should be randomized. Furthermore, all batching, mixing, fabrication, and testing should be in accordance with established specifications and methods, and the same personnel should be used for the same tasks throughout the trial batching. The purpose of such steps is to minimize the effects of extraneous variables on the results. Any anomalies or aberrations in procedure or in test results should be noted.

Analysis of results

The analysis of trial batch data from tests on fresh and hardened concrete uses established graphical and numerical techniques. A graphical overview of data using scatterplots and plots of raw data helps to identify general trends and effects as well as possible outlier points (resulting from recording errors or equipment malfunction, for example).

The next step is model fitting and validation. Standard analysis of variance (ANOVA) and linear regression techniques [5] are used to estimate model parameters. A full quadratic model will usually be fit first, and t-tests on the coefficients will indicate insignificant terms which can be eliminated from the model. If the trial batches are run in sequential blocks (see above), a preliminary analysis to assess the adequacy of linear models can be performed after the batches (and required tests) in the first block are completed. As stated previously, if linear models are sufficient, the second block of trial batches may not be necessary.

Once a model is chosen, the adequacy of the model is assessed by computing residuals and examining residual plots. The residuals are the deviations of the observed data from the fitted values, and they are estimates of the error terms, e_i, in the model (see Equation 2). The error terms are assumed to be random and normally distributed, and if this assumption holds, the residuals should exhibit similar properties. Plots of the residuals against batch order (run sequence), predicted values, and other parameters should be random and without structure.

Additional quantitative checks on the adequacy of the model can be made by calculating statistical measures such as the residual standard deviation and the predicted error sum of squares (PRESS). The residual standard deviation should be close to the replicate standard deviation calculated from the center points. The PRESS statistic is a measure of how well the model fits each point in the design, with a small PRESS statistic indicating a good fit.

An additional analysis that may be valuable at this stage is to examine contour plots of the response as a function of any two variables. The contour plots are similar to topographical maps and allow the user to see how the response varies over the ranges of the chosen variables. The contour plots allow visual identification of the best settings to achieve a particular response.

Optimization

The usual goal of mixture proportioning is to identify proportions that yield concretes meeting several performance criteria (that is, constraints on several responses) simultaneously, while minimizing cost. Cost is also a response since it can be calculated as a function of the independent variables (mixture proportions). In other cases, the goal may be to maximize or minimize a particular property (say, strength) irrespective of cost. Once suitable empirical models are chosen for each response, any response can be optimized with or without constraints on the other responses. Both graphical or numerical techniques can be used for this task, however graphical methods are best used when the number of responses (including cost) is three or less. The graphical methods involve overlaying contour plots of responses with the constraints indicated, creating a feasible region that meets all of the constraints. If desired, the feasible region can be superimposed on a cost contour plot to identify the point of minimum cost.

Numerical optimization is more widely applicable. A common approach to numerical optimization uses "desirability functions," d_i, which are defined for each response (including cost) [8]. These functions vary between zero and one and can be defined in several ways. Responses that are specified by minimum or maximum values, such as strength, have desirability functions that are step functions. For strength, $d_i = 1$ above the minimum value and $d_i = 0$ otherwise. Other responses, such as slump and fresh air content, are specified by ranges. Desirability functions for these responses could be two-sided step functions if all values were equally acceptable. Alternatively, if a target value is considered most desirable, the desirability function could vary from zero at the endpoints of the range to 1 at the target value (for example, a target slump of 100 mm within an acceptable range of 75 mm to 125 mm).

The optimal set of mixture proportions is the one that maximizes the geometric mean of all of the desirability functions. Numerical search techniques are used to identify this set. After the optimal mixture is determined, the predicted values of each response are calculated and checked to see that they meet the constraints and account for the uncertainty in the model. If necessary the constraints are modified to account for the uncertainty and a revised set of optimal proportions is identified using the procedure described above.

MECHANISTIC MODELS FOR APPLICATION TO MIXTURE PROPORTIONING

The goal of the NIST PHPCT program is "To enable the reliable application of high-performance concrete in buildings and the civil infrastructure by developing, demonstrating, and providing assistance in implementing a computer-integrated knowledge system, HYPERCON, incorporating verified multi-attribute models for prediction and optimization of the performance and life-cycle cost of HPC." Substantial progress towards this goal has been made and the growing suite of materials science-based models that has already been developed for that purpose is now available on the web in the form of an "electronic monograph" [9]. Among the models are ones for predicting a) microstructure development in hardening cement paste from information on the composition, sizes and shapes of cement particles[10]; b) the structures of mortars and concretes from which transport properties can be calculated [11]; and c) the service life of chloride-exposed steel-reinforced concrete [12]. Each of the existing models will be refined as new insights are gained into the phenomena they represent, and complementary models to complete the set needed for optimization of mixture proportions for specific applications, such as models for the flow of concentrated particulate dispersions[13] and for fire response of fiber-containing concretes [14], are being developed. When the models are combined into interoperable computer-integrated knowledge systems available on the internet, they will provide a coherent materials science base for many aspects of concrete technology.

As mechanistic computer models improve, it may one day be possible to predict concrete performance through computer simulation prior to batching any concrete. Although this idealization is well into the future, current, and near future, mechanistic and microstructural computer models, such as those being developed at NIST, may one day be used to develop mixture proportioning guidelines. Over small regions of parameter space, nearly any function can be approximated sufficiently well by a quadratic model. However, for mixture proportioning guidelines to be useful for the concrete industry, the model would have to accommodate a large parameter space. Over these wide variations in parameters, mechanistic models can give insight to the optimum functional form for the model equation.

The combined use of computer models and factorial experiment design for estimating the diffusion coefficient of concrete has already been demonstrated [11]. A combination of composite theory and the NIST microstructural model for cement paste hydration was used to calculate the chloride diffusivities for concrete mixtures. The results from the calculations were used to determine the parameters of the model equation. The resulting estimates of diffusivity were demonstrated to be "not unreasonable" by comparison to reported data. One could imagine expanding this to other properties. The near future may hold similar experiments for concrete properties such as viscosity and yield stress. Numerical techniques for simulation of dense solid suspensions have been used to estimate the viscosity of a matrix containing solid particles [13]. Further developments may include estimates of yield stress. Results from these simulations could then be compared to experimental measurements using equipment that has been developed only recently [15,16].

These statistical techniques may also prove to be powerful tools for future concrete mixture design and optimization. Future optimizations will benefit from reasonably accurate trial mixtures based upon computer models and existing model equations. One could imagine a future ACI 211 composed of an analytical equation that gives not only a prediction for the best trial mixture, but also an estimate of the uncertainty of the response. This estimated uncertainty could then be used to establish the required parameter space for the optimization using zero trial mixtures. One could then bypass the step of making a series of trial mixtures that are currently required for establishing a reasonable parameter space.

SUMMARY

As awareness of the potential of concrete to achieve higher performance grows, the problem of designing concrete to exploit the possibilities becomes more complex. Statistical design of experiments, such as the response surface approach described in the present paper, is a tool than can be of immediate help in increasing efficiency in selecting the optimum proportions for a high-performance concrete. For the future, prediction of performance of concrete using materials science-based models in conjunction with standardized databases of concrete material property data should make possible multi-attribute optimization. The optimization should be able to take into account several attributes at a time, including flow properties, strength development, dimensional changes, fire response, life-cycle cost, and environmental impact.

REFERENCES

1. RUSSELL, H G. ACI defines high-performance concrete. Concrete International, Vol 21, No 2, 1999, pp 56-57.

2. GARBOCZI, E J and BENTZ, D P. Computer simulation of the diffusivity of cement-based materials. Journal of Materials Science, Vol 27, 1992, pp 2083-2092.

3. FROHNSDORFF, G. Partnership for high-performance concrete technology. International Symposium on High-Performance and Reactive Powder Concretes, Eds. P-C Aïtcin and Y Delagrave, 1998, pp 51-73.

4. ACI Committee 211, Standard Practice for Selecting Proportions for Normal, Heavyweight, and Mass Concrete. ACI Manual of Concrete Practice, Vol 1, American Concrete Institute, 1996.

5. MYERS, R H and MONTGOMERY, D C. Response Surface Methodology, John Wiley & Sons, New York, 1995.

6. DE LARRARD, F and BELLOC, A. The influence of aggregate on the compressive strength of normal and high-strength concrete. ACI Materials Journal, Vol 94, No 5, 1997, pp 417-426.

7. BOX, G E P, HUNTER, W G and HUNTER J S. Statistics for Experimenters, John Wiley & Sons, New York, 1978.

8. DERRINGER, G and SUICH, R. Simultaneous optimization of several response variables. Journal of Quality Technology, Vol 12, 1980.

9. GARBOCZI, E J, BENTZ, D P, and SNYDER, K A. Modelling the Structure and Properties of Cement-Based Materials. http://ciks.cbt.nist.gov/garboczi

10. BENTZ, D P. Three-dimensional computer simulation of Portland cement hydration and microstructural development. Journal of the American Ceramics Society, Vol 80, 1997, pp 3-21.

11. BENTZ, D P, GARBOCZI, E J, and LAGERGREN E S. Multi-scale microstructural modeling of concrete diffusivity: Identification of significant variables. Cement, Concrete, and Aggregates, Vol 20, No 1, 1998, pp 129-139.

12. BENTZ, D P, CLIFTON, J R, and SNYDER, K A. Predicting service life of chloride-exposed steel-reinforced concrete. Concrete International, Vol 18, No 12, 1996, pp 42-47.

13. MARTYS, N S and MOUNTAIN, R D. Velocity Verlet algorithm for dissipative-particle-dynamics-based models of suspensions. Physical Review E, Vol 59, No 3, 1999, pp 3733-3736.

14. BENTZ, D P. Fibers, percolation, and spalling of high performance concrete. Submitted to ACI Materials Journal.

15. DE LARRARD, F, HU, C, SEDRAN, T, SZITKAR, J C, JOLY, M, CLAUX, F, and DERKX, F. A new rheometer for soft-to-fluid fresh concrete. ACI Materials Journal, Vol 94, No 3, 1997, pp 234-243.

16. FERRARIS, C F and DE LARRARD, F. Modified slump test to measure rheological parameters of fresh concrete. Cement, Concrete, and Aggregates, Vol 20, No 2, 1998, pp 241-247.

PARTICLE SIZE AND SHAPE ANALYSIS OF COARSE AGGREGATE USING DIGITAL IMAGE PROCESSING

C F Mora

A K H Kwan

H C Chan

University of Hong Kong

Hong Kong

ABSTRACT. An attempt of applying the digital image processing (DIP) technique to analyze the particle size distribution and shape characteristics of coarse aggregate is made. Several different types of aggregates have been analyzed, and their grading curves and shape parameters are compared to those obtained by conventional mechanical sieving and manual measurement. Since DIP produces only area gradation and uses a particle size definition different from the square sieve size used in mechanical sieving, direct comparison is not possible unless gradation basis and particle size definitions of the two methods are aligned. For this purpose, a simple method of converting area gradation to mass gradation is proposed. Moreover, a size correction factor is used to convert the particle sizes measured by DIP to equivalent square sieve sizes so that comparison between the DIP and mechanical sieving results can be made. The study demonstrates that DIP is a fast, convenient, versatile and accurate technique for particle size distribution and shape analysis of coarse aggregates.

Keywords: Aggregate, Digital image processing, Elongation index, Flakiness index, Grading

Mr Carlos F Mora is a Postgraduate Student studying for a Ph.D. degree at Department of Civil Engineering, The University of Hong Kong. After graduation from the University of Manitoba in Canada, he joined the University of Hong Kong and has since completed a research project on concrete preservatives. He is interested in all aspects of concrete technology, especially applications of digital image processing and artificial neural networks.

Dr Albert K H Kwan is a Senior Lecturer of Department of Civil Engineering, The University of Hong Kong. He obtained his doctorate from The University of Hong Kong and has acquired many years of practical experience before returning to academia. His research topics include concrete technology, tall building structures and earthquake resistant structures.

Prof H C Chan is a Professor and Head of Department of Civil Engineering, The University of Hong Kong. His main research interests include concrete technology, reinforced concrete structures and computer aided design of structures. He has published many journal papers and chapters of books, and served on many technical committees for the local government and engineering institute.

INTRODUCTION

Digital Image Processing (DIP) is a computerized technique by which a scene is captured, digitized and then processed so that information can be extracted from the objects depicted in it. In the past few years, DIP techniques have found widespread applications in many disciplines such as biology, medicine, meteorology, military and engineering [1]. Relatively, there have been very few applications of DIP in civil engineering [2,3]. DIP has been utilized to analyze the size, shape and spatial distribution of grains and pores in soil, study the microstructure of concrete, detect cracks in road pavement, measure structural deformations and evaluate traffic conditions [4]. However, it is believed that the applications previously mentioned are just a few of the possible applications.

In a research program designed to explore possible applications of DIP to concrete technology, the authors are investigating viable means of utilizing this technique to analyze the size and shape of aggregate particles in concrete. The study of size and shape of particles has been of great importance in the fields of particulate technology, soil mechanics, pavement construction and concrete technology. Early applications of DIP techniques to particle size and shape analysis have been tried by Barksdale et al [5], Li et al [6], Yue and Morin [7], and Kuo et al [8]. New methodologies are being developed every year, but it would be a few more years before all the pitfalls encountered can be overcome. The main obstacle in utilizing DIP techniques to analyze three dimensional objects is that DIP can only measure two dimensions. The third dimension, i.e. thickness, is not directly obtainable from DIP. Barksdale *et al* [5] have developed a method of estimating the thickness of particles by measuring the shadow length of each particle. However, the method fails if the spacing between the particles is not enough to accommodate the shadow projection. Another technique was developed by Kuo *et al* [8]. This technique is based on two orthogonal measurements of each particle which has to be arranged and adhered to an angle shaped transparent tray. But this method is more suitable for coarse rather than fine aggregates, since fine particles are very tedious and time consuming to be arranged one by one. In light of the difficulty of measuring the thickness from DIP output, the results have to be expressed in terms of area fractions rather than mass fractions; consequently the DIP results cannot be directly compared to those obtained by mechanical means.

DIP TECHNIQUE

Digital Image Processing has gradually developed and matured with the advancement of computer technology. Presently it is possible to perform DIP on a personal computer equipped with the appropriate hardware and software. The major steps in DIP can be separated into two categories, image acquisition and image processing.

Image acquisition is the process of capturing a scene and digitizing the image so obtained into a pixel array for storage in computer memory. The scene may be captured by means of specialized devices, such as charge coupled device (CCD) cameras or digital cameras. In addition, the image acquisition can also be performed through devices such as flat bed scanners or film scanners.

Image processing consists of two steps: manipulation of the image (strictly speaking, the pixel array) so as to identify the objects depicted in it and analysis of the pictorial information about the identified objects. The type of manipulation performed depends on the information required. Basically, it takes one image and returns another that has the required features enhanced (for instance, in the case of measuring the size of an object, the contrast between the object and the background is increased so as to increase the sharpness of the object boundary) and the unwanted features removed. Many specifically developed algorithms are available for analyzing the objects identified. Some examples are particle counting, geometric measurement and pattern recognition.

The DIP system used by the authors is the Quantimet 600S manufactured by Leica Cambridge Ltd. It incorporates a 3-chips CCD camera having a resolution of 736 × 574 pixels, a frame grabber with three A/D converters each of 8-bit resolution, a photographic stand fitted with light sources, a Pentium based computer and the necessary software for image analysis. A schematic representation of the system is shown in Figure 1. The following accessories are also fitted to the system for enhanced DIP capabilities: a film scanner, a MO disk drive, a CD-ROM writer drive and a high resolution color printer.

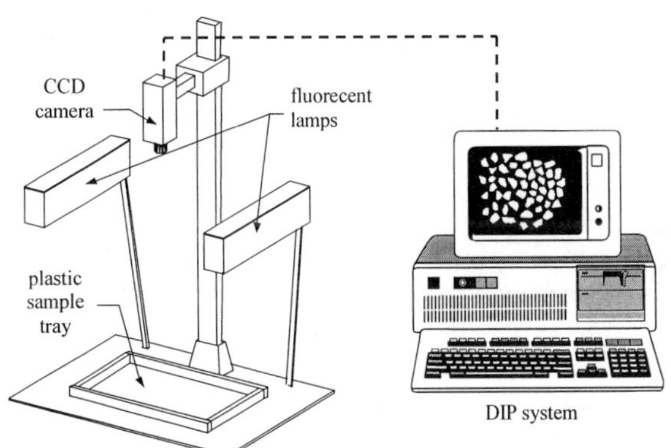

Figure 1 DIP system setup

PARTICLE SIZE DISTRIBUTION ANALYSIS BY MECHANICAL SIEVING

Mechanical sieving is the most commonly used method to determine the grading (i.e. particle size distribution) of concrete aggregates. The sieving operation divides a sample of aggregate into fractions each consisting of particles within specific size limits. However, after the sieving is done, not all particles retained on a sieve are larger than the sieve apertures. Particles slightly smaller than the aperture size may sometimes be stuck without passing through the sieve. Manual checking and brief hand sieving are thus required to make sure that all particles retained on a sieve are bigger than the sieve apertures. After sieving, the quantity of each fraction of particles is measured by weighing.

The results of mechanical sieving analysis are normally presented graphically in the form of grading charts. In the grading chart commonly used, the ordinate represents the cumulative percentage passing by mass of the aggregate (mass gradation) and the abscissa the sieve sizes plotted to a log scale. The following points should, however, be noted when interpreting the results of mechanical sieving:

(1) Particles passing through a sieve can actually have one dimension larger than the size of the sieve apertures. From Figure 2(a), it can be seen that an elongated particle having its length greater than the sieve size can pass through the sieve without any difficulties. Therefore, the sieve aperture size is a measure of the lateral dimensions of the particles only.

(2) A relatively flaky particle can pass through the sieve aperture, which is square in shape, diagonally as shown in Figure 2(b). Consequently, the breadth of a particle passing through a sieve can also be greater than the sieve size, although it has to be smaller than the diagonal length of the sieve aperture.

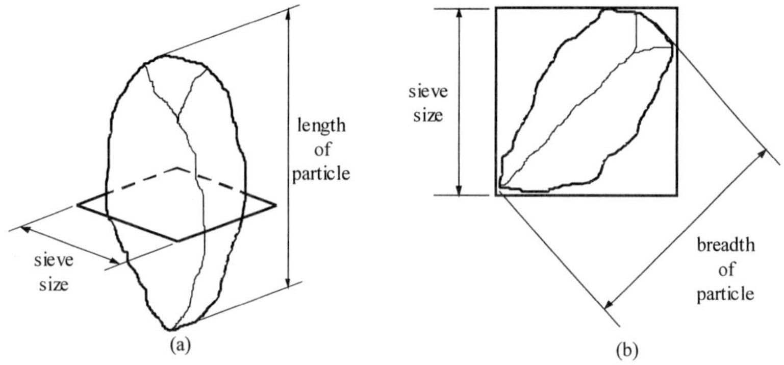

Figure 2 (a) An elongated particle passing through a square sieve aperture.
(b) Plan view of a flaky particle passing through a square sieve aperture.

PARTICLE SHAPE ANALYSIS BY MANUAL MEASUREMENT

Elongation and flakiness indexes are the most commonly used measures for gauging the general morphology of an aggregate sample. The procedures of their measurement [9] are based on rather arbitrary definitions that: (i) a particle is elongated when its length is larger than 1.8 times its mean sieve size, and (ii) a particle is flaky when its thickness is smaller than 0.6 times the mean sieve size.

Elongation and flakiness tests are performed after mechanical sieving, since they require the aggregate sample to be first separated into size fractions. The measurement of elongation index consists of manually gauging the length of each particle against a standard gauge for the particular size fraction and separating the elongated particles from each size fraction. Once all the fractions are tested, the quantity of elongated particles is measured by weighing and the elongation index is calculated as the ratio of the mass of elongated particles to the total mass of the sample expressed as a percentage. In similar way as described above, the flakiness index measurement is performed. The flakiness index is the percentage by mass of the aggregate particles that are regarded as flaky.

PARTICLE SIZE AND SHAPE ANALYSIS BY DIP

For comparison with the mechanical sieving and manual measurement results, the aggregate samples are first analyzed by the conventional methods and then by DIP. The schematic representation of the DIP setup is shown in Figure 1. When the aggregate particles are placed into the sample tray, they are carefully spread out so that they are not touching or overlapping each other or falling out of the boundary of the measurement area. After the aggregate particles have been properly positioned, the sample tray is placed on the photographic stand under the camera. If the captured image, which can be seen on the computer screen, is checked to be satisfactory, image analysis will be performed.

Having acquired a pixel image of the aggregate particles, image processing is performed to isolate the aggregate particles from the background. From the isolated aggregate particles, the following morphological parameters are measured: area, length and breadth of each particle. Area is defined as the projected area of the particle in its stable position. The length and breadth of a particle are defined as the length and breadth of the bounding rectangle that would enclose the particle area being analyzed, as shown in Figure 3.

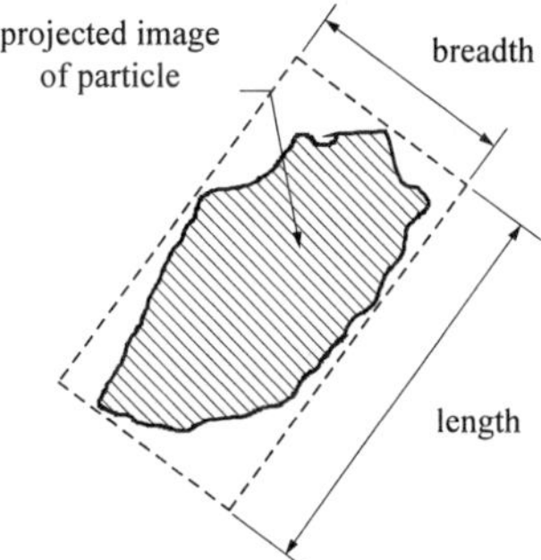

Figure 3 Definitions for length and breadth of a particle.

CONVERTING AREA GRADATION TO MASS GRADATION

In DIP, the volume or mass of the aggregate particles is not measured. In fact, since the image acquired is only a two-dimensional projection of the particles, even the thickness of the particles is not measured. Thus, aggregate fractions obtained by DIP can only be presented in terms of percentages by area of the aggregate on a stable horizontal surface (area gradation), as has been done by Yue and Morin [7].

On the other hand, in mechanical sieving, the amount of aggregate in each fraction is measured by weighing and hence the grading results are normally expressed as percentages by mass (mass gradation). Consequently, the DIP and mechanical sieving results are not directly comparable.

A simple method of converting the area gradation obtained by DIP to mass gradation so that the DIP results can be correlated to conventional mechanical sieving results and interpreted more easily has been proposed by the authors in a recent paper [10]. It is believed that aggregate particles from the same source have similar shape characteristics. Hence, the mean thickness of a particle may be estimated from the other dimensions of the particle as follows:

$$\text{mean thickness} = \lambda \times \text{breadth} \tag{1}$$

where λ is a parameter dependent on the flakiness of the particle. From this equation, the volume of the particle may be estimated by:

$$\text{volume} = \text{area} \times \text{mean thickness} = \lambda \times \text{area} \times \text{breadth} \tag{2}$$

Using this formula for the volume of a particle, the mass gradation is obtained as:

$$\%\text{ by mass passing a sieve} = \frac{\rho \times \lambda \times \sum_{i=1}^{p} (\text{area} \times \text{breadth})}{\rho \times \lambda \times \sum_{i=1}^{n} (\text{area} \times \text{breadth})} = \frac{\sum_{i=1}^{p} (\text{area} \times \text{breadth})}{\sum_{i=1}^{n} (\text{area} \times \text{breadth})} \tag{3}$$

where, ρ = density of aggregate

p = number of particles smaller than the sieve size

n = total number of particles in the sample

The values of λ and ρ are canceled out in the above equation and therefore the actual value of λ does not affect the mass gradation curve. Nevertheless, λ has a physical meaning and its value may be determined by:

$$\lambda = \frac{M}{\rho \times \sum_{i=1}^{n} (\text{area} \times \text{breadth})} \tag{4}$$

in which M is the total mass of the sample measured by weighing.

Having converted the area gradation to mass gradation, the grading results obtained by DIP can then be compared to those by mechanical sieving on the same grading chart, as shown in Figure 4(a).

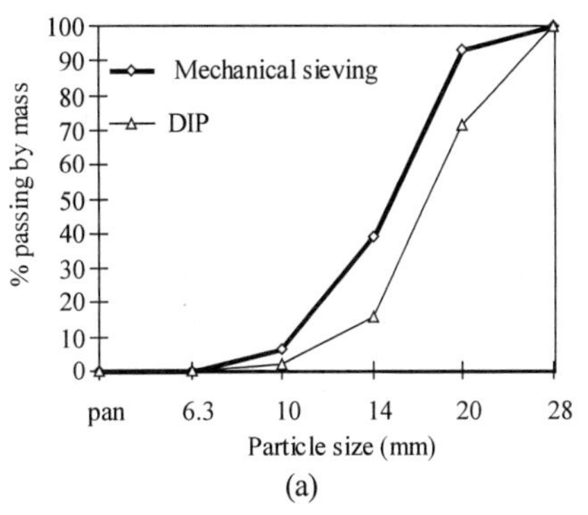

(a)

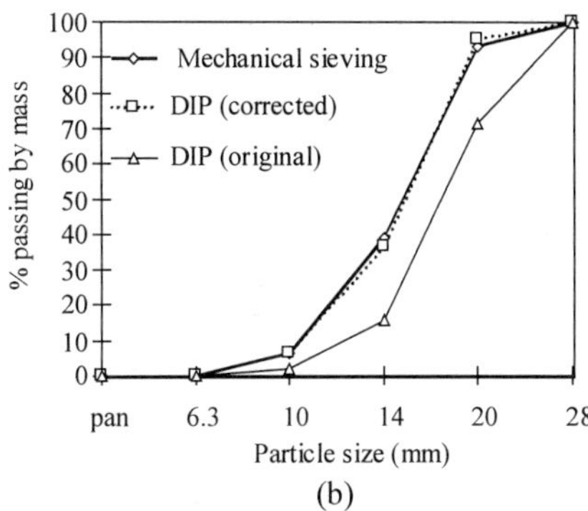

(b)

Figure 4 Grading curves of 20 mm granite aggregate: (a) before size correction and (b) after size correction

COMPARISON OF RESULTS

In mechanical sieving, the sizes of the particles are measured in terms of the sizes of the sieves that the particles pass through or are retained on. The particle sizes so determined are not the same as the breadth of the particles measured by DIP. In fact, the particle size measurement obtained by mechanical sieving depends also on the shape of the apertures of the sieves used. If the sieve apertures are circular, then the size of the sieve (diameter of the sieve apertures) that a particle can just pass through is very close to the breadth of the particle. But, if the sieve apertures are square, then the size of the sieve (width of the sieve aperture) that a particle can just pass through is generally only about 0.8 times the breadth of the particle. Hence, sieves with circular and square apertures give different particle size measurement results. According to Bernhardt [11], size measurement results obtained by sieves with circular apertures may be converted to equivalent square sieve sizes by:

$$\text{equivalent square sieve size} = C \times \text{circular sieve size} \tag{5}$$

where C is a size correction factor whose value is dependent on the shape of the cross-section of the particle but is generally within the range of 0.7 to 0.9. It is postulated herein that the breadth measured by DIP may also be converted to equivalent square sieve size using such a correction factor, as given below:

$$\text{equivalent square sieve size} = C \times \text{breadth} \tag{6}$$

Since the value of C depends on particle shape, it has to be determined for each type and source of aggregate. A trial and error process is needed to determine the value of C that would give the best agreement between the DIP and mechanical sieving results.

The values of C so determined for different aggregate samples together with the corresponding values of λ are given in Table 1. As expected, C falls between the range of 0.7 and 0.9.

After the correction factor is applied, the DIP sieve curve almost overlaps the mechanical sieve curve as can be seen from Figure 4(b). This indicates good agreement between the DIP and mechanical sieving results.

C is plotted against λ in Figure 5. Since both C and λ depend on the flakiness of the aggregate particles, they are inter-related. Using this graph, the value of C may be estimated from the corresponding value of λ.

Table 1 Size correction factor C and corresponding value of λ

AGGREGATE TYPE	AVERAGE OF 5 SAMPLES	
	C	λ
20 mm granite	0.82	0.32
10 mm granite	0.84	0.35
15 mm gravel	0.89	0.48
20 mm volcanic	0.81	0.31
10 mm volcanic	0.82	0.27

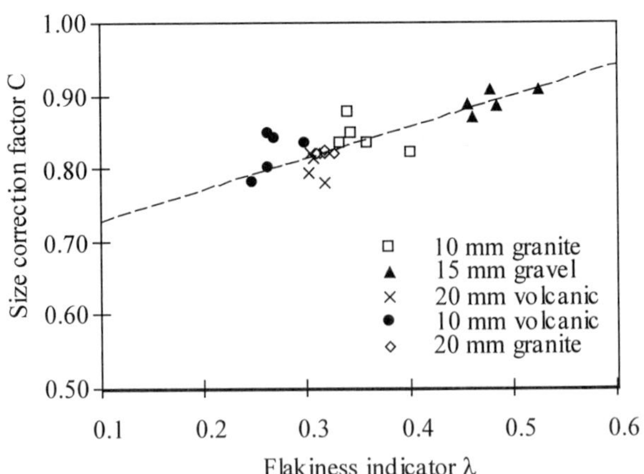

Figure 5 Relationship between C and λ

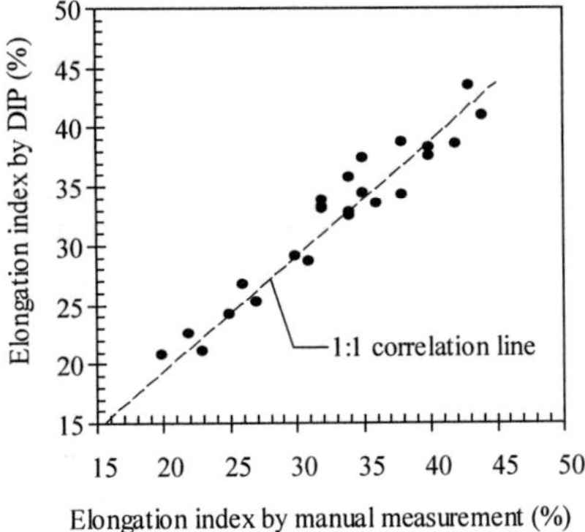

Figure 6 Elongation index results obtained by DIP and manual measurement.

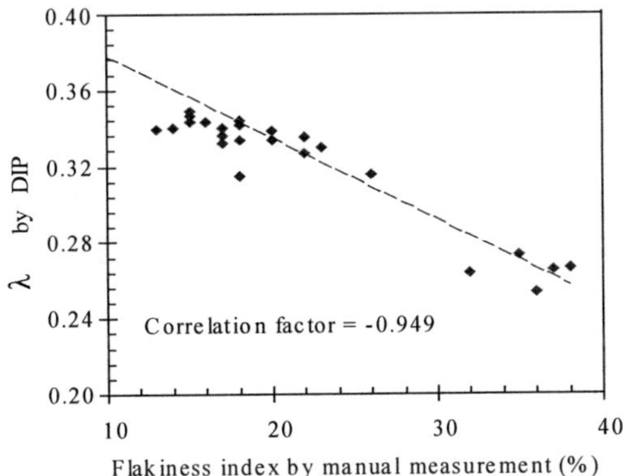

Figure 7 Comparison of λ to flakiness index obtained by manual measurement.

After converting to mass gradation and correcting the measured size to equivalent square sieve size, the elongation index obtained by DIP is compared to the corresponding value by manual measurement in Figure 6. Very good agreement between the results obtained by the two methods has been achieved.

The DIP method is unable to yield flakiness index directly. Nevertheless, there is a strong correlation between λ and the flakiness index obtained by manual measurement as shown in Figure 7, and thus λ may be taken as a measure of flakiness in its own right.

DISCUSSIONS

The DIP results have been converted to equivalent square sieve sizes and elongation indexes solely for the purpose of comparing with the results obtained by conventional methods. This helps to verify the accuracy of the DIP method.

In actual fact, the DIP results could have been interpreted directly without conversion to the conventional form. The breadth measured by DIP is a good measure of size in its own right. There is, strictly, no need to change breadth to equivalent square sieve size before plotting the grading curve. Moreover, with the use of DIP, it is no longer necessary to divide the particles into size fractions, since the size of each particle is measured. The size measurement results obtained by DIP can actually be plotted as a continuous grading curve.

DIP also has the advantage that it gives more information about the shape characteristics of the aggregate particles than the conventional elongation and flakiness tests. Theoretically, it should be possible to measure also the angularity and surface roughness from the DIP output. More research in this direction is recommended. In fact, the present definitions of elongation and flakiness indexes are designed to suit manual measurement. The length to breadth and mean thickness to breadth ratios which are directly obtainable from DIP output may be better measures of elongation and flakiness. As DIP becomes more popular, we may need to re-define the size and shape parameters.

CONCLUSIONS

1. A method for analyzing the size distribution and shape characteristics of coarse aggregate particles by digital image processing has been developed. Unlike other DIP methods which yield results only in terms of area gradation, the method developed herein has the capability of converting area gradation to mass gradation, provided all particles of the aggregate sample are from the same source.

2. After converting to mass gradation and equivalent square sieve sizes, the size and shape measurement results obtained by DIP are compared to those by conventional mechanical means. Very good agreement has been achieved, thus verifying the accuracy of the proposed DIP method.

3. Apart from being fast, convenient and versatile, the DIP method also has the advantage that it yields a lot more information than the mechanical procedures. There is the possibility that the DIP method would eventually replace the mechanical sieving method. In the long run, the size and shape parameters may need to be re-defined to suit.

ACKNOWLEDGMENTS

Part of the funding for the research work presented herein was provided by the Croucher Foundation of Hong Kong, whose kind support is gratefully acknowledged.

REFERENCES

1. IEEE. Proceedings of the 2nd IEEE Workshop on Applications of Computer Vision. Sarasota, December, 1994, 298pp.

2. MORA, C. F. AND KWAN, A. K. H. Applications of digital image processing in civil engineering. Proceedings of the XVI Jornadas Argentinas de Ingeniería Estructural, A. I. E., Buenos Aires, Argentina, September, 1998, published on CD ROM.

3. KWAN, A. K. H. AND CHAN, H. C. Applications of digital image processing technique to concrete technology. Proceedings of the Academic Exchange Workshop on Civil and Structural Engineering between Universities in Mainland China, Taiwan and Hong Kong, July, 1997, pp 100-110.

4. LEE, H. AND CHOU, E. Survey of image processing applications in civil engineering. Digital Image Processing: Techniques and Applications in Civil Engineering. A.S.C.E. Proceedings of the EF/NSF Conference, Hawaii, March, 1993, pp 203-210.

5. BARKSDALE, R. D., KEMP, M. A., SHEFFIELD, W. J. AND HUBBARD, J. L. Measurement of aggregate shape, surface, area and roughness. Transportation Research Record 1301, National Research Council, Washington D.C., 1991, pp107-116.

6. LI, L., CHAN, P., ZOLLINGER, D. G. AND LYTTON, R. L. Quantitative analysis of aggregate shape based on fractals. American Concrete Institute Materials Journal, Vol.90, No.4, July-August, 1993, pp 357-365.

7. YUE, Z. Q. AND MORIN, I. Digital image processing for aggregate orientation in asphalt concrete mixtures. Canadian Journal of Civil Engineering, Vol.23, 1996, pp 480-489.

8. KUO, C. Y., FROST, J. D., LAI, J. S. AND WANG, L. B. Three-dimensional image analysis of aggregate particles from orthogonal projections. Transportation Research Council, Washington D. C., 1996, pp 98-103.

9. NEVILLE, A. M. Properties of Concrete, 3 ed. Logman, 1988, 779 pp.

10. MORA, C. F., KWAN, A. K. H. AND CHAN, H. C. Particle size distribution analysis of coarse aggregate using digital image processing. Cement and Concrete Research, Vol. 28, No. 6, 1998, pp921-932.

11. BERNHARDT, C. Particle Size Analysis: Classification and Sedimentation Methods, Chapman & Hall, 1994, 428 pp.

INFLUENCE OF SILICA FUME AND DIFFERENT AGGREGATES ON THE ABRASION OF CONCRETE

A A Ramezanianpour

M R Ekhlassi

Amirkabir University of Technology

Iran

A Miyamoto

Yamaguchi University

Japan

ABSTRACT. Many concrete structures such as canals, spillways, industrial floors, and pavements are subjected to wear. It is very important to make a durable concrete which can resist against abrasion and erosion. The use of appropriate aggregate, low water cement ratio concrete mixtures, pozzolanic materials, finishing techniques and proper curing has shown a better performance concrete when exposed to abrasive conditions.

In this research the abrasion resistance of different concrete mixtures was assessed. Concrete mixtures were prepared with different types of aggregates and various percentages of silica fume. The abrasion resistance of concrete specimens were measured by water sandblasting method.

Results of the laboratory investigation show that the type of aggregate is the most significant factor enhancing the abrasion resistance of concretes. The use of silica fume as a cement replacement material improved the abrasion resistance of all concrete mixtures.

Keywords: Abrasion resistance, Aggregate, Silica fume, Compressive strength, Water sandblast.

A A Ramezanianpour is professor of concrete technology at the Amirkabir University of Technology and advisor for the BHRC, Tehran, Iran. He has been active in concrete research related with the durability for more than twenty years and has published 20 books and 70 papers. He is a member of ACI and IRCOLD.

M R Ekhlassi is a research fellow in civil Engineering, University of Amirkabir.

A Miyamoto is professor of concrete structures at the Yamaguchi University, Japan.

INTRODUCTION

Abrasion erosion damage results from the abrasive effects of waterborne silt, sand, gravel, rocks and other debris being circulated over a concrete surface during operation of a hydraulic structure. Spillway aprons, stilling basins, sluice ways and tunnel linings are particularly susceptible to abrasion erosion [1].

Concrete abrasion resistance is markedly influenced by a number of factors, including concrete strength, aggregate properties, surface finishing, and types of hardeners or toppings [2].

Apart from the aggregate properties and low water cement ratio etc, the use of silica fume can improve the abrasion resistance of concrete.

Numerous studies [3-6] have shown that concrete abrasion resistance is primarily dependent on the compressive strength of concrete; and some of studies [5,6] have indicated that compressive strength is the most important factor governing the abrasion resistance of concrete.

Regarding the test methods, researchers confirm that there is no simple test equipment to simulate all the actions to which a concrete floor may be exposed during service [7].

In this investigation the abrasion resistance of concretes containing various percentages of silica fume and different types of aggregates was evaluated. The compressive strength of concrete specimens was determined at various ages. For the assessment of abrasion resistance of concrete mixtures, the water-sandblasting method which is nearly similar to ASTM-C418 method was used.

TEST PROGRAM

Concrete Mixtures Incorporating Silica Fume

Concrete materials

Cement: ASTM Type 1 Portland cement was used. The chemical analysis of which is given in Table 1.

Silica fume: A newly produced silica fume from an Iranian source was incorporated in this study. The chemical analysis of the silica fume is presented in Table 1. Aggregates: The fine and coarse aggregates were local natural sand and crushed siliceous limestone, respectively.Superplasticizer: A sulphonated, melamine-formaldehyde condensate superplasticizer of German origin was used.

Mixture proportions

The concrete mixture proportions are given in Table 2. The design compressive strength of the control concrete mixture was 35 N/mm^2. Silica fume concrete mixtures were made in which 2,4,6,8 and 10 percent of the cement was replaced by silica fume. With the addition of a superplasticizer to some concrete mixtures, the water cementitious materials ratio was kept constant at 0.48 for all concrete mixtures. Attempt was made to produce concretes with a slump in the range of 60 to 100 mm.

Table 1 Chemical analysis of cementitious materials

	SiO_2 (%)	Al_2O_3 (%)	Fe_2O_3 (%)	Ca (%)	MgO (%)	SO_3 (%)	Na_2O (%)	K_2O (%)	L.O.I (%)
Type1 Cement	20	5	2.5	62.5	2.2	3.5	0.2	0.8	2.2
Silica fume	91.2	1	0.9	1.6	1.8	0.8	0.2	0.6	2

Table 2 Proportioning of concrete mixtures with various percentages of silica fume

MIX	CEMENT (kg)	AGGREGATE (kg)	WATER (kg)	SILICA FUME (kg)	$\frac{w}{c+scm}$	SP (l)	SLUMP (mm)
SF0	450	1710	215	0	0.48	0	60
SF2	444	1710	215	9	0.48	0	60
SF4	435	1710	215	18	0.48	0	60
SF6	425	1710	215	27	0.48	3	75
SF8	417	1710	215	36	0.48	8	95
SF10	408	1710	215	45	0.48	9	90

Casting and curing of test specimens

The test specimens were 150mm cubes both for compressive strength test and for the abrasion resistance test using water-sandblasting method.After casting, the moulded specimens were covered with water-saturated burlap and left in the casting room at 20± 2 ° C for 24 hours. Then they were demoulded and cured in fog room until the test ages.

Testing of specimens

The concrete specimens were tested in compression at 3,7,28 and 120 days. The test for abrasion resistance was carried out with water-sandblasting method nearly similar to ASTM. C418. In water-sandblasting method instead of air, water was used with sand to produce wear action. This test was carried out for 40 day specimens. The pressure on the nozzle was kept at 200 atm and the distance between nozzle and the surface of specimen was 75 mm.

Concrete Mixtures Containing Different Types of Aggregates

Concrete materials

Cement: ASTM Type 2 Portland cement was used. Aggregates: The fine and coarse aggregates were combination of conventional, granite and siliceous aggregates. Superplasticizer: A sulphonated, melamine-formaldehyde condensate superplasticizer of German origin was used.

Mixture proportions

The concrete mixture proportions are given in Table 3. The design compressive strength of the control concrete mixture was 30 N/mm^2. The granite and siliceous aggregates were mixed with conventional aggregates at the percentages of 0,40,60 and 100 to make 7 series of mixtures. With the addition of a superplasticizer to concrete mixtures, the water cement ratio was kept constant at 0.3 for all concrete mixes. The slump was in the range of 30 to 60 mm.

Casting and curing was as in the previous section.

Testing of specimens

The concrete specimens were tested in compression at 7 and 28 days. The test for abrasion resistance was carried out with water-sandblasting method. In this method, the pressure on the nozzle was kept at 150 atm and the distance between nozzle and the surface of specimen was 75 mm. The rate of flow of abrasive sand was 7200 ± 100 g/min.

Table 3 Proportioning of concrete mixtures containing different types of aggregates

MIX	AGG COMBINATION	CEMENT (kg)	AGGREGATE (kg)	WATER (kg)	$\frac{w}{c}$	SP (%)	SLUMP (mm)
C	%100 Conv.	400	1884	120	0.3	0.5	50
C6G4	%60 Con.+ %40 granite	400	1966	120	0.3	0.7	40
C4G6	%40 Conv.+ %60 granite	400	2008	120	0.3	0.7	40
G	%100 granite	400	2090	120	0.3	0.7	40
C6S4	%60 Conv.+ %40 siliceous	400	1926	120	0.3	0.7	40
C4S6	%40 Conv.+ %60 Siliceous	400	1946	120	0.3	0.7	40
S	%100 siliceous	400	1988	120	0.3	0.7	40

RESULTS AND DISCUSSION

Concrete Mixtures Incorporating Silica Fume

Compressive strength

The strength development of the control concrete mixture and mixtures containing silica fume are shown in Table 4 and Figure .1. The strength of concrete mixtures with and without silica fume increases at nearly the same rate with age up to 7 days . At longer ages concrete mixtures containing more silica fume show higher strength gain. This is clearly shown in Figure 1 where the concrete mixture incorporating 10 percent silica fume has the highest strength value at 120 days. The compressive strength of the concrete mixture with 10 percent silica fume is approximately 42 percent higher than that of the control concrete.

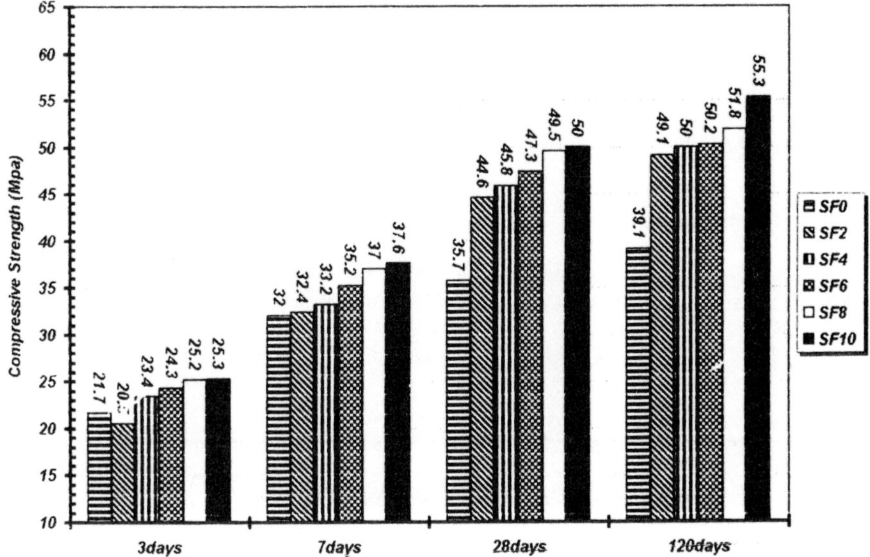

Figure 1 Compressive strength of concrete mixes with various percentages of silica fume

Table 4 Compressive strength of concrete mixtures incorporating silica fume

MIXTURE	SILICA FUME (%)	COMPRESSIVE STRENGTH (Mpa)			
		3 Days	7 Days	28 Days	120 Days
SF0	0	21.7	32.0	35.7	39.1
SF2	2	20.5	32.4	44.6	49.1
SF4	4	23.4	33.2	45.8	50.0
SF6	6	24.3	35.2	47.3	50.2
SF8	8	25.2	37.0	49.5	51.8
SF10	10	25.3	37.6	50.0	55.3

Abrasion resistance

The results of the resistance of concrete to abrasion using water-sandblasting method are given in Table 5 and Figure 2. It can be seen that concrete with higher replacement levels of the cement content by silica fume induced an decrease in the abrasion wear. However, this reduction in abrasion wear was only 12.5 percent when compared to the control concrete. The increase in abrasion resistance is attributed to the compressive strength and the transition zone of the concretes incorporating silica fume.

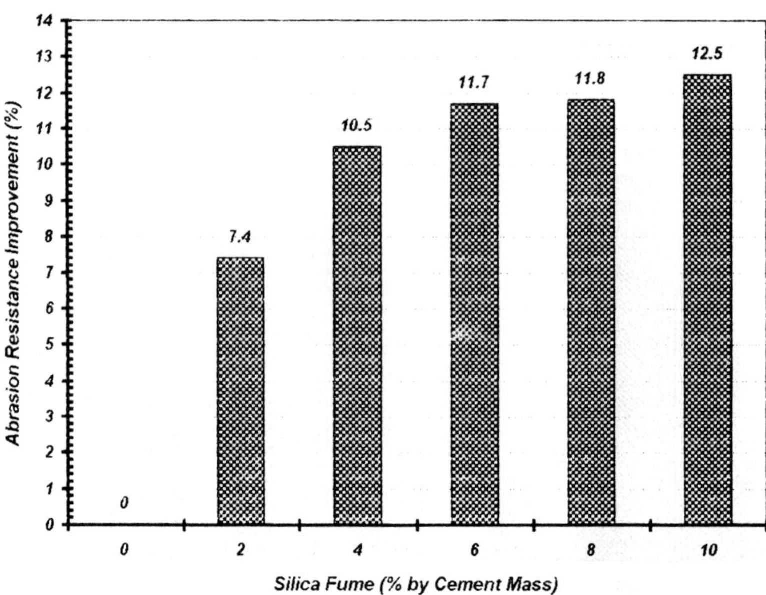

Figure 2 Abrasion resistance of concrete mixes with various percentages of silica fume

Table 5 Abrasion resistance of concrete mixtures incorporating silica fume

MIXTURE	SILICA FUME (%)	ABRASION RESISTANCE IMPROVEMENT (%)
SF0	0	0
SF2	2	7.4
SF4	4	10.5
SF6	6	11.7
SF8	8	11.8
SF10	10	12.5

Concrete Mixtures Containing Different Types of Aggregates

Compressive strength

The strength development characteristics of the concrete mixture, with different types of aggregates are shown in Table 6 and Figure 3. The concrete mixtures containing more granite aggregates show higher strength at 7 and 28 days. this is clearly shown in Figure 5 where the concrete mixture incorporating 100 percent granite aggregates has the highest strength value at any ages. Concrete mixtures containing siliceous aggregates have not shown any strength gain. (see Figure 3).

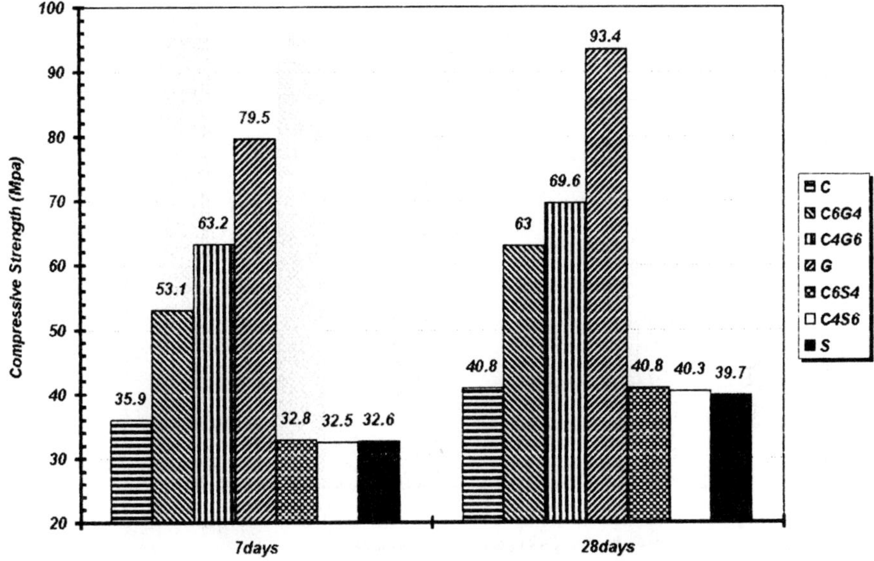

Figure 3 Compressive strength of concrete mixes with different types of aggregates

Table 6 Compressive strength of concrete mixtures containing different types of aggregates

MIXTURE	COMPRESSIVE STRENGTH (MPa)	
	7 days	28days
C	35.9	40.8
C6G4	53.1	63.0
C4G6	63.2	69.9
G	79.5	93.4
C6S4	32.8	40.8
C6S4	32.8	40.8
C4S6	32.5	40.3
S	32.6	39.7

Abrasion resistance

The results of the abrasion resistance of concretes using water-sandblasting method are given in Table 7 and Figure.4. In Figure.4, it can be clearly seen that the concretes with higher percentages of granite aggregates give higher abrasion resistance and lower abrasion wear. Abrasion resistance improvement in the mixture G is 35 percent and in the mixture S is 24.7 percent. It is also shown that although the compressive strength gain of the mixture S is insignificant but the abrasion resistance of this mixture obtained 24.7 percent improvement.

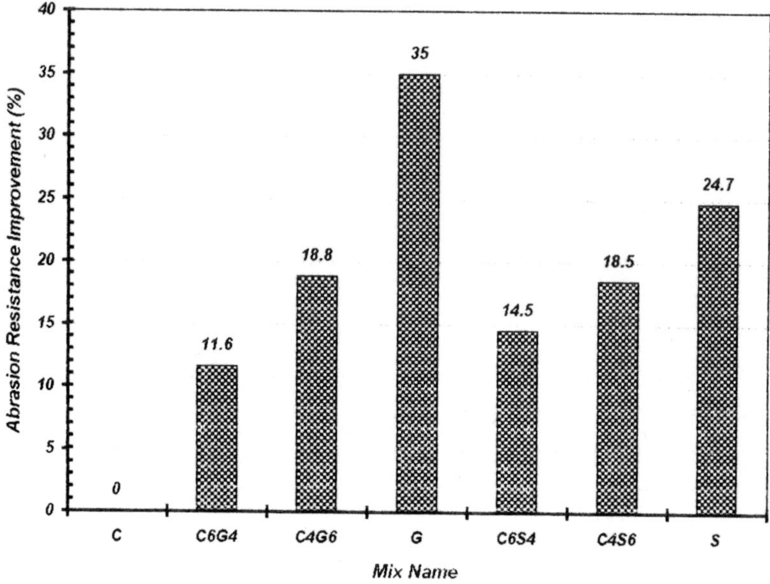

Figure 4 Abrasion resistance of concrete mixes with different aggregate types

Table 7 Abrasion resistance of concrete mixtures with different types of aggregates

MIXTURE	ABRASION RESISTANCE IMPROVEMENT (%)
C	0
C6G4	11.6
C4G6	18.8
G	35.0
C6S4	14.5
C4S6	18.5
S	24.7

CONCLUSIONS

For the concrete mixtures investigated, the following conclusions may be drawn.

1. Comparison of the results show that the type of aggregate is the most important factor affecting concrete abrasion resistance measuring by water-sandblasting method.

2. The highest abrasion resistance was observed in the mixture with 100 percent granite aggregate.

3. In silica fume concretes, the highest abrasion resistance was obtained in the mixture with 10 percent silica fume.

4. The water-sandblasting method is suitable for evaluating the relative resistance of concrete surfaces subjected to abrasive of waterborne particles.

5. The abrasion resistance of all concrete mixtures increases by increasing their compressive strength. However the increase in the abrasion resistance will not necessarily increase the compressive strength of concrete mixtures.

REFERENCES

1. ACI COMMITTE 210, “Erosion of Concrete in Hydraulic Structures”, A 210 R-87; ACI Materials Journal March-April 1987; 136-157.

2. NAIK, R, SINGH, S AND HOSSAIN, M. “Abrasion Resistance of High-Strength Concrete Made With Class C Fly Ash.” ACI Materials Journal, Nov-Dec 1995, PP 649-659.

3. ACI COMMITTE 201, “Guide to Durable Concrete”, ACI Manual of Concrete Practice 1988, ACI 201.2R-77

4. MEHTA, P K, “Concrete Structure, Properties and Materials”. Prentice Hall International, Series in Civil Engineering and Engineering Mechanics, 3rd Edition, 1986, 450 PP.

5. HADCHTI, K M, AND CARRASQUILLA, R L. “Abrasion Resistance and Scaling Resistance of Concrete Containing Fly Ash,” Research Report 481-3, Center for Transportation Research, Bureau of Engineering Research, University of Texas at Austin, Aug 1988, PP 185.

6. LAPLANTE, P, AIFCIN, P C, AND VEZIND. “Abrasion Resistance of Concrete”, Journal of Materials in Civil Engineering, V.3, No.1, Feb 1991, PP 19-30.

7. KETTLE, R, SADEGZADEH, M. “The Influence of Construction Procedures on Abrasion Resistance”, ACI-SP-100, Concrete Durability, 1987, Vol.2, PP 1385-1410.

STRENGTH COMPARISONS BETWEEN ROLLED SAND CONCRETE AND DUNE SAND CONCRETE

A Guettala

B Mezghiche

R Chebili

University of Biskra

Algeria

ABSTRACT. The objective of using sand concrete is to reduce costs especially in sand rich regions where transportation costs of aggregates over long distances makes any construction too expensive. This paper presents the mechanical characteristics of two types of sand concrete, a sand concrete based on rolled sand and a dune sand concrete, and compares these two types of sand concretes. It is also intended to determine the influence of the water/cement (0.4, 0.45, 0.5), cement/ sand (1/3, 1/4) and additives (plasticizers and thinners) on the mechanical characteristics: Compression and tension. It is observed that the cube strength rises with respect to time for the two types of sand concrete. The cube strength diminishes when increasing the water/ cement for rolled sand concrete and increases for the dune sand concrete. The influence of additives is quite noticeable; the rise in the cube strength is about 40 %. Taking into consideration these results, it is then recommended for the users the exploitation of rolled sand concrete with or without additives, however for the dune sand concrete, the use of additives is most important.

Keywords: Rolled sand concrete, Dune sand concrete, Comparison, Additives, Cube strength.

Mr A Guettala is Director of the Civil Engineering Institute, Lecturer, University of Biskra, (Algeria). He specialises in the permeation properties of concrete, and has performed researches on the durability of construction materials.

Dr B Mezghiche is a Lecturer in Construction Materials, Civil Engineering Institute, and Director of the Concrete Technolgiy Institute, University of Biskra (Algeria). He specialises in the use of binders and durability of condrete, published researches on Binders.

Dr R Chebili is The President of the Scientific Committee of the Architecture and Civil Engineering Institute, Lecturer, University of Biskra (Algeria). He specialises in materials and structural mechanics.

INTRODUCTION

Sand concrete is a construction material composed of sand, cement and a natural or industrial Filler. Sand concrete can replace successfully the traditional concrete because of its economical cost, its compressive strength that reaches (12-80 N/mm^2) [3], [4], and its high workability. Moreover, it can successfully be used where the reinforcement is overcrowded or where it is desired to have concrete surfaces that present a good appearance on removal of the formwork. It can also be used to produce bricks for construction. Sand concrete has been used for the first time in the third quarter of the nineteenth century by F. Coignet to construct the bearing wall Passy and F Coignet house in Saint-Denis in France. It has also been used in U.S.S.R in the beginning of the twentieth century to construct Kaliningrad harbor and Chernavskif bridge [5].

In Algeria where desert is about 80% of the total area, transportation of aggregates over long distances for construction can cause excessive cost to use the traditional concrete, also the abundance of sand in all this region can encourage use of sand concrete, mainly in Biskra (South East of Algeria) where the present research has been conducted.

This research is carried out in order to find the best content of sand concrete that can give a good workability and high compressive strength by using two different types of sand; rolled river sand and dune sand with cement of type C.P.A. 325 (Ordinary Portland Cement). Moreover, it is intended to determine the influence of water / cement (0.4, 0.45, 0.5), cement/sand (1/3,1/4) and lastly the influence of additives (plasticizers and thinners) on the mechanical characteristics : compression and tension. Finally results are compared for different type of sand.

CHARACTERISTICS OF USED MATERIALS

Sand

In the present study, two types of sand have been chosen (rolled river sand and dune sand). The later is highly abundant in the region of Biskra (South east of Algeria).

The density: The specific density is computed by using the pyknometre apparatus on a dry samples using sieves of 2 mm (ASTM D 845 5) and the apparent density is computed by using (ASTM C 71-29). The results are presented in Table, [1].

Table 1 The density of sand

TYPE OF SAND	ABSOLUTE DENSITY γs (Kg/m^3)	APPARENT DENSITY ρ (Kg/m^3)
River sand	2500	1550
Dune sand	2570	1500

Sand grading is performed using standard sieves (ASTM D422-63). Figure 1, shows curves representing the aggregate analysis of the different types of sands (rolled river sand and dune sand). It can be seen that river sand is gradually distributed, whereas dune sand is very fine.

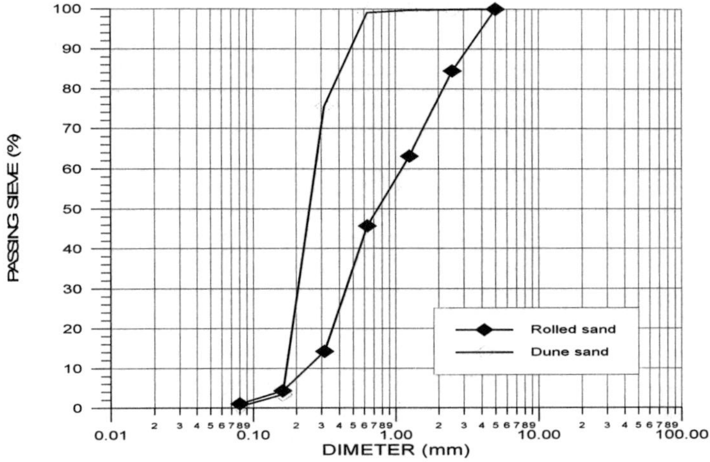

Figure 1 Aggregate analysis curves (rolled river sand and dune sand)

The sand equivalent values have been computed using the scale (NF P186598), and test results are shown in Table 2.

Table 2 Sand equivalent values

SAMPLE	SAND EQUIVALENT VALUE BY SIGHT	SAND EQUIVALENT VALUE BY TEST	SAND QUALITY
River sand	75.5	79.43	Clean sand
Dune sand	67.11	58.65	Dusty sand

Additives

The additives are used in small quantities not greater than 5% by weight of used cement in order to improve some of the characteristics of concrete. Some of these additives are produced by Granitex company in Algeria, among which we can mention plasticizers and thinners.

- Plasticizers are liquids that can easily be mixed in water with all kinds of cements. Their colour is brown and they have a density equal to 1.16 with pH = (7-8). The use of Plasticizers can improve the properties of sand concrete mixes by: reducing water mixes, increasing the cube strength, increasing the plasticity of concrete, produces a good workability and lastly reduces the time for setting.

- Thinners can improve the sand concrete characteristics by being stable and cohesive during handling and vibration. They increase the workability, the slump and the cube strength.

Cement

An ordinary Portland Cement manufactured in Algeria under the commercial name C.P.A. 325 was used, and tested to obtain the real strength using AFNOR recommendations [1]. The compressive strength for 28 days was tested and found to be equal to (388 N/mm^2).

QUALITY CONTROL

The quality control for workability of sand concrete has been conducted by using the Out Flow -Test [4]. The results are summarized in Table 3, function of the quality of sand the cement/ sand and the water / cement.

Table 3 Sand Concrete Workability

CEMENT / SAND	CONCRETE	WATER/ CEMENT	TYPE OF CONCRETE IN TERMS OF WORKABILITY		
			Without additives	1% plasticizers	1% thinners
1/3	River sand	0.4	Very cohesive concrete	very cohesive concrete	very cohesive concrete
		0.45	very cohesive concrete	cohesive concrete	plastic concrete
		0.5	Plastic concrete	mobile concrete	very mobile concrete
	Dune sand	0.4	very cohesive concrete	very cohesive concrete	very cohesive concrete
		0.45	very cohesive concrete	very cohesive concrete	very cohesive concrete
		0.5	very cohesive concrete	very cohesive concrete	very cohesive concrete
1/4	River sand	0.4	very cohesive concrete	very cohesive concrete	very cohesive concrete
		0.45	very cohesive concrete	cohesive concrete	Plastic cohesive concrete
		0.5	very cohesive concrete	cohesive concrete	Mobile concrete
	Dune sand	0.4	very cohesive concrete	very cohesive concrete	very cohesive concrete
		0.45	very cohesive concrete	cohesive concrete	cohesive concrete
		0.5	cohesive concrete	cohesive concrete	cohesive concrete

INFLUENCE OF CEMENT / SAND RATIO

Test results show that the cube strength increases as a function of time for all types of concrete sand, and it is also dependent on the percentage of cement/sand and the quality of sand. Moreover we can also conclude out of tests that the percentage of strength can easily reach 85% in river sand concrete case and 70 % in dune sand concrete. This is when the percentage of cement sand is 1/3. When the percentage of cement/sand is 1/4, the strength reaches respectively 75%, 65%. (These percentages are with respect to the strength during 28 days).

Cube Strength

In figures 2, 3, 4, 5, 6 and 7 the influence of water/cement is given against the cube strength. In these figures it is noticed that when this ratio is increased the cube strength is also increased in the case of dune sand. Whereas this later diminishes when it is the case of river sand, except when thinners are added with the cement/sand ratio 1/3, and then decrease again.

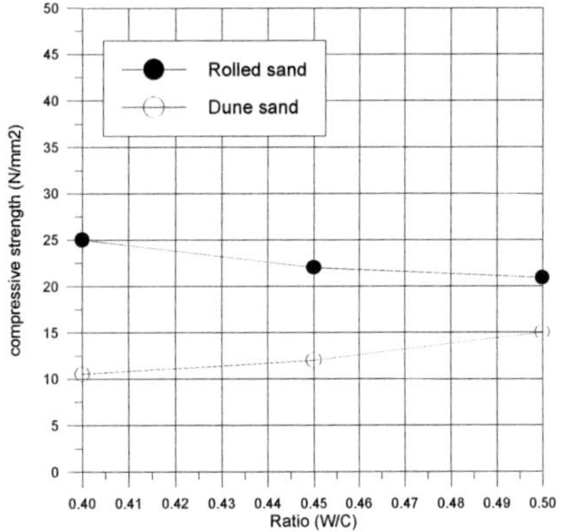

Figure 2 Influence of W/C on compressive strength with Cem/Sand = 1/4, no additives

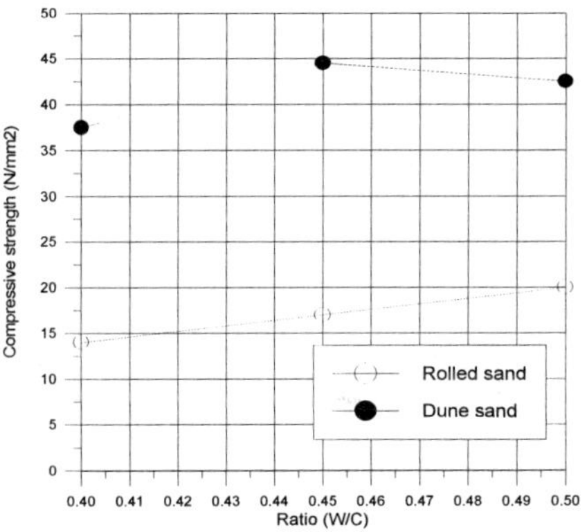

Figure 3 Influence of W/C on compressive strength with Cem/Sand = 1/3, no additives

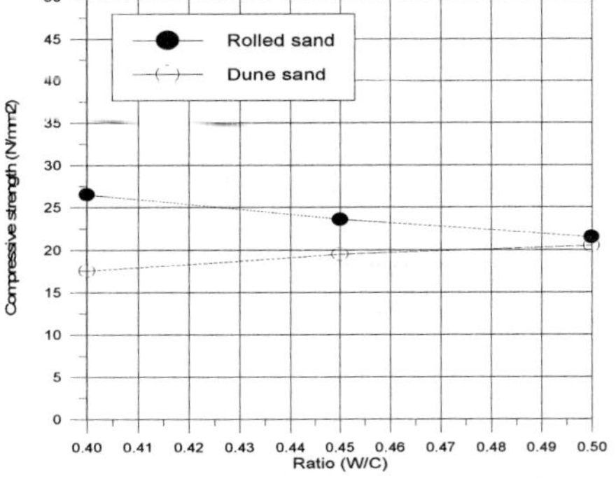

Figure 4 Influence of W/C on compressive strength with Cem/Sand = 1/4, 1% plasticiser

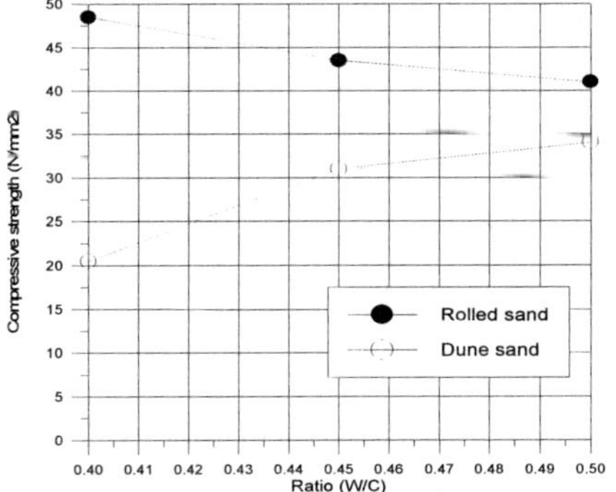

Figure 5 Influence of W/C on compressive strength with Cem/Sand = 1/3, 1% plasticiser

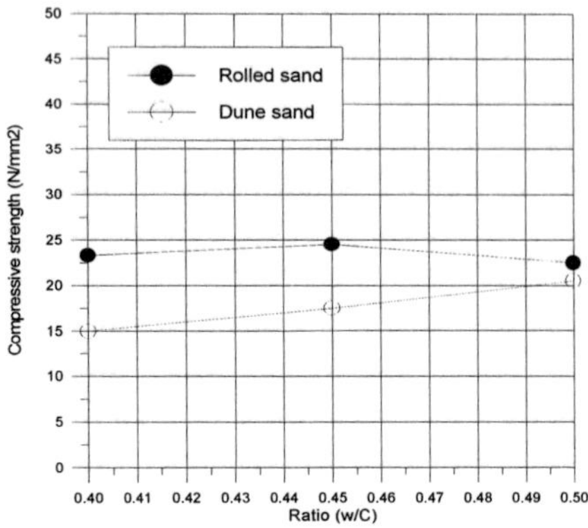

Figure 6 Influence of W/C on compressive strength with Cem/Sand = 1/4, 1% thinners

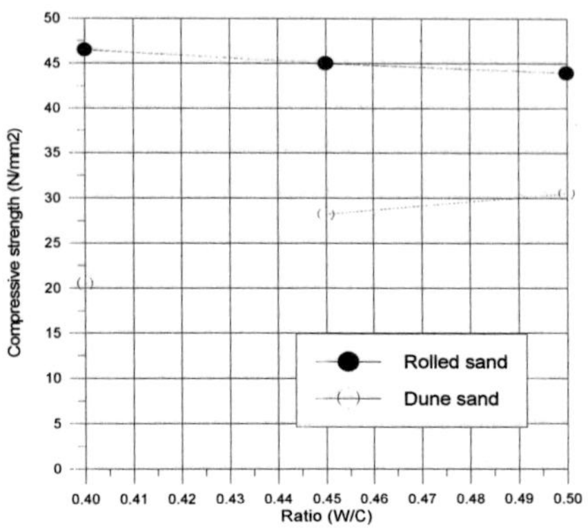

Figure 7 Influence of W/C on compressive strength with Cem/Sand = 1/3, 1% thinners

Strength in Tension (Flexure)

Tests studying the influence of water/cement ratio (0.4, 0.45 and 0.5) on the mechanical strength in tension shows that this later increases with increasing the (W/C) ratio in dune's sand concrete. This is apparently due to the fineness scale of sand = 1.22. Whereas in the case of river sand concrete the mechanical strength in tension decreases except when additives are not used, in this case the strength in tension increases then decreases because of the fineness of sand which is equal to 2.83. It is also noticed that for this case of sand, water is not needed in grater amount, that is way the strength decreases when additives are used, and increases then decreases again without additives, i.e. water/cement ratio = 0.45 is the optimal ratio for a better strength.

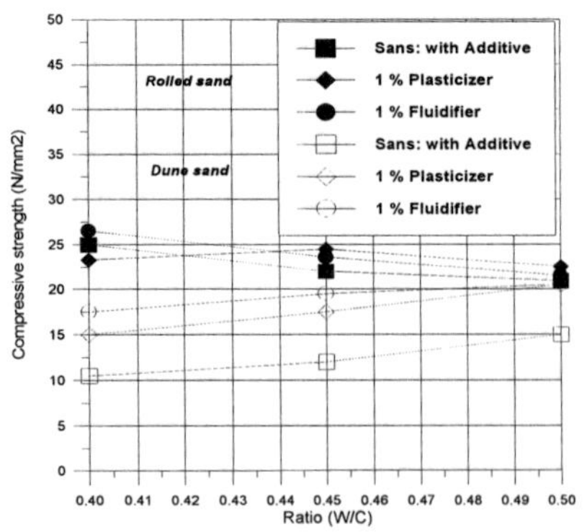

Figure 8 Influence of additives and W/C on compressive strength with Cem/Sand = 1/3

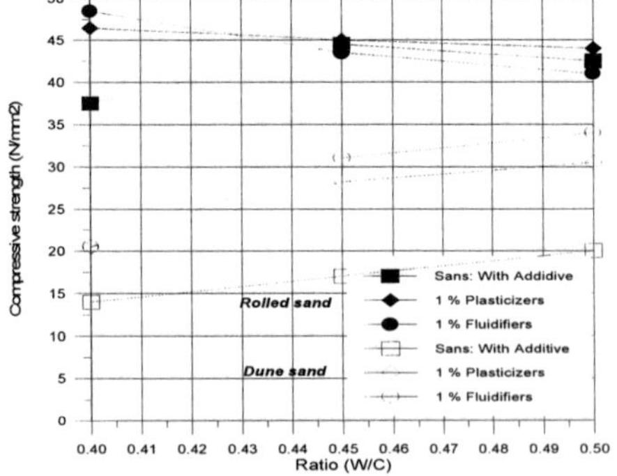

Figure 9 Influence of additives and W/C on compressive strength with Cem/Sand = 1/3

Influence of Additives

Concerning the effect of additives, the figures 8 and 9 show that the cube strength increases when 1% of additive is used (Thinners, Plasticizers), mainly in dune sand concrete. The percentage of the mechanical strength increases with 40% in case of cement/sand ratio = 1/4 and 50% in case of cement/sand ratio =1/3. For river sand concrete the effect of additive (Thinners) is almost negligible.

CONCLUSION

It has been shown through the experiments that sand concrete has almost equal strength to the ordinary concrete, mainly in river sand concrete. Increasing water/cement ratio increases the cube strength in dune sand concrete. It is almost the reverse of river sand concrete. The use of additives causes an increase in the strength up to 40% in dune sand concrete, while this increase is almost negligible in river sand concrete.

REFERENCES

1. AFNOR RECUEIL DE NORMES FRANÇAISE. Bâtiment béton et constituants de béton, Paris 1984.

2. ASTM. Annual Book of Standards, Philadelphia, 1968.

3. C.E.B.T.P. Synthese des Connaissances du Béton de Sable Operation 52 G 119 de Décembre, 1986.

4. CHANVILLARD, O. Basuyaux Une Méthode de Formulation des Bétons de Sable, Maniabilité et Résistance Fixées Bulletin des Laboratoires des Ponts et Chaussées No. 205, Septembre - Octobre 1996, Ref 4047, pp 49-63, Paris.

5. CHAUVIN J, J. Rapport Interne de Laboratoire Régional des Ponts et Chaussées de Bordeaux, Béton de Sable Ref. FAER 1.30.14.5 et 1.30. 24. 66, Janvier 1987.

6. LANCHON R. Cours de Laboratoire (1 et 2) Granulats Béton sols Editions Casteilla, Paris, 1988, pp 119.

THE ROLE OF WATER IN DETERMINING CONCRETE PERFORMANCE

P C Hewlett

British Board of Agrément

United Kingdom

ABSTRACT. Water is arguably the most important component in making durable concrete. It is, however, the least studied and understood at a fundamental level.

The paper reconsiders the role and nature of water and its disposition both initially and with time in concrete. The formation of residual porosity and its contribution to the various chemical processes resulting in beneficial and adverse changes are considered.

The relationship between degree of hydration and strength development particularly at low water:cement ratios is discussed along with the contribution from ancillary hydraulic binders.

If low water contents are used to begin with it is important that such levels are not reduced further during placing and curing. In this regard the newer water retaining admixtures are mentioned in the context of water's chemical manipulation.

Keywords: Water, Structure, Porosity, Durability, Strength, Binders, Admixtures

Professor Peter C Hewlett is the Director of the British Board of Agrément and Visiting Industrial Professor to the Department of Civil Engineering at the University of Dundee. He specialises in materials for construction and building both inorganic and organic and in particular of the enhanced performance of concrete by way of chemical modifications. He is President of the European Organisation for Technical Approvals and the UK Concrete Society.

INTRODUCTION

This paper is concerned with concrete performance, both initially and over time. It is assumed by way of good design and application that the resulting structure will be durable also. Design and construction should make the most of what is a well constituted and made material.

In Somerville's paper on engineering design and service life [1] it assumes a clear materials/systems specification can be given. Within the specification water is regarded as a macro component in the same way as the cement and aggregate. Is such a perception true?

Hydraulically active materials such as Portland cement, ggbs, pfa, microsilica and metakaolins depend upon water to produce the heavily hydrated structure building phases that change the cohesionless particulate mass to a solid. Water is vital. However, water is also responsible for causing accessible voidage in the hardened mass that permits self destructing degradation as well as bleeding, shrinkage all governing the quality of the concrete cover. These degradation processes are themselves dependent upon water that permit close chemical contact – a necessary precursor for chemical reaction to take place.

It is therefore somewhat perverse that water is both necessary and a nuisance. Water is also anomalous and perhaps the least understood component in concrete notwithstanding it comprises some 7-10% by concrete mass – 15-20% by volume. Anomalous in that the liquid itself has a higher density than its solid counterpart and an appreciable vapour pressure at normal ambient temperatures. Concrete made with soft water is stronger than that made with hard for the same cement content. These sentiments have been expressed before [2, 3].

Notwithstanding its significance, water remains relatively unstudied. However the advent of nuclear magnetic resonance based techniques may well permit a more scientific examination [4].

Water in concrete falls into three categories

1. Free water held by capillary forces. Such water is evaporable.
2. Adsorbed water held by surface forces. This water is also evaporable
3. Chemically bound water. This water is non evaporable

The presence of water whilst necessary during hydration and curing is, in part, responsible for later degradation reactions and may limit attempts to protect concrete using coatings and surface treatments [5].

For the moment let us assume that the concrete meets the designed for specification to begin with then we are concerned about retaining that performance and therefore regard has to be taken of the processes that cause concrete to deteriorate.

The issue of concrete deterioration is significant both technically and commercially. It has been estimated by Mays [6] that half a billion pounds is spent annually in the UK on the repair of concrete. Indeed corrosion costs Europe and the USA more than 3% of the GDP [6] with 70% of these losses being avoidable and 25% could be eliminated using existing technology.

The processes of deterioration are of course related but reciprocal to durability since deterioration means worsening or disintegrating. Durability on the other hand means 'lastingness, permanence and persistence' [7]. Both are controlled by the chemistry of the system.

Chemistry may be described 'as the branch of physical science concerned with composition, properties and reactions of substances' and a chemical as 'any substance used in or resulting from a reaction involving changes to atoms or molecules' [8].

The Issue of Scale

In linking together the chemistry with deterioration or durability we have to take regard of mechanisms and processes at a level of scale measured in nanometers (10^{-9}m or tens of angstroms) on the one hand whilst acknowledging that our practical concerns relate to scale measured in millimetres and centimetres when judging the practical significance of any deterioration in real articles and structures.

The scale factor of 10-100,000,000 causes difficulties in relating cause and effect, for instance the depassivation of steel caused by chloride ions (~0.3Nm ionic radius) and cracking resulting from corroding reinforcement measured in visible terms. This relationship is also exemplified in freeze/thaw action.

In addition to scale effects, water can take subtle forms. For instance according to Setzer [9] the pore water may be structured (adsorbed layers), prestructured (condensed) and bulk.

The structured water is contained in micro gel pores (less than 1nM) and consists of 1-3 molecular layers and remains unfrozen.

Prestructured/condensed water is contained in mesogel pores (1-30nM diameter) and a relative humidity of 50-98% is required for condensation to occur. Thus the freezing point of the water is depressed.

Bulk water is contained within the capillary pores (greater than 40Nm) and even these can be size classified, namely

40Nm – 1micron – micro capillary

1 micron – 30 microns – meso capillary

30 microns – 1Nm – macro capillary

The exact nature of this water and how it forms ice is complex and differs from what happens with 'free' water. The role of de-icing salts according to Setzer is dependent upon the manner in which the crystals form in the pore water.

Water transport occurs after ice has formed resulting in changed chemical potentials. The smaller pores become depleted and the larger pores fill progressively causing shrinkage and swelling respectively with consequential damage. The mechanisms are much subtler than previously thought.

The engineering outcome of the macro effect results initially from an accumulation of micro events. Thinking of the one whilst making engineering judgements of the other can cause problems.

The chemist or materials scientist will be wary about extrapolating data into the macro effect area and the engineer will not want to be hampered by the underlying detail particularly if the macro effect is reproducible.

How do you bring these two very different perceptions together? When concrete is made and placed it has usually been designed to have a function that may be measured in terms of specific characteristics, e.g. strength, permeability, density.

The retention of these characteristics with time is a measure of the durability and chemical changes that adversely affect these characteristics constitutes deterioration (Figure 1).

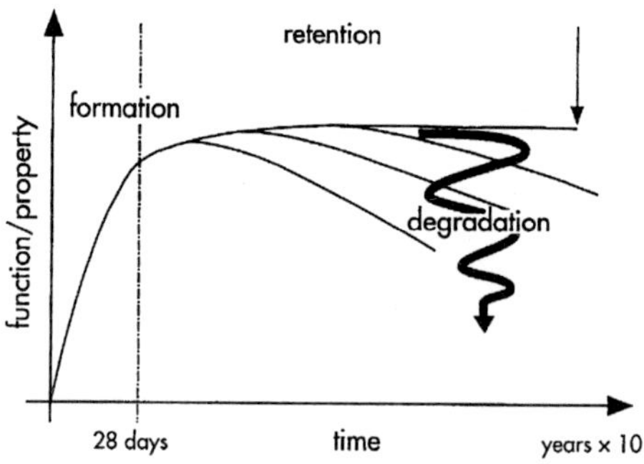

Figure 1 Time related trends

In a chemically inert and sterile atmosphere concrete as with other materials would seemingly last for ever. However, concrete is exposed to an environment that will impinge upon it in both negative and positive ways.

I am concerned in this paper about the negative effects that reduce or limit the lifespan of concrete in practical engineering terms and how these processes have their roots in the chemistry of Portland cement based concretes and the role of water.

The drive behind these changes is relentless in an endeavour to reach a low energy state or more correctly a low chemical potential in equilibrium with its surroundings.

Readily identified forms of degradation are listed in Table 1 below and identified as predominantly chemical or physical or both.

Table 1 The nature of various recognised forms of concrete downgrading

MODE OF DEGRADATION	UNDERLYING CAUSE
Sulphate attack	Chemical/physical
Freeze-thaw	Physical
Alkali-aggregate	Chemical/physical
Chloride ingress/metallic corrosion	Chemical/physical
Carbonation/Sulphation	Chemical
Acid/bacterial attack	Chemical
Soft water leeching	Chemical
Chelating chemicals	Chemical
Fire	Chemical/physical
Crystallisation	Chemical/physical
Cation exchange	Chemical/physical
Abrasion	Physical
Efflorescence	Physical

Chemical vs. Physical Changes

Distinguishing the chemical from the physical can be very difficult. It has been suggested by Bamforth [10] that present day Portland cements are less resistant to chloride ingress than those of 50 years ago. This is attributed to modern concretes not ageing beneficially (Figure 2). We are aware that the chemical composition of Portland cement has changed in that the tricalcium silicate to dicalcium silicate ratio has more or less reversed resulting in higher early strengths. Bamforth does not offer an explanation for the lack of property retention but does indicate a solution involving the use of calcium stearate as an integral waterproofing admixture – very old technology in a modern setting. A 50% reduction in chloride diffusion coefficient and about a 15% reduction in chloride levels are claimed under wetting and drying conditions. Others more by intuition than chemical knowledge have, by densifying the mix with ggbs, pfa microsilica and calcined metakaolin, offset shortfall in porosity/permeability (Table 2) [11]. Such trends have been recently corroborated [12] with diffusion coefficients increasing with w:c ratio. The benefit resulting from using normal water reducing agents and high range water reducing agents amount to 10-20% and 55% respectively.

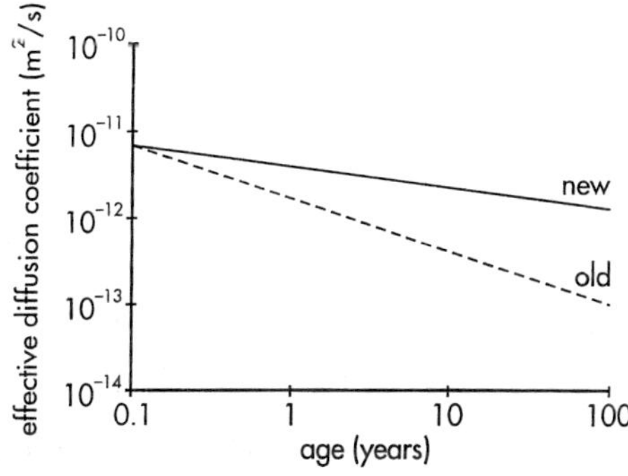

Figure 2 Influence of cement properties on effectivechloride diffusion coefficient

Concrete is no exception to other materials that are sometimes regarded as chemically stable such as dense polymers. The degrading agencies of water, oxygen, polluting gases, oxidising agents, exacerbated by impurities, defects and stresses resulting from shaping and forming all play their part [13 - 15]. Everything degrades, dust to dust , ashes to ashes!

It is apparent that nothing is permanent and the physical and chemical interactions are difficult to separate.

In order to obtain a practical view we may endeavour to simulate chemical degradation regimes and since the kinetics of negative change are often governed by diffusion and the chemical kinetics of the degradation reactions themselves both of which respond to temperature in line with an Arhenius relationship yields the prospect of accelerated testing if we know what the basic reactions are that govern degradation.

Such evaluations are commonplace in the aircraft component industry but notably absent within construction and in particular in the specific area of concrete.

Such evaluations may well feedback into compositional change. The loop is shown below:

Composition ⇨ Performance

⇧ ⇩

Modification ⇦ Evaluation

Another plastics example, resulting from these four interactions is the use of halogen substituted tetraglycidyl methylene dianiline resin in place of the unsubstituted resin, having a 40% reduction in water uptake relative to the non substituted polymer. Such a sequence applies, in principle, also to concrete.

Table 2 Diffusion resistance of cement paste types to chloride ions* [11]

CEMENT PASTE TYPE	DIFFUSIVITY / m^2s^{-1} x 10^{-13}	
	(1983)	(1989)
OPC	44.7	45.1
OPC/30% PFA	14.7	-
OPC/65% slag	4.1	-
OPC/40% slag	-	15.1
OPC/50% slag	-	10.2
OPC/70% slag	-	1.7
SRPC	100.0	100.5

*High diffusivity – high diffusion coefficient
– low diffusion resistance

THE NATURE OF CONCRETE

Concrete consists, primarily of hydrated cementitious materials, large and small aggregate with perhaps fillers and admixtures – a chemical compote.

The hydrated cementitious material consists primarily of calcium silicate hydrates, monosulphoaluminates and calcium hydroxide. Ordinary Portland cement yields some 20% of its mass as calcium hydroxide. The gel matrix is in some form of equilibrium with the pore water, itself containing dissolved salts, sodium, potassium, calcium hydroxides and sulphates.. If the concrete were allowed to dry out the saturated lime solution would precipitate out partially blocking the pores but leaving the system chemically quiescent.

Such an ideal situation is rare or at best transitory and concrete will interact with its surroundings that may contain:

1. more or less moisture than the concrete
2. chemicals within the surrounding water and/or gases within the atmosphere. For instance carbonation has an optimum relative humidity of 50-70% and can be predicted [16].

These other agencies will pass into/outfrom the concrete dependent upon absorption, diffusion, concentration gradient and chemical potential. However, for such mechanisms to work there must be a means of ingress/egress and water in some form must be available.

If the concrete had zero connected porosity and insufficient water available for the processes of dissolution and reaction to occur, then the degradation reactions would be much reduced and perhaps even stopped. Transport processes are at the root of degradation and service life [17].

Emphasis should be placed on achieving this state by chemically/physically changing the mix constituents or alternatively treating the hardened concrete to prevent the passage of moisture and water.

Whilst it may be tantalising to try and copy nature and create instant granite and marble that is an unlikely objective, notwithstanding these materials consisting of quartz, aluminosilicates and calcium/magnesium limestones respectively. Concrete is instant masonry and in geological terms is very immature. Such natural materials as granite and marble have resulted from degradation processes working over very long periods leaving the more stable compositions intact.

Let us consider the composition of concrete and the early physical chemistry that creates the starting point for any later degradation.

Cement Hydration

Of the four mineral phases in Portland cement the C_3S and C_3A phases react immediately on water contact and have been described previously [18]. Hydrolysis occurs preferentially on the grain surface and the number of sites determines how rapidly this process occurs. Calcium and hydroxyl ions pulse into solution and a layer of reaction product forms on the surface.

This layer shrouds the underlying grain giving rise to an induction period. However, this layer consisting of silicate ions, such as $H_3Si0_4^-$ and $H_4Si_20_7^{2-}$ has an amorphous structure. It is permeable to the diffusion of ionic species or water to and from the reaction sites. These events comprise phase 1 (Figure 3) [18].

Calcium and hydroxyl ions pass into solution and nuclei form by the assembly of atoms in the solution. In this stage whereas calcium ions can diffuse through the surface layer, that may act as a membrane. This we may regard as phase 2 (Figure 3).

The silicate ions beneath the layer cannot penetrate it and this increases the osmotic pressure. When the osmotic pressure is sufficient to rupture the surface layer at weak points, and the ionic concentration in solution becomes large enough, calcium hydroxide crystals and CSH gel are rapidly formed. This is phase 3. This marks the start of the acceleratory period of hydration.

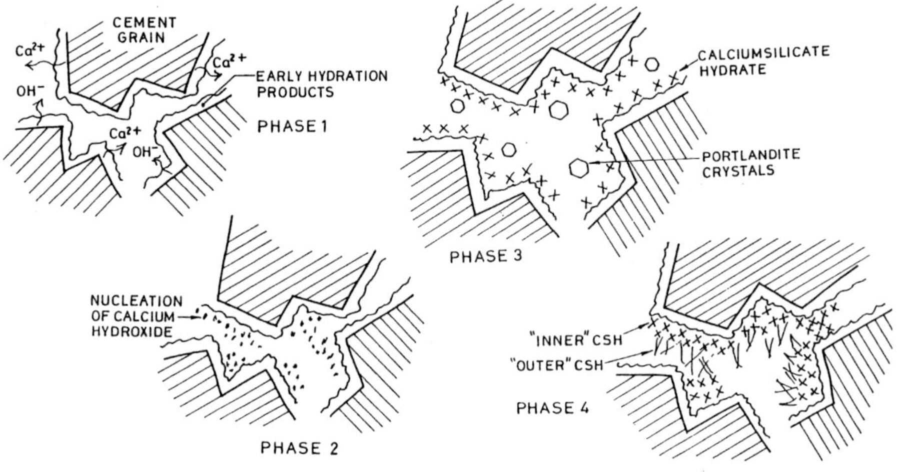

Figure 3 The four stages of Portland cement hydration [18]

Calcium hydroxide crystals may precipitate close to the grains where ionic concentrations are the highest or form in the pores away from the grains. However the growth of CSH is confined to outside the grain boundaries because of the difficulties in transporting silica. These 'outer products' act as barriers to ionic transport so that further hydration slows down. Later CSH gels called 'inner products' are formed beneath the 'outer products' and the hydration rate is controlled by water penetration and ionic diffusion. Further growth of the 'outer products' interweave with each other to form a porous solid matrix. This we may regard as phase 4.

Complicated though these reactions are, two relatively simple criteria have to be satisfied. Water must be present and intergranular space must be available to accommodate the crystalline growths.

The conventions refer to the gel:space ratio that in turn reflects the water:cement ratio used to make the concrete to begin with. I will come back to this point.

These recognisable stages of hydration are reflected in conduction calorimetry curves for instance, Figure 4 may also be split into 4 stages.

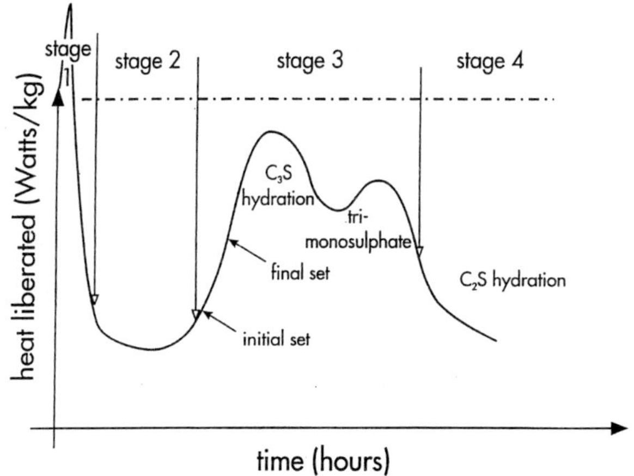

Figure 4 Heat evaluation vs time for Portland cement [18]

Stage 1 – rapid evolution of heat within the first few minutes. This peak results from:

a. The rehydration of calcium sulphate hemihydrate to give the dihydrate gypsum.

$$CaS0_4.0.5H_20 + 1.5H_20 \rightarrow CaS0_4.2H_20$$

b The hydration of free lime

$$Ca0 + H_20 \rightarrow Ca(0H)_2$$

c. The formation of ettringite

$$3Ca0.Al_20_3 + 3CaS0_4 + 31\text{-}32\ H_20 \rightarrow 3Ca0.Al_20_3.3CaS0_4.31\text{-}32\ H_20$$

d. The heat of wetting and solution

Stage 2 is an induction period that usually lasts from 30 minutes to 2 hours. During this stage the calcium ion concentration in solution rises slowly and nuclei of lime or calcium silicate hydrate form.

Stage 3 is called the acceleratory period. This stage contains the second major heat peak which is due to the rapid hydration of alite (C_3S). A large amount of CSH is formed on the surface of the cement grains, while crystals of calcium hydroxide precipitate either on the surface or in the pores. The initial and final set of the cement takes place during this period due to the rapid development of CSH. A third major peak may be observed when the C_3A content of the cement is over 12%. This peak indicates the hydration of C_3A and the transition of ettringite to the monosulphate ($3CaO.Al_2O_3.CaSO_4.31\text{-}32H_2O$).

Stage 4 is a deceleratory period and the final stage is a slow reaction period. Hydration of the cement (mainly Belite, C_2S) carries on slowly and is diffusion controlled.

Admixtures and Mineral Additions

The chemistry of these early stages is important and may well govern what happens later. For instance, the inclusion of chemical admixtures and/or reactive mineral additions that integrate into the early structure of the hydrated cement phase may well control ease of ingress and egress of water, reactive ions and gases without which degradation would be much reduced if not prevented.

The pictorial representation shown in Figure 5 taken from a recent paper by Nolan Basheer and Long [19] shows well the densifying contribution of pozzolanically reactive micro silica (refer later section).

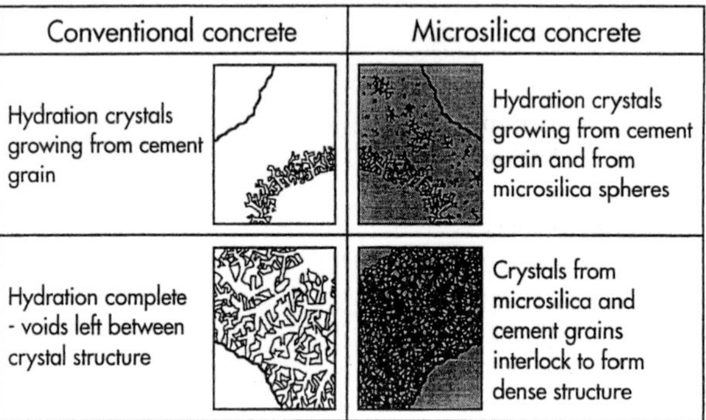

Figure 5 Comparison between hydration reactions [19]

ROLE OF WATER

Perhaps one of the most important and yet least dwelt upon chemicals affecting early structure and later responses of the cement paste phase is water.

Water controls the chemistry, the rheology and in large measure the physical nature of the hardened material.

Let us therefore consider this chemical a little closer.

As hydration proceeds, water is consumed that was in the mix to begin with until there is not enough to saturate the solid surfaces. The surfaces have also increased in area and the humidity falls creating a state of self desiccation that presumably slows down any further hydration. If our objective is to densify the mix Garboczi and Bentz [20] have shown that the degree of hydration also decreases as w:c ratio decreases. However 0.35 was the lowest ratio used (Figure 6). The gel pore space is independent of w:c ratio (1.5 – 2Nm). The initial space created between the grains into which the hydrates grow results directly from w:c ratio.

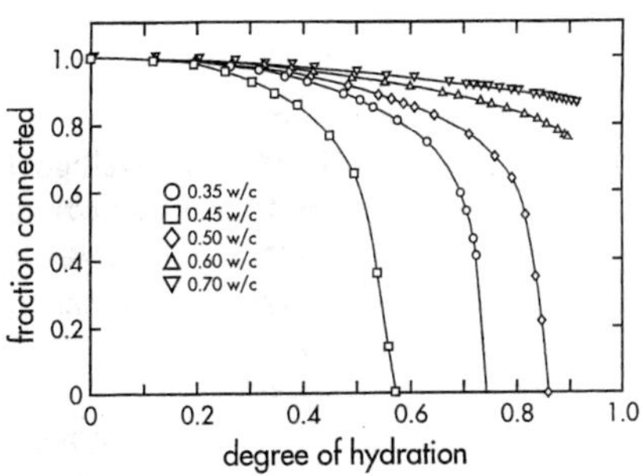

Figure 6 W/C ratio/hydration and connected voidage [13]

According to Neville [21] at w:c of 0.2 and below only 10% of the cement would be hydrated before capillaries cease to become continuous.

On the face of it there is a dilemma. The lower the w:c ratio, the lower the gel space and the less likely will complete hydration be.

How is it then that well dispersed mixes at very low w:c ratios can achieve very high strengths?

Neville also acknowledges [21] that such concretes can have higher strengths and suggests this may be due to the hydrated parts surrounding the unhydrated grain being thinner (Refer later to the reference of Goto and Uomoto [24].)

Strength, Porosity and Hydration

Strength, porosity and degree of hydration are all related to the water in the mix to begin with:

$$\text{Strength} = \text{strength}_o\,(1\text{-P})^n$$

for brittle materials [21]. If we densify the mix to the point where P=0 we should obtain maximum strength for that particular material. For concrete how do we minimise P whilst allowing the cement to hydrate? The void filling materials should of their very chemical nature be strength enhancing.

The familiar Abram's law could be considered to be a form of the porosity/strength law with the free water content (that not having reacted with the cement) causing the porosity plus any air [22].

Abram's law states:

$$\text{Strength} = \frac{K_1}{K_2 w/c} \qquad K_1 \text{ and } K_2 \text{ are constants}$$

Therefore strength increases as water cement ratio decreases and is valid over the water cement ratio range 0.3 – 1.20. What happens when water cement ratio is less than 0.3 assuming of course the concrete can be fully compacted?

Neville [21] quotes that Abram's obtained strengths of 280 MPa at w:c ratios of 0.08 but only by applying high pressures to consolidate the mix. In the presence of strongly dispersing admixtures such pressures may not be required and the early strength of cement hydration results from different and yet the same reaction products with the associated and inter layer water being dispersed differently.

The interplay of chemical dispersion and low water content is worth investigating at the Nm/micrometre level. The prospect of using nuclear magnetic resonance imaging/spectroscopy to establish the role of water in Portland cement hydration has improved if not, as yet, established itself [4, 23].

The nuclei ^{29}Si, ^{27}Al, ^{23}Na, and ^{1}H are all NMR active and occur in hydrated cement. A combination of magnetic resonance imaging (MRI), magic angle spinning (MAS), stray filled magnetic resonance (STRAFI) and NMR cryroporometry have already permitted monitoring of water ingress, pore size distribution and association between chloride ions and aluminium in forming Friedel's salt and the importance of calcium monosulphoaluminate in its formation. As such, variants of this technique may allow us to monitor the role of water at the Nm level.

What is the minimum water content that can be used and what happens to the gel:space ratio if that for hydrated paste is already at unity? Does minimising the capillary voidage adversely affect the growth of calcium silicate hydrates? Perhaps maximising density achieved using reactive fines can also mean maximising interlocking resulting from hydration products but resulting more from close packing less hydration is actually needed.

The work of Goto and Uomoto [24] on modelling hydration in dispersed systems based on a cube centred grain model would indicate this to be so.

Controlling Water Loss

If we are going to use low w:c ratios it is important that the water originally in the mix should not be reduced further by evaporation. Attempts to reduce vapour pressure by chemically 'spiking' the water to reduce its chemical potential have resulted in self-cure concretes [25]. Water soluble chemicals of the polyethylene glycol family associated with the water by way of hydrogen bonds (Figure 7) giving a vapour depressing effect greater than that from Raoult's law alone.

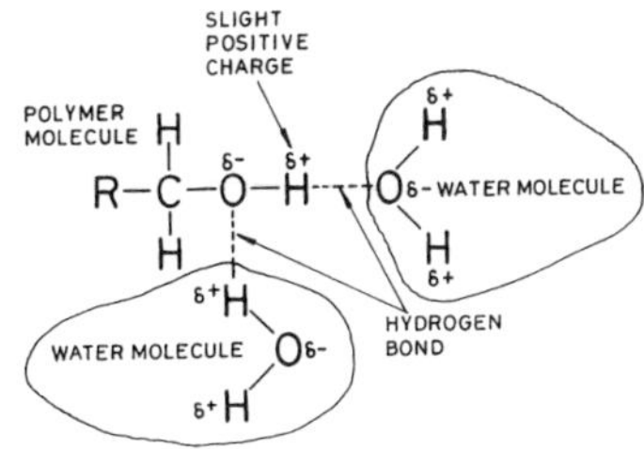

Figure 7 Hydrogen bonds between water molecules and an -OH group on a polymer molecule

Notwithstanding the previous comments, in general, considerably more water is added to concrete than is required to fully hydrate the cement. As a consequence such concretes are more easily penetrated by aggressive chemicals and degrade more readily. For the latter a w:c ratio of approximately 0.26 is generally considered sufficient. In practice w:c ratios of 0.4 – 0.6 are commonly used and are responsible for plastic and drying shrinkage.

As a creator of the means of ingress water should be minimised whilst not compromising strength or workability.

RILEM report 16 [26] (edited by H W Reinhardt) brings together a state of the art collection of papers dealing with chemical and physical cement microstructure and the role of water creating accessible voidage and its dependence on applied stress.

The means of reducing water contents to well below the 0.26 level are now commonplace and notwithstanding ratios of 0.2 or less strengths are enhanced even though the degree of hydration is presumably less than stoichiometric. However, if the reacting mass is densely packed the amount of hydration produced needed to achieve interlocking will also be reduced.

Water not only creates voidage it is the cause of dimensional change when lost from the mass. Why is it though that after setting and hardening moisture loss can cause long term contraction and shrinkage.

How are the contractile stresses created?

If the silicate network is intact why should water removal cause a volume reduction?

Surface tension plays a part for plastic and early drying shrinkage but is it true for long term shrinkage?

These matters are relevant because shrinkage causes tensile stresses and creates discontinuities that can then allow ingress by aggressive chemicals.

Again the chemical cause results in a physical effect.

Mobilising Cohesion

Strength/cohesion or the self attachedness results from interactions that may be mechanical as well as chemical. After all concrete is strong in compression but weak in tension.
Chemical forces are mobilised from the highest specific area that is generated from hydration yielding a microcrystalline mass.

The first type of cohesion is Van de Waals in nature and short range (attraction of solid surfaces).

The second type of cohesion results from chemical bonding. These forces are stronger and of longer range but fewer in number.

The generation of a multitude of very small particles would seem to be at the root of achieving strength (high resistance to degradation processes). Minimum water content, maximising density from a reactive mass would seem to be the basic tenets for reduced degradation.

Whilst changing the chemistry of the Portland cement itself is an option, indeed, but for other reasons, it has changed significantly over the last 50 years.

It is more likely that designer cements and bespoke compositions will result from the addition of reactive minerals that change the chemistry and physical form of the hardened mass.

Many reviews and publications exist that establish beyond reasonable doubt the technical merits of incorporating hydraulically active minerals that densify concrete making it stronger, less permeable and generally more durable.

One such recent publication by Malhotra and Mehta [27] and in particular Chapter 7, makes the emphatic point that 'to be durable, Portland cement concrete must be relatively impervious.'

Degradation processes such as carbonation, sulphation, alkali silica reaction, formation of thaumasite, freeze-thaw, chloride ingress and fire response depend upon water contained within the concrete to begin with and/or the influx of water during the concrete's working life.

Hydraulically Active Additions

The addition of hydraulically reactive cement supplements can substantially improve most properties when used optimally with water.

One particular group of pozzolanic materials of growing relevance are the calcined selected kaolins (aluminosilicates). These disperse readily and have high pozzolanic reactivity (Table 3) [28].

Their inclusion results in improved aggregate/paste bond; [28] and Figure 8.

Table 3 Reactivity of pozzolanic materials

MATERIAL	POZZOLANIC REACTIVITY (mg $Ca(OH)_2$ /g)
Blast furnace slag	40-106 (typical range)
Calcined paper waste	300 (typical value)
Condensed silica fume	427 – 581 (typical range)
Calcined bauxite	534 (typical value)
Pulverised fuel ash	196 – 875 (typical range)
Calcined metakaolin	1050 ± 100 (specification)

$AS_2 + 5CH + 5H \rightarrow$ Mixture of calcium aluminate/silicate hydrates $C_5AS_2H_5$ (average composition)

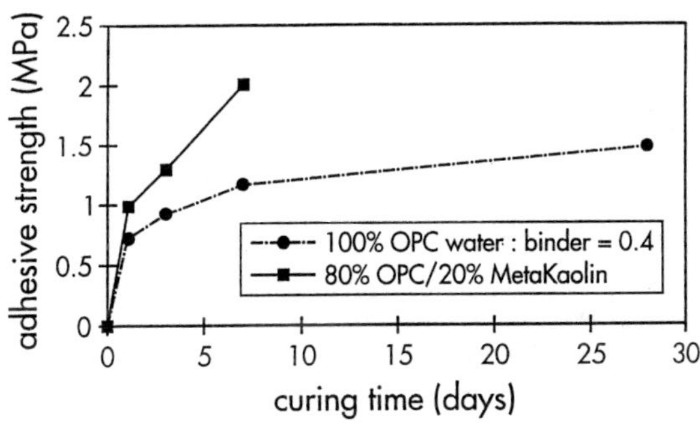

Figure 8 Effect of MetaKaolin on adhesive strength of past-limestone aggregate bond [28]

Reduction of pore sizes greater than 0.05 micrometres and in particular in the range 0.5 – 10 micrometres yielding improved permeability and absorption properties (Figure 9) as well as chloride diffusion (Table 4).

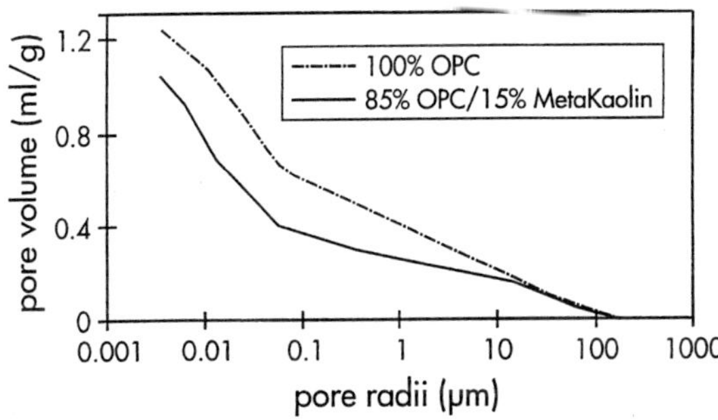

Figure 9 Effect of MetaKaolin on the pore size distribution of mortars [28]

Table 4 Chloride ion diffusion through mortars [28]

SAMPLE	WATER/BINDER RATIO	DIFFUSION COEFFICIENT (cm^2s^{-1} x 10^{-9})
100% OPC	0.6	220
	0.7	680
85% OPC/15% Metakaolin	0.6	18
	0.7	8

Key Generalised Items

1. All building fabrics degrade, it is mainly a question of rate and perception. To do so requires the presence of water both initially and later.
2. Improving the incipient chemical resistance of Portland cements is an unlikely prospect although further compositional changes may result from production needs.
3. A detailed explanation of the interplay between minimum voidage, reactive fines, water content and mechanical properties at very low water contents would assist optimising bespoke but durable concretes. The exact role of water, its retention and imbibition need fundamental study.
4. Chemical manipulation should be targeted at accommodating the inclusion of hydraulically active fines whilst using the minimum water content.
5. Densifying the mix may be coupled with imparting some long lasting water repellency either by chemical inclusion or post cure application.
6. Hydraulically active fines may lend themselves to chemical processing to impart particular durability characteristics.

REFERENCES

1. SOMERVILLE, G. Engineering design and service life: a framework for the future. Proceedings of STATS 21st Anniversary Conference 'Prediction of Concrete Durability' edited by J Glanville and A Neville, E & F N Spon, 16 November 1995, pp 58–75.

2 OWENS P L Concrete International, November 1989, Part 1, pp 68-74.

3. OWENS P L , Concrete International, December 1989, Part 2, pp 68-71.

4. HUNTER, G, JONES, M R AND HEWLETT, P C (Dundee University), HALSE, M R, AND STRANGE, J (University of Kent), MACDONALD, P J, GLOVER, P M AND MULHERON, M (Surrey University). Broadline MRI and MAS NMR characterisation of Portland cements: water and chloride mobility and control, joint final report EPSRC grants GR/K94874, GR/K71660, GR/K94881, September 1998.

5. Effects of moisture vapour transmission in delamination of coatings for concrete, Protective Coatings Europe, Vol 3, No 5, March 1988, pp 4-8.

6. MAYS, G (Editor). Durability of concrete structures, E & F N Spon 1992.

7. TEL Review April 1998, No. 61, pp 3.

8. Collins Dictionary and Thesaurus, Harper Collins, 1992, p 308.

9. SETZER, M J. Action of frost and de-icing chemicals - basic phenomena and testing, taken from Freeze-thaw durability of concrete, edited by J Marchand, M Pigeon and M Setzer, 1997, Chapter 1, pp 3-22. E & F N Spon.

10. BAMFORTH, P B. Materially affecting durability, Concrete Journal, Jan-Feb 1996, pp 21-22.

11. LAWRENCE, C D. Chloride ingress into concrete, BCA October 1989.

12. BAMFORTH, P B, PRICE, W F AND EMERSON, M. Contractor report 359, An international review of chloride ingress in structural concrete, TRL, 1997, pp162.

13. EURIN PH 'Degradation processes of organic building materials – a short review and some proposals for research' Durability of building materials Vol 1, 1982, pp 162-168. Elsevier Second International Conference Durability of building materials and components, Gaithersburg USA, 14-16 Sept 1981.

14. KATZ, S. Conservation of plastics – a race against time, Materials World, August 1995, pp 377-378.

15. MARTINS, R AND CAMPION, R. The effects of ageing on fibre reinforced plastics, Materials World, April 1996, pp 200-202.

16. BAMFORTH, P B. Guidance on the selection of measures for enhancing reinforced concrete durability, DoE/PIT contract C1 39/3/76 (cc 967), July 1997, Taywood Engineering Ltd, Technology Division.

17. KRUPP, J, HILSDORF, N G. RILEM Report No 12, Performance criteria for concrete durability, E& FN Spon ,1995, pp 280-293.

18. ZENG, S. Polymer modified cement: hydration, microstructure and diffusion properties, PhD thesis 1996, University of Aston, Birmingham.

19. NOLAN, E, BASHEER, P A M AND LONG, A E. Effects of three durability enhancing products on some physical properties of near surface concrete, Construction and Building Materials, 1995, Vol 9, No 9, pp 267-272.

20. GARBOCZI, E J AND BENTZ D P. Modelling of the micro structure and transport properties of concrete, Construction and Building Materials, Vol 10, No 5, 1996, pp 293-300.

21. NEVILLE, A M. Properties of concrete, 2nd edition, 1977, Pitman Publishing pp 687.

22. SEAR, L K A, DEWS, J, KITE, B, HARRIS, F C AND TROY J F. Abrams Law at higher water to cement ratios, Construction and Building Materials, Vol 10, No 3, 1996, pp 221-226.

23. HEWLETT, P C, HUNTER, G AND JONES, M R. Bridging the Gaps, Chemistry in Britain, January 1999, (in press).

24. GOTO, T AND UOMOTO, T. Strength development mechanisms of Portland cement pastes, Concrete Research and Technology, Vol 5, No 1, January 1994.

25. DHIR, R K, HEWLETT, P C AND DYER T D. Mechanisms of water retention in cement pastes containing a self curing agent, Magazine of Concrete Research, Vol 50, No 1, March 1998, pp 85-90.

26. RILEM Report 16. Penetration and permeability of concrete – barriers to organic and contaminating liquids, (Editor) H W Reinhardt, E& F N Spon, 1997, pp 332.

27. MALHOTRA, V M AND MEHTA, P K. Pozzolanic and cementitious materials, Advances in concrete technology, Vol 1, Gordon and Breach, 1996, pp 191.

28. ECC INTERNATIONAL, Commercial literature, Use of Metastar for the production of highly durable concretes and mortars, SBG 085, Second edition, November 1995.

ESTIMATION OF THERMAL PROPERTIES OF CONCRETE FROM IN-SITU MEASUREMENTS

N Nishida

K Ushioda

Nishimatsu Construction Company Limited

K Matsui

Tokyo Denki University

Japan

ABSTRACT. When thermal analysis is carried out, standard values of the important parameters are selected from previous data given in the literature or the average values of experiment conducted for that purpose. However, the properties of in-situ concrete may differ from properties of laboratory tests due to differences in the construction environment and casting conditions and the properties of in-situ concrete have to be used for a more reliable analysis. The authors present an inverse method for estimating thermal parameters from in-situ measurement data. For evaluating the validity of the method from field test data, the authors applied five sets of thermal data. It is found from the examination that a procedure based on the Gauss-Newton method serves as an effective tool for estimation of the thermal properties of concrete.

Keywords: In-situ concrete measurement, Inverse analysis, Thermal properties, Gauss-Newton method, Heat of hydration, Concrete slab

Dr Noriyuki Nishida is a Senior Research Engineer in the Nishimatsu Construction Company, Kanagawa, JAPAN. His research interests include thermal cracking, construction-aided observational systems, the properties of concrete at early ages, and self-compacting concrete.

Mr Katsushi Ushioda is a Research Engineer in the Nishimatsu Construction Company, Kanagawa, JAPAN. His main research interests include thermal stress analysis, the properties of concrete at early ages, and observational construction systems.

Professor Kunihito Matsui is Director of Department of Civil and Environmental Engineering, Tokyo Denki University, Saitama, JAPAN. He specializes in numerical computation for structural engineering, the durability of concrete and pavements. Professor Matsui has published widely and serves on many Technical Committees.

INTRODUCTION

It is well known that cracking of concrete poses a serious problem for concrete structures from the viewpoint of function and durability. Nonlinear temperature distribution in a concrete body due to the heat released by cement hydration causes cracking to occur at early age. Therefore, protection and control of cracks based on thermal stress analysis are important. Temperature distributions of a concrete body are affected by a number of parameters. When thermal analysis is carried out, the standard values[1 - 3] of these parameters are selected from previous data in the literature or average values from experiments conducted for that purpose. However, the properties of in-situ concrete may differ from those of laboratory tests, due to the difference in construction environments and casting conditions, and the properties of in-situ concrete have to be used for a more reliable analysis.

Considering the effective parameters from the above results, the inverse method is proposed for estimating thermal properties of concrete using in-situ measurements. Chikahisa and his coworkers[4] evaluated a convection heat transfer coefficient by using the simplex method. The authors[5, 6] proposed a method to estimate the values of five parameters (heat conductivity, specific heat of concrete, convection heat transfer coefficient and experimental constants of adiabatic temperature rise Q_∞ and γ) by using the Gauss-Newton method, based on the least square concept and construction site data instead of laboratory tests. However an inverse analysis is inherently unstable and requires special care to obtain reliable results[7, 8].

In this paper, the effectiveness of the inverse analysis algorithm for estimating thermal parameters from in-situ measurement data and for examining the results is demonstrated. For evaluating the validity of the method from field test data, the author applied five sets of thermal data consisting of two temperature regimes at the construction site. Based on the values of estimated parameters, the problems of the method and the models of thermal analysis have been evaluated.

THERMAL ANALYSIS OF CONCRETE

Temperature in a concrete body rises due to cement hydration and is transmitted to its ambient environment. A hypothetical model concrete-rock system is illustrated in Figure 1.

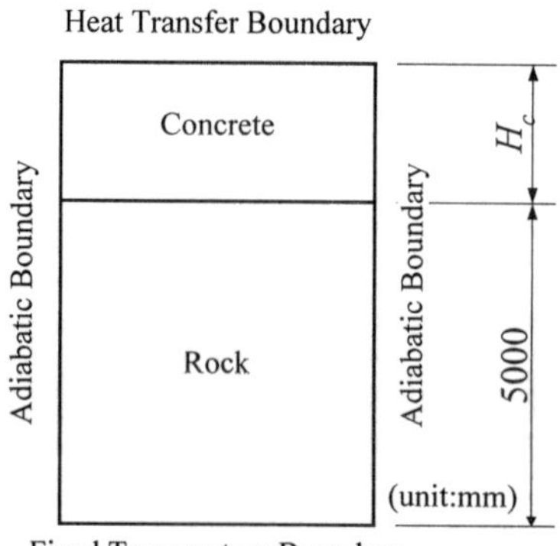

Figure 1 A hypothetical model concrete-rock system for thermal analysis

Taking into consideration the fact that the time increment Δt affects the stability and precision of numerical calculations, we transformed second order partial differential equations for non-stationary thermal conduction to first order simultaneous differential equations and solved by introducing an eigenvalue analysis. Specifically discretization of temperature with respect to the space x axis gives the following set of linear simultaneous differential equations:

$$[A]\left\{\frac{dT}{dt}\right\}+[B]\{T\}=\left(\rho_C C_C Q_\infty \gamma\, e^{-\gamma t}\right)\{F_1\}+\{F_2\} \quad (1)$$

$$\{T(0)\}=\{T_0\} \quad (2)$$

$[A]$:Matrix of thermal capacity, $[B]$:Matrix of thermal conduction, $\{T\}$:Temperature vector at node, $\{T_0\}$:Initial temperature vector, $\{F_1\}$:Heat flow flux vector for internal heat generation, $\{F_2\}$:Heat flow flux vector for heat conduction, ρ_C: Density of concrete, C_C: Specific heat of concrete, Q_∞,γ:Experimental constants referring to properties of adiabatic temperature rise.

METHOD OF INVERSE ANALYSIS

Since temperature distribution in a concrete body is a function of thermal properties $\boldsymbol{X}$ and time t, it can be expressed as $T_i(\boldsymbol{X},t)$ where subscript i refers to the location of i th thermal sensor. $\boldsymbol{X}$ is an unknown vector composed of five parameters X_1, X_2, X_3, X_4 and X_5 which represent the heat conductivity of concrete, specific heat of concrete, convection heat transfer coefficient from concrete to air, and two parameters in the hydration heat model.

The remaining parameters are considered as known. The five unknown parameters need to be determined so as to achieve good agreement between the computed and measured temperatures at thermal sensor locations by using the least square method. The temperature measurement is taken at N locations and let it be defined by $u_i(t)$, $(i = 1,\ldots, N)$. Then a least square functional can be defined as Equation (3) where t_0 and t_1 are the lower and upper limits of time used for the analysis. Unknown parameters X_i $(i = 1,\ldots, N)$ are determined so as to minimize Equation (3). Since it is a nonlinear least square problem, it requires an iterative computation to achieve the minimum of Equation (3).

In order to introduce the Gauss-Newton method, the following linear approximation Equation (4) will be utilized. Inserting Equation (4) to Equation (3), we want to determine ΔX_j such that the functional attains a minimum assuming $\boldsymbol{X}$ is given.

The necessary condition can be written as Equation (5). Equation (5) is a set of linear equations with respect to ΔX_j. The coefficient matrix is referred to as a normal matrix which often manifests the singular or nearly singular characteristics. Taking numerical instability into consideration, singular value decomposition is employed to solve for ΔX_j $(i = 1,\ldots, 5)$. The flow of computation is shown in Figure 2. In the figure the maximum allowable change on $|\Delta X_j|$ is set at 10% of X_j, because the linear approximation in Equation (4) has to hold.

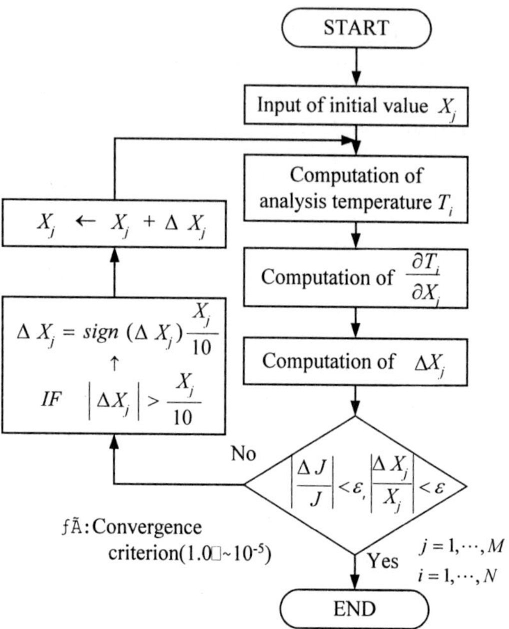

Figure 2 Flow of an inverse analysis

$$f(X) = \int_{t_2}^{t_1} \sum_{i=1}^{N} \{u_i(t) - T_i(X,t)\}^2 dt \tag{3}$$

$$T_i(X + \Delta X) = T_i(X) + \sum_{j=1}^{5} \left(\frac{\partial T_i}{\partial X_j} \right) \Delta X_j \tag{4}$$

$$\sum_{j=1}^{5} \left\{ \int_{t_0}^{t_1} \left(\sum_{i=1}^{N} \frac{\partial T_i}{\partial X_j} \frac{\partial T_i}{\partial X_k} \right) \right\} \Delta X_j = \int_{t_0}^{t_1} \sum_{i=1}^{N} \{u_i(t) - T_i(X,t)\} \left(\frac{\partial T_i}{\partial X_k} \right) dt \tag{5}$$

$(k = 1, \ldots, 5)$

MEASURED DATA OF CONSTRUCTION SITE

For evaluating the validity of the method from field test data, we applied five sets of thermal data consisting of two kinds of temperatures: slab-concrete body, and air at construction site. The measurement conditions of newly casted concrete are summarized in Table 1. Three or five thermal sensors are embedded in concrete and their locations are shown in Figure 3.

Five different heights of concrete are selected. The measuring duration is over one month after the concrete placing. The measured data of Type-C and E have noise. Thermocouples are used to measure temperature at several locations in a body of concrete.

Table 1 Measurement conditions of newly casted concrete

	TYPE OF CEMENT	*W/C* (%)	UNIT CEMENT CONTENT (kg/m^3)	DENSITY OF CONCRETE (kg/m^3)	MEASURE-MENT DURATION (DAYS)	HEIGHT OF CONCRETE (m)	NUMBER OF SENSORS	INITIAL CONCRETE TEMPERA-TURE (°C)	CURING TIME (DAYS)
A	OPC	56.0	284	2295		1.0	5	27.0	3
B	OPC	58.5	243	2326	60	1.5	3	12.6	6
C	OPC	57.1	280	2320	34	2.5	5	17.7	7
D	BB	57.1	251	2351	35	3.0	5	14.0	5
E	OPC	58.5	243	2326	27	3.5	5	16.2	6

PC: Portland cement
BB: Portland blast-furnace slag cement, the blend ratio of slag is 45%.

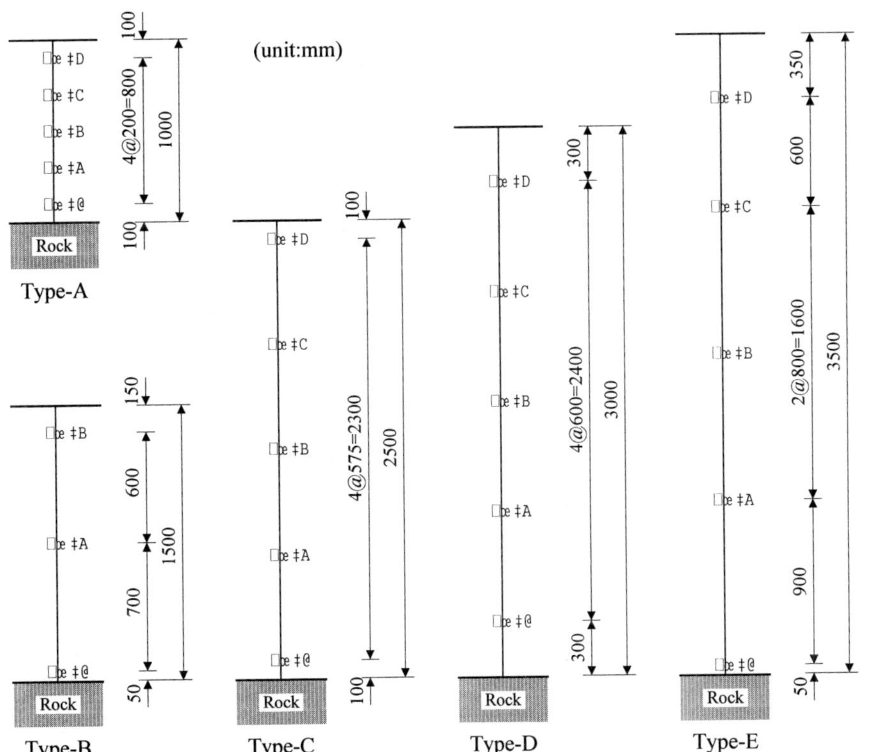

Figure 3 Location of thermal sensors

Table 2 Known parameters for thermal analysis

PARAMETERS	PRESCRIBED VALUES									
	TYPE-A		Type-B		Type-C		Type-D		Type-E	
	I	II	I	II	I	II	I	II	I	II
K_R: Heat Conductivity of Rock(W/m°C)	2.33	2.33	1.74	2.33	1.98	2.33	1.74	2.33	1.74	2.33
C_R: Specific Heat of Rock(J/kg°C)	795	795	2093	795	917	795	2093	795	2093	795
T_A: Air Temperature(°C)	Measured									
T_B: Boundary Rock Temperature(°C)	15.0	15.0	15.0	15.0	15.0	15.0	15.0	15.0	15.0	15.0
T_{C0}: Initial Concrete Temperature(°C)	27.0	27.0	12.6	12.6	17.7	17.7	14.0	14.0	16.2	16.2
T_{R0}: Initial Rock Temperature(°C)	Linear distribution of temperature between T_B and T_{C0}									
ρ_C: Mass Density of Concrete(kg/m^3)	2295	2295	2326	2326	2320	2320	2351	2351	2326	2326
ρ_R: Mass Density of Rock(kg/m^3)	2600	1800	1800	1800	2600	1800	2600	1800	1800	1800

T_B: Rock Temperature at Fixed Temperature Boundary(°C)

I: Properties of hard rock, II: Properties of soft rock

RESULTS OF INVERSE ANALYSIS

The known parameters for the thermal analysis are summarized in Table 2. These parameters are prescribed in references, which are obtained from laboratory or field tests. To compare the effect in difference of prescribed parameter values, I (properties of hard rock) and II (properties of soft rock) are used for estimation of thermal properties of concrete.

Unknown parameters K_C, C_C, α_C, Q_∞ and γ are initially assumed according to the JSCE design code[3]. Five thermal parameters are estimated from thermal data of in-situ concrete. The results of inverse analysis are presented in Table 3 along with their initial values.

Figure 4 and 5 show the comparison between the measured and the computed temperatures. It may be stated that both histories show very good agreement in the measured duration. If strong instability lies in an inverse problem, the computational process will show a slower convergence, will converge to different values, or will even diverge.

The results depend on the following conditions:

1) Number of sensors and their locations

2) Duration of measurement

3) Accuracy of prescribed parameters

4) Computation algorithm involved in the analysis.

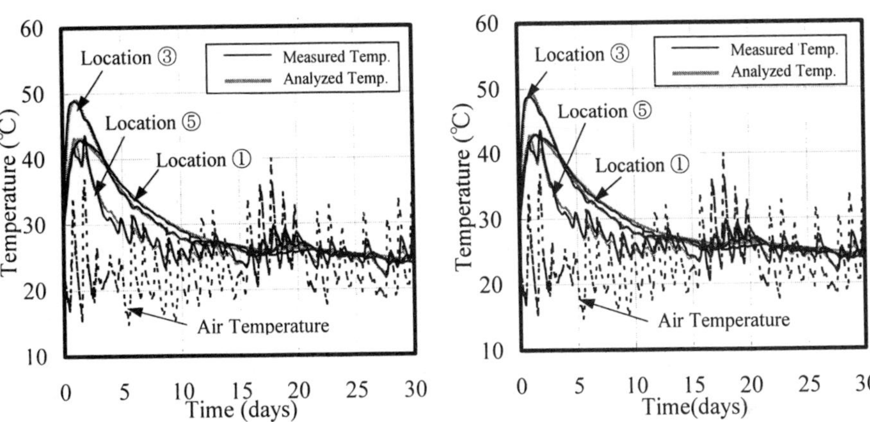

(a) Results of measured data within 3 days (b) Results of measured data within 45 days

Figure 4 Measured and analysed temperature history (Type A-I)

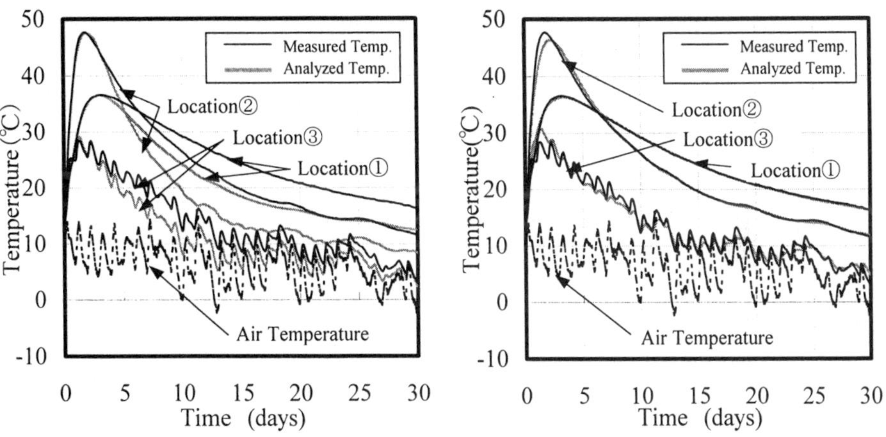

(a) Results of measured data within 3 days (b) Results of measured data within 30 days

Figure 5 Measured and analysed temperature history (Type B-I)

Duration of Measured Data

Four different lengths for measuring data are considered in each type, which are: 3, 7, 14 days and the measured maximum duration. The results of inverse analysis are presented in Table 3 along with their initial values. Change in the results after 14 days is not significant. These results are mean values in each term of data; however, slight changes within the length of measuring data, and Q_∞ particularly is different from the initial data. This is because thermal properties change depending on time and temperature of concrete body. Figure 4 demonstrates very good agreement regardless of the duration used for inverse analysis. However, Figure 5 shows different results, because identified parameter values are different depending on the duration.

Table 3 Results of inverse analysis from length of measuring data

	DURATION (days)	K_C (W/m°C)			C_C (J/kg°C)			α_C (W/m^2 °C)			Q_∞ (°C)			γ(1/DAY)		
		Initial value	Results I	Results II	Initial value	Results I	Results II	Initial value	Results I	Results II	Initial value	Results I	Results II	Initial value	Results I	Results II
	3		2.76	1.92		1478	910		10.4	7.1		30.2	29.7		1.956	1.694
	7		2.90	2.05		1398	947		11.1	7.7		31.5	29.8		1.839	1.693
A	14	2.67	3.06	2.17	1256	1440	972	12.8	10.7	7.4	43.3	31.0	29.4	1.344	1.900	1.732
	45		2.87	2.19		1448	961		11.9	7.4		31.3	29.7		1.860	1.704
	3		2.29	2.16		398	488		19.1	19.3		95.9	70.7		0.462	0.637
	7		2.28	2.26		222	966		15.3	17.4		200.0	48.6		0.242	0.867
B	14	2.67	2.66	2.38	1256	1419	1247	12.8	19.3	17.0	40.0	44.6	43.0	0.600	0.961	1.017
	60		2.49	2.30		1381	1278		18.7	17.3		44.5	42.3		0.976	1.052
	3		×	2.97		×	429		×	5.6		×	40.4		×	0.784
	7		×	2.67		×	1000		×	5.9		×	29.7		×	1.046
C	14	2.67	2.98	2.74	1256	1340	1223	12.8	7.2	6.7	44.0	29.8	28.6	0.919	1.109	1.090
	35		2.83	2.64		1134	1048		7.8	7.1		31.9	30.6		0.964	0.962
	3		3.91	2.91		1762	1232		36.2	23.6		53.8	53.0		0.887	0.881
	7		4.08	2.91		2018	1389		25.8	18.0		54.0	53.2		0.846	0.846
D	14	2.67	4.12	2.92	1256	2566	1802	12.8	32.1	22.3	41.0	52.5	51.5	0.588	0.876	0.879
	28		4.32	3.09		2637	1875		32.2	22.8		52.6	51.5		0.874	0.879
	3		×	2.48		×	612		×	1023		×	48.6		×	0.577
	7		1.88	2.36		419	779		21.1	70.2		52.6	44.7		0.487	0.587
E	14	2.67	2.00	2.38	1256	975	1186	12.8	72.8	99.9	39.9	41.6	40.2	0.740	0.608	0.627
	27		2.20	2.45		1227	1367		53.9	62.3		39.5	38.6		0.649	0.689

×: not converged

Noise in Measured Data

The quality of Type-C measured data is not good due to noise. Even if strong noise lies in measured data, the computed temperatures from identified parameters have to agree with the measured temperatures. Figure 6 illustrates the comparison between the measured and the computed temperatures in Type-C. The both histories show good agreement.

Influence of Prescribed Parameter Values of Rock

The results of Type-C-I and E-I diverge at 3 and 7 days in Table 3. However, the results using known parameters of C-II and E-II are estimated. If there are errors in the prescribed parameters of rock, the results of the inverse analysis may be different from their true values. To find the effect of errors on the five unknown parameters, 100 sets of prescribed parameters whose means are given in Table 2(Type-A-I, 45 days) and coefficients of variation are 10% are generated for the inverse analysis. The results are presented in Figure 7, which demonstrates rates of variation of unknown parameters. Positive sign indicates an increase of the identified results and negative sign is a decrease when prescribed parameter values are greater than those given in Table 2. It can be observed from Figure 7 that the effect of errors on the estimated results is small on Q_∞ and γ.

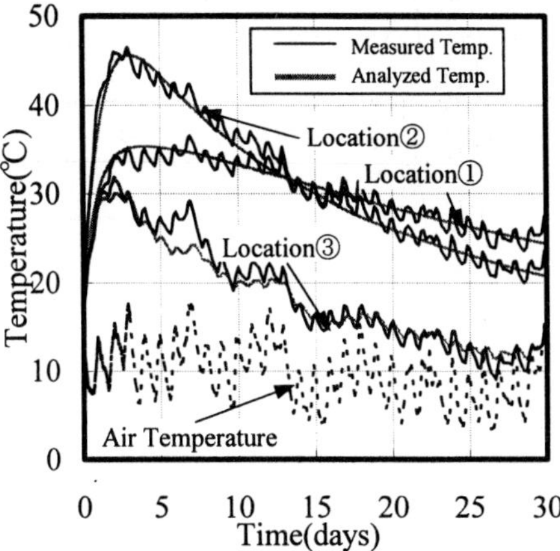

Figure 6 Measured and analysed temperature history
(Type C-I, measurement duration of 30 days with noise)

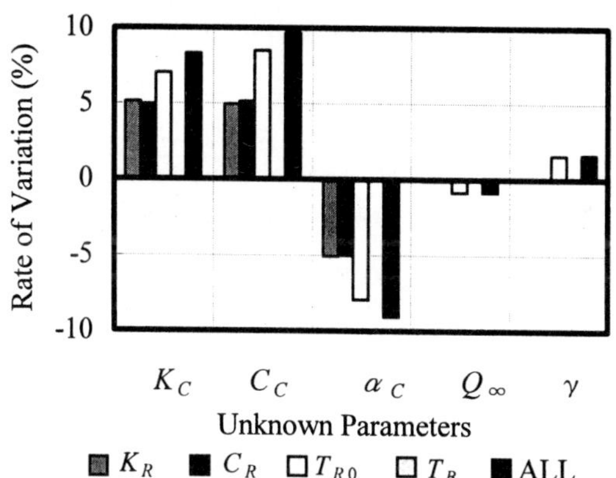

Figure 7 Sensitivities of unknown parameters (Type A-I, 45 days)

Convergence Rate and Number of Measuring Locations

An adequate number of sensors, their locations, and the status of convergency are investigated by carrying out inverse analysis using Type-A-I data in Figure 4.

The results are presented in Table 4 which shows that the number of sensors and their locations influence to identified results. All cases that meet the convergency requirement within 20 iterations include the location ① which is embedded close to the bottom of the concrete. Location ⑤ is also important to estimate the convection heat transfer coefficient since it is placed close to the convection heat transfer boundary. The convergency process of α_C is shown in Figure 8. There is no distinguishable difference in the convergency process between CASE 1 and 2. However, no convergence is observed in CASE 3, 6 and 8. According to our observation, three sensors (top, center, and bottom) are at least necessary.

Table 4 Convergent values of number of sensor locations (Type-A-I, 45 days)

CASE	LOCATIONS IN FIGURE 3	NUMBER OF ITERATION	K_C (W/m°C)	C_C (J/kg°C)	α_C (W/m^2°C)	Q_∞ (°C)	γ (1/DAY)
1	①②③④⑤	8	3.05	1435	10.6	29.7	1.892
2	①③⑤	7	3.03	1413	10.6	30.1	1.790
3	③④⑤	×	×	×	×	×	×
4	①②③	13	3.33	1186	6.9	31.9	1.746
5	①⑤	11	3.84	1168	11.5	29.1	1.576
6	③⑤	×	×	×	×	×	×
7	①③	15	3.31	1168	6.8	32.1	1.754
8	③	×	×	×	×	×	×

×: not converged

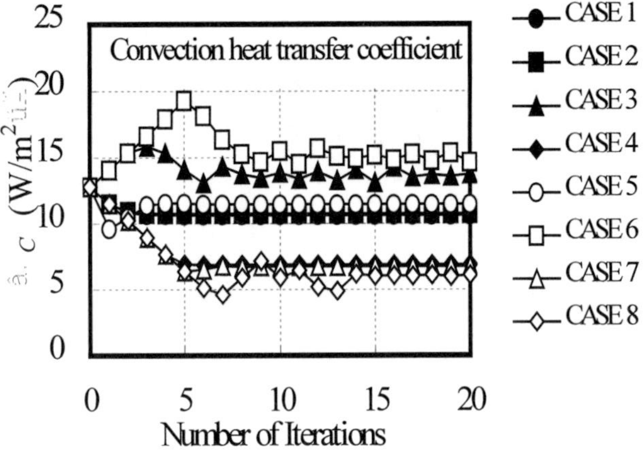

Figure 8 Convergence process(Type-A-I, 45 days)

CONCLUSIONS

An inverse analysis procedure has been developed based on the Gauss-Newton method to identify the five thermal properties of concrete from measured temperature at several locations in the concrete body and surrounding temperature. These limited preliminary studies resulted in the following conclusions:

1. An identification procedure based on the Gauss-Newton method serves as an effective tool for estimation of the thermal properties in concrete.

2. Stable convergence is observed when the duration of measured data is more than 14 days.

3. Errors in the values of prescribed parameters (heat conductivity, specific heat, fixed boundary temperature, and initial temperature of rock) give effect on the identified thermal properties of concrete.

4. The results change slightly with the measuring data length and the number of sensors and their locations.

5. It is desirable to have at least three measuring locations (top, center, and bottom) to identify the thermal parameters of new cast concrete from in-situ measurement.

ACKNOWLEDGMENTS

The authors deeply appreciate Professor Lewis A. Davis, Tokyo Denki University, for his valuable advice on English.

REFERENCES

1. ACI Committee 209. Prediction of Creep, Shrinkage, and Temperature Effects in Concrete Structures, Designing for Effects of Creep, Shrinkage, Temperature in Concrete Structures, SP-27, American Concrete Institute, Detroit, 1971, pp.51-93.

2. JCI. The Index on Controlling of Crack in Massive Concrete, 1986 (in Japanese).

3. JSCE. Standard Specification for Design and Construction of Concrete structures, Part of Construction, 1996 (in Japanese).

4. CHIKAHI, H, TSUZAKI, J, ARAI, T and SAKURAI, H. Estimation of heat transfer coefficients of mass concrete structures by means of back analysis, JSCE, J. of concrete Eng. and Pavements, No.541/V-17, 1992, pp.39-47 (in Japanese).

5. MATSUI, K, NISHIDA, N, DOBASHI, Y and USHIODA, K. An inverse method for estimation of thermal properties of mass concrete, Transactions of the Japan Concrete Institute, Vol.15, 1993, pp. 131-138.

6. NISHIDA, N, USHIODA, K, DOBASHI, Y and MATSUI, K. Estimation of Thermal Properties of Concrete on-site Measurements of Temperature, JSCE, J. of concrete Eng. and Pavements, No.544/V-32, Aug., 1996, pp.89-100 (in Japanese).

7. TARANTOLA, A. Inverse Problem Theory, Elsevier, 1987.

8. TANAKA, M. and BUI, H. D.(Eds.), Inverse Problems in Engineering Mechanics, Springer-Verlag, 1992.

INHOMOGENEOUS DISTRIBUTION OF MOISTURE CONTENT AND POROSITY IN CONCRETE

N Yuasa

Y Kasai

I Matsui

Nihon University

Japan

ABSTRACT. Inhomogeneity of structural concretes at the surface was studied in terms of pore structure and moisture distribution associated with drying immediately after demoulding. Lowering of the moisture content started at the surface and decreased, at the age of 28 days, influencing the outer 50 mm and being independent of the water-cement ratio. The surface had less bound water than the interior when drying started earlier and the water-cement ratio reduced. When the specimen was closer to the surface and subjected to drying at earlier stages, total pore volume and partial pore volume larger than 180 to 320 Å increased, which can be attributed to an interruption of hydration reactions.

Keywords: Structural concrete, Cover concrete, Moisture content, Pore structure, Drying, Inhomogeneity

Dr Noboru Yuasa is a Lecturer in the Department of Architecture and Architectural Engineering, College of Industrial Technology, Nihon University. His main interest is the durability of concrete structures covering both pore structure and moisture.

Dr Yoshio Kasai is an Emeritus Professor in the Department of Architecture and Architectural Engineering, College of Industrial Technology, Nihon University. His 45 years' research on concrete technology extends over a wide range including early-age properties, NDT of concrete structures, demolition and reuse of concrete and high-fluidity concrete.

Dr Isamu Matsui is a Professor in the Department of Architecture and Architectural Engineering, College of Industrial Technology, Nihon University. His main interest is the ergonomic aspect of building materials such as sensory evaluation of thermal, touching and aesthetic comfort.

INTRODUCTION

For practical reasons, formwork in concrete construction is removed at early ages, while hydration reactions are still proceeding. Because drying initiates from the surface of concrete, this results in a shortage of water necessary for hydration in regions near the surface. As a result, the surface concrete is likely to have reduced moisture and a coarser pore structure than those of the interior parts. These coupled phenomena may affect strength development and durability, and are closely related to the delamination-separation of finishing.

Though the effects of curing at early ages manifest as a function of depth from the surface, past studies [1-5] only consider relationships between curing conditions and resulting strength or durability as a whole, and the difference of moisture conditions and pore structure as functions of depth from the surface have never been extensively investigated.

This study deals with the effect of drying initiation ages on inhomogeneity of concrete in terms of moisture content and pore structure which are directly related to important properties such as strength and durability.

BACKGROUND

Among previous studies dealing with the distribution of moisture within the concrete structure, as a function of distance from the drying surface, Shiina [6] measured relative humidity inside concrete, and Tabata and others [7] used embedded electrodes but they simply showed the distribution of moisture conditions. Differences in pore structure along with the distance from the drying surface was studied by Chinou and others [8], in which it was pointed out that these became larger when the demoulding age was earlier and the water-cement ratio was larger. However, the specimens were cement pastes of 4x4x16 cm (drying area was 4x4 cm^2), the depth investigated was up to 37 mm from the drying surface and the moisture content was not measured.

This study develops previous investigations made by the authors [9, 10] and discusses changes in moisture content and pore structure as a cause of inhomogeneity, taking account of demoulding age, i.e. the age of drying initiation, and the effect of water-cement ratio.

AGE OF DRYING INITIATION AND INHOMOGENEITY

Experiments

Inhomogeneity of concrete subjected to drying was studied in a wide variety of mixture proportion and drying conditions. Water-cement ratio was 0.3, 0.4, 0.6 and 0.8, and the age of drying initiation was 1, 3, 7 and 28 days after mixing. The resulting variations of moisture content and pore structure were determined during 1 day to 1 year after mixing and at 28 days after mixing respectively.

Portland cement (density of 3.16 g/cm^3), river sand (surface-dry density of 2.62 g/cm^3 and F.M. of 2.83), river gravel (surface-dry density of 2.66 g/cm^3 and F.M. of 6.96) and chemical admixture were mixed at different water-cement ratio according to the mixture proportions in Table 1. Specimens were made with a standard metal form of 10x10x40 cm as shown in Figure 1. The compressive strength of these concretes, sealed cured at 20 °C, are shown in Table 1.

Table 1 Mixture proportions, air content, slump and mesured compressive strength of concrete

W/C	Unit water	Mass (kg/m^3)			Admixture(cc/m^3)		Air	Slump	Compressive strength (MPa)			
(%)	(kg/m^3)	Cement	Fine agr.	Coarse agr.	No.70	No.303A	(%)	(cm)	1day	3 days	7 days	28 days
0.3		616	545	976	-	6160	4.3	21.4	23.9	38.2	49.2	53.1
0.4	185	463	671	976	2315	-	4.6	15.6	12.1	24.5	33.6	40.6
0.6		308	838	939	770	-	4.7	21.1	4.86	13.3	19.4	28.8
0.8		231	865	976	577.5	-	4.5	20.3	2.81	7.37	11.7	15.6

After casting concrete, specimens were stored in an air-conditioned room at a temperature of 20°C and relative humidity of 60 %. Each end-side, 10 x10 cm, of the specimens was opened to indoor air for drying at the age of 1, 3 and 7 days and kept until the age of 1 year.

Moisture contents of the specimens were determined non-destructively by the ceramic moisture sensors developed by the authors [11] until the age of 1 year. The positions of the embedded sensors are shown in Figure 1.

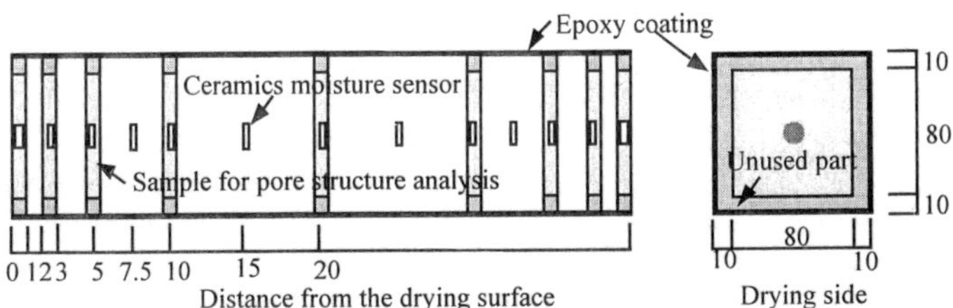

Figure 1 Dimension of the specimen

Bound water, and the non-evaporable water incorporated in the hydration products, were assumed to be lost on ignition at 600°C. As a measure of degree of hydration, the bound water content in a unit mass of hardened cement paste was defined. Specimens of 1 cm thickness were cut from the concrete prisms at depths of 0-1, 2-3, 4.5-5.5, 9.5-10.5, 19.5-20.5 from the drying surface, ground to a particle size between 2.5 to 5.0 mm, treated by acetone and then D-dryed.

The determination procedure of bound water content was to weigh the mass of a specimen before, W_0 (g), and after ,W_i (g), heating to 600°C for 1 hour. The heated specimens were then mixed with 10 % hydrochloric acid and stirred for 2 hours to dissolve cement components. Subsequently, the solution was heated at 600 °C for 1 hour, cooled in a desiccator and weighed as an insoluble residue W_{ns} (g).

The soluble component content (cement paste content in g/g) W_{Rs} and the bound water content W_{Rh} (g/g) can be calculated by equation (1) and equation (2) respectively.

$$W_{Rs} = (W_0 - W_{ns}) / W_0 \quad (1)$$

$$W_{Rh} = (W_0 - W_i) / W_0 \cdot W_{Rs} \quad (2)$$

Specimens used for bound water content determination were subjected to mercury porosimetry to determine pore structure ranging from 30 Å to 3.2 x 10^6 Å in radius. Effective pore volume V_{ep} (cc/g) that is the pore volume present in the cement paste in the specimen can be calculated from the measured pore volume V_{mp} (cc/g)using equation (3)[12],

$$V_{ep} = V_m \ / W_{Rs} \quad (3)$$

Results and Discussion

Variation of Moisture Content

Changes of moisture content with time are shown in Figure 2 with respect to drying initiation ages and water-cement ratio. All specimens showed rapid decrease in moisture content near the dying surface immediately after the initiation of drying. At the age of 28 days, the extent of the decrease in moisture content occurred over a depth of 50 mm from the surface and was independent of the water-cement ratio. Taking into account that hydration rate of available Portland cement reduces after 28 days, the hindered hydration zone can be estimated as 50 mm from the drying surface. When water-cement ratio is as high as 0.6 or 0.8 and the drying initiation age is as early as 1 day, the moisture content of concrete more than 50 mm from the drying surface changed significantly in relation to the distance from the surface.

Long-term moisture distribution, with little effects on drying initiation ages, resulted in an equilibrium moisture content which was a function of ambient temperature and relative humidity.

Specimens kept in a constant hygro-thermal condition, temperature of 20°C and relative humidity of 60 % for 1 year, resulted in a moisture content of 2.2 % at 5 mm from the drying surface irrespective of water-cement ratio and drying initiation ages. However, moisture content at the centre of the specimens, a 200 mm from the drying surface, resulted in larger value depending on water-cement ratio. It was 4.6 % and 5.5 % when water-cement ratio was 0.3 and 0.8 respectively.

Even before drying initiation, a decrease of moisture content uniformly from the surface to the concrete core was observed, which was more noticeable at low water-cement ratio and can be attributed to self desiccation. Self desiccation is an apparent drying phenomena caused by consumption of evaporable water by hydration. Therefore, the moisture content measured in this way is a consequence of evaporable water changes both by drying and hydration.

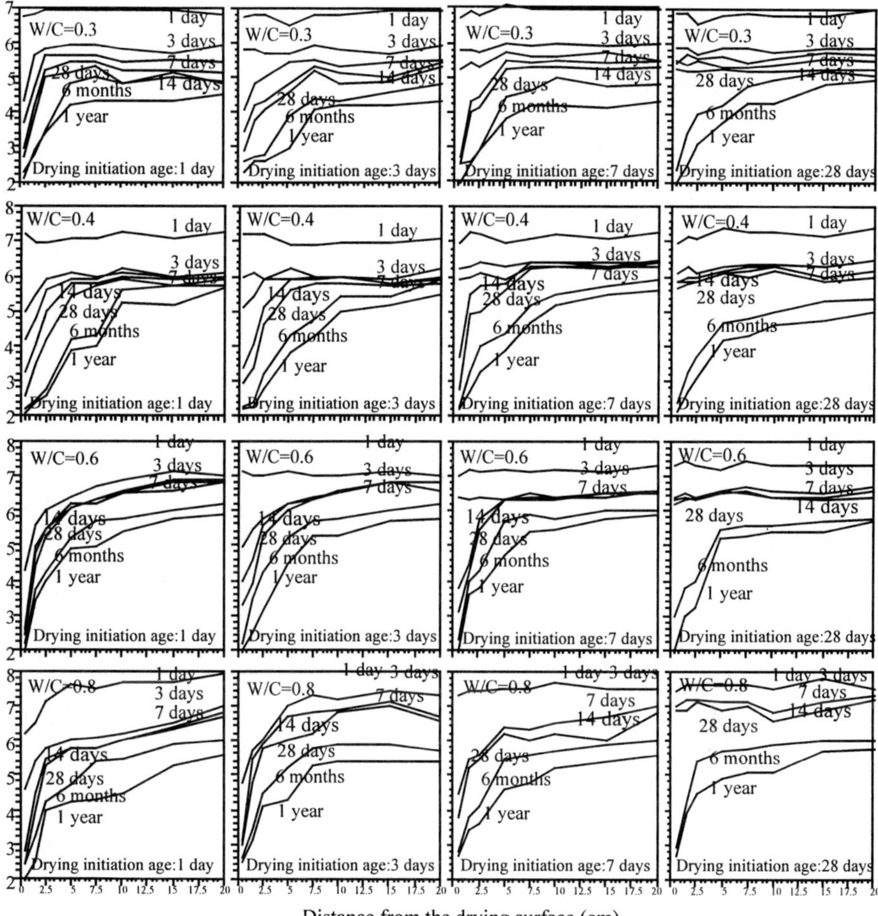

Figure 2 Effects of water-cement ratio and drying initiation age on moisture content

Distribution of Bound Water

Distribution of bound water at the age of 28 days is shown in Table 2. When the drying initiation age was earlier and the position was closer to the drying surface, the bound water content reduced. Taking undried specimens which were sealed until the age of 28 days as a reference, bound water contents of specimens dried from the age of 1 day with water-cement ratio of 0.3 was 70 % of that of the reference, and that with water-cement ratio 0.4 was 80 %. When the water-cement ratio was 0.8, the effect of drying initiation age on the bound water content was within the range of scatter of the test and no particular difference was observed. The effect of drying on the decrease of bound water was significant when water-cement ratio became smaller.

Table 2 Bound water content, total effective pore volume and median diameter

W/C	Drying initiation age (day)	distance from the surface(cm)	Bound water content (g/g)	Vep x10^{-4} (cc/g)	Median pore radius (Å)	W/C	Drying initiation age (day)	distance from the surface(cm)	Bound water content (g/g)	Vep x10^{-4} (cc/g)	Median pore radius (Å)
		0-1	0.154	1599	386			0-1	0.162	2208	637
		2-3	0.180	1272	299			2-3	0.187	1966	350
	1	4.5-5.5	0.201	1241	251		1	4.5-5.5	0.215	1674	238
		9.5-10.5	0.196	1180	236			9.5-10.5	0.214	1645	239
		19.5-20.5	0.190	1128	237			19.5-20.5	0.212	1576	235
		0-1	0.159	1400	364			0-1	0.172	2062	715
		2-3	0.183	1211	278			2-3	0.200	1716	385
	3	4.5-5.5	0.180	1177	247		3	4.5-5.5	0.217	1673	233
		9.5-10.5	0.178	1172	266			9.5-10.5	0.206	1609	228
		19.5-20.5	0.182	1151	228			19.5-20.5	0.206	1616	244
		0-1	0.171	1309	423			0-1	0.196	1744	444
		2-3	0.175	1297	257			2-3	0.215	1707	291
0.3	7	4.5-5.5	0.184	1261	248	0.4	7	4.5-5.5	0.220	1688	230
		9.5-10.5	0.183	1206	298			9.5-10.5	0.223	1659	240
		19.5-20.5	0.179	1177	245			19.5-20.5	0.213	1568	250
		0-1	0.184	1290	244			0-1	0.196	1531	328
		2-3	0.188	1210	240			2-3	0.213	1644	276
	sealed	4.5-5.5	0.183	1183	243		sealed	4.5-5.5	0.214	1620	233
		9.5-10.5	0.188	1285	266			9.5-10.5	0.211	1645	200
		19.5-20.5	0.187	1191	238			19.5-20.5	0.209	1627	217
		0-1	0.194	3644	1631			0-1	0.204	5053	2654
		2-3	0.211	2651	1535			2-3	0.197	4898	1536
	1	4.5-5.5	0.212	2541	561		1	4.5-5.5	0.211	4646	562
		9.5-10.5	0.215	2624	307			9.5-10.5	0.205	3691	350
		19.5-20.5	0.214	2631	358			19.5-20.5	0.197	3609	358
		0-1	0.186	3340	1740			0-1	0.211	4435	1740
		2-3	0.206	2646	867			2-3	0.199	4256	867
	3	4.5-5.5	0.207	2557	543		3	4.5-5.5	0.207	4214	372
		9.5-10.5	0.208	2588	317			9.5-10.5	0.199	3699	317
		19.5-20.5	0.208	2669	229			19.5-20.5	0.215	3661	329
		0-1	0.185	2669	1086			0-1	0.193	4055	1086
		2-3	0.204	2754	543			2-3	0.204	3880	619
0.6	7	4.5-5.5	0.207	2576	321	0.8	7	4.5-5.5	0.202	3794	321
		9.5-10.5	0.205	2575	275			9.5-10.5	0.204	3666	275
		19.5-20.5	0.204	2540	272			19.5-20.5	0.190	3654	272
		0-1	0.201	2691	203			0-1	0.193	3751	286
		2-3	0.203	2580	221			2-3	0.193	3779	282
	sealed	4.5-5.5	0.205	2840	244		sealed	4.5-5.5	0.196	3779	251
		9.5-10.5	0.198	2653	217			9.5-10.5	0.197	3682	290
		19.5-20.5	0.204	2653	229			19.5-20.5	0.201	3655	286

Pore Structure

The pore size distributions at the age of 28 days are shown in Figure 3. Samples near the drying surface showed pore size distribution shifting to larger pore side and had larger total pore volume, which was remarkable when the drying initiation became earlier and water-cement ratio became larger.

The total effective pore volume of all specimens is shown in Table 2. The relationship between distance from the drying surface and total effective pore volume is also shown. Past studies have shown that total effective pore volume is affected greatly by water-cement ratio, while in the present study, the total effective pore volume of specimens subjected to drying was larger than without drying when the drying initiation age became earlier and the position closer to the drying surface. The increase of total effective pore volume can be observed as a portion close to the drying surface. It was within 20 mm from the drying surface at water-cement ratio of 0.3, 50 mm at water-cement ratio of 0.4 and 0.6, and 100 mm at water-cement ratio of 0.8.

Total effective pore volume of specimens within 10 mm of the drying surface with water-cement ratio of 0.3 and drying initiation age of 1 day was nearly equal to that with water-cement ratio of 0.4 and without drying, and that water-cement ratio of 0.6 and drying initiation age of 1 day was nearly equal to that water-cement ratio of 0.8 without drying. These observations are significant when the relationship between curing conditions and water-cement ratio specified in the current standard practice is considered.

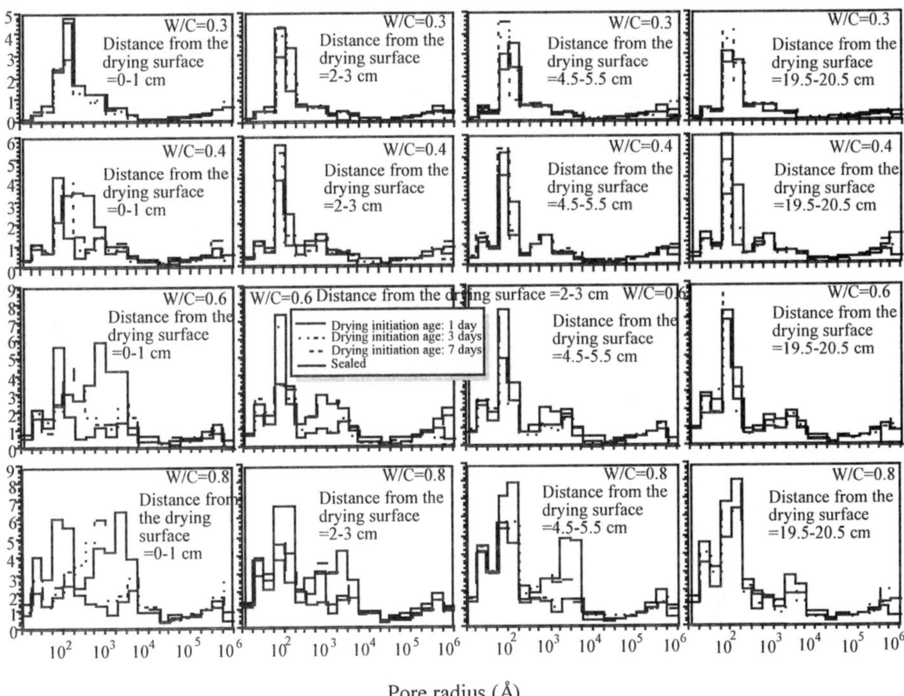

Figure 3 Effect of drying initiation age on pore size distribution

Table 2 shows the median pore radius, which is defined as a pore radius corresponding to 50 % of pore volume accumulated from the smaller side representing the deviation of distribution. It ranged from 200 to 300 Å for specimens without drying and was slightly affected by water-cement ratio, but when subjected to drying, the effect of the drying initiation age and distance from the drying surface was significant resulting in larger median pore radii as the drying effect became larger. The median pore radius shift greatly influenced by water-cement ratio.

The relative pore size distribution, a pore size distribution of dried concrete relative to undried concrete was considered, to examine changes of pore size due to drying. As shown in Figure 4, pore volume of dried concrete at a radius more than 180 Å or 320 Å increased while that of the other part decreased. This may be attributed to the interference of hydration due to drying resulting in the incomplete consolidation of pores at a radius more than 180 Å or 320 Å.

This tendency was significant when the drying initiation age was earlier and the position was closer to the drying surface, and the increasing pore volume extended over a larger pore radius. When water-cement ratio was large, this tendency was more conspicuous and reached a deeper portion from the drying surface.

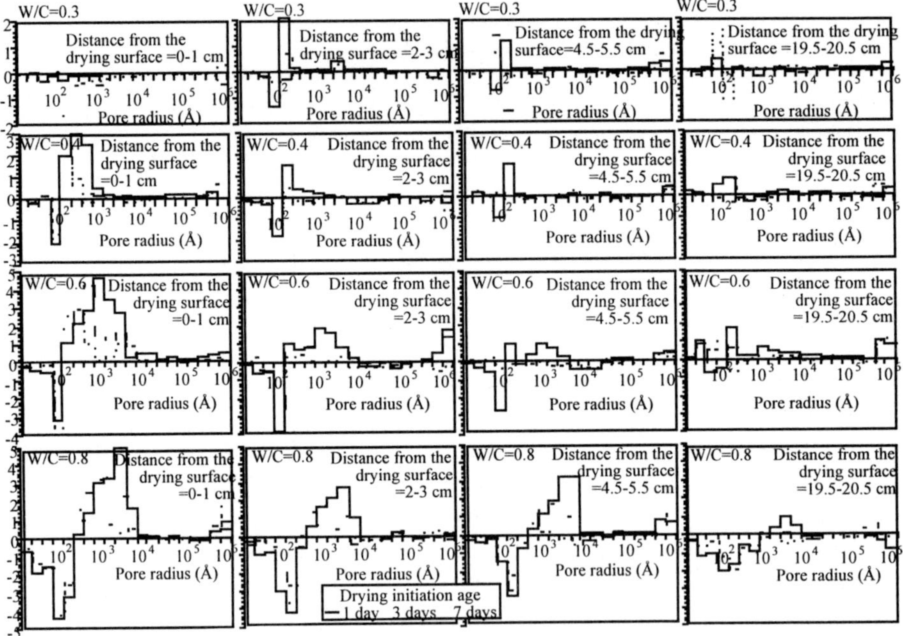

Figure 4 Increased or decreased pore volume due to drying

The degradation of reinforced concrete structures may be related to the increase in coarser pore volume, thereby, in order not to have coarser pores, the curing capable of preventing concrete at early ages from drying is necessary, especially at major water-cement ratio.

CONCLUSIONS

The inhomogeneity of moisture content and pore size distribution of concrete structures due to demoulding in the early ages has been discussed, and major findings of the work carried out, are as follows.

1. A decrease in moisture content started from the position near the drying surface immediately after drying. The decrease at the age of 28 days extended over 50 mm from the surface regardless of water-cement ratio. The long-term moisture distribution in specimens was independent of the drying initiation age and controlled by the distance from the surface and equilibrium moisture content, which is a function of ambient relative humidity and water-cement ratio.

2. The bound water content at a position near the drying surface was less than that of the interior part, when the drying initiation age was earlier. This decrease became significant when water-cement ratio was lower.

3. Interference of hydration due to drying resulted in an increase of pores with a radius more than 180 or 320 Å, in a decrease of the other part and in an increase of total effective pore volume when the drying initiation age was earlier and the position closer to the drying surface. This tendency became more noticeable at higher water-cement ratio.

4. The pore structure of concrete subjected to drying at early ages had a lasting effect. Coarser pores were likely to occur when the drying initiation age was earlier and the position closer to the drying surface.

REFERENCES

1. GILKEY, H J. The Moist Curing Concrete. Engineering News Record, 1937, Oct., pp 630-633.

2. PRICE, W H. Factors Influencing Concrete Strength. Journal of the American Concrete Institute, Feb., 1951, pp 418-432.

3. WATERS, T. The Effect of Allowing Concrete to Dry Before it has Fully Cured. Magazine of Concrete Research, 1955, Vol. 7, No. 20, pp 79-82.

4. IZUMI, I, et al. A Study on the Effects of Stripping Time of Sheathing and Initial Curing on Properties of Concrete in Structure. Journal of Structure and Construction Engineering, Architectural Institute of Japan, No. 449, 1993, pp 35-45.

5. SHIINA, K. Deformation and Internal Humidity of Concrete. Concrete Journal (Japan), Vol. 7, No. 6, 1969, pp 1-11.

6. TABATA, M, KOH, E AND KAMADA, E. Determination of Moisture Content by Electrode Method. Preprint of Annual Meeting of AIJ (Structure and Materials), 1976, pp 117-118.

7. CHINOU, S, HIRANO, T AND SHIIRE, T. Curing Conditions and Quality of Cover Concrete. Annual Report of Japanese Cement Association, No. 38, 1984, pp 266-269.

8. YUASA, N, KASAI, Y AND MATSUI, I. Quality of Cover Concrete: Change of Moisture Content and Pore Structure of Concrete Subjected to Drying. Prep. Annual Meeting of AIJ A, 1993, pp 449-450.

9. YUASA, N, KASAI, Y AND MATSUI, I. Quality of Cover Concrete: Change of Pore Structure of Cover Concrete Subjected to Drying. Prep. Annual Meeting of AIJ A, 1994, pp 199-200.

10. YUASA, N, KASAI, Y AND MATSUI, I. Measuring Method for Moisture Content in Hardened Concrete using Electric Properties of a Ceramic Sensor. Journal of Structure and Construction Engineering. Architectural Institute of Japan, No. 498, 1997, pp 13-20.

11. YOSHINO, T, KAMADA, E, TABATA, M AND YANAGI, T. Estimating of Concrete Strength Based on Pore Structure Consideration: Part I, Correlation between Compressive Strength and Pore Structure. Trans. Architectural Institute of Japan, No. 312, 1982, pp 9-17.

DEVELOPMENT OF HIGH DURABILITY CONCRETE BY USING PORTLAND CEMENT BASED BINDERS

V G Papadakis

Danish Technological Institute

Denmark

ABSTRACT. The effect of supplementary Portland cement-based binders, such as silica fume, fly ash, and slag, on concrete mechanical, physical, and chemical characteristics is briefly presented. Quantitative models for the estimation of the composition, the porosity, and the main durability characteristics of a concrete containing Portland cement-based binders are given. Using these expressions the estimation of the deterioration rate under carbonation or chloride penetration conditions is possible. The paper also reports the development of a high-durability concrete using low water-to-cement ratio, superplasticizer, and moderate quantities of additions (cementitious and pozzolanic fly ash and non-ferrous slag) which replace fine aggregates. Main mechanical properties and durability characteristics are presented and discussed. The conclusions from this work can contribute to a sustainable development of the concrete industry, improving service lifetime of the concrete structures and finding ecological solutions for disposal of industrial by-products.

Keywords: Carbonation, Chloride, Concrete, Durability, Fly ash, Modeling, Pozzolans, Slag.

Dr Vagelis G Papadakis is a chemical engineer, Ph.D., and Postdoctoral Researcher at the Danish Technological Institute (Building Technology, Concrete Centre, Taastrup, Denmark). His main research interest is in problems related to concrete durability (carbonation- ACI Wason Medal for Materials Research, chloride penetration, use of supplementary cementing materials).

INTRODUCTION

The good performance of concrete in service, including its durability, is equivalently a very important characteristic besides the usual required mechanical properties. Deterioration of concrete structures may be the result of a variety of mechanical, physical, or chemical processes. In reinforced concrete, the most serious deterioration mechanisms are those leading to corrosion of the reinforcement, which occurs after depassivation due to carbon dioxide or chloride-ion penetration [1,2]. Among the major problems facing the concrete engineering community at the end of this century is to improve concrete durability.

On the other hand, supplementary cementing materials have long been used as Portland cement additives or as active additions in concrete [2-5] due to economic and technological benefits. Among these materials are industrial by-products, such as fly ash from coal-burning electric power plants, slag from metallurgical furnaces, and silica fume from electric arc furnaces producing silicon and ferrosilicon alloys. These materials present cementitious properties either alone or in the presence of calcium hydroxide (pozzolanic activity), and thus improve the mechanical properties of concrete. Moreover, their rational use in concrete can retard the concrete deterioration mechanisms [6,7].

Contributions of practical or experimental character regarding the effect of these supplementary pozzolanic or cementitious materials on concrete properties abound in the literature [2,3,7]. Despite the wealth of practical information, efforts in the direction of description of the pozzolanic activity by chemical reactions and quantification of products and pore volume are very limited. This lack of theoretical approach has created a disagreement in the literature about the effect of these materials on concrete durability.

In the present work, quantitative models for the estimation of chemical and volumetric composition and the main durability characteristics of a concrete containing Portland cement-based binders are presented. These concepts are successfully applied in the development of a high-durability and high-strength concrete.

FUNDAMENTAL CONSIDERATIONS

Pozzolanic Activity and Products

The supplementary cementing materials (often referred all as pozzolans) can be analyzed in terms of oxides: CaO (C), SiO_2 (S), Al_2O_3 (A), Fe_2O_3 (F), SO_3 ($\bar{S}$), as the Portland cement, but in different proportions and compositions: They are richer in S, A, and F and poorer in C than the normal clinker. Moreover, a significant fraction of their constituent phases are in crystalline, non-reactive form. Their cementitious properties are due to the reaction of amorphous phases with calcium hydroxide (CH), and the formation of hydrated products with binding properties, similar to those produced during Portland cement hydration [3,7-9].Papadakis et al. [10] were the first to propose a general simplified scheme describing the pozzolanic activity in terms of chemical reactions, which was verified by an extended experimental investigation [11,12]. Thus, the main reactions in a high-gypsum system, in very simplified terms, are (using the cement technology notation):

$$2S + 3CH \rightarrow C_3S_2H_3 \quad (1)$$
$$A + C\bar{S}H_2 + 3CH + 7H \rightarrow C_4A\bar{S}H_{12} \quad (2)$$
$$C + H \rightarrow CH \quad (3)$$

As observed from Equation (1), the active silica of the pozzolan reacts with calcium hydroxide of low strength giving the higher strength component CSH. Calcium hydroxide originates either from the hydration of Portland cement or from the supplementary material itself; Equation (3). Due to this reaction and the reaction of the aluminate phase, Equation (2), an additional water binding takes place that reduces the total porosity. Therefore, if a pozzolan replaces aggregates and the water and cement content remain constant, a concrete of higher performance and durability than the control can be developed, due to a higher proportion in strength components and to a lower and more complicated pore volume.

Chemical and Volumetric Composition of Concrete

Let us consider as a base 1 m^3 of fresh concrete which is composed as follows:

C: kg cement / m^3 of concrete — ρ_C: cement density (kg/m^3)
P: kg pozzolan / m^3 of concrete — ρ_P: pozzolan density (kg/m^3)
A: kg aggregates / m^3 of concrete — ρ_A: aggregate density (kg/m^3)
W: kg water / m^3 of concrete — ρ_W: water density (kg/m^3)
D: kg admixtures / m^3 of concrete — ρ_D: admixture density (kg/m^3)
ε_{air}: m^3 of entrained or entrapped air / m^3 of concrete

The following balance equation should be fulfilled:

$$C/\rho_C + P/\rho_P + A/\rho_A + W/\rho_W + D/\rho_D + \varepsilon_{air} = 1 \quad (4)$$

Let us denote by $f_{i,c}$ and $f_{i,p}$ the mass fractions of the constituent i (i=C, S, A, F, $\bar{S}$) in the cement and pozzolan (silica fume, fly ash, or slag) respectively. Not all of the total mass of the oxide i in a pozzolan is active (only the glass phase). Let us denote by γ_i the weight fraction of the oxide i in the pozzolan which contributes to the pozzolanic reactions ("reactivity"). Taking into account the stoichiometry of the pozzolanic reactions and the Portland cement hydration as well, and using the molar weights and volumes of reactants and products, the amounts of the "finally" produced compounds and the concrete porosity (ε) can be determined (after "completion" of the hydration and pozzolanic activity). These amounts of CH, $C_3S_2H_3$ (CSH), in kg/m^3 of concrete, and ε are given as follows:

$$CH = \{1.321(f_{C,c} - 0.7f_{\bar{S},c}) - (1.851f_{S,c} + 2.182f_{A,c} + 1.392f_{F,c})\}\, C + \{1.321(f_{C,p} - 0.7f_{\bar{S},p}) - (1.851\gamma_S f_{S,p} + 2.182\gamma_A f_{A,p})\}\, P \quad (5)$$

$$CSH = 2.85(f_{S,c}\, C + \gamma_S f_{S,p}\, P) \quad (6)$$

$$\varepsilon = \varepsilon_{air} + W/\rho_w - \{0.249(f_{C,c} - 0.7f_{\bar{S},c}) + 0.191f_{S,c} + 1.118f_{A,c} - 0.357f_{F,c}\}(C/1000) - \{0.289(f_{C,p} - 0.7f_{\bar{S},p}) + 1.18\,\gamma_A f_{A,p}\}(P/1000) \quad (7)$$

When more than one pozzolan are used in the concrete mixture, then the term P in the above equations should be repeated for each pozzolan.

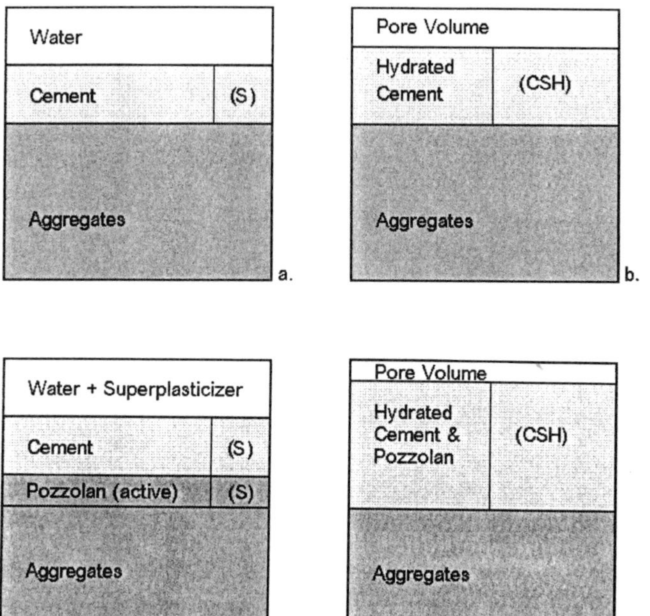

Figure 1 Schematic representation of the effect of pozzolan addition on hydrated products' content and pore volume. (a) fresh concrete without pozzolan; (b) hardened concrete without pozzolan; (c) fresh concrete with pozzolan replacing aggregates; (d) hardened concrete with pozzolan giving lower pore volume and higher hydrated product content

Application: For typical Portland cement: $f_{C,c}=0.63$, $f_{S,c}=0.21$, $f_{A,c}=0.05$, $f_{F,c}=0.035$, and $f_{\bar{S},c}=0.029$. For silica fume (SF): $f_{S,sf}=0.91$ and $\gamma_S=0.96$. For low-calcium fly ash (FL): $f_{S,fl}=0.54$, $f_{A,fl}=0.21$, and $\gamma_S=\gamma_A=0.82$ (high activity). For high-calcium fly ash (FH): $f_{S,fh}=0.39$, $f_{A,fh}=0.16$, $f_{C,fh}=0.23$, $f_{\bar{S},fh}=0.043$, and $\gamma_S=\gamma_A=0.71$. Thus, for the very common case of a concrete incorporating SF plus FL, the above equations are simplified:

$$CH = 0.26\,C - 1.62\,SF - 1.20\,FL \tag{8}$$

$$CSH = 0.60\,C + 2.49\,SF + 1.26\,FL \tag{9}$$

$$\varepsilon = \varepsilon_{air} + (W - 0.235\,C - 0.203\,FL) / 1000 \tag{10}$$

For the completion of the pozzolanic activity, the left-hand side of Equation (5) or (8) must be positive. Otherwise, there will not be enough lime solution to react with the entire quantity of the active S and A of the pozzolans. It is obvious from Equation (6) or (9) and illustrated in Figure 1, that when a pozzolan is used as an additive to concrete mixture replacing aggregates, the mass of the main strength component, CSH, always increases. On the other hand, when it replaces Portland cement, the CSH-content increases only if the content of the active silica in the pozzolan exceeds the silica content of the cement. This is usually the case in silica fume and low-calcium fly ash.

Rate of Concrete Carbonation

The evolution with time, t (s), of the carbonation depth, x_c (m), is given by the following analytical expression, based on fundamental reaction engineering modeling [13]:

$$x_c = \sqrt{\frac{2D_{e,CO2}(CO_2/100)t}{0.33CH + 0.21CSH}} \tag{11}$$

where, $D_{e,CO2}$ is the effective diffusivity of CO_2 in concrete (m^2/s), given by:

$$D_{e,CO2} = 1.64.10^{-6}\left(\frac{\varepsilon}{\frac{C}{\rho_C}+\frac{P}{\rho_P}+\frac{W}{\rho_W}}\right)^{1.8}(1-RH/100)^{2.2} \tag{12}$$

where, RH is the ambient relative humidity (%), and CO_2 the carbon dioxide content in the air at the concrete surface (%). Equations (11) and (12) are valid for $RH>50\%$ and $0.5<W/C<0.8$.

Rate of Chloride Penetration

It has been shown [14], that the most important parameter affecting the rate of chloride penetration in concrete is the effective diffusivity of Cl^-, $D_{e,Cl}$ (m^2/s). According to the parallel pore model the following semi-empirical equation was proposed (for NaCl, $0.5<W/C<0.7$):

$$D_{e,Cl} = \frac{2.4.10^{-10}}{\left(\frac{C}{\rho_C}+\frac{P}{\rho_P}+\frac{W}{\rho_W}\right)^2}\varepsilon^3 \tag{13}$$

As far as the steel bars have been depassivated, the corrosion progress depends on the relative availability of both water and oxygen. In this case, the oxygen diffusivity is given by an equation similar to Equation (12), where the term $1.64.10^{-6}$ is replaced by $1.92.10^{-6}$.

MECHANICAL TESTS

Materials and Specimen Preparation

A normal Portland cement and two supplementary cementing materials were used: a fly ash, categorized as cementitious and pozzolanic mineral admixture (10-20% CaO) [7], and a slag from a nickel production industry, categorized as mineral admixture of low reactivity [7]. The chemical composition of these materials is presented in Table 1. The by-products were ground to fulfill the EN 450 requirement limiting the amount retained on the sieve 45 μm to less than 40%. The mean particle diameter of all the materials was about 10 μm. Crushed aggregates were used with a typical grading and a maximum size of 3/8".

Table 1 Chemical analysis of materials and mixture proportions

OXIDE ANALYSIS (%)	C	S	A	F	$\bar{S}$	M	Resid.
Portland Cement	62.7	21.6	5.6	3.4	2.9	2.3	1.5
Fly Ash	16.0	49.2	16.1	10.4	2.8	3.6	1.9
Slag	5.2	36.3	10.8	40.5	-	3.0	4.2

CONTENT (kg/m^3)	C	W	A	P
Control	365	182.5	1825.0	-
F3	365	182.5	1770.2	54.8 (fly ash)
F6	365	182.5	1715.5	109.5 (fly ash)
F9	365	182.5	1660.8	164.2 (fly ash)
F9S (super/zer)	365	182.5	1660.8	164.2 (fly ash)
S3	365	182.5	1770.2	54.8 (slag)
S6	365	182.5	1715.5	109.5 (slag)
S9	365	182.5	1660.8	164.2 (slag)
S9S (super/zer)	365	182.5	1660.8	164.2 (slag)

The composition proportions were: water-to-cement ratio, W/C=0.5 and aggregate-to-cement ratio, A/C=5. With these proportions a mixture without supplementary binders was prepared ("control"), as well as three mixtures where 3, 6, and 9% of the aggregates had been replaced by fly ash (F3, F6, and F9, respectively) and other three mixtures where 3, 6, and 9% of the aggregates had been replaced by slag (S3, S6, and S9, respectively). It must be emphasized that in all these replacements all the other quantities remained the same, as shown in Table 1.

Two types of specimens were prepared: cylinders of 100 mm diameter and 150 mm height for compressive strength tests and beams of square cross-section 70 x 70 x 500 mm for flexural strength tests. Three specimens for every test were prepared. All the specimens after demoulding were placed into a water bath for six days and then they remained in room environment prior to their testing. The mixtures containing pozzolans presented lower slump than the control mixture, due to higher water adsorption from pozzolan particles during mixing. The mixtures F9 and S9 were particularly dry with almost zero slump, and as a final result their vibration was insufficient. For this reason new specimens of these mixtures were prepared, F9S and S9S, by using a high-range water-reducing admixture (ASTM C-494, type B, D and G) at a dosage of 0.5% of the cementitious constituents.

Mechanical Test Results

The results of compressive and flexural strength tests for all the specimens (mean values) are presented in Table 2, and comparative results are shown in Figure 2. Regarding the compressive strength, it is observed that both fly ash and slag gave higher strength than the control mixture, and among them fly ash was better. The increase of strength for the 90 days with 3% and 6% fly ash was about 13% and 15% respectively, whereas the corresponding percentages for 3% and 6% slag were 4% and 6% respectively. In the case of 9% fly ash or slag replacement without superplasticizers a dramatic reduction of the strength in both cases is observed, due to the inadequate slump that caused excessive porosity. But when superplasticizer was used (F9S and S9S) the compressive strength increased by 24% in the case of fly ash and 13% for the slag. It must also be emphasized that even in the case of F6 and S6 the requirement for superplasticizer was obvious and probably better performance could be achieved. By comparing also the results between 28 and 90 days the beneficial effect of the ageing on the strength of the mixtures containing pozzolans is clarified.

Table 2 Compressive and flexural strength results

SPECIMEN	COMPR. STRENGTH (MPa), 28 days	COMPR. STRENGTH (MPa), 90 days	FLEX. STRENGTH (MPa), 90 days
Control	45.6	46.7	6.66
F3	-	52.6	6.94
F6	51.9	53.9	7.12
F9	-	33.1	7.13
S9S	-	58.1	7.56
S3	-	48.8	7.04
S6	48.1	49.6	7.14
S9	-	43.9	7.21
S9S	-	52.9	7.71

Regarding the flexural strength results, the strength increase (90 days) for the mixtures F3 and F6 in relation to control mixture is 4% and 7% respectively, whereas for the mixtures S3 and S6 is 6% and 7% respectively. In the case of F9 or S9 the flexural strengths are slightly higher than those of F6 and S6 respectively and it seems that the inadequate vibration influences more the compressive strength. These strengths, however, are improved significantly when superplasticizer is used (mixtures F9S and S9S), where an important increase of 13.5% and 16% for fly ash and slag respectively, is achieved. It must also be emphasized that slag presented a better behaviour than fly ash in all cases.

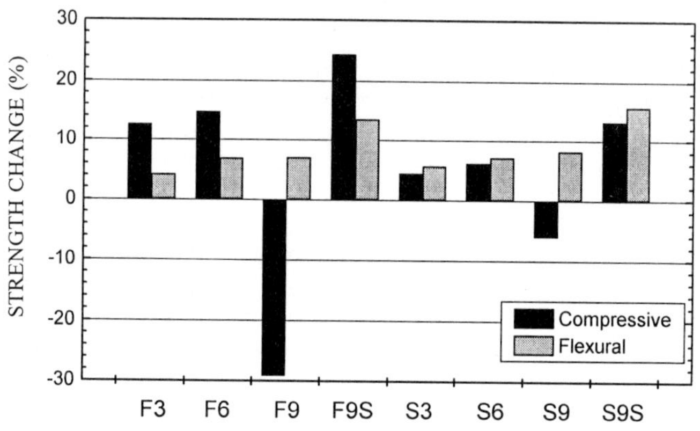

Figure 2 Change (%) of the compressive and flexural strength of pozzolanic mixtures in relation to control mixture

DURABILITY ASSESSMENT

The reliability of the fundamental expressions of durability characteristics, presented in the first part of this work, has been verified through experimental investigation in previous works [10-14]. Therefore, in the present work, the above expressions are used to estimate basic parameters for the durability of the mixtures presented, such as CSH and CH-content, concrete porosity, gas diffusivity, carbonation rate, chloride diffusivity, etc.Using the above equations and the composition parameters given in Table 1, the main durability characteristics were estimated and presented in Table 3. The only unknown parameters were the amorphous ratios γ_i of the oxides in the pozzolan which can be determined using either X-ray diffraction or chemical criteria [3,7]. In a previous work [11], it was found that CaO is almost completely active, i.e. $\gamma_C=1$, and for this particular fly ash, $\gamma_S=\gamma_A=0.5$ (moderate reactivity) and for this slag, $\gamma_S=\gamma_A=0.2$ (low reactivity). In both cases, the ferrous phase can be considered as non-reactive ($\gamma_F=0$). For the calculations, ε_{air} was considered zero. As shown in Table 3, the CSH-content increases with the replacement of aggregates by a pozzolan, in a higher degree for fly ash than for slag. On the other hand, the CH-content decreases in both cases. The concrete porosity decreases also significantly in the case of fly ash, giving much lower values for effective diffusivity of CO_2, O_2 and chloride ions than those in the control mixture. Due to these reductions, significant decreases in the propagation of carbonation and chloride penetration are observed. For the mixture F9, the carbonation depth decreases by 34% and the chloride diffusivity by 70%. In Figure 3a, the main mechanical strength characteristics for the fly ash mixtures, such as compressive and flexural strength, CSH-content and concrete porosity, are illustrated. The agreement between strength increase, CSH increase and porosity decrease is obvious. On the other hand, pure durability characteristics, such as carbonation depth and diffusivities of C1- ions and O_2 are given in Figure 3b. The beneficial effect of fly ash on concrete durability, as it is added in the mixture replacing aggregates, is also obvious.

Table 3 Main durability characteristics of pozzolanic mixtures

MIX	CSH (kg/m^3)	CH (kg/m^3)	ε	(1) $D_{e,CO2}$ $x10^8(m^2/s)$	(2) x_c (mm)	(3) $D_{e,Cl-}$ $x10^{12}(m^2/s)$	(1) $D_{e,O2}$ $x10^8(m^2/s)$
Control	219.0	94.9	0.09	2.12	20.8	2.41	2.44
F3	257.4	70.5	0.08	1.62	18.2	1.64	1.90
F6	295.7	46.0	0.08	1.23	15.8	1.11	1.44
F9	334.1	21.6	0.07	0.92	13.7	0.73	1.08
S3	230.3	88.7	0.09	1.84	19.3	2.01	2.15
S6	241.7	82.5	0.09	1.61	18.0	1.69	1.88
S9	253.0	76.3	0.09	1.40	16.8	1.42	1.64

(1) RH=65%
(2) RH=65%, CO_2=0.05%, t=50yr
(3) NaCl diffusion in saturated concrete

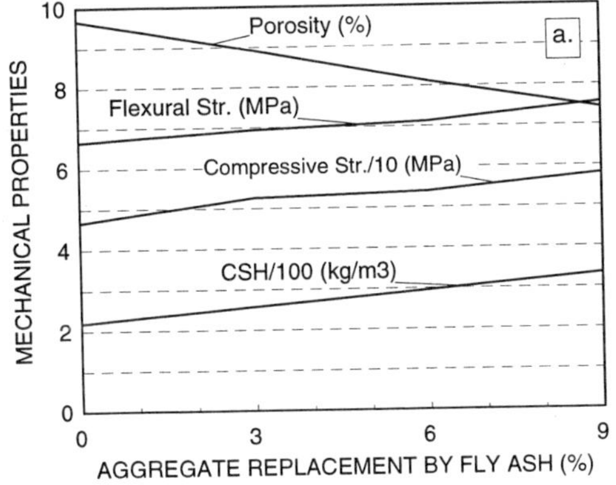

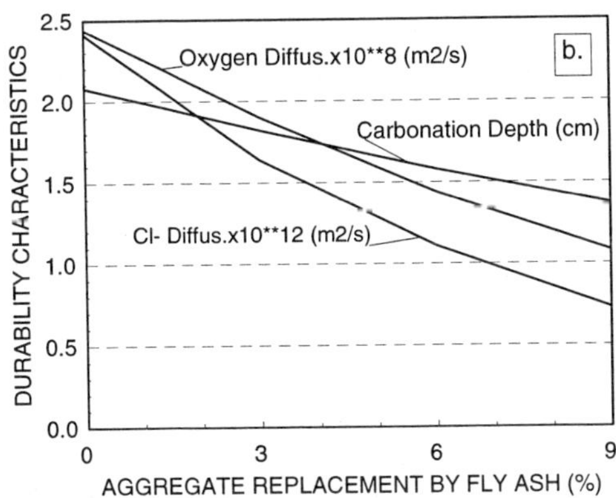

Figure 3 Effect of replacement of aggregates by fly ash on (a) mechanical properties and (b) durability characteristics of concrete

CONCLUSIONS

1. A quantitative yet practical model predicting the final chemical composition, porosity, and durability characteristics of a pozzolanic concrete is presented. Using the model expressions, the rates of carbonation and chloride penetration can be estimated.

2. These models and the experimental results show that when a pozzolan is used as an additive to the concrete mixture and the comparisons are made for the same water-to-normal Portland cement ratio, both strength and durability increase appreciably.

3. Among all mixtures were examined that with W/C=0.5, (A+P)/C=5, fly ash to replace 9% of aggregates, and superplasticizer, presented the best behaviour in both mechanical properties and durability characteristics giving a high durability and performance concrete.

ACKNOWLEDGEMENTS

The European Commission, DG XII (TMR Programme), and the Danish Technological Institute provided financial support for this work.

REFERENCES

1. CEB. Durable Concrete Structures - CEB Design Guide. Bulletin d'Information 182, Lausanne 1989. 268p.

2. NEVILLE, A M. Properties of Concrete. 4th ed. Longman, Essex 1995, pp 62-107, pp 649-723.

3. SERSALE, R. Aspects of the chemistry of additions. in: Advances in Cement Technology, Ed. S N Ghosh, Pergamon Press, New York 1983, pp 537-567.

4. MASSAZZA, F. Pozzolanic cements. Cem. Concr. Comp., Vol.15, 1993, pp 185-214.

5. MEHTA, P K. Role of pozzolanic and cementitious material in sustainable development of the concrete industry. Proc. 6th Int. Conf. on the Use of Fly Ash, Silica Fume, Slag, and Natural Pozzolans in Concrete, Ed. V M Malhotra, ACI SP-178, Bangkok 1998, pp 1-20.

6. DHIR, R K, EL-MOHR, M A K AND DYER, T D. Developing chloride resisting concrete using PFA. Cem. and Concr. Res., Vol.27, No.11, 1997, pp 1633-1639.

7. MEHTA, P K. Pozzolanic and cementitious by-products in concrete - Another look. Proc. 3rd Int. Conf. on the Use of Fly Ash, Silica Fume, Slag, and Natural Pozzolans in Concrete, Ed. V M Malhotra, ACI SP-114, Trondheim 1989, pp 1-43.

8. AITCIN, P C, AUTEFAGE, F,CARLES-GIBERGUES, A AND VAQUIER, A. Comparative study of the cementitious properties of different fly ashes. Proc. 2nd Int. Conf. on the Use of Fly Ash, Silica Fume, Slag and Natural Pozzolans in Concrete, Ed. V M Malhotra, ACI SP-91, Madrid 1986, pp 91-113.

9. URHAN, S. Alkali silica and pozzolanic reactions in concrete. Part 1: Interpretation of published results and a hypothesis concerning the mechanism. Cem. and Concr. Res., Vol.17, No.1, 1987, pp 141-152.

10. PAPADAKIS, V G, FARDIS, M N AND VAYENAS, C G. Hydration and carbonation of pozzolanic cements. ACI Mat. J., Vol.89, No.2, 1992, pp 119-130.

11. PAPADAKIS, V G. Supplementary Cementing Materials in Concrete- Activity, Durability, and Planning. Technical Report, Danish Technological Institute, Taastrup, Denmark, 1999.

12. PAPADAKIS, V G. Experimental investigation and theoretical modeling of silica fume activity in concrete. In press. Cem. and Concr. Res., 1999.

13. PAPADAKIS, V G, FARDIS, M N AND VAYENAS, C G. Effect of composition, environmental factors and cement-lime mortar coating on concrete carbonation. Materials and Structures, Vol.25, 1992, pp 293-304.

14. PAPADAKIS, V G, ROUMELIOTIS, A P, FARDIS, M N AND VAYENAS, C G. Mathematical modeling of chloride effect on concrete durability and protection measures. In Concrete Repair, Rehabilitation and Protection, Ed. R K Dhir and M R Jones, E. & F.N. SPON, London 1996, pp. 165-174.

CORROSION RESISTANCE INCREASE OF CEMENT BINDERS MODIFIED BY WASTE MINERAL ADDITIVES

J Jasiczak

A Lowińska-Kluge

Poznan University of Technology

Poland

ABSTRACT. The authors of this paper have for many years been engaged in scientific research of the effect of aggressive environments on concrete structures. The collected information directed the authors attention to the reasons for concrete failure, as well as the necessity of concrete protection. The exploration of materials for improving the concrete microstructure lead to the discovery of new additive AG (activated granules), and the use of different materials available in Poland namely silica fumes and fly ash. As a result of the complexity of the research programme in relation to other similar works, the assessment of the role of binders in mortar testing was required. The results of mortar stored in acid, sulphate and ammonium environment were analysed, and the influence of mortar composition on cement stone microstructure change was determined. The outcome of this research work is useful for designers and contractors building new waste and water treatment plants in Poland.

Keywords: Corrosion, Durability, Ordinary Portland cement, Blast furnace slag cement, Condensed silica fume, Fly ash, Microscopic and X-ray analysis

Professor Dr Jozef Jasiczak is Vice-Dean of the Faculty of Civil Engineering Poznan University of Technology, Poland. He specialises in the use of statistical methods for quality control ready mix production and the durability and protection of concrete structures working in the aggressive environment. He is a member of Buildings Materials Association Polish Academy of Science. Professor Jasiczak has published on many International Conferences Proceedings (Brussels'95, Dundee'96, Hyderabad'97, Düsseldorf'97, Tromsφ'98).

Dr Aldona Lowińska-Kluge is a Lecturer in Institute of Structural Engineering Poznan University of Technology. Her main research interests include the permeability properties of concrete, the prediction and modelling of concrete degradation processes. She discovered new mineral additive AG [3]. She published also in many International Conferences Proceedings (Hyderabad'97, Düsseldorf'97, Tromsφ'98).

INTRODUCTION

In Poland, during the last few years, the construction of waste-water treatments plants develops rapidly. For these plants the key technological structures from concrete are realised. In the result, the problem of cement selection (ordinary, or supplemented by additives), as well as of technology for concrete mixture preparation arises. The addition of superplasticisers and binder supplement by mineral additives of waste nature, are preferred. The very important problem is also the assurance of produced structures durability, assuming concrete protection. Investigations carried out by this paper Authors should to determine the influence of different aggressive environments (acid, sulphate and ammonium) on durability of concrete made from ordinary cement, from cement with slag admixture, or from cement supplemented on building site by fly ash or microsilica.

EXPERIMENTAL DETAILS

Materials

Portland cements, ordinary (OPC), blast furnace slag (BFS) and ordinary with additives fly ash (FA) and condensed silica fume (CSF) were used. A fine quartz sand of SiO_2 content ≥ 98% was used as a filler. Properties of binders will be presented on the conference. Using binders (OPC and BFS) are from the serial productions cement's industry, CSF is produced in so-called " submerged – arc electric furnaces, FA is the ash from coal power station. All these materials were studied by the authors and results presented in the Conference.

Mixture Proportions

Proportions of mortars used in Table 1 are shown.

Table 1 Mix proportions of cement mortars. Specimens 0,04×0,04×0,16 m

NO.	COMPONENTS	QUANTITY OF COMPONENTS, kg			
		OPC	BFS	FA	CSF
1.	Cement	0,45	0,45	0,27	0,45
2.	Sand	1,35	1,35	1,35	1,305
3.	Water	0,225	0,225	0,225	0,225
4.	FA	-	-	0,18	-
5.	CSF	-	-	-	0,045

Curing Environments

Two based curing conditions were used: clean water and different corrosion environments. The corrosion environments were simulated by:

- acid - anhydrous solution of HCl, pH = 4,
- sulphate - anhydrous solution of Na_2SO_4, ions concentration of SO_4^{-2} about 3750 mg/l,
- ammonium - anhydrous solution of NH_4Cl, ions concentration of NH_4^+ about 600 mg/l.

RESULTS AND COMPARISONS

Influence of Aggressive Environments on Concrete Strength

The mortars worked out according to data from Table 1 by 28 of days in water, and afterward by 12 of months in water (first half of batch) or in aggressive environment (second half of batch) were kept. After 28 of days, as well as after 1, 3, 6 and 12 months of samples curing, the mortars bending and compressive strength were assessed. Initial and final results of investigations in the Table 2 are shown.

Table 2 Mortar compressive strength and corrosion resistance coefficient values

MORTAR TYPE	f_C 28 DAYS	CHANGE IN MORTAR COMPRESSIVE STRENGTH FOR MORTARS STORED FOR 12 MONTHS (N/mm²)				CR
Storage Environment	H_2O	H_2O	HCl	NH_4^+	SO_4^{-2}	f_{cl} / f_{H2O}
PC	42.6	49.3-59.3	53.0-44.1	53.7-45.6	52.6-49.1	1.075-0.744
BFS	45.9	51.1-65.2	51.1-65.2	53.1-50.8	49.5-57.8	1.076-0.770
FA	23.6	31.4-45.0	31.4-45.0	37.1-36.4	36.2-41.1	1.242-0.836
CSF	71.0	78.2-106.1	78.2-106.1	79.8-89.1	79.6-91.6	1.031-0.807

Corrosive Resistance Coefficient

Comparing the strength of mortar after curing in aggressive environment against the strength of the same mortar after curing in water, the image of corrosion progress was obtained (see Figure 1).

This relations are described by thy corrosive resistance coefficient CR. Corrosion progress by coefficient CR value measured is different for different binders and different aggressive environments. Comparisons for selected environment in the last column of Table 2 are presented.

Rentgenostructural Investigations

Investigations on powder preparations, obtained from mortars after curing by 12 of months in aggressive and comparative water environments, were carried out. For identification of compounds existing in mortars, the software's X-DATA and X-VIEW were used. Diffractogramm examples for aggressive ammonium environment on the Figure 2 are shown.

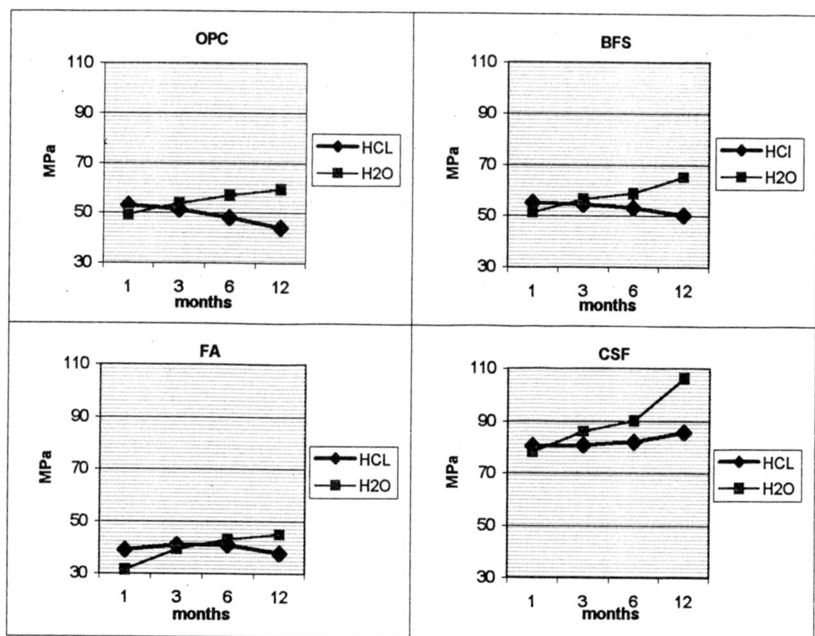

Figure 1 Changes strength of mortars curing in acid and water environments

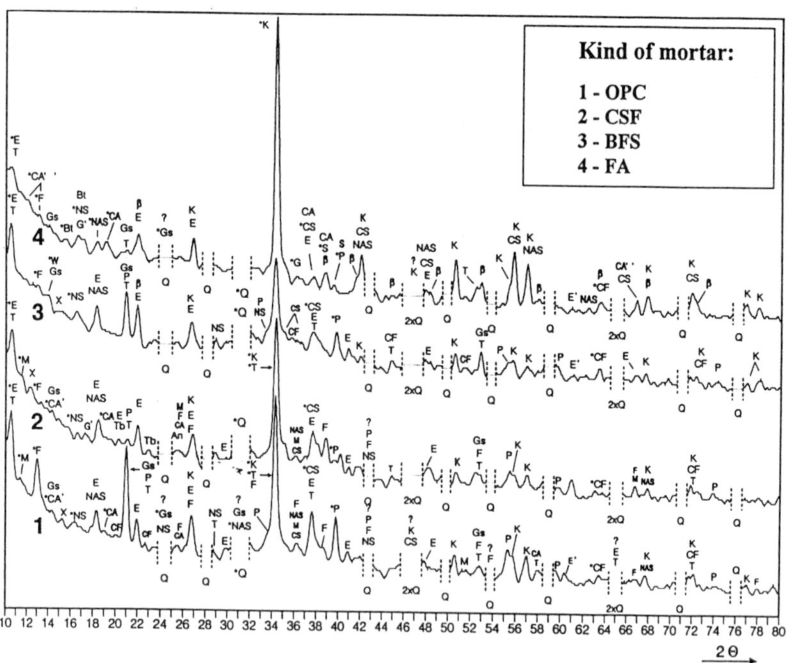

Figure 2 Comparison of diffractograms tested mortars kept by 12 months in ammonium environment

Microstructure Investigations

Microstructure images of 4 kinds of mortars after curing in water, and of 3 kinds from aggressive environments, using scanning microscope JSM 50 A JEOL equipped with roentgen microprobe were obtained. Mentioned images show the tested sample areas with material changes, provoked by different environments. Comparisons on the Figure 3 are shown.

CONCLUSIONS

1. Corrosive resistance of cement binders from the same clinker obtained, is diversified for different environment kinds. OPC cement is not resistant on action of applicated aggressive environments. The addition of milling slag (BSF) to the clinker, improves the resistance against acid and ammonium, and brakes sulphate corrosion as well.

   The addition of 40% FA to the OPC cement is favourable in the case of acid and ammonium corrosion, however is not so positive for sulphate corrosion. The addition of 10% CSF visibly increases the mortars strength, and brakes the corrosion process.

2. Mortars microstructure investigations confirm the presence of diversified phases of cement stone, according to environment kinds. In the environment of clean water, the portlandite, ettringite, tobermorite phases and additive derivatives (ash grains for FA and additional tobermorite phases for CSF, improving the tightness) are formed.

   For the other samples of mortars, the products of binders reaction with aggressive environment were found. Samples OPC in all cases show the microstructure destruction, as well as appearance of long- and short-fibre ettringite (sulphate corrosion).

   In the remaining mortars, the salts typical for environments (e.g. Friedel's salt for ammonium environment) or the compounds of lower expansively (monosulfates instead of swelling ettringite), were formed. Full qualitative and quantitative analysis by rentgenostructural method can be assured.

3. Investigations of mortars mechanical and chemical properties carried out confirmed, that bathing of microsilica as the additive for strength and chemical resistance improvement, is useful. By bathing of milling slag and fly ash to OPC, the improvement of binder chemical resistance and braking of corrosion process can be obtained as well.

   Application of such treatments is the shortest way to concrete structures exploited in aggressive environments durability assurance.

ACKNOWLEDGEMENTS

The authors would like to express their appreciation for the research grants made by Polish Committee of Research Works (No 11-552/99/DS).

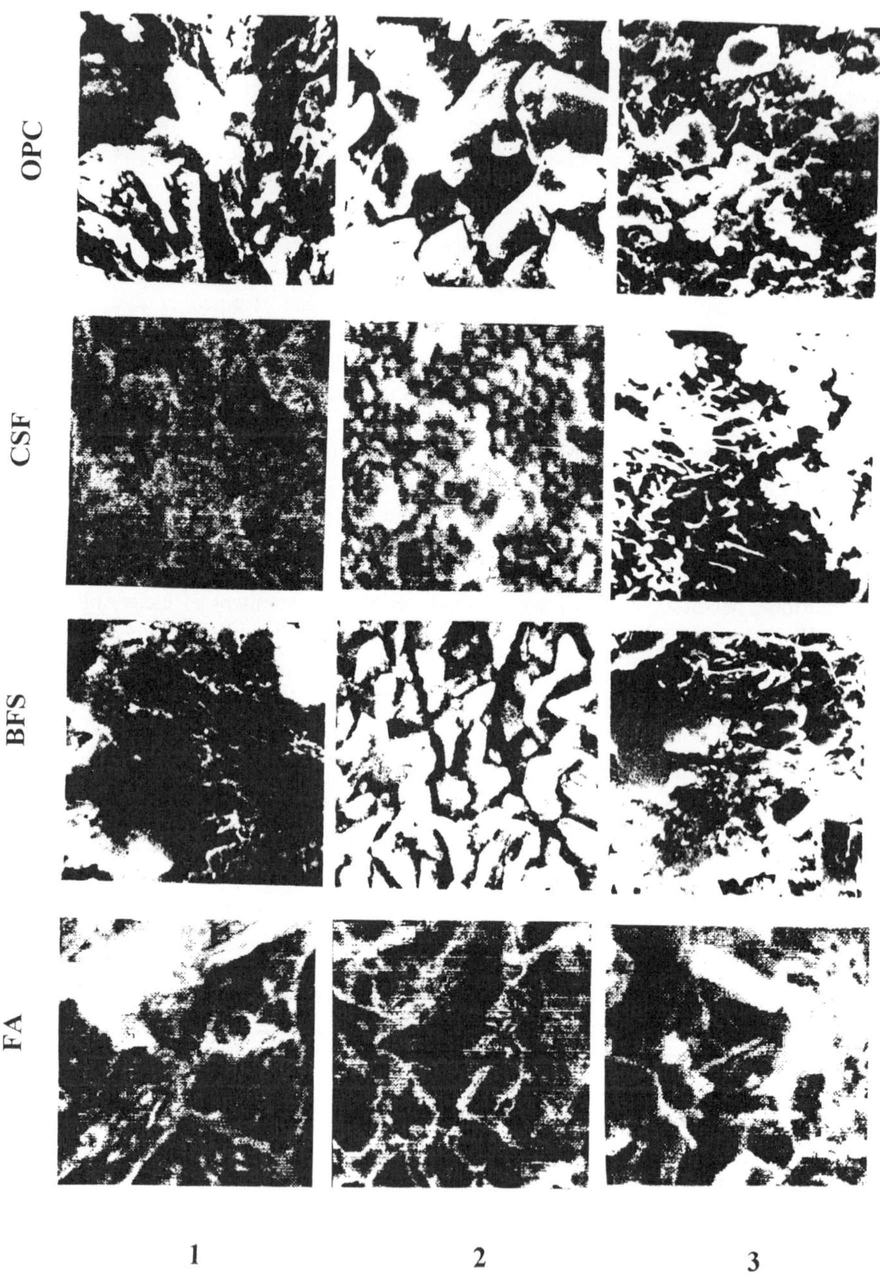

Figure 3 Microscopic photos four kind of mortars kept in:
1-acid, 2-ammonium, 3-sulphate environment

REFERENCES

1. JASICZAK, J. Reconstruction of the Concrete Sanitary Collectors Damaged by Hydrogen Sulphide Aggression. Proceedings. Int. Conference. Concrete In The Service of Mankind. Infrastructure and Utilities. Dundee, 1996, pp. 235 – 244.

2. MALHOTRA, V.M. RAMACHANDRAN, V.S., FELDMAN, R.F., AICTIN, P.C. Condensed Silica Fume in Concrete. CRC Press, Inc. Boca Raton, Florida, 1987. Pp. 221.

3. LOWINSKA-KLUGE, A. Patent No. 166057. Patent Office, Poland, 1995.

PROPERTIES OF CONCRETE CONTAINING ARTIFICIAL ZEOLITE MADE BY CHEMICAL CONVERSION OF FLY ASH

I Ujike

Ehime University

Japan

ABSTRACT. The artificial zeolite converted from fly ash is characterized by high cation exchange capacity, high absorptivity and high catalytic activity. It has been found that the newer performance is added to concrete by using the zeolite. This paper investigates the effects of incorporation of zeolite in concrete and on its hardened properties. Any decrease of slump due to the incorporation of zeolite is controlled by the use of superplasticizer. Compressive strength of concrete with water binder ratio of 50% and 65% is hardly affected by the incorporation of zeolite, although that of concrete with water binder ratio of 40% decreases with the increase of zeolite replacement ratio. The air permeability and chloride penetration are lowered by the use of zeolite.

Keywords: Artificial zeolite, Mix proportions, Compressive strength, Drying shrinkage, Permeability, Chloride penetration, Pore size distribution

Dr Isao Ujike is Associate Professor in Department of Civil and Environmental Engineering, Ehime University, Ehime, Japan. His research interests include the permeation properties of concrete, evaluation of pore structure and micro cracking in concrete and time-dependent behavior of reinforced concrete.

INTRODUCTION

The coal has come to be noticed again as energy resources from view point of the exhaustion of petroleum and safety of nuclear power. However, the disposal of fly ash discharged from thermal power station using coal for fuel is at stake. Much of fly ash has been disposed of as soil wastes for reclamation materials though same of that has been recycled as admixture for concrete. In future, it may become difficult to secure the reclaimed site with increasing the discharged fly ash. The conversion of fly ash into zeolite-like materials is one of techniques for recycling fly ash effectively[1]. The zeolite converted from fly ash is called artificial zeolite and is distinguished from natural zeolite or conventionally synthesized zeolite. The artificial zeolite is characterized by high cation exchange capacity, high absorptivity and high catalysis compared to natural zeolite and is cheaper than conventionally synthesized zeolite[1].

It has been reported that the addition of zeolite in concrete gives the effect of preventing the expansion due to alkali aggregate reaction[2]. Furthermore, studies have shown that performance such as humidity controlling, absorption of nitrogen oxides (NOx) and water purification is added to concrete by using zeolite[3][4][5]. However, in order to use the artificial zeolite more actively as admixture for concrete, it is essential to be aware of fundamental properties of concrete using artificial zeolite.

This study investigates the effects of incorporation of artificial zeolite on mix proportion, compressive strength, drying shrinkage and permeability and chloride ingress examined.

EXPERIMENTAL PROCEDURES

Materials

Portland cement was used. Sea sand with specific gravity of 2.55, fineness modulus of 2.68 and crashed gravel with maximum size of 20mm, specific gravity of 2.62 and fines modulus of 6.52 were used as fine and coarse aggregate, respectively. Two types of chemical admixtures were used, superplasticizer composed of polycarboxylic ether and air entraining agent composed of alkylarrylsulfonic compound.

Commercially available artificial zeolite with sodium ions was used in the experiments except for chloride penetration test. Conversion of fly ash into zeolite is carried out as follows[1]: The slurry is made by adding 2 mol/*l* NaOH water solution to fly ash. The slurry is heated at about 100°C for five hours. After the completion of reaction, the contents are washed with water to eliminate the excess of NaOH and then are air-dried. About 60% to 80% of fly ash is converted into zeolite with sodium ions. The artificial zeolite consists mainly of faujasite and phillipsite. The zeolite is crystallized on particle surface of fly ash as shown in Figure 1. The physical properties and chemical composition of artificial zeolite given in Table 1.

Chloride penetration test was carried out using zeolite with aluminum ions. The zeolite with sodium ions is stirred with 0.05 mol/*l* $AlCl_3$ water solution, followed by washing with water and air-drying. The zeolite with aluminum ions is made by cation exchange capacity of zeolite itself.

Table 1 Chemical compositions and physical property of artificial zeolite

CHEMICAL COMPOSITION (%)									SPECIFIC GRAVITY
SiO_2	Al_2O_3	Fe_2O_3	CaO	MgO	Na_2O	K_2O	TiO_2	Ig.loss	
41.9	19.4	3.1	1.6	0.6	5.1	0.9	1.0	25.3	2.38

Table 2 Mixture proportions of concrete and mortar

W/(C+Ze) (%)	Ze/(C+Ze) (%)	s/a (%)	UNIT WEIGHT (kg/m^3)					SP (g/m^3)	AE (g/m^3)
			Water	Cement	Zeolite	Aggregate			
						Fine	Coarse		
40	0	44	170	425	0	720	948	3400	17
40	10	44	170	383	42	709	940	6400	34
40	20	44	170	340	85	704	934	9400	64
40	30	44	170	297	127	700	927	19100	11
50	0	46	170	340	0	777	951	3400	14
50	10	46	170	306	34	774	946	5780	18
50	20	46	170	272	68	770	941	7480	25
50	30	46	170	238	102	766	936	9520	31
54	10	47	185	306	34	782	908	3400	71
59	20	47	200	272	68	767	865	3400	95
62	30	48	210	238	102	750	847	3400	150
65	0	49	170	262	0	860	931	5240	10
65	10	49	170	236	26	856	928	4980	26
65	20	49	170	209	52	854	924	5240	34
65	30	49	170	183	78	852	920	6550	47
65	0	---	272	419	0	1373	---	2930	17
65	10	---	272	375	42	1364	---	5030	34
65	20	---	272	333	83	1356	---	8800	54
65	30	---	272	291	125	1347	---	12570	75

Ze:Artificial zeolite SP:Superplasticizer AE:Air entraining agent

Mix Proportions

Table 2 shows the mix proportions of concrete and mortar used in the tests. The mixtures of concrete were proportioned to have a slump of 8"2cm and air content of 4.5"1%. Water binder ratio (W/(C+Ze)) were changed from 40 to 65%. Replacement ratio of zeolite were varied in three ways, 10, 20 and 30%.

Test Procedures

For each concrete, three sets of cylindrical specimens (10φ x 20cm) were cast, and tested in compressive strength and elastic modulus. Specimens were cured in water at 20 °C for the period of 7, 28 and 91 days.

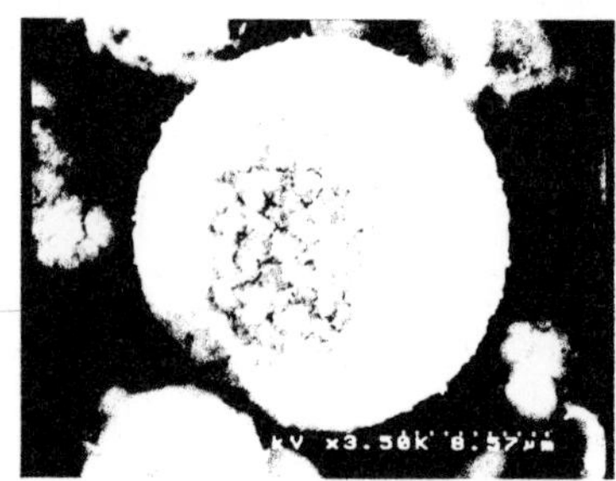

Figure 1 SEM photograph of artificial zeolite

Drying shrinkage was measured using two sets of concrete prisms (10x10x20cm). The age when drying start was 14days. Gauge points were put on top and bottom surface of prism and concrete shrinkage was measured by dial gauge with minimum reading of 1/1000mm. Test was carried out in the room maintained at 20°C and 70%R.H.

Air permeability test was carried out. Two sets of concrete prisms (15x15x5cm) were cast, and cured in water for the period of 14 days. After that, specimens were cured in air until drying rate of 3%. Drying rate is ratio of volume of evaporated water to volume of specimen. Air pressure of 0.2N/mm^2 was applied to specimen.

For chloride penetration test, Two sets of mortar prisms (4x8x16cm) were cast, cured in water for the period of 14 days and was immersed in NaCl water solution with concentration of 10%. Specimens were covered with epoxy resin adhesives except for one surface (4x8cm). After immersion of 1 month, samples were taken from each 1.5cm from unsealed surface in order. Total chloride content and water soluble chloride content were measured in accordance with the method of Japan Concrete Institute[6].

Mortar specimen cured in water for 14 days is crushed into grain of about 5mm and was dried at 105 °C until reaching constant weight. After pretreating, the pore size distribution of mortar was measured by the mercury penetration method. The mercury penetration test was carried out at least two times for each mixture.

Tests mentioned above were not conducted for concrete containing fly ash. Because, the fly ash as raw material corresponding to artificial zeolite used in this study was not obtained.

RESULTS AND DISCUSSION

Mix Proportions

Figure 2 shows effects of zeolite replacement ratio on unit water content and on dosage of air entraining agent. The unit water content and dosage of air entraining agent increase almost linearly with the increase of zeolite replacement ratio. The moisture absorption of zeolite is contributed to the increase of unit water content. Furthermore, in general, fly ash replacement mixtures result in better flowability than mixtures without fly ash due to effect of spherical particles of fly ash. Although the artificial zeolite is made from fly ash, the advantageous effect on reduced unit water content is not expected for artificial zeolite. Because, the particle surface of fly ash become uneven by crystallization of zeolite as shown in Figure 1 previously. Furthermore, the increase of dosage of air entraining agent is caused by carbon in artificial zeolite. The carbon contained in fly ash remains in artificial zeolite.

Any decrease of slump due to the incorporation of artificial zeolite is controlled by use of superplasticizer in this study. The dosages of superplasticizer to obtain the required slump, provided that unit water content is constant, are shown in Figure 3. The dosage of superplasticizer is expressed in percentage of unit binder content. The dosage of superplasticizer in water binder ratio of 65% is almost the same except in the case of zeolite replacement ratio of 30%. For water binder ratio of 40% and 50%, the dosage of superplasticizer increases with the increase of zeolite replacement ratio and there is no significant difference in dosage of superplasticizer. But, in the case of water binder ratio of 40% and zeolite replacement ratio of 30%, the concrete with required slump is not obtained. The consistency of concrete changes with the increase of dosage of superplasticizer discontiniously. That is, the slump of concrete added superplasticizer up to 4% is below 5cm. And when the dosage of superplasticizer is more than 4%, concrete comes to have high cohesion and slump reaches 20cm or more.

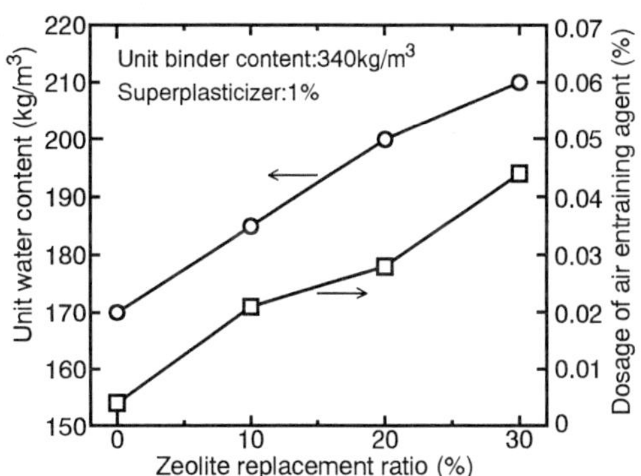

Figure 2 Effects of zeolite repolacement ratio on unit water content and dosage of AE agent

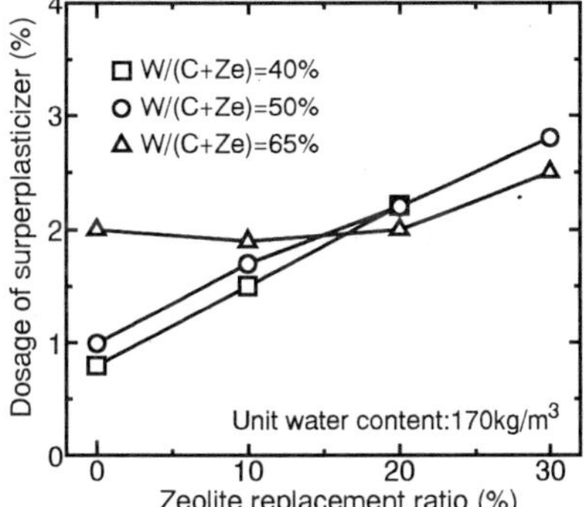

Figure 3 Effect of zeolite replacement ratio on dosage of superplasticizer

Compressive Strength

Relations between compressive strength and zeolite replacement ratio are shown in Figure 4. Compressive strength of concrete with water binder ratio of 40% decrease with the increase of zeolite replacement ratio independent of curing period in water. When water binder ratio is 50%, compressive strength of concrete cured in water for 7 days also decrease with the increase of zeolite replacement ratio. But, compressive strengths of concrete containing zeolite after 28 days and 91 days almost the same as that of concrete without zeolite. In the case of water binder ratio of 65%, though compressive strength decreases by incorporation of zeolite, the difference in zeolite replacement ratio hardly affects the compressive strength. The effect of incorporation of zeolite on elastic modulus is almost the same tendency as that on compressive strength, though the results are not illustrated.

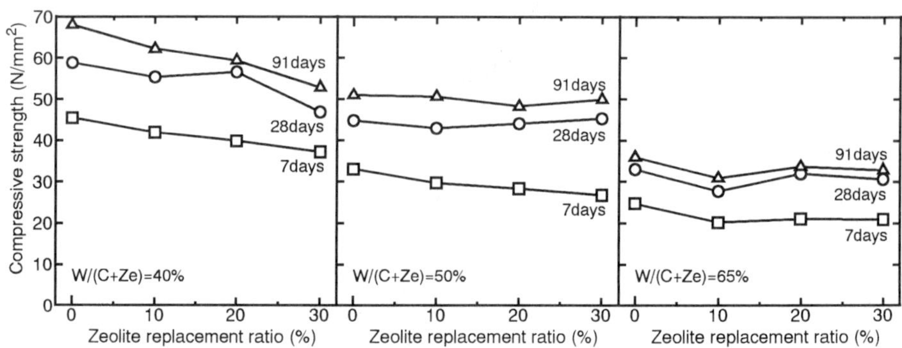

Figure 4 Relation between compressive strength and zeolite replacement ratio

The development of compressive strength with time is shown in Figure 5. The rate of compressive strength development of concrete with zeolite is approximately the same as that of concrete without zeolite. The artificial zeolite made from fly ash as raw material is considered to have at least pozzolanic activity because about 60% to 80% of fly ash is converted into zeolite.

Drying Shrinkage

Figure 6 shows the time dependent change of concrete shrinkage strain. The incorporation of zeolite hardly affects the development of drying shrinkage. Figure 7 shows time dependent change of loss of water. The loss of water from concrete increases with the increase of zeolite replacement ratio. This increase may be due to moisture absorption and desorption of zeolite. Therefore, drying shrinkage of concrete with zeolite is lower than that of concrete without zeolite when compared at the same level of water loss. As the concrete incorporating zeolite has humidity controlling function, it is necessary to test the drying shrinkage under varying humidity condition like actual environment.

Air permeability

Figure 8 shows effect of zeolite replacement ratio on air permeability coefficient. The air permeability coefficient decreases by incorporation of zeolite. Partial replacement of cement by zeolite above 20% has the effect of lowering the air permeability. This tendency corresponds to the measured result of pore size distribution by the mercury penetration method, to be shown later.

Chloride penetration

Figure 9 shows the total and water soluble chloride content in the surface layer (0 to 15mm) of mortar specimen immersed for 1 month. The total and water soluble chloride content decreases with the increase of zeolite replacement ratio. These reduction due to the incorporation of zeolite is reflected in the change in pore structure of mortar. Figure 10 compares the pore volume and the pore size distribution of mortar incorporating zeolite with those of mortar without zeolite.

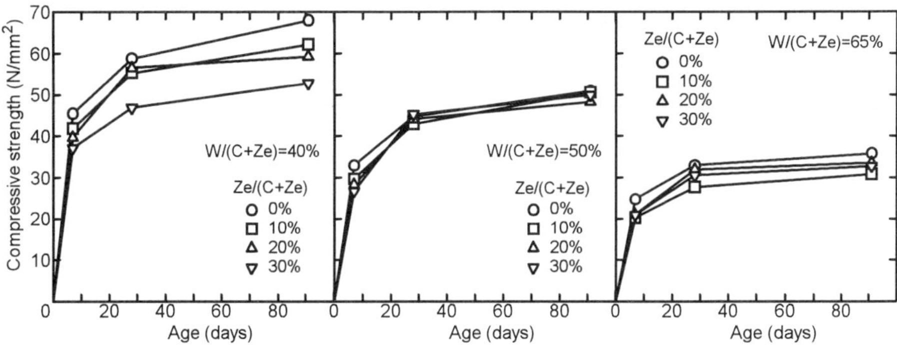

Figure 5 Relation between compressive strength and age

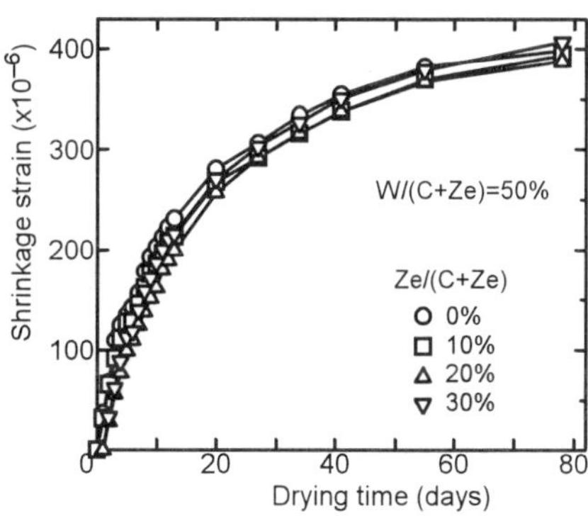

Figure 6 Development of drying shrinkage

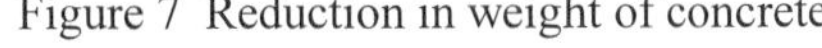

Figure 7 Reduction in weight of concrete

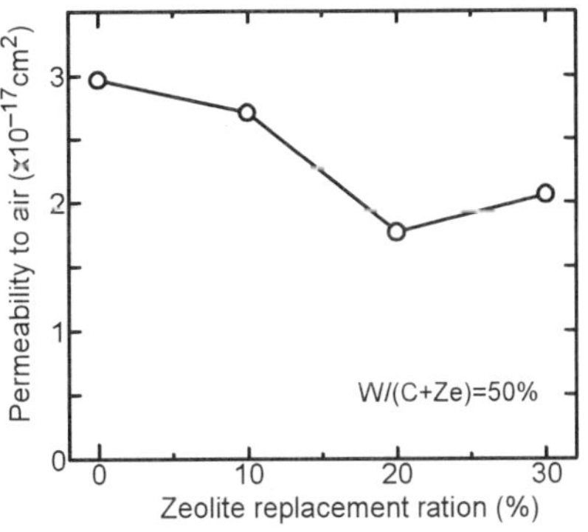

Figure 8 Effect of zeolite replacement ratio on air permeability coefficient

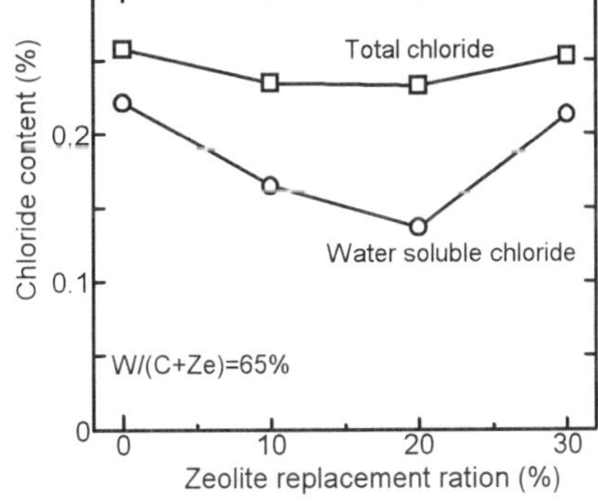

Figure 9 Effect of zeolite replacement ratio on chloride content

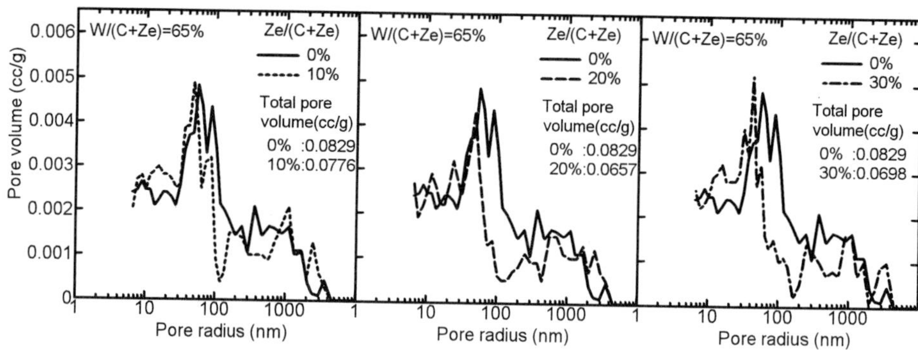

Figure 10 Pore size distribution

By incorporation of zeolite, the pore volume ranging in size from about 50nm to 1000nm decreases and the volume of pore with size of about 50nm or less increases. Consequently the total pore volume decreases. It is noticeable in the case of zeolite replacement ratio of 20% and 30%. However, the decrease of water soluble chloride content due to incorporation of zeolite is not explained only by the change in pore structure. The difference between total and water soluble chloride content indicates the fixed chloride content. The fixed chloride contents of mortar with zeolite replacement ratio of 20% and 30% is greater than those of mortar with zeolite replacement ratio of 0% and 10%. This may be explained by the chloride binding capacity produced by the zeolite. It is supposed that aluminum ions emitted from zeolite by the cation excahneg capacity may form an insoluble complex, calcium chloro-aluminate hydrate. The effect of partial replacement of cement with zeolite on chloride penetration should be further investigated.

CONCLUSIONS

The results obtained in this study can be summarized as follows:

1. The unit water content and dosage of air entraining agent increase almost linearly with the increase of zeolite replacement ratio when slump and air content are maintained at the same value as concrete without zeolite.

2. When unit water content is constant, any decrease of slump due to the incorporation of artificial zeolite is controlled by use of superplasticizer.

3. Compressive strength of concrete with water binder ratio of 40% decrease with the increase of zeolite replacement ratio. In the case of water binder ratio of 50% and 65%, the difference in zeolite replacement ratio hardly affects the compressive strength.

4. There is no significant difference in the development of drying shrinkage between concrete with zeolite and without zeolite.

5. The partial replacement of cement with zeolite makes the pore structure fine and reduces the air permeability coefficient and the penetration of chloride.

ACKNOWLEDGMENT

The author gratefully acknowledges funding for this study from Shikoku Construction Public Utility Association and Japan Cement Association.

REFERENCES

1. HENMI, T. Synthesis of hydroxy-sodalite ("zeolite") from waste coal ash. Soil Science and Plant Nutrition, 1984, Vol.33, No.3, pp.517-521.

2. WANG, Z Y, SATO, H, NAGAOKA, S and NAKANO, K. Studies on effectiveness on natural zeolite for preventing expansion due to alkai-aggregate reaction. CAJ Proceedings of Cement & Concrete, 1990, No.44, pp.470-475.

3. SAGAE, A. Some characteristics and utilizations of zeolite mixed humidity controlling concrete panel. Proceedings of the 26th Symposium on Thermal Environment AIJ, Tokyo, 1996, pp.45-52.

4. TAMAI, M. NO*x* absorbing concrete. Concrete Journal, 1998, Vol.36, No.1, pp.33-36.

5. HAYASHI, M, MIZUGUCHI, H, UEDA, T and MASUDA, S. Effects of blast furnace slag and artificial zeolite on water purification of porous concrete. Proceedings of the 4th Annual Conference of the JSCE Shikoku Chapter, Tokushima, 1998, pp.390-391.

6. JAPAN CONCRETE INSTITUTE. The proposal of standards for testing and protection of corrosion of reinforced concrete structures

THE IMPORTANCE OF MECHANICAL BOND BETWEEN CEMENT PASTE AND AGGREGATE

N Buch

J Early

Michigan State University

United States of America

ABSTRACT. The bond between cement and aggregate results from some combination of mechanical interlocking between cement hydration products and the aggregate surface, and chemical bond resulting from a reaction between the cement paste and aggregate. Bond between aggregate and cement is an important factor in the strength of concrete, especially the flexural strength, but the nature of bond is not fully understood, nor is there a standard test procedure to measure bond. It has been hypothesized that aggregate type and gradation, aggregate texture, surface treatment of aggregate, mixing process and mineral admixtures affect the properties of bond. There is in general, an agreement that the interface between cement and aggregate is the "weak link" in concrete, since it has more porosity than bulk paste. This paper investigates and summarizes the factors that affect the morphological, mechanical and chemical nature of the cement-aggregate bond .

Keywords: Transition zone, Interfacial bond, Surface texture, Aggregate, Mechanical bond, Aggregate interlocking.

Dr Neeraj Buch is an assistant professor in the department of Civil and Environmental Engineering at Michigan State University in East Lansing, Michigan, U.S.A. His research interests are in the design and analysis of rigid pavements, design and analysis of rehabilitation alternatives for rigid pavements, and fiber reinforced concrete properties. He is a member of several professional societies in the field of transportation and concrete materials.

Mr Jason Early is a graduate research assistant in the department of Civil and Environmental Engineering at Michigan State University in East Lansing, Michigan, U.S.A. He is pursuing a Masters degree in Civil Engineering and is expected to complete the requirements by December, 1998.

INTRODUCTION

It is routine to consider portland cement concrete (PCC) as a material consisting of three phases: hardened cement paste (HCP), aggregate and the interfacial transition zone (ITZ). Clearly, the strength of the concrete depends on the intrinsic strength of the HCP and possibly of the aggregate and upon the strength of the bond between HCP and the aggregate. An investigation of the stress-strain properties of aggregate, HCP and concrete reveals that even though the aggregate and HCP individually exhibit linear elastic behavior, the composite concrete shows signs of non-linearity, particularly at high stress levels. This indicates that the interface between the two constituent media, namely the ITZ, contributes to the apparent ductility of the concrete [1].

Investigation of the interfacial bond between aggregate particles and the cement paste indicates that the nature of the hydrated cement paste at the interface is different from the bulk cement paste. There is general agreement that the interface between cement and aggregate is the "weak link" in concrete, since it has more porosity. It has been found that the interface porosity is relatively high and decreases towards the bulk paste [2]. It is in this weak zone that micro-cracks initiate and propagate around the aggregates, which implies that the fracture resistance of the interface is lower than that of the bulk paste. Figure 1a. illustrates the nature and path of crack propagation at 12 hours of age, whereas Figure 1b. illustrates the nature and path of crack propagation after 28-days of curing because the mortar-aggregate bond is stronger and the crack travels through the aggregate

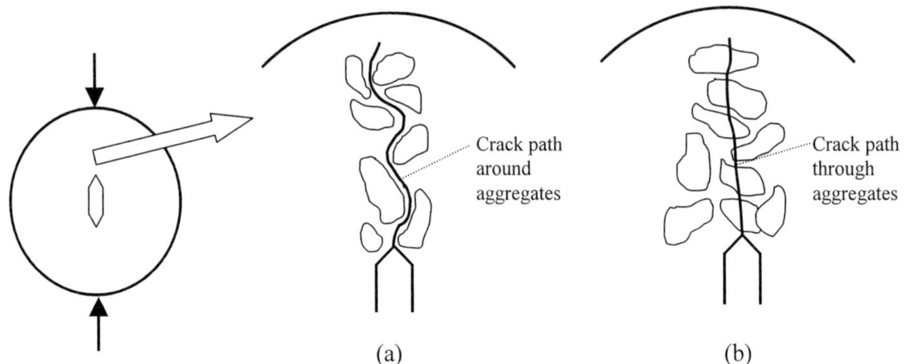

Figure 1 Schematic of crack paths, (a) 12 hours; (b) 28-days

The structure of the interfacial compounds have been shown to be dependent upon the physical and chemical properties of the aggregate type and gradation, aggregate texture and surface treatment and mixture design parameters. Strubl et. al. in 1980 [3], presented a comprehensive summary of the research findings on the cement paste-aggregate bond published up to that time. The summary stated that the bond between cement and aggregate results from some combination of mechanical interlocking between cement hydration products and the aggregate surface, and chemical bond resulting from a reaction between the cement paste and aggregate.

The objective of this paper is to summarize the morphological, and mechanical nature of the cement-aggregate bond and its impact on concrete performance.

MORPHOLOGICAL NATURE OF BOND

Study of the morphological nature of the aggregate-cement paste bond typically involves the examination of the cement hydration products at the interface under a microscope. Farran (1956) [4] hypothesized that bond strength developed due to the chemical reactions between aggregate and cement paste. Scrivener and Pratt (1986) [5] reported that "relative movement of the sand and cement grains during mixing, and possibly settling of aggregates before the cement paste sets, may lead to regions of low paste density around grains and to areas of localized bleeding at the aggregate-cement paste interface." It is known from literature that the interface region is in general much different from the bulk paste in terms of morphology, composition and density.

Many interface models have been published in the literature, and the general consensus is that the thickness of the interfacial region is approximately 40-50 μm, with the major difference in characteristics from the bulk paste occurring within the first 20 μm. Interestingly, the weakest part of the interface of this interfacial region does not lie at the physical interface, but 5-10 μm away from it within the paste fraction [6].

MECHANICAL NATURE OF BOND

The magnitude of the bond strength varies widely according to the rock type and surface roughness of the aggregate. Some researchers have used this as indirect evidence of chemical reaction between the cement and rock types. It has also been hypothesized that variation bond strength can be explained by different roughness factors of different aggregates. Regardless, of whether the bond is primarily due to either mechanical interlocking or chemical reaction, the true surface area of the aggregate available for bonding is an important aspect of bonding. The true surface area entails the size, shape and surface texture of aggregate particles.

Methods of Improving Mechanical Bond

There are no standard test methods to measure or quantify strength of the interfacial zone. Some of the indirect methods of measuring bond are: (1) "push out" test, in which cement mortar is cast against a aggregate prism and the interface is tested in shear, (2) modified indirect tensile test, in which a predetermined notch is cast, the specimen is tested in the indirect mode and fracture face is analyzed, and (3) volumetric surface texture analysis (VST), through this test process the micro- and macro-texture of the fracture face can analyzed and also the crack path can be determined. The subsequent sections of this paper will discuss methods of improving the mechanical bond between cement paste and aggregate.

Use of improved materials

Scholer (1967) [7] suggested that the stress level for initiation of micro-level cracks in concrete is primarily a function of the mortar strength. Hence, one way to improve interfacial bond strength would be to improve the strength of the cement mortar. However, this needs to be compatible with the other functional needs of the resulting concrete mixture design. For example, if a higher cement content is used to increase mortar strength in concrete, the shrinkage strains in concrete may also increase, thereby reducing any benefit that may have been garnered from increased mortar strength. It can also be argued that the use of "premium" aggregates, which improve bond strength between aggregate and cement paste, should be used. However, this is governed by the availability of these aggregates locally since a major part of the cost of aggregate is transportation.

Processing methods

The surface texture and roughness of the aggregate has an influence on the interfacial strength. The use of crushed aggregates, sand blasted aggregates and aggregates coated with calcium hydroxide are some of the methods by which bond strength can be enhanced. Figures 2 -4 illustrate the impact of aggregate processing (surface treatment) and type of aggregate on the indirect tensile strength.

Table 1 summarizes VST results of concrete cores from real life pavements in Michigan. There is a striking difference in the micro- and macro structure of the fracture surfaces and also in the paths of the cracks. The visual assessments of surface texture are found to be mostly in agreement with the VSTR values. That is, higher VSTR values generally correspond to "rougher" texture assessments. Results also indicate a relationship between the mode of concrete fracture and surface texture. Cracks, which propagated through the aggregates, were generally found to have smoother surface texture (and lower VSTR's), whereas cracks propagating around the aggregates were associated with rougher texture (and higher VSTR's).

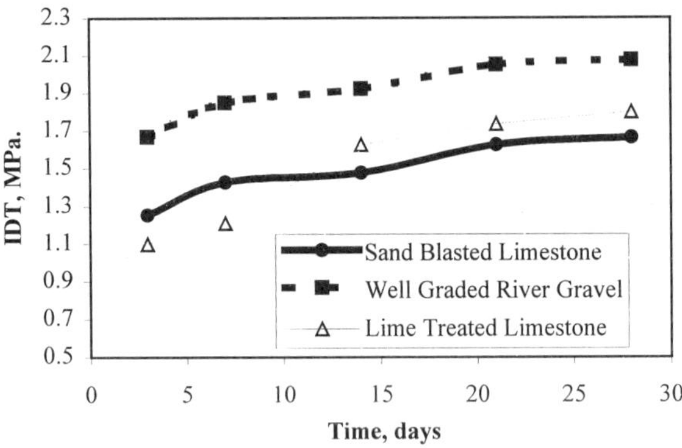

Figure 2 Indirect tensile strength as a function of aggregate treatment and type

There is a marginal improvement in the indirect tensile strength of sand blasted limestone concrete mixtures. The softer limestone is subjected to "roughening" under the sand blasting process and thereby improving the mortar-aggregate bond.

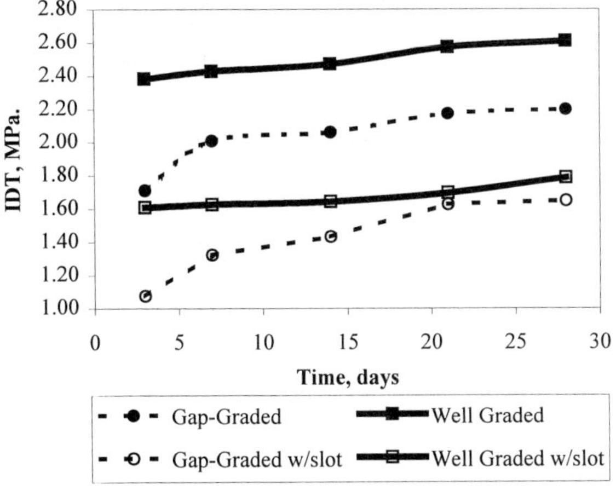

Figure 3 Impact of gradation on indirect tensile strength

Use of admixtures

Mehta and Monteiro (1986) (8) reported that an increase in bond strength resulted by the addition of silica fume. This is because less free water (bleed water) at the interface during specimen preparation was available. It has also been hypothesized that grain refinement can lead bond strength improvement.

However, it is not clear whether this primarily a chemical effect (due to the pozzolanic reaction) or a filler effect due to particle size and shape. Bentur and Goldman (9) have shown that effects similar to those of silica fume can be obtained with carbon black, which is completely non-reactive, suggesting that the filler effect play, a major role. Grain refinement replaces large crystals with smaller ones in the interfacial zone. Grain refinement can be achieved by the addition of silica fume or with use of carbonate aggregates.

IMPACT OF BOND ON PROPERTIES OF CONCRETE

The interfacial zone, generally the "weakest link" is considered the strength-limiting phase in concrete. It is because of this weak link that concrete fails at a considerably lower stress level than the strength of either of the two main components. Studies on normal strength concrete have in general shown that, as the cement-aggregate bond strength increases, the strength of concrete also increases, whether in tension, compression, or flexure.

The magnitude of this improvement is in the range of 10%-40%, improvements in tensile strength being higher than those in compressive strength [10]

Table 1 Volumetric surface texture test results

SPECIMEN NAME	AGGREGATE TYPE	MICROTEXTURE VSTR (cm^3/cm^2)	MICROTEXTURE	MODE OF FRACTURE (T)HROUGH / (A)ROUND
CARB-1	Carbonate	0.0678	Smooth	90%T
CARB-2	Carbonate	0.0767	Smooth	99%T
CARB-3	Carbonate	0.0958	Smooth	Poor Visability[a]
CARB-4	Carbonate	0.0691	Smooth	98%A
CARB-5	Carbonate	0.0429	Smooth-Moderate	99%T
NG-1	Natural Gravel	0.1624	Rough	95%A
NG-2	Natural Gravel	0.1238	Rough	60%A
NG-3	Natural Gravel	0.2498	Moderate	90%A
NG-4	Natural Gravel	0.1406	Moderate	80%A
NG-5	Natural Gravel	0.0550	Moderate	98%A
RCY-1	Recycled Concrete	0.1426	Rough	80%T
RCY-2	Recycled Concrete	0.0419	Smooth	99%T
RCY-3	Recycled Concrete	0.1699	Moderate	85%T
RCY-4	Recycled Concrete	0.0878	Moderate	90%T
RCY-5	Recycled Concrete	0.0635	Smooth	95%T
SLAG-1	Slag	0.0663	Smooth	70%T
SLAG-2	Slag	0.0781	Smooth	99%T
SLAG-3	Slag	0.0659	Smooth-Moderate	95%A

[a] Poor visability - was unable to tell whether the fractures went through or around the aggregate particles.

However, in order to study the effects of the interfacial properties on concrete fracture, strength measurements are inappropriate, instead, we should perhaps investigate fracture parameters, such as fracture toughness, stress intensity, or fracture energy. These parameters assist in explaining the process of crack propagation leading to failure.

Figures 4 and 5 illustrate the impact of aggregate type and aggregate surface treatment on the stress intensity of the concrete respectively. Sand blasted CLS resulted in higher stress intensity when compared to the sand blasted SRG. The softer limestone allowed the sand grains to indent the surface resulting in a rougher surface and a better bond between the cement paste and the aggregate.

Probably, the harder SRG aggregate was more resistant to impregnation of sand. The influence of aggregate treatment was also analyzed using the ANOVA analysis. The results of which are summarized in Tables 2a and 2b.

Table 2a Influence of aggregate treatment on K_{IC}(ANOVA-Analysis)

GROUPS	COUNT	SUM	AVERAGE	VARIANCE
No Treatment	5	2.57	0.514	0.00063
Lime Treated	5	2.78	0.556	0.00053
Sand Blasted	5	2.61	0.522	0.00072

Table 2b ANOVA statistics

SOURCE OF VARIATION	F-STATISTICS	P-VALUE	F-CRITICAL
Between Groups	3.968	0.047559	3.885
Within Groups	-	-	-
Total	-	-	-

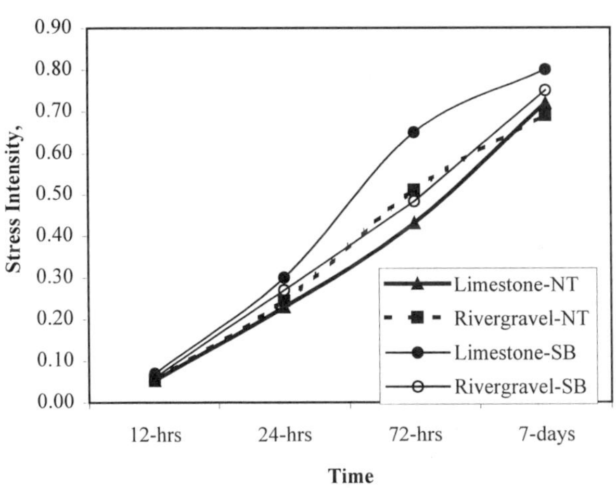

Figure 4 Stress Intensity as a function of aggregate type and treatment

It can be seen from Figure 5 that for SRG and CLS, the gap graded mixture yields higher critical stress intensity irrespective of the aggregate size when compared to the companion well-graded mixture. The analysis of variance (ANOVA) at a significance level of 95% revealed that both aggregate type and gradation have a significant impact on the stress intensity factor of the concrete mixture. Tables 3a and 3b summarize the ANOVA analysis results. Also, the ratio of K_{IC}(GG)/K_{IC}(WG) decreased with decreasing aggregate size.

The results of the above mentioned ratio are summarized in Table 4. This ratio indicates that for both the aggregate types, gap-graded(GG) concrete mixtures exhibit a higher stress intensity when compared with the well-graded(WG) concrete mixtures. The characteristics of the ITZ also influence the durability of the concrete mixture. The corrosion rate in pre-stressed and reinforced concrete elements is influenced by the permeability of concrete. This is due to the presence of micro-cracks in the transition zone between the aggregate and steel. The effect of water/cement ratio on the concrete permeability and strength is in general dependent on the relationship between the porosity of the hcp and water/cement ratio. The previous discussion about the influence of structure and properties of the ITZ on concrete illustrates the fact that it is more appropriate to think in terms of the effect of water/cement ratio on the concrete mixture as a whole.

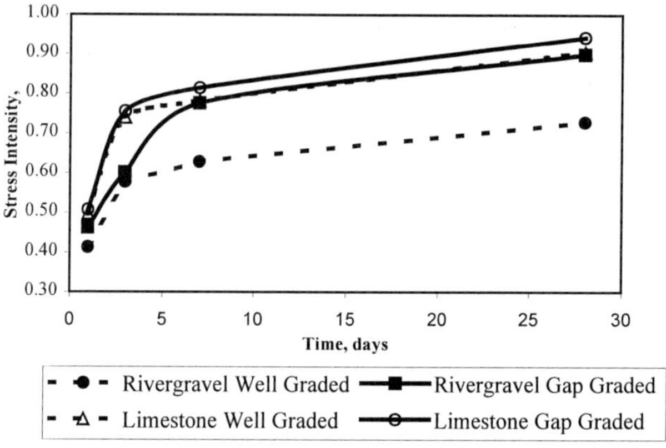

Figure 5 Impact of aggregate type and gradation on stress intensity

This is due to the fact that aggregates characteristics, such as, size, shape and texture have an impact on the mechanical interaction between cement mortar, water/cement ratio and the ITZ. In general, everything else being the same, the larger the aggregate the higher will be the local water/cement ratio in the transition zone, and, consequently, the weaker and more porous will be the concrete.

Table 3a Influence of aggregate gradation on K_{IC}(ANOVA-Analysis)

GROUPS	COUNT	SUM	AVERAGE	VARIANCE
37.5-WG	5	2.57	0.514	0.00063
37.5-GG	5	2.96	0.592	0.00087
25-WG	5	2.31	0.462	0.00032
25-GG	5	2.58	0.516	0.00043

Table 3b ANOVA statistics

SOURCE OF VARIATION	F-STATISTICS	P-VALUE	F-CRITICAL
Between Groups	25.469	2.48E-06	3.238
Within Groups	-	-	-
Total	-	-	-

Table 4 Ratio of $K_{IC}(GG)/K_{IC}(WG)$

AGGREGATE TYPE	MAX AGGREGATE SIZE	1-DAY RATIO	7-DAY RATIO
Rivergravel	37.5 mm	1.12	1.23
	25 mm	1.10	1.19
Limestone	37.5 mm	1.02	1.04
	25 mm	0.95	1.03

CONCLUSION

By now, the micro-structural characteristics of the ITZ are well documented and understood. However it is evident from the above discussion that there are still considerable problems in determining with any confidence the mechanical properties of ITZ and in assessing how changes in the ITZ relate to changes in concrete properties.

REFERENCES

1. NEVILLE, A M, AND BROOKS, J J. Concrete Technology, Longman Scientific and Technical Publishers, London, and John Wiley and Sons, New York, 1987.
2. MITSUI, K, LI, Z, LANGE, D A, AND SHAH, S P, Relationship between Microstructure and Mechanical Properties of the Paste-Aggregate Interface, ACI Materials Journal, Vol. 91, No. 1, January-February 1994, pp. 30-39.
3. STRUBL, L, SKALNY, J, AND MINDESS, S. A Review of Cement-Aggregate Bond, Cement Concrete Research, Vol. 10, pp. 277-286.
4. FARRAN, J. Contribution Mineralogique a l'etude de l' adherence entre les Constituants Hydrates Des Ciments et les Materiaux Enrobes, Rev. Mater. Constr. Trav. Publics, Ed. C., pp. 490-491, 155-172, 191-209, 1956.
5. SCRIVENER, K L, AND PRATT, P L. A Preliminary Study of the Microstructure of Cement/Sand Bond in Mortars, Proceedings, 8th International Congress on the Chemistry of Cement, Vol. III, pp. 466-471, 1986.
6. MINDESS, S. Mechanical Properties of the Interfacial Transition Zone: A Review, ACI SP 156-1, pp. 1-9.
7. SCHOLER, C F. The Role of Mortar-Aggregate Bond Strength in the Strength of Concrete, Highway Research Record, 210, National Research Council, pp. 108-117, 1967.
8. MONTEIRO, P J M, AND MEHTA, P K. Improvement of the Aggregate-Cement Paste Transition Zone by Grain Refinement of Hydration Products, Proceedings, 8th International Congress on the Chemistry of Cement, Vol. III, pp. 433-437, 1986.
9. BENTUR, A. Microstructure, Interfacial Effects, and Micromechanics of Cementitious Composites, Ceramic Transactions: Advances in Cementitious Materials, Vol. 16, pp. 523-549, 1991.
10. MINDESS, S. and Alexander, M., Mechanical Phenomena at Cement/Aggregate Interfaces, Material Science of Concrete Volume IV, Edited by Jan Skalny and Sidney Mindess, American Ceramic Society, pp. 263-282, 1995.

STUDY ON THE PROGRESS OF DETERIORATION OF SODIUM SULFATE USING MORTAR SPECIMENS

M Hironaga

Central Research Institute of Electric Power

H Sasaki

Hazama Corporation

T Endo

Tohoku Gakuin University

Japan

ABSTRACT. The research herein described was carried out by immersing mortar specimens in sodium sulfate solutions having different concentrations (5%, 10%) for 1 year, in order to determine the degree of sodium sulfate's infiltration into the mortar and establish relations between ettringite formation and the period of immersion. A new method of analysis using an X-ray micro-analyser and observing in detail the infiltration of sulfate ion, etc. into the mortar and the formation of ettringite is proposed.

The authors work indicated that the formation depth of ettringite was proportional to the square root of the period of immersion and the degree of deterioration (gravimetric change, change in flexural oscillation, etc.).

Keywords: Mortar, Chemical-resistance, Sulfate corrosion, Ettringite, Deterioration, EPMA.

Mr Michihiko Hironaga is a Member of Back End Project Natural Barrier Team, Abiko Laboratory, Central Research Institute of Electric Power.

Mr Hajime Sasaki works at the Technical Laboratory, of the Hazama Corp.

Dr Takao Endo works in the Civil Engineering Dept., Faculty of Engineering, Tohoku Gakuin University.

INTRODUCTION

Gypsum is formed when sodium sulfate acts on concrete structures and reacts with calcium hydroxide produced from cement hydration. The gypsum formed by such reaction acts with 3CaO Al_2O_3 or 3CaO Al_2O_3 $6H_2O$ in cement. Consequently, ettringite is formed and a large expansion pressure may result.

It is known that the expansion pressure of this ettringite generates cracks in concrete potentially affecting durability.

Previous research [1-5], has established the deterioration mechanisms associated with sulfate attack: longitudinal expansion, gravimetric loss, reduced compressive strength, etc., of mortar and concrete due to sodium sulfate. In cases where it is difficult to take countermeasure to prevent such effects, maintenance and repair of the structures must be carried out over long periods and it is necessary to establish an appropriate covering depth, etc.

This paper, based on the results of measurements and analyses described above, discusses the relations between infiltration and deterioration due to sodium sulfate as well as the relations between the formation of ettringite and the period of immersion.

EXPERIMENTAL DETAILS

Mixing Conditions

The mortar under test had a cement-to-sand ratio of 1:2 with water content (W/C) of 65%, in accordance with specified mortar mixes in construction.

Specimen Shape / Size and Curing Method

Immediately after preparing the mix a flow test was conducted and the specified flow value confirmed. The specimen was then placed in a mold of 40 x 40 x 160 mm. On the second day after placement, the specimen was demoulded and water curing carried out for 28 days. Exposure tests commenced thereafter. Meanwhile, only the mortar specimens for chemical analysis were subjected to tar-epoxy base coating (3000#H), so that erosion could take place from one side alone on the troweled surface which might have a rough texture due to the effect of bleeding.

Chemical Immersion Test Method

In this study, the mortar specimens were immersed in sodium sulfate solutions (still solutions) having concentrations of 0%, 5% and 10%, and their longitudinal length change, gravimetric weight change, relative dynamic modulus of elasticity and bending strength were measured at specified ages.

Measurement Items

Measurements were made conforming to the following standards every month after sodium sulfate immersion:

- Bending strength (in conformity with JIS R 5201)

- Compression strength (in conformity with JIS A 1108)

Method of Analysis by EPMA

Following the required period of immersion specimens were cut to 40 x 40 x 10 mm size by means of a diamond cutter and dried in a vacuum desiccator. The analytical surface was mirror-finished and carbon vaporized.

The analysis was conducted using an X-ray micro-analyser JMX-8621MX with the following analytical conditions: Acceleration voltage: 15kV; irradiation current: 2.0 x 10^{-6}A; measuring time: 30 ms; mesh size: 20 nm, the number of meshes: 512 x 512.

TEST RESULTS

Bending Strength Ratio

The results of measurements are shown in Figure 1. There was tendency for bonding strength ratio increases just after the immersion irrespective of solution concentrations. However, the specimens immersed in solutions of 5% and 10% concentration exhibited reductions following six months of immersion which were greater for the latter.

Compression Strength Ratio

The results of measurement are shown in Figure 2. There was tendency for increases in compressive strength ratio just after the immersion irrespective of solution concentrations. However, again the specimens immersed in solutions of 5% and 10% concentration exhibited reductions following 6 months of immersion, with the latter having the greater reductions.

Visual Observations

Signs of deterioration such as cracking were not observed during the first 3 months of immersion. By six months, the presence of micro cracks were observed, centering around the corner of the specimen. Such cracks were found spreading in the axial direction, but their propagation over the specimen surface was not confirmed. After that, no remarkable crack growth was confirmed on and after the ninth month.

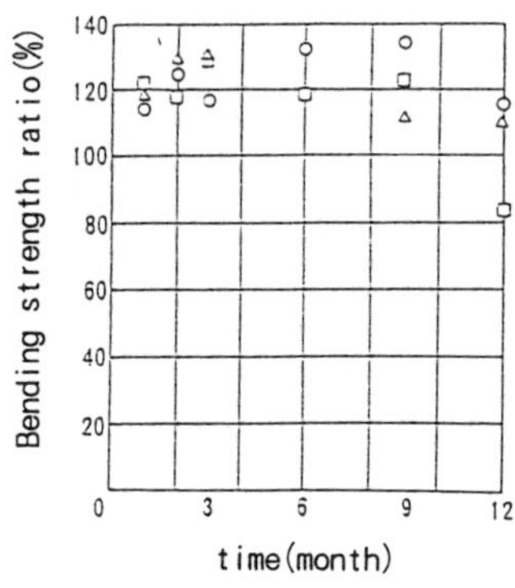

Figure 1 Bending strength ratio

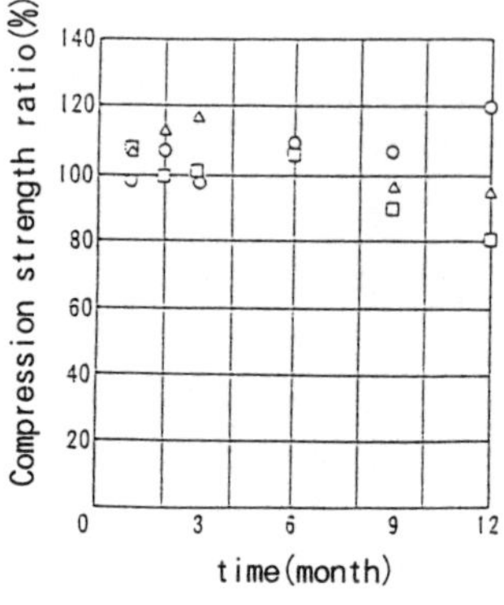

Figure 2 Compression strength ratio

Results of Analysis by EPMA

The infiltration of S into the mortar specimen at each age, analysed by EPMA, is shown in Figures 3 through 4. Shown on top of the figure is the surface of the specimen and the surface affected by erosion due to sodium sulfate. The colour bar shown at the right end in the figure represents weight concentration % when infiltrated S was converted into SO_3, indicating that the brighter the colour in the upper part, the higher the weight concentration % of S.

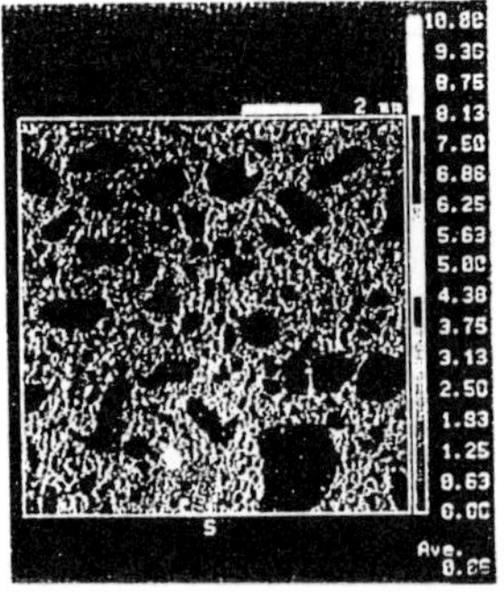

Figure 3 Distribution of ions before the immersion

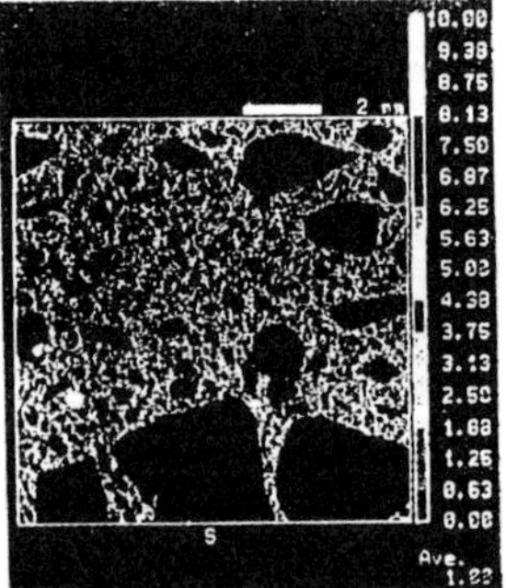

Figure 4 Distribution of ions 1 month after the immersion in a 5% solution

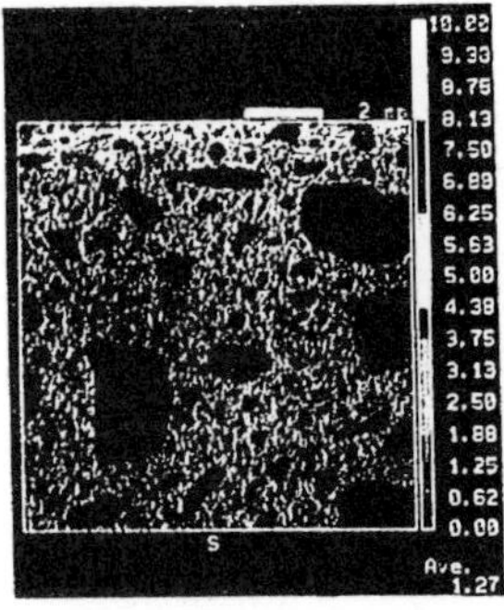

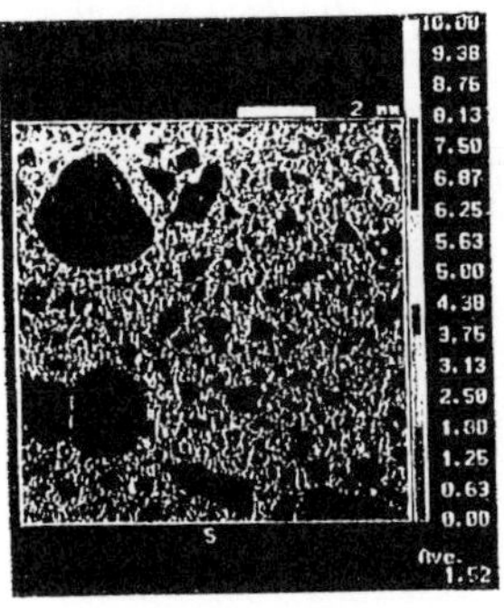

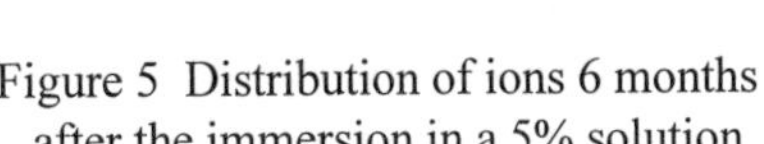
Figure 5 Distribution of ions 6 months after the immersion in a 5% solution

Figure 6 Distribution of ions 12 months after the immersion in a 5% solution

TEST RESULTS

Relation Between Specimen Deterioration and Sodium Sulfate Infiltration Based on the Chemical Immersion Test and Chemical Analysis

In this paper relations between the infiltration of sodium sulfate and various test parameters have been studied. Changes were observed in some of these on the ninth month of immersion including longitudinal expansion, bending strength ratio and compression strength ratio. From this, it can be considered that these reflect the deterioration of the mortar specimens relatively sensitively. On the other hand, only minor changes in the relative dynamic modulus of elasticity were obtained suggesting that the changes do not influence this parameter.Thus, it is after the sixth month of immersion when bending strength ratio, etc. are influenced by sodium sulfate infiltration. According to the result of chemical analysis by EPMA, however, the infiltration of S is confirmed by 1 month of immersion in both 5% and 10% solutions, and it is evident that more than a half of the specimen is infiltrated by 3 months of immersion, though no change is observed in the behaviour of a longitudinal change, etc., at this time.The authors considered that these phenomena indicate that even if sodium sulfate infiltrates mortar specimen, no ettringite is formed immediately and the reaction takes place in steps until ettringite is produced and deterioration occurs.

Study on the Identification of Ettringite Formation Steps

The chemical analysis by EPMA indicates only the infiltration of S into the specimen. This analysis indicates diffusion of S is underway and when its reaction takes place. Although the infiltration of S can be identified from these analytical results, no immediate conclusions can be drawn on the formation of ettringite which is considered to contribute to the deterioration.

That is to say. it is necessary to study a step which becomes rate-limiting to the deterioration in the various reaction steps, clarify the reaction domain of that step and then to study the progress of deterioration due to sodium sulfate. Otherwise, no study can be made of the deterioration rate. Shown in Figures 7 and 8 are changes in the depth of domains whose SO_3 gravimetric concentrations are 3%, 5% and 7%, during the respective immersion periods, with time.

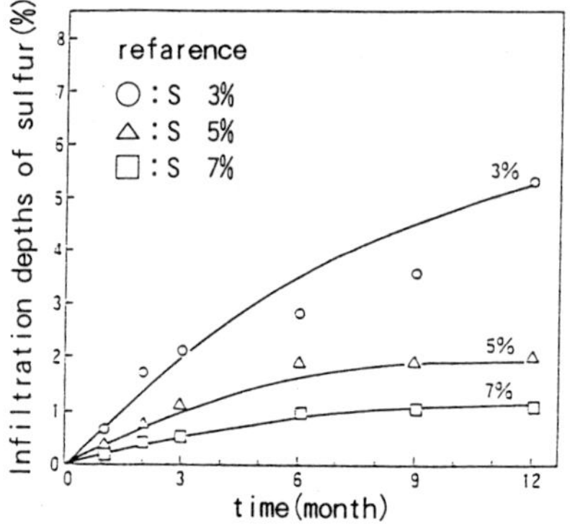

Figure 7 Infiltration depths of sulfur (5% solution) by gravimetric concentration

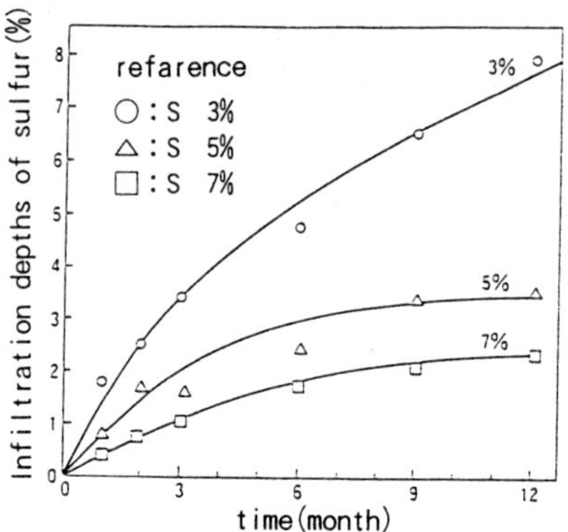

Figure 8 Infiltration depths of sulfur (10% solution) by gravimetric concentration

Considering these results from the steps of reaction, the domain represented by 3% in the gravimetric concentration is possibly attributable to the indication before diffusion and reaction in the first and second steps. On the other hand, the domain over 5% in the gravimetric concentration indicates reaction in the third step, that is to say, possibly causing the formation of mono-sulfate and ettringite and reaching a steady state by 6 months of immersion, which also agree well with the bending strength ratio, etc., decline by 6 months.

CONCLUSIONS

The condition of the deterioration of mortar specimens, the infiltration of sodium sulfate into the specimens and the minerals produced were examined by the immersion test and EPMA, and the relationship between the formation of ettringite and period of immersion studied, thus enabling a hypothesis to be formulated regarding the progress of deterioration.

In addition, although the infiltration of sulfur, etc., into the specimen from relatively early stages was confirmed by EPMA, no sign of deterioration was recognised in the mortar specimens.

It was also indicated that the formation depth of ettringite is in proportion to the square root of the immersion period(Figure 9).

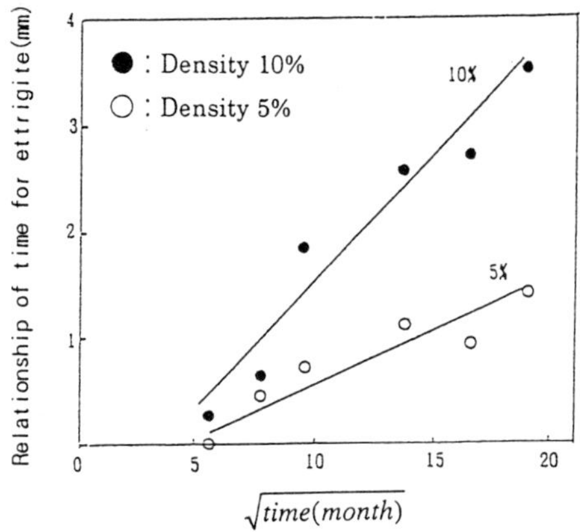

Figure 9 Relationship of time for ettringite

REFERENCES

1. K KISHITANI AND N NISIZAWA. "The Series of Durability for Concrete Structure Chemical Attack", Gihoudou.

2. H MATUSITA. "The Study of Durabiliy for Concrete by Sulfate", No. 7, JCI Reports, 1985.

3. A YODA. "The Study of Concrete for Chemical Resistance", No. 4, Report of Japan Construction Institute, 1984.

4. H IKENAGA. "The Durability of Concrete in Acid and Sulfate of Different Density and Kind". Report of Cement, 1983.

5. A YODA. "The Study of Concrete for Chemical Resistance", No. 3, Report of Japan Construction Institute, 1983.

INCREASING THE STRENGTH OF CONCRETE WITH VIBRATION-ACTION TECHNOLOGY

R Hela

L Bodnarova

Technical University Brno

Czech Republic

ABSTRACT. The contribution summarises the findings from an extensive investigation aimed at the verification of the possibility to influence the resultant physical and chemical properties of concretes with the aid of vibro-activation. During vibro-activation we act upon the freshly placed concrete mix with repetitive vibration cycles. The first cycle is started after a stipulated maturing period and the following cycles are repeated with stillstand intervals in accordance with a selected algorithm of vibro-activation. The influence of vibro-activation was verified on concrete mixes of different compositions and varying consistences with a constant cement dose and on concrete with differing cement doses and a constant water - cement ratio. The results have shown that an expendiently selected vibro-action cycle allows the attainment of higher short - term as well as long - term strengths of the concrete and that also further properties of the conctere are improved, and possibly also cement savings can be attained while maintaining the same resultant strengths. Mentioned is also the importance of using a suitable vibro- activation algorithm (with factual strength results for various periods of maturing).

Keywords: Concrete, Compacting of concrete, Vibration-action technology, Algorithm of vibration-action technology.

Dr Rudolf Hela is a Research/Teaching Assistant in Department of Building Materials Technology of the Faculty of Civil Engineering at Technical University, Brno, Czech Republic. He is specialized in the field of concrete technology and is involved in problems connected with redevelopments of reinforced concrete structures including diagnostics and construction-technical survey before starting the redevelopment works.. He is member of the Association for concrete structures redevelopment, technical committee of the Association of concrete producers, and Technical standardization committee CSNI.

Lenka Bodnarova is a Research/Teaching Assistant in Department of Building Materials Technology of the Faculty of Civil Engineering at TU Brno, Czech Republic. She specialises on of the interaction of High Speed Water Jet and concrete, possibility of the exploitation of High Speed Water Jet technology in redevelopments of concrete structures.

INTRODUCTION

The production of concrete building sections for civil, industrial or engineering buildings has its irreplaceable position in the present building industry. The production of these precast elements consists of several technological sections. The compacting of the deposited concrete mix in shaping moulds belongs among the most important ones. The compacting is most frequently performed by means of vibration. The goal of the compacting of the concrete mix is to reach the maximum consistence of the cement bond and the mixture of aggregates and in this way to reach good physical-mechanical properties of hardened concrete economically. The required property in the production of prefab concrete sections is also to reach quickly a high beginning strength because of an early dimantling and in this way to reach a higher turn over of expensive shaping moulds. One of the possibilities how to reach this effect is the exploitation of controlled vibration-action technology of the cement bond in fresh concrete during its additional compacting. This technology enables in a controlled and planned way with a certain exact intention to influence the degree of the compacting of concrete, the character of its porous structure, the microstructure of concrete, an increase of short term and long term compressive strength of concrete, it enables to reach higher volume weights but also to improve the cohesion of concrete with steel reinforcement in reinforcing elements, to improve the water-tightness and resistance against the effects of frost and those of chemical de-freezing ingredients. The improvement of physical-mechanical properties of concrete is reached without intervention in the structure of concrete mixes especially without demands on higher portions of cements. The economic demands of input raw materials for the production of the concrete mix are in this way not increased [1], [2]. In actual fact has been verified under practical conditions the possibility of saving concrete in manufacture of prestressed bridge beams with a length of 16 m, volume of the concrete 8,5 m^3. There have been attained savings of 60 kg/m^3 cement CEM I 42,5 R upon maintaining the required 28 - day strengths of 55 N/mm^2.

THE PRINCIPLE OF THE METHOD

The controlled vibration-action technology is in reality an intentional, planned effecting by means of cyclic dynamic impacts (vibration) the fresh concrete deposited in the shaping mould or scaffolding in suitable time intervals fixed ahead by a constructed algorithm which exploits the laws in stiffening of cement bond. Frankly speaking in question is the internal activation of the cement bond by means of a controlled mechanical effect which at the same time provides the concrete mix with additional compacting and decreases its porosity. The favourable impact of the technology of vibration-action technology on the quality of concretes may be explained so that after its first compacting (mostly by means of vibration) there occurs in further vibration cycles a reduction of the size of pores, but also the total porosity of the cement bond. Due to mutual vibration of the grains of the aggregate coated with the cement bond there occurs the friction of the surface coats of cement grains being dissolved. Water more easily penetrates to the till not wetted cores of cement grains- in this way the area and the quantity of the active hydration material in the cement bond are increased - in principle there occurs an additional grinding of cement inside the concrete. In this way the measured surface of cement is increased which has the same impact as the application of a cement having been more finely ground. The production of more finely ground cements directly in the cement works is expensive and increases the consumption of primary energy needed for additional grinding of cement.

The principle of vibration-action technology consists in purposeful, planned alternation of vibration with the period of calm on the principle of an algorithm constructed in advance. The determination of individual time periods of the cycles depends on many factors. The basis of the construction of algorithm according to which we may control the process of vibration-action technology, is the determination of the beginning of stiffening of the concrete mix being compacted. The beginning of stiffening depends for instance on the water coefficient (w), on the sort and quantity of cement, on the sort of aggregates having been used, on the type of the admixes and ingredients, on the consistence of the concrete mix on temperature during which the concrete mix is produced, compacted and during which there occurs its stiffening.
The algorithm of vibration-action technology consists of several rest and activation intervals for instance according to the following diagram:

$$DVA = V + O + AC_1 + I + AC_2 + I + AC_3 + ... I + AC_n$$

where, DVA = the period of vibration-action technology
V = the period of basic vibration
O = the period of maturing of fresh concrete in the mould
AC = activation (vibration) cycle
I = the rest interval between AC.

The number of activation intervals and their value O and I depends on concrete conditions of production and the sort of concrete mix having been used. In case of the application of an unsuitable algorithm of vibration-action technology mainly if the total time of the cycle surpasses as far as behind the beginning of stiffening of concrete, there may occur a considerable deterioration of the properties of concretes having been produced. If we act with the activators of vibration on the concrete mix at the time when the hardening of the cement bond starts there occurs a mechanical destroying of the starting adhesions on the contacts of individual grains which causes an expressive reduction of strength of concretes (Figure 1, maturing time 5 hour). The importance of finding a suitable vibro-activation cycle, especially the concrete selection of the maturing period O, after which we start to vibro-activate the concrete mix, is shown in Figure 1.

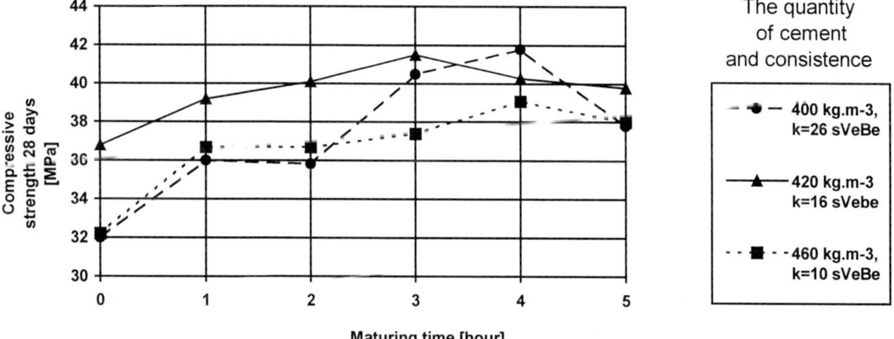

Figure 1 Influence of maturing time on the strengths after 28 days of normal maturing

RESULTS OF EXPERIMENTAL TESTS

During the experimental programme tests were carried out on specimens of thickness 150 mm. During the course of experiments there was verified the influence of vibration-action technology on concrete mixes of different compositions (Table 1).

Table 1 Composition of the concrete mixes (aggregate - proportioned by volume)

CEMENT CEM I 42,5R	400-460 kg/m^3
Agg. 0-4 mm	38 %
Agg. 4-8 mm	22%
Agg. 8-16 mm	40%

For each verified series were produced comparing samples which were compacted just during the basic period of vibration. The physical-mechanical properties were found out at the age of 1, 7, 28 and 150 days. The influence of vibration-action technology was evaluated on the basic physical-mechanical properties of hardened concretes, i.e. in the volume weight, compressive strength of concrete, the tensile strength of concrete under bending and the dynamic modulus of elasticity in compression and in shear.

Further on we verified the integral effect of concrete with the inserted steel reinforcement, waterproof of concrete and the resistance of the surfaces against the influence of water and chemical de-freezing materials.

During the investigation of the influence on the microstructure of concrete we have investigated the porosity of the hardened cement stone by the method of pressure mercury porosimetry. We have carried out the investigation by the electronic modular grid microscope, by differential thermic analysis and by the x-ray diffraction analysis with the goal to find out the qualitative changes of minerals arisen during the hydration of cement [3].

Compressive Strengths of Concrete

The compressive strengths after 24 hours have reached an increase as much as by 25 % in comparison with the concretes just vibrated. It is interesting and important that even the strengths after 28 days reach an increase as much as 20%. The long term strengths (after 150 days of normal maturing) preserved a similar trend as after 28 days.

Figure 2 and Table 2 present a comparison of 1- day and 28-day compressive strengths of concretes that have only been vibrated and concretes vibro-activated by the most effective vibroactivation mode (see Figure 1).

Table 2 Comparison of 1- day and 28-day compressive strengths

CONCRETE	1-DAY STRENGTH [MPA]		28-DAY STRENGTH [MPA]		MIX PROPORTIONS			
	V	VA	V	VA	Cem I 42,5 R	Aggregate (% volume)		
						0-4	4-8	8-16
BS 1	6,3	8,7	31,5	39,0	400 kg/m^3	38	22	40
BS 2	5,8	8,5	36,4	42,2	420 kg/m^3	38	22	40
BS 3	6,8	8,7	32,2	40,0	460 kg/m^3	38	22	40

* V - only vibration, VA - the most effective vibroactivation

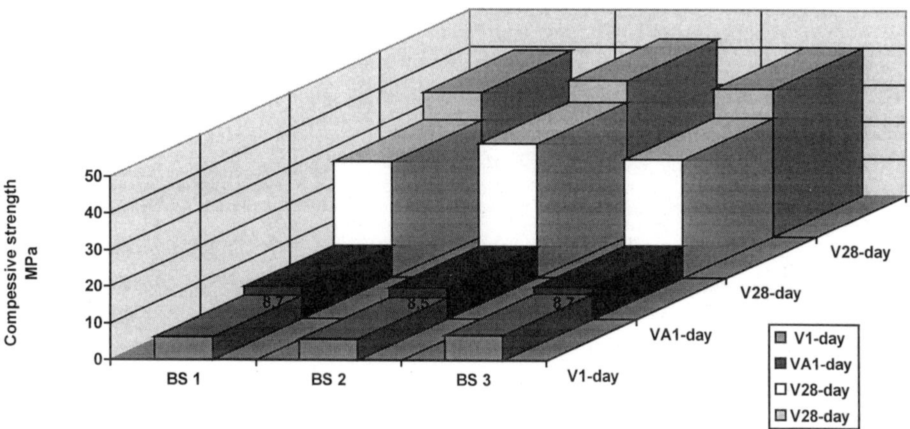

Figure 2 Comparison of strengths after 28 days, * V - only vibration, VA - vibroactivation

The Cohesion of Reinforcing Steel with Concrete

There has been investigated the influence of various vibro-activation algorithms onto the cohesion of steel with the concrete. The vibration algorithm was the following:

$$DVA = V + O + AC_1 + I + AC_2 + I + AC_3 + ... I + Ac_7$$

where: V = 2 min, O = 1 - 5 hours, I = 15 min, AC = 1 min

The following table 2 summarised the results of the tests aimed at monitoring of the influence of the number of activation cycles AC onto the cohesion of steel with the concrete.

Table 2 The adhesion of steel with the concrete, maturing time O = 2 hours

NUMBER OF ACTIVATION CYCLES	FORCE REQUIRED FOR PULLING OUT A REINFORCING BAR [%]
0 (Concrete mix without vibroactivation)	0
4	22
5	28
6	39
7	45

The cohesion of steel with the concrete was determined in terms of the force required for pulling out a reinforcing bar dia. 14 mm from a concrete cube with an edge of 150 mm. The results have again shown a better cohesion in the application of vibro-activationin comparison with the concretes having been just vibrated. In practice it means a considerable possibility of reducing the batches of cement in concrete mixes while maintaining the required strengths or a possibility of increasing the strength of concretes at a constant batch of cement.

Verification of the Influence of the Change of the Parameters of Vibration

In the comparison of the strengths of concretes vibrated and vibroactivated (there were used various values of frequencies of vibration at practically congruent values of acceleration from 26 to 30 m.s-2 at a constant time of vibration) there is demonstrated a considerable influence of frequency on the strength of concretes. And this is demonstrated both in case of the short time strengths after 1 and 7 days and also in case of long term strengths after 28 days. The differences in strengths are more important in the short time strengths, e.g. after 1 day of normal maturing the concrete vibrated by the frequency just 33 Hz has reached the strength 4,0 MPa and the same concrete vibrated with the frequency 66 Hz has reached the strength 10,4 MPa which is 2,6 times higher value. In case of the concretes vibroactivated by the frequency 66 Hz is against the frequency 33 Hz an increase of one day strengths even 2,9 times higher, after 28 of normal maturing is this increase 1,9 times higher [4], [5].

SUMMARY

A favourable influence of vibroactivation appears by the reduction of the contents of air pores in the cement bond, the character and the distribution of the newly arisen pores are changed, the character of the microstructure of concrete is changed as well, there is obvious a higher share of tricalciumsilicate ($3CaO.SiO_2$) in comparison with the concretes having been just vibrated. An expressively positive influence is then on the physical-mechanical properties of concrete, especially on the compressive strength of concrete and in the final consequence also on the durability of concrete. An important parameter influencing the effectiveness of vibroactivation is the frequency of activators of vibration which may positively influence the improvement of properties of concrete even at a congruent algorithm of vibroactivation and the consistence of the concrete mix [4], [5].

ACKNOWLEDGEMENTS

This work was supported by The Ministry of Education of Czech Republic (project PG 98547).

REFERENCES

1. AVRAM, I AND VOINA, L. L´influence de la revibration de pate cement durcil et des betóns, Rev. Des. Mat. de constr. No 619 a

2. MELUZIN, O. Concrete technology, VUT Brno, 1994

3. POWERS, T. C. Vibrated Concrete, Proc. Coner. ACI

4. HELA, R. The Laws of dynamic methods of Compacting, Brno 1989

5. HELA, R. The Importance of Compacting of Concrete Mixes, Brno 1986

6. UNÈÍK, S AND PÁNIS, R AND PAVLÍK, V. Monosti zniovania nákladov pøi výrobe transportbeónu. In. Stavby, 1997, 45, 2, pp. 77 - 77

THEME TWO:

PERFORMANCE SPECIFICATION AND DURABILITY BY INTENT

Keynote Paper

REPAIR PHILOSOPHY FOR CONCRETE STRUCTURES

K Tuutti

Skanska Group

Sweden

ABSTRACT. The need to carry out repairs of concrete structures is increasing exponentially. The service life of structures varies by different occasions and it courting dangerous to apply repair methods without an analysis of the consequences. This report discusses the milestones in durability research in Sweden. Lack of knowledge in this area particularly evident as regard the corrosion of steel in concrete. A repair philosophy is established in this paper demonstrating that the mechanisms behind concrete deterioration must be analysed and identified in the design procedure of a repair system. Two case studies are described: a frost resistance problem and a reinforcement corrosion problem. In both cases a coal fibre composite system has been used.

Keywords: Durability, Frost resistance, Steel corrosion, Acid attack, Moisture analysis, Repair philosophy, Coal fibre composite.

Dr Kyösti Tuutti is Director of Research & Development in the Skanska Group. He is also a visiting professor at Lund Institute of Technology, specialised on durability of building materials and service life predictions.

INTRODUCTION

Deterioration mechanisms of concrete structures have been of great interest the past two decades. Thousands of reports have been published on the subject of corrosion of steel in concrete and other mechanisms such as frost resistance, alkali-silicon reactions, acid attack, etc. However, the main purpose of all research in this area, which is to summarise all parameters into a precise formula describing the service life of a concrete structure, with a significance of 10 years, is still lacking. What is the reason for this when billion of dollars are being spent in research. Probably one factor is the complexity of the problem. There is not just one parameter, for an example the chloride profile in the material that will have an impact on service life. A structure or a concrete material does not produce a homogeneous electrochemical condition for the reinforcement. Microenvironments are established due to variations in moisture condition and the existence of imperfections such as voids and cracks. Different acids will decompose the concrete in various ways where reaction products may prevent dissolution of compounds, etc.

However, we are able to observe the degradation of concrete structures and therefore we need more of knowledge about these processes. Repair of concrete structures is considerably more complex than the design and construction of new structures. Deterioration mechanisms or the origin of the problem must be known before any design or repair method is selected. Both expertise in material and structural technologies must be included in the design process of a repair methodology. Careful considerations must be given to a range of questions, see Figure 1.

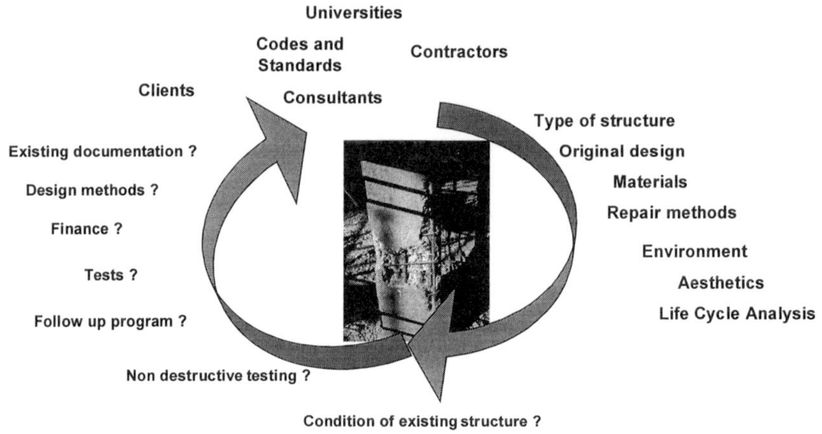

Figure 1 Schematic sketch of partners involved in the process and the aspects that must be considered, Täljsten [1]

SIGNIFICANT STAGES OF DURABILITY RESEARCH IN SWEDEN

Swedish researchers, professor Bergström [2] and professor Fagerlund [3], were pioneers in the specific task of service life predictions of building materials. A focus on such a goal created a paradigm shift in the planning and design of research activities.

Prioritisation of parameters that must be studied suddenly gave a approach. Existing data produced by colleagues in other countries was used due to the lack of time to make all necessary parameter studies. In the beginning of 1970, several researchers directed by these far-sighted excellencies were able to visualise a holistic view of the service life issue. However, several excellent research programs were initiated which have generated very valuable results for the durability area. A typical case in point is the research in moisture properties for different materials at Lund Institute of Technology.

Studies of moisture conditions in different qualities of concrete by professor Bergström and professor Fagerlund have been of great importance in the understanding of different deterioration mechanisms. This research was established with the aim of understanding the behaviour of building systems used in Sweden, especially the risk of poor indoor conditions. During that time contractors in Sweden introduced a change in the production process of buildings and the choice of materials. Impermeable plastic carpets with layers of organic compounds were placed on floors. In buildings in where moisture conditions were high problems could raise. This was called the "Sick building syndrome" which could be experienced on occupants. Unexpectedly the knowledge gained in regard to predicting moisture conditions inside a concrete structure became one of the milestones in the durability research.

The durability of concrete can be discussed in terms of four different mechanisms, the frost resistance, corrosion of steel reinforcement, alkali silica reactions and acid attack. In all four deterioration processes water has a major impact. The water in the pore system can freeze and cause spalling and interior damage in cold environments. Water is needed in the reactions between alkali and silica in the ASR process. Today it is also possible to predict the moisture conditions in different materials if fluctuations occur, see Figure 2

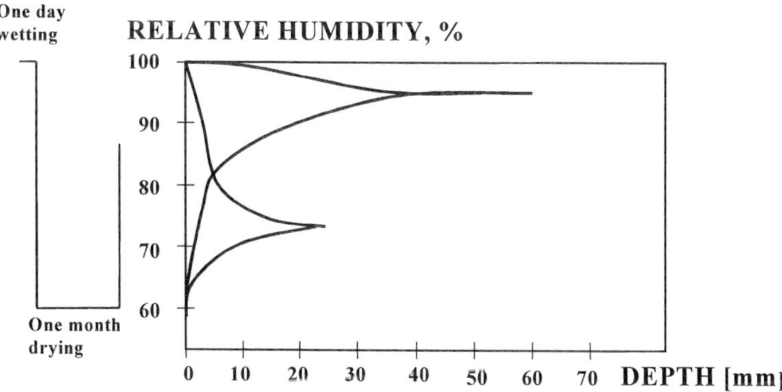

Figure 2 Demonstration of the moisture variation in two different concretes, W/C 0.40 respective 0.60, with an environmental cycling of 1 day capillary suction and 1 month drying in RH = 60%, Arfvidsson and Hedenblad [4]

The term critical degree of saturation and standardised frost resistance test procedures are other milestones in the frost resistance field.

Corrosion of reinforcement in concrete structures is of a more global interest and therefore more thoroughly studied. Diffusion processes, mathematical modelling and the relevance of different material parameters are well known. However, one important parameter, the corrosion threshold value, must be further studied before we can expect a breakthrough in this area.

Theoretical calculations and modelling of the time of initiation in a chloride rich environment demonstrate the lack of knowledge regarding the vital parameter, the threshold value of chloride ion concentration, which changes the passive stage to an active corrosion stage. A demonstration of this can be seen in Figure 3. Normally the reader believes that the sample with the lowest chloride concentration is preferable. Figure 3 demonstrates that the case can be just the opposite.

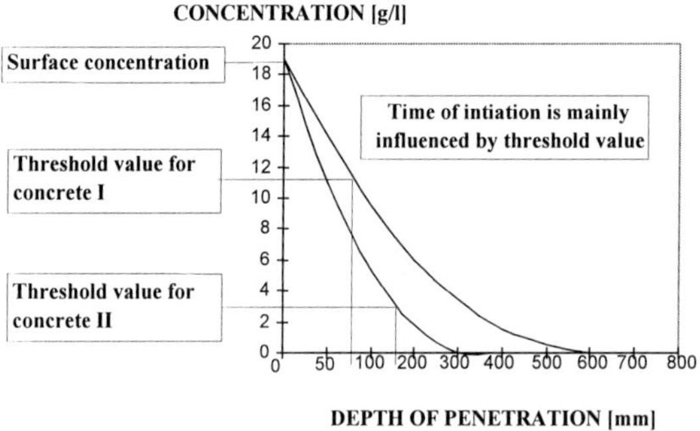

Figure 3 Schematic sketch of the time of initiation for two different concretes as a function of depth, Tuutti [5]

In the Swedish "High Performance Concrete" project and the research done mainly by Karin Pettersson, CBI and Paul Sandberg, Cementa, and reported by Fidjestöl, Jørgensen, Pettersson, Sandberg, Tuutti [6], several threshold values have been measured for different concrete qualities surrounded by typical natural chloride contaminated environments.

The highest threshold values were found in the most constant conditions, for example concrete immersed in seawater. Additions that decrease the hydroxide concentration result in a decrease of the threshold value. A surrounding environment with a fluctuating moisture condition, such as a splash zone, decreases the homogeneity.

This has a positive effect in the constitution of corrosion cells and will reduce the threshold value, see Figure 4.

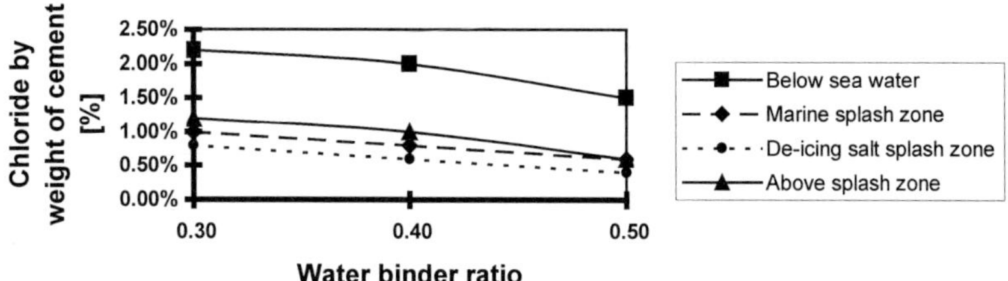

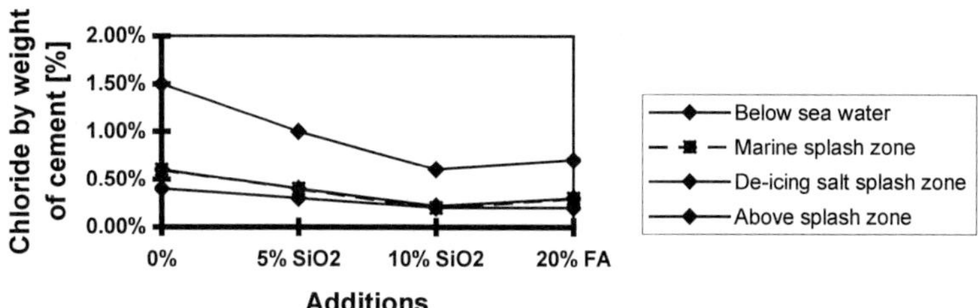

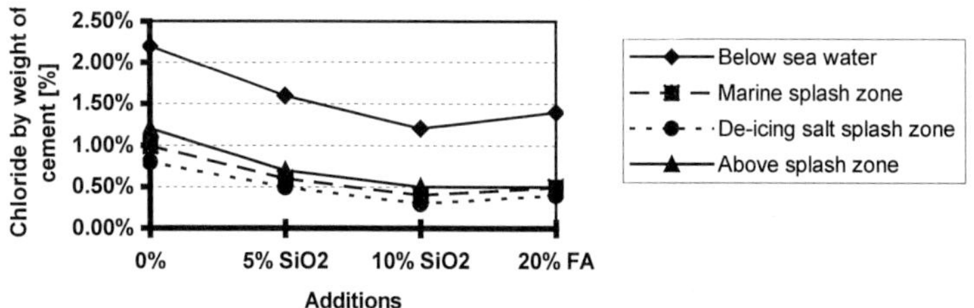

Figure 4 Influence of water binder ratio and mineral additives on the chloride threshold value in uncracked concrete at various exposure conditions, Fidjestöl, Jørgensen, Pettersson, Sandberg, Tuutti [6]

An interesting point often discussed in the literature is the effect of different parameters. The initiation time of a steel corrosion process has been calculated with the figures given above, on the assumption that the surrounding is a marine environment. The results are presented in Table 1 as relative figures to a mix with W/C = 0.40 and 100% PC. The surface concentration has been estimated to 36 g Cl^-/litre which is about twice as high as the normal concentration in seawater. The field exposure tests indicate that the concentration in the concrete pores is higher than in the surrounding environment.

Table 1 Calculated relative corrosion initiation periods for different compositions

CONCRETE COMPOSITION	W/B = 0.50	W/B = 0.40	W/B = 0.30
100% PC	0.35	1.00	3.20
95% PC + 5% SiO_2	0.25	0.75	4.50
90% PC + 10% SiO_2	0.15	0.75	1.25
80% PC + 20% PFA	0.10	0.60	0.90

The results presented in Table 1 indicate that the effect of silica fume is ideal in a high performance concrete. The effect of PFA is negative for all compositions. Use of 100% PC is optimum for all compositions with W/B > 0.40. Thus, the Swedish tests indicate that the reduction of chloride permeability, by additions of mineral additives, is not as effective as the high threshold value in PC concrete. However, the most effective parameter, inducing a long initiation period is the use of a low water binder ratio.

Nowadays we often are discussing the effect of the most incredible products. These products are normally characterised as waste materials. The owner of such a material will naturally try to increase its value by selling it to the concrete industry. Therefore the normal procedure is to make comparison tests, in a laboratory, of a few of the parameters.

The parameters chosen are those which appear to produce good behaviour for specific properties only. Several studies have demonstrated that a laboratory comparison procedure, which is not calibrated to a long-term field exposure, could not be used in service life predictions.

Unfortunately most of the research in the corrosion area consists of small-scale tests in laboratories. Field tests indicate that it is very difficult to extrapolate laboratory results to practical conditions. Yet an another milestone in this work is the consciousness that a natural exposure combined with precise measurements of material properties and electrochemical conditions simplify the understanding and provide opportunities to quantify them.

Finally, the service lives of concrete structures thus not end at the moment an active stage propagates the deterioration. Research on high performance concrete of different compositions indicates that in some extremely dense materials, the rate of corrosion will be limited to values, which we normally classify as a passive stage. Also cracks in such materials will be more tolerable due to the difficulty to forming an electrolytic macro cell.

Acid attacks are even more complex due to the formation of by-products that could retard diffusion processes and the dissolution of the matrix. An interesting demonstration of such an effect was presented by Rombén [7]. Chemical attack by hydrochloric acid on concrete is initially influenced by the dissolution of CaO.

After about 4 years the second stage appears in which the dissolution of Fe_2O_3 increases the rate of chemical attack, see Figure 5.

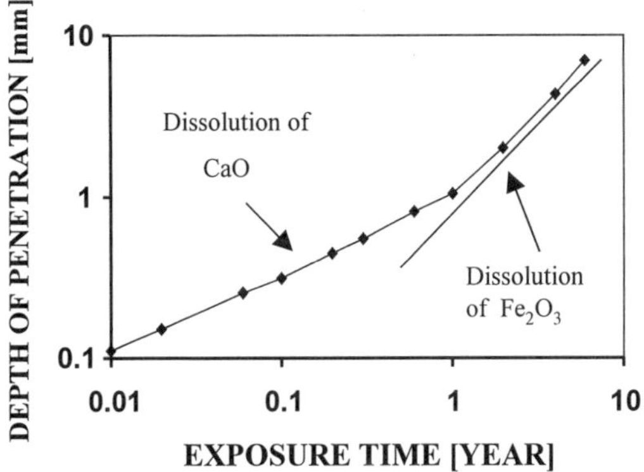

Figure 5 Experimental demonstration of different dissolution mechanisms for concrete exposed to hydrochloric acid. Rombén [7]

PHILOSOPHY IN THE PROCESS - REPAIR OF CONCRETE STRUCTURES

Structures that are in a immediate need of repair indicate that the material properties have changed due to the impact of the environment. Time of service verifies the suitability of a material or a combination of materials for a specific use. An important issue is the possibility of restoring the structure to its former condition. It is not certain that a repair improve the durability properties. Several practical cases have shown faster deterioration after repair action.

The philosophy of a repair design procedure is to use excellencies of both material science and design. The team must identify actual deterioration mechanisms, load bearing capacities, and possible impact from different repair technologies. In other words, they have to understand the behaviour and properties of different materials and how different parameters affect service life. Generally it is obvious the repair of concrete is much more difficult compared with steel, timber, brick, etc. The problem is related to the concrete pore structure that is partially filled with water. An impermeable seal protects the material against aggressive substances but can also saturate parts of the material. This enlarges the volume of the material and also accelerates the deterioration that is influenced by a high water content. Therefore, use of permeable materials will be preferable to concrete repair systems.

A typical example is the repair of reinforcement corrosion damaged concrete. The initiation of the corrosion process is either a pH-neutralisation of the material or a high concentration of aggressive ions, such as chlorides. In the case of a chloride initiation we normally replace the concrete around the steel with new concrete. Other options are the use of electrochemical methods such as cathodic protection, washing out the chlorides from the concrete. However, concrete must also be frost resistant in countries where the temperature alternates below freezing point. An accurate selection of materials is vital especially if the new material is less permeable than the old concrete. In this specific example it is also difficult to assess how much of the concrete should be replaced.

Repair of corroded areas only may lead to the establishment of new anode areas resulting in a rapid deterioration. The repair strategy must try to attain physical, chemical and electrochemical homogeneity. Durable materials that will not have an impact on interface areas should be used.

A CASE STUDY – STRENGTHENING WITH COAL FIBRE COMPOSITE

Structures are not always lacking in durability properties. Grain silos that were designed in Sweden during the late 1960s lack about 50% of necessary reinforcement. Contributory factors were inadequate theoretical knowledge of loads during the emptying process of such structures and negligence in the quality control of the construction process. Today several silos are heavily cracked and some collapses have been observed. These silos need some type of load bearing strengthening systems. Use of coal fibre composites is a possible technique for this purpose, see Figure 6.

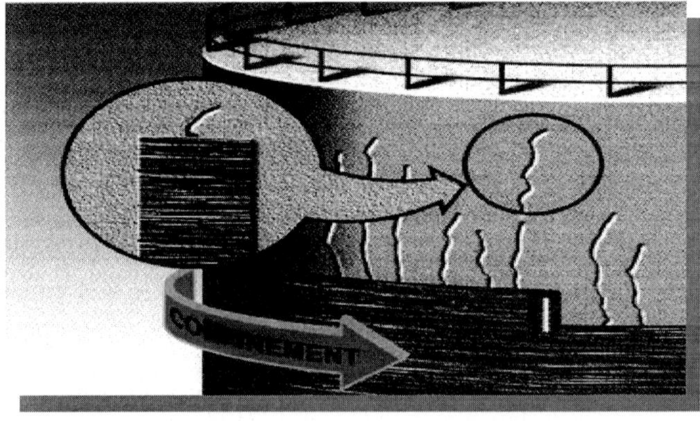

Figure 6 Silos, heavily cracked due to overloading can be repaired by coal fibre composites. The nordic climatic conditions necessitate moisture design. In particular calculations of the degree of saturation at the interface concrete versus the dense repair material, Täljsten [1]

Several types of silos ranging from individual cylindrical buildings to a package of several half-cylinders. A cylindrical structure will be surrounded with the new material without structural problems. The package type of the half-cylinders can not be completely surrounded if the confinement is placed on the outside. Transmission of tensile forces will occur in the ends of each string, see Figure 7.

As a result of practical unsuccessful attempts experience of water impenetrable coatings on concrete surfaces, laboratory tests were made to simulate this case. The project team could foresee an accumulation of water behind the coating that could disintegrate the concrete. The challenge was to study possible coal fibre composite repair systems associated to the frost resistance.

Figure 7 Anchorage of strengthening materials is a problem at interfaces marked with arrow

Concrete cylinders with different W/C were fully or partially covered with a coal fibre composite and tested in a climatic chamber with alternating temperature. The samples were either conditioned to reach a specific degree of water saturation or placed in contact with free water during the tests. The tests demonstrated that the crucial factor for frost deterioration was the critical degree of water saturation achieved regardless of the concrete was sealed or not. However, the suction of water was retarded whether the mantle surfaces were completely sealed.

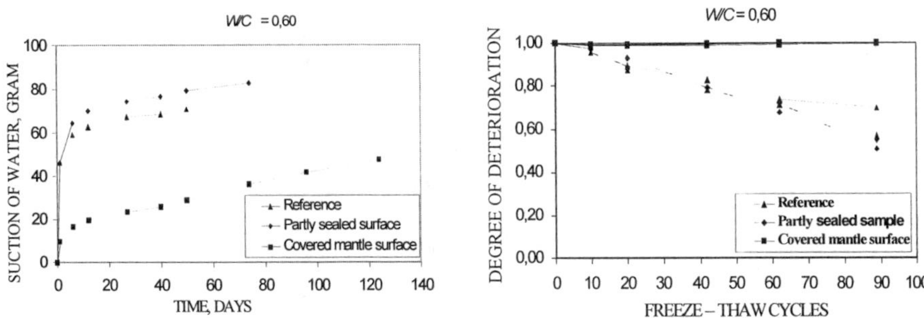

Figure 8 Covered mantle surface indicates a low suction process that has a direct impact on the frost resistance of concrete, Hassanzadeh [8]

The laboratory tests indicate that frost damage will occur if the concrete becomes water saturated. Repair strategy with regard to these findings was to keep the concrete surface as water permeable as possible for evaporation water but to repel rain and condensed water as much as possible. A sectional strengthening by using coal fibre strings and an impregnation of the mantle surface was recommended.

The idea was also to prevent a slow accumulation of water from the inner parts, even if the stored materials were such a low RH, that a drying of the concrete should occur. All water in the concrete must be able the possibility to evaporate between the strengthening strings.

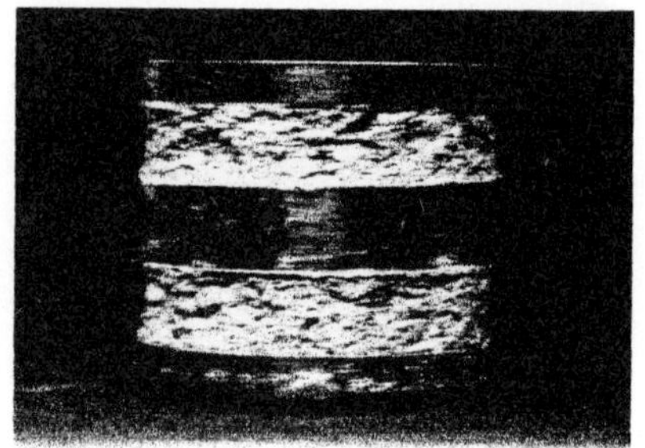

Figure 9 Typical result from frost resistance tests the concrete on the left is not frost resistant and the seal prevents scaling, Hassanzadeh [8]

The laboratory tests indicate that frost damage will occur if the concrete becomes water saturated. Repair strategy with regard to these findings was to keep the concrete surface as water permeable as possible for evaporation water but to repel rain and condensed water as much as possible. A sectional strengthening by using coal fibre strings and an impregnation of the mantle surface was recommended. The idea was also to prevent a slow accumulation of water from the inner parts, even if the stored materials were such a low RH, that a drying of the concrete should occur. All water in the concrete must be able the possibility to evaporate between the strengthening strings.

A CASE STUDY – REPAIR OF CORROSION DAMAGED CONCRETE

Reinforcement corrosion and severe spalling of the concrete cover damaged a chimney. This structure was heated up by the energy flux of the smoke, which prevents frost resistance problems. However, the concrete quality was mediocre and could not reduce the corrosion rate of reinforcement. The alternatives were to repair the structure with a method that could prevent further corrosion or to use a method whereby a further corrosion was of no consequence. A corrosion protection could be achieved by removing carbonated concrete cover and replacing it for an example with shotcrete. Such a drastic repair action was costly and did not meet the client's demand of aesthetics. The most attractive alternative was the strengthening of the structure with coal fibre composite in such a way that the load bearing capacity could be taken over by the repair system and the non reinforced concrete behind. The idea was to let the corrosion process at the reinforcement bars propagate freely. That process would generate an increase of volume around the anode areas, and this in turn means a post tensioning of the strengthening system, which could be acceptable.

Both these case studies demonstrate intelligent ways of serving the market with reconditioning of old concrete structures. We have identified considerable risks associated with applying repair methods without an analysis of the consequences.

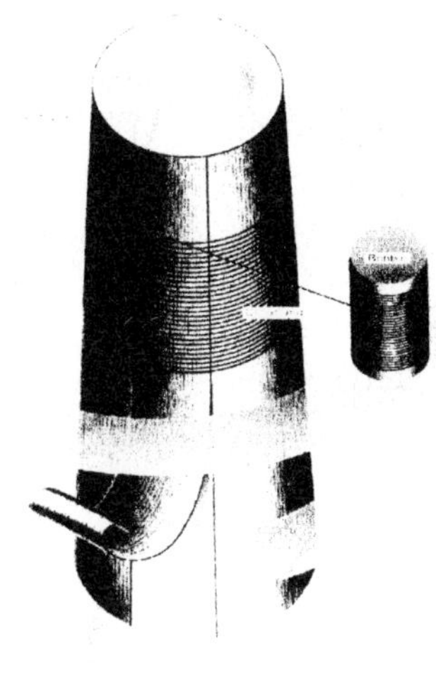

Figure 10 A repaired and strengthened chimney with a coal fibre composite, Täljsten [1]

REFERENCES

1. TÄLJSTEN, B. Personal communication, Skanska AB (1998).
2. BERGSTRÖM, S-G. Studies on concrete technology, dedicated to professor Sven G. Bergström on his 60th Anniversary, Svedish Cement and Research Institute, (1979).
3. FAGERLUND, G. The critical degree of saturation method of assessing the freeze/thaw resistance of concrete, Materials and Structures 10 : 58, (1977).
4. ARFVIDSSON AND HEDENBLAD. Calculation of moisture variations in concrete surfaces, Lund Institute of Technology, Building Materials & Building Physics, (1991).
5. TUUTTI, K. Assessment of service life of concrete structures, Selected Research Studies from Scandinavia, Lund Institute of Technology, TVBM-3078, (1997).
6. FIDJESTÖL, JÖRGENSEN, PETTERSSON, SANDBERG AND TUUTTI, Corrosion of steel in high performance concrete, Summary of the Swedish National High Performance Concrete Programme, Cement and Research Institute, (1998).
7. ROMBÉN, L. Aspects on testing methods for acid attacks on concrete – further experiments, Cement and Research Institute, Research report 9:79, (1979).
8. HASSANZADEH, M. Frost resistance of coal fibre strengthened concrete structures, Lund Institute of Technology, Building Materials, (1998).

THE EFFECT OF CURING ON SURFACE CHLORIDE CONCENTRATION

D W Law

Liverpool University

A N Fried

Kingston University

United Kingdom

ABSTRACT. Curing is regarded as being of major importance in the production of high quality concrete. However, there has been some suggestion that well cured concrete can result in high surface concentrations of chloride ions. This paper reports a set of experiments on three Concrete mixes made using Portland Cement(PC), of nominal strengths 10, 15 and 40 N/mm^2, which were subjected to a range of curing regimes. The specimens were tested for cube and tensile strength in accordance with the British Standard BS:1881: Parts 108, 117 and 124. The specimens were also exposed to wet/dry cycling with 1 M Sodium Chloride, with the chloride profile subsequently being determined from dust drillings. The strength data indicted that for weak concrete's crushed at 28 days, shorter curing times appear more beneficial than a longer curing period. This is probably due to the moisture within the concrete softening the products of hydration rather than weaker concrete resulting. With stronger concrete, longer curing was beneficial to strength as expected. In general there was a good correlation between strength and surface chloride levels, with weaker concrete having higher surface chloride levels. These findings indicated that there was minimal evidence that better curing resulted in higher surface chloride levels.

Keywords: Curing, Durability, Cube strength, Tensile strength, Chloride, Diffusion

Dr David W Law is a Post-Doctoral Research Assistant at Liverpool University, UK. His main research interests are in the areas of steel reinforcement corrosion and the durability of reinforced concrete structures, with particular emphasis on the use of electrochemical methods.

Dr Anton N Fried is a Lecturer in Civil Engineering at Kingston University, UK. His main areas of research are the properties of concrete and masonry.

INTRODUCTION

Curing is regarded as being of major importance in the production of high quality steel reinforced concrete [1-4]. The main benefits of curing are to assist strength development, to improve the durability of concrete against chemical attack and to prevent surface cracking. The importance of curing is to enable the full hydration of cement by water. Through this process the initially water filled pores are blocked by the formation of the hydration products which increase concrete strength and reduce its permeability. The hydration reaction does not occur instantaneously, but progresses with time, thus if water is lost from the concrete during this reaction, the development of the concrete can be impaired. Problems associated with inadequate curing are a reduced final strength and an increased permeability, which will allow chloride ions to progress to any reinforcing steel present in a shorter time period [5,6]. The rapid drying of the concrete can also result in the cracking of the concrete surface. This can be unsightly, in the long term render the concrete susceptible to freeze/thaw attack and enable the easy ingress of aggressive agents such as chlorides.

In order to overcome these problems the adequate curing of concrete by limiting surface water loss and so ensure full hydration is recommended. A number of different curing methods have been developed which generally fall into two groups. Those that prevent moisture loss from the surface ; polythene sheeting, curing membranes and permeable formwork and those that keep moisture in close contact with the concrete surface ; ponding, damp sand, damp Hessian. The second group are the most efficient, but more labour intensive and difficult to monitor.

However, there have been suggestions that well cured concrete can result in higher surface levels of chlorides being developed, due to the increased levels of C-S-H gel which are formed due to improved hydration [7]. Previous research has shown that higher surface levels of chloride have been developed for mixes with higher cement contents [8]. It would be expected that those mixes with higher cement content would have higher C-S-H gel content.

High surface chloride levels are a cause for concern as the total chloride content at the surface effects the rate at which the chloride migrates through the concrete, based on Fick's Second Law of Diffusion [9]

$$C_x = C_s \left(1 - \operatorname{erf} \frac{x}{2(D_c t)^{1/2}} \right) \qquad (1)$$

C_x - total chloride level at depth x
C_s - total chloride level at surface
erf - error function
D_c - chloride diffusion coefficient
t - time

From Equation (1) it can be seen that a higher surface level of chloride will result in a higher chloride content at depth x. However, the curing method will also affect the chloride diffusion coefficient. The better the curing the denser the concrete and the lower the diffusion coefficient. Thus if the curing increases the surface level of chloride this may be offset by the reduced diffusion coefficient of the bulk concrete.

EXPERIMENTAL

Curing and Chloride Ingress

In order to investigate the effects of different curing methods on the initial build up of surface chloride content, four different curing methods have been studied, together with an uncured control. The curing methods adopted were curing tank (CT), curing room (CR), plastic sheeting (PS) and a range of curing membranes. Specimens were subjected to 3, 7 and 28 days curing for the curing tank, curing room and plastic sheeting. The procedures adopted cover a range of curing regimes used in the laboratory and on site [10,11].

In order to monitor the quality of the curing, tensile and cube strength tests were undertaken on each set of specimens at 28 days. The cube strength tests were conducted in accordance with BS:1881: Part 108:1983. The tensile strength tests were conducted in accordance with BS:1881: Part 117:1983.

Chloride exposure was conducted on 150 mm cubes with a 10 mm rebate on the top surface, Figure 1. The exposure regime was ponding with 1M NaCl solution every two weeks for an initial two month period. This was then changed to ponding with a 1 M NaCl solution every four weeks. The chloride profile was determined after 120 days exposure by taking dust drillings from the blocks. The drillings were taken at four random points on the surface and combined for analysis. The analysis was conducted in accordance with BS:1881: Part 124, using electrochemical chloride determination.

Test were conducted on two concrete mixes, nominally 10 and 40 Nmm^{-2} for the curing tank, curing room and plastic sheeting. The curing membranes were tested on 15 and 40 N/mm^2 concrete mixes. In addition a 20 N/mm^2 concrete was used for strength tests only. The mix designs are given in Table 1.

Curing Methods

As mentioned previously, four different curing regimes were used in the experiments, together with an uncured control.

Curing tank

Specimens of the 10, 20 and 40 N/mm^2 concrete mixes were subjected to 3, 7 and 28 days in a curing tank. The tank was filled with tap water and maintained at 20°C throughout the curing period. The specimens were demoulded after 24 hours and fully immersed in the curing tank throughout the curing period.

Those specimens removed after 3 and 7 days curing were then stored in the laboratory until testing. Chloride ponding commenced the day after the specimens were removed from the curing tank.

Curing room

Specimens of 10, 20 and 40 N/mm^2 concrete mixes were subjected to 3, 7 and 28 days in a curing room. The curing room was maintained at 25°C and 90 % relative humidity. The specimens were demoulded after 24 hours and immediately placed in the curing room. Those specimens removed after 3 and 7 days were placed in the laboratory for storage until testing. Chloride ponding commenced the day after the specimens were removed from the curing room.

Plastic sheeting

Specimens of 10, 20 and 40 N/mm^2 concrete mixes were subjected to 3, 7 and 28 days under plastic sheeting. The specimens were demoulded after 24 hours and placed under close fitting plastic sheeting in the laboratory. Those specimens removed from the plastic after 3 and 7 days were placed in the laboratory for storage until testing. Chloride ponding commenced the day after the specimens were removed from under the plastic sheeting.

Table 1 Mix design by weight

CONSTITUENT	15 N/mm^2 (CURING MEMBRANES)	10 N/mm^2 (CT, CR, PS)	20 N/mm^2 (CT, CR, PS)	40 N/mm^2 (ALL SPECIMENS)
Cement	257	144	277	460
Water	180	195	180	180
Fine Aggregate	485	727	671	570
Coarse Aggregate	1454	1293	1247	1160

Curing membranes

Curing membranes minimise the loss of moisture by forming a film over the surface of the concrete which is relatively impermeable to water and water vapour. Thus moisture is retained within the concrete and good curing is achieved.

Two commercially available curing compounds denoted Compound A and B whose properties are summarised in Table 2 were used in the testing. These compounds were applied directly after striking and six hours after striking in accordance with the manufactures instructions. Following the application of the compounds the specimens were stored in the laboratory until testing. Chloride ponding commenced 24 hours after the application of the curing compound.

Table 2 Curing compound properties

COMPOUND	SPECIFIC GRAVITY (20°C)	CURING EFFICIENCY (%)
A	0.87	90
B	1	75

Control

Control tests were conducted on each of the mix designs. These specimens were demoulded after 24 hours and stored in the laboratory. Chloride ponding commenced the day after the specimens were demoulded, Figure 1.

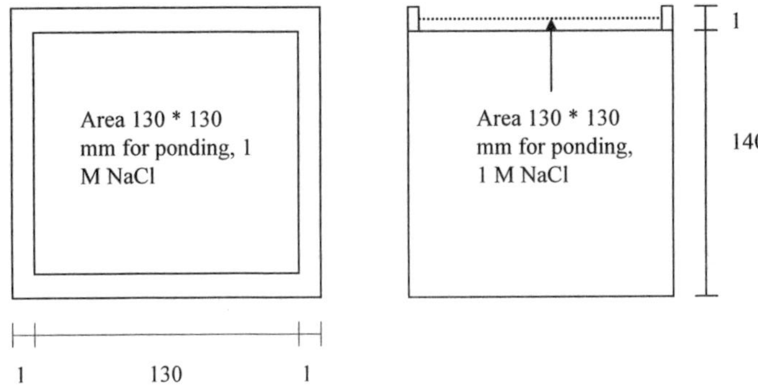

Figure 1 Ponding specimen for chloride exposure

Strength Tests

For the 10 and 40 N/mm^2 and with each variation of curing, three, 150 mm cubes and three, 150 mm diameter x 300 long cylinders were manufactured to enable the cube and tensile splitting strengths of the concrete to be determined. For the 20 N/mm^2 concrete, 100 mm cubes for variation of curing were produced.

RESULTS AND DISCUSSION

The 28 day cube strength test results and 28 day tensile strength results are given in Tables 3-7. The chloride profiles are given in Figures 2-5. The chloride data is presented as the % chloride by mass of sample.

Table 3 Cube and tensile strength data, curing membrane 15 N/mm^2

CURING REGIME	CUBE STRENGTH (N/mm^2)	TENSILE STRENGTH (N/mm^2)
A (0 hours)	29.7	2.9
A (6 hours)	30.9	2.8
B (0 hours)	30.5	3.0
B (6 hours)	28.9	2.5
Air	27.85	2.3

Table 4 Cube and tensile strength data, curing membrane 40 N/mm^2

CURING REGIME	CUBE STRENGTH (N/mm^2)	TENSILE STRENGTH (N/mm^2)
A (0 hours)	62.4	3.8
A (6 hours)	60.7	3.5
B (0 hours)	61.1	4.0
B (6 hours)	59.5	3.8
Air	54.6	3.2

Table 5 Cube and tensile strength data, curing room, curing tank and plastic sheeting, 10 N/mm^2

CURING REGIME	CUBE STRENGTH (N/mm^2)	TENSILE STRENGTH (N/mm^2)
Curing Tank (3 days)	9.78	1.27
Curing Tank (7 days)	7.42	1.13
Curing Tank (28 days)	5.78	0.95
Curing Room (3 days)	9.09	1.06
Curing Room (7 days)	6.93	0.99
Curing Tank (28 days)	6.20	0.82
Plastic Sheeting (3 days)	8.67	1.06
Plastic Sheeting (7 days)	7.07	1.06
Plastic Sheeting (28 days)	7.53	1.01
Air	5.71	0.94

Table 6 Cube and tensile strength data, curing room, curing tank and plastic sheeting, 40 N/mm^2

CURING REGIME	CUBE STRENGTH (N/mm^2)	TENSILE STRENGTH (N/mm^2)
Curing Tank (3 days)	55.0	2.68
Curing Tank (7 days)	58.5	2.83
Curing Tank (28 days)	60.4	2.63
Curing Room (3 days)	53.6	2.66
Curing Room (7 days)	58.9	2.89
Curing Tank (28 days)	58.0	3.00
Plastic Sheeting (3 days)	53.1	2.56
Plastic Sheeting (7 days)	55.2	2.65
Plastic Sheeting (28 days)	55.8	2.62
Air	52.3	2.50

Table 7 Cube strength data, curing room, curing tank and plastic sheeting, 20 N/mm^2

CURING REGIME	CUBE STRENGTH 24 HOUR SOAKED (N/mm^2)	CUBE STRENGTH AS CURED (N/mm^2)
Curing Tank (3 days)	35.1	38.6
Curing Tank (7 days)	35.2	40.2
Curing Tank (28 days)	35.9	36.1
Curing Room (3 days)	30.8	35.8
Curing Room (7 days)	31.6	39.0
Curing Tank (28 days)	33.1	41.5
Plastic Sheeting (3 days)	32.3	33.4
Plastic Sheeting (7 days)	31.1	42.1
Plastic Sheeting (28 days)	29.9	43.3
Air	29.3	40.6

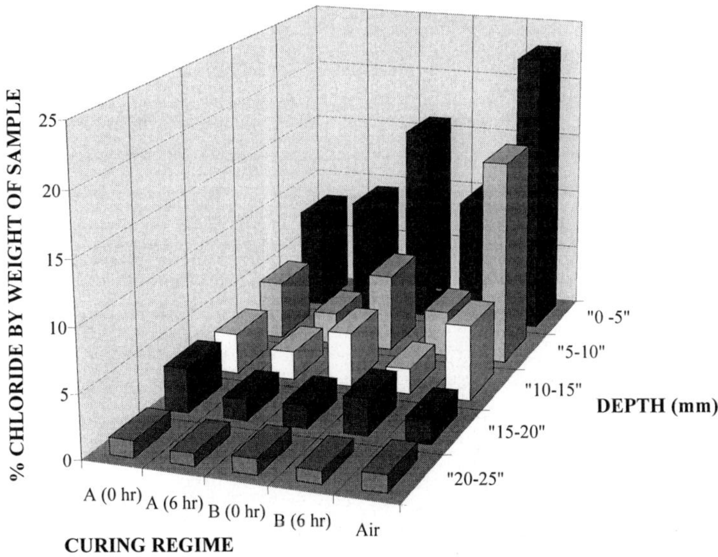

Figure 2 Chloride profiles curing membranes, 15 N/mm^2 Mix

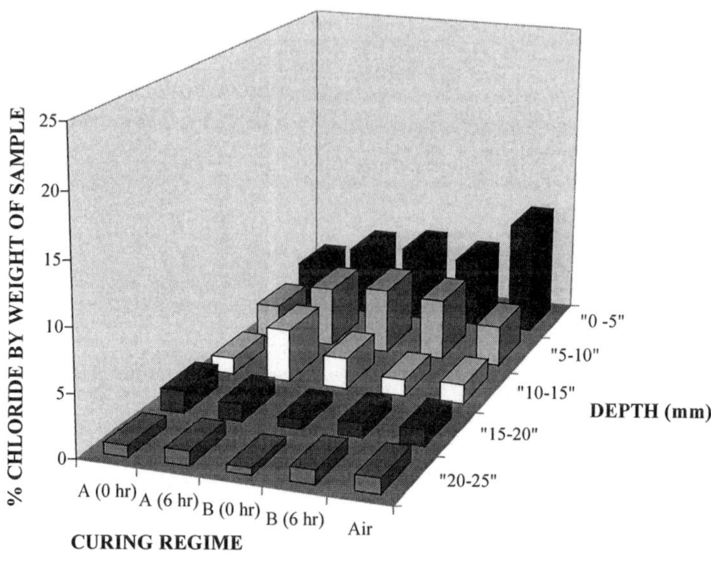

Figure 3 Chloride profiles curing membranes, 40 N/mm^2 mix

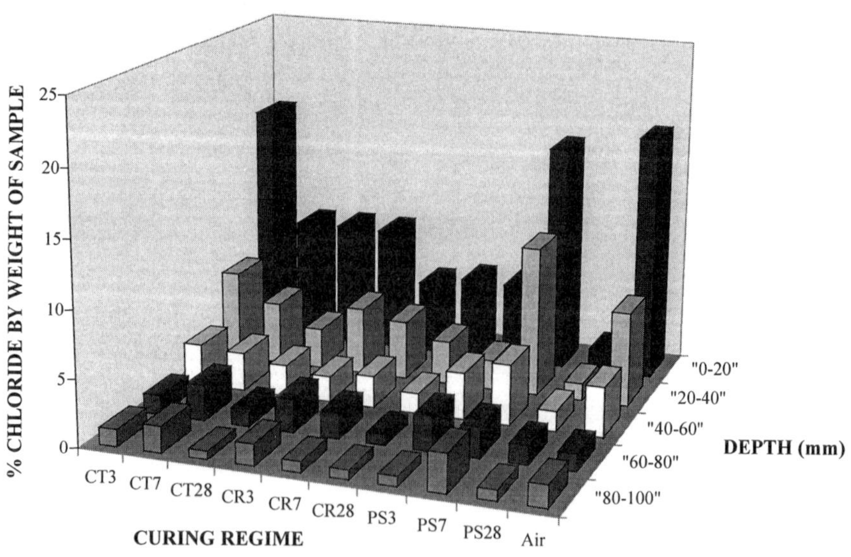

Figure 4 Chloride profiles curing tank, curing room, plastic sheeting, 10 N/mm^2 mix

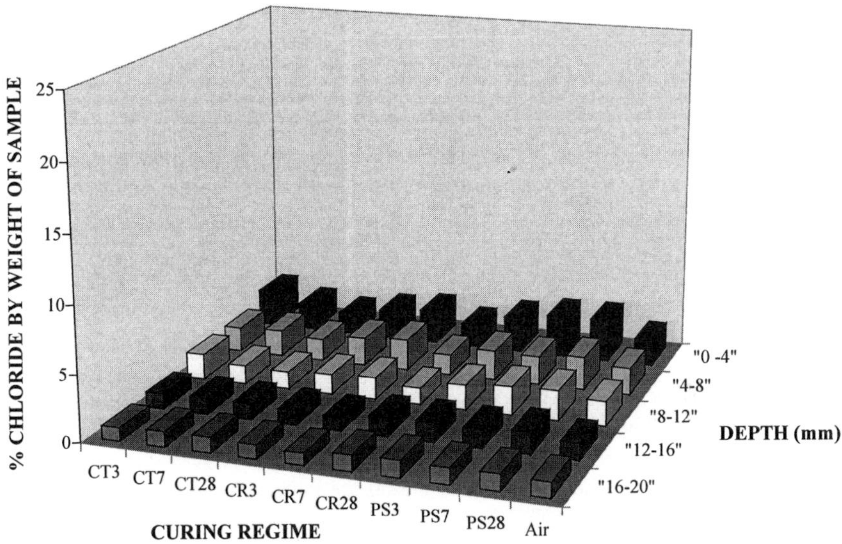

Figure 5 Chloride profiles curing tank, curing room, plastic sheeting, 40 N/mm^2 mix

Effect of Curing on Concrete Strength

From the data in Table 3, it is evident that with weaker concrete painted with a curing membrane, the strength generally decreases if there are delays in applying the compound. The exception occurred when compound A was not applied for six hours to compressive specimens. Here there was a 4% strength loss. From Table 4, the 40 N/mm^2 concrete covered with curing membrane, delays in painting the membrane on the concrete resulted in a loss of both tensile and cube strength as expected.

The cube strength of the control specimens of the two 40 N/mm^2 mixes (See Tables 4 and 6) are similar, these being 54.6 and 52.3 N/mm^2. The corresponding tensile strengths are however 3.20 and 2.50 N/mm^2. However, the cube strength of concrete cured using the curing membranes was higher in all instances (except for that immersed in the curing tank for 28 days) than that cured in the curing tank, the curing room or under polythene. No exceptions occurred when comparing the tensile strengths of the 40 N/mm^2 concrete. These findings indicate that the concrete's were reasonably similar (as the cube strengths of the controls indicate) but that curing by membrane produces stronger concrete.

From analysis of the data in Table 5, with the 10 N/mm^2 concrete cured for 3, 7 or 28 days in water, in the curing room or under polythene and with its uncured control, the following is evident. (a). Cube strength decreases as the length of curing increases with the exception of concrete cured under polythene where there was an increase in strength for concrete cured for 28 as opposed to 7 days. The strength decrease when curing increased from 3 to 7 days was 24, 24 and 19% for concrete cured under water, in the curing room and under polythene respectively. The corresponding decrease was 22, 11 and -6.5% when curing was increased from 7 to 28 days. This finding is unexpected and indicates a reduction in strength with improved curing. (b). Tensile strength decreased when cured for 7 as opposed to three days with fully submerged specimens (11%) and those in the curing room (7%) but no change to tensile strength occurred when the length of close cover curing was increased. The strength further decreased when curing was increased to 28 days. Drops of 16, 17 and 5% occurred for specimens in the curing tank, curing room and under polythene respectively when curing increased from 7 to 28 days. Again this finding indicates a reduction in strength with improved curing.

From Table 6, with the 40 N/mm^2 concrete cured for 3, 7 or 28 days in water, in the curing room or under polythene and with its uncured control, the following is evident. (a). Cube strength increases as the length of curing increases with the exception of concrete cured in the curing room where there was a drop in strength for concrete cured for 28 as opposed to 7 days. The strength increase when curing increased from 3 to 7 days was 6.3, 9.8 and 4.0% for concrete cured under water, in the curing room and under polythene respectively. The corresponding numbers are 3.2, -1.5 and 1.1% when curing is increased from 7 to 28 days. This clearly indicates that the early days of curing are important but that lengthening curing from 7 to 28 days is not as significant. (b). Tensile strength increases when cured for 7 as opposed to three days in all instances but a drop in strength occurred when the length of curing increased from 7 to 28 days for concrete in the curing tank and under polythene. The strength increase when curing increased from 3 to 7 days is 5.6, 8.6 and 3.5% for concrete cured under water, in the curing room and under polythene respectively. The corresponding numbers are -7.0, 3.8 and -1.1% when curing is increased from 7 to 28 days. When comparing the results from tables 5 and 6 curing weaker concrete's appears to be detrimental. However as curing (i.e. the maintaining of a moist environment) and testing occurred on the same day it is possible that the moisture content of the concrete when tested may affect its strength, which if significant has a greater impact on weaker concrete's[12]. Hence, the 28 day strengths for both mixes may have been depressed compared to the 3 and 7 day cured specimens. This may account for the slight falls in strength observed in some specimens for the 40 N/mm^2 mix at 28 days compared to the 7 day data. However, this would not account for the fall in strengths observed for the 3 and 7 day values of the 10 N/mm^2, or indeed account for all of the observed decrease for the 7 to 28 day values. This would indicate that for very weak concrete curing may be detrimental.

For the 3, 7 and 28 days regimes the ranking in the cube and tensile strengths for the curing tank, curing room and plastic sheeting was generally consistent for both the weak and strong concrete. However, there was considerable variation in the performance of the different curing period employed in each mix. In general concrete strengths reduced slightly as the curing regimes changed from full immersion, to curing room to polythene wrapping and obviously to air cured

Overall there was greater variability in weaker than the stronger mixes, though the data still indicated that the control specimens have the lowest cube and tensile strength data for all four sets of specimens, there being a single exception of the 28 day curing tank 10 N/mm^2 mix. Thus the data confirms the beneficial effects of curing on concrete strength development.

In order to examine if increases in moisture content in the concrete at the time of testing affected the strength data determined a 20 Nmm^{-2} concrete was tested both, as cured, and with 24 hours soaking prior to testing. The 28 day cure specimens were removed from the curing regime 24 hours early and either left to dry for 24 hours, as cured, or soaked for 24 hours. The 3 and 7 day specimens were removed at 3 and 7 days, with the soaked specimens being immersed in water for 24 hours before testing, the as cured being left to dry. The results from Table 7 indicate that the effect of soaking prior to testing reduces the cube strength independent of curing regime. In general the difference in strengths was most marked for concrete cured for 7 and 28 days, with the obvious exception of the 28 day, as cured specimen, exposed to the curing tank regime.

Effect of Curing on Chloride Ingress

The chloride profiles of both curing methods displayed chloride profiles with two to three times higher surface chloride levels in the weak concrete than the strong concrete. The surface levels of the 10 and 15 N/mm^2 mixes were similar, but the chloride ions had progressed to a significantly greater depth in the 10 N/mm^2 mix. Chloride levels greater than 1 % by mass of sample were detected at depths of 80-100 mm into the concrete. For the 15 and 40 N/mm^2 mixes the chloride ions have only reached the 1 % levels in the 12-20 mm depth range. This is attributed to the 10 N/mm^2 concrete being highly permeable and has enabled the chloride ions to move freely through the open pore structure of the concrete. This open pore structure will result in the concrete having a high chloride diffusion coefficient and as such surface chloride levels have not had to build up before the chloride ions have been able to diffuse through the concrete, Equation (1).

Analysis of those specimens treated with the curing membranes showed the control uncured concrete had the highest surface levels of chloride for both mixes. The curing Compound B with zero hours exposure is the next highest again for both mixes, followed by Compound A with six hours exposure, Compound B with six hours exposure and Compound A with zero hours exposure.

The results for the 3, 7 and 28 day curing regimes showed no regime or duration performing consistently better or worse for both the 10 and 40 N/mm^2 mixes. The uncured control gave the highest surface chloride levels for the 10 N/mm^2 mix, though the curing tank 3 day and the plastic sheeting 7 day values were of a similar level. In the 40 N/mm^2 mix all of the specimens gave comparable results, with the uncured control being one of the lower values.

The 3, 7 and 28 day 40 N/mm^2 mix gave consistently lower surface chloride levels than the curing membrane 40 N/mm^2 mix. This is in direct contrast to the curing performance where the curing membrane 40 N/mm^2 mix gave higher strengths than the 3, 7 and 28 day mix. However, this data is not supported by the observed relationship between strength and surface chloride content within the two mixes. In the 3, 7 and 28 day mix the curing tank regime gave the highest strengths but did not display any appreciably higher surface chloride levels than any of the other regimes. The uncured control specimens gave the lowest strength data and had the highest surface chloride levels for three of the four mixes studied. Also the 40 N/mm^2 mix has significantly less chloride than the 15 N/mm^2 mix. Thus overall the data would indicate that improved curing does not result in increased levels of chloride at the concrete surface.

CONCLUSIONS

1. Increasing the length of curing resulted in a loss of strength for the very weak 10 Nmm^{-2} concrete, but with stronger concrete a strength increase occurred.

2. An increase in the moisture content, soaking, prior to testing reduced the cube strength

3. Delays in applying curing compounds resulted in reduced 28 day strengths.

4. In very weak concrete the very high permeability results in the rapid build up of chloride ions at depths where steel reinforcement may be located.

5. There is little evidence for curing increasing the surface levels of chloride. Overall the data shows that curing reduced the surface levels in three of the four mixes studied and had no detrimental effect in the fourth.

REFERENCES

1. NEVILLE, A M., Properties of Concrete, Pitman, London, 3rd Edition, 1981

2. WATERS, T., The effect of allowing concrete to dry before it is fully cured, Magazine of Concrete Research, 1955, July, pp79-82

3. ACI COMMITTEE 308, Recommended practice for curing concrete, ACI-308, American Concrete Institute, Detroit, 1981

4. TAN, K AND GJORV, O E., Performance of Concrete under different curing conditions, Cement and Concrete Research, 1996, Volume 26, No. 3, pp355-361

5. GOTO, S AND ROY, D M., The effect of W/C ratio and curing temperature on the permeability of hardened cement paste, Cement and Concrete Research, 1981, Volume 11, pp575-579

6. PARROTT L J., Water absorption, chloride ingress and reinforcement corrosion in cover concrete: some effects of cement and curing, Procedings Corrosion of Reinforced Concrete in Construction, Cambridge, 1996, pp146-155

7. BAMFORTH P B, LAW D W AND CHAPMAN-ANDREWS J F., Deterioration prevention in reinforced concrete structures subject to hostile environments, BRITE PROJECT P-1552-1-85, March 1992

8. GJORV O E AND VENNESLAND O, Diffusion of chloride ions from seawater into concrete, Cement and Concrete Research, 1979, Volume 9, pp229-238

9. CRANK J, The mathematics of diffusion, The Clarendon Press, Oxford, 1956

10. NOLAN E, ALI M A, BASHEER P A M AND MARSH B K., Testing the effectiveness of commonly used site curing regimes, Materials and Structures, 1997, Volume 30, Jan-Feb, pp53-60

11. WANG J AND BLACK A, Performance evaluation of curing membranes, Procedings 3rd International RILEM Conference, Concrete in Hot Climates, pp199-205

12. POPOVICS S, Effect of curing method and final moisture condition on compressive strength of concrete, ACI Materials Journal, 1986, Volume 83, July-August, pp650-657

DETERMINATION OF THE EFFECTIVE CHLORIDE DIFFUSION COEFFICIENT IN CONCRETE VIA A GAS DIFFUSION TECHNIQUE

A Sharif

K F Loughlin

A K Azad **B C M Nawaz**

King Fahd University of Petroleum and Minerals

Saudi Arabia

ABSTRACT. A new experimental technique based on gas diffusion through a thin sample specimen is presented to evaluate the effective chloride diffusion in concrete. The counter-diffusion of two gases helium and nitrogen through a concrete disc is used to establish the porosity-to-tortuosity ratio (ε/τ) of the concrete. From measurements of the porosity via porosimetry, the tortuosity of the concrete can be determined. The effective chloride diffusion in concrete is a function of (ε/τ) ratio and the diffusion coefficient of chloride ion in water which is a known value. The test takes 3 to 4 hours to complete. Excellent results were obtained when correlated with the conventional diffusional measurement techniques for concrete with different water / cement ratios and cement content.

Keywords: Diffusion, Effective diffusion coefficient, Porosity, Static diffusion cell, Stefan-Maxwell equations, Tortuosity.

Alfarabi Sharif is the Director, Center for Engineering Research and Professor of Civil Engineering at King Fahd University of Petroleum and Minerals (KFUPM), Dhahran, Saudi Arabia. He received his Ph.D. from Washington University, St. Louis, Missouri, in 1982. His research interests include composite structures, prestressed concrete, durability, assessment and repair of concrete structures.

Kevin F Loughlin Associate Professor of Chemical Engineering at KFUPM, graduated from the National University of Ireland with a BSc. and MEngSc. in Chemical Engineering and obtained his Ph.D. from the University of New Brunswick, Canada, in 1970. His research interests are adsorption and diffusion in molecular sieves, zeolites and porous media.

Abul K Azad Professor in the Department of Civil Engineering at KFUPM. He obtained his DEng. Degree from Concordia University, Montreal, Canada, in 1973 and has been engaged in teaching, research and industrial work. His research interests include structural optimization, damage assessment and durability of concrete.

C M Nawaz is a graduate student at KFUPM. He received his M.Sc. from KFUPM in 1995.

INTRODUCTION

The corrosion of steel reinforcement in concrete and its damaging impact have generated significant research interest in the area of chloride penetration or diffusion in concrete. The transport phenomenon of the chloride ions through concrete has been modeled by numerous researchers in the past by assuming diffusion through porous media. Fick's second law of diffusion has been extensively used to find the effective diffusion coefficient D_e of chloride ions in concrete from data generated on laboratory samples. The main problem is that the evaluation of D_e using conventional concentration difference type tests is slow [1].

Previous attempts to rapidly determine the effective diffusion coefficient D_e for pastes have met with limited success. Goto and Roy [2] applied potential difference of up to 2 V (dc) across thin paste specimens (< 5 mm thick), assuming that this small perturbance did not change the diffusion of chloride ions significantly. Whiting [3] carried out accelerated chloride ingress tests on a bridge deck, in the laboratory and in situ, using an applied potential differences of 60-80 V (dc), which was adopted by AASHTO standards as the chloride penetration test [4]. These results were correlated later by Andrade and Sanjuan [5] and found that the D_e from Whiting tests were different from the D_e obtained from the normal diffusion tests. Dhir et al. [6] applied a potential difference of 10 V (dc) across specimen and also conducted conventional non-perturbative tests. A close relationship between the D_e values from both tests was established allowing D_e to be estimated within a week.

This paper reports the development of a simple rapid test to determine the effective diffusion coefficient of chloride ion in concrete. The test is based on gas diffusion through concrete, from which the porosity to tortuosity ratio of concrete is evaluated based on the Stefan-Maxwell equations relating the concentration gradients, molecular fluxes, and the molecular and Knudsen diffusion concepts [7] in a porous medium. This then enables the effective diffusion of chloride ions in concrete to be evaluated. The results of this test are correlated with the conventional ponding techniques for evaluating the chloride diffusion coefficient [8] to assess the validity of the experimental method.

SPECIMEN PREPARATION

In preparing the concrete specimens Type V sulphate resisting Portland cement, prewashed limestone aggregate, and prewashed beach sand were used. The grading of coarse aggregates presented in Table 1 was selected to conform with ASTM C-33. The specific gravity and absorption of coarse and fine aggregates are presented in Table 2. Concrete cylinders 5 cm (2 in.) in diameter and 15 cm (6 in.) long were cast. The specimens were demolded 24 hr after casting and continuously covered with wet burlap/towel to cure at lab temperature of 18 to 20 C (64 to 68 F) for 2 weeks. Later, 1 cm (0.39 in.) thick concrete discs were sliced out from the top, bottom, and middle of these cylinders. A total of 18 disc specimens has been tested for the different mix concrete design. The prepared specimens were stored and tested at laboratory conditions of 18 to 20 C (64 to 68 F) temperatures. The parameters considered in the concrete mix design were water-cement ratio 0.4 and 0.55 and cement content 300, 350, and 400 kg/m^3 (18.7, 21.8, and 25.0 lb/ft^3). The coarse-to-fine aggregate ratio was kept constant at 1.5 for all mixes.

Table 1 Grading of course aggregates

SIZE, * IN.	PERCENT WEIGHT RETAINED	PERCENT CUMULATIVE WEIGHT RETAINED	PERCENT PASSING	ASTM C-33 NO. 7
3/4	0	0	100	100
1/2	10	10	90	90-100
3/8	45	55	45	40-70
3/16	45	100	0	0-15
3/32	0	100	0	0-5

*1 in. = 2.54 cm

Table 2 Absorption and specific gravity

SIZE, * IN.	ABSORPTION, PERCENT	BULK SPECIFIC GRAVITY
Fine aggregate	1.562	2.55
3/16 in	1.621	2.607
3/8 in.	1.477	2.606
1/2 in.	1.034	2.643

*1 in. = 2.54 cm

APPARATUS SETUP AND MEASUREMENTS

The apparatus consists simply of two diffusion cells [(a) and (b)] separated by a thin concrete disc (1 cm thick) as shown schematically in Figure 1. To facilitate holding in the molds, the concrete discs were glued to PVC rings with epoxy. The complete setup consists of the two diffusion cells, a reference cell, gas flow lines, a detector and a recorder as schematically presented in Figure 2.

To prepare for the experimental runs, the apparatus is loaded with the two gases as follows. The reference cell is purged with helium gas for approximately 5 min. During this purging, the outlet of the reference cell is connected to a gas burette at atmospheric pressure, and the flow rate is adjusted to approximately 50 to 100 cm^3/min. When the 5 min are up, first the inlet valve is closed to prevent pressurization above atmospheric pressure, then the outlet valve is closed.

Next the cell is purged in a similar manner, except that nitrogen is passed through Cell (a) and helium through Cell (b). Both outlets are connected to atmospheric gas burettes and the flow rates are adjusted to 50 to 100 cm^3/min. It is important that the flow rate be maintained at a minuscule value of 50 to 100 cm^3/min, otherwise the thermistor would be initialized using a forced convective flow rather than the natural convective flow which will be used during the experiment. Further, the flow rate must be large enough to prevent contamination of each cell by diffusion through the discs. After purging for 5 min, the inlet cell valves are closed first (to prevent pressurization) and the outlet valves immediately after. The initial cell voltage difference is noted on the recorder. The gases diffuse counter-currently through the concrete specimen. The concentration of nitrogen, x_2, in Chamber (b) is recorded continuously with respect to the reference cell which is filled with helium. Initially, at $t = 0$, the concentration of helium, x_1, in Chamber (b) is zero. Thermistors are affixed to the diffusion and reference cells to detect any change in electrical resistance with respect to the reference cell. The thermistors' resistance varies due to the change of thermal conductivity of the gas phase as it varies with time. As the gaseous diffusion progresses, the concentration in the diffusion cells changes, leading to a change in the electrical resistance.

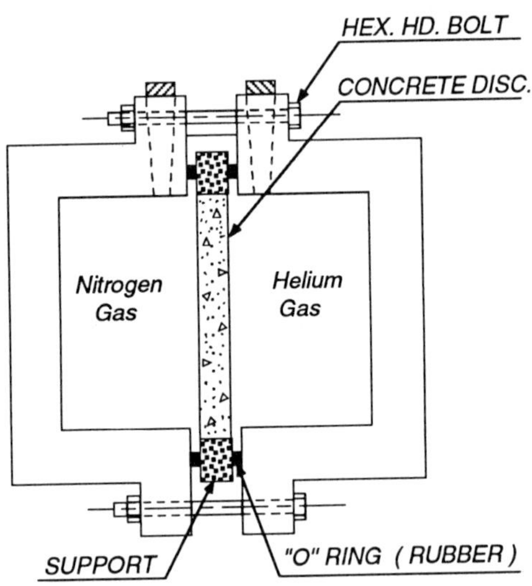

Figure 1 Sketch showing diffusion cell

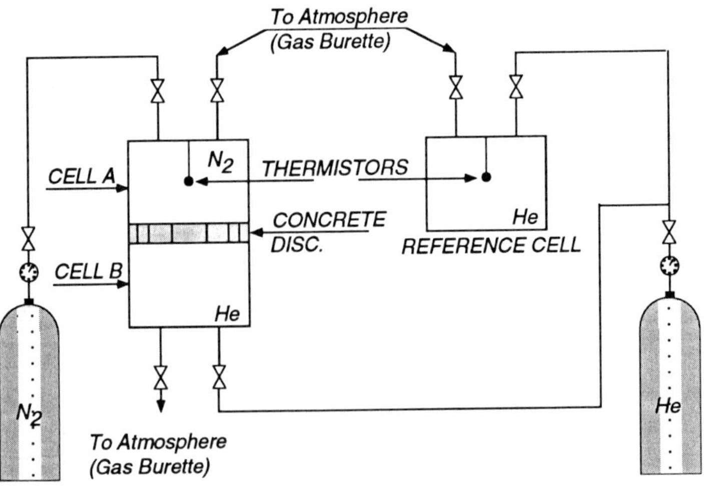

Figure 2 Static diffusion: values are for evacuation and preloading – closed during experiment

The test measurement output is recorded continuously on a chart recorder representing the concentration x_1 in Chamber (b) with respect to the reference cell. The concentration difference given in mV is plotted versus the distance the chart has traveled in cm as shown in Figure 3. Knowing the speed at which the chart is advancing, the distance can be converted to time units. Initially, at $t = 0$, Chamber (b) contains only nitrogen, i.e. $x_1 = 0$, which corresponds to a potential difference of 1.0 mV. Similarly at the time when the system attains a steady state, i.e. $x_1 = 0.5$, corresponds to the smallest potential difference as may be observed from Figure 3. A straight-line graph is plotted from these two points for the potential difference versus the helium concentration x_1 as shown in Figure 4. In an earlier study Holborow [9] established the validity of this calibration technique. This is the same principle used to measure concentration via a thermal conductivity cell in a gas chromatographic experiment i.e. the change in resistance of the thermistor varies linearly with concentration. From Figures 3 and 4 the helium concentration x_1 in Chamber (b) can be evaluated at any time t.

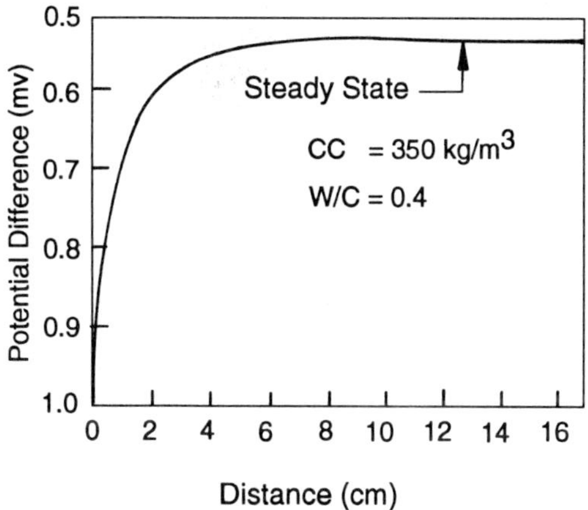

Figure 3 Chart obtained from gas diffusion test

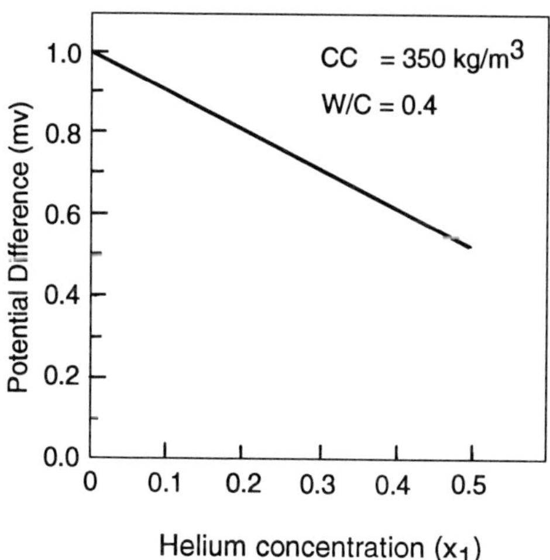

Figure 4 Plot of potential difference vs. helium concentration (x_1)

GAS DIFFUSION IN CONCRETE

Gas diffusion in a porous media at a constant pressure could be of molecular diffusion, Knudsen diffusion, or a combination of the two depending on the size of pores [7]. Molecular diffusion takes place if the pores of the medium are relatively large. In molecular diffusion, molecular collisions with each other predominate as compared to that with the pore walls. Hirschfelder et al. [10] presented the following equation for the molecular diffusion coefficient which has been derived from the molecular theory of gases

$$D_{12} = \frac{0.001858T^{3/2}\left[\frac{1}{M_1} + \frac{1}{M_2}\right]^{1/2}}{P\sigma_{12}^2\,\Omega_D} \tag{1}$$

where the subscripts 1 and 2 represent two different gases diffusing in the medium. Knudsen diffusion is applicable when the pores of the medium are small and the molecules collide with the pore walls much more frequently than with each other. The Knudsen diffusion coefficient in a straight cylindrical pore is given by the following expression.

$$D_K = 9700\,r_e\sqrt{T/m} \tag{2}$$

In concrete as in porous catalysts or adsorbents where small and large size pores exist, the diffusion process is a combination of molecular diffusion and Knudsen diffusion. However, the voidage in concrete is much lower than in porous adsorbents or catalysts.

The static gas diffusion cell is selected to measure the transport properties through the thin concrete specimens. An essential consideration in the static diffusion cell is to estimate the gas side film resistance adjacent to the surface of each side of the specimen if any confidence is to be placed in the results. The transport of gas from one cell to the other involves passage through three resistances: 1) the resistance of the film on one side, 2) the resistance of the concrete specimen, and 3) the resistance of the film on the second side. If the resistance is just due to the concrete, the (ε/τ) parameter can easily be determined. Otherwise the resistances of the gas film must be considered.

To measure both the gas film resistances and concrete resistances, a dynamic apparatus was used similar to that employed by Dogu and Smith [11]. This experiment basically involves injecting a pulse into a system where a carrier gas flows across one side of a concrete specimen, and the peak response detected at the other side is analyzed. When this test was performed, no peak was detected on the other side of the specimen even though many test variations were employed. The reason is that all the resistance to transport resides within the concrete specimen, and consequently the film resistances are negligible in comparison to the diffusion resistances and may be neglected. Accordingly, the dynamic test was replaced by a static test in which the film resistances are assumed negligible.

The porosity-to-tortuosity ratio (ε/τ) is then evaluated using Equation (3) and (4) [8].

$$\theta = \left[-\frac{1}{2}\left\{1+\frac{D_{12}}{D_{1K}}\right\}In(1-2x_1)+\frac{1}{4}\left\{1-\frac{D_{2K}}{D_{1K}}\right\}In(1-2x_2)+\frac{1}{2}\left\{1-\frac{D_{2K}}{D_{1K}}\right\}x_1\right] \tag{3}$$

$$\theta = \frac{AD_{12}}{VL} t \frac{\varepsilon}{\tau} \tag{4}$$

The variables in Equation (3) are x_1, the mole fraction of the gas (helium) in Chamber (b) at any time, and θ defined by Equation (4). Once x_1 is evaluated experimentally at any time t, the value of θ can be evaluated from Equation (3). Then the ratio of porosity of tortuosity ε/τ, which is a constant for a porous medium, can be calculated from Equation (4) at any time t.

The porosity-to-tortuosity ratio (ε/τ) is an important characteristic of any porous medium. Once this ratio is evaluated, the effective diffusivity D_e of any liquid fluid in concrete can be evaluated using Equation (5)

$$D_e = \frac{\varepsilon}{\tau} D \tag{5}$$

where D is the diffusion of chloride ions in water and D_e is the effective diffusivity of chloride ions in concrete.

RESULTS AND DISCUSSION

The molecular and Knudsen diffusion coefficients for helium and nitrogen for the different concrete are evaluated using Equations (1) and (2). The molecular diffusion coefficient D_{12} for the two gases is independent of the porous medium and its computed value is 0.485 cm^2/sec at 19°C.

To evaluate the Knudsen diffusivities, the mean pore radius must be established. The mean pore radius of the pores and the total porosity of the concrete specimens were determined by mercury intrusion porosimetry according to ASTM D 4284. Intrusion was performed up to 6000 psia. For the test, a value of surface tension γ equal to 485 dynes and contact angle equal to 140 deg was used. In addition to the applied pressure and the total pore volume recorded, the approximate pore diameter ranges, the pore volume, and the surface area are also obtained. The total porosity ε and the effective pore radius r_e in Angstroms are presented in Table 3. The results indicate that the porosity ε increases with the water-cement ratio. No discernible trends could be ascertained for either the pore radius or the cement content. Using the mean pore radius values from Table 3, the Knudsen diffusion coefficients for each concrete for helium and nitrogen were evaluated and are tabulated in Table 4.

A table of θ values was calculated for each x_1 value between 0.000 and 0.499 in steps of 0.001 by solving Equation (3) using a computer program. From the experimental data of x_1 versus time, the equivalent θ value for each x_1 point was determined using the mass balance equation, and since the time and other parameters in Equation (1) are known, the values of (ε/τ) are determined. Theoretically, the value of (ε/τ) for a particular concrete should be a constant. However, the values of (ε/τ) ratio are found to vary slightly due to experimental measurements. Therefore, the average value and standard deviation of (ε/τ) are calculated for each particular concrete. The resulting (ε/τ) values and mean standard deviation values are tabulated in Table 5. As the ε values are also known (see Table 3), the tortuosity values can also be calculated and these are tabulated in Column 4 of Table 5. The reported values of (ε/τ) are approximately a factor of 10 smaller than would be observed for a porous catalyst or adsorbent. This is attributable to the lower ε values which are of the order of 0.4 and to the larger tortuosity values which are of the order of 2 or 3 for adsorbents or catalysts. The smaller the value of tortuosity the more open the concrete structure. From the tortuosity values in Table 5, it is evident that higher w/c contents and lower cc values give very open concrete specimens. The consequence is that, for these concretes, the expected values of the effective diffusion coefficient of chloride ions should be large. The opposite should be expected low w/c and high cc concretes; for these the anticipated chloride ion diffusivity is expected to be much lower.

Table 3 Mercury intrusion test results*

w/c	CEMENT CONTENT, (kg/m^3)					
	300		350		400	
	ε	Å	ε	Å	ε	Å
0.4	7.57	224	7.39	127	8.39	177
0.55	8.3	122	10.64	135	8.52	175

*porosity in percent; pore radius in Å; 1 kg/m^3 = 16.02 lb/ft^3

Table 4 Knudsen diffusion coefficient values

w/c	D_K, cm^2/sec					
	HELIUM GAS			NITROGEN GAS		
	CEMENT CONTENT, kg/m^3					
	300	350	400	300	350	400
0.4	0.185	0.104	0.146	0.070	0.040	0.055
0.55	0.101	0.112	0.144	0.038	0.042	0.055

*1 kg/m^3 = 16.02 lb/ft^3; 1 in.^{2} = 6.45 cm^2

To ascertain the fit of the average (ε/τ) values relative to the experimental values, a plot of ℓn $(1 - 2x_l)$ vs. time is developed to verify the procedure. A typical plot is presented in Figure 5 for w/c = 0.4 and cement content = 350 kg/m^3 and as may be observed the fit is excellent. Finally, the effective diffusion coefficient D_e for chloride ions in concrete is evaluated using Equation (2). The average value of (ε/τ), the standard deviation (SD) and D_e are listed in Table 5 for each concrete mix. As may be observed, the calculated values follow the trends discussed earlier. The largest D_e values (33.14 x 10^{-8} and 37.33^{-8} cm^2) are reported for high water-cement values, low cement content concretes, and conversely the lowest D_e values (9.7 x $^{-8}$ and 8.82 x 10^{-8} cm^2/sec) are reported for low water-cement ratio and high cement content concretes. The effective diffusion coefficient D_e obtained using the gas diffusion technique is compared to that measured using the conventional ponding technique [8]. The specimens in the conventional ponding technique were ponded with sodium chloride solutions 80 days after casting. The ponded solution was changed daily for the specimens to maintain constant chloride concentration. The effective chloride ion diffusion coefficient from the conventional ponding technique was obtained using Fick's second law solved for the appropriate boundary conditions. Concrete slab specimens used for the ponding technique and concrete cylinders used for gas diffusion test were cast together for each concrete mix. A comparison of the effective chloride diffusion coefficient D_e obtained from each test is listed in Table 6 for the different concrete mixes. The ponding technique and the gas diffusion technique reveal excellent agreement.

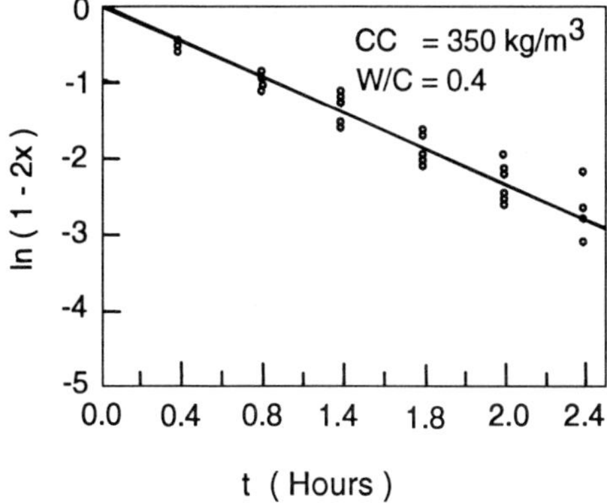

Figure 5 Fit for $\ell n\,(1 - 2x_1)$ vs. t

Table 5 Gas diffusion results

cc, kg/m^3	w/c	ε/τ	τ	SD	D_e, 10^8 cm^2/sec
300	0.40	0.0090	8.4	0.0021	11.34
	0.55	0.0263	3.16	0.0042	33.14
350	0.40	0.0077	9.60	0.0011	9.70
	0.55	0.0192	5.54	0.0046	24.19
400	0.40	0.0070	11.99	0.0023	8.82
	0.55	0.0110	7.75	0.0023	13.86

1 kg/m^3 = 16.02 lb/ft^3

Table 6 Comparison of effective chloride diffusion coefficient of D_e values evaluated from ponding and gas diffusion tests

cc, kg/m^3	w/c	D_e x 10^8, cm^2/sec CHLORIDE PONDING	GAS DIFFUSION
300	0.40	10.32	11.34
	0.55	33.70	33.14
350	0.40	8.28	9.70
	0.55	28.10	24.19
400	0.40	7.63	8.82
	0.55	16.46	13.86

1 kg/m^3 = 16.02 lb/ft^3; 1 in.^{2} = 6.45 cm^2

CONCLUSIONS

The gas diffusion technique has revealed excellent results for estimating the effective chloride diffusion coefficient in concrete. The method is fast and practically applicable since it takes 3 to 4 hr for its completion. Further testing is needed to include different types of concrete mixes to ascertain the applicability of this method.

ACKNOWLEDGEMENTS

The authors gratefully acknowledge the support of the King Fahd University of Petroleum and Minerals for the project CE/Corrosion/167 under which this study was conducted. The porosity tests were carried out by Imperial Chemical Industries, UK., using a micromeritics 9220 Mercury Porosimeter.

NOTATIONS

A = cross sectional area of concrete disc specimen (cm^2)
D = diffusion coefficient of NaCl in water at 19 °C (1.26 x 10^{-5} cm^2/sec)
D_K = Knudsen diffusion coefficient (cm^2/sec)
D_{1K} = Knudsen diffusion coefficient for helium
D_{2K} = Knudsen diffusion coefficient for nitrogen
D_{12} = molecular diffusion coefficient of gas one helium to gas two nitrogen (cm^2/sec)
L = thickness of concrete disc (cm)
M_i = molecular weight for gas i
P = absolute pressure
r_e = mean pore radius (cm or Å)
T = absolute temperature in Kelvin
t = time

V	=	volume of chamber (cm^3)
x_I	=	more fraction of helium in chamber "a" at any time t
ε	=	porosity of concrete
τ	=	tortuosity factor to account for the sinuosity of the pores along the path of diffusion
Ω_D	=	"collision integral" for molecular diffusion, a dimensionless parameter defined as kT/ε_{12} where k is the Boltzmann constant and ε_{12} is the energy of molecular interaction for gases one and two, which is defined as $\varepsilon_{12} = \varepsilon_1\varepsilon_2$

REFERENCES

1. PAGE, C. L., SHORT, N. R AND TARROS, A. Diffusion of Chloride Ions in Hardened Cement Pastes. Cement and Concrete Research, V. 11, No. 3, 1981, pp.395-406.

2. GOTO, S. AND ROY, D. M. Diffusion of lons through Hardened Cement Pastes. Cement and Concrete Research, V. 11, No. 51, 1981, pp. 751 -759.

3. WHITING, D. Rapid Measurements of Chloride Permeability of Concrete. Public Roads, V. 45, No. 3, Dec. 1981, pp. 101 - 112.

4. AMERICAN ASSOCIATION OF HIGHWAY AND TRANSPORT OFFICIALS. Standard Method of Test for Rapid Determination of the Chloride Permeability of Concrete. AASHTO T 277-83, Washington, D.C., 1983.

5. ANDRADE, C., AND SANJUAN, M. A. Experimental Procedure for the Calculation of Diffusion Coefficients in Concrete from Migration Tests. Advances in Cement Research, V. 6, No. 23, 1994, pp. 127-134.

6. DHIR, R. K.; JONES, M. R.; AHMAD, H. E.; AND SENEVERATUNE, A. M. Rapid Estimation of Chloride Diffusion in Concrete. Magazine of Concrete Research, V. 42, No. 152, Sept. 1990, pp. 177-185.

7. SATTERFIELD, C. N. Mass Transfer in Heterogeneous Catalysis. Addison-Wesley, 1971.

8. NAVAZ, C. M. Chloride Diffusion in Concrete/Prediction of the Onset of Corrosion in Reinforced Concrete Structures. MSc. Thesis, King Fahd University of Petroleum and Minerals, Dhahran, Saudi Arabia, Dec.1994.

9. HOLBOROW, K. A. Multicomponent Sorption Equilibria of Gases in 5A Zeolite. Ph.D. Dissertation, University of New Brunswick, 1975.

10. HIRSCHFELDER, J. O., CURTISS, C. F. AND BIRD R. B. Molecular Theory of Gases and Liquids. John Wiley and Sons, New York, 1954.

11. DOGU, G. AND SMITH, J. M. Rate Parameters from Dynamic Experiments with Single Catalyst Pellets. Chem. Eng. Sci., V. 31, 1976, pp. 123-135.

THE INFLUENCE OF SURFACE ABSORPTION ON SULFATE ATTACK

J P Camps

R Jauberthie

F Rendell

Université de Rennes

France

ABSTRACT. An assessment of the behaviour of an Ordinary Portland Cement mortar to sulfate conditions has been studied. The mortars was subjected to a six month immersion exposure test in three environments : water, sulfuric acid and ammonium sulfate. These tested included a study of the behaviour of stressed mortar subjected to sulfate attack. The durability of the material strength, capillarity and sample swelling were noted through the period of exposure. At the end of the exposure strength changes were determined by testing in flexure. The result indicate that the sorptivity of the mortar is sensitive to the condition of cure and the type of mortar skin in contact with the fluid. The samples subjected to sulfuric acid exhibited less loss of durability than those exposed to ammonium sulfate.

Keywords: Sulfate, Cure, Sorptivity, Capillarity, Durability, Portland cement.

Dr J P Camps is Maître de Conférences, teaching at the Institut Universitaire de Technologie, Département de Génie Civil, Université de Rennes 1, France. His research area developed at INSA's GTMa lab involves the durability and protection of building materials and the behaviour of clay cement mixes.

Dr R Jauberthie is Maître de Conférences, Département Génie Civil, Institut National des Sciences Appliquées, Rennes, France. His main research interests include the properties of different phases in hydrated cements, the durability and the protection of concretes.

Mr Franck Rendell Independant Engineer specialising in durability of concrete. Researcher Département Génie Civil, Institut National des Sciences Appliquées, Rennes, France. Previous research has included a study of macro cell corrosion of reinforcement in concrete. Technical author currently writing a book concerning evaluation in the environmental engineering.

INTRODUCTION

Within a wastewater environment there is an extensive use of concrete as a construction material, this investigation involves a study of its behaviour thus enabling an assessment of durability. Concretes under these conditions, will be subjected to a series of aggressive vectors of attack, in general the principle cause of attack will be sulfuric acid that is produced chemically and bio-chemically in the aerated zones above water level [1] . Similarly, this type of environment is generally rich in ammoniate compounds of which ammonium sulfate [2] will have the most deleterious effect on concrete. The interesting difference between these two forms of chemical attack is that in the case of sulfuric acid, the low pH will tend to dissolve the portlandite thus opening the matrix of the material to attack. However the surface deposit of gypsum is thought to provide a protective skin to the concrete [3]. The influence of stress on a concrete will have the effect of inducing micro-cracking within the matrix of the material and thus opening further paths for chemical attack [4]. The use of capillarity to measure the surface sorptivity of the material is seen as having a considerable value in assessing the changes in the near surface pore structure of the mortar.

The use of techniques such as capillarity has been used to examine the surface durability of materials [5] [6], however the techniques have generally been criticised for producing inconsistent results ; this being due in many cases to the sensitivity of the method to the test technique and more importantly the initial moisture content of the sample. The current method of reporting capillarity is to determine the slope of a plot of mass of water absorbed per unit area against the square root of time. Justification for the form of this relationship has been justified on the grounds of diffusion equation, capillary rise and by pure empirical observation [7].

Under chemical attack the properties of the surface layer of the material are extremely important, effects such as carbonation have been shown to have a beneficial effect in improving the initial durability of a cementitious material to chemical attack. The surface layer of a concrete has been shown to have properties that are not representative of the core material [8][9], high cement content and low aggregate content, thus implying that the surface layer is highly susceptible to the conditions of cure.

The aim of this study is to make a comparison of the degradation of an ordinary portland cement mortar under the action of ammonium sulfate and sulfuric acid, and to correlate the loss of durability to the sorption assessed by capillarity.

SAMPLE PREPARATION AND METHODS OF TEST

Characterisation of the Material

The samples used in the test programme consisted of a 16cm x 4cm x 4cm prisms fabricated using a standard cement mortar mix. The aggregate used in the mix was Standard Sand conforming to ISO 679 and the method of fabrication was carried out in accordance to EN 196. The mix details were as follows:

Water / Cement Ratio	0.5
Aggregate / Cement Ratio	3.0
Cement Content	500 kg/m^3

The cement type used was an Ordinary Portland Cement – CPA CEM 1 52.5 CP2 de SAINT-PIERRE-LA-COUR. An analysis of the cement type is summarised in Table 1.

Table 1 Oxide analysis of the cement used in the tests

Ins	SiO_2t	Al_2O_3	Fe_2O_3	CaO	MgO	K_2Ot	Na_2Ot	SO_3	P.F.	CaOfree
0.26	20.15	5.18	2.76	65.13	0.69	0.99	0.17	2.85	1.51	1.31

To characterise the resistance of the cement to sulfate attack the Aggressivity Modulus (Grün), the lime standard (Kühl) and the Hartmann and Mangotich ODF [10] were calculated for each of the cements. The C_3A content was calculated using the Bogue and Dahl Equation.

Table 2 Characterisation of the cement type

AGGRESSIVITY MODULUS	LIME STANDARD	HARTMANN AND MANGOTICH ODF	C_3A CONTENT %
0.32	101.7	3.03	9.1

Curing Regimes

Three cure regimes were used in the programme. The curing regime consisted of two stages, an initial period in a humidity chamber, relative humidity 100% temperature 20°C, samples were then transferred to an air conditioned chamber, relative humidity 55% temperature 20°C for the second phase of the regime. The total period of cure was 45 days the regimes of cure are :

Regime A 1 day at 100% RH - transfer to the 55% RH
Regime B 7 days at 100% RH – transfer to the 55% RH
Regime C 28 days at 100% RH - transfer to the 55% RH

Characterisation of the Mortar

At the end of the period of cure the strength, porosity and capillarity of the samples were measured. These results are reported in Table 3.

Table 3 The characterisation of the mortar at the end of the 42 day period of cure

	DENSITY	CHANGE IN MASS	SHRINKAGE	POROSITY	FLEXURAL STRENGTH	COMPRESSIVE STRENGTH
	kg/m^3	G	Mm/m	%	kN	kN
1 day cure	2233	-21.88	0.65	16.2	3.45	681
7 day cure	2252	-15.75	0.57	16.0	4.25	749
28 day cure	2291	-7.96	0.31	15.7	4.85	800

Capillarity Testing

The experimental method used for assessing the capillarity of the samples consisted of measurement of sample mass gain over a period of one hour. Samples were weighed and then placed on a saturated sand bed covered with a layer of filter paper. The depth of water surrounding the sample was not allowed to exceed 1mm. The sample was removed from the bed, excess water lightly wiped off and the sample weighed. This process was carried out at time intervals that increased exponentially. Care was taken to cover the sample whilst on the sand bed to minimise evaporation losses from the mortar. The sample was then immersed in water for a period of 48 hours, re-weighed to obtain the saturated weight, it was then oven dried at 105°C for 3 days to obtain the dry weight.

Exposure Tests

Throughout the period of cure the mass and the shrinkage were monitored for each sample type.

The following stage in the programme consisted of the comparison of the characteristic of the mortar throughout a 6 month exposure to a controlled environment. Four curing environments were used in the test :

Stored in air at 20°C and 55%RH
Stored in tap water at 20°C
Stored in sulfuric acid solution, 30 g/l SO_4
Stored in ammonium sulfate, 30 g/l SO_4

Throughout the period of exposure the mass and linear dimension of the samples were measured at approximately monthly intervals. After 150 days the samples stored in the sulfate solutions were transferred to a water solution for one month and then transferred to the air conditioned chamber for another month to allow stabilisation of there moisture contents.

It was noted that during the test period the samples in the sulfuric acid solution were covered in a soft white amorphous deposit, the degree of surface deterioration to the samples was very evident. A strong smell of ammonia gas was noted when the samples in ammonium sulfate were examined, large needle crystal forms grew on the surface and there was no visible evidence of material loss.

The capillarity of the samples was measured prior to exposure and at the conclusion of the test, i.e. after washing and drying.

A second set of samples were subjected to a load in flexure of 50% of the failure load plus immersion in sulfate solutions as described above. The creep behaviour was monitored and the sample strength obtained at the conclusion of the test.

Analysis of Capillarity Results

The quantification of capillarity is based on the measurement of the slope of a plot of mass gain per unit area against the square root of time; where the area refers to the surface in contact with the fluid. A typical capillarity result and statistical analysis are as follows:

Capillarity $g/cm^2/min^{½}$	0.00327
Standard error of the slope	0.00012
Regression coefficient	0.987

A recognised weakness in methods of measuring sorptivity is that the results are dependent on the initial sample moisture content, wherever possible samples were tested at close to 55% RH. However to minimise the effect of moisture content variation a mathematical correction method was devised for correction of results to a standard oven dry condition. The method of correction is based on the assumption that the mass change of a sample is proportional to the square root of time.

To quantify the accuracy of the method sets of control tests were carried out. From an analysis of the results of these tests it was noted that the influence of the variation in material between samples introduced the greatest error. Standard error when comparing repeated observations on the same sample was 10%, compared with a standard error of 20% when comparing two different samples.

RESULTS OF THE CAPILLARITY TESTING

Influence of Cure, Type of Liquid and Nature of Mortar Skin

Capillarity was assessed for one group of samples prior to the immersion tests. The variation of capillarity with duration of cure at 100% RH clearly shows that the sorptivity of the mortar decreases with increased period of cure, i.e. from 0.03 $g/cm^2/min^{½}$ at 1 day cure to 0.005 $g/cm^2/min^{½}$ at 28 day cure.

The transport of liquids through a media by a process of capillary suction implies that the phenomena is dependent on the viscosity and surface tension of the liquid. (Kropp) The initial process of fluid uptake will also be a function of the reactivity of the liquid with the material. To test this relationship capillarity tests were carried out for a range of solution types and concentrations; in each test direct comparisons of capillarity were made between the sulfate solution and water. The resulting of capillarities were found to be in the ratios shown in Table 4.

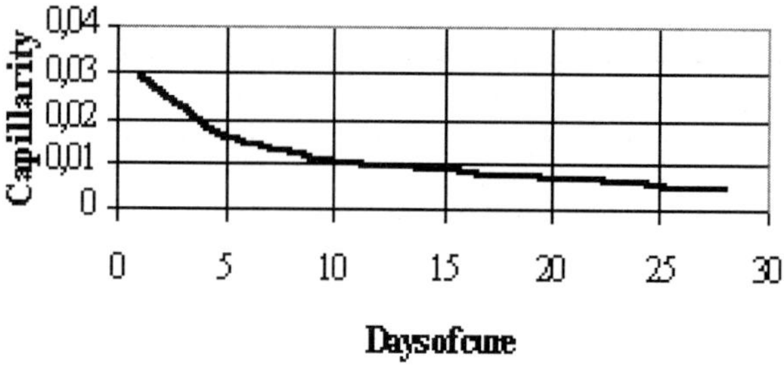

Figure 1 The influence of the period of cure at 100% RH on the capillarity

Table 4 Ratio between water capillarity / sulfate capillarity

TAP WATER	SULFURIC ACID 1.5 g/l	SULFURIC ACID 30 g/l	AMMONIUM SULFA 1.5 g/l	AMMONIUM SULFA 30 g/l	SODIUM SULFA 30 g/l
1.00	0.91	0.48	0.93	0.90	1.63

A series of capillarity tests were carried out to examine the variation in the sorptivity between various types of surface. Four surfaces were examined, the surface formed by the mould, the top surface, a fracture face and a surface produced by a saw cut into the heart of the mortar. The ratio of capillarities was observed to be as follows:

Fracture	1.4
Mould	1.0
Saw Cut	1.0
Trowelled surface	1.6

Influence of Exposure to Capillarity

The capillarity of the samples subjected to the immersion test was measured before and after the period of exposure. It should be noted that the final capillarity was measured after the washing process. All quoted results indicate a comparison of the same sample and therefore increasing the significance of differences between readings. Table 5 sets out a variation in the capillarity of samples over the 6 months exposure period.

Sample Swelling

The change in linear dimension was measured throughout the period of immersion. It can be seen from Figure 2 that the sample in ammonium sulfate started to shrink after removal from the solution, at 150 days, whereas the sample in sulfuric acid remained virtually unchanged.

Table 5 The influence of immersion of the capillarity of the samples

	SULFURIC ACID 30g/l				AMMONIUM SULFATE 30g/l			
Cure	Before	After	Difference		Before	After	Difference	
1 day	0.0138	0.0198	0.0061	36%	0.0134	0.0133	-0.0001	-1%
28 days	0.0186	0.0178	-0.0008	-4%	0.0126	0.0105	-0.0021	-18%

difference indicates the increase in capillarity over the period of test.

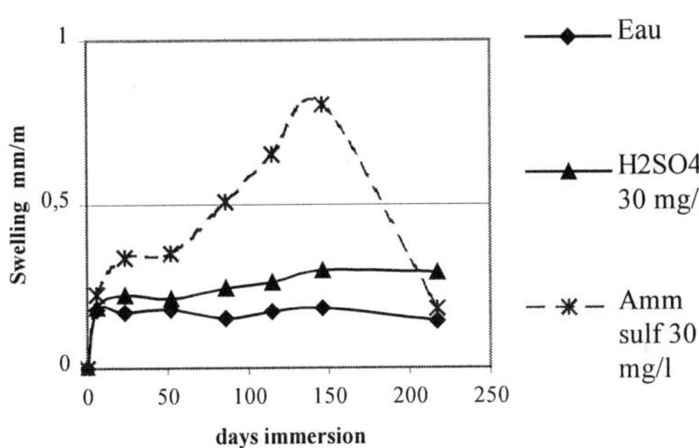

Figure 2 The variation in the sample length of the samples over the period of immersion

Loss of Mass

The sample weights were monitored throughout the period of exposure. The period of cure did not have a significant effect on the overall mass loss, i.e. prior to immersion, to after washing and drying. The general trend was for a mass loss of 22 g (3.4%) for the samples in sulfuric acid and 2 g (0.3%) for samples immersed in ammonium sulfate.

Sample Strength

To assess the impact of exposure on the strength of the mortar the samples were tested at the end of the programme, the 16 cm x 4 cm x 4 cm prisms were tested in flexure and then the halves were tested in compression. Table 6 sets out the sample strengths recorded at the end of the exposure test.

Table 6 The influence of exposure on the strength of the samples

CURE	AIR	WATER	SULFURIC ACID		AMMONIUM SULFATE	
1 day	3.7	5.0	4.8	97.2%	5.1	102%
28 days	4.2	5.5	5.7	105%	5.6	103%

Flexural strength kN unstressed samples

CURE	AIR	WATER	SULFURIC ACID		AMMONIUM SULFATE	
28 days	5.2	5.9	5.2	88%	2.5	45%

Flexural strength kN; samples stressed in flexure at 50% of the failure load

DISCUSSION

There is a good agreement between the material characteristics and the condition of cure; the increase in capillarity with poor curing is reflected in the increase in porosity. This opening of the pore structure for poorly cured concrete potentially permits a higher transport of aggressive agents into the concrete. The situation is compounded because the incomplete hydration in a poorly cured concrete will also be associated with lower strength. These results reinforce the observations from several field studies that condition of cure is one of the dominant parameters effecting durability.

The capillarity results for the different types of surface confirm the observation by Kreijger et.al. that the surface type has an important influence on durability. During the immersion tests in sulfuric acid, initial deterioration and gas evolution were observed on the trowelled surface of the sample; this is possibly due to the high portlandite content that is usually associated with laitance.

The difference in the modification to the material characteristic due to immersion in sulfuric acid and ammonium sulfate are noticeable.

Capillarity

The measurement of the final capillarity after the washing process enabled a quantification of the sorptivity of the inner layer of the material. In the case of sulfuric acid it is clear that for poorly cured samples there is an increase in capillarity which is to be expected because of the high dissolution of portlandite resulting from acid ingress into an open mortar structure. For the well cured mortar the change in capillarity is insignificant. The poorly cured samples immersed in ammonium sulfate showed no change in capillarity, therefore one can speculate very little change in pore structure, whereas the well cured mortar experienced a reduction in permeability.

Mass Loss

Mortar samples in sulfuric acid exhibits a mass loss of ten times greater than those immersed in ammonium sulfate. The heavy mass loss in sulfuric acid was observed to be a surface degradation.

Swelling Behaviour

The behaviour of the control sample in water and the sample in sulfuric acid are similar whereas there is a marked difference in the ammonium sulfate exposure. The samples immersed in ammonium sulfate experience a strong swelling which was reversed on dilution of the solution ; this shrinkage is probably due to the re-dissolution of the expansive sulfate products due to the change in equilibrium within the material. This reversal was not noted in the samples in sulfuric acid.

Strength Changes

The effect on material strength is not clearly apparent, however in a second phase of this work stressed samples were subjected to the same environmental conditions. Work by Gérard [11] indicated a 100 fold increase in concrete samples in tension at a load level increased from 40% to 70% ultimate strength. Under stressed conditions the samples in ammonium sulfate suffered highly significant reduction in strength in comparison with those in sulfuric acid

It is therefore suggested that the actions exhibited in the two environments are as follows : In a sulfuric acid environment the pore structure is initially opened due to the dissolution of the portlandite and then filled with precipitated gypsum, the resulting layer effectively blocks further migration of sulfates ; this effect does not occur in an ammonium sulfate environment. Impact of the Gypsum Barrier: The barrier effect can be seen in the capillarity results from the washed samples. In the case of the 28 day cured sample immersed in sulfuric acid, the mortar capillarity remained unchanged, indicating no modification to structure behind the barrier.

The overall reduction in capillarity for similar samples in ammonium sulfate reflects the pore blocking by reaction products thus indicating a greater depth of sulfate penetration. The effect of the barrier is also seen in the swelling and shrinkage behaviour of the samples. Samples in sulfuric acid are relatively insensitive to the solutions whereas those in ammonium sulfate respond rapidly to the presence and dilution of the sulfate solution ; this indicates an unhindered migration of the sulfate into the matrix of the material, thus confirming a relatively rapid response of the mortar to the environment. The large reduction in flexural strength of samples in ammonium sulfate reflects the inflow of sulfate ions to the core of the mortar thus changing the internal properties. The limited change in strength exhibited by the sample in sulfuric acid possibly reflects the protective nature of the gypsum layer.

CONCLUSIONS

1. The method of capillarity testing as a tool for the assessment of durability was found to be effective, the results were consistent with other observations.

2. The influence of cure has a dominant effect on the capillarity of the mortar, this is reflected in the variation in porosity and strength.

3. Samples immersed in sulfuric acid solution experienced a 3% loss of mass, however the deterioration mechanism is confined to the surface, the gypsum layer forms a protective barrier to the matrix of the material.

4. In contrast, the samples immersed in ammonium sulfate, where no protective barrier is formed, the loss of material was very light (0.3%) but the mortar was highly sensitive to swelling and strength change. This indicates a change in the internal properties of the material due to the migration of sulfate ions into the core of the mortar.

REFERENCES

1. AZIZ, M A, KOE, L C C. Durability of concrete sewers in aggressive subsoils and groundwater conditions ,Geotechnical aspects of restoration works. ed Balasubraman et al, Roterdam,1990, pp 299-310.

2. BICZOK, J. Concrete corrosion and concrete protection. Akademiai Jiadl, 8th edition, Budapesta, 1972.

3. JAUBERTHIE, R, LANOS, M, TEMINI, M, and CAMPS, J P., Bétons en environnement acide : dégradation ou protection? Congrès Universitaire de Génie Civil, Reims, France, 1998, pp 73-80.

4. NÄGELE, E. New and powerful method for the evaluation of multi parameter corrosion tests.,Cement and Concrete Research,,Vol 25 No 6, 1995, pp.1209-1217.

5. GUMMERSON, R J, HALL, C, HOFF, W D."Water movement in porous building materials – II.", Building and Environment, Vol 15, 1980, pp. 101-108.

6. EMERSON, M. Mechanisms of water absorption by concrete.,Protection of Concrete, Ed R.K. Dhir J.W. Green,,E&FN Spon, London, 1990, pp 689-700.

7. KROPP, J, HILSDORF, H K, GRUBE, H, ANDRADE, C. "Transport mechanisms and definitions.", RILEM REPORT 12, Performance criteria for concrete durability Ed J.Kropp H.K. Hilsdorf,E&FN Spon, London, 1995, pp 4-13.

8. MEHTA, P K, MANMOHAN, D. Pore size distribution and permeability of hardened cement paste. Proc. of the 7th Int. Congress on the Chemistry of Cement, 1980, pp 1-6.

9. KREIJER, P C. The skin of concrete: composition and properties., Materiaux et Constructions RILEM Vol 17, April 1984, pp 275-283.

10. HARTMANN, C, MANGOTICH, E. "A method for predicting sulphate durability of concrete", Bryant and Katherine Mather Symposium on Concrete Durability, American Concrete Institute, SP-100, 1987, pp 2027-2040.

11. GÉRARD, B, BREYSSE, D, AMMOUCHE, A, HOUDUSSE, O, DIDRY, O. Cracking and permeability of concrete in tension.,Materials and Structures,Vol 29, April 1996, pp 141-151.

EVALUATION ON CHLORIDE INDUCED DAMAGE FACTOR USING ACTUAL IN-SITU DATA

M Matsushima

T Tsutsumi

Tokyo Electric Power Service Company

Japan

ABSTRACT. To accurately evaluate chloride induced deterioration, consideration must be given to the effects of regional differences, the locations of structures and other variables in addition to actual structure deterioration data. Deterioration data accumulated over more than 20 years for reinforced concrete structures within Tokyo Bay was collected and then used to study the factors governing the surface chloride ion density, the diffusion coefficient, and the reinforcement corrosion rate, which are important parameters in relation to chloride induced corrosion. The main factors affecting chloride ingress were member location and concrete compressive strength.

Keywords: Diffusion coefficient, Chloride ion density, Data analysis, Sampling data in-situ, Reinforcement corrosion rate

Dr Manabu Matsushima is a senior research engineer in the Research and Development Department of Tokyo Electric Power Service Company, JAPAN. He received his Doctor of Engineering Degree from Tokyo Denki University in 1994. His research interest is the application of reliability theory to concrete members in RC structures. He is a member of JSCE, JCI, and SOFT.

Dr Tomoaki Tsutsumi is a Senior Research Engineer at the Engineering Research Center of Tokyo Electric Power Co., Ltd. He received his Doctor of Engineering Degree from Tokyo Metropolitan University in 1997. His research interests include deterioration of RC structures and inspection methods for damaged RC structures. He is a member of JSCE and JCI.

INTRODUCTION

This study, concerning factors influencing chloride induced deterioration in Tokyo Bay, was performed by collecting, organizing, and analyzing inspection and repair records accumulated over more than 20 years for RC structures, including landing piers for petroleum and coal and revetments, constructed along the seashore of Tokyo Bay. In line with earlier research [1], the chloride induced deterioration process was categorized as the incubation period up to the point where reinforcement is started by the diffusion of chloride ions and the development period which begins with reinforcement corrosion followed by the appearance of cracks in the reinforcement axially caused by pressure from the expansion of the reinforcement. The governing parameters during the incubation period are the surface chloride ion density and its diffusion coefficient, while the governing parameter during the development period is the reinforcement corrosion rate. This study was a qualitative study of factors influencing these parameters.

FACTORS INFLUENCING CHLORIDE INDUCED DETERIORATION

Generally, chloride ion diffusion is, as a diffusion phenomena conforming to the concentration slope, evaluated by Fick's diffusion equation as follows.

$$C_C(x,t) = C_0\left\{1 - \mathrm{erf}\left(\frac{x}{2\sqrt{D_C \cdot t}}\right)\right\}$$

Where,

$C_c(x,t)$: Chloride ion density in concrete at a depth from the concrete surface x at time t from the beginning of diffusion. (kg/m^3)
C_0: Chloride ion density on the concrete surface (kg/m^3)
erf(): Error function.
D_c: Equivalent diffusion coefficient of chloride ions. (cm^2/sec)
x: Depth from the concrete surface. (cm)

It is possible to represent the chloride diffusion process based on the equivalent diffusion coefficient D_c and the surface chloride ion density C_0 in above equation. These values, namely the equivalent diffusion coefficient and the surface chloride ion density, are believed to vary according to environmental conditions, the concrete quality, etc. For this reason, attempts have been made to clarify the characteristics of the factors which govern the equivalent diffusion coefficient and the surface chloride ion density [2]. Figure 1 presents the relationships between factors influencing the diffusion of chloride ions. As it shows, the equivalent diffusion coefficient is influenced by mix proportion factors such as cement type, casting method and other execution conditions.

The surface chloride ion density is governed primarily by environmental conditions: distance from the sea and so on. Many researchers have pointed out that the equivalent diffusion coefficient and the surface chloride ion density are effected by mix proportion conditions, execution conditions, and by environmental conditions [3]. but many of these studies are limited to laboratory tests and outdoor exposure experiments, and there are relatively few study results of actual structures. Therefore, for this study, five items which can actually be evaluated in site, as shown below, were set as influential factors in place of mix proportion or execution conditions, and a study was performed of their relationship with the surface chloride ion density and the equivalent diffusion coefficient.

Takeda et. al. [4] exposed specimens in a splash zone, and both under the surface and in the air above the surface of the ocean to measure chloride ion diffusion. These tests revealed that the surface chloride ion density on the specimens was highest for those to the splash zone, less for those exposed under the surface of the ocean, and least for those in the air above the surface. The installation location of structures is categorized as shown in Figure 2.

Concrete structures can basically be classified into beams, columns, walls, and slabs. A survey of prestressed concrete beam deterioration performed by the former National Railway Company revealed large surface chloride ion density on the bottom surfaces of beams, but only a few on the bottom surfaces of slabs. [5] This study referred to these results to determine the categorization shown in Figure 3: one based on shape of member and installation location. Beams and columns are not differentiated because it is assumed that columns, being located below deck slabs and beams of bridges, are exposed to sea water splash conditions as severe as those which effect beams. because there are differences in the way that surfaces of walls and slabs are washed by rain water, essentially they should be in different categories.

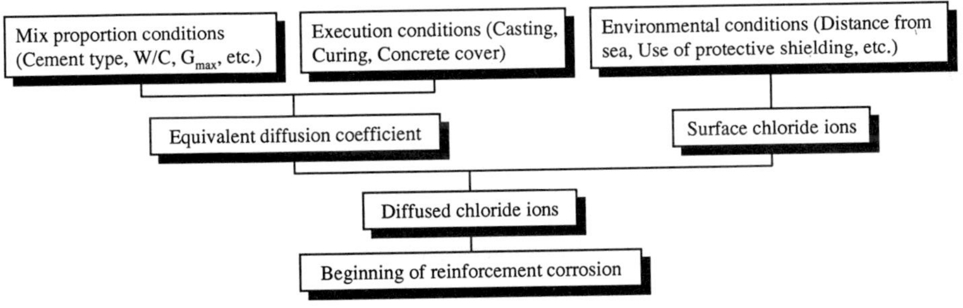

Figure 1 Relationships of factors influencing the diffusion of chloride ions

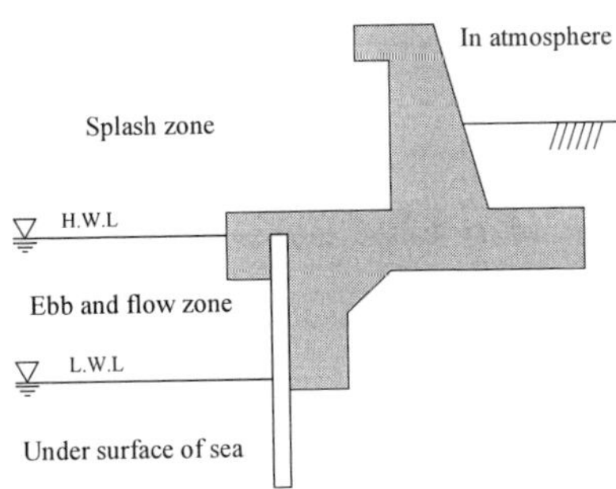

Figure 2 Location of structure

However, because it would have been difficult to distinguish between the two using the data provided for this study, data for the two members, which have similar simple shapes, is combined, despite the lack of precision of this approach. To eliminate the effects of wind direction, when analyzing the material, only data for members located in the same direction towards the seashore was included.

Surface chloride ion density is governed by the extent to which the member is exposed to flying salt water splash. Although conducted not in the same region where the data for this study was obtained, Kashino et. al.[5], measured flying chloride ions in the atmosphere above solid land, discovering that the difference in the chloride ion density adhering to parts protected from the spray by shielding was about 1:0.4 within 100 m of the ocean. Therefore, both the surfaces on the seashore side and opposite side (below referred to as rear side) of structures were studied. However, because most of the structures targeted by this study are port facilities, all sides of beams and columns were located in splash zones. Consequently, because it is difficult to distinguish from the seashore side and rear side of structures, their effect is excluded from this study. As in the member category case, to eliminate the effects of wind direction on this factor, the study categorized the walls and slabs as seashore side and rear side. Walls and slabs are categorized as shown in Figure 3.

Based on laboratory experiments, one of the authors [6] represents the chloride ion diffusion properties as the corrosion current value to obtain a negative correlation with the compressive strength. It is known that compressive strength is described by a function of the water/cement ratio, and the diffusion properties of chlorides are also influenced by the water/cement ratio [2]. Therefore, in this study, compressive strength was considered a factor which substitutes for the water/cement ratio, and its relationship with the equivalent diffusion coefficient was studied.

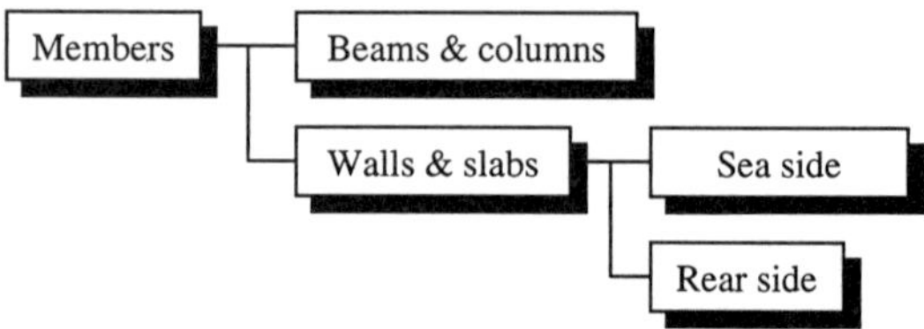

Figure 3 Member and member surface categories

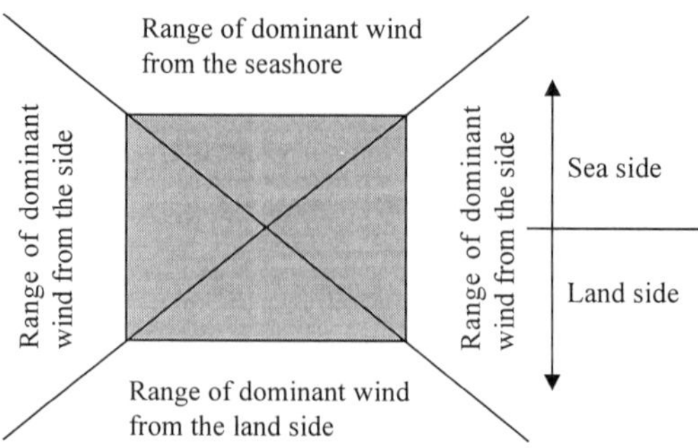

Figure 4 Relationship between structure and dominant wind direction

In order to study the effects of the dominant wind direction on marine structures in Tokyo Bay, the relationship between dominant wind direction and location of structures was considered under three categories as shown in Figure 4. Because this factor is assumed to influence the surface chloride ion density, its effects on the equivalent diffusion coefficient were not examined.

ANALYSIS OF RESULTS

Maintenance engineers for existing concrete structures have to supplement daily inspections of the structures with periodical inspections performed once every six months to determine if repair work is necessary. If necessary, this includes the extraction of a concrete core. The data used for this study included diffusion coefficient and surface chloride ion density data obtained from the density distribution of chloride ions in the depth direction of concrete cores obtained during such inspections. The chloride ion density data used in this paper was obtained using the method of potentiometric titration.

Figure 5 shows the distribution of surface chloride ion density. Although the surface chloride ion density tends to be higher under the surface of the sea and in tidal zone than it is in the splash zone, there is insufficient data to permit a clear theory regarding these tendencies.

Figure 6 shows the distribution of surface chloride ion density and Figure 7 shows the distribution of the equivalent diffusion coefficient. Assuming that data actually measured in-situ represents the state of the structure's concrete, basically, the data obtained should be used without modification. But, because it includes data considered idiosyncratic, featuring severe localized deformation, the method proposed by Grubbs [7] was used to dismiss abnormal data. The dismissal limit values obtained were n=26 for the surface chloride ion density, n = 27 for the equivalent diffusion coefficient, and 5% for the significance level. Both figures reveal a great difference between the surface chloride ion density, which ranges from 1.0 kg/m^3 to 21.0 kg/m^3. And while there is less data for walls than for beams, the distribution range is identical for both members. Although data for the equivalent diffusion coefficient of beams and walls varies, their distributions are almost identical, with both distributed between 1.0×10^{-8} and 1.0×10^{-7}cm^2/sec. A W-examination [8] was performed to determine whether the data within the range of this survey should be clearly categorized based on member category or whether it is within the range of simple scattering. The level of significance of the test was set at 5%. The results are shown in Table 1. This table reveals that the surface chloride ion density and equivalent diffusion coefficient were part of the same population in the case of both walls and beams, leading to the conclusion that the member category is not influential.

Figure 8 shows the distribution of surface chloride ion density, while Figure 9 shows the distribution of equivalent diffusion coefficients. Despite considerable scattering of the surface chloride ion density, there are differences in the distribution ranges obtained for the seaside and the rear side; with more surface chloride ions observed on the seaside than the rear side. The seaside/rear side ratio is 1: 0.6. The ranges of the equivalent diffusion coefficient data on the seashore and rear sides are almost identical, with the distribution pattern represented by a lognormal distribution. As in the case of concrete specimens used for laboratory tests, if there were no cracks or other defects, scattering of the equivalent diffusion coefficient was limited to scattering at the time the specimens were prepared and had a normal distribution, but within the range of this data, it was a lognormal distribution.

This is a result of the fact that actual structure data includes larger values as a result of execution scattering, fine cracks or other defects.

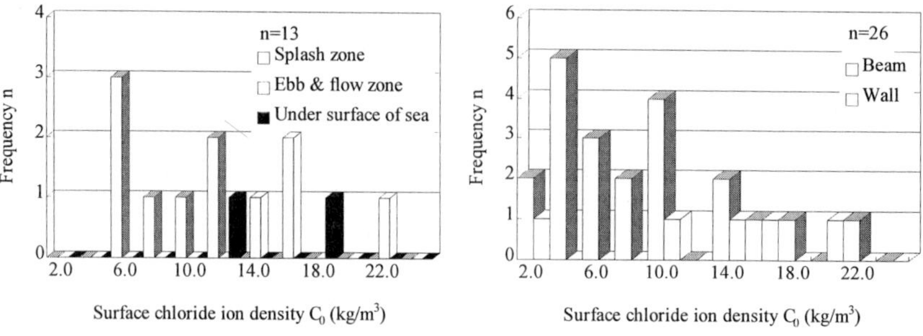

Figure 5 Surface chloride ion density

Figure 6 Surface chloride ion density

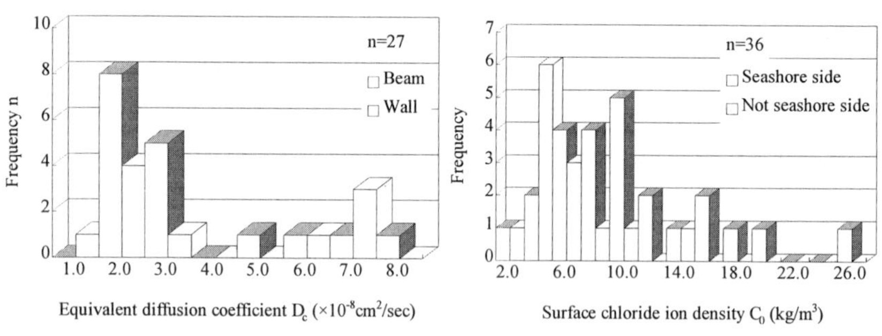

Figure 7 Equivalent diffusion coefficient

Figure 8 Surface chloride ion density

Table 1 Results of W Examination

NULL HYPOTHESIS	ALTERNATIVE HYPOTHESIS	NUMBER OF DATA		RESULTS
		m	n	
Surface chloride ion density on beams and walls in the same population	Smaller surface chloride ion density on walls	5	21	Null hypothesis
Equivalent diffusion coefficient for beams and walls in the same population	Lower equivalent diffusion coefficient on walls	9	18	Null hypothesis

Note) Data number m: Data indicating the null hypothesis. n: Data in another direction

Figure 10 presents the relationship between compressive strength and equivalent diffusion coefficient. The specimens used for compressive strength were taken close to the specimens used for the diffusion coefficient measurements. As the same figure shows, the measured values are widely scattered and there is no clear correlation between the compressive strength and the equivalent diffusion coefficient. Laboratory tests performed by one of the authors [6] has revealed a negative correlation between the compressive strength and equivalent diffusion coefficient, but the study results of actual structures have not shown as clear a relationship as that revealed by the results of laboratory tests. There are a number of reasons for this discrepancy. The structures studied have been in use for more than 20 years, fine cracking has likely occurred under the service load on the structures, the structures covered by the data differ, and the effects of variances in construction quality are dominant.

Figure 11 shows the distribution of surface chloride ion density. Here, the dominant wind direction is set based on year-round data obtained from wind vanes and anemometers installed near the structures. This figure shows that the distribution range of the data is almost identical on the seaside, sides at right angles to the seashore, and from the mountainside, indicating that it is not effected by the dominant wind direction. This is because the data was obtained inside Tokyo Bay.

Before the above results were summarized, beam members were analyzed in order to study the characteristics of the relationship between surface chloride ion density and elapsed years. The results are presented in Figure 12. This figure shows that although there is scattering and bias in the data, no clear relationship with elapsed years is evident. Based on the above study results, the surface chloride ion density was limited to splash zone and to seaside data, while the equivalent diffusion coefficient was limited to splash zone data to summarize the distribution of the surface chloride ion density and the equivalent diffusion coefficient of reinforced concrete structures in Tokyo Bay. The results are shown in Figures 13 and 14 respectively. The results of an χ-square test show that a normal distribution is appropriate for the surface chloride ion density and a lognormal distribution is suitable for the equivalent diffusion coefficient. This is because while the surface chloride ion density is influenced by the environment where the structure stands, the equivalent diffusion coefficient is, as stated above, influenced by data including latent defects in the concrete, and high value data is dominant. The distribution for the surface chloride ion density is a mean value of 8.84 $kg/m^3$with a standard deviation of 5.78 kg/m^3 (obtained using 67 data). The distribution for the equivalent diffusion coefficient is a mean value of 1.73×10^{-8} cm^2/sec with a standard deviation of 1.59×10^{-8} cm^2/sec (obtained using 66 data).

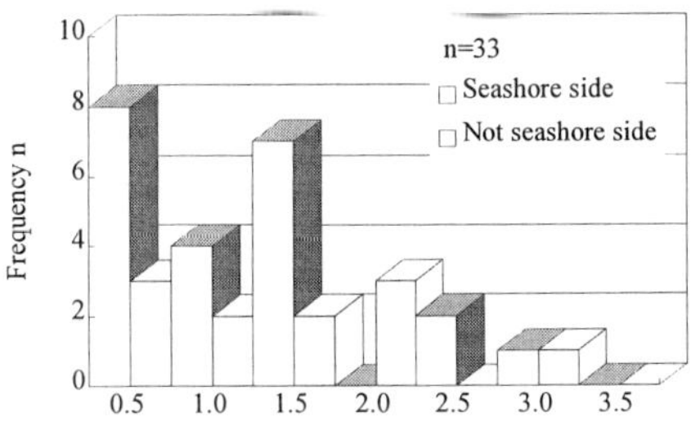

Figure 9 Equivalent diffusion coefficient

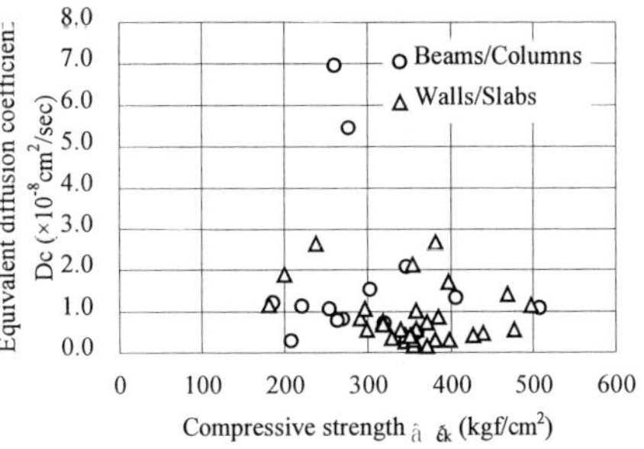

Figure 10 Compressive strength and equivalent diffusion coefficient

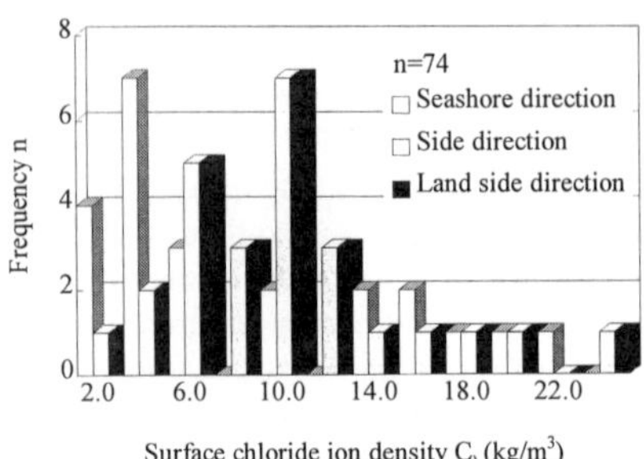

Figure 11 Surface chloride ion density

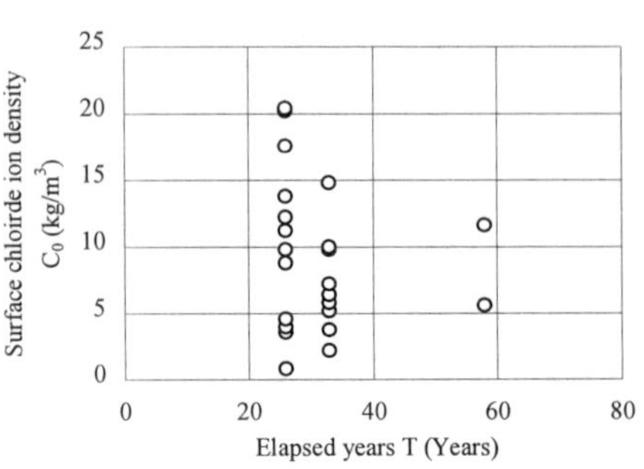

Figure 12 Surface chloride ion density and elapsed years

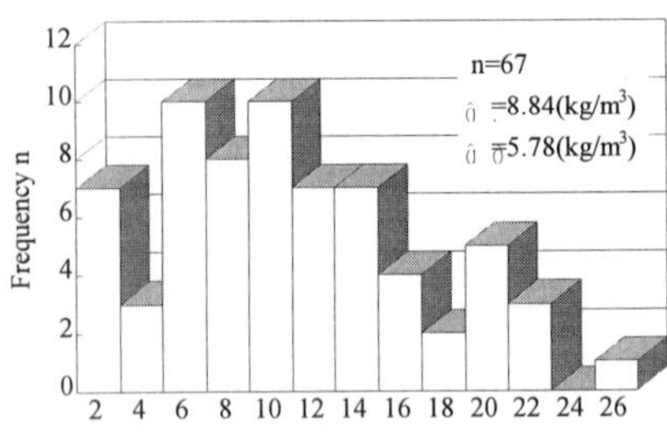

Figure 13 Surface chloride ion density

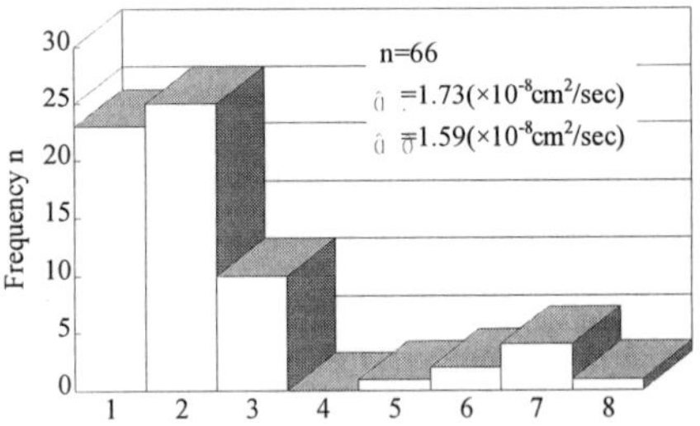

Figure 14 Equivalent diffusion coefficient

REFERENCES

1. MIYAGAWA, T. Durability and estimation on life span of concrete structures in the chloride atmosphere, Symposium on durability and estimation on life span of concrete structures, JCI, pp.47-54, 1988.4 (in Japanese).

2. MASUDA, M, TOMOZAWA, M., YASUDAS, M, HARA, K. Experimental study on diffusion speed of chloride ion into concrete, The 10th annual meeting of JCI, 10-2, pp.493-498, 1988 (in Japanese).

3. OHTA, M, SASAKI, S, SAKAI, T, TAKAHASHI, Y. Chloride diffusion into concrete at seaside, The 13th annual meeting of JCI, 13-1, pp.589-594, 1991 (in Japanese).

4. TAKEDA, M, SAKOTA, K, TOGAWA, S. Estimation on chloride diffusion of various concrete based on outdoor exposure tests, The 13th annual meeting of JCI, 13-1, pp.595-600, 1991 (in Japanese).

5. OHTSUKI, N, KASHINO. N, KATAWAKI, K, KOBAYASHI, A, MIYAGAWA, T. Durability design of concrete structures-chloride induced deterioration-, Gihou-do, 1986.4 (in Japanese).

6. TSUTSUMI, T, YAMAMOTO, S, MISURA and MOTOHASHI, K. Effects of composition and age on chloride permeability of concrete, The 6th annual meeting on durability of materials and components, 1993.

7. JAPANESE STANDARDS' ASSOCIATION, JIS handbook -Quality control-, pp.653-659, 1984 (in Japanese).

8. WONNACOTT, T H, AND J R: Introductory Statistics-2nd Ed., John Wiley & Sons Inc., 1969

NON AND PARTIALLY DESTRUCTIVE TESTING OF FORTY CONCRETE BRIDGES

S M A Tajalli

S R Rigden

Queen Mary and Westfield College London

United Kingdom

ABSTRACT. This paper reports the methodology of research used to relate rate of deterioration of concrete bridges to an assessment of quality of build and exposure condition from information collected from long term bridge inspection records held by bridge owners together with the results from partially destructive and non destructive tests on a representative sample of these bridges. The bridges tested were carefully selected and scattered throughout the Greater London area in order to cover different field situations.

In order to understand and compare the individual examples given of the rate of deterioration of each fault noted it would help considerably if the cause of each defect could be determined. A series of site tests and laboratory tests of the concrete in the close vicinity of each fault was therefore carried out. The assessment of exposure included measuring relative humidity (RH) and air permeability (AP) in the cover concrete, using the method developed by the BCA was correlated to the historical records of the defects of more than 40 bridges.

High air permeability in the bridges tested was found to be related to low humidity and low humidity was related to high penetration of carbonation. Conversely high humidity is associated with low penetration of carbonation therefore where high humidity and defective concrete co-existed, this was likely to be due to causes other than carbonation, such as salt attack. However, seepage was noted in over one fifth of the structures audited and this usually occurred at the expansion joints and this was the cause of the majority of the corroded parts of the deteriorated structures.

Keywords: Bridges deterioration, Partially destructive testing, Carbonation, Air permeability, Relative humidity

S M A Tajalli PhD Graduate of the Department of Civil Engineering, Queen Mary & Westfield College, University of London

S R Rigden Senior Lecturer of the Department of Civil Engineering, Queen Mary & Westfield College, University of London

INTRODUCTION

Wide variations in the depths and rates of carbonation in concrete bridges within apparently sound concrete have been noted by the authors. The Maunsell Report [1], recommends that further work is needed on the interpretation of the data obtained from permeability results and their relationship to the durability of concrete. This paper discusses the major features of a research programme to investigate the quality of build, air permeability, AP and relative humidity, RH of cover concrete, from which it was hoped to address the questions raised in the Maunsell Report[1]; leading to more accurate durability predictions in concrete structures. Carbonated concrete requires the presence of moisture before corrosion of reinforcing steel can be induced. It is, therefore, important that the RH within the cover concrete is noted as well as the permeability. A method that enabled measurement of both permeability and RH of the cover concrete was therefore needed.
The parameters included in assessment of the quality of build were: cover depth, Schmidt hammer, carbonation depth, uniformity of cores, cement paste, aggregate grading, voids, voids sizes and voids shape. Some of the individual results will be presented in this paper.

SITE TESTING

The testing of degradation of real models of concrete investigated over true periods of time to gain the required data is both expensive and time consuming, taking decades rather than years. The alternative was to study records of defect histories which cover long occasions of particular experience which were detailed enough to establish a comprehensive picture. The cause of each defect though was mostly unknown as was the quality and mix design of the concrete concerned. To obtain information in relation to these factors a series of non and partially destructive tests were carried out on a representative sample of bridges. Samples obtained from these were taken to the laboratory for further testing. The aim was to be able to combine information relating to quality of build such as cover to reinforcement and concrete quality together with the information obtained from the records to develop more sophisticated degradation models.

From the locations that were visited 46 structures were found to be suitable for testing. The site testing was carried out by the authors assisted by a Queen Mary and Westfield College Civil Engineering Department laboratory technician and began with a general visual survey similar to the inspections carried out by the LUL. The tests carried out included cover survey, Schmidt hammer reading, core extraction for measurement of carbonation and strength, visual inspection of the quality of build, drill powder for chloride content measurement and permeability and relative humidity monitoring by the use of inserts. On some of these structures more than one member was tested resulting in 57 members being tested in total. At least three cores were drilled from each of the members, 200 cores were attempted with 162 cores being successfully extracted from the chosen members. Access to the structures was often difficult and for safety reasons it was important that these tests were carried out at least two meters from the track[2]. The policy of the London Underground is to avoid delays and stoppages to the normal services. The repairs, maintenance and inspections are therefore carried out while the trains are running and when the services are coming to rest. Underground stations and a number of bridge structures had to be tested during "Engineering Hours" when trains were not running.

This usually restricts work to four hours from midnight to four o'clock in the morning. The station supervisor through the line manager had to be informed prior to arriving for the site tests. A suitable electrical connected had to be arranged since generators were not allowed in stations or near tracks in the vicinity of stations. The site tests were planned and carried out to comply with BS6089[3] with transportability of equipment, accessibility, safety and inconvenience to the general public also being considered.

BACKGROUND TO RELATIVE HUMIDITY (RH) AND PERMEABILITY

There has been an increasing interest in the properties of concrete structures other than compressive strength when considering problems of degradation and serviceability. A number of different techniques have been used by various researchers to measure the permeability of the cover concrete but it is disappointing to find that there has been relatively little work on permeability testing of concrete in-situ, either carried out on site or on specimens removed from the site (Concrete Society Technical Report 31, 87)[4].

The humidity condition of concrete is a critical determinant of permeability, with high humidity leading to reduced permeability. It has therefore been recommended that samples should be removed and tested in the laboratory where they could be conditioned to standard humidity states. There are limitations on measuring the permeability of in-situ concrete and there is little quantitative correlation data available, comparing permeability results and long term degradation. The generation of such in-situ permeability data has been recommended by the Concrete Society Technical Report 31[4], as the first step to achieving a relation between permeability and long term performance. More research is needed to quantify the time taken for corrosion to occur under various conditions with varying permeability in the cover concrete. High carbonation rates will occur with high permeability concrete, but this will not result in corrosion if the humidity of the concrete is low (Parrott 90)[5]. It has been shown by Dinku et al[6] that compressive strength alone could not be used to assess the durability properties of concrete. Their investigation indicated that the permeability coefficient in relation to RH should be used to assess quality of construction, but they also expressed their concern on the reliability of permeability testing when RH was higher than 75%.

BRITISH CEMENT ASSOCIATION METHOD OF MEASURING RH & AP

The permeability and humidity measurements were made using the method developed by Parrott at the British Cement Association[5, 7] in which air permeability (AP) and relative humidity (RH) of the surface concrete are measured.

168 of the British Cement Association[5, 7] designed permeability plugs were inserted into 40 bridges in the London area. 35mm deep by 20mm diameter holes were drilled into the concrete and stainless steel inserts were then sealed into these holes using silicon or other sealing material, with a tightly screwed cap that sealed the opening of the insert. Three inserts were placed at each location, located in a triangular arrangement wherever possible, the inserts being 100mm away from each other. Permeability readings were then taken and the results were then analysed and correlated against carbonation depths measured by drilling cores in vicinity of the inserts.

The inserts were left in place for a short period of time to allow the inside of the hole and the concrete to reach temperature and moisture equilibrium. A form was designed to cover all the possible details that could be collected at the time of monitoring (Figure 1).

Permeability and RH readings were then taken soon after installation of the inserts and repeated every three-months in order to try to determine seasonal effects.

The apparatus consisted of: a probe, timer, pressure meter connected to a safety valve, air pump and five containers of salt solutions. The probe was used to measure the humidity and temperature. The containers were partially filled with highly viscous solutions of certain salts producing controlled humidities which varied under different temperatures as seen in Figure 1. The full response of the probe takes 30 seconds when the RH is under 80%, but takes longer when the RH is higher than 80%. Temperature also needed to be taken prior to measuring humidity, since the temperature should not fluctuate by more than ±1 C whilst taking the reading.

Care had to be taken using this apparatus on site, since a probe that is sensitive to ±0.1 C temperature and humidity ±0.1% will be sensitive to other factors such as rain, saturated insert, sunlight, dust and small particles of salts scattered around the containers. To overcome these problems, various methods were used, such as working under an umbrella when it was raining or was very sunny. To obtain the air permeability of the concrete the cavity was pressurised to 0.75 bar using an air pump and the time taken for the pressure to drop from 0.6 (P1) bar to 0.4 (P2) bar was noted. Permeation through the cover concrete and around the joint between the concrete and the insert could be seen as patches of fine or large bubbles on the concrete surface if liquid soap solution has previously been applied to the surface of the adjacent concrete.

SITE TESTING RESULTS

Figure 2a shows the cover to the reinforcement as measured by a covermeter. This figure gives the cover in five ranges and it is seen that the greatest number of measured depths (32%) were found to be in a range of 25-35 millimetres. Lack of cover was clear in a number of structures and as Figure 2a shows 17.7% of members were found to have 10 mm or less concrete cover to the reinforcement surrounding the extracted cores. Three quarters of the structures tested were made of reinforced concrete materials whilst a number of cores were extracted from mass concrete construction, see Figure 2b. In the cases where the covermeter did not detect any reinforcement within 120 mm of the surface of the concrete, these were also defined as 'mass concrete'. The third most popular type of construction within the LUL system was found to be concrete encased RSJ's, which constituted 17% of the members that were tested.

The results from the Schmidt hammer, found by the authors, were very variable but in-spite of this variation most of the rebound numbers from the Schmidt hammer trials were found to indicate relatively high strength concrete, and the 30-40 range was obtained from more than 45% of the collected data as seen in Figure 2c.

The registered depth of carbonation was related to the age of the bridge and the carbonation growth was calculated in millimetres per year, and this is shown in Figure 2d. Concrete that had suffered from a carbonation rate of 0.3-0.5mm/year was found in 28% of the total number of cores obtained. The level of carbonation that had penetrated into concrete was generally found to be uneven and the maximum penetration only is shown in Figure 3a. It is seen that 10-20 mm depth of carbonation was found to be the commonest.

Date: 3/8/94	Operative R & A	Structure Overbridge
Code/Type: TO683	Year Built: 1949	Nearest Stn.: SW
Construction: R.F.	Element: Column	General Con.1..**2**..3..

Rain(mm) 1=1, 2=2, 3=3, 4=4, 5=5, 6>5, 7=none   Bold: represent conditions

Weather: Sunny **1** . Cloudyb 2 Rain 3 Wind 4	Surface:Dry **1**..Wet 2..,Exposure:1..**2**.. 3

Humidity Calibration & Readings used in: RHc = RHs + (Rc - Rs) x K

Atmosph. RH-Temp	25^{oc}, 20, 15, 10, 5^{oc} (RHs)	Saturated S./Field Rs
51 25.1 ^{oc}	32.7..33.0..33.4..33.8..34.1..	MgCl2, (32.7)
	58.2..59.6..61.0..62.4..63.8..	NaBr, (50.7)
	80.2..80.6..81.1..81.9..82.7..	(NH4)2SO4, (62.5)
	90.3..90.7..91.0..91.4..91.8..	BaCl2, (68.1)
	97.0..97.3..97.7..98.0..98.5..	K2SO4, (72.7)

Insert Readings for RH, (RH_{c1}=58.95, RH_{c2}=64.17, RH_{c1}=61.93) Average = 61.68

No/Cov	Rc	Temp	At. RH-Temp	RHs	Rs	Temp
1-	53.4	24.2 ^{oc}		59.6	53	25.3 ^{oc}
2-	54.1	24.4 ^{oc}	24.4 ^{oc}	59.6	50.9	25.9 ^{oc}
3-	52.6	24.6 ^{oc}		59.6	50.6	25.8 ^{oc}

Permeability K=c/t * (P1-P2)/(P1+P2) where c=252 * 10^{-16}. But if P1=0.6 & P2=0.4,

K=50.4 * 10^{-16}/t (1) & K=0.84*10^{-16}*(0.6-P2)/(0.6+P2) when t= 300

No.	1st Reading (t)	2nd Reading (t)	3rd Reading (t)	Ave (Sec)
1-	..1.,.39 ,...61	...1.,..46.,..34.	...1,.47.,.19.	103.713
2-	......,..25.,..21.	......,..34.,..02.	.....,..33,.91.	31.047
3-	...1.,..17.,..27.	...1.,..22.,..27.	..1.,..18,.90.	79.52

Permeability using equ (1): K_{c1}=0.485x10^{-16}, K_{c1}=1.6234x10^{-1}, K_{c1}=0.6338x10^{-16}, ($K_{cAve.}$ – 0.9144x10^{-16})   Reinforcement Depth, Comment & Visual Inspection

Figure 1 A typical example of a repaired member of a reinforced concrete overbridge showing relative humidity and air permeability readings

The age distribution of the cores is shown in figure 3b, it is seen that 37.3% of them were drilled from different structures that had been in service for the last 60-70 years.

The field data[8] was used to obtain defect observations of those members subjected to site testing and the last observed fault was then divided by its age taken at the time when it was inspected to indicate the rate of the member deterioration, (see Figure 3c). The categorised five rates of deterioration indicates that these structures mostly suffered from a slow rate of deterioration in the range 0.1-0.3 (Fault/Year).

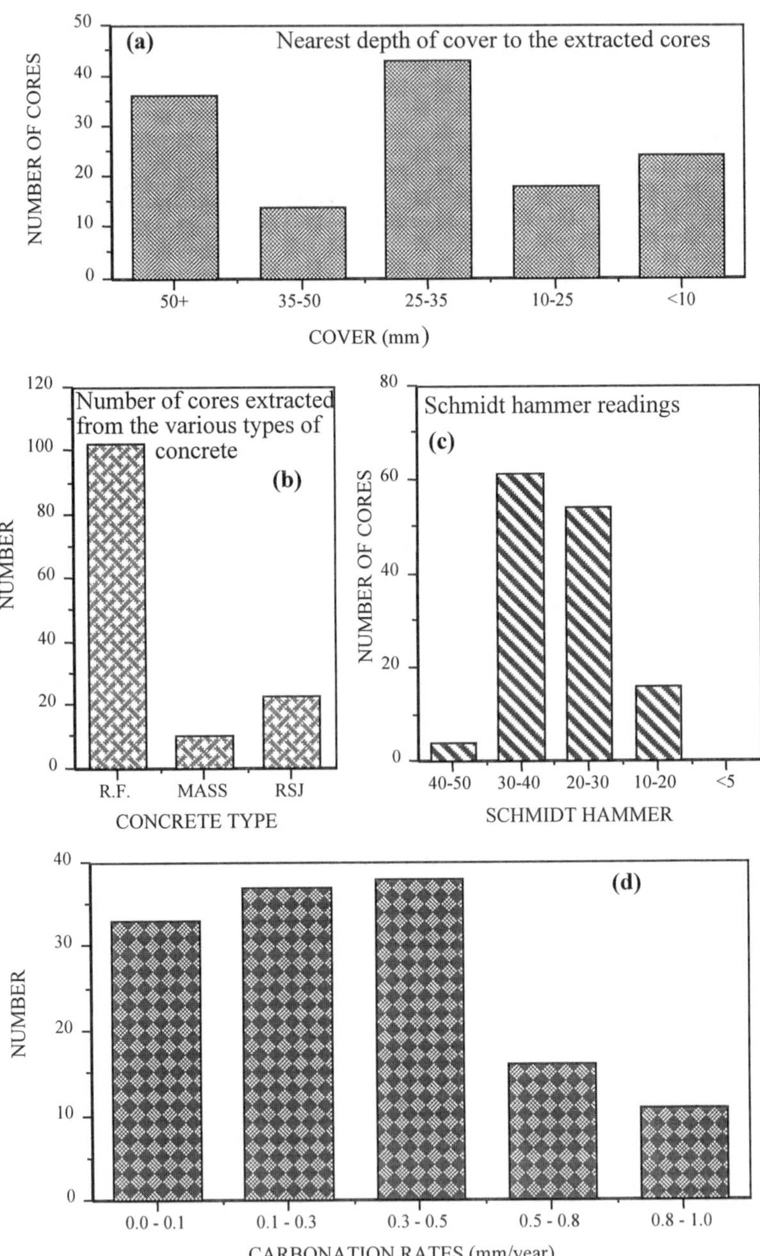

Figure 2 Distribution of results from the 40 bridges site tesing (a) cover (b) concrete type, (c) schmidt hammer and (d) carbonation rate

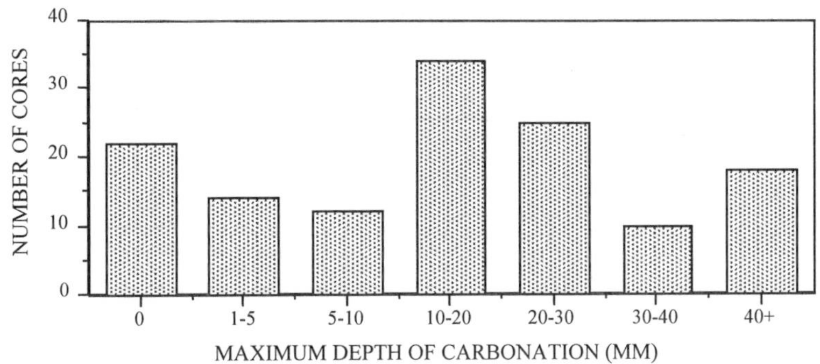

Figure 3a Number of in-situ cores v carbonation depth

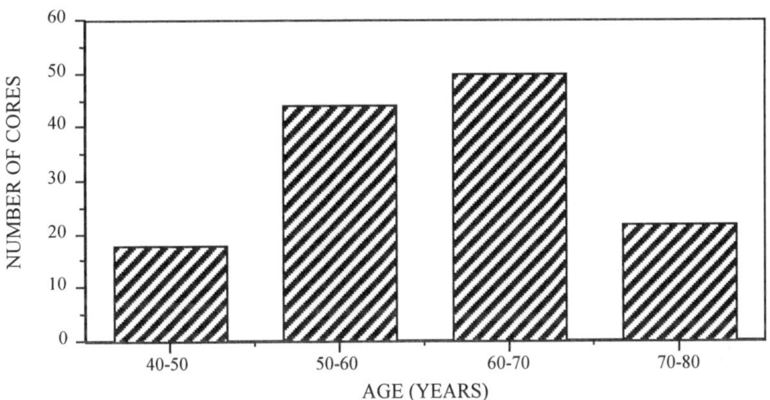

Figure 3b Age distribution of extracted cores

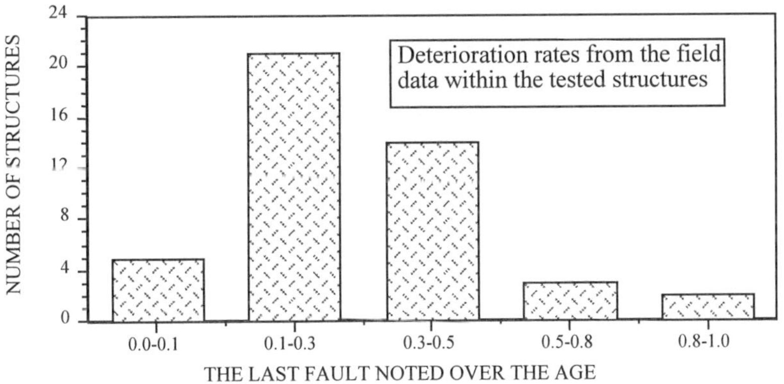

Figure 3c Deterioration rates (Fault/Year)

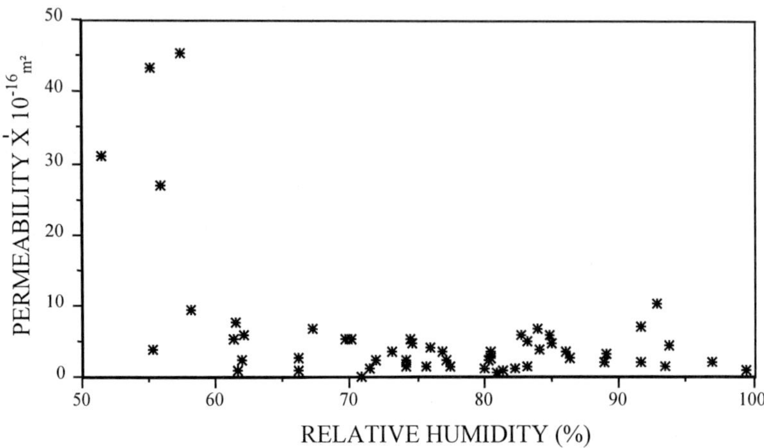

Figure 4 Plot of air permeability against relative humidity of concrete members from 40 bridges subjected to exposure 1 & 2

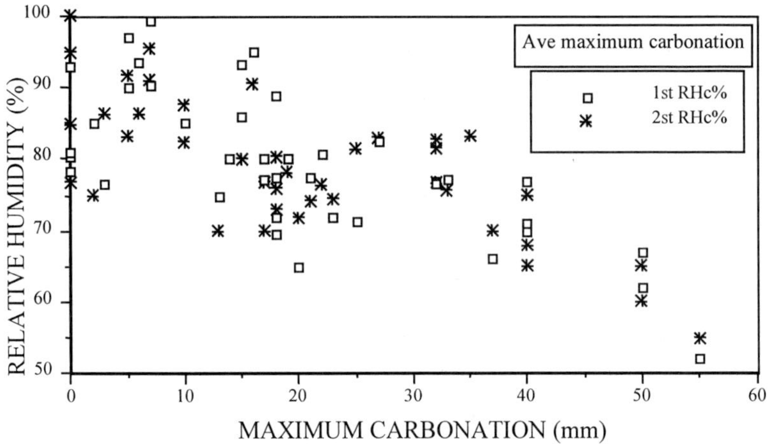

Figure 5 Relative humidities of 40 bridge members subjected to site testing against the carbonation depth

However the averaged results from a set of 3 inserts are shown in Figure 4, and these shows that the air permeability can be quite high when the humidity of the tested area stays below 60%. The results of RH from a set of three inserts were then averaged to correlate them to the averaged maximum depth of carbonation obtained from three cores. Figure 5 shows the results obtained in different seasons with a set of inserts being read twice and highlights that high humidity is related to low penetration of carbonation.The RH of cover concrete was generally found to be in excess of 70% and related to an air permeability range of less than 10 x 10^{-16} m^2, see figure 4. Figure 5 shows a region of approximately 25 mm penetration of carbonation within the 70% RH. It is then likely that this lower depth of carbonation where defects were still existed combined with other causes of corrosion than only carbonation.

The spray from the running traffic containing de-icing salts and the contaminated water seeping through the defective joints could therefore be another factor in causing corrosion. The drilled powder taken from forty bridges subjected to site testing are therefore under investigation for chloride content.

The occurrences of seepage would then become an important study, reported by Rigden et al[8], since this excess water can easily be drained by properly designed structure. The records[8], however, showed that only 4% of the members have suffered from corrosion, while more than 21% have suffered from seepage. The underestimated incidence of lack of cover and consequent corrosion is therefore under investigation by correlating permeability and quality of build against rates of deterioration.

CONCLUSIONS

To establish a measure of quality of build of each structure, cores and samples were taken from 40 bridges together with the data about carbonation depth and air permeability related to the relative humidity from 168 BCA designed plugs inserted in these structures.

At least three cores were drilled from each of the members, 200 cores were attempted with 162 cores being successfully extracted from the chosen members. Wide variations of depths of carbonation in 57 members tested were therefore found. Concrete that had suffered from a carbonation rate of 0.3-0.5mm/year was found in 28% of the total number of cores obtained. The age distribution of the cores has shown that 37.3% of them were drilled from different structures that had been in service for the last 60-70 years.

The rates of deterioration were found to vary considerably but seem to be showing reasonable correlation with the general quality of construction and the degree of exposure experienced by the element over its life. This latter condition may well have changed during the element's life time, which makes this correlation particularly difficult to predict. However, More than 21% of the structures suffered from seepage, while only 4% of the members reported as being suffered from corrosion. High humidity was also related to low penetration of carbonation and above 60% relative humidity to low air permeability. The analysis of chloride contents of the samples extracted from 40 bridges tested are under investigation.

ACKNOWLEDGEMENTS

The authors would like to express their profound gratitude to Dr. E. Burley, former head of the Civil Engineering Department for supervising the research. The assistance of the staff of the Bridge Section-London Underground Ltd- particularly the chief bridge engineer, Mr G. Bessant is gratefully acknowledged. The assistance of A. Abu-Tair, Roger Nelson and the late Gil Schubert for the in-situ testing is greatly appreciated. Thc assistance of Dr L. Parrott of the BCA is also greatly appreciated.

REFERENCES

1. DEPARTMENT OF TRANSPORT, DOT (1989), 'The performance of concrete in bridges', A survey of 200 Highway bridges, H MS O, London, April.

2. LONDON UNDERGROUND LTD, Track Safety for Track Accustomed Person UK.

3. BS 6089, (1981), Assessment of concrete strength in existing structures', Parts 4, The Institution of British Standard.

4. TECHNICAL REPORT NO 31 (1987), ' Permeability testing of site concrete - A review of methods and experience', The Concrete Society

5. PARROTT L J (1990),' Assessing carbonated in concrete structures', Interl. (5th) conference, Durability Building Materials & Components, B.S.C. Nov., pp 575-586

6. DINKU A & REIHARDT H W (1996), 'The influence of storage conditions on the gas permeability and carbonation of concrete', CRRP, Pro. Dundee, June, pp 195-204.

7. PARROTT L J (1991),' Measurement of air permeability and relative humidity in cover concrete', British Cement Association, January 1991.

8. RIGDEN S R, TAJALLI S M A, BURLEY E. ABU-TAIR A.I., 'Service Life Prediction of Concrete Bridges', Department of Engineering, QMW ; International Congress, Concrete Repair, Rehabilitation and Protection, Dundee June 1996.

FIELD INVESTIGATION OF CHLORIDE DIFFUSION IN CONCRETE USING THE RAPID CHLORIDE PERMEABILITY TEST

W Akili

University of Qatar

Qatar

ABSTRACT. Service life of reinforced concrete structures is often governed by the rate of chloride diffusion through the cover. To estimate the time it takes chloride ions to reach reinforcement, the coefficient of chloride diffusion needs to be determined or assumed. Since chloride diffusion in concrete is an extremely slow process, conventional testing may take as long as 6 months to a year to finish.

To reduce testing time, accelerated test methods have emerged. Acceleration of chloride ions through concrete is achieved by the application of an electric voltage gradient across the thickness of the test specimen. The most prominent amongst accelerated test methods is the Chloride Ion Penetration test, referred to here as CP, and designated as T277 and C1202 by AASHTO and ASTM.

This paper addresses the application of the CP test as a field control method, making use of an extensive data base derived from a construction site on the coast of Qatar. Tests were carried out on vibrated and unvibrated (tremmied) concrete cores extracted randomly during construction. The data was statistically analyzed and correlations between CP test results and degree of consolidation of the concrete were developed. The accuracy and reliability of CP results appear to be a function of the homogeneity of the concrete matrix and whether a uniform chloride front can be maintained during testing. While the test is undoubtedly useful in evaluating concrete resistance to chloride ingress in general, caution must be exercised in interpreting results obtained from the field; particularly so when the concrete has not been properly consolidated.

Keywords: Chloride diffusion, Chloride ion penetration (CP) test, Electric voltage gradient, Vibrated and unvibrated concrete, Degree of consolidation.

Professor Waddah Akili is Professor of Civil Engineering at the University of Qatar, Doha, Qatar. His main areas of specialization are Geotechnical Engineering and Engineering Materials, with interest in degradation of construction materials under Arabian Gulf conditions. He has published widely, and is a member of several professional international societies and technical committees concerned with performance and degradation of materials.

INTRODUCTION

High-quality impermeable concrete cover of reinforcing steel remains the most practical approach to preventing the ingress of chloride ions into the bulk of the concrete. To assess the potential resistance of concretes to chlorides in a reasonably short period of time, researchers have relied on the technique that utilizes voltage gradient to force the chloride ions to migrate more rapidly through the pores of the concrete under test.

This approaches was developed into a test method by Whiting and was later adopted as an AASHTO standard test method [1-3]. However, there have been conflicting reports on whether this test can reasonably predict the resistance of insitu concrete to chloride permeability; and therefore a certain degree of uncertainty regarding its application on a day to day basis has arisen. [4-6].

Whiting's method – referred to here as the CP test – has recently been imposed as a quality control measure on large-scale industrial projects along the coast of Qatar. The environment along the Qatar coast, similar to other Gulf neighbouring states, is extremely severe and very hostile to concrete [7,8].

The paper addresses the application of the CP test in Qatar, and highlights problems that have arisen as a consequence of its use as an acceptance or rejection test of hardened concrete in the field. The paper make use of several hundred CP test results performed on selected cores during the construction of a major coastal project. The test data were examined and correlated in an attempt to arrive at a sound technical argument on when and how the CP test can be of use.

THE CP TEST

The AASHTO/ASTM test method (referred to as the CP test), requires monitoring the amount of electrical current passing through the disk-shaped standard test specimen (51 mm thick × 102 mm nominal diam.), when an electric potential difference of 60V dc is maintained across the two ends of the specimen during a 6 hour period. One end of the specimen is immersed in a sodium chloride solution and the other in a sodium hydroxide solution. The total charge passed, in coulombs, has been found to be related to the resistance of the specimen to ion penetration [2,3]. Total charge passed is determined by integration of the current-time curve over the 6 hour test period; the higher the charge passed, in coulombs, the higher the potential for chloride ion penetration into concrete. At the start of test, temperatures of the specimen, applied voltage cell and solutions were always maintained at 20 to 25°C.

Table 1 shows the correlation between charge passed and the potential of concrete to chloride ion penetration [2,3]. The concrete is believed to be highly conductive when the total charge passed is over 4000 coulombs. Various researchers have examined the use of the rapid chloride permeability test, and a consensus appear to have emerged on the following inherent limitations [5,6]. (i) The rise in sample temperature brought about by the passage of the current will induce changes in the porewater structure of the sample, resulting in overestimates of its chloride permeability. (ii) The test tends to measure all the ions present rather than chloride ions only. (iii) The use of aggregates that contain iron compounds – such as Gabbro – will indicate higher coulombs as compared to limestone aggregates.

(iv) The diffusion of ions during testing is affected by the quality of the concrete core i.e., the presence of: cracks, cavities, surface irregularities, etc. can adversely affect test results. (v) Results reported in coulombs are not meaningful to an engineer who wishes to estimate the structural life against chloride diffusion, unless appropriate correlations have been established between the test results and chloride diffusion in the field. (vi) The test method has been shown to give results that sometimes correlate very poorly with standard (unaccaelerated) tests. (vii) Lack of prior experience with this test, and use of non-standard equipment (since shelf items are not yet available) will undoubtedly contribute to inaccuracies of test results.

Table 1 Chloride ion penetration based on electric charge passed*

CHARGE PASSED (Coulombs)	POTENTIAL CHLORIDE ION PENETRATION
> 4,000	High
2,000 - 4,000	Moderate
1,000 - 2,000	Low
100 - 1,000	Very low
< 100	Negligible

*as per CP test write up (AASHTO: T277/ASTM: C 1202-94)

PROJECT CONCRETE

The contract specifications drawn for concrete works have provided for the production and placement of high quality concrete for: bored tremmied (unvibrated) piles, and for structural (normal) concrete used throughout the project. Microsilica in the range of 5 to 10% by weight of total cementatious materials was specified. Water cement ratio was 0.4. Coarse aggregates were imported Gabbro. The total chloride content arising from all ingredients in a mix, including: cement, water and admixtures shall not exceed 0.15% in 95% of all specified tests, and no single test result would show more than 0.3% chloride ions by weight of cement in the mix. Ordinary Portland cement, conforming to BS 12 and BS 4550, was specified and used throughout the works.

The mix design used throughout, was to achieve a target slump of 150 mm of wet mixing at 0.4 water cement ratio. Typical composition and weights were: 360 kgs of OPC, 26 to 40 kgs of microsilica, 780 kgs of crushed sand, 310 kgs of 14 mm nominal coarse aggregates, 930 kgs of 20mm nominal coarse aggregates, 154 litres of water and 7.5 litres of a plastisizer.

Typical results of: concrete cube strengths, cube density and CP tests, carried out under laboratory-controlled conditions, are shown in Table 2. The trial results shown (Table 2) are well within the acceptable tolerance laid down in BS 1881 for strength and density. The CP test results shown (Table 2) also indicate compliance with a single-operator precision [3]. ASTM C1202-94 states that "the results of two properly conducted tests by the same operator on concrete samples from the same batch should not differ by more than 35%". All four CP test results shown (Table 2) are well within this tolerance. Additionally, the mean CP value of all four trials does meet the specified upper value of 1000 coulombs.

It should be pointed out that, although the CP test results on laboratory trial mixes do conform to specifications by meeting the 1000 coulomb upper limit; tests run on samples from field cores - as shown later - would not necessarily meet this requirement, particularly when concrete is unvibrated.

Table 2 Compressive strengths*, density* and chloride permeability (CP)* of trial mixes

TRIAL NO.	STRENGTH, N/mm^2 3 days	7 days	28 days	DENSITY, kg/m^3	CP coulombs
# 1	28.5	34.5	47.5	2530	887
# 2	27.5	33.0	47.0	2510	934
# 3	27.0	34.0	47.5	2507	862
# 4	29.5	34.5	44.0	2520	1006
Mean value	(28.1)	(34.0)	(47.8)	(2517)	(922)

*Avg. of 3 cubes
Note: Mix details: w/c = .40; slump = 150 mm; opc = 360 kg/m^3: 26 to 40 kgs/m^{3}of microsilica.

ANALYSIS OF CP TEST RESULTS

The Analysis of CP tests performed on hardened concrete cores, that were at least 28 days old, were examined, sorted out and subjected to statistical analysis. Equal weight was assigned to each test result irrespective of the measured value. The outcome is presented under two headlines: structural concrete and piling concrete.

The climate conditions on site, typical of coastal arid environment, are very severe. High temperature during the months of June through September, could reach 50 ±2^{O}C with alternating high and low humidity. In the winter time (December – February), high temperature of 25 ± 2^{O}C are recorded, and often coupled with long durations of very high humidity.

Structural Concrete

Over two hundred test results, at the rate of one test per day, were performed on cores from various structural concrete elements of the project. Figure 1 shows the CP values versus degree of consolidation for each test core. Degree of consolidation is defined here as: measured core density at CP test time over 28 days cube density. The degree of consolidation, a function of mix characteristics and placement conditions - is an indirect measure of the morphology of core's pore structure.

The higher the degree of consolidation the higher the resistance to chloride transport into the bulk of the hardened concrete. The assumption being made is that chloride ions penetrate from outside through the surface by diffusion. Most modelling of transport mechanisms assumes the concrete to be homogeneous and crack-free.

Unfortunately this is not the case, and cracking is a feature in concrete sections. Expansive forces due to concrete deterioration and loading are always at work causing cracking and/or widening of existing cracks.

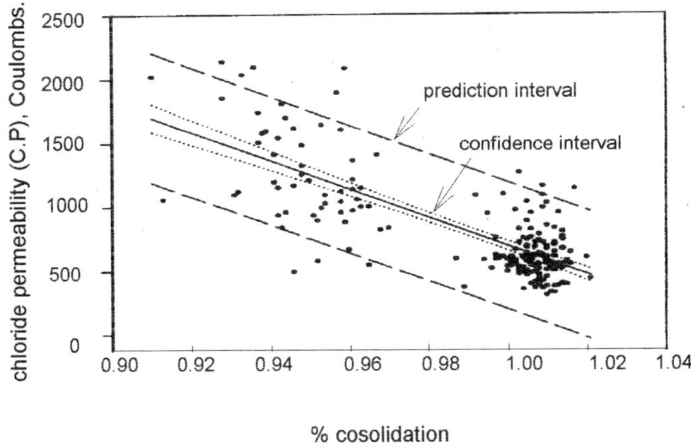

Figure 1 Effect of degree consolidation on chloride permeability (CP) in structural concrete

Figure 2 shows CP test results versus degree of consolidation in a bar chart format. Figure 3 shows a bar chart presentation of number of tests (attempts) versus the different consolidation ranges attained. As shown in Figure 3, majority of tests have exhibited a degree of consolidation higher than 100%.

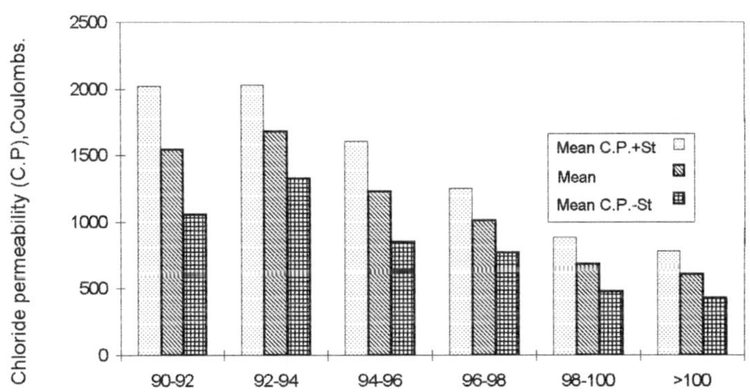

Figure 2 Chloride permeability (CP) versus % consolidation in structural concrete

Based on the above, the following general trends appear to have emerged: (i) Degree of consolidation has a relatively strong effect on CP test value, i.e., the higher the degree of consolidation the lower the CP value.

(ii) Measured CP values have ranged from a high of 2138 to a low of 302 with a mean of 792.5 coulombs. (iii) 66% of all cores tested have attained a degree of consolidation of over 100%, and 30% of all cores, have attained 96% and higher consolidation. (iv) Out of 202 tests, five tests only have exhibited CP values higher than the threshold value of 1000 coulombs.

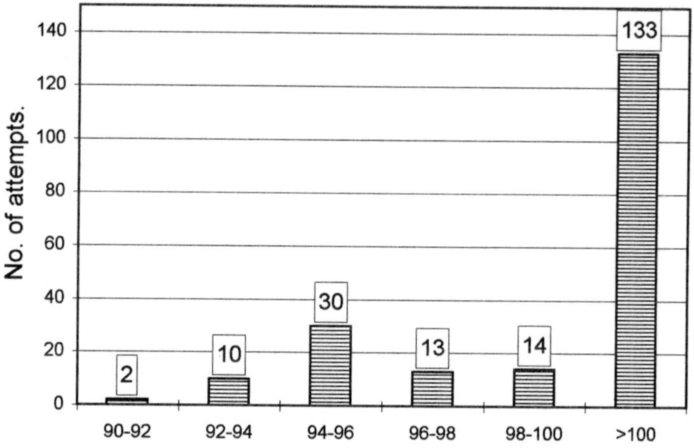

Figure 3 Number of CP tests (attempts) versus percent consolidation in structural concrete.

Piling Concrete

During piling works, which preceded structural works, 154 CP tests were carried out, at the rate of one a day, on cores extracted from the top part of selected piles. The tremmied (unvibrated) piling concrete had the same properties and composition as that of structural (vibrated) concrete. Figure 4 shows CP results versus degree of consolidation.

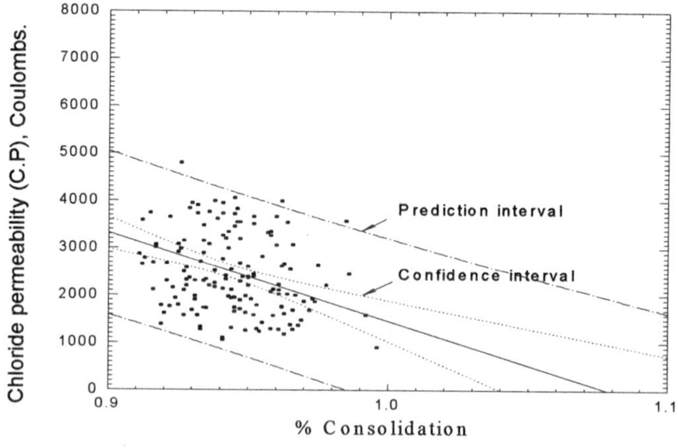

Figure 4 Effect of degree of consolidation on chloride permeability (CP) in piling concrete

Figure 5 is a bar chart presentation of measured CP values versus degree of consolidation. Figure 6 presents number of tests (attempts) per consolidation range in the form of a bar chart.

Based on the analysis of all test data under piling concrete, and unlike structural concrete, the CP values appear to be only moderately affected by the increase in percent consolidation within the range of 90 to 100%. Test results on cores with lower degree of consolidation than 90% exhibited relatively high CP values. This is to imply that other causal factors appear to be responsible for this noted rise in the CP values. It is conceivable that ions, including chloride ions, can move during testing of low consolidation unvibrated cores, through minute channels rather than continuous micropres. Measured CP values in this type of concrete (piling concrete) have ranged from a high of 5800 to a low of 900 with a mean value of 2518 coulombs. It is interesting to note that only 3% of all 154 cores in this category have attained a degree of consolidation of 98% and higher. Out of 154 CP tests, six have exhibited CP values higher than 4000 coulombs and 119 tests were higher than 2000 coulombs; only one test (out of 154) had lower CP value than the threshold of 1000 coulombs.

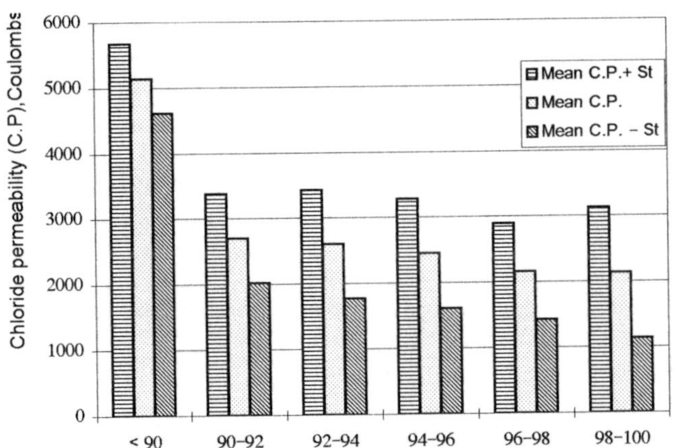

Figure 5 Chloride permeability (CP) versus percent consolidation in piling concrete

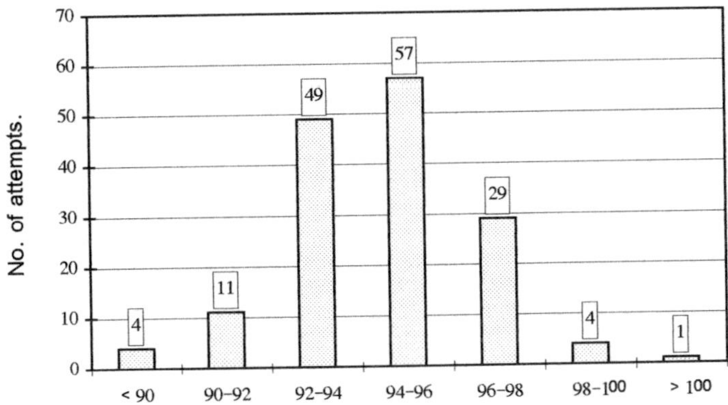

Figure 6 Number of CP tests (attempts) versus percent consolidation in piling concrete

DISCUSSION

As expected, vibrated concrete elements (structural concrete) have exhibited markedly lower CP test values in comparison to unvibrated trimmed concrete (piling concrete). Vibrations do not only densify fresh concrete, but also assist in homogenizing the mix before set time, thus reducing the formation of undesirable flaws in the form of microcracks and voids. Cracks and voids, when present in concrete, tend to expedite chloride ion movement through the matrix, i.e., the ions could take a short cut by moving through these flaws rather than by diffusion through the porestructure.

Nearly all CP test results on structural concrete have met the limit of 1000 coulombs set by the specifications. And in accordance with table 1- drawn from test write up - concrete that meets the 1000 coulomb limit is presumed to posses high resistance to chloride ion penetration. On the other hand, CP results run on piling concrete have been considerably higher, and have ranged between 1000 and 4000 coulombs; implying that tremmied piling concrete is less resistant to chloride ion penetration (Table 1).

The CP results as presented (Figures 1 to 6) do not account fully for the positive contribution of the microsilica towards ion diffusion in hardened concrete. This is because 6 months to one year is normally required for microsilica in concrete to reach a degree of maturity where it can develop its intricate microstructure [9]. Therefore, the CP test results performed at 28 days of casting or shortly after, do not reflect the positive role that microsilica imparts in delaying the migration of cations, and in particular, chloride ions. Limited number of CP tests performed nine months after casting (data not shown here), have exhibited much reduced CP results in comparison to present (Figures 1- 6) 28 days measurement. Difference was two to three fold lower, indicative of significantly higher resistance to chloride ion ingress. If a 100% downward correction is applied onto present results (Figures 1- 6) to account for the microsilica contribution, the projected values would drop appreciably to bring the number of tests (that would meet the threshold of 1000 coulombs) in piling concrete to over 50% of all such tests. Whether or not this is appropriate, does require further investigation.

In addition to the many shortcomings and limitations of the CP test [4-6], its utility as a field measure depends also on: the coring process, how representative are cores of conditions on site? , and the care and precision exercised during the test. Prior to performing the CP test on insitu concrete, the values given in Table 1 – including the threshold limit of 1000 coulombs – should be checked to see whether appropriate for materials and conditions on site. In fact, a broad spectrum of CP measurements need to be conducted during mix design stage, preceding any CP field testing, to establish correlations between: (i) mix variables and charge passed (in coulombs); (ii) effect of consolidation of selected mix on CP results, (iii) effect of curing time on charge passed, particularly when microsilica is used; and (iv) CP readings (in coulombs) against actual chloride permeability measurements from conventional diffusion test methods.

While the technique and underlying principles behind the CP test are sound and appropriate, further work is needed to define acceptance limits and to arrive at statistical schemes for acceptance and rejection of concrete in the field.

CONCLUSIONS

Analysis of several hundred CP test results (AASHTO 277/ASTM C1202) performed on insitu microsilica concrete, intended to measure hardened concrete resistance to chloride ion ingress, has shown the following trends:

1. Results of CP tests on structural (vibrated) concrete correlate well with the degree of consolidation in the field, and results of all such tests have fallen below the threshold limit of 1000 coulombs, set by specification, indicative of high resistance to chloride ingress.

2. Conversely, CP tests performed on tremmied (unvibrated) piling concrete, correlate poorly with the degree of consolidation, and majority of results have fallen above the threshold limit of 1000 coulombs, indicative of lesser resistance to chloride ion ingress. This is attributed largely to the inhomogeneity of tremmied concretes and the presence of flaws – in the form of cracks – that tend to expedite ion movement through the matrix rather than by diffusion through micropores.

3. The CP test, although useful for: design, research and general guidance; has not been properly tuned as a field acceptance test on a day to day basis.

REFERENCES

1. WHITING, D AND MITCHELL,T. A history of the Rapid Chloride Permeability Test. Transportation Research Record No. 1335 (preprint) Jan. 1992, pp 55-62.

2. AMERICAN ASSOICATION OF STATE HIGHWAY AND TRANSPORTATION OFFICIALS. Standard test method for rapid determination of the chloride permeability of concrete. AASHTO T 277-83. Washington, D.C., 1983.

3. ASTM. Standard test method for Electrical Indication of Concrete's Ability to Resist Ion Penetration. ASTM C1202-94, 1994, pp 620-625.

4. SHA'AT A A H, BASHEER,P A M, LONG, A E, AND MONTGOMRY, F R. Reliability of the Accelerated Chloride Migration Test as a measure of chloride diffusivity in concrete. Proc. Inter. Conf. Corrosion and Corrosion Protection of Steel in Concrete. Ed. R N Swamy, 1994, Vol. 1, pp 446-460.

5. ZHANG,J-Z. AND BUENFELD,N R. Development of the Accelerated Chloride Ion Diffusion (Acid) test. Proc. Int. Conf. Corrosion and Corrosion of Steel in Concrete. Ed. R N Swamy, 1994, Vol. 1, pp 395-401.

6. FELDMAN, R F, CHAN, G W, BROUSSEAU, R J AND TUMDIDAJSKI P J. Investigation of the Rapid Chloride Permeability Test. Am. Concrete Inst. Materials Journal, Vol. 91, 1994, pp 246-255.

7. AKILI,W. Reinforced concrete structures on the Arabian Gulf coast: a durability assessment. Proc. Int. Congress. Concrete in the Service of Mankind. Eds. R K Dhir and N A Henderson, 1996, pp 1–10.

8. AKILI,W, MANSER,I, AND COGAN, K. Deterioration of reinforced concrete deep water wharves in Qatar: condition causes and repair. Proc. Int. Conf. Corrosion and Corrosion Protection of Steel in Concrete. Ed. R N Swamy, 1994, Vol. 1, pp 105-115.

9. RSHEEDUZZAFAR AND S.E. HUSSAIN. Durability mechanisms of blended cement concrete. Proc. 4th Int. Conf. Deterioration and Repair of Reinforced Concrete in the Arabian Gulf. Ed. G L MaCmillan. Bahrain, 1993, pp 909-925.

VISUAL ASSESSMENT OF CONCRETE FOOTINGS IN INDUSTRIAL PLANTS

O E K Daoud

Dar Al-Handasah Consultants

I A Ibrahim

University of Cairo

Egypt

ABSTRACT. More than 1000 footings in an old oil refinery in the Arabian Gulf were assessed by a condition survey specially developed for this purpose. The exposure conditions during operation, the observed distresses, their degree, and the overall condition of the footings were recorded in a data collection sheet. The footings were categorized according to the observed distresses by applying a classification method employing a weighting system to each distress and degree of damage. Poor detailing of the joint between the concrete and the equipment chassis and the use of cement mortar with high w/c ratio in sealing anchor bolts were found to be responsible for many of the steel corrosion distresses. Cyclic wetting by seawater or combined attack from cyclic wetting with other pollutants affected 75% of the footings and these footings suffered 75% of the total reported distresses. The most frequent distresses reported were cracking, delamination, and spalling and together they represented 60% of the total.

Keywords: Reinforced concrete, Corrosion, Damage assessment, Condition survey, Oil refineries, Wetting and drying, Seawater attack, Cracking, Footings

Dr Osama E K Daoud is the Head of Management of Construction and Contracts Department, Dar Al-Handasah Consultants (*Shair and Partners*), Cairo, Egypt. He has a particular interest in concrete durability problems in the Middle East, concrete technology, design and construction quality control, fracture mechanics applications to reinforced concrete, performance of building materials and construction management. He has published more than twenty-five papers in these fields.

Dr Iman A Ibrahim is an assistant Professor in the Department of Civil Engineering, University of Cairo, Egypt. She is interested in the fields of concrete technology and the mechanics of failure in composite materials.

INTRODUCTION

Design for durability has become a growing trend in many important sea-front structures. Concrete in industrial facilities has been less fortunate. The distresses occurring in such facilities, their frequency and causes are not yet fully understood. Daoud and Ibrahim [1] found that concrete exposed to the environment in an oil refinery looses 25% of its strength in 15 to 40 years. Moreover, severe deterioration of other physical properties was also recorded. Previous knowledge of the exposure conditions, the resulting distresses and their frequencies are important information in the design of concrete structures in such facilities. O'Connor [2] and Novokshchenov [3] reported an alarming case of concrete deterioration at two industrial facilities in the Arabian Gulf. The use of seawater in concrete mixing and curing was proved to be the main cause of deterioration. Concrete footings are used to support equipment in oil refineries. Damage assessment for these footings is a major problem facing maintenance staff. Most of the available literature provides assessment methods for pavements, bridges and buildings [4]. In this study a survey of more than 1000 footings in a 45 years old oil refinery has been conducted. A data collection sheet was used to list the exposure conditions, the resulting distresses, their degree and the overall condition of the footings. A classification technique based on a point system was developed for the damage assessment. Such information would enable the designers to take into account the environmental effects expected, hence reducing the deterioration occurring.

UNIT DESCRIPTIONS, HISTORIES AND DESIGN DATA

Although the refinery contains more than 63 production units, this study is limited to the concrete footings of seven units. A full description of the information available in the design documents is given in the report by Daoud, Qazweeni and Higab [5].

The seven units were all constructed between 1949 and 1978. The oldest unit, the Washery (Wash) has 55 footings and pedestals. Caustic Soda and Lead were used to wash Kerosene. No design data were found for this unit. The Products Loading Pump House (PLPH), the Hydrobon Platforming (PLAT), and the Crude Distillation #3 (CDU3) were all built in 1957. They have, 61, 138 and 238 footings, respectively. The PLPH and PLAT were put out of operation since 1986 and 1982, respectively. Both received no maintenance during their service life. The Fractionation Units #1 and 2 (FU 1 and FU2) were built in 1962 and 1965, respectively. Both had been subjected to concrete maintenance work, involving jacketing to the footings and exposed pedestals. The newest unit, the Bitumen Plant (BIT), was constructed on the Gulf shore. Its environment is heavily contaminated with SO_2 and H_2S. All the units were constructed from concrete mix with Cement:Sand:Gravel ratio of 1:2:4 and with cylinder compressive strength of 21MPa. Ordinary Portland Cement was used in all units except BIT, which was built using sulfate resisting cement (SRC). PLAT, CDU3 and BIT were subject daily to floor washing using seawater to wash off oil residues. Only footings in BIT were protected by polythene sheet placed under the footings and self-adhesive, self-sealing bitumen based membrane on the buried vertical surface. The rest of the units had no protection on exposed or buried parts of the footings.

From the above it is clear that no provisions were taken by the designers to protect the concrete from environmental factors. Exposed concrete elements received no protection while buried concrete received very moderate protection in one unit.

Information available on the design requirements was incomplete and no information on the construction phase was found. This is typical for projects in the Middle East. In all cases the minimum cement dose was not specified and the mix proportions quoted were those used in normal residential dwellings in the 40's and 50's. The maximum w/c ratio was not specified. In the newest unit, built in 1978, the cement specified was SRC, which was commonly used in those days in the Gulf States. This practice was responsible for many failures in the area, due to the weakness of this cement in resisting the dual attack of chlorides and sulfates because of its low C_3A content [6]. Low C_3A would allow more free chloride ions in the concrete to attack steel bars. The designers were not aware that the concrete would be subjected to a daily wash by seawater.

METHODOLOGY OF FIELD SURVEY AND VISUAL INSPECTION

The field survey involved visual inspection of the pedestals/footings, followed by a general evaluation of their condition. This data was used later to select the locations for further investigations such as core cutting or powder sampling. The results from these samples could be found in ref. [1]. It was first necessary to develop a visual inspection method since no recognized method was available in the literature for such elements. Because a large portion of the footing is underground, visual inspection is limited to a small portion; this situation creates a problem in categorizing the footing's condition. Also, the exposed portions of the footings (especially those observed in the refinery) are subjected mainly to compression stresses throughout their service life, rather than flexure or tensile stresses. Accordingly, the visual distresses recognized by known methods for evaluating beams and slabs are not applicable here. Furthermore, due to the huge number of footings/pedestals involved (1078 footings with 1276 pedestals), the procedure has to be adaptable to computerization so that the accumulated data can be entered and retrieved easily by a database program. Taking account of the above constraints, a visual inspection data sheet was developed along with the suggested methodology for its use.

The visual inspection methodology for footings and pedestals focused on identifying the degree of distresses observed on the footings and the exposure conditions for each footing. This information was then used to determine the frequency of each distress and its relationship with the exposure type. The visual survey used a special field inspection form shown in Figure 1 and described as follows. Thirteen types of distress were listed in the inspection form and marked for presence or absence. Degrees of distress were also recorded for five types selected to assess the actual conditions of the footings and the associated repair methods. The definition of each distress and its degree were identified and explained to the condition survey team. These can be found in detail elsewhere [7]. In order to minimize variability of results by different operatives, the team inspected the first unit together and cross-checked their results before inspecting the remaining units.

The overall condition for each footing/pedestal was graded as good, medium or poor. This was a subjective elementary evaluation towards categorization for coring purposes, and repair methods. It was determined according to the seriousness of the damage observed. Footings with localized or no surface defects were assessed as good; those needing extensive patching and crack sealing, as medium; and those requiring partial or total replacement, as poor. Later on this assessment was used as a guideline in the developing of the point system used to quantify the damage in each footing.

FIELD INSPECTION FORM

UNIT NAME:
FOUNDATION/ STRUCTURE No.:
PEDESTAL NO.:
FOUNDATION/STRUCTURE CODE No.:

Exposed Length (cm):
Exposed Perimeter:

Equipment Type:
Equipment Code No.

TYPES OF DISTRESSES **Degree of Distress**

TYPES OF DISTRESSES	Degree of Distress		
1. Spalling	❒ H	❒ M	❒ L
2. Surface cracking	❒ H	❒ M	❒ L
3. Steel Corrosion	❒ H	❒ M	❒ L
4. Disintegration	❒ H	❒ M	❒ L
5. Efflorescence	❒ Y	❒ N	
6. Delamination	❒ H	❒ M	❒ L
7. Crazing	❒ Y	❒ N	
8. Exposed Steel	❒ Y	❒ N	
9. Stains	❒ Y	❒ N	
10. Segregation	❒ Y	❒ N	
11. Honeycombing	❒ Y	❒ N	
12. Repairs & Patching	❒ Y	❒ N	
13. Others	❒ Y	❒ N	

DEGREE OF IMPORTANCE
❒ H ❒ M ❒ L

METHODS OF REPAIR
❒ 1. Removal of loose material & patch
❒ 2. Injecting cracks + surface coat
❒ 3. Surface coat
❒ 4. Total Replacement
❒ 5. Partial replacement
❒ 6. Steel sandblasting/replacement
❒ 7. Repair top grout layer

EXPOSURE TYPE

EXPOSURE TYPE		
1. Salt water	❒ Y	❒ N
2. Chemicals	❒ Y	❒ N
3. Gases	❒ Y	❒ N
4. Heat	❒ Y	❒ N
5. Vibrations	❒ Y	❒ N
6. Wetting & Drying	❒ Y	❒ N
7. Steam	❒ Y	❒ N

Type:...........................
Type:...........................

SAMPLES FOR:

Chemicals	Cores	Photos
❒	❒	❒

INSPECTED BY:.......................
DATE:..

OVERALL CONDITION

Poor	Medium	Good
❒	❒	❒

Figure 1 Data collection sheet used in field inspection

Exposure type was recorded to help in analyzing the results and to obtain better understanding of the causes of deterioration. Seven types were listed:

1. ***Salt water:*** Salt water is used to cool the equipment, and may cover the element during operation or maintenance due to leaks or splashes. In many units seawater was frequently used to wash floor slabs.
2. ***Chemicals:*** Leaking pipes that transport chemicals can harm concrete. When frequent maintenance necessitates emptying vessels and storage tanks, chemical residues splashed onto concrete elements: these were also recorded.
3. ***Gases:*** Gases, such as H_2S and SO_2, pollute the refinery environment in different concentrations. Footings directly exposed to gases and gas type were identified.
4. ***Heat:*** Heat can cause differential thermal expansion, leading to cracking.
5. ***Vibration:*** Footings carrying pumps and motors are subjected to high-frequency vibrations that can contribute to internal cracking.
6. ***Wetting and drying:*** Cyclic wetting resulting from daily washing or leaking joints cause alternate expansion/contraction, and hydration/dehydration, which can cause mechanical damage when salt crystals form and dissolve within the paste pores.
7. ***Steam.*** Footings directly exposed to steam nozzles erode and disintegrate.

Other information such as the degree of importance and the method of repair were included to help defining the optimum repair strategy with minimum impact on the unit operation.

METHODOLOGY FOR FOOTINGS CLASSIFICATION

A method was developed to classify, quantitatively, the damage to the pedestals according to the distresses observed and their degree. The overall classification made by the field survey team was used to test this method. The method assigns weighting points for each distress and degree as a measure of the damage resulting from the distress relative to other distresses. For example, the effect of cracking, disintegration, or delamination was considered to be higher than spalling and corrosion since the former is expected to be more detrimental to the concrete and would involve more serious maintenance works. Spalling and corrosion are expected to be more localized and hence would require less extensive repair. Weighting systems were developed and tested against the overall condition assessment done during the site survey. Upon applying the point system to the surveyed pedestals, a quantitative evaluation of the damage was possible by adding points resulting from the distresses observed in each pedestal. The resulting figure representing the total damage was named the Footing Classification Number (FCN). The higher the FCN, the lower the quality of the concrete in the footing. The resultant FCN was compared with the overall condition and the following limits representing the three concrete conditions, poor, medium and good, were identified:

Good condition	$0 \leq \text{FCN} \leq 3$
Medium condition	$4 \leq \text{FCN} \leq 21$
Low condition	$22 \leq \text{FCN}$.

The degree of confidence in the weighting points was determined by finding the percentage of the pedestals that produced FCN within the above limits to the number of footings of the same category classified visually during the survey. The higher this percentage, the higher the degree of confidence in the system. Several trials were made by modifying the weighting points in order to reach the maximum confidence ratios.

RESULTS AND DISCUSSION

Effect of Exposure Type on Frequency of Distresses

Table 1 shows combinations of the exposure types reported in the refinery with the resulting frequency of distresses. These combinations are not cumulative, i.e., the combination of wetting and drying with gases does not include the individual observations from cyclic wetting and gases, but the recorded distresses from footings exposed to these two in combination only. Only major combinations, which account for more than 93% of the total recorded distresses, are shown in the table. It is clear from Table 1 that wetting and drying is a dominant cause of deterioration in the refinery. Footings exposed only to wetting cycles suffered 27% of the total number of distresses; and when cyclic wetting is combined with other exposures, this is increased to 75%. Heat, gases, and vibrations were found to play a minor role in the damage recorded, either when considered individually or when combined with cyclic wetting. Distresses due to exposure to gases, heat, or vibrations were 0.2%, 1.2% and 0.3%, respectively of the total recorded. Chemicals produced 2.3% of the distresses when considered individually. However, when chemicals were combined with cyclic wetting or seawater, they produced almost 10% of the distresses. Salt water alone or combined with gases or chemicals caused 22% of the distresses observed. This indicates the detrimental results from washing the floors by seawater.

Table 1 Frequency of Distresses at Different Exposure Conditions

	TYPES OF DISTRESSES													
Exposure type	spall	crack	Corr	disintg	Delamn	efflor	crazg	exp st	stains	seg.	hn.com	total	%	# pedls expsd.	Distress/ pedestal
wetting & drying	69	178	28	28	91	48	41	26	34	0	4	547	26.7	265	2.06
wet & dry+salt	21	32	9	13	28	5	8	4	5	3	3	131	6.4	41	3.20
wet & dry+chemicals	18	20	6	7	18	18	2	6	15	3	2	115	5.6	59	1.95
wet & dry+vibrations	13	18	1	0	23	11	1	1	0	0	0	68	3.3	29	2.34
wet & dry+heat	0	1	0	0	1	0	4	0	1	0	0	7	0.3	5	1.40
wet & dry+gases	0	3	0	0	2	0	0	0	0	0	0	5	0.2	4	1.25
wet & dry+salt+chem	13	15	8	8	14	19	5	8	8	2	0	100	4.9	28	3.57
wet & dry+salt+gas	18	69	2	17	63	8	11	2	16	2	3	211	10.3	123	1.72
wet & dry+salt+vibr.	3	2	0	0	4	1	1	0	2	0	0	13	0.6	8	1.63
wet & dry+chem+vibr.	12	20	2	5	14	8	0	1	8	0	0	70	3.4	43	1.63
wet & dry+chem+gas	0	2	0	0	2	0	0	0	3	0	0	7	0.3	4	1.75
wet & dry+vibr.+gas	0	1	0	0	1	0	0	0	0	0	0	2	0.1	1	2.00
wet & dry+salt+chem+vibr.	4	4	1	1	3	2	1	1	0	0	0	17	0.8	5	3.40
wet & dry+salt+chem+heat	0	4	1	0	2	5	1	0	2	0	0	15	0.7	13	1.15
wet & dry+chem+vibr.+gas	4	4	1	1	2	0	0	0	2	0	0	14	0.7	6	2.33
wet & dry+salt+steam+gas	1	3	0	0	4	0	1	0	1	0	0	10	0.5	6	1.67
wet & dry+salt+heat+gas	1	1	0	1	0	0	0	0	1	0	0	4	0.2	2	2.00
wet & dry+heat+vibr.+steam	0	1	0	0	0	0	0	0	0	0	0	1	0.0	1	1.00
wet & dry+salt+chem+gas	14	25	0	8	24	4	1	0	9	0	1	86	4.2	71	1.21
wet & dry+salt+chem+gas+heat	2	6	1	3	3	0	1	2	2	0	0	20	1.0	9	2.22
wet & dry+salt+chem+gas+vibr.	18	21	1	1	10	6	0	0	9	0	0	66	3.2	31	2.13
wt & dy+salt+chem+gas+steam	2	0	0	0	1	0	0	0	0	0	0	3	0.1	2	1.50
wt & dy+salt+chem+heat+steam	0	1	0	0	1	0	1	0	0	0	0	3	0.1	1	3.00
w&d+salt+chem+ht+stm+vibr+gas	2	3	1	0	1	0	0	0	3	0	0	10	0.5	3	3.33
gas	1	1	0	0	1	0	0	0	0	0	0	3	0.1	1	3.00
chemicals	3	10	0	0	12	10	2	0	9	0	0	46	2.2	25	1.84
heat	7	10	2	1	5	0	0	0	0	0	0	25	1.2	17	1.47
vibrations	1	2	1	1	1	0	0	1	0	0	0	7	0.3	2	3.50
heat+vibrations	0	1	0	0	0	0	0	0	0	0	0	1	0.0	1	1.00
chemicals+vibrations	4	17	2	2	10	3	9	0	3	0	0	50	2.4	21	2.38
salt+cem+vibr.	1	1	1	1	1	1	0	1	1	0	0	8	0.4	1	8.00
None	43	24	27	40	44	11	4	26	29	0	0	248	12.1	186	1.33
Total	275	500	95	138	386	160	94	79	163	10	13	1913	93.5	1014	1.89
*Total as a % of **1913** given above*	14	26	5	7	20	8	5	4	9	1	1	100			

From the above, it is clear that concrete in such industrial facilities should be designed with all the necessary precautions used in water structures in the splash zone.

Table 1 shows that cracking and delamination represent almost half the observed distresses. Spalling alone was observed in 14% of the damage obtained. This is a high figure considering that the spalls were due to environmental attack and not mechanical loading. Distresses resulting directly from steel corrosion (spalling, delamination, corrosion, and exposed steel) accounted for 41.7% of the total distresses. Cracking in reinforced concrete may result from environmental attack or flexural stresses. Since the latter did not exist in the exposed part of footings under study, it would be reasonable to assume that most of the cracks are due to sulfate attack or steel corrosion. Hence, the total damage due to steel corrosion may have been up to 70% of the distresses reported. Novokshchenov [3] suggested that the corrosion process in a liquid gas/sulfur plant, which suffers similar pollution as an oil refinery, consist of two principal stages; 1, onset of rusting due to the presence of CL^- and 2, acceleration of rusting due to the synergistic effect of CL^- and industrial pollutants including SO_2 and H_2S.

Corroded steel bars that have more than 100 mm concrete cover were observed. This was attributed to the penetration of water along the anchor bolt. Epoxy mortars were not common in the 1940's and 50's, hence cement mortars were used to seal anchor bolts used in fixing machinery or steel frames. Cement mortars were normally made using high water/cement ratio to increase flowability. This would result in a permeable paste that facilitates water penetration to surrounding steel bars causing the observed corrosion. Large spalls were observed due to corrosion of equipment chassis. The poor construction detail of the joint between the chassis and the concrete footing allowed water penetration causing corrosion to the steel chassis and hence expansion. It is clear that protecting the concrete against water ingress would noticeably improve its long-term performance. This may be more effective than using other precautions such as increasing concrete cover or using epoxy coated bars.

Table 1 shows that pedestals not exposed directly to any attacks suffered 12.1% of the total distresses. This is high percentage considering that the pedestals were apparently not exposed to detrimental factors. Many of these pedestals suffered from exposed steel due to inadequate concrete cover. The number of pedestals suffering from each exposure type is shown for comparison. The highest frequency occurred for pedestals exposed to cyclic wetting alone, 265 pedestals. Pedestals exposed to seawater attack, with and without gases, and chemicals represented more than 20% of the total pedestals (1066). This shows that more than 75% of the pedestals were exposed to fresh or seawater in combination with other pollutants. It is interesting to note that this is almost the same percent of distresses recorded for those pedestals when compared to the total. The number of distresses per pedestal for each exposure type is shown. This is an indicative figure, however it shows the degree of severity for each exposure type. The average was 1.9 distress per pedestal, based on the exposure types included in the table. Based on this rate, salty water appears more aggressive than fresh water. The former produced 3.2 distress per pedestal, while the later produced 2.1.

Effect of Unit Operational Conditions on Frequency of Distresses

Table 2 shows the frequency of each distress type in all units and the percent of these distresses to the total of each type. PLAT suffered the most, followed by FU2, Bitumen and CDU3. The least number of distresses was recorded for the Washery.

Table 2 Frequency and Percentage of Occurrence for Distresses in Each Unit

Distress	PRODUCTION UNITS														
	Bitumen		CDU#3		FU#1		PLPH		Washery		PLAT		FU#2		Total
	No.	%	No.	%	No.	%	No.	%	No.	%	No.	%	No.	%	No.
spalling (high)	5	5.2	13	13.5	13	13.5	15	15.6	3	3.1	41	42.7	6	6.3	96
spalling (mid.)	7	13.7	11	21.6	8	15.7	2	3.9	3	5.9	12	23.5	8	15.7	51
spalling (low)	31	21.8	28	19.7	10	7.0	22	15.5	6	4.2	28	19.7	17	12.0	142
cracking (high)	11	8.1	18	13.3	15	11.1	13	9.6	7	5.2	50	37.0	21	15.6	135
cracking (mid.)	32	20.6	32	20.6	11	7.1	10	6.5	3	1.9	31	20.0	36	23.2	155
cracking (low)	66	22.8	45	15.5	31	10.7	10	3.4	7	2.4	48	16.6	83	28.6	290
corrosion (high)	0	0.0	1	1.8	5	8.9	6	10.7	0	0.0	41	73.2	3	5.4	56
corrosion (mid.)	2	8.0	0	0.0	3	12.0	2	8.0	0	0.0	7	28.0	11	44.0	25
corrosion (low)	6	33.3	2	11.1	0	0.0	1	5.6	0	0.0	3	16.7	6	33.3	18
disintegration (high)	0	0.0	3	10.3	2	6.9	4	13.8	2	6.9	18	62.1	0	0.0	29
disintegration (mid.)	1	2.0	6	12.2	1	2.0	3	6.1	1	2.0	28	57.1	9	18.4	49
disintegration (low)	17	26.2	16	24.6	6	9.2	2	3.1	0	0.0	15	23.1	9	13.8	65
delamination (high)	14	12.7	26	23.6	9	8.2	9	8.2	4	3.6	31	28.2	17	15.5	110
delamination (mid.)	24	22.4	25	23.4	9	8.4	6	5.6	8	7.5	11	10.3	24	22.4	107
delamination (low)	37	20.7	37	20.7	18	10.1	16	8.9	4	2.2	41	22.9	26	14.5	179
efflorescence	8	4.9	26	15.9	14	8.5	0	0.0	22	13.4	34	20.7	60	36.6	164
crazing	14	14.7	1	1.1	1	1.1	0	0.0	0	0.0	33	34.7	46	48.4	95
exposed steel	5	6.1	1	1.2	8	9.8	9	11.0	0	0.0	40	48.8	19	23.2	82
staining	48	27.9	27	15.7	11	6.4	7	4.1	5	2.9	58	33.7	16	9.3	172
segregation	0	0.0	7	53.8	2	15.4	1	7.7	0	0.0	1	7.7	2	15.4	13
honey combing	2	14.3	4	28.6	2	14.3	1	7.1	1	7.1	4	28.6	0	0.0	14
Total distresses	330		329		179		139		76		575		419		2047
No. of Pedestals	225		222		125		61		40		197		194		1064
distress/pedestal	1.467		1.482		1.432		2.279		1.900		2.919		2.160		1.924

However, this may not be representative since the size of the unit and the number of footings involved must be considered. Table 2 also shows the ratio of the total number of distresses recorded, 2047 to the number of inspected pedestals, 1064 (two pedestals were ignored due to incomplete information). This ratio, which is indicative of the damage, has an overall average of 1.9 distresses per pedestal. Units having higher values will indicate more severe damage than the average. The opposite is true for units having lower values. It is interesting to note that PLAT and PLPH produced the highest damage per pedestal, 2.9 and 2.3, respectively. These two had no maintenance during their operational life and were out of operation for 4 to 8 years. This may explain the high ratio of damage reported. Unit FU2 produced slightly higher ratio than the average. It is worth mentioning that this unit was subjected to harsh operating environment and chlorine spray for years. The Washery produced an average damage ratio while the bitumen plant, CDU3 and FU1, were below average.

Table 2 also shows that the highest occurrence of severe distresses; such as high spalling, cracking, corrosion, disintegration, delamination, as well as exposed steel, staining, and honeycombing, occurred in the PLAT. The highest occurrence of medium degree distresses occurred in FU2, while low degree damages occurred the most in the bitumen plant (BIT). This supports the earlier remark that the lack of maintenance in the PLAT affected the damage observed. BIT, which was exposed to seawater washing and located near the Gulf shore, produced the highest occurrence of low-degree corrosion, equal to the FU2, which is 20 years older. It also produced the highest occurrence of low-degree disintegration and spalling. The bitumen plant was unique in its highly polluted atmosphere, which caused high staining, cracking, and delamination. The Washery unit suffered the most from efflorescence. This was due to its exposure to salts and chemical attack from the caustic soda and related chemical operations in the unit. The PLPH showed a high degree of damage related to steel corrosion, though the unit was operated apparently in dry conditions. This occurred basically in the footings supporting the steel shed covering the unit.

The unit was close to the Gulf shore, which exposes it to chloride attack. CDU3 suffered the same occurrence ratios as the PLAT for medium-degree spalling, cracking, and low-degree spalling, cracking, disintegration, and delamination. However, it suffered relatively few observations of high-degree distresses. These results are interesting since both units are located in the same area, have the same age and apparently made of concrete with the same specifications. The only major difference in their operational conditions was that the PLAT has not received any maintenance during its service life and was out of operation for 8 years while CDU3 is still in operation.

Footing Classification Number

Table 3 shows trials done to determine the weightings for the distresses observed and Figure 2 shows the ratio between the number of footings falling within each group as identified in limits given earlier and the actual number of footings classified within the group visually. This ratio is identified as the degree of confidence. A gradual improvement was obtained in each trial. Trial #3 agreed with the overall classification by the surveying team with degrees of confidence of 91%, 84% and 86% for pedestals classified as poor, medium and good, respectively. In other words, 91% of the footings classified visually to be in poor conditions produced FCN values greater than 22. The same is applied to other values. Such an agreement was considered to be sufficient.

Table 3 shows that by reducing the weightings for low delamination, disintegration, and efflorescence and increasing the weightings for high and medium cracking, better confidence limits were obtained, particularly for medium and poor quality pedestals. The weightings shown in trial #3 reflect the relative damage resulting from each distress. For example, the damage from medium disintegration, delamination, and cracking is more severe than high spalling and corrosion. This agrees with the definitions of these distresses, since disintegration and delamination involve greater area of the footing than spalling and corrosion. Efflorescence causes twice as much damage as crazing, staining, segregation, honey combing, and exposed steel. Efflorescence indicates attack by salt, which may be

Table 3 Weighting points for different trials

	DEGREE				
Distress	High	Medium	Low	Existing	not exist.
		Trial #1			
Spalling	6	4	2	NA	0
Cracking	6	4	2	NA	0
Corrosion	6	4	2	NA	0
Disintegration	10	8	6	NA	0
Efflorescence	NA	NA	NA	4	0
Delamination	10	8	6	NA	0
Crazing	NA	NA	NA	1	0
Exposed steel	NA	NA	NA	1	0
Staining	NA	NA	NA	1	0
Segregation	NA	NA	NA	1	0
Honey combing	NA	NA	NA	1	0
		Trial #2			
Spalling	6	4	2	NA	0
Cracking	10	8	2	NA	0
Corrosion	6	4	2	NA	0
Disintegration	10	8	6	NA	0
Efflorescence	NA	NA	NA	4	0
Delamination	10	8	2	NA	0
Crazing	NA	NA	NA	1	0
Exposed steel	NA	NA	NA	1	0
Staining	NA	NA	NA	1	0
Segregation	NA	NA	NA	1	0
Honey combing	NA	NA	NA	1	0
		Trial #3			
Spalling	6	4	2	NA	0
Cracking	10	8	2	NA	0
Corrosion	6	4	1	NA	0
Disintegration	10	8	2	NA	0
Efflorescence	NA	NA	NA	2	0
Delamination	10	8	2	NA	0
Crazing	NA	NA	NA	1	0
Exposed steel	NA	NA	NA	1	0
Staining	NA	NA	NA	1	0
Segregation	NA	NA	NA	1	0
Honey combing	NA	NA	NA	1	0

NA not applicable

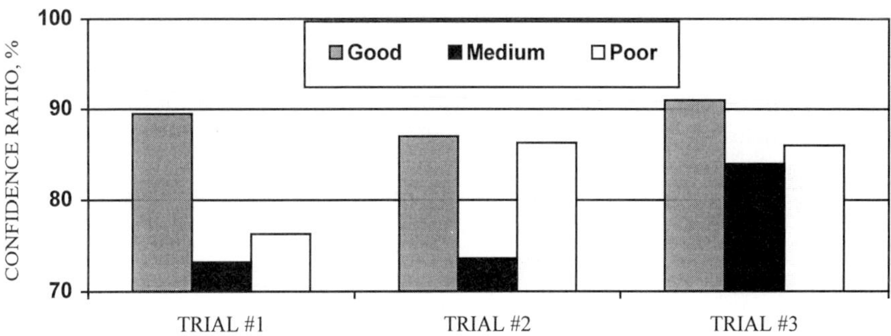

Figure 2 Variations in confidence ratios for each trial of weighing points

damaging. In the same manner, high degree cracking causes ten fold as much damage as crazing, and so on. Based on this data, priority should be given to repair of footings with distresses having high weightings to minimize cost and time for repair work and prolong the useful life of the concrete footings.

CONCLUSIONS

A method was developed to assess concrete footings in industrial plants quantitatively. Concrete footings in oil refineries are exposed primarily to cyclic wetting, and should be designed for such exposure. The effects of heat, vibrations, chemicals, and gases, though harmful to concrete, are not major factors in its deterioration. Protection from water attack should be provided during the design stage by the use of dense concrete coated with a protective layer to stop water ingress. The joints between the equipment and the footings supporting them should be carefully detailed to avoid water penetration. The use of seawater in washing the refinery floor, made of reinforced concrete, is detrimental and should be stopped. This may be common in the Arabian Gulf due to the unavailability of fresh water. Routine inspection and repair works are very important for keeping the degree of distresses, within controllable limits. In this project, the units that were put out of operation gave greater distress than units continuously maintained.

REFERENCES

1. DAOUD, O E K AND IBRAHIM, I A. Impact of environmental and operating condition in oil refineries on concrete properties. ACI Mat. Jour, 1996, Vol. 93, No.4, pp. 307-318.

2. O'CONNOR, J P. Middle eastern concrete deterioration: Unusual case history. J. of Performance of Constructed Facilities, ASCE, 1994, Vol. 8, No. 3, pp. 201-212.

3. NOVOKSHCHENOV, V. Deterioration of reinforced concrete in the marine industrial environment of the Arabian Gulf-A case study. Mats. and Strucs., 1995, Vol. 28, No. 181, pp. 392-400.

4. ACI COMMITTEE 201, "Guide for making a condition survey of concrete in service", ACI 201.1R-68, Revised 1984, Manual of Concrete Practice.

5. DAOUD, O K, AL-QAZWEENI, J AND HIGAB, K. Causes of deterioration and repair measures in concrete foundations at a Kuwaiti Refinery. Progress report, 1990, KISR 3468, Kuwait Inst. for Scientific Res., Kuwait.

6. AL-QAZWEENI G AND DAOUD O K. Concrete deterioration in a 20 years old structure in Kuwait, Cem. and Conc. Res., 1991, Vol. 21, No. 6, pp. 1155-1164.

7. DAOUD, O E K, AND IBRAHIM, I A. Condition Survey and Damage Assessment of Concrete Footings in Industrial Facilities. J. of Mat. in Civil Eng., ASCE, 1997, Vol. 9, No. 4, pp. 161-170.

STRUCTURAL ASSESSMENT OF DETERIORATED REINFORCED CONCRETE BEAMS UNDER MARINE ENVIRONMENTS FOR MORE THAN 20 YEARS

H Yokota

T Fukute

H Hamada **A Mikami**

Port and Harbour Reasearch Institute

Japan

ABSTRACT. This study, which forms a part of long term exposure test programme of concrete members under marine environments, deals with physical and chemical examination of deteriorated reinforced concrete (RC) beams after 23 years of exposure. The two types of beams were tested to obtain the degree of deterioration and mechanical behaviours. The relationship between the materials degradation and overall performance of the beams was examined and a structural assessment of the deteriorated beams was discussed. From the results and discussion, it is concluded that degradation of concrete and the decrement in load carrying capacity of the beams did not significantly appear up to 23 years of exposure.

Keywords: Chloride, Deterioration, Marine environment, Reinforced concrete (RC) beam, Corrosion of bars, Load carrying capacity.

Dr Hiroshi Yokota is a Director of Structural Mechanics Laboratory, Port and Harbour Research Institute, Ministry of Transport, Yokosuka, Japan. He specialises on structural behaviours and design of RC, PC, and steel-concrete composites. His recent research interests include the effect of deterioration of materials to overall performance of concrete structures.

Dr Tsutomu Fukute is a Director of Planning and Design Standard Division, Port and Harbour Research Institute. His main research interests include recycling of concrete, durability of marine concrete, high flowable concrete, and airport pavements. Dr Fukute has published papers widely.

Dr Hidenori Hamada is a Director of Materials Laboratory, Port and Harbour Research Institute. His main research interests include long-term durability, maintenance of concrete structures, and development of new construction materials.

Mr Akira Mikami is a Research Engineer in Structural Mechanics Laboratory, Port and Harbour Research Institute. His main research interests include structural design of concrete port structures and non-destructive test to deteriorated concrete structures.

INTRODUCTION

Among reinforced concrete (RC) marine structures, many examples of heavy deterioration and degradation have been collected particularly because of chloride attacking by severe sea water splashing or sea breezing. For ensuring the reliability of existing structures for a required period of service life, an evaluation of deteriorated structures, especially a structural assessment is very important [1]. Reliable service life predictions can also be made by evaluating the durability performance of the materials in that environment. However, the effect of the deterioration condition, such as crack formation, corrosion of steel bars, etc. to the load carrying capacity of deteriorated members has not been made clear.

Port and Harbour Research Institute (PHRI) has been conducting several series of exposure tests in marine environment including RC beams described in this present paper. Many test beams with various research parameters including crack condition, crack width, cover depth of embedded steel bars, cement types, etc. have been exposed to marine environment for a certain period. Up to the exposure time of 10 years, the test results have been discussed [2]. Following this former work, loading tests of deteriorated RC beams were executed after 23 years of exposure and the relationship between the degree of materials degradation and load carrying capacity of beams was examined. Together with the former test results, the trend of deterioration with exposure time in marine environments was discussed.

EXPERIMENTAL PROCEDURES

Test Beams

All five test beams have the identical length of 2400 mm, but their cross sections and bar arrangements are varied according to the type of beam designated 'S' and 'L' as shown in Figure 1. Beams S and L were reinforced with 13 mm and 16 mm diameter deformed bars with a yield strength of 363 N/mm^2, respectively. Deformed bars of 6 mm in diameter were used in both types of beams as stirrups with a spacing of 100 mm.

All the beams were made with normal concrete without any surface coatings or mineral admixtures. Ordinary Portland cement was used and its unit weight was 300 kg/m^3. The water-to-cement ratio and the air content of the concrete were designed to be 0.68 and 4%, respectively. River sand and crushed stone with the maximum size of 20 mm were used.

After placing of concrete, the beams were moisture cured for 1 day and then demoulded, followed by curing in air until the start of exposure. Concrete cylinders of 150 mm diameter and 300 mm high were also made for obtaining compressive strength of concrete.

Exposure Conditions

All the beams and cylinders have been exposed to real marine environments at the Port of Sakata which is located north-west of Japan (38°56' N, 139°47' E) facing to the Sea of Japan. While the annual average temperature there is around 11.9°C, the minimum temperature between December and March reaches below zero mostly every day, which could bring freezing and thawing action. Furthermore in winter, the daily maximum wind velocity is more than 25 m/s, which produces much splashing there.

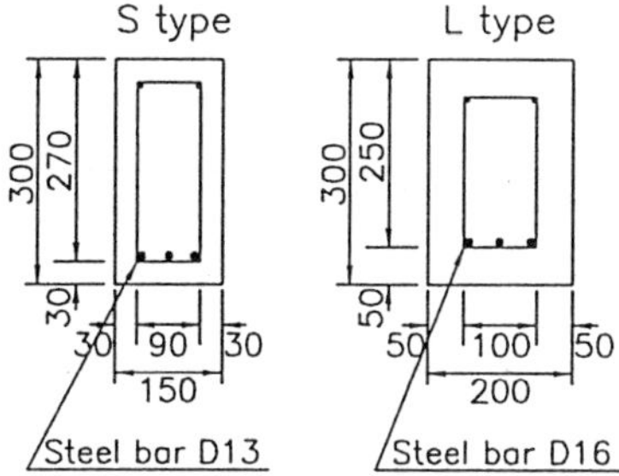

Figure 1 Cross section of test beams

The beams and cylinders were placed in tidal zone just in front of a caisson type quaywall two months after placing of concrete, where the beams have been subject to wet and dry condition alternately due to the tidal action. After 23 years of exposure, the beams and the cylinders were taken out for laboratory tests.

Laboratory Tests

At the loading test, the beams were simply supported with a span of 2100 mm. Load was applied symmetrically at two points 700mm apart (see Figure 2 in which a triangle symbol and a closed circle represent a support and a load applying point respectively). The applied load was monotonically increased up to the maximum crack width of 0.25 mm. Since it was not possible to measure strains of embedded bars, the crack width was used to estimate strain of the bars. According to the formula in JSCE Standard Specification, tensile stresses of the bars in Beams S and L were 363 N/mm^2 (equal to the yield stress) and 230 N/mm^2 respectively at the maximum crack width of 0.25 mm. Once the maximum crack width reached 0.25 mm, the load was applied repeatedly controlled by the midspan deflection.

After the loading test, all of the embedded reinforcing bars were taken from the concrete and their corroded areas were observed. The chloride analysis was carried out according to JCI-SC4. Hot nitric acid solution was used as solvent. The chloride content is then expressed as acid-soluble chloride content. The samples were taken from three different portions of the beam: near the surface, around the tensile bar, and the centre of cross section of the beam.

RESULTS AND DISCUSSION

Deterioration of Beams and Corrosion of Embedded Bars

Figure 2 shows the appearance of Beam S1 before the loading tests after cleaned up with sandblasting. Slight deterioration appeared partly on the surface of the beam: rust stains, uncovered aggregates, and rough surfaces. At the scaled parts, air bubbles of 3 to 10 mm in diameter appeared. The maximum width of cracks was about 2 mm. Other beams showed almost the same degree of deterioration. Together with the results of 20 years' autopsy, it is considered that heavy deterioration such as spalling of the cover concrete did not appear on the outer surface of the beams up to 23 years of exposure to marine environment.

Table 1 Corroded area of reinforcing bars (%)

	GRADE	1	2	3	4	TOTAL
S1	Tension bars	7	9	23	1	40
	Comp. bars	9	22	37	23	91
	Stirrups	14	11	16	7	48
S2	Tension bars	8	13	43	5	68
	Comp. bars	21	23	56	0	100
	Stirrups	15	12	12	0	39
L1	Tension bars	3	10	39	0	52
	Comp. bars	6	6	43	1	56
	Stirrups	2	4	7	10	23
L2	Tension bars	13	6	20	0	38
	Comp. bars	44	25	12	0	80
	Stirrups	8	8	4	0	20
L3	Tension bars	21	18	53	0	92
	Comp. bars	31	20	47	0	98
	Stirrups	4	14	0	0	18

Grade
1: Slight corrosion; brown coloured dot rust on the surface
2: Medium corrosion; brown coloured very thin corrosion partially
3: Severe corrosion; black coloured product on the surface
4: Heavy corrosion; reduction of the cross sectional area of bars

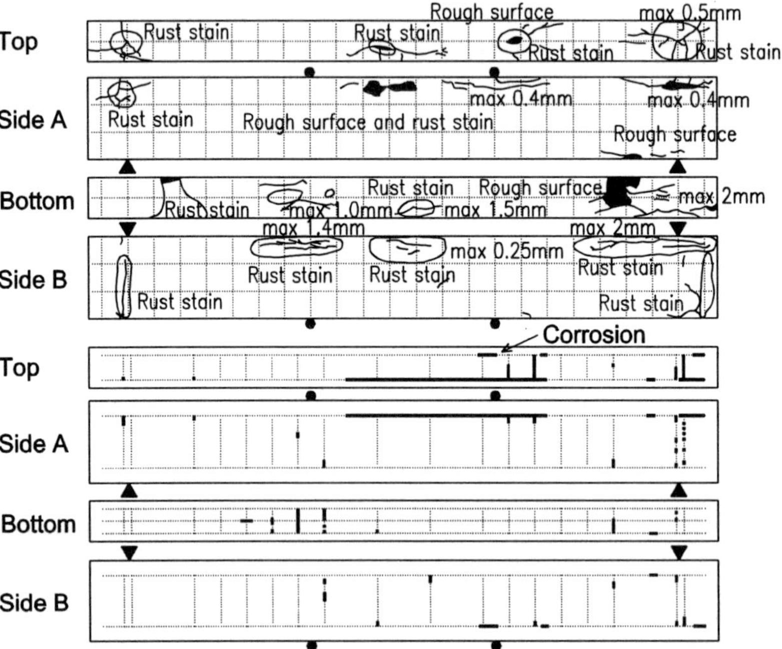

Figure 2 Appearance of beams and corrosion of the embedded bars

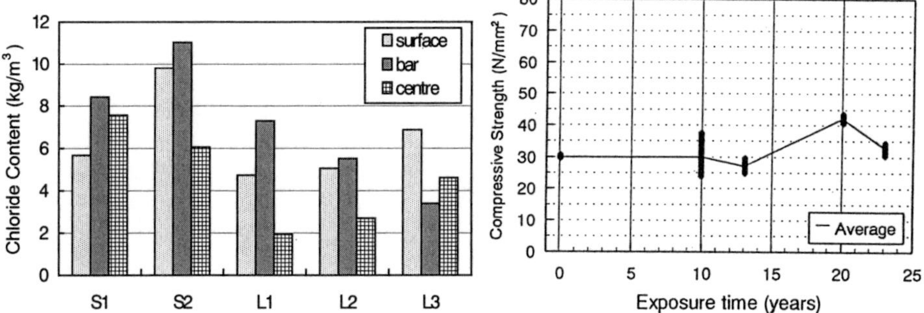

Figure 3 Chloride ion content

Figure 4 Compressive strength of concrete

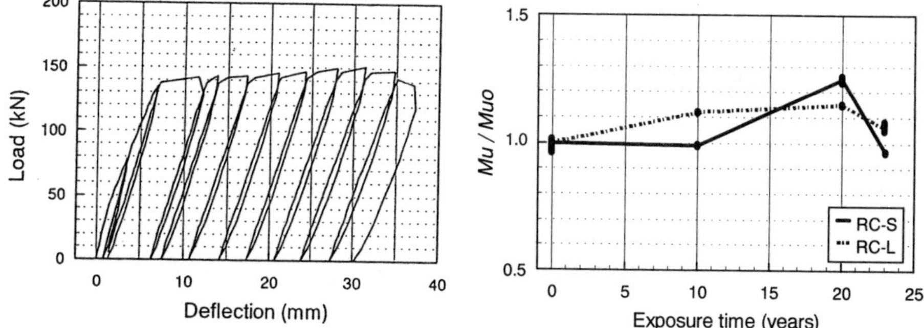

Figure 5 Load vs midspan deflection

Figure 6 Changes in ultimate flexural moment

Corrosion due to chloride attack of embedded bars occurred at almost the same position as the visibly deteriorated part. Table 1 summarises the corroded area of bars in terms of the ratio to the total area of bars.

It shows that most of the corrosion was categorised among Grades 1 to 3. Heavy corrosion was significantly observed only in compression bars of S1 and stirrups of L1. Those bars were located slightly above the highest water level. Thus, it is considered that longer drying period can accelerate the degree of corrosion.

Chloride Ion Content

Figure 3 shows the chloride ion content at the three parts of the beams. The amount of chloride ion was rather large compared to the value of 1.2 kg/m^3 which is considered to initiate the bar corrosion. Although the large amount of chloride ion has penetrated into the beams, the corrosion of bars did not occur on their all surfaces. Thus, corrosion of embedded bars did not start only with the factor of the amount of chloride ion in concrete.

Compressive Strength of Concrete and Load Carrying Capacity of Beams

The change in compressive strength of concrete against the exposure time is shown in Figure 4. The compressive strength has been obtained in average of 2-5 cylinder specimens. The compressive strength has showed no significant changes and not decreased below the design compressive strength of 24 N/mm^2.

The relationship between the applied load and the midspan deflection of Beam L3 is shown in Figure 5. A crack was initiated at the pure bending span at the beginning of the load application, and then the number of cracks increased. Flexural cracks were dominant and crushing of concrete occurred at ultimate load in all the beams. Thus, the failure mode of the beams was a tension controlled flexural failure. The load-deflection relationship also implied the flexural failure. That is, up to the probable yield of bars, the load-deflection curve was almost linear, and after that, further load carrying capacity was not recorded in spite of the hardening of bars. The reason behind that is not clear, but the failure of bond between the bars and the concrete might occur due to partial corrosion.

According to the JSCE Standard Specification, the calculated ultimate load of L beams is 141.4 kN on condition that each safety factor was set to be 1.0. The ultimate load was slightly larger than the calculated ones; thus, it will be possible to estimate the load carrying capacity up to 23 years with the conventional calculation method. This was also confirmed with S beams. Although much corrosion was observed in the bars, no significant decreasing in ultimate load did not occur. This is because the heavy corrosion of bars occurred mainly in compression bars and stirrups. It is also concluded that slight and medium corrosions do not have much effect on structural performance of the beams.

Figure 6 shows the changes in ultimate flexural moment of the beams against the exposure time. In this figure, the ultimate flexural moment (M_u) is expressed as the ratio to that before the start of exposure (M_{uo}). After 20 years of exposure, these ratios slightly decreased as the exposure time increased. However, a conclusion that load carrying capacity of the beams decreased after 20 years of exposure cannot be derived from the present test results because of the limited number of beams. Further tests for longer period of exposure will be required.

CONCLUSIONS

1. There was no significant deterioration of concrete after 23 years of exposure. The concrete beams showed good to excellent performance after 23 years of exposure from the viewpoint of mechanical behaviours, though corrosion of bars occurred at the large amount of areas.

2. The load carrying capacity of the exposed beams became slightly inferior to the former results of 10 years of exposure. The ultimate flexural moment after 23 years of exposure, however, can be estimated with the calculation method according to the JSCE Standard Specification.

REFERENCES

1. SCHUPACK, M AND O'NEILS, E F. Observations and testing of post-tensioned beams exposed to severe weathering for 33 years. American Concrete Institute, SP-163, 1996, pp.383-407.

2. HAMADA, H, OTSUKI, N, AND HARAMO, M. Durabilities of concrete beams under marine environments exposed in Port of Sakata or Kagoshima (after 10 years' exposure), Technical Note of the Port and Harbour Research Institute, No.614, June 1988.

RECOMMENDED METHOD FOR EARLIER INSPECTION OF CONCRETE QUALITY BY NON-DESTRUCTUVE TESTING

T Soshiroda

K Voraputhaporn

Shibaura Institute of Technology

Japan

ABSTRACT. In the previous report, it is concluded that the optimum age for ultrasonic method to evaluate quality of concrete is 24-h and the one for rebound hardness method is 3-day practically, and a method combined 1-day pulse velocity with 3-day rebound index is recommended for the earlier inspection of concrete quality in structures. This study deals with the earlier estimation of 28-day strength by the recommended method. The estimation equation has been established by multiple regression analysis of test results on 27 mixes of normal concrete that had a range of 20-65 MPa strengths. After three years applying the equation to a series of test results the mean error of the estimation was 9% approximately. This method can provide in steps early and versatile information about the potential quality of concrete in structures without destructive testing.

Keywords: Concrete in structures, Quality inspection, Earlier evaluation, Nondestructive testing, Optimum age for testing, Ultrasonic method, Rebound hardness method, Combined method, Estimation of 28-day strength

Professor Tomozo Soshiroda is Leader of the Building Materials and Construction Engineering, Graduate School, Shibaura Institute of Technology, Tokyo, Japan. He specializes in the quality evaluation methods of concrete in structures, with particular reference to nondestructive testing and earlier evaluation. Professor Soshiroda has published many papers related to the concrete technology, especially on the anisotropy of concrete, and served on many Technological Committees.

Mr Kongkij Voraputhaporn is a graduate student, Shibaura Institute of Technology. He studies on the application of nondestructive testing to the quality inspection of concrete in situ.

INTRODUCTION

For the quality control of concrete in structures the two following methods are very advantageous provided they are reliable in reasonable extent: (1) nondestructive testing, (2) earlier evaluation testing. Besides, when a method can serve both of the above two, "*It kills two birds with one stone*".

Elvery and Ibrahim has shown the possibility of prediction of 28-day strength from the ultrasonic pulse velocity at earlier age[1]. Meanwhile it is known that the rate of development of pulse velocity slows down more rapidly than the rate of development of compressive strength as the age of concrete increases [1] [2].

From these facts, the investigation to obtain the optimum age of concrete for nondestructive testing was undertaken. As the results it was reported previously [3] that the optimum age for measuring the ultrasonic pulse velocity to evaluate quality of concrete is 24-hour and the one for measuring the rebound index is 3-day practically, and the earlier the better. It was required to be "the optimum age" that satisfied the sensitivity to the changes in concrete quality and the stable correlation to 28-day strength of concrete. As one of the conclusion in the previous report, a nondestructive method combined 1-day pulse velocity with 3-day rebound index is recommended for the earlier inspection of concrete quality in structures.

In the presented paper experimental results obtained during 1994~1995 and 1998 are discussed to confirm the validity of the recommended method and to establish an estimation equation of 28-day potential strength of concrete for reference.

EXPERIMENT

Experimental Outline

Experimental outline is shown in Table 1.

Table 1 Experimental outline

SERIES	W/C, %	STRENGTH, MPa	TESTING ITEMS AND AGE, day		
		range	Vp	R	F
1994	60.0, 65.0, 65.5, 70.0	28.8 ~ 40.9	1, 3, 28	3, 28	28
1995	30.0, 42.0, 67.9, 70.0	21.7 ~ 64.9			
1998	27.6, 28.0, 30.1, 54.7, 55.6, 62.5, 71.4, 83.8	14.0 ~ 81.1			

*Abbreviations:Vp = Pulse velocity R = Rebound index F = Compressive strength
**Materials used : Ordinary Portland Cement, River sand and River gravel

Testing

The velocity measurements were made by using a "Pundit" ultrasonic apparatus. The rebound index measurements were made by using a "Schmidt Hammer/type NR" with a test anvil.

Five cubes (150 mm) for each mix were used for all tests and all of them were cured in the water at 20±1°C.

Procedure of Discussion

Experimental results are discussed according to Figure 1.

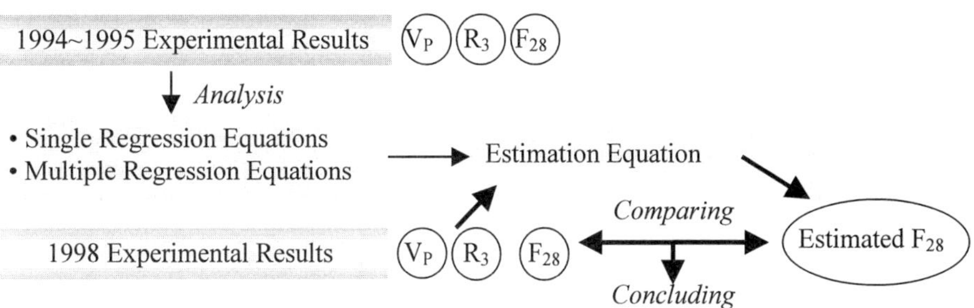

Figure 1 Procedure of discussion of results

RESULTS AND DISCUSSION

Regression Analysis of Test Results (1994 ~ 1995)

The test results obtained in 1994 ~ 1995 were analyzed for single regression and multiple regression. The analysis results are shown in Table 2. Pulse velocity and rebound index at any age give good correlation to 28-day strength. Pulse velocity gives less correlation than rebound index, especially at 28-day. Combined method at early age gives better correlation than the one at 28-day.

The Relationship Between Strength at 28-Day and Pulse Velocity at 1-Day

The results are shown in Figure 2. The regression line and equation obtained from 1994~1995 data are shown and 1998 data are plotted in the figure. Besides, as reference, the curve after Elvery *et al.* [1] and 1991 data [4] are shown here. The regression line agrees approximately with Elvery's curve and the points plotted from the other two sources tend to gather near by the two lines. In the range of low velocity (low strength) the regression line ('94~'95) tends to be lower than Elvery's curve, because the former is higher than the latter in the aggregate/cement ratio[1].

In the case of the high strength concrete, it seems that the velocity has a limit like as "the ceiling". These facts suggest that the regression line should be chosen from curves in the wider range of strengths.

Table 2 Analysis results of series 1994~1995

NDT METHOD		N	REGRESSION EQUATION	INDICATOR		
				Correlation coefficient	Standard deviation, MPa	Mean error, %
Pulse velocity	1-day	27	$F_{28}=44.52V_{P1}-126.83$	0.912	4.91	12.65
	28-day	27	$F_{28}=54.18V_{P28}-206.27$	0.725	8.23	18.71
Rebound index	3-day	27	$F_{28}=1.61R_3-1.37$	0.957	3.46	6.88
	28-day	27	$F_{28}=1.47R_{28}-16.85$	0.962	3.28	8.02
Combined Method	1+3-day*	27	$F_{28}=14.60V_{P1}+1.16R_3-44.45$	0.967	3.03	5.93
	28-day	27	$F_{28}=0.63V_{P28}+1.46R_{28}-19.31$	0.962	3.28	8.01

Suffixes indicate ages   * 1-day pulse velocity and 3-day rebound number

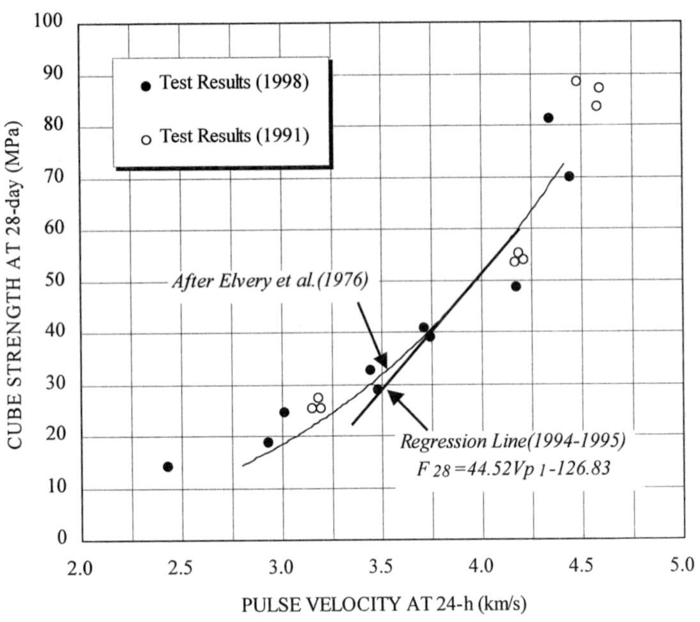

Figure 2 Relationship Between Strength at 28-days and Pulse Velocity at 24 hours

The Relationship Between Strength at 28-day and Rebound Index at 3-day

Figure 3 shows the regression line and equation obtained from 1994~1995 data are shown and 1998 data are plotted in the figure. The points plotted from 1998 data tend to agree approximately with the regression line in the range of 30~70 MPa strengths, but tend to lower in the range of strengths less than 20 MPa. The new regression line obtained from data in wider range of strengths should be required to get better precision.

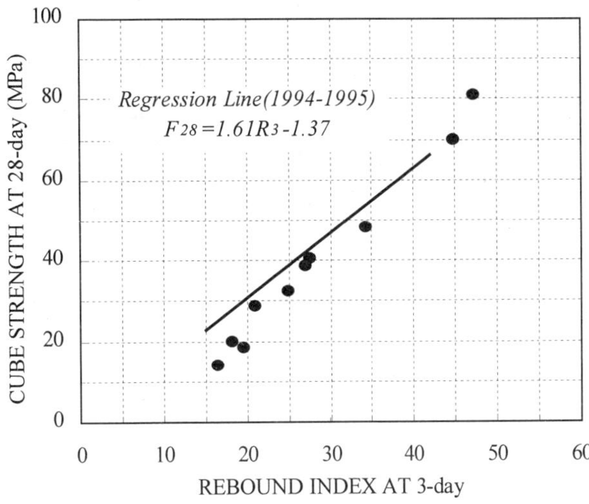

Figure 3 Relationship between strength at 28-day and rebound index at 3-day

28-day strength Estimation from Nondestructive Testing Measurement at Early-Age

Using the multiple regression equation, 28-day strength was estimated from the 1-day pulse velocity and the 3-day rebound index obtained in 1998. The results comparing with the actual 28-day strength are shown in Figure 4 (A).

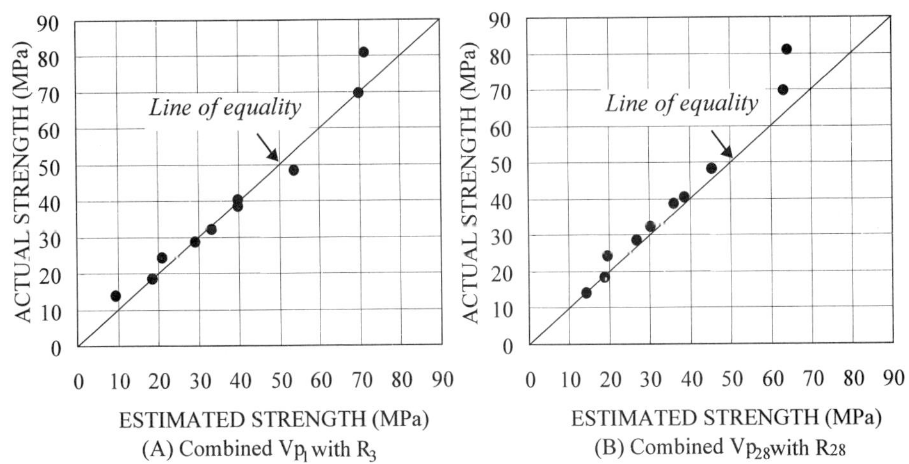

Figure 4 Comparison of estimated 28-day strength with actual 28-day strength

Most of points plotted tend to agree approximately with the line of equality. It is found that the use of the combined method improves the precision to estimate 28-day strength comparing with each single method respectively. One higher strength point has about 15% error. The mean error is about 9%.

For comparing, the results of the estimation of 28-day strength from 28-day pulse velocity and rebound index are shown in Figure 4 (B). Most of points plotted also tend to agree approximately with the line of equality except two higher strength points.

CONCLUSIONS

1. The earlier estimation of 28-day strength by the nondestructive testing combined 24-hour pulse velocity with 3-day rebound index is possible with reasonable accuracy. This method can provide in steps early and versatile information about the potential quality of concrete in structures without destructive testing[6].

2. For practical use it is recommended that the estimate equation is established by the results from preliminary test. The equations shown in Table 2 are to be referred.

ACKNOWLEDGEMENTS

The authors would like to express their appreciation to Mr. Hidetoshi Sonohara and Miss Ayako Kimura for their cooperation in a part of this research.

REFERENCES

1. ELVERY, R H AND IBRAHIM, L A M. Ultrasonic assessment of concrete strength at early ages, Mag. of Concr. Res., Vol 28, No.97, 1976, pp 181~190.

2. ACI COMMITTEE 228. In-Place Methods for Determination of strength of Concrete, Materials Journal of American Concr. Inst., Vol 85, No.5, 1988, pp 451.

3. SOSHIRODA, T AND VORAPUTHAPORN, K. Optimum testing age for nondestructive evaluation of concrete quality Approach to earlier inspection, Proceedings of the 7th International Expertcentrum Conference on Non-destructive Testing and Experimental Stress Analysis of Concrete Structures, Košice, Slovakia, 1998, pp 42~47.

4. SOSHIRODA, T AND SONOHARA, H. Earlier Evaluation of Concrete Quality by Non-destructive Testing, Proceedings of the 3rd Beijing International Symposium on Cement and Concrete, Vol 2, 1993, pp 859-862.

5. SOSHIRODA, T AND KIMURA, A. Recommended method for earlier inspection of concrete quality by nondestructive testing, Conc. Res. and Techn., Vol 8, No.1, 1998, pp 11~19.

6. SOSHIRODA, T. Ultrasonic pulse velocity as an indicator of concrete quality, Proceedings of the Second International Conference on Concrete under Severe Conditions, Tromsø, Norway, Vol 3, 1998, pp 2093~2102.

NEW ULTRA-LOW PERMEABLE CONCRETE

D Ball

David Ball Group plc

United Kingdom

ABSTRACT. Aggressive media are transported into concrete by permeation, diffusion, adsorption and capillary suction. Recent research investigated concrete for two principle properties; water permeability and sorptivity.

The paper examines wetting and drying, sorptivity, wick action, pressure induced flow and water permeability in particular. The relevant transport mechanisms which influence the durability of concrete have been reviewed and include diffusion of gasses and water, diffusion of ions permeation of water under pressure and capillary suction. Close correlation has been established between these processes and the rate of carbonation and depth of chloride penetration.

Detailed microstructural analysis of modified and unmodified cement pastes have been examined. Reduction of water transport through concrete by sorptivity and pressure induced flow can be achieved, significantly improving durability.

Keywords: Porosity, Permeability, Sorptivity, Pressure induced flow, Capllarity, Particle packing, Hydration.

Mr David Ball is a Member of the Concrete Society, a Fellow of the Institute of Management and a Fellow of the RSA. He is a Council Member of the Cement Admixtures Association and formerly Assistant Engineer at Middle Level Commissioners, March, Cambs. He began his career at Sir William Halcrow and Partners, Consulting Engineers and he has worked both in the UK and overseas.

INTRODUCTION

The durability of concrete is principally related to the ability of water (with or without aggressive ions), oxygen and carbon dioxide to penetrate the concrete pore structure [1]. Aggressive media are transported into concrete by mechanisms such as permeability, diffusion, absorption and capillary suction. The work investigated concretes for two principle properties, namely water permeability and sorptivity.

Permeability is influenced by the general nature of the pore system in the cement paste component of concrete as well as the interfacial zone with aggregate particles which is often the location of microcracking [1]. Waterproofing admixtures impart a hydrophobicity to treated concrete. They are designed to limit the ingress of water into concrete through a variety of mechanisms. Water repellents which may be applied to the surface of concrete are based on silicone resins, whereas waterproofing membranes, such as bitumen coatings/rubber latex produce a tough coating when applied to concrete.

Admixtures to reduce the penetrability of concrete have been widely available since the early 1900's. Their use has been reviewed [2], however, most of the work undertaken in this area was performed in the 1920's and 1930's [3]. A review of the mechanisms of moisture movement in concrete is necessary for the influence of water repelling and waterproofing compounds to be elucidated.

The degree of saturation of pores within a concrete governs to a large extent the movement of water and the potential for ionic transport through that concrete.

The main mechanisms of water transportation through concrete are:

- absorption
- wick action
- pressure induced flow
- water vapour diffusion

WETTING AND DRYING

Wetting of concrete is normally as a result of water absorption, however, when conditions of high humidity prevail pores will fill with water through vapour diffusion. Conversely the reverse occurs when relative humidity conditions outside the concrete are lower. Repeated wetting and drying can induce salt crystallisation and loss of absorbed moisture which can cause microcracking and ultimately an increase in moisture movement.

Change in the pore structure of concrete due to drying can lead to irreversible shrinkage and it is generally thought that the smaller pores, which hold water more strongly (Table 1) are those that collapse last when concrete drying takes place.

Table 1 Relative humidity below which pore drying occurs [15]

PORE RADIUS ()	RELATIVE HUMIDITY (%)
10000	99.9
1000	99.0
100	89.9
10	34.8

Concrete containing pozzolans can have an increased tendency to be subject to pore changes as a result of drying [4]. The pores most affected are those of >100Å and these are of significance to water transport through concrete [5].

Water will absorb into unsaturated concrete on contact. The pores of importance in absorption range from approximately 1μm to 10Å by virtue of their capacity for capillarity, however it should be noted that much larger pores also support capillarity action.

The suction force acting within any single capillary can be defined by:

$$F = 2r\acute{O} \cos \O$$

Where : Ó is the surface tension of the fluid in N/m^2 and Ø is the angle of contact of the fluid with the pore wall.

The suction force is positive as long as Ø is less than 90°, however, maximum capillary rise is time dependent and does not occur instantaneously.

SORPTIVITY

Sorptivity is often used to characterise the ability of a concrete to absorb and transmit water by capillarity. The initial absorption rate can be given by the mass of liquid absorbed per unit area of sample and a function of contact time. Early absorption is governed by the square root of time [6, 19]. Providing aspects of the testing procedure are kept constant, e.g. specimen geometry, orientation etc. sorptivity S ($mm/min^{1/2}$) can be calculated from:

$$S = slope/Az$$

where: A is the specimen cross sectional area and where the slope is taken from a line plot of specimen weight gain against time $^{1/2}$. Z is obtained from z= (M_{sat}-M_o/AL) where z= effective empty porosity, M_{sat} is the mass of the specimen at saturation, and M_o is the mass of the specimen at the start of the test.

The main factors that determine the sorptivity of concrete are temperature, presence of chemical impurities in permeating water, moisture content, porosity, pore size distribution and connectivity of pore space. Capillary suction is greatly influenced by the moisture state within a concrete [6] and pore structure characteristics such radius, tortuosity and continuity of capillaries will influence the rate of water absorption of a concrete [10].

For example, as initial water content increases sorptivity decreases and under normal test conditions the initial water state is a very important parameter to control.

During the process of absorption, water is also absorbed onto the surface of pores within a concrete. The thickness of this absorbed layer is about 26Å and this should be taken into consideration when calculating the effective pore size [5].

The theory and practical importance of the property of sorptivity as a universal measure of the ability of a concrete to absorb and transmit water by capillarity has been thoroughly reviewed [6]. This showed that hydraulic diffusivity (D) can be regarded as the fundamental water absorption and transmission property of a porous material which underlies sorptivity. However, determining S is easier and as capillary absorption of the large majority of porous construction materials conform to $t^{1/2}$ it is a convenient means to compare the sorptive behaviour of these materials.

As sorptivity is influenced by the initial water content of the sample to be measured, the degree of compaction and temperature, the experimental procedures used in this work were designed to limit these effects.

WICK ACTION

Wick action is the transport of water from the saturated face of a concrete element to a drying face. It can be considered as a combination of absorption and water vapour diffusion where water vapour leaving a drying face creates a condition for water to migrate from a saturated face. The flow of water through concrete due to steady state wick action is a simple function of concrete element thickness. Ingress of soluble salts into concrete can be rapidly facilitated by wick action. Water transport by wick action has been described [7], and where chloride ingress is concerned [8] and is not considered further in this paper.

PRESSURE INDUCED FLOW

Permeability is the transport property that relates to the ease with which water will flow through a material under the action of differential pressure. The coefficient of water permeability (K_v m/s) can be determined according to D'Arcy [9] from:

$$K_v = d^2 v/2ht$$

where: d is the depth of penetration, h is the applied head of water and t is the duration of the test. K_v is then not a materials characteristic but describes the flow of water [10] and assumes that under the particular conditions of measurement the influence of sorptivity is negligible.

WATER PERMEABILITY

Liquids and gasses will flow through a concrete as a result of a pressure head. The coefficient of permeability can be used to describe the permeation of liquids and gasses due to a pressure head when the viscosity of the liquid is considered and laminar flow is assumed.

In this case the coefficient of permeability becomes a materials characteristic and permeability is a flow property.

It is established that as w/c ratio decreases and the degree of hydration increases then concrete permeability decreases [11]. For example as w/c ratio rises from 0.5 to 0.7 the increased fraction of pores within the 750Å to 1.32μm range can dramatically increase the amount of water permeating a concrete for a given head of water.

The intrinsic permeability of materials may be described, where k / m^2 is defined by:

$$k = K_V \mu / pg$$

where: μ = viscosity of fluid, p = density of fluid and g = gravity and the permeability of a material can be described for any given fluid or gas provided that the permeability coefficient of any other fluid is known.

DURABILITY OF CONCRETE

The relevant transport mechanisms that influence the durability of concrete have been reviewed [10] and include:

- diffusion of gasses and water etc. through empty pores, microcracks and interfaces
- diffusion of ions or solvation of gasses in pore solution
- permeation of water or aqueous solutions under pressure
- capillary suction of water or aqueous solutions through empty/non-saturated pores

A close correlation with the following processes has been established experimentally [16] and incorporates:-

- rate of carbonation
- weight loss duc to freeze/thaw
- abrasion resistance
- depth of chloride penetration
- weight gain due to sulphate attack

A number of generic and proprietry materials and products have been examined for their ability to improve water pressure resistance (K_v) and absorption (S) in concretes. The effect on compressive strength was also determined (UCS) and the results are as shown in Table 2. The control concrete results in all cases is shown as 1.00.

A detailed commentary on the Table is not included in this paper except to draw attention to the difficulty that most of the mechanisms have in improving all three characteristics of K_v value, S value and compressive (UCS) strength. Some were good at achieving one, and occasionally two features but failed to show significant differences across all three parameters simultaneously. A reduction in compressive strength cannot be acceptable where high durability is paramount. Only three of the mechanisms showed any improvement in strength and all except one failed to show any significant improvement in the key durability factors of resistance to hydostatic pressure, low absorption and increased compressive strengths. Proprietry products were added exactly as the manufacturers instructions.

Table 2 Selection of potential generic and proprietary waterproofing materials [5]

MATERIAL	DOSE (%W/W)	K	UCS	S
Control	-	1.00	1.00	1.00
GGBFS	40	1.43	0.91	1.02
PFA	30	0.85	0.84	1.02
Silica Fume	5	0.55	1.08	1.00
Ca-bentonite	2	1.43	0.85	0.98
Limestone dust	20	0.27	1.10	0.95
Hydrated lime	5	0.34	0.95	0.96
Mortar sand	5	0.67	0.84	0.93
Sodium Stearate	0.2	0.44	0.88	0.76
Calcium stearate	1	1.73	0..98	0.32
Aluminium stearate	0.5	0.85	0.84	0.63
Butyl stearate	2	0.80	0.83	0.22
Caprylic acid	0.5	0.42	0.53	0.29
Soyabean oil	0.2	4.22	0.48	0.51
Mineral oil	0.5	0.91	0.72	0.77
Tar emulsion	10	0.57	0.86	0.91
EVA	5	1.57	0.61	0.75
SBR	5	0.61	0.67	0.64
Diethyl ethanolamine	13	0.49	0.77	0.93
Dimethyl ethanolamine	23	0.31	0.97	1.02
Sodium silicate	1	0.46	0.94	0.98
Aluminium powder	0.2	0.84	0.79	0.99
'T'	30**	1.28	0.83	0.39
'U'	1	0.20	0.93	0.95
'V'	9*	0.34	0.90	0.97
'X'	6*	0.33	0.88	0.94
Modified concrete	2	0.33	1.40	0.13

* Added per m³ of concrete
** Added as 30L/m³ at a w/c of 0.35
Where control is 1.00, K is pressure induced flow, S is sorptivity and UCS, compressive strength

As the internal moisture state of a concrete will have an influence on the results obtained, samples were conditioned after curing in a controlled RH in specially constructed chambers. Conditioning by drying at 105°C in an oven can be employed but this has several major disadvantages [5]. In this work, therefore, conditioning in a controlled humidity (55 or 75RH as appropriate) was employed until satisfactory conditioning was deemed to have taken place.

Samples were not tested until a stable or near stable internal moisture equilibrium had been established i.e. when the weight of samples (of approximately 1kg) became constant or a weight loss of <0.1g/day (or 0.01% of the sample mass) was recorded.

Each batch of samples was manufactured with its own control for comparison. Control values are shown as 1.0 The comparative results are as shown in Table 3.

WATER TRANSPORT TESTS

Sorptivity

Sorptivity testing involved conditioning 100mm diameter x 50mm deep cylindrical specimens at a particular RH (75% RH) then placing one of the flat surfaces in contact with water and monitoring weight gain with time. A plot of weight gain versus square root of time often produces a bilinear plot. The initial gradient of the earlier line, expressed in terms of unit cross sectional area of concrete are used to derive S in units of mm/min $^{1/2}$. For comparative purposes with their respective control samples the value S is also determined as the rate of weight gain due to sorption in grams per square root of time (min).

Pressure Induced Flow

The water permeability testing of concrete also involved conditioning of 100 mm cylindrical samples at 75% RH. After conditioning, these samples were loaded into stainless steel cells and enveloped in latex sleeves which, when compressed, provide a watertight jacket around the specimen. A water pressure of 3 bar was applied to the cast face of each specimen. The weight gain and depth of water penetration, after 4 days as measured from the water front visible on breaking samples were used to calculate K_v in m/s.

Chloride Diffusion

This method entailed immersing water saturated concrete cylinders in saturated sodium chloride solution for a period of one month. The ingress of chlorides into the body of the cylinder was determined from powdered samples obtained at pre-determined (2mm) depth intervals. In practice, however, salts normally enter a concrete through a combination of diffusion and absorption of a salt solution [10]. This combination of mechanisms provides a faster route of penetration than diffusion alone. However, in this test diffusion of chlorides alone was measured as test samples were vacuum saturated with water prior to testing.

Samples were extracted every 2mm by a milling machine. The powder obtained was then acid digested and the chloride content determined by an auto potentiometric titration technique. The profile was used to determine the effective chloride diffusion coefficient using a least squares method of best fit to the theoretical chloride profile with the surface concentration and effective diffusion coefficient being the regressor variables.

GAS DIFFUSION

Oxygen Diffusivity

The test procedure involved sealing cylindrical samples in latex sleeves in a steel jacket. Oxygen of known temperature, pressure and flow rate was passed over one side of the specimen and nitrogen gas over the other at the sample temperature and pressure. The amount of oxygen entering the nitrogen gas stream as a result of diffusion through the sample was analysed using a Zirconia oxygen analyser. The data obtained was then used to calculate the diffusion coefficient according to Fick's first law.

$$F = -D \frac{dc}{dx}$$

where D = diffusion coeff (m^2/s)
C = concentration (g/m^3)
X = distance (m)

Carbonation Method

This method is based on that of [12]. Cylinders were wrapped in tape with the top face exposed and placed in a carbonation chamber. The chamber had a controlled humidity and carbon dioxide content of 80% and 4%, respectively. The temperature of testing was controlled at 25°C. After 28 days exposure cylinders were removed and split. The depth of carbonation was measured using phenolphthalein (0.1 % soln., 50:50 methanol/ deionised water) from the top, exposed, surface of the sample. The rate of carbonation is primarily controlled by diffusion of the gas within the pore structure of the concrete [13] which in turn relates to the quality of concrete and its internal moisture state.

STRENGTH

Compressive Strength

Compressive strength testing was carried out in accordance with [17] on 100mm cubes. Durability of concrete is a function of strength, low permeability, extremely slow rates of carbonation and chloride ion ingress [10] so it is important to astablish that no sacrifice in strength occurs in achieving low permeability.

Examining the modified concrete for the properties described above indicated results which clearly show that the modified concrete mix has substantially improved durability properties and is much more corrosive resistant over un-modified concrete. Micrograph analysis shows increased hydration of modified concrete (see Figures 1- 4).

Table 3 Comparison of results

PARAMETER	CONTROL CONCRETE W/C RATIO 0.45	MODIFIED CONCRETE (2% W/W OF CEMENT CONTENT) W/C RATIO 0.38	COMMENTS
Slump	55mm	55mm	
Sorptivity (s)(mm/min½)	1.00	0.13	Sorptivity reduced to 13% of the control
Pressure induced flow (KK) ($g.m^2 min^{-1}$)	1.00	0.33	A 3 fold improvement over the control
Oxygen diffusion ($m^2 s^{-1} \times 10^{-8}$)	1.00	0.71	Approx 30% reduction over the control
Rate of oxygen transmission ($ml.m^{-2} s^{-1}$)	14.3	17.5	Approx 18% reduction
Chloride diffusion coeff ($m^2. s^{-1} \times 10^{-12}$)	1.0	0.2	Factor of 5 reduction in rate of transmission
Chloride content @ 5mm (mg/kg)	7833	2703	65% improvement @ 5mm depth
Chloride content @ 10 mm (mg/kg)	4750	698	85% improvement @ 10mm depth
Carbonation mm depth	4.0	<1.0	Factor of 3 improvement
Compressive strength 7 days (N/mm^2) 28 days	33 43	52 64	57% increase 48% increase

MICROSTRUCTURAL EXAMINATION OF NEW LOW PERMEABLE CONCRETE

The development of a dense microstructure within concrete is fundamental for good durability. Admixtures that interfere with hydration processes whether by accelerating or retarding setting or by modifying the chemistry of hydration products can significantly influence durability performance. As previously discussed, waterproofing admixtures modify concrete by altering pore structure and or by lining pores with hydrophobic coatings. The latter are liable to be only a few molecules thick and this will make them very difficult to characterise by normal methods of microstructural examination.

However, other changes, such as in the spatial distribution of pores within a modified system are appropriate for microstructural analysis.

The microstructure of selected samples was examined using a Jeol 540 Low Vacuum scanning electron microscope. Samples were prepared by freeze drying followed by resin impregnation under vacum. The surface of samples were then polished to ¼ micron using diamond paste. Specimens were examined by backscattered electron imaging (BSE) at a pressure of 20pa.

Modified concrete, produced during this programme of work, was stored for microstructural examination. The oldest concretes were over a year old when examined. Other concretes of approximately 6 months age and made to constant water/cement ratio were also available for study, only after equilibrium/stability had been reached, ie: constant weight.

An examination of concrete microstructure is facilitated by the use of back scattered electron imaging (BSE) techniques when mineral phases with a high mean atomic number (such as unhydrated cement) produce a larger number of backscattered electrons and appear as lighter shades of grey. Typically, the hydration phases present lower mean atomic numbers, and are seen in darker grey shades whereas porosity is black. The use of a grey-scale imaging facility enables specific features of concern to be identified and studied.

Figure 1 shows a BSE image (x100) of a control concrete produced during this work. The bright anhydrous cement grains can be seen with a rim of hydrated material within the original boundary of the grain. This paramorphic material is what is often referred to as inner hydration product. The outer hydration products (occupying the original water filled porosity) are composed of C-S-H gel (darker grey) and portlandite (mid grey). Aggregates are the large mid grey features with distinct boundaries associated with porosity at the interfacial zone [14].

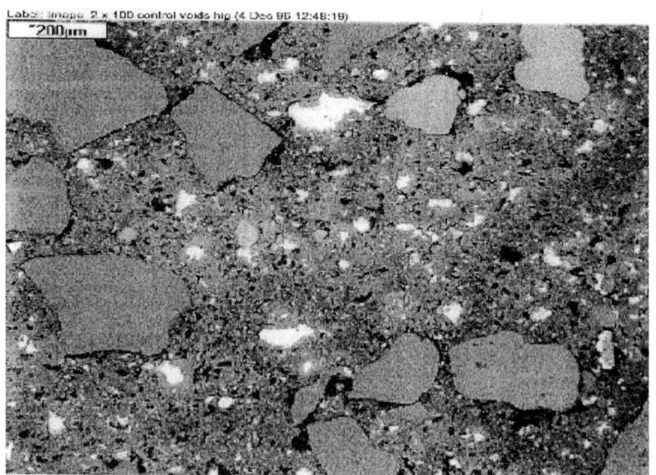

Figure 1 BSE image (x100) of control concrete

The capillary pores of >1 micron within the matrix are numerous and distinct. The increase in porosity at the aggregate/paste interface presents a distinct feature of importance to concrete durability performance, [18] and may be observed in the micrograph. However, it should be noted that at the magnifications used here not all the capillary porosity can be observed.

Figure 2, is an image which shows the spatial distribution of capillary porosity within the control concrete described above. The porosity has been highlighted through grey scale thresholding and in this field occupied 8 % of the field area.

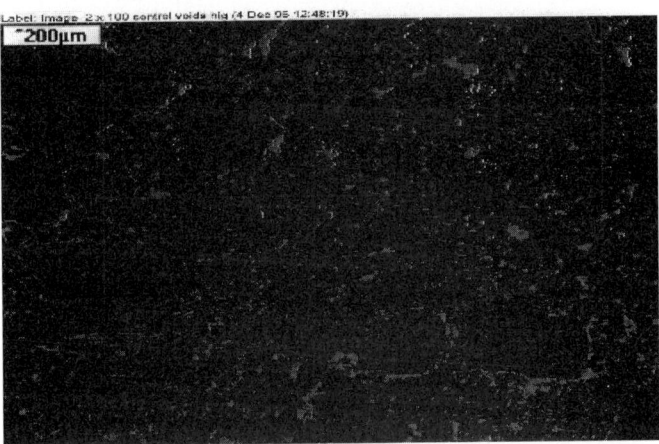

Figure 2 Image of control concrete showing spatial distribution of capillary porosity

Figure 3 shows a typical BSI image of modified concrete taken at approximately x350

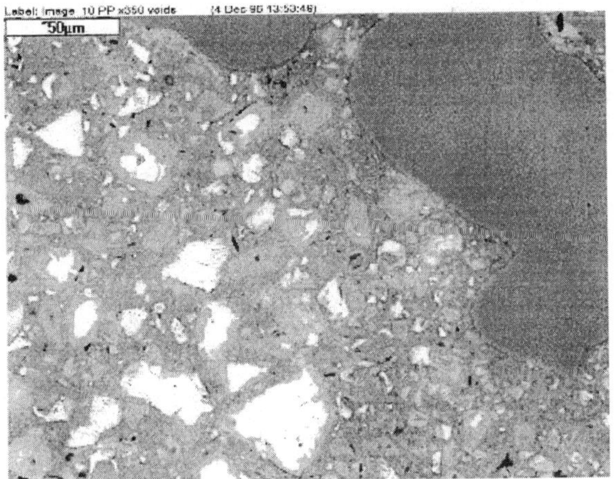

Figure 3 BSI image of modified concrete magnitude approximately x350

Figure 4 is the ‘reversed out’ image of figure 3 showing greatly reduced permeability.

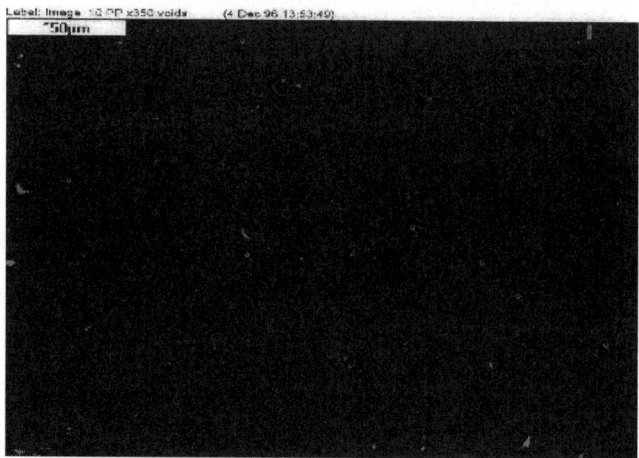

Figure 4 Reversed image of Figure 3 showing reduced permeability

Ultra-low-permeable (modified) concrete was seen to be distinct when compared to the control concrete for a number of reasons. The effects of reduction in w/c were apparent in that the area fraction occupied by capillary porosity visible at this magnification was much reduced at <1% in comparison with the control value of 6%. In Figures 3 and 1 the spatial distribution of capillary porosity in modified concrete and the control concrete is highlighted, respectively. An examination of the two Figures clearly illustrates the reduced large capillary porosity that results from the modified concretes and a reduction in w/c ratio in the region of 10% - 15%.

In addition, the pozzolanic activity which is responsible for preventing the formation of calcium hydroxide (portlandite) by promoting C-S-H has the effect of further infilling porosity, which was particularly apparent at the aggregate paste interfaces observed in, for example, Figure 4. The increased production of gel phases, which have a lower density may also contribute to a reduction in porosity in ultra-low, permeable concrete.

CONCLUSIONS

The main conclusions of the work are:

1. Low-permeable, modified concrete has a distinct microstructure. In lower w/c concrete, portlandite is replaced by C-S-H gel, the packing of cement hydrates in the region of the aggregate/cement interface is improved and the capillary porosity greatly reduced.

2. Water transport through concrete is a result of distinct mechanisms. By measuring sorptivity (S) and pressure induced flow (K_v), the relative performance of concretes can be compared.

3. The reduction of water transport through concrete by sorptivity and pressure induced flow can be facilitated by a single dose modifier to concrete.

4. Modified concrete was capable of significantly reducing (by a factor of 3) the sorptivity of concrete. Resistance to oxygen diffusivity was improved and a significant reduction (by a factor of 5) in chloride diffusion coefficient was recorded.

5. The modified concrete was generally better performing than other concretes (plasticisers, waterproofers, etc) and appeared to increase early strength development, as shown for example in Tables 2 & 3.

REFERENCES

1. NEVILLE, A M. Properties of Concrete, (4th. Edition), (1995), Longman, UK.

2. ACI COMMITTEE 212. Chemical Admixtures for Concrete, ACI Mat. J., 86, pp 297-327, (1989).

3. JUMPER, C H. Tests on Integral and Surface Waterproofing for Concrete. J. Am. Conc. Inst., 48, pp 209–241, (1931).

4. DAY, R L AND MARSH, B K. Measurements of Porosity in Blended Cement Pastes. Cem. Concr. Res. 18, 1, pp 63-73, (1988).

5. AL ISA, M A H I. Admixtures to Reduce Chloride Ingress into Concrete. PhD Thesis. Department of Civil Engineering, Imperial College, London, (1994).

6. HALL, C. Water Sorptivity of Mortars and Concretes: a Review, Mag. Concr. Res., 41, 147, pp 51-61, (1989).

7. BUENFELD, N R. Measuring and Modelling Transport in Concrete for Life Prediction in Concrete Structures, STATS Conference, Chapman and Hall, 16, November 1995.

8. BUENFELD, N R, SHURAFA-DAOUDI, M T AND MCLOUGHLIN, I M. Chloride Transport due to Wick Action in Concrete, RILEM International Workshop on Chloride Penetration into Concrete, Domaine Saint Paul, St-Remy-les Chevreuse, France, October, 15-18, 1995.

9. VALENTA, O. The Permeability of Concrete in Aggressive Conditions. Proc. 10th. Int. Cong. on Large Dams (Montreal), pp 103-117, (1970).

10. KROPP, J, HILSDORF, H G, ANDRADE, C AND NILSSON, L O. Transport Mechanisms in Concrete, in: Performance Criteria for Concrete Durability. RILEM Report 12, (Eds Kropp J and Hilsdorf H K), E and F Spon, London, (1995).

11. MEHTA, P K AND MANMOHAN, C. Pore Distribution and Permeability of Hardened Cement Pastes, 7th Int. Symp. on the Chemistry of Cement, (Paris), Vol. III, pp Vii-1 to Vii-5, (1980).

12. ROBERTS, M H. Carbonation of Concrete Made with Dense Natural Aggregates. BRE Information Paper IP 6/81, BRE, Garston, Watford, UK, (1981).

13. YING-YU, L AND QUI-DONG, W. The Mechanism of Carbonation of Mortars and the Dependence of Carbonation on Pore Structure. Proc. Katherine and Bryant Mather International Conference on Concrete Durability, ACI Special Publication 100, Volume II, (Ed Scanlon, J M), pp 1915-1943, 1987.

14. SCRIVENER, K L, CRUMBIE, A K AND PRATT, P L. A Study of the Interfacial Region Between Paste and Aggregate in Concrete. In: Bonding in Cementitious Composites (Eds Mindness, S and Shah, S P), Mat. Res. Soc. Symp. Proc. 114, Pittsburgh, Pa. (1988).

15. CZERNIN, W. Cement Chemistry for Civil Engineers, George Goodwin Ltd. (Second Edition), p 91, (1980).

16. KROPP, J. Summary and Conclusions in: Performance Criteria for Concrete Durability. RILEM Report 12, (Eds Kropp J and Hilsdorf H K), E and F Spon, London, pp 280–293, (1995).

17. British Standards 1881. Testing of Concrete, Part 116, Determination of Compressive Strength of Test Specimens, (1988).

18. CRUMBIE, A K, SCRIVENER, K L AND PRATT, P L. The Relationship Between Porosity and Permeability at the Surface Layer of Concrete and the Ingress of Aggressive Ions. Mat. Res. Soc. Symp. Proc. 115, Pittsburgh, Pa. (1989).

A GENERAL MODEL OF CONCRETE BEHAVIOUR FOR TIME-DEPENDENT LOADING

A V Zabegayev

Moscow State University of Civil Engineering

Russia

ABSTRACT. A general model of concrete, adequately reflecting all main features of its behaviour under time-dependent (instant, short-term standard, low cycling, multi-cycling, long-term) loading is proposed. The model is based on a presentation of concrete as three-level hierarchical system and allows for changes (pore system intersection by microcracks, mass transfer etc.) in real structure of concrete at micro-level under the loads. It is shown why 1iquidgaseous phase governs the time-dependent processes in concrete and their description for axial compression is given on the base of simple differential relationships. An example of RC beam analysis, implementing the model is also given. A good coincidence of the theoretical and corresponding test results has been obtained.

Keywords: Concrete, Hierarchical system, Time-depending loads, Pore system intersection, Microcracks, Mass-transfer, Reinforced concrete structures, Analysis.

Professor Alexander V Zabegayev is Pro-Rector and Head of Department of Reinforced Concrete Structures, Moscow State Univensity of Civil Engineering. He is Emeritus Scientist of Russian Federation (Governmental Award), Member of Presidium of National Concrete Association, Fellow in the Institution of Civil Engineers (UK), Member of SECED (UK). He specialises in the reinforced concrete dynamic testing and analysis, deformational mechanics of concrete. He is a co-author of the most popular (in the former USSR) text-book on reinforced concrete structures. Professor Zabegayev is also known as author of a number of books and over 90 papers, partly published abroad. He also serves on many international organisations, dealing with structural engineering, safety and education.

INTRODUCTION

The research in concrete, having been carried out during last years [1], aims a profound investigation of its structure and nature of the processes, governing its behaviour at various stages of its life. Joint efforts of specialists in chemistry of cement and concrete, material science, concrete technology have revealed to great degree the links between structure and properties of concrete. However, designers can not effectively use these achievements because of an absence of a physical model, adequately reflecting the structure and, as a consequence, typical features of concrete performance under loading. Apparently, the model may not be a conservative system with constant parameters, as concrete can represent absolutely different physical (and, in some cases chemical) media at various stages of its existence (hydration, cracking, instant and long-term resistance etc.). Eight major requirements to meet by the model, describing behaviour under time-dependent loads, were formulated by Kesler (1962). According to the author's information, no models satisfying ad these requirements have been put forward. It may be explained by the fact that the attempts undergone have been based on rather rough assumptions, referring mostly to macrostructure of concrete.

In this paper a mechanical model of normal-weight concrete, satisfying the above-mentioned requirements, is proposed. It is based on modem data about concrete structure and describes from a general point of view the response to venous time-dependent loads: long-term, low- and multi-cycling, short-term, instant). Having been originally elaborated for software analysis, the model turns out to be also acceptable for the traditional one; a relevant example is given at the end of the paper.

ASSUMPTIONS

Considering concrete the hierarchical system [2] leads to the conclusion that at meso-level it represents a two-phase material with aggregate grains embedded into cement matrix. Initial microcracks in the interface between the aggregate and the matrix is the most essential feature of the media.

At micro-level the matrix is usually considered as an environment, containing solid, liquid and gaseous phases. Notice, that this approach has to be corrected if light-weight or high-performance concrete is under consideration.

Axial compressive loading initially causes closing of open pores, related to the initial concave block of < <stress-strain> > diagram, usually neglected in the analysis. Further deformation (within the range of stresses up to 0,4-0,6 of the peak one, standard loading) is mainly linked with elastic slipping of the initial microcracks shores. On exceeding the above mentioned level of loading the initial cracks start grow, first within the interface, then into the matrix in parallel with the acting force. At stresses of 0,75-0,85 of the peak ones an interconnected network of microcracks forms, followed by macrocracks, visible at the peak stress. A character of failure (longitudinal, shear macrocracks) depends on both inner and scale factors.

It is important to emphasize that the cracks, growing through the matrix, intersect pores and capillaries, containing water and vapour; an increasing pressure, caused by the load, promotes their leakage (mass-transfer). On the other hand there are evidences that the opening microcracks intensify water and moisture absorbency [3]. A special experimental study, modelling processes of the mass-transfer was undergone by the author [41].

According to other experimental data [5], resistance of normal-weight mean-strength concrete, related to the slipping of microcrack shores, essentially exceeds the resistance of uncracked matrix, which, in turn, is obviously greater than that with cracks.

Comparative experiments with dried and normal specimens under standard static and instant loading witness that concrete time-dependent response mainly depends on the liquid gaseous phase, first of all, presence and migration of free water (see, f. e.[6]).

GENERAL MODEL

The elasto-visco-brittle model, shown in Figure 1, allowing for above mentioned and some other relevant phenomena, was elaborated by the author to tie the existing data on concrete time depending loading 17]. Parameter E_{sl} characterises the concrete resistance refuted to rnicrocrack slipping in the interface; E_{mc} denotes the cracked matrix resistance, while E_m = E_m - E_{mc}, where E_m is a similar parameter for the uncracked matrix; b is a coefficient of viscosity. Apparently, two springs in the upper element jointly model solid uncracked matrix while visco element simulates the liquid-gaseous phase.

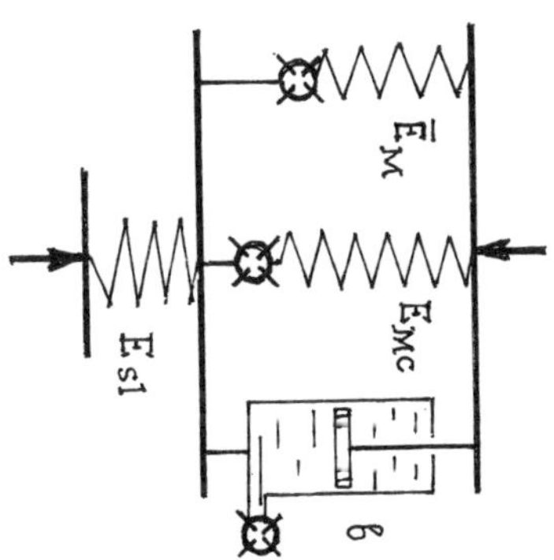

Figure 1 Model of concrete

When axial compressive stresses attain level of 0,4-0,6 of the standard peak one, the left spring fails; that corresponds to the beginning of crack grow in the matrix as well as pressing liquid gaseous phase out of pores and capillaries (the outlet valve in Figure 1 fails).

On attaining level of 0,75-0,85 of the peak by axial stresses the second spring of the upper element also fails. This level corresponds to the microcrack network forming in the matrix, i.e. the beginning of actual specimen failure. This level is reasonable to denote as time-independent strength of concrete (see Figure 2).

Greater resistance can be referred to as but a temporary one, related mostly to deformation of the visco element. Under instant loading the liquid performs incompressible and greater stress can be sustained within a short time. If the stress is constant but exceeds the limit, marked in Figure 2, the strain grows in time, finally attaining its ultimate value. These phenomena are known as strain-rate effect and creep, respectively, determining two extreme time-dependent regimes of loading. Presumably, if the model is correct, it would adequately describe the concrete response to other time-dependent loading, first of ad, cycling loading.

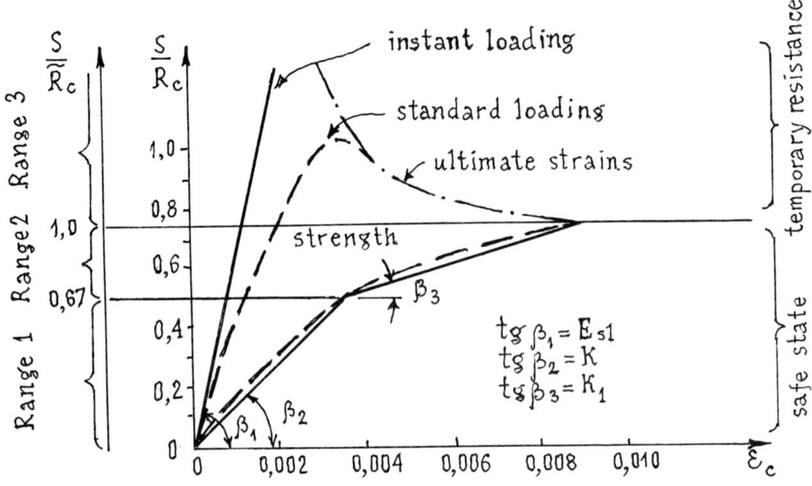

Figure 2 Relative stress - strain diagram

PARAMETERS OF THE MODEL

Denoting time-independent strength R_c, one can get 0,67 R_c for the stress corresponding to the beginning of crack growing. As the liquid phase behaves as an incompressible one under instant loading, the upper element's strain is blocked in this case, so the lower spring only can deform. It is well known that this regime of loading is used for determination of initial modulus of elasticity of concrete, E_c. Thus, it is naturally to assume $E_{sl} = E_c$.

E_m can easily be derived from ultimate linear creep data (under stress < 0,67 Rc). Really, the following relationship is valid for any stress from this range (Range 1 in Figure 2):

$$1/K = 1/E_{sl} + 1/E_m, \tag{1}$$

where K = parameter of generalized stiffness of a specimen, obtained as secant modulus of the graph of ultimate linear creep. E_m can be derived from (1) as follows:

$$E_m = K E_{sl} / (E_{sl} - K). \tag{2}$$

Similarly, one can get for Range 2 (Figure 2):

$$E_{mc} = K_1 E_{sl} /(E_{sl} - K_1), \tag{3}$$

where Kl = parameter of generalized stiffness, similar to K, but obtained from non-linear creep data.

Determination of the coefficient of viscosity can be accomplished now on the base of general relationships and experimental data and is given below.

GENERAL RELATIONSHIPS

Let us consider increasing loading, sustained load and unloading as the most general set of regimes of time-dependent loading, assuming the increasing loading to be linear in time:

$$S(t) = S_0 t / \tau_1, \tag{4}$$

where S = current stress caused by the load; S_0 = the peak stress; τ_1 = period of loading.

Range 1 Equilibrium equation (inertia is neglected)

$$S(t) = E_m \varepsilon_1 + b \dot{\varepsilon}_1 = E_{sl} / \varepsilon_2 \tag{5}$$

leads to the solutions as follows:

for the increasing load

$$\varepsilon_1 = S_0 [t - b(1 - exp(-E_m t / b)) / E_m] / E_m \tau_1 ; \tag{6}$$

$$\varepsilon_2 = S_0 t / E_{sl} \tau_1; \tag{7}$$

$$\varepsilon_c = \varepsilon_1 + \varepsilon_2, \tag{8}$$

here ε_1 and ε_2 are strains of the upper element and the lower spring, respectively;

for the sustained load (S = So = const)

$$\varepsilon_1 = (\bar{\varepsilon}_1 - S_0 / E_m) exp(-E_m(t - \tau_1) / b) + S_0 / E_m, \tag{9}$$

where $\acute{\varepsilon}_1$ is obtained from (6), substituting t = τ_1; ε_2 is identified by (7);

- for the unloading, assuming it also to be linear as follows:

$$S(t) = S_0 [1 - (t - {}_1) / \tau_2], \tag{10}$$

where τ_2 = period of unloading, one can obtain

$$\varepsilon_1 = [\tilde{\varepsilon}_1 - S_0(\tau_2 + b / E_m) / E_m\tau_2] \exp(-E_m (t - \tau_1) / b) + \\ + S_0 (\tau_1 + \tau_2 - t + b / E_m) / E_m\tau_2; \tag{11}$$

$$\varepsilon_2 = S_0 [(1 - (t - \tau_1) / \tau_2] / E_{sl}. \tag{12}$$

As it follows from (6), when $t \Rightarrow \infty$, $\tau_1 \Rightarrow \infty$, ε_1 aspires S_0 / E_m It is also clear from (9), that ε_1 aspires the same value when $t \Rightarrow \infty$, and $\tau_1 = 0$, $\acute{\varepsilon}_1 = 0$ (instantly applied sustained load). This relates to the well-known fact, that final concrete strain at the same stress does not depend upon period of loading.

According to (11), when $t = \tau_2$ (end of the unloading),

$$\varepsilon_1(\tau_2) = [\tilde{\varepsilon}_1 - S_0 (\tau_2 + b / E_m) / E_m\tau_2] \exp(- E_m (\tau_2 - \tau_1) / b) + \\ + S_0 (\tau_1 + b / E_m) \neq 0, \tag{13}$$

since a residual deformation exists and a shape of unloading block of stress-strain diagram depends on the unloading rate or, in other words, on the liquid-gaseous phase. When the unloading is fast, the block asymptotically approaches a straight line with a slope, identified by the initial modulus, Es˜' usually observed in tests. Further strain recovery takes place due to the deformation of visco element, caused by the unfolding upper element springs (matrix) pressure. Corresponding equilibrium equation will be

$$E_m \varepsilon_1 + b \dot{\varepsilon}_1 = 0, \tag{14}$$

and its solution can be obtained as follows:

$$\varepsilon_1 = \varepsilon_1 (\tau_2) \exp(-E_m(t - \tau_2) / b). \tag{15}$$

Notice, if $t \Rightarrow \infty$, $\varepsilon_1 \Rightarrow 0$

After n+1 cycle of multi-cycling loading one can get instead of (6):

$$\varepsilon_1 = [\varepsilon_{1n} - S_0 (t_n - b / E_m) / E_m\tau_1] \exp(-E_m (t - t_n) / b) + \\ + S_0 (t - b / E_m) / E_m\tau_1, \tag{16}$$

where ε_{1n} = strain, attained at the end of unloading of the n-th cycle; t_n = time of the end of the n-th cycle. It can be shown from (16) that current strain is a function of loading and unloading rates. The greater the rates, the more cycles are needed to attain maximal strain, since the inelastic strain per cycle is small. The approach proposed also adequately reflects other features of concrete behaviour under cycling loading; this reaffirms the model.

Under instant loading the upper element is blocked and $\varepsilon_c = \varepsilon_2$, as from (6), (7) and (8).

The mean value of coefficient b can be obtained now from experimental creep data and relationship (17), derived from (9):

$$b = - E_m(t - \tau_1) / \ln [(\varepsilon_1 - S_0 / E_m)/(\bar{\varepsilon}_1 - S_0 / E_m)]. \tag{17}$$

Range 2. Similar procedures can easily be used to obtain governing relationships at this stage. However, the mass-transfer, caused by liquid-gaseous phase leakage into cracks, essentially reduces the viscosity immanent this stage. A special study of this factor has just been finished [8]. It may be preliminary assumed in this paper, that the outlet cross-sectional area and area of cracks increase proportionally; as a result stress S may be assumed proportional liquid drop rate in the visco element. Experiments prove practical validity of this assumption. Thus, coefficient b becomes constant and, being denoted b1 for this range, may be determined as follows:

$$b_1 = - E_{mc}(t -\tau_1)/ \ln [(\varepsilon_1 - S_0 / E_{mc} + C)/(\bar{\varepsilon}_1 - S_0 / E_{mc} + C)], \tag{18}$$

where $C = 0,67\tilde{R}_c (E_m - E_{mc}) / E_m E_{mc}$.

Range 3. A temporary resistance of concrete only is considered at this stage. The rnicrocrack network reduces dramatically a matrix ability to resist further loading, so the latter factor may be neglected in the analysis.

The equilibrium equation will be

$$S = b_2 \dot{\varepsilon}_1 = E_{sl} \varepsilon_2, \tag{19}$$

so one can get for the increasing load:

$$\varepsilon_1 = \varepsilon_1 (t_2) + S_0 (t^2 - t_2^2) / 2 b_2 \tau_1, \tag{20}$$

as well as for the sustained load

$$\varepsilon_1 = \varepsilon_1 (t_3) + S_0 (t - t_3) / b_2. \tag{21}$$

Where t_2 and t_3 = times when R_c is attained and the increasing loading ceases, respectively; b_2 = coefficient of viscosity for Range 3

Ultimate strain can be easily obtained now from test results under standard rate of loading, as in a number of standards this value is codified (see, f. e., [9]). As it follows from (20), the smaller period of the sustained load, the greater stress would be applied to attain the ultimate strain. This corresponds to strain-rate effect in real concrete. Coefficient of viscosity values (b, bl, b2) were determined for middle- and high-strength mature concrete and a special procedure of their calculation was elaborated LX]. The model proposed adequately reflects other properties of concrete, f. e. relaxation, as well as concrete behaviour under other regimes of loading (uniform strain rate, etc.) within ascending block of stress-strain diagram.

APPLICATION OF THE MODEL

Applicability of the model to the analysis of structures is illustrated here by an example of a beam, instantly loaded and then left under the sustained load. Assuming concrete to perform within Range 1, time-dependent stresses in compressed concrete and tensile reinforcement may be determined in traditional way:

$$S_c\,(t) = 2M\,/\,b\,x(t)\,[d - x(t)/3] = 2M\,/\,b\,d^2\tilde{x}\,(1 - \tilde{x}/3); \qquad (22)$$

$$S_s\,(t) = M\,/\,A_s\,[d - x(t)/3] = M\,/\,n\,b\,d^2(1 - \tilde{x}/3); \qquad (23)$$

where M= bending moment; b and d = cross-sectional width and effective depth, respectively; A' = cross-sectional area of longitudinal reinforcement; n = As / b d; x(t) = x (t) I d. Plain cross-section law gives

$$\varepsilon_c / \varepsilon_s = \tilde{x}\,/\,(1 - \tilde{x}). \qquad (24)$$

Concrete strain at a time t for this case can be derived from (8) (7) and (9):

$$\varepsilon_c = S_c\,\{[E_m + E_{sl}\,Q\,(t)]\,/\,E_m\,E_{sl}\}, \qquad (25)$$

Where

$$Q\,(t) = 1 - exp\,(-E_m\,t\,/\,b). \qquad (26)$$

Corresponding strain in the longitudinal reinforcement will be

$$\varepsilon_s\,(t) = S_s\,(t)\,/\,E_s. \qquad (27)$$

Substituting (22) into (25), (23) into (27) and then both results into (24), one can get the equation for defining x at any predetermined moment t

$$\tilde{x}\,(t) + \tilde{x}\,(t)\,r\,(t) + r\,(t) = 0, \qquad (28)$$

where r (t) = 2 n Es [Em + Es' Q (t)]/ E_{sl} E_m

Further procedures are absolutely traditional; for instance, the edge stress in compressed concrete and corresponding stress in longitudinal reinforcement may be obtained from (22) and (23), while bending stiffness B of the critical cross-section will be:

$$B\,(t) = E_s\,A_s\,d^2(1 - \tilde{x}/3)\,(1 - \tilde{x}). \qquad (29)$$

Comparison of analytical beam deflections calculated by means of the proposed procedure with a number of experimental results has shown their good coincidence (the discrepancy does not exceed 7-10%). However, the model may obviously be more effective in numerical approaches (f. e., FEM) and relevant software.

CONCLUSIONS

Concrete structures can be more explicitly reflected in physical models than it has been assumed. The model proposed illustrates such feasibility and demonstrates a greater influence of liquid-gaseous phase than was first thought. Further improvement of the model is being connected with an adequate description of the process governing the interaction between deforming matrix and liquid-gaseous phase transfer, including the meniscus formed in capillaries, intersected by microcracks [4], etc. Relevant description should also consider the roles of free and absorbed water within the liquid-gaseous phase; an essential step forward in this term has recently been made in [8]. The model can also easily be extrapolated on concrete softening.

ACKNOWLEDGEMENTS

The author conveys his profound gratitude to professors Rolf Lenshow and Ivar Hooland, Norwegian Technical Higher School, University of Trondheirn, for fruitful discussions on an original idea of this paper.

REFERENCES

1. DHIR, R K et al. Concrete In The Service of Mankind, Proceedings of the International Conference held at the University of Dundee, Scotland, UK, E & FN Spon, London, 1996.

2. WITTNANN, I H. Structure of Concrete with Respect to Crack Formation, Fracture Mechanics of Concrete, Elsevier, 1953, 43-74.

3. ROSSI, P. Influence of Cracking in the Presence of Free Water on the Mechanical Behaviour of Concrete, Magaane of Concrete Research, 1991, 43, No.154, 54-57.

4. ZABEGAYEV, A V, TAMRAZYAN, A G. On the Irdluence of Inner Moisture on Concrete Deformations, Beton i Zhelezobeton (Concrete and Reinforced Concrete), 1997, No. 1, 5-7.

5. MINDESS, S. The Application of Fracture Mechanics to Cement and Concrete: A Historical Review, Fracture Mechanics of Concrete, Elsevier, 1983, 1-30.

6. REINHARDT, H W. et al. Joint Investigation of Concrete at High Rates of Loading, Materials and Structures, 1990, 23, 213-216.

7. ZABEGAYEV, A V. On Elaboration of General Model of Concrete Behaviour, Beton i Zhelezobeton, 1994, No 6, 23-26.

8. TAMRAZYAN, A G. Improvement of Methods of Analysis of Reinforced Concrete Structures on the Base of Structural Theory of Concrete, D.Sc. Thesis, Moscow State University of Civil Engineering, 1998, 420.

9. NORGES STANDARDISERINGFORBUND. Concrete structures. Design rules. NS 3473E, 1992, 78.

CHLORIDE INGRESS IN MARINE CONCRETE EXPOSED TO THERMAL CYCLES

A Taheri

K van Breugel

Delft University of Technology

Netherlands

ABSTRACT. Data on chloride penetration into concrete exposed to a simulated aggressive marine environment are presented. Concrete specimens, large beams and small cubes, are subjected to 90 complete exposure cycles of wetting and drying, plus heating and cooling. The applied exposure condition consists of a drying period of 42 hrs, followed by a wetting phase of 6 hrs with salt water containing 5% NaCl. The drying phase, itself, is a thermal regime characterised by a temperature swing from 20 °C to 60 °C within a period of 12 hrs. This simulates, with some acceleration, the aggressive marine environmental condition in hot regions, with varying daily temperatures including, direct solar radiation. Totally 315 temperature cycles and 90 cycles of wetting/drying were applied to specimens in this experiment. It was observed that temperature and humidity variations promote chloride penetration into marine concrete significantly. This particular study shows that alternating thermal loads applied to large beams, did not cause significant microcracks. The thermally-induced microcracks, however, had only minor effects on chloride penetration in "restrained" concrete beams, compared to the relatively "stress free" small specimens. The effect of two other significant parameters, i.e. type of curing and type of cement, on chloride ingress rate is also investigated.

Keywords: Chloride, Cracking, Curing, Heating/cooling, Marine concrete, Portland cement, Restraint, Slag cement, Temperature, Thermal stresses, Wetting/drying.

Dr Ali Taheri is a senior researcher at Delft University of Technology. He obtained his PhD degree from the same university in 1998 following a four year research programme on concrete technology with special emphasis on durability related issues. He has been involved in design and execution of several marine projects in Iran since 1983. His research interests are durability of concrete, temperature effects on concrete and the evaluation of deterioration processes in concrete marine structures such as bridges, harbors and key walls.

Dr Klaas van Breugel is a senior researcher / lecturer at Delft University of Technology. Main topics are modeling of concrete behavior at early ages and the structural design of concrete structures for environmental protection.

INTRODUCTION

The environmental conditions marine concrete structures are exposed to are considered among the most severe natural exposure conditions. These are considered even more severe in areas, like the Persian Gulf, where solar radiation causes concrete temperatures of up to 60°C and higher [1]. These high temperature promote the rate of chloride penetration into the concrete and hence the chloride-induced corrosion of the reinforcing steel. The effect of temperature is considered to be two-fold. On the one hand temperature-induced strains may cause microcracking of the concrete in cases where movements are restrained. In real concrete structures which are exposed to daily thermal cycles, the surface layer of the structure will always experience a certain degree of restraint and is, therefore, prone to cracking. These cracks, generally microcracks, are believed to increase flow channels into concrete and hence the rate of penetration of chloride ions. On the other hand the fact that penetration processes, for example diffusion, increase at higher temperatures, this will also promote the rate of chloride penetration. In the splash zone of marine concrete structures, the alternating drying and wetting of concrete may further contribute to the build up of high chloride contents in the concrete.

Among the different environmental factors which determine the rate of chloride penetration, the effect of temperature-induced microcracking has often been considered to be one of the most severe ones [2,3]. In order to prove whether alternating thermal loads will indeed cause the expected microcracks, one should consider large concrete structures in which thermal stresses occur. In most laboratory tests dealing with temperature effects on chloride penetration small specimen are used. These small specimens, however, are almost free to deform when the temperature of the concrete varies. Hence only minor thermal stresses will occur, too small to create microcracks.

To generate thermal compressive and tensile stresses in a small concrete specimen, exposed to thermal cycles, the specimen should be placed in a frame in which the thermal deformations are prevented from occuring in either a passive or an active way. Although this is possible, several complicating experimental difficulties have to be tackled. A very efficient and practice-oriented way of testing is through the use of large concrete elements, in which thermal eigen stresses can occur when these elements are thermally loaded from one side. The principle of this test method will be explained here, together with the chloride penetration data observed after 1 and 6 month exposure.

EXPERIMENTAL DETAILS

Three 400 x 750 x 6000 mm^3 beams and a large number of 150 mm^3 cubes were made for this experiment. Specimens were subjected to wetting/drying and heating/cooling cycles to simulate the complex splash/tidal zone of marine structures under severe environmental conditions. A general view of the test set-up and a cross section of the beams are shown in Figure 1.

Full size beams were selected for this experiment to provide the restrained conditions occurring in a natural way as in practice. Large deep beams and small cubes represent the "restrained" and "unrestrained" conditions for the imposed strains, respectively. Three beams and their accompanying cubes, shown in Figure 1, are identified with "Set 1", "Set 2" and "Set 3". The first and the third sets are made with ordinary Portland cement and the second set with blast furnace slag cement. All three beams and cubes are subjected to alternate wetting and drying cycles with salt water. The first and second sets were also subjected to temperature variation. Salt water can get onto the surface of all three beams and remain during the wetting period inside the "dikes" installed on the top of the beams.

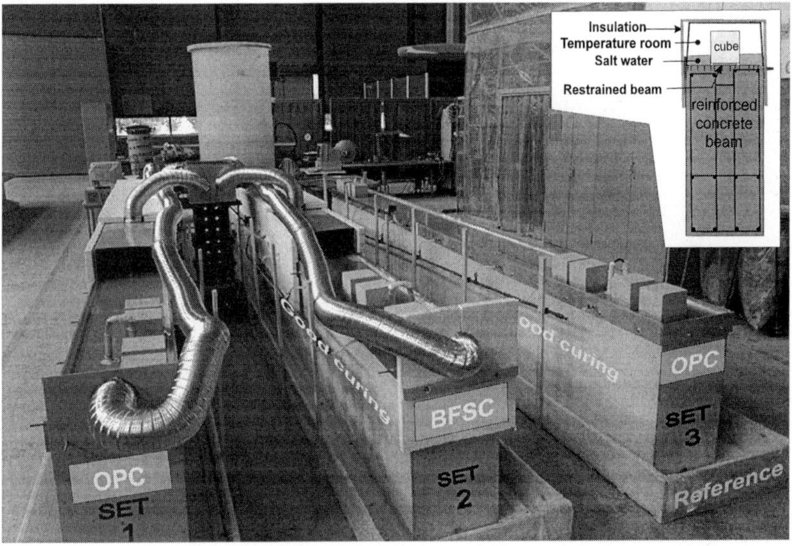

Figure 1 Overall view of the experimental set-up - The temperature rooms at the top of the beams are totally closed and sealed during the test

Figure 2 shows a top view of the beam, together with a coding system for the different "specimens". Each "specimen", either beam section or cube, is named by a 3-digit code, explaining the restraint condition, cement type, exposure condition and curing regimen. For example C2P refers to the cube (unrestrained) made with slag cement and exposed to alternating cycles of wetting and drying plus heating and cooling and cured with temperature of 38 °C (poor condition). B2P represents the specimen with the same conditions but for the beam section (restrained) and so on.

Materials

Two types of concrete were used in this investigation, namely normal Portland cement concrete and blast furnace slag cement concrete, with the latter about 70% of the Portland cement replaced by slag. The cement (or binder) content was 320 kg/m^3 in either mixes. A high W/binder (0.58) was selected in order to enhance and facilitate ingress of chlorides in the relatively short duration of the test.

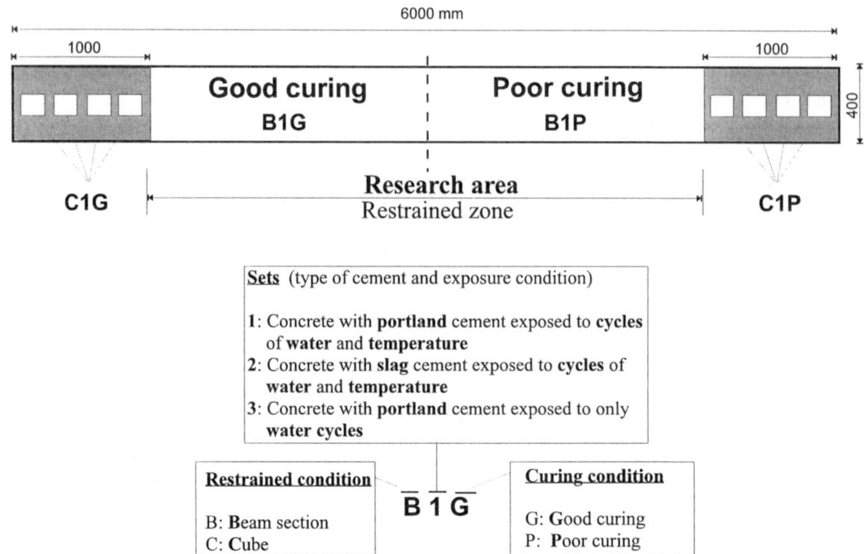

Figure 2 Top view of a beam used in the experiment and coding system – The research area of the beam is illustrated

The surfaces of all the beams and cubes were covered for one day to prevent the plastic shrinkage during the hardening process. Two curing regimens were imposed upon the specimens, 1) good curing and 2) poor curing. One half of each beam, as well as corresponding cubes, were cured in the normal condition, i.e. room temperature and humidity, for 14 days. The other halves were exposed to the controlled environment with a temperature of 38 °C and relative humidity of 50%, also for 14 days, to simulate the elevated temperature curing conditions (Figure 2), as might occur in tropical regions. All the beams and cubes were stored in open air for about six weeks to ensure good curing before exposure to the test regime.

Exposure Conditions

At the age of 56 days, the first two sets of beams and cubes were subjected to salt water and temperature variations and the third set only to wetting and drying cycles with constant temperature. The applied exposure condition consists of a drying period of 42 hrs followed by a wetting phase of 6 hrs with salt water of 5% NaCl content. The drying period itself is a step function of temperature with a period of 12 hrs changing from 20 to 60° C. Consequently a complete cycle of exposure is 48 hrs. The third beam and cubes are acting as the reference set (Figure 1). During the test the control rooms were closed and fully covered with insulation layers down to 300 mm below the beam surface. More information about the features of this experiment is given in [4].

RESULTS AND DISCUSSION

Cores of 50 mm in diameter were taken from the concrete specimens and slices of 6 mm thick were cut from the cores for chemical analysis to determine the chloride concentrations along the depth of specimens. Two drilling programs, i.e. after one month and six months of exposure, were performed. The chloride data, on the basis of the two measurements for all three sets of beams and cubes, were collected. Those are the total chloride content, free and bound ions, by % weight of cement in the concrete. In the following, the effect of restrained conditions as well as moisture and temperature fluctuations on the rate of chloride penetration into concrete beams, for both measurements, are discussed. The influence of curing condition and cement type on performance of concrete specimens, with regard the chloride ingress, is also presented. Further information about the effect of various parameter studies on chloride penetration can be found in other paper [5].

Effect of Restrained Condition

The chloride profiles of the beam and cubes for the first set, for both measurements, i.e. after 18 and 78 complete cycles of exposure, are presented in Figure 3. It is seen that after one month slightly more chloride have penetrated into the beam sections compared to the cubes. Observations in this phase of experiment indicated that, because of the restrained condition, some microcracks occurred at the surface of the beam, while the cubes with the same exposure condition remained uncracked. These microcracks, however, were only developed at the very near surface of the beams. Therefor they did not significantly increase the permeability and hence the ingress of chlorides into the concrete beams.

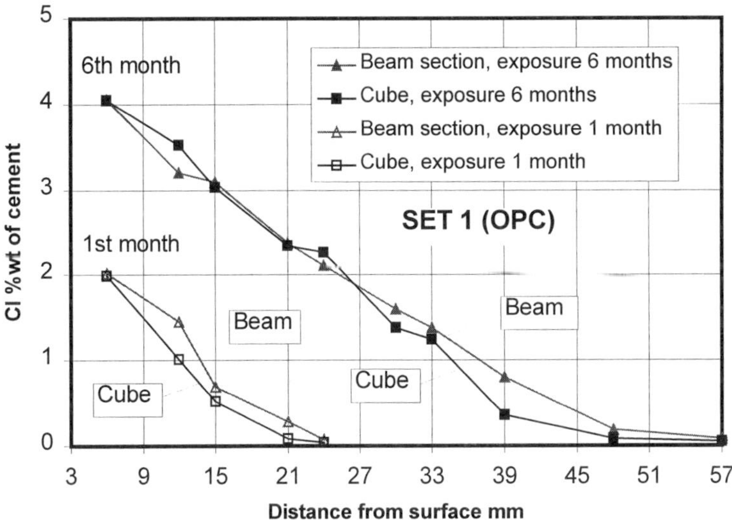

Figure 3 Chloride penetration in concrete beams (restrained) and cubes (unrestrained) in the first set of specimens made with Portland cement, after 1 & 6 months of test – Good curing

It was thought that further exposure to the simulated environment would cause more propagation of the cracks leading to more chloride ingress. But crack investigations, using the florescent method, showed that although the cracks were increased in width at the end of experiment, but they have not penetrated deep into concrete. So, no significant increase of chloride content was found in concrete beams due to microcracks at the beam surfaces in this experiment. It is clearly shown in Figure 3 that the chloride concentration in the restrained beam at the end of the test is almost similar to the cubes which were relatively free of stress.

It could be also possible that the thermally induced cracks have been sealed off due to self-healing effect. In an investigation, Jacobsen et al found a critical width for crack healing being about 0.2 mm [6]. The cracks occurred in our experiment were less than 0.2 mm in width. It might also be due to the swelling of the concrete during the heating period causing the cracks to be narrower and prevent further ingress of chlorides. The swelling of concrete exposed to seawater has been reported by Bijen [7].

Effect of Thermal Cycles

The chloride data of the first and third beam, B1G and B3G, which are cured under good conditions, are presented in Figure 4, for both measurements. It is seen that chloride penetration process has been significantly promoted on the first beam due to temperature rise and fall cycles. This effect is more evident in measurements after six months. The chloride ions have penetrated down to 60 mm in B1G which is exposed to temperature cycles while in the third beam, i.e. B3G, the penetration depth is not more than 24 mm with much lower chloride concentration.

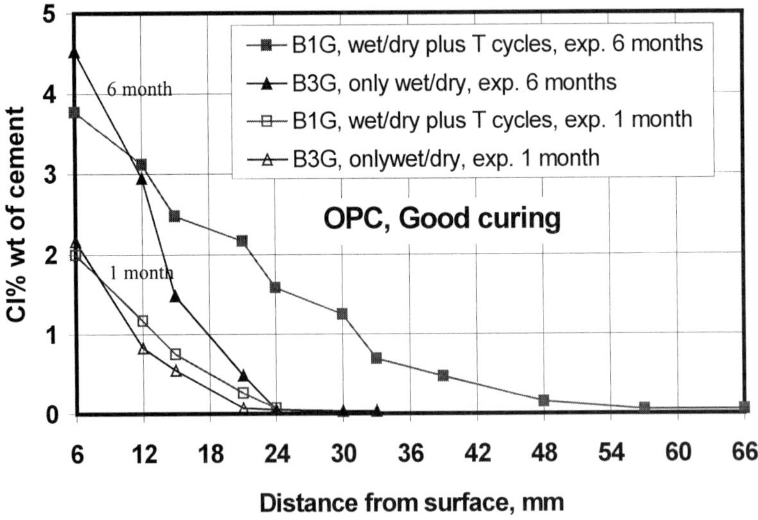

Figure 4 Chloride penetration into concrete beams exposed to temperature cycles (set 1) and constant room temperature (set 3), for two measurements - Good curing condition

Progress of the chloride penetrations over time in this figure shows that for the specimen which had not been submitted to temperature, B3G, the penetration depth has not much increased. The concentration at the upper layers, however, has considerably increased. This might be attributed to the capillary absorption due to wetting and drying cycles. But for the specimen which is subjected to both thermal and hygral cycles, B1G, the two profiles are almost parallel to each other and penetration has reached down to 60 mm. This is certainly not due to only capillary suction which normally takes place at the near surface layers. That is rather the thermal cycles which have caused the deep profiles. It could, therefore, be explained that temperature fluctuations have controlled the chloride transport process rather than capillary suction which is the very nature of wetting and drying exposure. The chloride contents in the surface of specimens with thermal cycles found to be significantly lower than those without temperature cycles.

In specimens with temperature variations, however, the drying out of the concrete surface is much more than specimens with constant temperature. Therefore the capillary suction is expected to be more at the surface which appears not to be the case in this experiment. This is in agreement with findings of other investigations. Chloride profiles from sea water exposed concrete structures in hot countries is reported to have shown much lower chloride content at the concrete surface [8]. This point should therefore be considered when the simplified Fick's law is used for prediction of chloride ingress into concrete structures.

Effect of Slag Addition

The performance of concrete with and without slag with regard the chloride penetration is illustrated in Figure 5. This graph compares the data, at the end of experiment, of the first and second beams made with Portland and slag cement, respectively. The poor resistance of concrete with Portland cement compared to relatively excellent performance of slag concrete in this simulated aggressive marine environment is demonstrated.

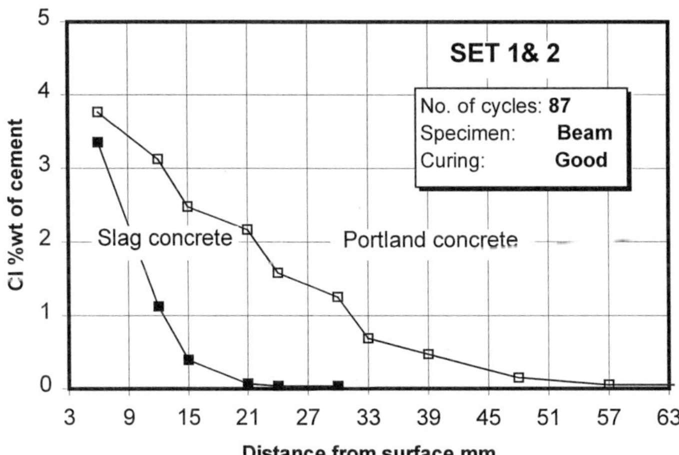

Figure 5 Chloride penetration after 87 cycles of exposure into beam sections with good curing for concrete with Portland and slag cement

This pattern of penetration, i.e. substantial increase in chloride resistance of slag concrete, was found for all the specimens (cubes and beams) regardless of curing conditions. The findings, however, are in good agreement with those of Gjørv [9] and Bijen [10]. They found good performance of concrete with slag cement compared with Portland cement.

Effect of Curing

In Figure 6 the chloride profiles after six months of exposure are plotted for OPC and BFSC specimens with both good and poor curing conditions. An increase in chloride ingress into specimens cured under poor condition is observed for Portland concrete whereas in slag concrete no significant increase is found. The effect of elevated temperature curing was more pronounced in the first period of experiment in Portland concrete. However, it is well known from literature that the capillary porosity and permeability of concrete is increased if it is exposed to a high temperature during hydration and curing process [11]. These data show that concrete with slag cement has relatively good resistance to chloride ions in such severe environment regardless of the applied curing conditions, at least in the conditions considered in this particular test.

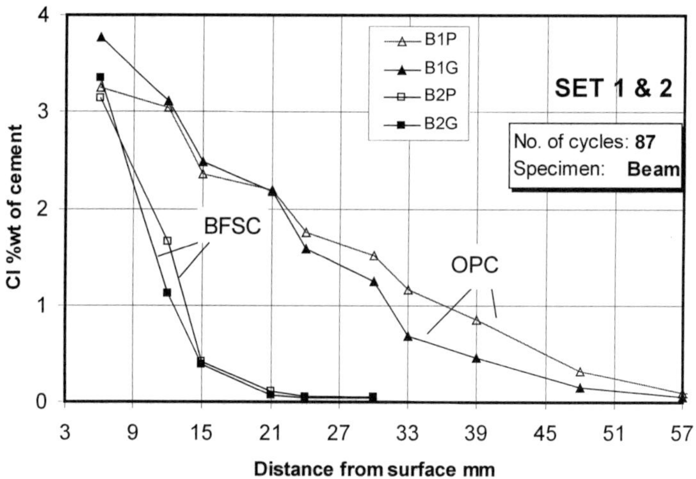

Figure 6 Chloride profiles of the beam sections made with Portland cement and slag cement – after six months of exposure

CONCLUSIONS

Concrete specimens in this investigation are loaded in alternate compression and tension through varying temperature. It was found that aggressive environmental conditions, i.e. wetting/drying cycles and temperature variations, have generally a significant effect on the ingress of chloride ions in marine concrete structures compared to normal environments. Furthermore, the following conclusions were made:

1. No significant microcracks were occurred at the surface of the concrete beams due to the imposed exposure condition. Comparison of large elements and small specimens, both exposed to temperature and hygral variations, showed that minor microcracking in the surface layers due to restraint of temperature-induced deformations have minor or no effect on chloride penetration rate in large specimens. The results of this study, with the particular test and boundary conditions, show that the cracks occurred at the concrete surfaces are not deep enough to accelerate the chloride penetration process. This does not mean, however, that in case of other thermal loading regimes microcracking will always be of minor importance.

2. Temperature cycles played a dominant role on promoting chloride ingress in marine concrete. The pronounced effect of temperature cycles, however, was found to be attributable to the higher temperature than to the formation of minor microcracks.

3. Slag cement concretes performed better than Portland cement concretes in the simulated aggressive marine environment.

4. Elevated temperature curing has a substantial effect on promoting the chloride transport in marine concrete, especially concrete with Portland cement.

REFERENCES

1. FOOKES, P G. Concrete in the Middle East - Past, present and future: a brief review. Concrete, 1993, pp 14-20.

2. AHN, W, et al. Accelerated durability testing of marine reinforced concrete under fatigue loading. Proc. Third Int. Conf. on Performance of Concrete in Marine Environment, supplementary papers, Ed. P.K. Mehta. St. Andrews-by-The-Sea, 1996, pp 191-202.

3. MEHTA, P.K. AND BREMNER, T W. Concrete in the marine environment - Some lessons for the future. Proc. Third Int. Conf. on Performance of Concrete in Marine Environment, Odd E. Gjørv Symposium on Concrete for Marine Environment, Ed. P.K. Mehta, St. Andrews-by-The-Sea, 1996, pp 175-190.

4. TAHERI, A. An experimental approach to study durability of concrete in marine environment under cyclic thermal loading. Progress in Concrete Research, Section of Concrete Structures, Faculty of Civil Engineering, Delft University of Technology, 1996, **5**, pp 65-75.

5. TAHERI, A AND VAN BREUGEL, K. Chloride penetration in concrete structures in aggressive marine environment - Experimental simulation. Proc. 7th Int. Conf. on Structural Faults and Repairs, Edinburgh, Scotland, 1997, pp 325-334.

6. JACOBSEN, S, et al. Concrete cracks: Durability and self-healing – A review. Proc. Second Int. Conf. on Concrete under Severe Conditions – Environment and Loading, CONSEC '98, Tromso, Norway, 1998, pp 217-231.

7. BIJEN, J M J M AND VAN DER WEGEN, R G. Swelling of concrete in deep seawater. Proc. Third Int. Conf. on Durability of Concrete, Nice, France, 1994, pp 389-407.

8. POLLOCK, D J. Concrete durability tests using the (Persian) Gulf environment. Proc. First Int. Conf. on Deterioration and Repair of Reinforced Concrete in the (Persian) Gulf. Bahrain, 1985, pp 427-441.

9. GJØRV, O E. Diffusion of chloride ions from seawater into concrete. Cement and Concrete Research, 1979, pp 229-238.

10. BIJEN, J M J M. Durability aspects of the King Fahd causeway. Concrete in Hot Climates. Ed., M.J. Walker, E & F N Spon: London, 1989.

11. CLARK, B A, et al. Electron-Optical evaluation of concrete cured at elevated temperatures. Proc. Int. Symposium on How to Produce Durable Concrete in Hot Climates, Ed. C. MacInnis. San Juan, Puerto Rico, ACI SP-139, 1992, pp 41-60.

BENCHMARKING THE CARBONATION OF HARDENED CONCRETE

M R Jones

R K Dhir

M D Newlands

A M O Abbas

University of Dundee

United Kingdom

ABSTRACT. This paper describes a study carried out to determine the potential for benchmarking the relative performance of hardened concrete in a carbonating environment. The development of an existing 1 year simulated natural carbonation test method is described. To conform to the test environment limits, active control over temperature, relative humidity and CO_2 concentration were required through the use of a climate controlled storage room and in-house CO_2 controller system. A novel three mix normalisation procedure is also described. The 1 year carbonation depths for a series of mixes of various binder types were ranked against a PC/PFA 30% Reference mix. Simulated natural carbonation depths were also compared to 20 weeks accelerated carbonation depths, showing the potential for using an accelerated test method to assess relative performance. The three mix normalisation procedure allowed a direct comparison between concrete mixes and showed the possibility of trade off between binder content, cover depth and carbonation resistance.

Keywords: Simulated natural carbonation, Accelerated carbonation, Active environmental control, Normalisation, Reference mix, Benchmarking.

M R Jones is a chartered civil engineer and senior lecturer in the Concrete Technology Unit in the Department of Civil Engineering at the University of Dundee. His research focuses mainly on binder technology, concrete durability and repair and maintenance.

Professor R K Dhir is the Director of the Concrete Technology Unit and Professor of Concrete Technology at the University of Dundee. He is a member of numerous national and international technical committees and has published extensively on many aspects of concrete technology, binder science, durability and construction methods.

M D Newlands is a research student studying the development of performance specifications for carbonation resistance of concrete and long term prediction of concrete durability in the Concrete Technology Unit at the University of Dundee.

A M O Abbas is currently undertaking research into the development of performance specifications for carbonation resistance of concrete in the Concrete Technology Unit at the University of Dundee.

INTRODUCTION

During the 1960's the types of concrete typically specified and delivered were comparatively simple. Prescription mixes, usually of a standard 28 day strength of around 20MPa with PC the main binder were commonplace. Specialised mixes were frequently concretes containing RHPC or SRPC [1]. Demands on ready-mix concrete suppliers have changed considerably over the past 35 years. With the large variety of constituent materials available today a typical ready-mix supplier can be faced with potentially 1000's of combinations of binders, aggregates and admixtures. For example, in Europe alone there are now 25 different binders available for use in structural concrete as specified in ENV 197-1[2].

This abundance of relatively new constituent materials has obvious implications for specifying concrete with a view to long-term durability. As the construction industry moves towards performance based specification, the long-term durability of a concrete mix is the key factor in specification. Comparing the long-term performance of concrete mixes against a known reference criterion, for example a concrete mix of known performance, has the potential to allow a trade off between a number of design parameters including strength, cover depth and resistance to deleterious agents such as atmospheric CO_2 or chlorides and thereby allows greater flexibility to designers and specifiers. Sufficient knowledge of long-term durability performance can also allow ready-mix concrete suppliers to tailor mixes in terms of material costs whilst having confidence that the mix will still perform satisfactorily, protecting embedded steel from corrosion in any environment.

SIMULATED NATURAL AND ACCELERATED CARBONATION TESTING

The most common cause of steel corrosion is through the process of carbonation[3]. Atmospheric CO_2 gas reacts with the pore fluids of the concrete reducing the pH. In the presence of moisture this can lead to the corrosion of embedded steel through the breakdown of the passive oxide layer protecting the steel. The long-term carbonation resistance of many binders and combinations of binders is relatively unknown and in order to benchmark concrete mixes against a reference criterion, a reliable and repeatable test method is required so that one can have confidence in the performance of the reference mix and relative performances of the test mixes

There are two common methods for testing the carbonation resistance of concrete. Simulated natural carbonation involves storing the concrete in an atmosphere with a small partial pressure of CO_2 of typically 0.03% to 0.04% by volume of air. In an attempt to simulate the effects of natural atmospheric CO_2, this test philosophy has the disadvantage of being somewhat time consuming as carbonation is a relatively slow process with some studies reporting that achievement of significant carbonation depths can take up to 20 years [4].

Accelerated carbonation testing subjects concrete to a much higher partial pressure of CO_2, in many cases between 4% and 100% by volume of air. This method has the distinct advantage that the rate of carbonation is increased considerably thus test periods are comparatively short. However, it is felt by many that accelerated carbonation is not a true reflection of the actual carbonation resistance of concrete mixes due to the effects of higher partial pressures of CO_2 on the physical and chemical properties of concrete [5]. It is for this reason that test results from specimens exposed to CO_2 levels closer to natural atmospheric levels are more widely accepted.

Simulated Natural Carbonation Testing

In natural exposure, concrete is exposed to a variety of micro-climates depending on its position within a structure. The three main exposure conditions which are concerned with carbonation-induced corrosion can be defined as:

1. Indoors

2. Outdoors sheltered from rain

3. Outdoors unsheltered from rain

The Committee for European Normalisation (CEN) have produced a draft test philosophy for simulated natural carbonation testing [6]. Concrete specimens are exposed to an atmosphere of 0.03%-0.04% (equivalent to 350±50ppm) CO_2 by volume, a temperature of 20±2°C and a relative humidity of 65±5%. In an attempt to replicate the 3 exposure conditions above, the test method has 3 storage conditions subjecting the concrete to varying degrees of moisture:

- **Class 1:** Continuous storage at 350±50 ppm CO_2, 20±2 °C and 65±5 %RH

- **Class 2:** As Class 1 but test specimens immersed in water for 6 hours every 28 days

- **Class 3:** As Class 1 but test specimens are immersed in water for 6 hours every 7 days

Depths of carbonation are measured at various test ages up to 1 year on concrete prisms by following recommendations of RILEM CPC-18 [7] where testing is by phenolphthalein indicator. The depth of carbonation of the test concrete is then compared to a concrete of established performance, the Reference mix.

The draft test method was identified as having a high statistical variability by CEN, thus the Concrete Technology Unit are developing the recommendations of this draft as part of an extensive study into the carbonation of hardened concrete involving a combination of binders available in the UK.

Conformance of test storage conditions with CEN Limits

In order to have confidence in test results for benchmarking purposes, it was necessary to conform to the test storage conditions and a number of environments were tested for variations in temperature, relative humidity and CO_2 concentration.

S1 - Normal laboratory environment

Concrete test specimens were exposed in the normal laboratory environment as this was simple. The environmental conditions were monitored over a period of one month and the three environmental parameters (temperature, relative humidity and CO_2 concentration) fluctuated significantly out with the test limits, Figure 1a, b and c.

S2 - Sealed controlled climate storage room

A room of volume 9.5m^3 with controlled temperature and RH was filled with dummy concrete specimens giving an exposed surface area of approximately 125m^2. With the room sealed for one month, CO_2 concentration was monitored. Figure 1a shows that the CO_2 concentration was, on average very low as it had depleted rapidly over a short period of time due to the carbonation reaction with the concrete. Temperature and relative humidity conformed to the test limits.

S3 - Controlled climate storage room with door slightly ajar

The door of the storage room was kept ajar to allow exchange of fresh air, however this increased the mean temperature, Figure 1b, and lowered the mean relative humidity, Figure 1c. The CO_2 concentration also increased when the laboratory air entered the storage room.

S4 - Controlled climate storage room with door opened for 24 hours every 48 hours

The door of the storage room was kept ajar for a 24 hour period every 48 hours in an attempt to minimise fluctuations. Although the mean CO_2 and temperature was now within the test limits, Figure 1a and 1b, the relative humidity was not.

From the four trial environments it was decided that active control was required on all three environmental parameters in order to conform to the test limits.

S5 - Controlled climate storage room with retro fitted CO_2 injection system

Proprietary controller and monitoring systems were prohibitively expensive, thus an in-house CO_2 controller and injection system, Figure 2, was retro fitted to the climate room at a cost of approximately £2500 (excluding both UK VAT and labour costs). Standard table fans were placed within the storage room to provide a turbulent atmosphere and prevent the formation of locally depressed partial pressures of CO_2.

The performance of the active control system in the storage room was monitored over a period of 1 month and the environmental parameters were kept within the test limits, Figure 1a, b, c.

Normalisation of Test Data

Given the relatively long period of the test and that depths of carbonation were likely to be small at 1 year, it was felt necessary to formulate and adopt a normalisation procedure to offset the inevitable variability in the production of test specimens at the initiation of the test. In comparing the performance of a test concrete with that of a Reference mix, the basis on which the mixes were compared, eg concrete grade, must be similar if a true comparison of performance is to be made.

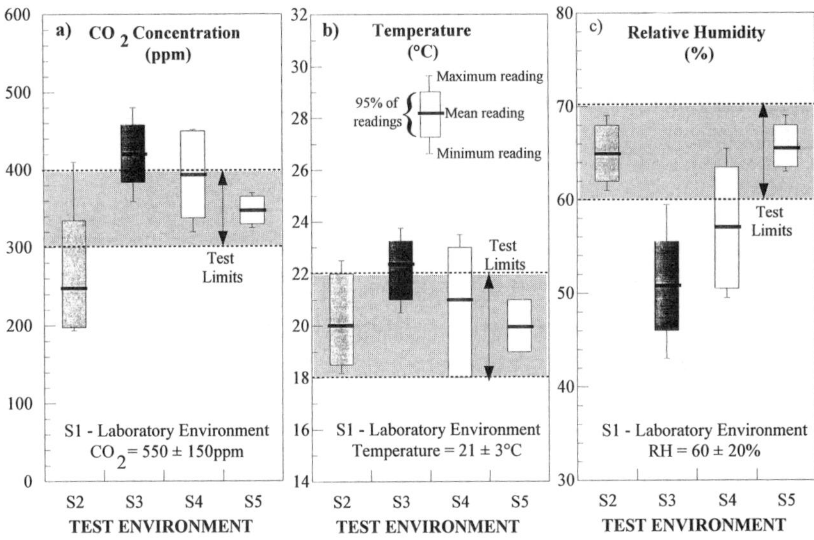

Figure 1 Box-whisker plots showing variation of a) CO_2 concentration, b) temperature and c) relative humidity for various test environments in the test chamber for a period of 1 month

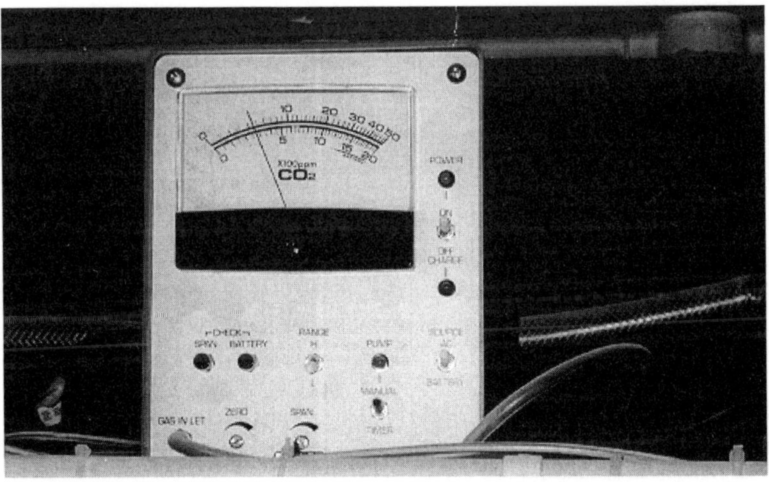

Figure 2 CO_2 monitoring/controller and injection system

The three mix normalisation procedure was as follows:

1. Cast one primary mix designed to give the required test concrete grade.

2. Cast two secondary mixes with ±8% of the primary mix binder content. These mixes have the same free water as the primary mix and are volumetrically adjusted by altering the fine aggregate content.
3. Plot the 28 day standard compressive strength against binder content for the three mixes. Obtain the exact binder content required to achieve the test concrete grade, Figure 3a.

4. Plot depth of carbonation against binder content for the three mixes. Obtain the exact carbonation depth for the required test concrete grade by reading from the binder content axis, Figure 3b.

The example shown in Figure 3 gives the results of a PC mix with 30% replacement by weight of PFA, the primary mix being designed for a standard 28 day compressive strength of 37 N/mm².

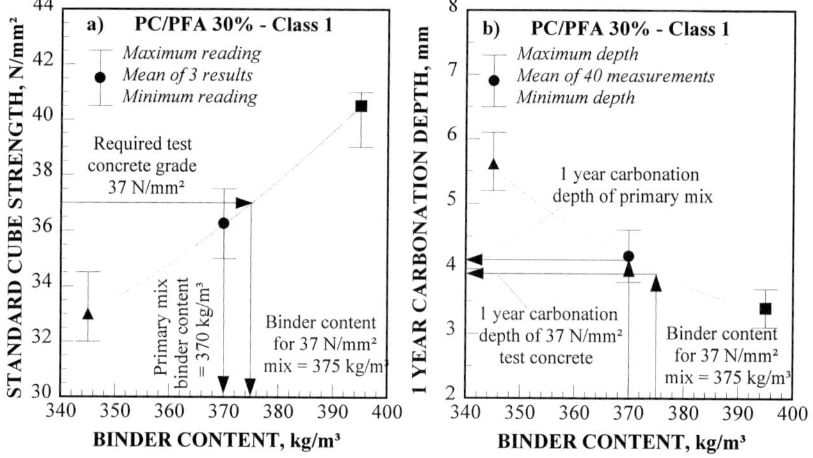

Figure 3 Adopted normalisation procedure, a) standard 28 day compressive strength is plotted against binder content and b) 1 year carbonation depth is plotted against binder content for primary and secondary mixes

The primary mix has an actual 28 day compressive strength of 36 N/mm², thus an additional 5 kg/m³ is required to achieve 37 N/mm². From Figure 3b the corresponding carbonation depth at 1 year is interpolated for exactly 37 N/mm². Although the difference in actual 1 year carbonation depth of the primary mix and the 1 year carbonation depth of the required test concrete grade is very small, if one were to project these depths on a simple $\sqrt{t}$ relationship, the difference would increase substantially with time.

A limited series of 10 repeated mixes of PC/PFA 30% were tested with and without the normalisation procedure. Table 1 shows a fivefold reduction in the coefficient of variation (V%) when using the normalisation procedure. A Chi-squared Test also showed that at a 95% confidence level there was a significant difference between the 10 mixes without using the normalisation procedure. However, when adopting the normalisation procedure the difference between the results became insignificant at the same confidence level.

Table 1 Statistical analysis of 10 repeated PC/PFA 30% mixes with and without the normalisation procedure

	SIMULATED NATURAL CARBONATION TEST	
	With Normalisation(1)	Without Normalisation(1)
Maximum, mm	4.0	6.0
Minimum, mm	3.5	3.0
Mean, mm	4.0	4.0
Standard Deviation, mm	0.2	1.3
V, %	6.0	31.0
Chi-squared Test(2)	2.56	7.60

(1) Test series included full control over storage conditions.
(2) At 95% confidence level, results are insignificant if Chi-squared test is less than 2.73.

Accelerated Carbonation Testing

Accelerated carbonation testing was also used in this study and involved the use of a chamber previously developed at Dundee University [8]. Specimens were exposed to an atmosphere of 4% CO_2 by volume, a temperature of 20 ± 2°C and a relative humidity of 55 ± 5%. A single exposure class with no cyclic wetting and drying was adopted and results compared to Class 1 exposure in the simulated natural carbonation test. Specimens were exposed for a period of 20 weeks and the normalisation procedure detailed earlier adopted.

BENCHMARKING RELATIVE CONCRETE PERFORMANCE

A series of 10 mix combinations constituting Portland cement (PC) from 2 sources, pulverized fuel ash (PFA), ground granulated blast furnace slag (GGBS), metakaolin (MK) and condensed silica fume (CSF) at various replacement levels were tested. The mixes were subjected to both simulated natural and accelerated carbonation testing. The adopted Reference mix, PC/PFA 30% was chosen as it was deemed to have a recognised satisfactory performance and is a mix commonly used for structural purposes.

Most urban structures now require a concrete grade of between 30-40 N/mm² [1], thus in the experimental programme the primary mix was designed to achieve a characteristic strength of 37 N/mm² after 28 days of standard water curing. Furthermore, this cube strength corresponds also to 30 N/mm² cylinder strength used widely in Europe [9].

The relative performance of all mixes to the reference mix, normalised to 37 N/mm² for both simulated natural and accelerated carbonation testing, is shown in Figure 4.

Across the three exposure classes in the simulated natural carbonation test it is clear that the comparative rankings of the binders vary with Class 1 exposure different from Classes 2 and 3. GGBS mixes show improved performance compared to the Reference mix as the degree of pore saturation increases, whereas metakaolin seems less sensitive to this effect.

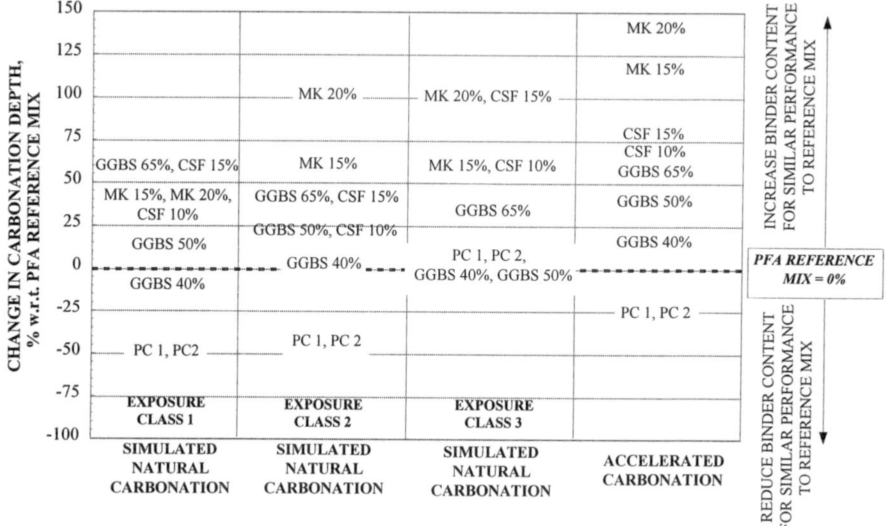

Figure 4 Comparison of carbonation depths of test mixes and PFA Reference mix for 1 year simulated natural and 20 weeks accelerated carbonation tests at 37 N/mm²

This phenomenon has been seen in other published studies [10] where under cyclic wetting and drying, mixes of similar grade performed comparably as pore saturation levels increased.

Figure 4 shows the relative performance of the mixes after 20 weeks exposure to accelerated carbonation. A direct comparison to the simulated natural carbonation, Class 1 exposure shows the ranking of the test mixes varies slightly. However, a correlation coefficient of 0.84 was found when comparing 1 year simulated natural carbonation depth with 20 week accelerated carbonation depth indicating the potential to use accelerated carbonation testing as a means of ranking the relative performance of concrete mixes. By testing the concrete at various ages up to 1 year, a relationship between depth of carbonation and time can be established for the concrete. From this relationship carbonation depths beyond 1 year may be predicted by:

$$d_c = k\, t^n$$

Where			
	d_c	=	depth of carbonation, mm
	k	=	rate of carbonation, mm/year (k depends on binder type)
	t	=	time, years
	n	=	constant dependant on the level of pore saturation.

Adjusting Test Mix Properties to Achieve Reference Mix Performance

Using the three mix normalisation procedure has the potential to allow the test mix proportions to be manipulated so the mix achieves similar performance to that of the Reference mix. Figure 5 gives an example of a CSF 15% mix in simulated natural carbonation, Class 1 exposure. The actual performance at 37 N/mm² is shown to be 6mm thus to achieve a 1 year carbonation depth of 4mm the binder content would have to be increased from 285 kg/m³ to 315 kg/m³.

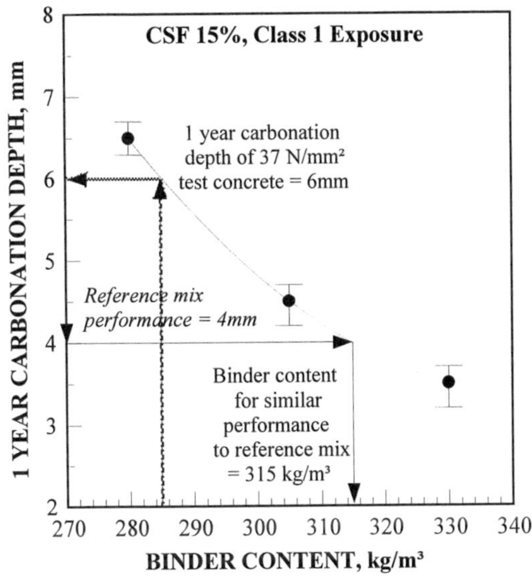

Figure 5 Example of achieving similar performance to the Reference mix by adjusting the binder content of the CSF 15% test mix

The establishment of a rate of carbonation for the three mixes in the normalisation procedure can also allow a trade off between carbonation resistance and cover depth. Rather than increasing the binder content of a mix, a nominal increase in cover depth would have the same effect. This is of particular significance in mixes which out-perform the Reference mix. In large structures reducing the cover depth may be more economically significant than reducing the binder content.

Practical Implications

The novel test method developed allows the relative performance of existing and new materials to be compared under strict laboratory conditions in a relatively short time period. Variations previously encountered [6] have been markedly reduced, increasing confidence in the results. The method is a step towards the development of an explicit design procedure for concrete durability and can allow designers to work alongside concrete suppliers to optimise material characteristics in a given environment.

CONCLUSIONS

In order to fully conform to the simulated natural carbonation test limits, full active control over temperature, relative humidity and CO_2 concentration is required. The application of a three mix normalisation procedure significantly reduced initial variations usually experienced in producing similar concrete, allowing a unique comparison of concretes on a similar grade basis.

In benchmarking the performance of various mixes against a Reference mix, accelerated carbonation methods have the potential to allow a reasonably rapid judgement of relevant carbonation resistance. However, the simulated natural carbonation test has the potential to allow long-term predictions of absolute carbonation performance.

By establishing the rate of carbonation of a concrete of unknown performance and benchmarking this against concrete of established performance, there is the potential to allow trade off between mix proportions, cover depth and carbonation resistance.

ACKNOWLEDGMENTS

The authors would like to acknowledge the support provided for the project by the UK Department of the Environment, Transport and the Regions, Ash Resources Ltd, Castle Cement Ltd, Cementitious Slag Makers Association, ECC International Ltd, PowerGen (PFA Sales) plc and Rugby Cement Ltd.

REFERENCES

1. BROWN, B, What the ready-mixed concrete industry has to offer, Concrete, Vol.31, No.2, February 1997, pp 14-18

2. ENV 197-1: 1992, Cement: Composition, specification and conformity criteria - Part 1: Common Cements.

3. PARROT, L.J., Some effects of cement and curing upon carbonation and reinforcement corrosion, Materials and Structures, 29, 1996, pp 164-173

4. PARROT, L J. A review of carbonation in reinforced concrete. C&CA/BRE Report C/1-0987. British Cement Association, July 1987, 45 pp.

5. BUENFELD, N R, HASSANEIN, N M, JONES, A J. An Artificial Neural Network for Predicting Carbonation Depth in Concrete Structures. 2nd ASCE Monograph on ANNs in Civil Engineering, 1997, Ch. 4

6. CEN, Measurement of the carbonation depth of hardened concrete. CEN Report CR12793, 17th January, 1997

7. RILEM Committee C56, Measurement of hardened concrete carbonation depth. CPC-18, Materials and Structures, 21(126), 1988, pp453-455.

8. DHIR, R K, JONES, M R, MUNDAY, J G L. A practical approach to studying carbonation of concrete., Concrete, Vol 19, No 10, October 1985, pp 32-34

9. HARRISON, T A. European standards, mañana? Concrete, Vol 31, No 2, February 1997, pp.24-26.

10. CONCRETE SOCIETY. The Use of GGBS and PFA in Concrete. Technical Report No. 40, 1992, 142pp.

DURABILITY OF CONCRETE STRUCTURES AND SPECIFICATION

C Bob

University Politechnica of Timisoara

Romania

ABSTRACT. The study, which represents part of the author's preoccupations concerning the durability of the concrete structures, deals with the influence of concrete strength on depth of the concrete cover. A quantitative model of reinforcement corrosion for both initiation period and corrosion process periods is presented in the work. The author's model is compared with other theoretical / experimental models which take into account the concrete compressive strength as a factor influencing the average value of carbonation depth. On the other hand, the corrosion process rate is presented as a function on its main influencing factors.

As a result of such studies an important conclusion for design is drawn: the depth of the concrete cover for the durability of the reinforced structures depends on concrete strength, environmental parameters and service lifetime requirements. A comparison between the author's model and European Norms is presented.

Keywords : Reinforcement corrosion, Concrete carbonation, Chloride penetration, Concrete cover, Durability, Service life, Codes, Specifications.

Professor Corneliu Bob is currently professor of Building Materials and Reinforced Concrete Structures, University "Politehnica"of Timişoara, Romania. He is also head of the National Research Institute INCERC-Timişoara branch. Professor Bob has published many papers and some books on various aspects of Civil Engineering : new types of concrete; durability and protection of constructions ; analysis of reinforced concrete structures.

INTRODUCTION

In order to make quantified statements as to the service life of a concrete member or a structure it is necessary to know the factors that affect service life. The durability aspects of concern in relation to failure or deterioration include chemical attacks, reinforcement corrosion, freeze – thaw bursting, alkali-aggregate reactions, fatigue, erosion. From these only the first four are really important [8]. The durability of concrete structures depends both on the resistance of the concrete against physical and chemical attack and on its ability to protect embedded steel reinforcement against corrosion. World-wide, the most common damage is the corrosion of the reinforcement adjacent to the exposed surface.

If concrete is able to withstand satisfactorily the environmental and working conditions to which it is exposed during the intended lifetime, the following factors have to be taken into account by the designer: choice of suitable constituents and concrete composition; mixing, placing and compacting of the fresh concrete; curing of the concrete. The durability requirements in ENV 206 for the fresh and hardened concrete related to the environmental exposure are: water-cement ratio, minimum quantity of cement, air content, frost resistant of aggregates, types of cement for plain and reinforced concrete, and concrete cover to reinforcement.

A QUANTITATIVE MODEL OF REINFORCEMENT CORROSION

A numerical calculation method for both initial period (time until deterioration start) and corrosion process period (time of deterioration) is presented.

The author of this paper has suggested a formula (based on Fick's Law) for the average value of the depth of carbonation [1], [2] and for the chloride ion penetration [3] as factors of initial period.

The formula, proposed by the author in 1986, takes into account: the binding capacity by coefficient **c** of type of cement, environmental conditions by coefficient **k**, surface concentration by coefficient **d** and permeation properties by concrete compressive strength $\mathbf{f_c}$

$$x = \frac{150ckd}{f_c}\sqrt{t} \qquad (1)$$

x- average depth of carbonation or chloride penetration, mm;
$\mathbf{f_c}$ – concrete compressive strength at time **t** , N/mm2
t- time of CO_2 or/and Cl^- action, years.

Numerical values of **c, k** and **d** are presented in Table 1.

The compressive strength, used in formula (1), can be established with nondestructive methods or other experimental procedures; the strength values are also used for the structural analysis of the buildings.

Table 1 Numerical values of influence coefficients

CARBONATION PROCESS

c – Cement Type

Cement	**I** 52.5	**I** 42.5	**II/A**	**II/B**	**III/A**
C	0.8	1.0	1.2	1.4	2.0

k – Environmental Conditions

Environmental conditions		Indoor	Outdoor protected	Outdoor average	Wet concrete
RH	%	<60	70-75	80-85	>90
	k	1.0	0.7	0.5	0.3

d – Concentration of CO_2

	%	0.03	0.10
CO_2	g/m³	0.36	1.20
D		1.00	2.00

CHLORIDE ION PENETRATION

c– Cement Type

Cement	**I**	**II/A**	**II/B**	**III/B**
c	1.0	0.9	0.75	0.67

k – Environmental Conditions

Environmental conditions	Value of **k** = $\mathbf{k_1} \times \mathbf{k_2}$					
Temp	°C	0-5	5-15	15-25	25-35	35-45
	$\mathbf{k_1}$	0.67	0.75	1.00	1.25	1.50
RH	%	50		0.85		100
	$\mathbf{k_2}$	0.75		1.00		0.75

d – Concentration of Chloride Ions

% of surface concentration	in front 0%	20	50	65	85
d	2	1	0.5	0.33	0.16

Note: % of surface concentration represents critical chloride concentration (around 0.2 % by weight of cement content for carbonated concrete and 0.4 % for noncarbonated concrete) from chloride environment (surface concentration)

There is also possible to use the compressive strength of concrete at 28 days, f_{c28}, instead of actual compressive strength, $\mathbf{f_c}$ at time **t**; an appropriate formula for increasing the concrete strength in time can be used :

$$f_c = 0.69 f_{c28} \log t_i \qquad (2)$$

where t_i is the concrete age in days and formula (2) is valid if $t_i \in (28; 360)$.

A rational application of formula (1) by using concrete compressive strength at 28 days is obtained by considering: the compressive strength f_{c28} for the first year and the compressive strength f_c for the next (t-1) years of service life of the concrete structure. For this assumption the formula (1) will become (t_i=360):

$$x = \frac{150ckd}{f_{c28}}(1 + \frac{\sqrt{t-1}}{1.766}) \qquad (3)$$

Other formulas for the average depth of carbonation, based on the concrete compressive strength, have been proposed:

Parrott [6], 1987:

$$x = \sqrt{521t}\exp(-0.05 f_{c28}) \qquad (4)$$

Duval [5], 1992:

$$x = (\frac{1}{2.1\sqrt{f_{c28}}} - 0.06)\sqrt{365t} \qquad (5)$$

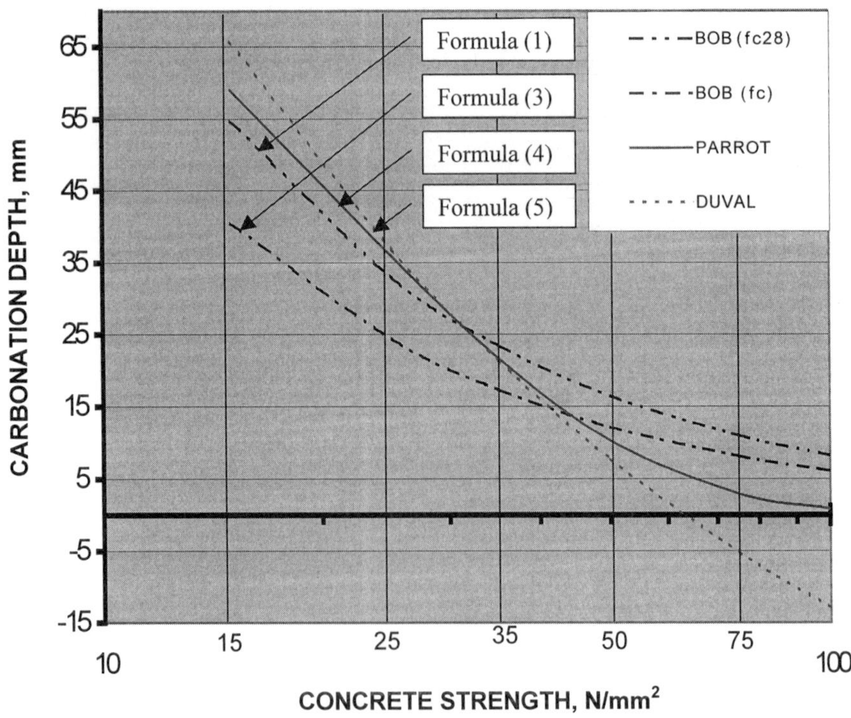

Figure 1 Carbonation depth as a function of concrete strength

In Figure 1 the carbonation depth as a function of concrete strength is presented in accordance with formulas presented. The author's formula have been used for two cases: formula (3) which takes into account the compressive strength $\mathbf{f_c}$ at time $\mathbf{t}$; formula (1) where $\mathbf{f_{c28}}$ was used instead of $\mathbf{f_c}$. The curves presented in Figure 1 were established for : CO_2 = 0.03 %, relative humidity RH < 60%; Portland cement Type I 42.5; service life $\mathbf{t}$ = 30 years.

From Figure 1 it can be observed that the curves related to the author's formulas take into account more properly the compressive strength

The corrosion process rate of the reinforcement in accordance with the author's studies, is presented in Figure 2.

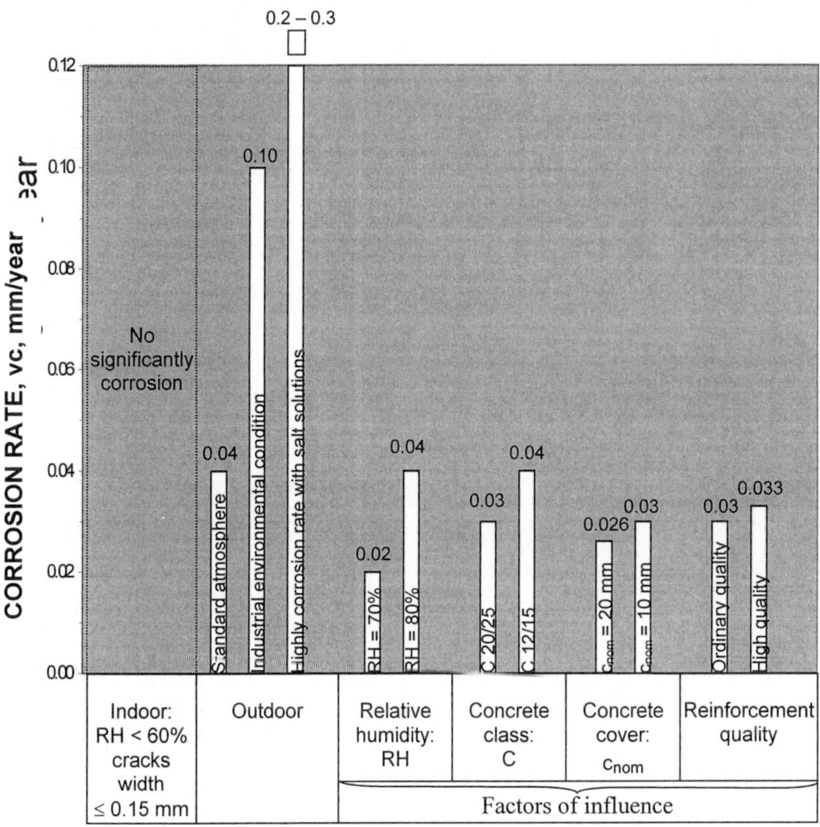

Figure 2 Corrosion process rate of reinforcement

COVER TO REINFORCEMENT AND CONCRETE QUALITY FOR DURABILITY

From the "Quantitative Model "there is possible to points out the importance of two parameters for the durability of the reinforced concrete structures under normal conditions : the concrete cover and concrete strength.

The concrete cover, in mm, and concrete quality for durability are presented in Table 2 : the minimum or nominal concrete covers specified in European norms as well as in accordance with the author 's model are presented [1, 2, 7].

The minimum concrete covers specified in EC 2 for reinforcement and prestressing steel are presented. The nominal covers and concrete strength classes, according to Concise Eurocode [4], should not be less than the value given in Table 2 for the appropriate exposure class. For prestressing steel, the concrete cover to a pre-tensioned tendon or to a duct containing a post-tensioned tendon, should be 5 mm greater than values given in Table 2 from Concise Eurocode; on the other hand, the minimum cover should be not less than twice the tendon size or the duct diameter.

The values calculated from the author's model are in accordance with environmental conditions (exposure class 1 to 4) for different concrete strength as well as with design value of service life. For exposure class 2 b – humid environment with frost – the calculated values of the concrete cover have taken into account a diminished value of concrete strength by 20%. The recommended values are established in function of the calculated data but each value is a multiple of 5 mm. For other environmental conditions, concrete strength and service life time, the calculated and recommended values of minimum concrete cover can be established from the author's model.

The nominal values c_{nom} of the concrete cover are equal with minimum value c_{min} (Table 2) plus tolerance. According to EC 2, the tolerances are 5 to 10 mm for in situ cast concrete and 0 to 5 mm for precast elements.

Other reasons for larger covers are:

- Ensuring bond strength
  $c_{min} \geq \phi$, where ϕ is the reinforcement diameter;

- Use of large aggregate size with $d_g > 32$ mm
  $c_{min} \geq \phi + 5$ mm ;

- Concrete elements on
  foundations, $c_{min} \geq 75$ mm; soils, $c_{min} \geq 40$ mm;

- Prestressed concrete with
  round section reinforcement, $c_{min} \geq 2\phi$; profiled reinforcement, $c_{min} \geq 3\phi$

- Ensuring fire protection - where the nominal concrete cover exceeds 70 mm, special skin reinforcement will be required.

Table 2 Concrete cover

METHOD OF CALCULATION			EXPOSURE CLASS					
			1	2a	2b	3	4 a	4 b
Minimum cover (mm), from **Eurocode 2** for:	Reinforce – ment		15	20	25	40	40	40
	Prestressing steel		25	30	35	50	50	50
Nominal cover(mm) to reinforcement from **Concise Eurocode** for concrete class:	C 25/30		20	-	-	-	-	-
	C 30/37		20	35	-	-	-	-
	C 35/45		20	35	35	40	40	40
	C 40/50		20	30	30	35	35	35
	C 45/55		20	30	30	35	35	35
Minim cover (mm) to reinforcement from **Author's Model** for concrete class, and environmental conditions; t = 100 years	C 25/30	C	-	18.75*	22.5*	44.62**	59.63**	59.63**
		R	15	20	25	45	60	60
	C 30/37	C	-	16.67*	20.0*	36.80**	51.83**	51.83**
		R	15	20	25	40	50	50
	C 35/45	C	-	15.0*	18.0*	31.32**	45.66**	45.66**
		R	15	15	20	30	45	45
	C 40/50	C	-	12.5*	15.0*	23.90**	35.85**	35.85**
		R	15	15	20	25	35	35
	C 45/55	C	-	10.71*	12.9*	19.80**	29.70**	29.70**
		R	15	15	15	20	30	30
	RH, %		< 60	75 - 85		75 - 85	75 - 85	
	Cl^-		0	0		5 g/l	10 g/l	

Notes : C – calculated values ; R – recommended values
* - depth of carbonation ; ** - chloride ions (Cl^-) penetration

CONCLUSIONS

1. The quantitative model of reinforcement corrosion for both time until deterioration starts-initiation period-and time of deterioration-corrosion process period – is an analytical tool for providing a diagnostic guide for the control of concrete structures.

2. The use of the compressive strength as an influencing parameter for durability, particularly carbonation, was chosen for the authors model since it reflects the quality and content of the cement, the water – cement ratio, the aggregate characteristics, the casting conditions. Moreover, the concrete compressive strength is the major criterion when assessing the quality of a concrete class for the design of a new concrete structure as well as for judging of a concrete structure which has to be renovated.

3. As a result of the author's studies an important conclusion for design is drawn: the influence of the strength and depth of the concrete cover for the durability of the reinforced concrete structures is critical.

REFERENCES

1. BOB, C. The model of corrosion of the reinforcement in concrete. (in Romanian), Proceedings of the Symposium ICCPDC Timişoara, 1986.

2. BOB, C. Some aspects concerning corrosion of reinforcement. Proceedings of the International Conference "The Protection of Concrete", Dundee, Sept.1990, E & F.N. SPON, London.

3. BOB, C. Probabilistic assessments of reinforcement corrosion in existing structures. Concrete in the Service of Mankind, E & F.N. SPON, London, 1996

4. BEEBY, A W, AND NARAYAANON, R S. Concise Eurocode for the design of concrete buildings. Crowthorne, BCA, 1993

5. DUVAL, R. La durabilité des armatures et du béton d'enrobage. La Durabilité des Bétons, Presses de l'Ecole Nationale des Ponts et Chaussées, Paris 1992

6. PARROTT, J. A review of carbonation in reinforced concrete. Cement and Concrete Association, Ed. Wrexham Springs, Slough, July 1987.

7. COMMISSION OF THE EUROPEAN COMMUNITIES. Eurocode No 2 Design of concrete structures. Final Text, 1991

8. IABSE CONFERENCE DAVOS 1992, Structural Eurocodes, Report

9. ROMANIAN CODE STAS 10107/0-90. Design and detailing of concrete reinforced concrete and prestressed concrete structures. (in Romania).

THEME THREE:

WHOLE LIFE COST AND DURABILITY AUDIT

Keynote Paper

DURABILITY DESIGN OF CONCRETE STRUCTURES MINIMISING TOTAL LIFE CYCLE COSTS – CONSIDERATIONS AND EXAMPLES

M Geiker

COWI

Denmark

ABSTRACT. With recent and ongoing major Danish construction works as a basis, e.g. the Metro in Copenhagen, important design considerations are dealt with. These design considerations include the need, initially in the design process:

- to identify environmental exposure
- for the owner to select and define the design service life
- for the owner to select a maintenance strategy

Based on an evaluation of possible deterioration and acting transport mechanisms as well as constructability and requirements to materials and structural detailing (e.g. cover, supplementary protection), the application of the so-called multi-barrier protection strategy is dealt with.

Keywords: Durability design, Total life span costs, Verification.

Dr Mette Geiker is Chief Engineer in Concrete Technology with COWI, Denmark. She has experience in the specification of concrete for special purposes and assessment of service life of new and existing structures, in the investigation, assessment and repair evaluation of deteriorated reinforced concrete structures, especially structures affected by chloride initiated reinforcement corrosion, alkali silica reactions and freeze/thaw attack, and in testing of materials. Dr. Geiker is vice chairman of *fib* Commission 8, Concrete, and member of RILEM TC 116 & TC 160, ACI TC 231 & TC 365, and the Danish Concrete Council, and past president of the Danish Concrete Society and the ACI Copenhagen Chapter.

INTRODUCTION

Based on his year-long experience, de Sitter has introduced the Law of Fives: "One pound spent in phase A equals five pounds in phase B, 25 pounds in phase C, and 125 pounds in phase D [1]. Where the phases are:

A Design, construction, and curing
B Initiation processes are under way, but propagation phases of damages has not begun yet
C Propagating deterioration has just begun
D Advanced state of propagation with extended damage occurring

De Sitter concludes that it is high time that engineers working in scientific institutions, engineering consultants, and contractors take up this challenge and concentrate on phase A and B.

A more homogeneous level of safety with regard to degradation can be obtained by applying reliability based durability designs. A report on systematic attempts to introduce both a general theory of structural reliability and existing calculation models for common degradation processes into structural design has recently been published, [2]. Furthermore, a performance based durability design methodology is being developed, [3]. The methodology is based on principles of reliability analysis in contrast to the present design method which to a large extent is empirical (deem-to-satisfy rules). This new design methodology will be based on realistic and sufficiently accurate environmental and material models, capable of predicting the future behaviour of a concrete structure. Reliability based design will not be dealt with further in this paper.

MULTI-BARRIER APPROACH

To obtain a high reliability, a so-called multi-stage barrier design has been formulated. The strategy includes the following, [4].

Identification of:

- environmental exposure
- critical structural zones
- transport and deterioration mechanisms
- determining parameters

Selection of the optimal combination of:

- barriers
- future protective means
- appropriate maintenance strategy.

Service life design hence starts with identifying types and aggressiveness in the environment, and to differentiate them on the structure elements.

Example - Reinforcement Corrosion, Barriers to Consider

Concrete structures in car parks, balcony accesses, road bridges, harbours, tunnels, and swimming pools are exposed to aggressive water containing chlorides. When chlorides in sufficient amounts reach the reinforcement, it starts to corrode, and the service life of the structure decreases.

Reinforcement corrosion can be prevented both in the design phase and in the operation phase. Traditional repair, however, must often be repeated each 5-15 years. Many means can hinder the corrosion process either in the design phase or during operation.

Structural detailing

The concrete cover over the reinforcement is of outmost importance for the durability. As a rule of thumb, double thickness of the cover causes four times longer service life. Limitation of joints and construction joints and simplification of structural detailing decrease the risk of early deterioration. At the same time, the structural detailing must allow for correct placement, compaction and curing of the concrete.

Dense concrete

As well as the execution (e.g. compaction, moisture and temperature curing) the constituent materials and the mix composition have significant influence on the imperviousness of the concrete.

Barriers on the concrete surface

The outer barriers can be flashing, membranes, surface protection, and impregnation.

Surface coating of the reinforcement

The reinforcement can be protected by metallic surface treatment, such as hot galvanised, or a non-metallic coating as epoxy. Hot galvanised steel is protected due to the higher tendency of zinc to corrode. Epoxy-coatning creates a barrier between the reinforcement and the concrete, if it is free from flaws and defects.

Alternative reinforcement materials

Examples of alternative materials for reinforcement are fibre reinforced plastic materials and stainless steel, both non-corrosive.

Bars of fibre-reinforced plastic consist of parallel fibres (glass, aramid or carbon) in a matrix of epoxy or polyester resin. The fibres provide strength and stiffness, whereas the matrix protects the fibres and distribute strains.

Stainless steel is alloy steel with chromium and nickel or molybdæn (other constituents may be added). The alloying increases the corrosion resistance. The most used alloy steel is AISI 316, whereas duplex alloy steel is used in very aggressive environments.

Cathodic protection

Cathodic protection hinders or controls reinforcement corrosion by introducing an electrical current from an anode through the concrete to the reinforcement. The reinforcement is thus polarised and corrosion inhibited. At the same time negative ions, e.g. chlorides, are rejected.

Electro-chemical chloride extraction

Chloride extraction is based on the same principle as cathodic protection, but much larger currents are applied, causing negative ions to migrate from the reinforcement to the concrete surface, where they are collected and removed.

The ingress of aggressive substances should be verified by means of corrosion sensors. Hereby supplementary corrosion protection can be initiated at the optimal point, and the total costs of repair and maintenance reduced.

RELIABILITY UPDATING

To facilitate the operation and maintenance planning, reliability updating of service life prediction is undertaken by incorporating inherent uncertainties due to:

- inaccurate deterioration models
- inaccurate testing methods
- insufficient information

and by carrying out a probabilistic treatment of information to ensure:

- consistent treatment of uncertainties
- assessment of the influence of parameters
- sensitivity analysis to identify critical parameters
- evaluation of benefits of additional tests and inspections.

Thus, by repeated investigations and incorporating reliability updated and improved knowledge of the deterioration phenomena and their time dependency, a reliable forecast of future deterioration and means to evaluate the consequences of alternatives can be obtained.

These achievements are to be used to improve the traditional O&M procedures, including dynamic inspection frequency and targeted maintenance procedures, to meet the owner's specific service life requirements.

CASE STORY 1 - THE GREAT BELT LINK

The USD 4 billion Great Belt projects have included one of the longest suspension bridge spans in the world; incorporated the most extensive pre-casting of a major bridge with elements weighing up to 7,400 tonnes placed by floating crane; and a tunnel driven through probably the most difficult ground conditions ever encountered with a Tunnel Boring Machine. The railway link was opened in 1997, and the roadway link a year later.

The owner's requirements regarding durability complied:

- 100 years of service life
- minimal total life cycle costs
- improved structural and performance reliability

The 100-years perspective of the design resulted in very strict requirements to all phases of the design, construction, and operation processes, particularly as the structure is exposed to relatively aggressive environments.

The main considered deterioration mechanisms were:

- freeze/thaw damage
- alkali-silica reactions and sulphate attack
- reinforcement corrosion

A detailed study of available models describing the foreseeable critical deterioration mechanisms in a quantified way revealed little to help the design. Only limited practical means of calculating the transport of aggressive substances into concrete was available.

Figure 1 The Great Belt Link, Railway Tunnel

Multi-Barrier Strategy

To obtain a high reliability, the multi-stage barrier design was applied, [4.5]. The concentrations of the most important aggressive substances were:

- chloride ions in the soil and in sea water: 19 000 ppm
- sulphate ions in the soil: 2 500 ppm

These concentrations are not very high compared to the concentrations found in other parts of the world, and the temperature levels are also limited. However, in a 100-year perspective and for structures subjected to evaporative effects, this exposure is still serious.

The types and number of protective barriers have been adjusted to the relative degrees of aggressiveness in the environment, taking into account the consequences of deterioration and the feasibility, and to ease maintenance and repair works in the future.

Figure 2 The Great Belt Link, West Bridge

The following barriers were selected, based on a 1st order service life prediction:

Tunnel:

- annular grout
- dense concrete (w/c = 0.35, three powder mix of cement, fly ash and microsilica)
- appropriate cover
- provision for future cathodic protection
- epoxy coating of reinforcement

Bridges:

- dense concrete (w/c = 0.35, three powder mix of cement, fly ash and microsilica)
- appropriate cover
- provision for future cathodic protection
- Permeable Formwork Liner (PFL) or silane impregnations.

Besides cover, structural detailing was considered with regard to service life.

Concrete Mix

The concrete mix designs were evolved from national and international experience within high quality concretes, but governed by Danish knowledge and experience, and available materials. Selected requirements to the concrete are summarised in [6].

Figure 3 The Great Belt Link, East Bridge

Curing

To ensure the highest possible integrity and low penetrability of the outer concrete layer when exposed to saline environments, the curing time was related to the type of concrete. Curing also included control of differential temperatures during hardening. The theoretical treatment of this problem has been a research and development key area in Denmark for more than 25 years, and much knowledge and experience are available. The general requirement is that the concrete must remain crack-free during hardening

Usually this should be achieved if differential temperatures within a concrete pour are maintained below 18°C, and if the differential temperature across a construction joint is maintained below 12°C. However, in several major castings these requirements proved insufficient to ensure crack-free concrete. Detailed 2D, and sometimes 3D, finite element calculations, incorporating the early age transient concrete properties determined on the actual concrete mix, were required. The results enabled the contractor to adjust the initial concrete temperature, the rate and timing of internal cooling with cast-in cooling pipes, the time of stripping the form, and the external weather conditions, for cracking to be avoided or minimised. These efforts were very successful once the complexity of the system was fully understood. The availability of sophisticated FEM programmes was essential for this exercise.

Improved Maintenance Planning

The operation and maintenance strategy, selected by the Owner to provide a high reliability and low total life cycle costs, includes, [7]:

- routine inspections
- intensified inspections at hot-points such as exposure and non-conformances
- smart structures monitored with durability sensors
- reliability based service life updating
- pro-active preventive and corrective maintenance.

Service Life Updating

A realistic service life forecast of concrete structures needs verification of the design parameters as the design quality in itself is wishful thinking. The real quality is determined during execution, and performance is only detectable later. Service life prediction therefore requires:

- in-situ testing of actual quality
- applicable deterioration models.

Monitoring

Durability monitoring is part of the long-term operation and maintenance procedures, and constitutes an integral part of the service life design of the Great Belt Link. The monitoring undertaken is based on hypotheses of deterioration and available data. The following mechanisms are considered:

- chloride corrosion of reinforcement
- freeze/thaw attack

The monitoring includes the following actions:

- durability monitoring of in-situ performance through Corrosion Monitoring System, CMS, and measurement of chloride and moisture profiles
- determination of chloride threshold for initiation of corrosion and critical degree of saturation (frost)

450 corrosion sensors (6-step corrosion-cells) have been installed. The principle of the corrosion sensors is described in [8]. The selected areas for monitoring are tidal, splash and atmospheric zones, and construction joints at the West Bridge and the East Bridge; and intelligent segments at the East Tunnel.

The sensors provide information on:

- time-to-corrosion directly
- potentials
- resistivities (moisture level)
- temperature
- rate of corrosion

CASE STORY 2 - THE COPENHAGEN METRO

The first deep underground mass transit system in Copenhagen, the Metro, is now under construction, with structures both below and above the ground. The owner's main requirements to the concrete works were 100 years' service life, a design-and-construct contract – i.e. primarily performance requirements – and use of the European standards.

It was the first time Eurocodes were to be used for an all-Danish project. At the beginning of 1995, the relevant European standards were: ENV 1992-1-1, ENV 206, and ENV xxx, Execution of concrete structures. These standards did, however, not provide the necessary requirements and guidelines: ENV 1992-1-1 is based on 50 years' service life and does not include structures exposed to penetrating saline water, and neither ENV 206 nor ENV xxx were ready. Thus, several other standard systems and local Danish test methods had to be included, as well as valuable expereince of the use was available from the current Swedish-Danish fixed link.

The project development as well as the current status of the Copenhagen Metro are described in [9 and 10].

Figure 4 The Copenhagen Metro

Environmental Exposure

The civil works comprise the following main types of concrete structures: tunnels (bored, NATM, and cut-and-cover); deep stations; ramps; viaducts; above-ground stations. These structures will be subjected to one or more of the following exposure conditions: ground water, occasionally with chlorides and sulfates; industrially contaminated soil; freeze/thaw action; de-icing salts.

Durability

The design life for the permanent works is 100 years. For some components such as edge beams, flexible joints, bridge deck membranes, a design life of less than 100 years is accepted, provided that foreseen remedial measures are described.

Precautions against the influence of the environment and foreseen use should be taken by: appropriate detailing, including reinforcement layout; placing structural parts at the furthest possible distance from exposure (e.g. traffic); use of a minimum 1% slope on horizontal surfaces to allow water to drain; limiting the influence of design and execution (including construction joints) on the risk of early cracking due to temperature and shrinkage.The contractor should select the constituent materials, the mix composition and the structural detailing (e.g. cover) reflecting the exposure conditions. Four exposure classes were specified:

E Extra:
Exposure to penetrating water containing chloride or ponding water with de-icing salt.

A Aggressive:
Humid indoor and outdoor environment where alkalis and chlorides have access to the concrete surface. Frost action is possible.

M Moderate:
Structures indoor and outdoor in humid environment. Frost action possible.

P Passive:
Dry environment with no chloride action.

When selecting the environmental class for each structural component, attention had to be paid to the special environmental influences to which the structural part may be exposed, also during construction. Furthermore, the risk of accumulation of aggressive substances by evaporation from walls, exposed to ground water or due to splashing with contaminated water should be taken into consideration in the design phases.

Constituents, Concrete Composition and Structural Detailing

In a design-and-construct contract, the contractor is responsible for the detailed design to meet the structural and durability requirements. Thus, the owner's requirements should preferably be stated as performance requirements. In 1995, reliable and verified performance test methods and acceptance criteria were not available. For example, the negative effect of carbonation on the porosity of slag concrete was not – and still is not – included in frost testing. Also, the effects of items such as concrete composition, age, and moisture were not included in the testing and modelling of resistance to chloride ingress. The owner requested that the project should not be a full-scale, real-time durability experiment and thus a set of requirements and guidelines were given to:

Materials

- Constituent materials
- Concrete composition (guidelines)
- Concrete properties

Execution

- Cover (guidelines)
- Early age cracking

Quality assurance

- Pretesting
  - Concrete composition
  - Full-scale trial casting
- Production testing.

The contractor was free to select other concrete compositions and covers, provided the suitability and the long-term performance were documented.

Exposure classes for the regions covered by the CEN-member states are defined in ENV 1992-1-1 and divided into five main classes and nine sub-classes. Requirements for concrete constituents, composition and properties are given in ENV 206:1989.

During the revision of ENV 206, an unsuccessful attempt was carried out to agree on performance requirements. Substituting prescriptive ('detailed') requirements were suggested but unfortunately not agreed upon

In view of this lack of agreement, the relationships given in ENV 206:1989 between exposure classes and requirements should be used only as guidelines.

Constituent Materials and Concrete Composition

In 1995, the available European standards on binders included the cement standard ENV 197 and the standard for fly ash EN 450. DS 427 is a translation of ENV 197 but with fewer cement types. European standards were not available for silica fume, chemical admixtures, water, and aggregates.

Cement types CEM I and CEM II/A-L as well as fly ash and silica fume are well known under Danish exposure conditions and are allowed for all exposure classes, except CEM II/A-L which is allowed for exposure classes M and P, only. Other cement types might be used if their suitability and long-term performance in the actual structure and environment were documented.

For exposure classes E, A and M, a maximum silica fume content of 5.0% and a maximum fly ash content in the range 10-20% were proposed. As mentioned above, other concrete types could be selected, if their suitability and long-term performance in the actual environment could be documented.

Provision for Cathodic Protection of Reinforcement

To provide provision for future cathodic protection – and earthing – requirements were given for electrical connection between reinforcement steel.

Durability Monitoring

Corrosion sensors should be installed at selected positions to facilitate future operation and maintenance planning by verifying the rate of ingress of chlorides and the threshold chloride content for chloride-induced corrosion.

Pre-Testing and Production Control

Constituent materials

Apart from test methods for chromate content and heat of hydration, test methods for cement were covered by EN 196. Test methods for fly ash were covered by EN 450, EN 451 and EN 196, except for the available alkali content (EN 196-21 test for total alkali content, for which no requirement is given).

The European standards only to a small extent covered test methods for silica fume. Remaining test methods were covered by ASTM, DIN and test methods prepared by the Danish Technological Institute. A European standard for water for concrete was not available, so national test methods had to be applied. European test methods for alkali content (pr EN 480-12) and chloride content (pr EN 480-10) in admixtures were available. Requirements in accordance with pr EN 934-2 and -6 concerning consistence retention and workability consistency should be complied with.

No European standards were available for fine and coarse aggregates. Instead, ASTM, BS, CSA (Canadian Standards Association), NT Build (Nord Test Build), and the Danish Technological Institute test methods were used.

Concrete

European standards were not available. For air content, density, initial setting time and bleeding, test methods from NT Build were used, while ISO standards were used to test workability. The air stability of the fresh concrete is important to all concrete, exposed to frost in a saturated or near- saturated condition.

Transportation, secondary admixture dosage, pumping, placing, compaction, and finishing may all influence the air content and air void system in the hardened concrete. Thus, it was considered necessary to measure the air stability by casting 1m3 blocks using similar methods as planned for construction.

European standards were not available for the test of hardening and hardened concrete. Thus, methods of NT Build, ASTM, SS (Swedish Standards), Danish Technological Institute, and AEC Consulting Engineers were used (see Table 1).

Table 1 requirements for hardening and hardened concrete

PROPERTY		TEST METHOD
Compressive strength	28 days (declared value only)	NT Build 203
Heat development	Q, T, α (declared value only)	NT Build 388
Air void structure (frost resistance)	< 0.35 mm > 0.35 mm Spacing factor	ASTM C 457
Frost resistance	m56 m56/m28, m112/m56	SS 137244
	Dilation after 3, 6, 12, 24 weeks moist curing	ASTM C 671
Chloride ingress	Diffusion coefficient	NT Build ?
	Coulombs passed or electrical resistivity (declared value only)	ASTM C 1202 or resistivity test
Petrographic analysis	Impregnated plane sections Thin section	TI-B 3 and 5

Many of the Metro structures will be exposed to moisture as well as freezing and thawing, causing a risk of either surface scaling or internal cracking. No European standard for testing risk of freeze/thaw damage was available. Generally, requirements to the air void structure are given as substitute requirements for frost resistance.

Such requirements are also given to the concrete for the exposed parts of the Metro, supplemented with requirements to performance testing during pretesting and – in the case of non-conforming air void structure or significant changes in petrographic test results – during production. The performance testing includes testing for surface scaling (SS 127244) and internal cracking (ASTM C671). The duration of the performance test methods is between two and six months, which unfortunately causes a significant delay in the final decision on approval or rejection.

Performance testing of resistance to chloride ingress is carried out by both an immersion test (APM 302, now NT Build 443) and a migration test (ASTM C1202 or a simple resistivity test). The immersion test provides information about the acceptance of the concrete during pretesting, whereas the migration test, being much more rapid, may be applied for production testing, the acceptance criteria being declared based on correlated immersion and migration testing. As both the porosity and the conductivity of the pore liquid affect the migration test results, the migration test should be supplemented by an immersion test once a year.

Repair works

Experience from other major construction projects shows the need for repairs to be carried out during construction. Relevant requirements for constituent materials, composition, production, fresh, hardening and hardened concrete properties must be applied. Specific requirements and test methods for repair were not covered by European standards. Concerning patch repair, national standards as SS, ASTM and Danish methods were used. Injection of cracks with epoxy materials should follow the German guidelines, ZTV.

Pre-testing, including full-scale trial casting

In major construction projects, the contractor often has to face many problems with the first large construction element or first large section of the structure. To provide a first-hand documentation of the suitability of the selected constituents and concrete composition, pre-testing is performed by casting and testing 1m3 concrete blocks.

Following the preliminary approval, full-scale trial castings were carried out to provide a sound basis for detailed planning of the execution and a final approval of the concrete mix and construction procedures. At the full-scale trial castings, all requirements to the constituent materials and the concrete mix and requirements to other properties should be documented to be fulfilled simultaneously, using the construction methods proposed in a method statement for construction. The full-scale trial casting should be of minimum $15m^3$ for each mix design.

CONCLUDING REMARKS

Total life costs can be minimised by limiting extensive deterioration repair works. Thus, effort should be made establish a service life design of high reliability and a system for verification of the design and early warning of possible deviations.

To obtain a high reliability, a multi-stage barrier design has been formulated and used, among others, on two large Danish transportation projects. A more uniform safety level is expected obtained when reliability based durability design is introduced as a common tool.

Understanding of transport and deterioration mechanisms and quantitative prediction of concrete's service life as well as the related performance test methods have not yet reached a stage which allows for concrete works specifications stating performance requirements, only.

REFERENCES

1. CEB-RILEM INTERNATIONAL WORKSHOP. Durability of concrete structures: Workshop Report by Rostam, S. COWI, 1984. Also available as CEB Bulletin of information no. 152, 1984

2. SARJA, A AND VESIKARI, E. Durability design of concrete structures, Chapman & Hall, 1996.

3. ROSTAM, S AND FABER, M. Probabilistic performance based durability design of concrete structures (DuraCrete). Nordic Concrete Research, Research Projects 1996, pp. 316-317

4. INTERNAL REPORT ON THE DURABILITY DESIGN FOR THE GREAT BELT LINK TUNNEL. COWI, 1989. Concept published in, by Rostam, S.: Durability design - the European approach. ACI Concrete International, July 1993

5. VINCENTSEN, L J AND HENRIKSEN, K R.: The Great Belt Link - Built to last, ACI Concrete International, July 1992, pp. 30-33

6. KJÆR, U, SØRENSEN, B AND GEIKER, M.: Chloride resistant concrete - theory and practice. International conference on concrete across borders, Odense, 1994, pp. 227-237

7. GEIKER, M, ROSTAM, S AND VINCENTSEN, L J.: Next generation bridge management systems. International conference on repair of concrete structures, Svolvær, Norway, 1997, pp. 24-32

8. SCHIESSL, P AND RAUPACH, M.: Monitoring system for the corrosion risk of steel in concrete structures. ACI Concrete International, July 1992, pp.52-55

9. JACOBSEN, A S.: The Copenhagen Metro project. Concrete, Vol. 32, No. 7, 1998, pp. 22-23

10. ODGAARD, A.: The Copenhagen Metro - Status of the Works. Concrete, Vol. 32, No. 7, 1998, pp. 27-28

DAMAGE ASSESSMENT AND LIFE-SPAN PREDICTION OF CEMENT CONCRETE PAVEMENTS IN NEPAL

M P Aryal

Tribhuvan University

Nepal

ABSTRACT. The paper consists of three parts. In the first part, method of quantitative evaluation of secondary stresses in the hardened concrete under various conditions of temperature and humidity has been proposed. Prediction of the life-span of concrete road pavements considering these stresses in various road sectors of Nepal has been carried out in the second part of the paper. Modulus of rupture of hardened concrete has been adopted as the determining strength parameter to predict the durability of concrete under the conditions of uneven drying-wetting, heating-cooling. Additional safety factors have been worked out and proposed for designing concrete road pavement in different climatic conditions. This approach, which has been proposed under Nepal's typical climatic condition, is equally adaptable to different climatic conditions.

Keywords: Hardened cement concrete, Secondary stresses, Drying-wetting, Heating-cooling, Design truck, Life-span, Safety factor.

Dr Mohan Prasad Aryal is an Associate Professor, Tribhuvan University, Institute of Engineering, Pulchowk Campus (College), Nepal. He specialises in the field of concrete structures. The author conducts lecture classes, guides project and research works for undergraduate and postgraduate students in Civil Engineering in Pulchowk Campus.

INTRODUCTION

Cement concrete in Nepal has still been limited mainly in buildings, bridges, irrigation and other small structures. But it is receiving popularity in many other structures. Different types of pavement structures like bus terminals, bus parks and some of the rural and urban roads in these days are being designed and constructed based on this material. Being the most common and easily available material in the country, its use is becoming very common from public to private sectors as well as from small to large-scale projects. Furthermore, concrete, being able to incorporate safely millions of tons of waste materials, proves itself much more environmentally friendly and obviously the future material in construction industry. However, it has been observed by various researchers that durability of hardened cement concrete working under the conditions of drying-wetting, heating-cooling decreases by about 5 to 10 times compared to its design life-span. This is the major aspect, which has been focused in this paper.

CLIMATIC CONDITIONS IN NEPAL

The unique feature of Nepal is the sharp contrast in its landform. In the north-south direction, the altitude varies from less than 100 metres in the south to above 8,000 metres in the north within a stretch of less than 200 kms. This variation in topographical and altitude features, has fostered a diversity in climatic conditions.

The average relative humidity of the atmosphere at 17:40 hrs in different altitudes in Nepal (Kathmandu, Biratnagar, Pokhara and Nepaljung) is about 64%. The minimum value of the relative humidity in these places come to be 56%, 41%, 52% and 40% and rainfall on the same months are 38 mm, 7 mm, 22 mm and 6 mm respectively [1]. Similarly, the temperature variation is also from 5° C to 30° C.

TRANSPORTATION SYSTEM

Automobile road is the major means of transportation system in Nepal. Except in Kathmandu and the terai region, which in terms of area consists of only about 17% of the total territory of the country, no major airports in the hilly regions have been constructed. Furthermore, airways is far more costlier and not affordable to the common people.

Road networks have also not been widely spread all over the country. There are about 10,000 kms total road network in which only 6,000 kms consist of metalled portion. Development of transportation system in the country with the use of local and future material like concrete, considering the specific climatic conditions, is one of the major challenges for infrastructure development in the country.

WORKING SCHEME OF ROAD PAVEMENT

The annual average relative humidity of the atmosphere and the moisture content of the ground below concrete pavement slab play an important role in creating secondary stresses in the pavement slab. The data on climatic conditions mentioned above show that earth and sub-base remain in wet condition because of the rain and capillary action, whereas the relative humidity becomes low. Concrete pavement laid on the ground in such conditions undergo uneven drying. Base of the pavement possesses the moisture equal to that of the earth, whereas its moisture at the exposed surface tends to be equal to that of the atmosphere. Uneven drying causes uneven shrinkage, which ultimately creates secondary stresses. Similar types of secondary stresses are also developed due to temperature variation in the pavement slab when the temperature of the saturated layer is greater to that of the atmosphere.

Secondary Stresses

With the assumption of the above mentioned working scheme of the concrete road pavement along with the assumption of the linear regime of moisture and temperature distribution, following mathematical relationships have been developed by the author [2] to calculate the secondary stresses in the road pavement slab due to moisture and temperature gradients:

For linear regime of moisture gradient:

$$\sigma_{x(y)} = -\beta.E(W_s - W_y) + \frac{\beta.E}{h}[(h-h_o)(W_s - C_1 - 0.5C_2h_{o)}] + \frac{\beta.E}{h^3}.Y[hh_o - h_0^2)6(W_s - C_1) - C_2(h^3 - 3h^2h_0 + 6hh_0^2 - 4h_0^3)] \quad (1)$$

For linear regime of temperature gradient:

$$\sigma_{x(y)} = -\beta_T E.(T_s - T_y) + \frac{\beta_T.E}{h}[(h-h_o)(T_s - Q_1 - 0.5Q_2h_{o)}] + \frac{\beta_T.E}{h^3}.Y[hh_o - h_0^2)6(T_s - Q_1) - Q_2(h^3 - 3h^2h_0 + 6hh_0^2 - 4h_0^3)] \quad (2)$$

where,

β - coefficient of shrinkage;
β_T - coefficient of thermal expansion / contraction;
E - modulus of elasticity of concrete;
W_s - moisture content at the bottom layer of the slab in saturated state (100%);
T_s - temperature of the bottom saturated layer of the slab;
W_y - change of moisture content along the depth (Y - axis) of the slab;
W_T - change of temperature along the depth (Y - axis) of the slab,

$$W_y = C_1 + C_2.Y \quad (3)$$
$$W_T = Q_1 + Q_2.Y \quad (4)$$

In which C_1, C_2, Q_1, Q_2 - constants

Relationship for secondary stresses for non-linear regime of moisture distribution for a particular case (i. e. h = 300 mm; h_0 = 100 mm) has also been developed [3] as follows:

$$\sigma_{x(y)} = -\beta E[(6+1.307Y_w+0.02Y_w^2+0.00293Y_w^3) + 10.36 + 1.25\ Y_\sigma] \quad (5)$$

In which, $(-h/2 + h_0) \leq Y_w \leq h/2$ (6)

$-h/2 \leq Y_\sigma \leq h/2$ (7)

Y_w ,Y_σ - have been measured from neutral axis of the slab element, upward positive and downward negative

h - total depth of concrete element (slab);

h_0 - depth of saturated portion of concrete element (slab) at bottom;

Coefficient of shrinkage and thermal expansion / contraction of the concrete pavement element has been assumed to be 0.00003 and 0.000009. These are the constant parameters that were obtained in the experiment [3] for the concrete grades which are widely used in Nepal and are very much comparable with the data in relevant literatures [4].

Secondary stresses due to moisture gradient have been calculated in two different slab thickness (i.e. 300 mm and 200 mm) under linear regime of moisture distribution. For this the relative humidity of the atmosphere has been adopted 60%. Similarly, stresses under linear and non-linear regimes of moisture distribution have also been obtained with the help of the above mentioned relationships (Eq. 1, Eq. 5). Data show that the maximum stresses at the bottom fibre of the concrete slab are 5.84 MPa and 5.54 MPa under linear and non-linear regimes of moisture distribution respectively. (Refer Table 1). For linear regime of moisture distribution the thickness of the saturated layer of the concrete element (h_o) has been taken one third of its total depth, whereas for non-linear regime, actual data obtained from the experiments [3], have been used. Similarly, it has also been observed that due to moisture gradient stresses at the bottom layer of the pavement element with the thickness of 300 mm and 200 mm are 5.84 MPa. (Refer Table 1). Based on this data, no variation in the secondary stresses due to change in thickness have been adopted within the range of the pavement slab thickness.

Equation 2 has been used to calculate secondary stresses due to temperature gradient for the same scheme of work of the pavement slab.

Table 1 Secondary stresses due to shrinkage

β	E, MPa	H= 300, h_0 =100; MOISTURE DISTRIBUTION: LINEAR.		h=200; h_0=67 MOISTURE DISTRIBUTION: LINEAR		h=300; h_0=100 ; MOISTURE DISTRIBUTION: NON-LINEAR	
		Y_y(mm)	$\sigma_{x(y)}$(MPa)	Y_y(mm)	$\sigma_{x(y)}$(Mpa)	Y_y(mm)	$\sigma_{x(y)}$(MPa)
0.00003	22000	0	5.84	0	5.84	0	5.54
0.00003	22000	50	0.97	33	1.04	100	1.42
0.00003	22000	100	-3.89	67	-3.91	100	-2.71
0.00003	22000	150	-2.19	100	-2.21	150	-2.87
0.00003	22000	200	-0.49	133	-0.51	200	-2.38
0.00003	22000	250	-1.22	167	1.24	250	0.76
0.00003	22000	300	2.92	200	2.94	300	7.19

Relative humidity (R.H.) = 60 %; h = 300 mm and 200 mm; h_0 = 100 mm and 67 mm; moisture distribution: linear and non-linear

Y_y – starts from bottom layer and ends to top layer of concrete slab. Refer on columns 3, 5 and 7. (+) Tension; (-) Compression.

Primary Stresses

The reason of primary stresses in the road pavement is the external load from the traffic. Traffic data collected from the department of roads in Nepal have been used and analysed. Collected data were the two-way traffic data counted at specific location of roads in different road sectors in the country. As most of the roads are two-laned, 50% traffic on the opposite lane has also been added in the lane under consideration for the calculation of the design truck. Average projection of the traffic volume for a period of 15 years has been made assuming that the average life-span of the road pavement is 30 years. Based on the average growth of population and past traffic increment rate, annual increment of the traffic volume has been adopted to be 2% per year. Based on various literatures [5], it has been assumed that the vehicle load less than two tons in gross weight does not provide any impact in the damage process of the pavement / structure. With this assumption, no data of vehicles less than two tons in gross weight have been considered.

The distribution of traffic load has been represented by a fatigue design truck. The gross weight (W_F) of the design truck has been calculated from the traffic data in such a way that the number of cycles to failure for the fatigue design truck is the same as that of the total number of cycles to failure for different trucks in the distribution. The following relationship [5] has been used to calculate the fatigue design truck:

$$W_F = \sum_{i=1}^{i=n} (\alpha_i W_i^3)^{1/3} \tag{8}$$

where, α_i - fraction of trucks with a weight W_i.

W_i - average load of each type of the vehicles in the distribution.

Traffic data have been presented in Table 2.

In order to make the calculation easier without exaggerating the design consideration, simplified method has been adopted for the assessment of primary stresses. This method is basically based on limit state method of design of concrete structures.

According to the limit state design conception, the total safety factor (material and load) without considering the size effect comes to be 2.25 (i.e. 1.5 x 1.5). With this total safety factor, the level of stress developed in the concrete road pavement slab, due to external loading, comes to be 44.4% of its characteristic strength (f_{ck}). This will be the maximum load (stress) that comes to the pavement during the movement of design truck. The minimum load (stress) on the slab becomes only its own weight and this has been assumed to be about 20% of the total primary load (stress).

In this way the range of primary design load (stress) in the concrete road pavement has been fixed to be 44.4% and 10% of its characteristic strength. These values are within the endurance limit of the concrete material.

MODULUS OF RUPTURE

Modulus of rupture of the concrete pavement in this study has been adopted based on the data given in the literatures. Some experimental studies have also been made by the author in order to verify the values of the modulus of rupture given in various literatures and / or codes. Local materials have been used for that purpose. It has also been observed from the experiments that for the adopted scheme of work of the concrete pavement slab, which is under saturation at the base, where it attains maximum value of tensile stress due to bending, the modulus of rupture decreases by about 12% to 18% compared to the specimen in room temperature (25°C-30°C) with the relative humidity 65% to 80%. This reduction due to water saturation has also been incorporated while calculating the modulus of rupture of the concrete slab. Based on the literature and experimental data, modulus of rupture of the concrete pavement slab has been obtained from the following relationship:

$$f_{rup} = 0.84 \times 0.70 \times \sqrt{f_{ck}} \quad (9)$$

But level of maximum and minimum stresses in a cycle under external loading have been adopted without the reduction coefficient due to saturation as it is not considered in actual design practice. Concrete with characteristic strengths (f_{ck}) of 15, 20, 25 and 30 MPa are very common in the construction industries in Nepal. Studies of the pavement has been carried out on these common types of concrete grades. The grades have been identified as M15, M20, M25 and M30.

SECONDARY STRESSES IN VARIOUS REGIONS

The territory of the country has been divided into three different regions, i.e. eastern, central and western and the average values of secondary stresses in the year due to moisture and temperature gradients have been used for these regions. The average value of temperature stresses for eastern, central and western regions have been found to be 0.19, 0.30 and 0.30 MPa respectively. The minimum value of temperature stress has been adopted to be zero, as compressive strength, being not critical, has not been considered. No reduction, due to creep, has been made for temperature stresses.

Similarly, average shrinkage stresses with the consideration of 85% creep have been found to be 0.59 MPa in eastern region and 0.77 MPa in central and western regions respectively. The average shrinkage stress variation have been found to be 0.10 MPa in eastern region and 0.34 MPa in central and western regions respectively.

Maximum values of secondary stresses due to shrinkage and temperature are time dependent parameters. For example, the minimum relative humidity of the atmosphere happens to occur in the day, during which the shrinkage stress happens to be maximum. But, the minimum temperature in general happens to be at night or in the morning, during which the temperature stress is maximum. As the phase difference between maximum values of shrinkage and temperature stresses for the adopted scheme of work of the concrete pavement is about 12 hours, the two stresses can not be combined to get maximum values of secondary stresses. Shrinkage stress has been found to be the dominating. So while adopting the secondary stresses either 100% of the shrinkage stress or 70% of the maximum combined stresses, whichever is greater, has been adopted for damage assessment and life span prediction.

FATIGUE CYCLES CALCULATION

The number of cycles (N) of loading and unloading with and without considering the effect of secondary stresses have been calculated using two different equations (Eq. 10 & Eq. 11), developed for 50% and 95% confidence levels by Aas - Jacobsen and J. M. Hanson with some modification in equation (10) based on the results of the relationship proposed by Japanese Society of Civil Engineers [6]. The equations are as follows:

$$\text{Log N} = \frac{1-(f_{max}/f'_c)}{\beta(1-(f_{min}/f_{max}))} \qquad (10)$$

$$\text{Log N} = \frac{1-(f_{max}/f'_c)}{\beta(1-(f_{min}/f_{max}))} - 1.9 \qquad (11)$$

where, f_{max}, f_{min} -maximum and minimum stresses in the cycle, f'_c - modulus of rupture calculated using Equation 9.

DAMAGE ASSESSMENT AND LIFE-SPAN CALCULATION

Miner's hypothesis has been adopted to determine the accumulation of fatigue damage and life-span prediction of the concrete pavement under varying stresses. According to the Miner's hypothesis the concrete pavement slab will not fail if

$$\sum_{i=1}^{i=n} \frac{N_i}{N_{fi}} \qquad (12)$$

where, N_i - number of constant amplitude cycles at stress level "i"

N_{fi} - number of cycles that will cause failure at that stress lenel "i" &

"k" - number of stress levels.

Based on the calculations it has been observed that the concrete pavement can withstand some hundreds to some thousands cycles of loading - unloading and the life-span of the concrete pavement under the condition of uneven drying and wetting has been decreased drastically. With the higher grade of concrete the effect of secondary stresses have been found decreasing.

In order to design the concrete road pavement for a specific life-span under such conditions, additional load safety factor has been proposed. Additional safety factors for different grades of concrete for a life-span of 30 years with 50% confidence level has been presented in Table 2.

Table 2 Effect of primary and secondary stresses and additional safety factor in concrete pavement design

REGIONS	STATION	PRESENT TRAFFIC PER DAY	PROJECTED TRAFFIC/DAY AFTER 15 YEARS	ADDITIONAL SAFETY FACTOR FOR 30 YEARS LIFE-SPAN IN VARIOUS CONCRETE GRADES FOR 50% CONFIDENCE LEVEL			
				M 15	M 20	M 25	M 30
Eastern	Charali	53	70	1.35	1.25	1.20	1.15
	Lahan	77	102	1.40	1.30	1.20	1.15
	Banepa	236	311	1.65	1.50	1.40	1.30
Central	Bharatpur	311	410	1.70	1.50	1.40	1.35
	Panchkhal	52	69	1.50	1.35	1.25	1.20
	Kurintar	240	317	1.65	1.50	1.40	1.30
	Bardghat	129	170	1.60	1.40	1.35	1.25
Western	Butwal	278	367	1.70	1.50	1.40	1.35
	Lamahi	54	71	1.50	1.35	1.25	1.20

CONCLUSIONS

1. Relationship between moisture gradient and shrinkage stress has been developed. The proposed general relationship with the consideration of creep effect can directly be used in calculating secondary stresses in concrete road pavements in different road sectors. Similar to the shrinkage stress, relationship proposed for temperature stress can also be used.

2. Secondary stresses in concrete pavement slab under linear and non-linear regimes of moisture distribution and linear regime of temperature distribution, under particular scheme of work, have been presented. Experimental verifications of these relationships have also been observed. But these have not been presented in this paper.

3. Traffic data collected from different road sectors of Nepal have been analysed and presented in the form of fatigue design truck.

4. Damage assessment of concrete pavement in different road sectors of Nepal have been made based on Miner's hypothesis and life-span of different grades of concrete pavements have been calculated.

5. In order to design the concrete road pavement for a life-span of 30 years with 50% and 95% confidence levels, under the conditions of uneven drying-wetting, heating-cooling, additional load safety factors for different grades of concrete, have been proposed for different road sectors of Nepal.

REFERENCES

1. MINISTRY OF WATER RESOURCES, DEPARTMENT OF IRRIGATION, HYDROLOGY AND METEOROLOGY. Climatological Records of Nepal, Kathmandu, 1986, Vol. 1, pp 187.

2. ARYAL, M P. Durability of hardened cement concrete under the condition of uneven shrinkage and swelling. Proceedings of the Fifth International Conference on Concrete Engineering and Technology, Kuala Lumpur, 5-7 May, 1997, pp 251-261.

3. ARYAL, M P. Research report on defectiveness of cement concrete under the condition of uneven shrinkage and swelling. Tribhuvan University, Institute of Engineering, Kathmandu, August, 1998, pp 104.

4. MEHTA, P K., PAULO, J M, MONTEIR,O. Concrete, Microstructure, Properties, and Materials, Indian edition, Indian Concrete Institute,1997, pp 548.

5. BS 5400 : Part 10, Steel, Concrete and Composite Bridges, Code of Practice for Fatigue, 1980, pp 55.

6. HANSON, J M. Design for fatigue. Handbook of Structural Concrete edited by Kong F.K., Evans R.H. et al, 1983, pp.16.1 – 16.35.

DETERIORATION OF REINFROCED CONCRETE STRUCTURES IN A MARINE ENVIRONMENT IN BRAZIL: A CASE STUDY

J Andrade

D Dal Molin

Federal University at Rio Grande do Sul

Brazil

ABSTRACT. Nowadays the corrosion of reinforcement is the most harmful damage that can occur in reinforced concrete structures. The case study is one analysis form utilized for to verify the damages evolution when the structure is exposed in environment. In a residential building located at saline environment in Brazil was made measures of carbonation depth, chloride ions and to apply an new assessment method for to establish quantitatives boundaries for structures degradation. The results showed that the carbonation depth was smaller than concrete cover, but the amount of chloride ions in structural elements was biggest than the threshold value. The structure deterioration degree showed which the structure have a critic level of degradation, where is necessary the immediate intervention on structure for to restore the serviceability levels. So, this work show make real the possibility of to verify the damages evolution in a structure, using simple inspection procedures with a numeric analysis.

Keywords: Durability of structures, Corrosion of reinforcement, Marine environment, Case study, Life cycle analysis, Numerical modelling.

MSc Jairo Andrade is a PhD researcher in the Federal University at Rio Grande do Sul, Brazil. His main research interest is concrete durability, in particular durability modelling and applications of reliability theory in durability of reinforced concrete structures.

Professor Denise Dal Molin is a PhD at the Federal University at Rio Grande do Sul, Brazil. Her main research interests are concrete durability, admixture effects on concrete properties and high performance concrete.

INTRODUCTION

Nowadays a great amount of damage that happens in reinforced concrete structures is observed. Many countries around the world are located in very agressives environments, like the Norway Coasts [1] and the Arabian Gulf [2]. In the United States resources at an order of millions of dollars were spent over the last years in inspections and rehabilitation programs, mainly in the railway structures [3]. So, many studies are conducted with the purpose of identifyng the degradation forms and providing informations to minimize this kind of occurence in new structures. In this way, the case studies are extremely important, having been done in high scale all around the world [4,5,6], when significant information is given about the construction and climatic effects in many degradations forms. In Brazil, a high temperated country in which the most important cities lay near the sea, analysis as this one were conducted by many authors, who did inspections in industrial instalations that presented a high degree of agressivity because of the chemical product emission [7], and analysis in residential buildings that presented corrosion of reinforcement at advanced level [8]. So, this paper made an analysis about the degradation problems in a structure located in a saline environment and tried to make considerations about the source of damages as well as to quantify the structure deterioration degree [9].

Structure Description

The building was located in an urban area, 3 km distant from the sea, in a crossing of two streets of high traffic. It's structure has a garage pavement and 13 floors. It's composed by reinforced concrete elements (columns, beams and slabs), with a triangular form in the plant. The structure has a residential use, and didn't suffer any change by it's occupants.

Inspection Methodology

Visual Inspections

A preliminary and detailed inspection was made on the structure. Every structural element was analysed, in a methodology described in some references [9,10], when this stage of work was responsable by the correct assessment of the damages. The presence and physical characteristics (eg.: width, length) of cracks, spalling, deflections and others forms of damage's evidence were the measurable parameters employed for to provide informations for to establish the structure deterioration degree. Unfortunately, some information was not avaliable, such as construction details and concrete properties, and the problem of access imposed some restrictions on the amount of investigation in this case.

Carbonation Depth

Measures of the carbonation depth in the columns located in garage and ground level were made, since they were in areas with highest concentration of carbon dioxide (CO_2) of the environment. The analysis was made in four columns in the structure, changing the sample points in function of the building orientation.

Chloride Ions

Concrete samples were removed from four points of the structure in three differents deepers. In each one was collected a concrete sample and sent to the laboratory for determination of chloride in mass (corresponding to the amount of element in the cement weight).

Deterioration Degree of Structure

Nowadays, every information about deterioration in reinforced concrete structures is exposed in qualitative forms, without any quantitative classification about the damage evolution and the right moment to conduct the rehabilitation works. A new methodology was developed in University of Brasília, Brazil, for maintenance of reinforced concrete structures [11], with the purpose to establish a quantitative classification for degradation degree to individual elements, family of elements (columns, slabs, beams) and the whole structure, taking in account the more expressive parameters for damage's definition, your evolution and the environmental influence in this process.

However, a very important consideration might be made: it's very difficult to make any kind of prognostic about the life cycle of structures. The big amount of factors that have a significant influence in many deterioration processes doesn't interact separatedly, but has a holistic (or sinergic) effect. This sinergism made difficult the reliability of all deterioration models proposed [9]. So, this methodology is a starting point for to infer the possibility of damages classification based in simplificated procedures, in which is necessary more detailed analysis to validate completely the program.

The bulding is divided in families of structural elements, in order to identify groups that have similar structurals characteristics. After this, a matrix is made for each structural element, exposed to the possible damages to the family, with a damage consideration factor. This factor, varying from 1 to 10, shows the relative importance of a specific damage in a family, whereas the serviceability of the element, as showed in Table 1

Table 1 Family of structural elements, damages and consideration factors (F_p) [11]

COLUMNS	
Damage	F_p
Geometry deflection	8
Foundation problems	10
Foundation infiltration	6
Segregation	6
Efflorescence	5
Crushing evidence	10
Insufficient concrete cover	6
Cracking	10
Carbonation	7
Chloride ions	10

BEAMS	
Damage	F_p
Segregation	4
Efflorescence	5
Desagregation	7
Insufficient cover	6
Excessive deflection	10
Cracks	10
Carbonation	7
Infiltration	6

The same damage can have different consideration factors, depending on the family in which they are inclosed, and the damage's consequences in each one. So, the next step is to establish the evolution of the damage, through the damage intensity factor. This factor takes in account the present stage of element and it's deterioration evolution, as showed in Table 1.

Table 1 Damage intensity factor for some damages (F_i) [11]

DAMAGE	F_p	INTENSITY FACTOR OF DAMAGE (F_i)
Segregation	6	1 → Superficial and little significant in element geometry; 2 → significant in element geometry; 3 → significant in element geometry, with reinforcement exposure; 4 → loss of cross section of element.
Insuficient cover	6	1 → less than detailed in Codes and/or Standards, without allowed the reinforcement localization; 2 → less than detailed in Codes and/or Standards, allowed the reinforcement localization and/or exposed reinforcement in small distances; 3 → deficient with exposed reinforcement in large distances
Cracks	10	1 → width less than maximum foreseen in Standards; 2 → stabilized, with width until 40% over the Standards limits; 3 → excessive width stabilized; 4 → excessive width not stabilized.
Carbonation	7	1 → superficial signs; 2 → existent, without to reach the reinforcement; 3 → reaching the reinforcement in wet environment; 4 → at large extentions, reaching the reinforcement in wet environment.
Chloride ions	10	2 → in internal elements with dry environments; 3 → in external elements with dry environments; 4 → in wet environment.

This factor considers some aspects related to funcionality, aesthectics and structural safety of element, with the environmental influences in function of exposure conditions and the protection systems adopted.

The damage degree (D) for a structural element is tied to the damage consideration factor and the intensity of damage, as shown in Figure 1.

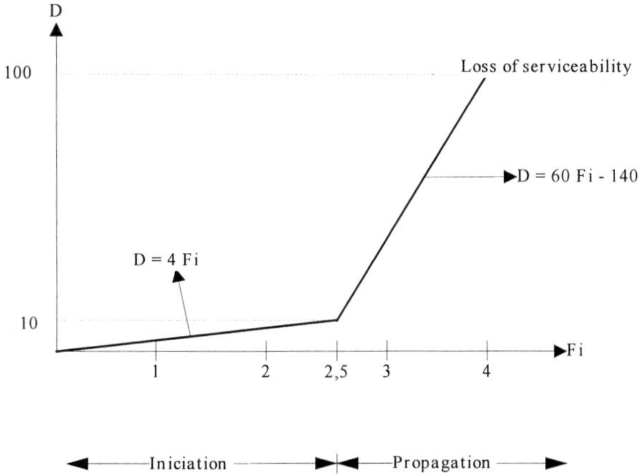

Figure 1 Damage degree (D) *versus* damage intensity factor (F_i) [11]

The first part of the graphic corresponds to the damage's iniciation stage, starting at 0, in which there is no degradation evidence, until the point at (2,5;10). For the propagation stage, that corresponds at D > 10, the damage's effect is more evident, and can take to an unserviceably state. For Fp = 10, which represents the more unfavourable condition of element, the degree is:

$$D = 4\ F_i \text{ for } F_i \leq 2 \tag{1}$$

$$D = 60\ F_i - 140 \text{ for } F_i \geq 3 \tag{2}$$

Damages with consideration factors smaller than 10, the damage degree is obtained by the expressions:

$$D = 0{,}4\ F_i\ .\ F_p \text{ for } F_i \leq 2 \tag{3}$$

$$D = (6\ F_i - 14)\ F_p \text{ for } F_i \geq 3 \tag{4}$$

Equations 1 and 3 try to model the initiation phase, take into consideration the intensity factor of damage. In accordance with Table 1, when the damage is in initiation phase, the value of F_i is less than 2. The same analogy is made for the Equations 2 and 4, what considers the propagation phase for to establish the damage degree. These parameters was adopted for to provide quantitatives boundaries, based in damage's evidence, for each stage of damage.
The element deterioration degree (G_{de}) is obtained by:

$$G_{de} = D_{máx} \text{ for } m \leq 2 \tag{5}$$

$$G_{de} = D_{máx} + \frac{\sum_{i=1}^{m-1} D_{(i)}}{m-1} \quad \text{for } m > 2 \qquad (6)$$

when:

$D_{(i)}$ = damage degree
m = number of damage in element.

For two different damages in the same element, the deterioration degree corresponds to the greatest damage, in order to avoid any simplification (as a medium value) that can take to mistakes in the expression of damage's reality. The next step is to calculate the deterioration of family elements, G_{df}, as shown below.

$$G_{df} = \frac{\sum_{i=1}^{n} G_{de(i)}}{n} \qquad (7)$$

where:

n = numbers of elements in a family with $G_{de} \geq 15$

For the G_{df} determination, is established that only the most expressive damages, the ones with $F_i \geq 2,5$. This procedure was adopted to avoid that the value of damages with no great importance influences in the final result. Finally, the structure deterioration degree is calculated, in funcion of the degradation of the elements family and of the factor of structural relevance, as shown in Table 2.

$$G_d = \frac{\sum_{i=1}^{k} F_{r(i)} . G_{df(i)}}{\sum_{i=1}^{k} F_{r(i)}} \qquad (8)$$

where:

k = numbers of families with elements present in the strucuture;
F_r = structural relevance factor of each family;
G_{df} = deterioration degree of the family.

Table 2 Relevancy factor (F_r) for each family of elements [11]

ELEMENTS FAMILIES	F_r
Not structurals concrete elements	1,0
Upper reservoir	2,0
Stairs, lower reservoir, secondaries slabs	3,0
Slabs, foundations, secundaries beams and colums	4,0
Mains colums and beams	5,0

RESULTS AND DISCUSSION

Visual Inspections

Based in several detailed inspections on the structure, it was observed that there is a great amount of damages. At the garage level the Gerber beams have deterioration in an advanced stage, with cracks that vary from a few milimeters until 1,5 centimeters. These problems happen because of the expansion forces originated in a corrosive process, that leads to the crushing of the columns top by innadequate stress distribution.

Failures in waterproof systems in roof gutters were observed in contour beams. In Brazil, this type of protection systems has a warranty time of nearly 5 years, when, over this period, the product must be replaced. This fault can lead to concrete leaching, with the ocurrence of stalactites of calcium carbonate in beams deep.

The most dangerous damage observed in structural elements happens in the columns, in different intensities. From the garage level until the top of columns the cracking of concrete cover is evident, with extensively variable width over the elements.

The insuficient concrete cover was observed in many points of the analysed structure. In any construction, mainly the ones located in saline environment, this quality parameter of construction must be reached, in order to maximize the physical and chemical protection at the reinforcement. Poor quality concrete, with varying resistence between 15 and 18 MPa, and insuficient concrete cover can lead to the occurence of corrosion of the reinforcement, decreasing the life service of the structure.

Carbonation Depth

The results of the carbonation depth measures is shown in Table 3.

Table 3 Carbonation depth in some elements

STRUCTURAL ELEMENT	CONCRETE COVER (mm)	CARBONATION DEPTH (mm)
Column P9	24	5
Column P10	28	22
Column P18	24	12
Column P30	15	5

The improvement of CO_2 level in the concrete elements didn't reach the reinforcement in any analysed point. In the garage pavement, where the concentration of CO_2 is biggest than the environment one in function of the vehicle discharge, the carbonation depth is smaller than the concrete cover.

Chloride ions

The results of laboratorial analysis of this element vary in a range of 0,455 at 1,526% in cement weight. This data was plotted in a graphic, as shown in Figure 2.

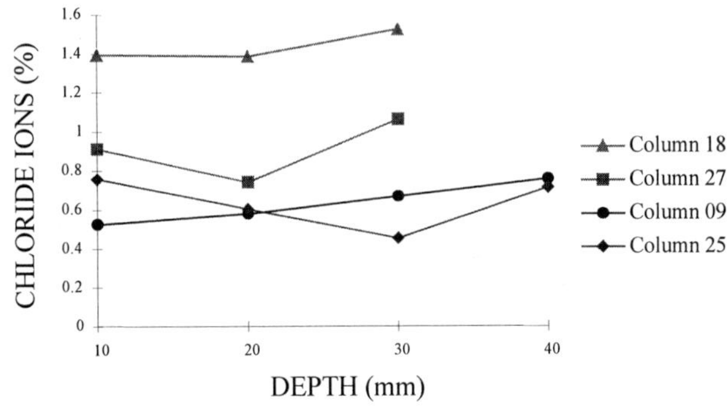

Figure 2 Penetration curves of chloride ions in structural elements

In all points the amount of chloride ions was biggest than the reccomended values (0,4% in cement weight). However, we doesn't have a clear tendency about the possible source of this element in structural members. When the chloride comes from only the environmental action (saline wind), the amount of Cl ions is bigger in the surface of the structural element, and decreases with the increase the depth of the same.

However, in this specific case, the graphic configuration indicates possible contamination of construction materials used in construction process. The chloride ions were deposited over the sand and incorporated to the concrete mass. That occured during the long interruptions in the construction phase of building.

Other possible source of this element is the use of accelerators, that have one or more calcium chloride ($CaCl_2$) forms as the active ingredient in it's chemical composition. One more source for this occurrence is a very common procedure in Brazil of washing the columns ceramic coating. This process is done using muriatic acid (a commercial form of chloridric acid), that has a big amount of chloride ions. That's done without any kind of care, because the users don't have any idea about the harmful effects in the durability of the structure.

Analysis of Structure Deterioration Degree

The structural elements inspected were the columns (in number of 18), the garage and frontal beams (16) and Gerber beams (10) at the garage, totalizing 44 elements.

The damages shown in one column are exposed in
Table *4*, in which was observed that the most dangerous damages are due to the presence of a high amount of chloride and excessive cracking.

Table 4 Example of damages in column P09

DAMAGES	Fp	Fi	D
Segregation	6	3	24
Corrosion spot	7	2	5,6
Cracking	10	4	100
Carbonation	7	2	5,6
Chloride ions	10	4	100

The deterioration degree (G_{de}) for this element was calculated, having been obtained the value of 123,2. The Table 5 shows some limits for this parameter. These numerics values can't be considered as absolute, but only as indicative in the moment of taking a decision about the recuperation services.

Table 5 Levels of element deterioration [11]

LEVEL OF DETERIORATION	G_{de}	NECESSARIES MEASURES
Low	0 - 15	acceptable level
Medium	15 - 50	periodic observation
High	50 - 80	detailed periodic observation
Critical	> 80	intervention immediate for reestablish the funcionality

So, observed that the structural element has a G_{de} greater than the critical level reccomended, it's showed the immediate need of an intervention in this specific element. The next step is to determinate the structure deterioration degree, as showed in Table 6.

Table 6 Structure deterioration degree (G_d)

ELEMENT FAMILY	G_{df}	F_r	$G_{df} \times F_r$
Columns	109,48	5	547,4
Beams	67,12	5	335,63
Gerber beams	112,39	4	449,56
	Total	14	1332,59
			$G_d = 95,2$

Thus, observed that the value obtained is high, compared with the ones showed at Table 7, it is necessary the immediate start of the recuperation tasks.

Table 7 Levels of structure deterioration [11]

LEVEL OF DETERIORATION	G_d	NECESSARIES MEASURES
Low	0 - 15	acceptable level
Medium	15 - 40	periodic observation
High	40 - 60	detailed periodic observation
Critical	> 60	intervention immediate for reestablish the funcionality

The basis for the corrective actions explained above was developed take into account some works where the methodology proposed is applied. The expert opinion in this kind of work have a great importance for to establish these limits values. The model proposed is tested in many structures in Brazil and the results obtained showed a great correlation between the damages observed in practice and the numerical limits reached.

FINAL CONSIDERATIONS

In Brazil a lack of information about the many degradations processes in reinforced concrete structures is observed. The sinergism existent among the saline environment, high temperatures, relative humidity and the poor quality control of concrete production process (workmanship, materials and procedures inadequate) can lead to damages in reinforced concrete structures [9].

So, a case study was conducted in a reinforced concrete structure in order to identify the sources, the occurrence forms and the factors that affect the most important damage in these structures: the reinforcement corrosion. The inspection works have a critical importance in a correct diagnostics of the problem, because it can define the efficiency of a rehabilitation task.

The visual analysis showed a holistic effect in many damages occurrences that lead to the increasing of the deteriorating level. Concrete leaching, high humidity and insufficient concrete cover were also problems that can also lead to the maximization of the corrosion process. The carbonation depth was smaller than the concrete cover. But many analysis points have been made for a greater reliability of data, although the same was collected in areas with a big concentration of CO_2 (garage pavement).

The amount of chloride ions in the structural elements was highest than the accepted limits proposed by many authors (0,4% by cement weight).

Despite the difficulty in order to determinate the service life of many kinds of structures, it is possible to estimate the deterioration level from single procedures, with a satisfactory degree of reliability. This approach has a fundamental importance because it establishes numerical limits for a correct evaluation of deterioration process, and indicates the best moment for the rehabilitation works [9].

REFERENCES

1. MEHTA, P K. (Ed.) Odd E. Gjorv Symposium on Concrete for Marine Structures. Proceedings. New Brunswick, 1996. 279 pp.

2. AL-AMOUNDI, O S. Durability of Reinforced Concrete in Agressive Sabhka Environments. ACI Materials Journal, v. 92, nº 3, 1995. pp 236-245.

3. STEWART, M G, ROSOWSKY, D B. Time-Dependent Reliability of Deteriorating Reinforced Concrete Bridge Decks. Structural Safety, 20, 1998. pp. 91-109.

4. WEST, R. A Case Study to Illustrate Design and Construction Factors Affeting the Durability of Concrete Structures. Design Life of Buildings. Thomas Telford, Londres, 1985. pp. 25-33.

5. SHALABY, H M, DAOUD, O K. Case Studies of Deterioration on Coastal Concrete Structures in Two Oil Refineries in the Arabian Gulf Region. Cement and Concrete Research, v. 20, nº 6, 1990. pp. 975-985.

6. NOVOKSHCHENOV, V. Deterioration of Reinforced Concrete in the Marine Industrial Environment of the Arabian Gulf - A Case Study. Materials and Structures, v. 28, nº 181, ago, 1995. pp. 392-400.

7. FIGUEIREDO, E, ANDRES, P. Damages in Industrual Plants in Brasil: A case Study. Symposium of Structures Rehabilitation. Proceedings. Porto Alegre UFRGS, Brazil (in Portuguese). pp. 283-301.

8. CASCUDO, O, REPETTE, W. Damages Induced for Chloride Ions in Reinforced Concrete Structures in Residencial buildings: a case study. 37ª Annual Meeting of Brazilian Institute of Concrete, Goiânia, Brazil (in Portuguese). Proceedings 1995. pp. 219-231.

9. ANDRADE, J. Durability of Reinforced Concrete Structures: Analysis of Damages In Northeast Region of Brazil. M.Sc. Dissertation (in Portuguese). Federal University of Rio Grande do Sul, Brazil, 1997. 148 pp.

10. MAYS, G. (Ed.) Durability of Concrete Structures: Investigation, Repair, Protection. London. E & FN Spon, 1992. 269pp.

11. CASTRO, E K. Developement of Methodology for Maintenance of Reinforced Concrete Structures. M.Sc. Dissertation (in Portuguese) University of Brasília, Brazil 1994. 185 pp.

COST-OPTIMISED MAINTENANCE OF RC HIGHWAY STRUCTURES

D P Rowe

M B Roberts

R P Gardiner

Maunsell Ltd

United Kingdom

ABSTRACT. The increasing use of risk-based structural assessment methods demands a correspondingly detailed approach to the consideration of repair strategies. A management methodology is required which benefits from improved information availability whilst retaining a strategic overview. This paper discusses such a methodology for the structural maintenance of reinforced concrete highway structures.

The paper develops a process for optimising repair strategies. The process combines automated whole-life cost evaluation with the interactive incorporation of complex site constraints, typically found with real structures. In particular, the significant impact of temporary support systems on repair contract costs and programmes is addressed. The methodology is illustrated by the development and evaluation of two fictitious strategies for a representative motorway viaduct.

Keywords: Highways, Bridge, Maintenance, Management, Whole-life, Cost optimisation, Deterioration, Contract, Economic, Strategy

David P Rowe is a chartered civil engineer with 8 years experience of transport planning, design and construction at Maunsell Ltd. His interest in life cycle costing is currently being pursued as a cornerstone of the Agile Construction Initiative at the University of Bath.

Dr Mark B Roberts is a chartered civil engineer and is currently Team Leader for several projects looking at optimised maintenance strategies for the Highways Agency. He joined Maunsell in 1991 after completing his PhD research. His technical experience includes deterioration modelling, reliability analysis, carbon-fibre and steel plate bonding, remote monitoring technology and bridge management systems.

Richard P Gardiner is an experienced Technician within the Highways Division of Maunsell Ltd. He joined Maunsell in 1987 where he has worked on the design of major highway schemes utilising his experience of MOSS. He has worked on the supervision of major highway refurbishment contracts and has recently worked on the optimisation of highway maintenance schemes.

INTRODUCTION

Infrastructure managers are faced with increasing demands for improvement in service levels without a commensurate increase in resources [1]. Management strategies need to be developed that maintain infrastructure with minimal disruption to the services they provide. The traditional approach to the maintenance of structures has sought to repair all deterioration with a first "hit", the constraint to timing of the work being availability of financial resources. The large number of assets that now need to be maintained, combined with a history of maintenance backlog, means that this approach is no longer tenable – a method for rationalisation must be sought.

The need for rationalisation has caused a shift in emphasis towards strategic maintenance planning built on an improved understanding of the structures being managed. Strategic planning promotes a more intelligent prioritisation of maintenance activities; matching work programmes to the requirements of individual structures or parts of structures. This type of approach is being adopted for the maintenance of many highway structures [2].

A step change in the assessment of reinforced concrete highway structures is being promoted via the joint application of reliability analysis and deterioration modelling [3],[4]. These techniques quantify structural deterioration in terms of risk variation with advancing time. These risks can then be managed through appropriately timed structural maintenance.

Maintenance activities need to be programmed to ensure structures operate at a safe level of risk at a minimised optimised cost commensurate with available resources. This paper describes the development of a generic methodology to address this need supported by an illustration for a sample group of reinforced concrete viaduct structures.

METHODOLOGY

Overview

Bridge management has previously been described entirely in terms of cost algorithms [5] However, there are many issues for which a financial proxy is difficult to generate and might even be considered inappropriate [6] The methodology described here adopts the latter premise by proposing a policy-led multi-stage interactive process.

A set of work programmes or strategy is initially developed to satisfy policy objectives. Policy objectives provide the constraints or boundary conditions for each strategy and are typically defined by a mix of financial and non-financial terms; including, for example, environmental, safety, political and social consequences of work programme options. Constraints may be "hard" with a rigid definition of what is acceptable or "soft" where there are many degrees of acceptability. The most influential hard constraint for the maintenance of highway structures is safety [7]. The safety criterion is represented by the latest time to maintenance intervention, that is, the time yet to elapse before structural deterioration will result in an unacceptable level of risk. Some restorative action must therefore occur within the time to intervention.

The strategy is evaluated through a multi-stage financial appraisal. The cost model uses a variable or marginal approach, achieved by separating the time-related contractor overhead from individual maintenance activities. This overhead is re-introduced as a single project cost following determination of the duration of project programme.

The Financial Appraisal

Structures exist in many forms. They are built up from components that are often not easily maintained or replaced in isolation and whose performance is intimately linked. Experience and judgement has identified groups of components or parts of structures which lend themselves to being concurrently repaired, and for whom further separation would not be feasible in practical terms. For example, the repair of a bridge deck would need to be co-ordinated with the replacement of waterproofing and resurfacing operations. The smallest part of a structure likely to be repaired on a single occasion may be termed an element.

The repair of an element typically comprises an initial repair action followed by subsequent repair actions. Taking the example of a cathodic protection system, the initial repair action would be its installation followed by subsequent repair actions to restore the anode and electrical system. The cost of undertaking an initial repair action is first evaluated as if procured as an isolated contract. Naturally separate procurement of repairs is unlikely to provide the most cost-effective strategy, but it is necessary to provide a consistent starting point for evaluation and comparison of more complex strategies.

The initial strategy of separately procured repair actions is termed the reference strategy. The reference strategy data provides the base data for the evaluation of further strategy options. These strategy options are developed by collecting elements into contracts and applying cost savings which accrue through grouping the elements into a single contract thus:

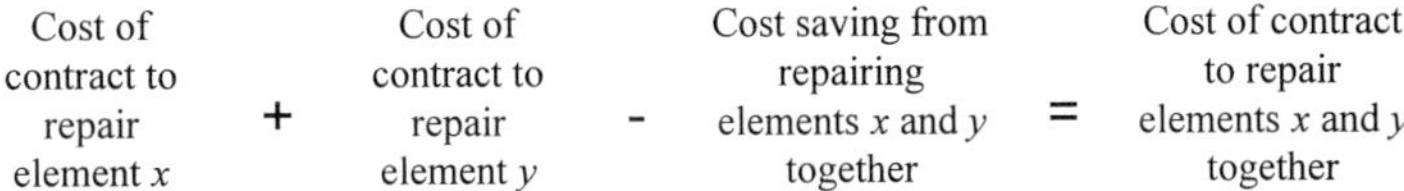

Evaluation of cost savings is derived from the combination of two hypotheses which have been developed from the following observations:

- Structural maintenance work is characterised by sequential activities;
- Repair and maintenance contracts tend to operate on a low turnover.

The sequential feature of structural maintenance work is promoted by onerous restrictions on the scope for weakening the structure during the course of repairs, particularly where the structure supports live carriageway during the repair contract. This is illustrated in Figure 1 by the example contract programme for the cathodic protection of a reinforced concrete support structure. Activities in the repair contract are generally seen to be in series with one activity finishing before the next can start. The contract programme shows activities separated into three broad categories:

- Repair action factors (the notional time to undertake repair activities);
- Programme factors (adjustments to the notional times for repair activities brought about by the specific site circumstances and contract requirements);
- Contract overhead factors (time-related overheads applied to the repair action and programme factors).

The relationship between the factors may be expressed as:

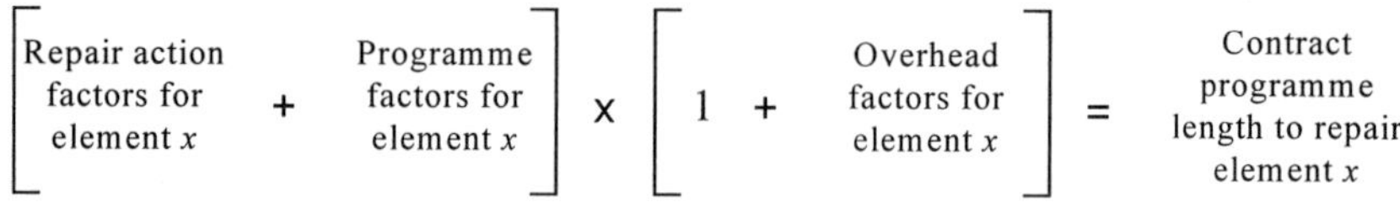

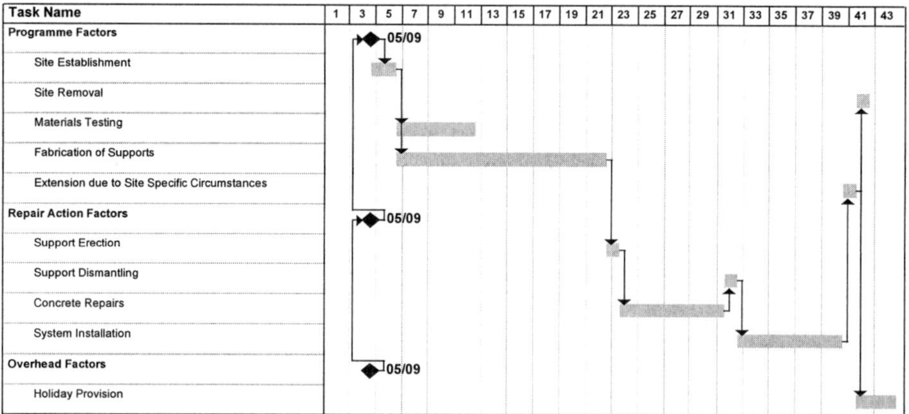

Figure 1 Example contract programme

These categories are then separated further. The example contract programme in Figure 1 shows the repair activities expressed in terms of temporary support erection and dismantling, concrete repairs and system installation. System installation represents application of the anode and electrical system for cathodic protection of the concrete. The degree of contract breakdown is a compromise:

- it has to be sufficient to yield reliable estimates;
- and it has to be brief to enable practical application of the model.

Repair contracts tend to operate on a low turnover due to long programmes, which are necessary to accommodate the sequential nature of repair work. Consideration of the influence of long programmes has provided a causal link to the variation in contract costs. The programme shown in Figure 1 has a duration of 40 weeks during which time perhaps only £0.25m would be spent. Consequently overheads prove highly significant and cannot simply be related to contract value. This principle is illustrated by the relationship:

$$function\begin{bmatrix}\text{Contract}\\\text{programme}\\\text{length to repair}\\\text{element } x\end{bmatrix} + function\begin{bmatrix}\text{Contract cost}\\\text{to repair}\\\text{element } x\end{bmatrix} = \begin{matrix}\text{Contract cost}\\\text{overhead}\\\text{to repair}\\\text{element } x\end{matrix}$$

The *function of contract programme length to repair an element x* is effectively the minimum feasible time-related overhead to maintain a site. A representative value has been determined from the lowest practical operating levels of site establishment, head office overhead, contract management and supervision costs.

The combination of low turnover and sequential working supports the assumption that, if work activities are well planned, the length of programme should not be constrained due to the availability of resources. The combined repair activities will tend to occur concurrently and a critical path then emerges for the group of elements. The overhead value is used to convert the accrued time savings into cost savings.

Many structural repair options require the installation of temporary works of considerable complexity, particularly if the supported carriageway is to remain open to traffic. The procurement, installation, storage and flexibility for re-use of temporary works represents a strategic challenge and is central to the financial optimisation process.

Figure 2 shows the cost breakdown for the cathodic protection of a partially delaminated concrete structure with and without the fabrication, installation and removal of temporary support. The high proportion of costs associated with temporary support fabrication promotes the case for a strategic approach to temporary works planning, in conjunction with the planning of repair works.

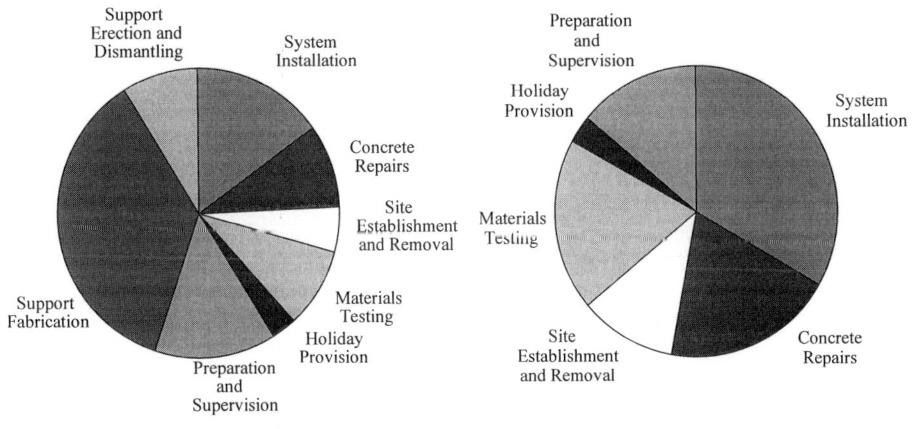

Figure 2 Cost breakdown for cathodic protection

Relationships between compatible temporary works and support structures can be developed, such that the strategic planning of temporary works and repairs may be combined. Subsequent repair actions would typically be required following an initial repair action. It is considered unrealistic and unnecessarily complex to apply the detailed proposals for costing of contract options for these subsequent contracts. This is justified by the following observations:

- The anticipated subsequent repair action costs are typically small compared with the initial repair costs for all repair options, particularly when the costs are discounted;
- The interval and cost of the initial repair action are based on assumed deterioration levels.

It is therefore considered reasonable to assume that subsequent repair and maintenance actions would be organised into efficient contracts, as some uncertainty will not result in significant variation in the financial evaluation. This allows evaluation of subsequent repair actions by the simple application of unit rates.

ANALYSIS OF EXAMPLE VIADUCT

Setting The Scene

The example section of reinforced concrete viaduct comprises thirty six supports constructed using a typical column and crossbeam arrangement with an abutment at each end. The arrangement of piers is shown in Figure 3. Repair options have already been implemented on twenty-three of the piers, whilst the remaining thirteen have yet to be allocated a repair option. Several site specific problems are included in the analysis. In particular, the viaduct crosses a canal, a major road and, for part of its length, is situated immediately over a minor road.

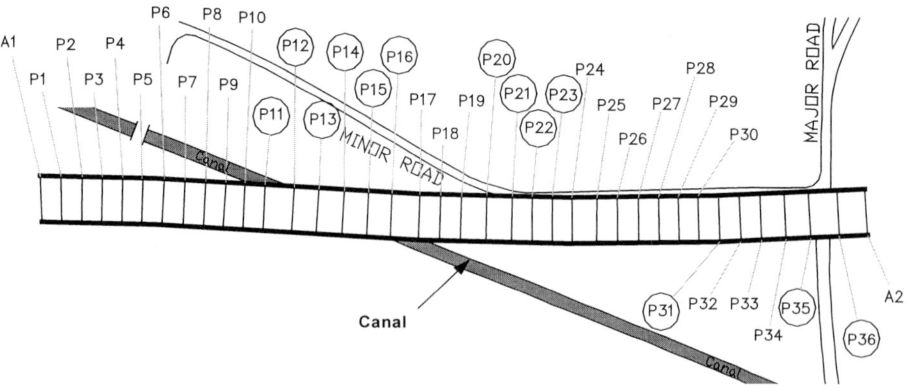

Figure 3 General arrangement of example viaduct

Repair of the piers is governed by the following summary of constraints:

P11 to P16	Installation of temporary support structures would close the canal. The canal can be closed but only outside the leisure season for a maximum of 30 weeks. With concurrent repairs to some or all of these piers, temporary support structures must be erected from P11 and P16 inwards and from P13/14 outwards due to land access restrictions.
P11 to P21	Temporary support structures are difficult to erect due to low headroom.
P31 & P35	Installation of temporary support structures would close the minor road.
P35 & P36	The close proximity of the road would require lane closures.

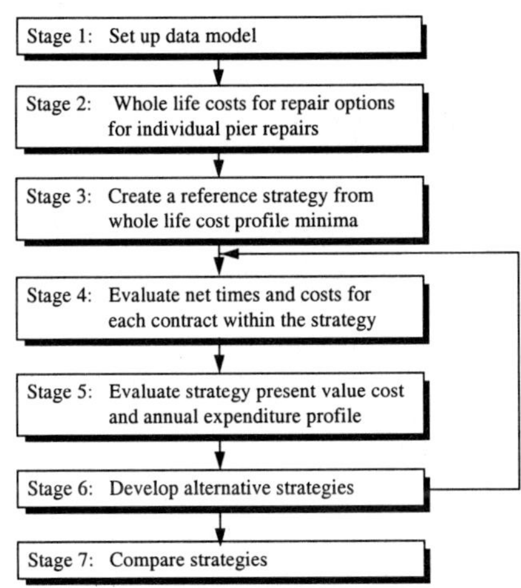

Figure 4 Flow chart of analysis procedure

Stage 1: Set Up Data Model

The analysis follows the flow chart shown in Figure 4. Algorithms have been developed to link time and cost parameters, derived from historical repair contracts, to representative geometric parameters to form the repair action factors, programme factors and overhead factors. The algorithms have been developed from the principles outlined in the discussion of methodology. The example provides a pilot application for the methodology – detailed discussion of the validity of data and algorithms used will be the subject of further papers.

Stage 2: Whole Life Costs for Repair Options For Individual Pier Repairs

Figure 5 shows a series of present value cost profiles for Pier P13. Cost profiles are produced for the following repair options, each of which comprises an initial repair action followed by subsequent repair actions:

- Cathodic protection – repair of delaminated concrete prior to the application of an impressed current anode and subsequent maintenance of the anode system;
- Concrete replacement – replacement of chloride contaminated concrete;
- Element replacement – demolition and renewal of entire element.

Each cost/year point on the cost profile represents the discounted whole-life cost for the profiled option if the initial repair action were implemented in that year. The example uses a 100 year time horizon for the calculation of whole-life costs. The whole-life cost would include repair, maintenance and operation of the system after the initial repair action. For example, if cathodic protection is applied in eight years time then, with a maintenance cycle time of 15 years, subsequent repair costs would arise from maintaining the anode system at years 23, 38, 53, 68, 83 and 98. Additionally, an annual monitoring cost would be applied from and including year 9.

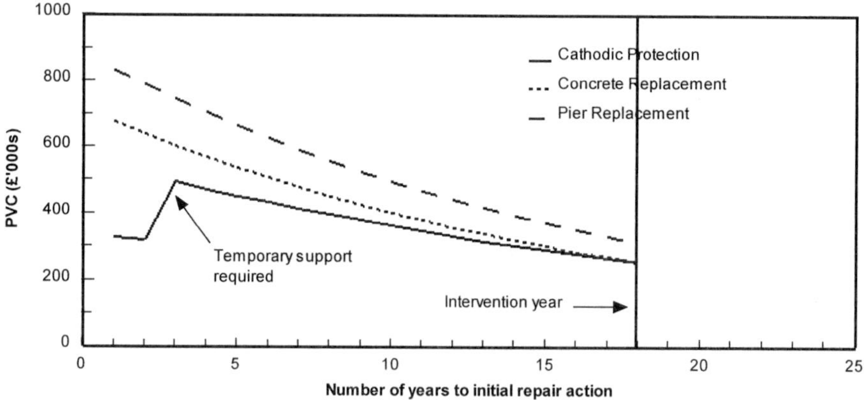

Figure 5 Present value costs for repair options

The whole-life cost would also include a do-minimum monitoring and minimum safety treatment prior to the initial repair action. It can be seen from Figure 5 that the discounted whole-life cost for implementing cathodic protection in 8 year's time for Pier 13 is £400k. Using this rationale, cost profiles are produced for each of the piers. Although the cost profiles differ in detail, they have several common characteristics:

- The provision of temporary support for the cathodic protection repair option results in a large step in the cost profile.
- Element replacement is typically the most expensive repair option at any time within the evaluation period; except for piers with a high proportion of delaminated concrete.
- All cost profiles have negative gradients with the exception of the step in the cathodic protection profile - unfeasibly high deterioration is needed to give a positive gradient. This illustrates the often-quoted problem of cost discounting outstripping deterioration.
- Where an intervention date has been identified within the evaluation period, the lowest Present Value Cost (PVC) is typically found at this time.

The profiles might initially suggest that the obvious time to repair is at or approaching the time to intervention. But the profiles assume each pier is repaired in isolation. The two competing strategies put forward in the example demonstrate that a well-devised repair programme can improve whole-life cost by repairing well within times to intervention.

Stage 3: Create a Reference Strategy from Whole Life Cost Profile Minima

Observation from the cost profile that the most cost-effective repair of individual piers occurs at the end of the time to intervention provides the starting point for a reference strategy. Contracts for repair are developed based on repairing each element at the end of the intervention period.

Stage 4: Evaluate Net Times and Costs for Each Contract within the Strategy

Table 1 shows the summary tabulated output of the reference strategy. The data are grouped chronologically by year of initial repair action. Comparison with the intervention year will show these data to be the same. The repair option is referenced by means of a number – "2" represents cathodic protection. An explanation of the table is provided for piers 12, 13, 16 and 21 repaired in 2015. The section is highlighted in Table 1. The first four rows provide data for each individual pier. The fifth row provides a summary of the data assuming the piers are repaired in the same contract.

Repair action factors are shown in columns 4 to 8. Column 4 headed "float" identifies the programme float for repairs to each pier with respect to the pier with the longest repair programme. The critical pier is that with zero float – in this case pier 12. It is assumed that repairs can be undertaken concurrently and therefore the totals in row five of columns 5 to 7 headed "time" are simply the values from pier 12. It is worth noting that the concurrency principle could be invalidated by site-specific conditions. Column 8 headed "cost" are evaluated as those costs which are dependent only on the quantity of work and hence there is no scope for cost rationalisation. They are thus simply added to provide the totals in row five.

Programme factors are shown in columns 9 to 14. Cost and time have been aggregated for piers 12, 13, 16, and 21 allowing for considerations of cost rationalisation and concurrency. The time and cost aggregations for fabrication of temporary supports are notable (columns 13 and 14). The rationale for these columns cannot be considered in terms of an individual contract. The example assumes some scope for re-use of supports, but interchangeability is restricted between different element geometries. The repair strategy begins with a stock of supports which increases with time as supports are fabricated for contracts. The computer analysis incorporates a dynamic database that accounts for this changing stock.

Overhead factors are shown in columns 15 to 17. In the highlighted example of piers 12, 13, 16, and 21, column 15 shows an addition to the programme critical path for contractors holidays. This is a pro rata allowance calculated on the time components from each of the repair action and programme factors. The calculation is based on the summary values shown in row 5. This time value is then added to the time components from each of the repair action and programme factors to provide the overall contract period shown in column 18:

$$10.4 + 1.7 + 29.5 + 2.0 + 7.0 + 20.0 + 4.2 = 74.8 \text{ weeks}$$

Column 16 shows the overhead cost. This is calculated as a pro rata cost based on the 74.8 weeks overall programme length. The pro rata cost factor is described in the methodology as the "minimum feasible time-related overhead which would be needed to maintain a site".

Column 17 shows the contract preparation and supervision cost. It is a pro rata allowance calculated on the cost components from each of the repair action and programme factors plus the overhead cost. The calculation is based on the summary values shown in row 5. This cost value is then added to the cost components from each of the repair action and programme factors and the overhead cost to provide the overall contract cost shown in column 19:

$$565.5 + 25.0 + 78.0 + 340.2 + 280.7 + 198.0 = £1487.5\text{k}$$

Table 1 Strategy based on latest intervention dates

				Repair Actions				Programme Factors						Overhead Factors			Total	
				Install System	Install Support	Conc Repair		Site & Materials		Traffic and Environmental		Fabrication		Contractor's Hols	O/head	Prep/ Supvn	Capital Cost and Contract Duration	
Beam Ref	Repair Option	Intvn Year	Float (Wks)	Time (Wks)	Time (Wks)	Time (Wks)	Cost (£'000)	Time (Wks)	Cost (£'000)	Time (Wks)	Cost (£'000)	Time (Wks)	Cost (£'000)	Time (Wks)	Cost (£'000)	Cost (£'000)	Time (Wks)	Cost (£'000)
Contract 1		Repair Year		2000														
P22	2	2000	28.3	8.8	2.0	9.5	92.5	2.0	25.0	0.0	0.0	15.9	172.8	2.3	151.8	73.2	40.5	515.4
P35	2	2000	0.0	9.1	2.0	37.5	243.0	2.0	25.0	0.0	370.6	15.9	172.8	4.0	264.4	126.5	70.5	919.7
				9.1	2.0	37.5	335.4	2.0	25.0	0.0	370.6	20.0	345.6	4.2	280.6	168.0	74.8	1242.7
Contract 2		Repair Year		2003														
P23	2	2003	0.0	8.8	2.0	10.5	110.0	2.0	25.0	0.0	0.0	15.9	172.8	2.4	155.9	76.7	41.6	540.3
P31	3	2003	5.1	0.0	1.7	14.5	95.6	2.0	25.0	0.0	32.4	15.0	86.4	2.0	131.8	59.3	35.2	414.3
				8.8	2.0	10.5	205.6	2.0	25.0	0.0	32.4	15.0	86.4	2.3	152.0	80.1	40.5	565.4
Contract 3		Repair Year		2005														
P14	2	2005	0.0	9.1	2.0	21.5	170.8	2.0	25.0	2.0	58.8	15.9	172.8	3.2	208.7	102.9	55.7	739.0
				9.1	2.0	21.5	170.8	8.0	25.0	2.0	58.8	0.0	0.0	2.6	169.3	70.3	45.1	494.2
Contract 4		Repair Year		2007														
P36	2	2007	0.0	9.1	2.0	23.5	204.0	2.0	25.0	0.0	207.6	15.9	172.8	3.2	208.8	104.3	55.7	749.4
				9.1	2.0	23.5	204.0	8.0	25.0	0.0	207.6	0.0	0.0	2.6	169.3	71.8	45.2	504.7
Contract 5		Repair Year		2008														
P11	2	2008	0.0	9.1	2.0	14.5	154.6	2.0	25.0	2.0	41.3	15.9	172.8	2.7	180.9	93.7	48.3	668.3
				9.1	2.0	14.5	154.6	8.0	25.0	2.0	41.3	0.0	0.0	2.1	141.5	60.5	37.7	422.9
Contract 6		**Repair Year**		**2015**														
P12	**2**	**2015**	**0.0**	**10.4**	**1.7**	**29.5**	**227.4**	**2.0**	**25.0**	**2.0**	**78.0**	**15.3**	**113.4**	**3.7**	**242.0**	**110.4**	**64.5**	**796.2**
P13	**2**	**2015**	**13.6**	**9.4**	**1.7**	**17.0**	**148.6**	**2.0**	**25.0**	**2.0**	**46.7**	**15.3**	**113.4**	**2.8**	**188.2**	**85.8**	**50.2**	**607.6**
P16	**2**	**2015**	**17.6**	**10.6**	**1.7**	**11.8**	**117.6**	**2.0**	**25.0**	**2.0**	**33.6**	**15.3**	**113.4**	**2.6**	**172.1**	**76.4**	**45.9**	**538.0**
P21	**2**	**2015**	**24.0**	**7.8**	**1.6**	**8.1**	**71.9**	**2.0**	**25.0**	**1.0**	**0.0**	**15.0**	**86.4**	**2.1**	**141.3**	**54.4**	**37.7**	**379.1**
				10.4	**1.7**	**29.5**	**565.5**	**2.0**	**25.0**	**7.0**	**78.0**	**20.0**	**340.2**	**4.2**	**280.7**	**198.0**	**74.9**	**1487.5**
Contract 7		Repair Year		2020														
P15	2	2020	0.0	9.4	1.7	22.0	166.6	2.0	25.0	2.0	59.2	15.3	113.4	3.1	208.0	93.3	55.5	665.6
				9.4	1.7	22.0	166.6	8.0	25.0	2.0	59.2	0.0	0.0	2.6	171.2	70.0	45.7	492.0
Contract 8		Repair Year		2025														
P20	2	2025	0.0	7.8	1.6	8.4	77.5	2.0	25.0	2.0	0.0	15.0	86.4	2.2	146.2	56.1	39.0	391.3
				7.8	1.6	8.4	77.5	8.0	25.0	2.0	0.0	0.0	0.0	1.7	110.6	36.2	29.5	249.4

Stage 5: Evaluate Strategy PVC and Annual Expenditure Profile

The expenditure profile with time is shown in Figure 6. The reference strategy offers a present value cost of £3392k. Undiscounted expenditure amounts to £5459k.

Stage 6: Develop Alternative Strategies

An alternative strategy was developed based on the policy objective of promoting minimal disruption to third parties. This was achieved through contemporary repair of piers in close proximity to the roads and canal. In particular, the piers spanning the canal have been grouped in a single contract in 2005. This minimises the canal possessions required.

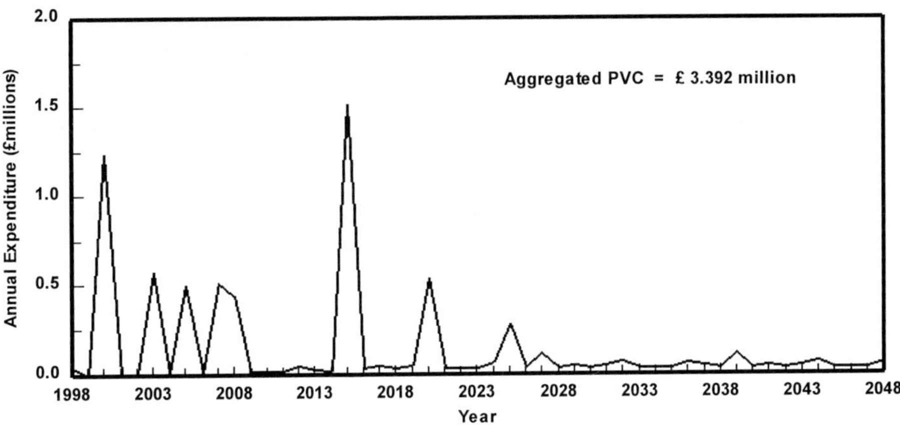

Figure 6 Expenditure profile based on latest intervention dates

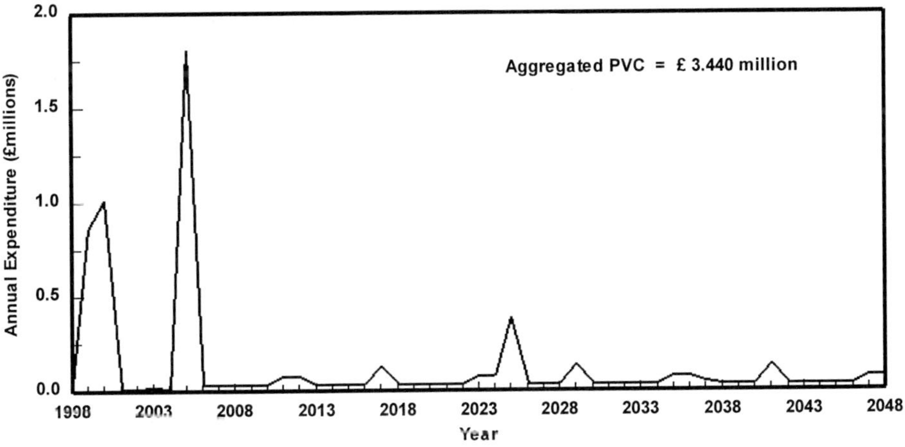

Figure 7 Expenditure profile based on minimum disruption to third parties

Stage 4 has been implemented on this alternative strategy to yield the results shown in the summary table in Table 2. Stage 5 has been implemented to produce the expenditure profile with time that is shown in Figure 7. The alternative strategy offers a present value cost of £3440k. Undiscounted expenditure amounts to £4009k.

Table 2 Strategy based on minimum disruption to third parties

				Repair Actions				Programme Factors						Overhead Factors			Total	
				Install System	Install Support	Conc Repair		Site & Materials		Traffic and Environmental		Fabrication		Contractor's Hols	O/head	Prep/ Supvn	Capital Cost and Contract Duration	
Beam Ref	Repair Option	Intvn Year	Float (Wks)	Time (Wks)	Time (Wks)	Time (Wks)	Cost (£'000)	Time (Wks)	Cost (£'000)	Time (Wks)	Cost (£'000)	Time (Wks)	Cost (£'000)	Time (Wks)	Cost (£'000)	Cost (£'000)	Time (Wks)	Cost (£'000)
Contract 1		Repair Year		1999														
P21	2	2015	5.6	7.8	0.0	6.1	32.1	8.0	25.0	1.0	0.0	0.0	0.0	1.4	91.2	25.2	24.3	173.6
P22	2	2000	0.0	8.8	2.0	8.8	86.1	2.0	25.0	0.0	0.0	15.9	172.8	2.2	148.9	71.7	39.7	504.5
P23	2	2003	0.2	8.8	2.0	8.5	89.0	2.0	25.0	0.0	0.0	15.9	172.8	2.2	147.9	72.0	39.4	506.7
			0.0	8.8	2.0	8.8	207.2	2.0	25.0	1.0	0.0	20.0	345.6	2.6	169.0	119.5	45.1	866.4
Contract 2		Repair Year		2000														
P31	2	2003	31.4	7.8	1.4	8.0	66.7	2.0	25.0	0.0	18.8	14.4	34.2	2.0	133.7	45.5	35.6	314.5
P35	2	2000	0.0	9.1	2.0	37.5	243.0	2.0	25.0	0.0	370.6	15.9	172.8	4.0	264.4	126.5	70.5	919.7
P36	2	2007	21.0	9.1	2.0	16.5	151.5	2.0	25.0	0.0	165.6	15.9	172.8	2.7	180.9	91.2	48.2	649.0
			0.0	9.1	2.0	37.5	461.1	2.0	25.0	0.0	370.6	14.4	34.2	3.9	258.3	137.5	68.9	1004.2
Contract 3		Repair Year		2005														
P11	2	2008	8.5	9.1	2.0	13.0	138.9	2.0	25.0	2.0	37.5	15.9	172.8	2.6	175.0	89.9	46.7	639.1
P12	2	2015	1.0	10.4	1.7	19.5	152.4	2.0	25.0	2.0	53.0	15.3	113.4	3.1	202.3	89.4	53.9	635.5
P13	2	2015	9.5	9.4	1.7	12.0	111.1	2.0	25.0	2.0	34.2	15.3	113.4	2.5	168.3	74.8	44.9	526.8
P14	2	2005	0.0	9.1	2.0	21.5	170.8	2.0	25.0	2.0	58.8	15.9	172.8	3.2	208.7	102.9	55.7	739.0
P15	2	2020	7.0	9.4	1.7	14.5	110.3	2.0	25.0	2.0	40.5	15.3	113.4	2.7	178.2	77.3	47.5	544.7
P16	2	2015	11.0	10.6	1.7	9.3	91.3	2.0	25.0	2.0	27.3	15.3	113.4	2.4	162.2	69.6	43.3	488.8
			0.0	9.1	2.0	21.5	774.8	2.0	25.0	2.0	60.0	24.0	453.6	3.6	240.8	235.1	64.2	1789.4
Contract 4		Repair Year		2025														
P20	2	2025	0.0	7.8	1.6	8.4	77.5	2.0	25.0	2.0	0.0	15.0	86.4	2.2	146.2	56.1	39.0	391.3
			0.0	7.8	1.6	8.4	77.5	2.0	25.0	2.0	0.0	14.6	52.2	2.2	144.7	50.4	38.6	349.8

Stage 7: Compare Strategies

The principal observations in drawing a comparison of these alternatives are:

- The strategies show very similar PVC outcomes, the reference strategy being just 1% better than the alternative strategy. But the alternative strategy requires 26.5% less cash expenditure than the reference strategy.

- The alternative strategy requires £1.8m to be available in year 7 when all piers crossing the canal (except P17 which is not anticipated to need structural repair) are included in a single notional contract. This compares with the maximum expenditure under the reference strategy of £1.5m in year 2. The small number of piers when considered for repair over a period of several decades leads to significant expenditure spikes. In practice, the importance of these spikes will depend on the breadth of the strategy.

The reference strategy requires four separate possessions of the canal, three periods of closure for the minor road and two periods of lane closures for the major road. The alternative strategy rationalises this into a single possession for the canal, a single period of lane closure for the major road and two closures of the minor road.

CONCLUSIONS

The methodology provides a framework for the evaluation of reinforced concrete highway structures. The approach identifies generic relationships. These relationships have been developed using historical contract data. The relationships have been incorporated as algorithms into pilot computer software.

The pilot software has demonstrated a more rapid and integrated evaluation of repair strategies than has previously been possible. Hence it has provided a practical means to assist the optimisation of repair strategies based on whole-life costs. The improved evaluation process provides the potential to readily examine the impact of policy objectives and assists in the development of long term strategies tailored to the developing knowledge of structures.

REFERENCES

1. LEADBETTER, A. (1996) Can we afford the bridge rebuild cost? Assessment and demolition of structures conf., Inst. Civ. Engrs., Sep.

2. LOUDON, N. (1995) Structure management plans and whole life costing. Bridge Management conference, June, London.

3. CROPPER, D; JONES, A; ROBERTS, M. (1998) A risk based maintenance strategy for the Midland Links motorways viaducts. Bridge Management conference, June, London.

4. FRANGOPOL, D; ESTES, A; AUGUSTI, G; CIAMPOLI, M. (1997) Optimal bridge management based on lifetime reliability and life-cycle cost. Proc. Of the Intl. Workshop on Optimal Performance of Civil Infrastructure Systems, April, Portland, Oregon.

5. DE BRITO, J; FRANCO, F. (1994) Bridge management policy using cost analysis. Proc., Instn Civ. Engrs Structs & Bldgs, Vol. 104, Nov. pp. 431-439.

6. MISHAN, E. (1998) Cost Benefit Analysis, Fourth edition, 1994 reprint, Routledge, ISBN 0-415-10922-1. Part VI.

7. HAYNES, L. (1997) Introduction. Safety of Bridges. Thomas Telford. ISBN 0-7277-2591-2. Part 1, pp. 3-6.

ASBESTOS CEMENT PIPES IN AGGRESSIVE GROUND WATER: A CASE STUDY

P Dux

University of Queensland

Australia

ABSTRACT. This paper presents a case study of deterioration of asbestos cement sewerage pipes from attack by sulphuric acid. The pipe network was only eight years old and served an estate that had been built largely on reclaimed swampland. While the symptoms of attack were typical, featuring extensive deterioration above the average waterline and very little deterioration below, the source of the dissolved sulphides leading to hydrogen sulphide generation and subsequent acid formation around the crown of the pipes was not obvious. The pipe network was relatively new and there was little opportunity for stagnant sewage conditions to develop. Regular flushing with alkali to eliminate or at least substantially limit the amount of dissolved sulphide existing as hydrogen sulphide had been undertaken once deterioration was discovered. Despite the above factors, decay continued at an alarming rate. It was postulated that the pipes were substandard. The investigation described in this paper confirmed via petrographic analysis that the attack was from sulphuric acid. It was shown that the pipes were not defective and that the source of dissolved sulphide was ground water which entered the pipe network through numerous fractures. Prior to the investigation, the intrusion of ground water had been viewed favourably as effecting a continuous flushing of the system. The paper discusses the methodology of the investigation, the phenomena leading to deterioration and makes recommendations concerning the management of such networks.

Keywords: Asbestos cement pipes, Acid attack, Aggressive ground water, Dissolved sulphides, Sulphuric acid

Dr Peter Dux is a Reader in Civil Engineering at the University of Queensland, Australia. His areas of research and consulting include concrete technology, concrete structures and stability of steel structures. He was awarded the 1995 T Y Lin prize by the ASCE, for research into prestressed concrete. He is Vice-President of the Queensland branch of the Concrete Institute of Australia.

INTRODUCTION

Asbestos cement pipes are formed from sheet comprising Portland cement and ground sand mortar, and asbestos fibre. The sheet is rolled to the required shape, compressed and autoclaved. Pipes for sewerage are dipped in bitumen, all surfaces being coated. Manufacture of asbestos cement pipes ceased in Australia well over a decade ago, smaller diameter sewerage pipe networks being now constructed mainly in plastic. However, the sewerage systems of many municipalities contain extensive lengths of asbestos cement pipes which generally give excellent service. The coastal municipality in which the investigation described in this paper took place, has many hundreds of kilometres of asbestos cement sewerage pipes. The replacement cost is about $A50000 per kilometre.

The coastline to the North of the municipality is generally emergent with swamps and numerous mangrove islands. In many places, urban waterfront development has extended Northward, involving the reclamation of swampland. One such new suburb was sewered in the mid-eighties with asbestos cement pipes, around the time that manufacture ceased. In the following years it became known that ground movement had led to numerous fractures and a video survey of the affected network was conducted. Investigation of the failure of some domestic sewerage services had revealed severe deterioration of pipes. Because of the high ground water inflow and the relatively few homes connected to the local network, conditions did not exist for acid attack associated with domestic sewage. Other measures such as occasional alkali dosing were undertaken. Continued deterioration led to the postulation that the pipes were most likely defective. It was also held that the influx of ground water effected continuous flushing and hence served to prolong the life of the defective pipes.

Treated wastewater from the system was used to irrigate parklands and a golf course. Concern arose over the increasing salt content of this water and its likely effect on grasses. The decision was taken to replace the local network with plastic pipes. It was just after the beginning of this work that the investigation described herein took place, one possible outcome being litigation by the local authority against the pipe manufacturer.

STAGES OF INVESTIGATION

The investigation had three main stages:

1. A field survey involving
   - the development of a classification system for severity of decay
   - detailed logging of the condition of pipes as they were removed, in relation to position in the network and depth below water table
   - the making of a photographic record
   - collection of other data.

2. Pipe material analyses including petrographic analysis to identify residual material in affected pipes and to estimate the composition of sound material, and analysis of sound pipe material for calcium oxide content and density.

3. Chemical analysis of ground water with particular emphasis on the dissolved sulphide content.

FIELD INVESTIGATION

Field monitoring took place until a continuous record for about 1 km of pipes had been made. A five-point classification system was developed ranging from category 1 representing pipes unaffected or with, at most, part of the internal bitumen lining removed, to category 5 representing pipes deteriorated to the point of possible collapse. Figure 1 shows a rough cross-section of a category 5 pipe. Salient points from the figure which apply to all affected pipes are:

- No deterioration has occurred at the outer surface where the bitumen layer remains intact.
- No deterioration has occurred in the region of the invert, over that part of the inner surface which is typically submerged (Fig. 1 shows some scale build-up at the invert).
- Above the average waterline, the pipe has decayed leaving what appears to be a mat of asbestos fibre backed by the remaining sound material. The worst decay is not at the crown but more to the sides.

Figure 1 Cross-section of pipe, category 5

Crude field tests for pH at the reaction front were conducted with multi-test litmus paper at the crown, the expanded mat of asbestos having been removed. Typical results were in the range 4.0 – 4.5. Similar tests on ground water indicated a pH of around 7.0. It was advised that the pH of sewage in unaffected networks typically ranged from 6.5 to 7.5 without alkali treatment.

Table 1 presents some condition data representing pipes between manholes spaced at around 90 m. The consistent trend is for pipe condition to worsen with depth, sometimes gradually, sometimes suddenly across a manhole. The reduced levels in Table 1 are with respect to mean sea level. Actual ground surface level is a few metres higher.

Table 1 Sample field data

MANHOLE NO.	R.L. (metres)	CONDITION
M1	0.0	
M2	-0.5	M1 – M2 Category 1
M3	-1.0	M2 – M3 Category 2 – 3
M4	-1.5	M3 – M4 Category 5
M5	0.0	
M6	-0.65	M5 – M6 Category 1
M7	-1.25	M6 – M7 Category 3 - 4
M8	-2.0	M7 – M8 Category 5
M9	-1.5	
M10	-2.0	M9 – M10 Category 4 – 5
M11	-2.2	M10 – M11 Category 4 – 5

Excavations were de-watered by well points, with perforated spears up to 6 m in length, the continuous flow of collected ground water being then discharged into the stormwater system. The water smelt of hydrogen sulphide. While this indicated the presence of some dissolved sulphide, this gas can be detected by smell at very low concentrations. Samples of the water were collected for analysis using a fixing agent to prevent the further release of gas. Ground water entering the network through broken pipes could be seen from some manholes. Figure 2 taken with high-speed film shows a spray of very turbulent water.

Figure 2 Ground water entering pipe at break

MATERIAL ANALYSES

Petrographic Analysis

Petrographic analysis involved examination under microscope of thin sections taken from pipes of different condition. Samples were impregnated with coloured resin before sectioning. The following points were confirmed:

- Material in the region of the invert was sound.
- In the damaged upper zones, the amorphous cement paste (ie. the various components of hydrated cement) had been completely destroyed.
- Common in residual products were silica, gypsum and asbestos fibre.

It is well known [1] that the acids destroy cement hydrates creating salts of the acid. Sulphuric acid attack produces calcium sulphate which hydrates to produce gypsum as a precipitate. Whereas aggressive agents such as sulphates attack components of the hydrated cement, acids typically attack all products of hydration.

Estimates of the composition of sound asbestos cement were made by two counts each of 100 widely spaced points on thin sections of sound pipe. The results appear in Table 2 as percentages of the total point count.

Table 2 Approximate composition of sound asbestos cement based on point counts

SAMPLE NO	1	2
Cement hydrate	55	62
Quartz	24	22
Asbestos	21	16

The above compositions identify siliceous sand as the aggregate fraction and a mortar rich in cement.

Chemical Analysis for CaO and Density Analysis

Eight samples from variously affected pipes were tested for calcium oxide content. The cement used in manufacture was Type A or Portland cement, known to have a typical CaO content of 64.5% by mass. Because the aggregate had been identified as crushed quartz (Table 2), the cement content of samples could be reasonably inferred from the CaO contents. The results of testing and estimated cement content as percentage of mass of sample are presented in Table 3. The results indicate a mortar component rich in cement with negligible variation between samples.

Densities were measured by two standard processes for three samples of sound material taken from pipes of categories 1, 3 and 5. Results in Table 4 show the material to be of high density. A general conclusion from the material analyses is that they revealed nothing to suggest sub-standard manufacture.

Table 3 Measured calcium oxide contents and inferred cement contents (% by mass of sample)

SAMPLE NO	1	2	3	4	5	6	7	8
CaO content	26.2	25.4	25.9	26.6	26.5	26.9	27.0	25.9
Cement content	40.6	39.4	40.2	41.2	41.1	41.7	41.9	40.2

Table 4 Densities of sound asbestos cement samples (kg/m^3)

METHOD	SAMPLE 1	SAMPLE 2	SAMPLE 3
Water displacement	2568	2440	2587
Helium displacement	2657	2698	2585

CHEMICAL ANALYSIS OF GROUNDWATER

Ground water samples for analysis were taken at the outlet to the trench de-watering system. It was clear that some dissolved sulphides were immediately lost to the atmosphere as hydrogen sulphide gas. Because the de-watering spears could draw water over their installed depth of 6 metres, the samples represented a mix of ground water to this depth.

It is likely that ground water drawn from depths close to the pipes would be diluted by fresher water from higher levels. Some results of analysis are presented in Table 5.

Table 5 Selected results from chemical analyses of ground water

PRINCIPAL CATIONS (ppm)		PRINCIPAL ANIONS (ppm)		OTHER PROPERTIES (ppm)	
Na^+	3600	HCO_3^-	480	Bio. O_2 Demand	7
K^+	125	Cl^-	6000	Chem. O_2 demand	140
Ca^{++}	200	SO_4^{--}	910	Sulphide S^{--}	5.8
Mg^{++}	405				

It is clear that the water is not sea water; for example, the chloride ion concentration is around one third that of sea water. Results of particular interest are the pH of 7.2 at 22°C (about the average water temperature) and the sulphide concentration of 5.8 mg/l, confirmed to be principally dissolved sulphide. While a concentration of 5.8 ppm might seem small, in terms of potential for hydrogen sulphide release and subsequent conversion to sulphuric acid, this concentration is very high [2].

DISCUSSION OF RESULTS

Acid Attack in Sewerage Systems

The evidence from the investigation points strongly to sulphuric acid attack as the cause of the deterioration. Acid attack in sewerage systems usually stems from the generation of hydrogen sulphide gas and its subsequent oxidation. The generation of gas from sewage depends initially on the presence of anaerobic bacteria in the slime layer below water level and on the presence of conditions which permit the bacterial reduction of sulphate drawn from the sewage, to sulphide. If the dissolved oxygen concentration is low, the sulphide diffuses through the slime layer to enter the sewage. The anaerobic bacteria can exist in an environment of pH 5.5 - 9.0 with the optimum pH range being 7.5 - 8.0 [2].

Within the sewage, the sulphide exists as a combination of HS^- and dissolved H_2S, the relative proportions depending on the pH of the sewage [2]. At a pH of 7.2 and at normal temperatures around 40 % the dissolved sulphide exists as H_2S and is available for release into the atmosphere of the pipe. Aerobic bacteria around the moist, exposed interior surface oxidise the gas into acid.

The difficulty with the notion that sulphides originated from sulphate reduction within the pipe in this instance is that the attack was very severe in a very short time span. The sewage had small designed retention time and was sourced from a small, primarily domestic catchment. Much older pipe networks within the municipality with higher flows of sewage and longer retention times exhibited no significant decay, the difference being that these sound networks did not suffer from significant ground water intrusion.

This suggests either that sulphides were introduced into the system from a source other than sewage or that the pipes were so severely substandard as to offer little resistance to what would otherwise have been minor attack. As discussed earlier, tests of the pipe material indicate that the latter possibility was remote. In contrast, chemical analysis also discussed earlier identified the source of the dissolved sulphides as being the intruding ground water.

Origin of Sulphides in Ground Water

The most likely origin of dissolved sulphides is the reduction of sulphate ions to sulphide ions by anaerobic bacteria during the decomposition of organic matter under oxygen deficient conditions. A second considered source is from acid sulphate soil, the origin of sulphide being the oxidation of iron pyrite. However, the region has no history of problems with such soils.

Rate of Deterioration

Parameters relating to the deterioration of the pipe material such as the history and distribution of pipe defects, the rate of inflow of ground water and its chemical properties with time, could not be determined. Hence, prediction of the rate of destruction of pipe material is imprecise. From discussions with officers of the municipality and from inspection of leaks during the replacement program, the rate of intrusion was high. That is, ground water constituted a substantial portion of the total sewage pumped through the local station.

Not all network branches were badly fractured, hence some parts must have sustained very high rates of inflow. These parts could be expected to show most deterioration.

The following table presents calculated average rates of regression of the inside surface of three pipes within the network. Predictions are based on the method in Reference 2. The source of dissolved sulphide is assumed to be the ground water alone. Properties are as in Table 5.

Table 6 Predicted rates of deterioration of three pipes

CONDITIONS	PIPE 1	PIPE 2	PIPE 3
Diameter (mm)	225.0	150.0	150.0
Slope (%)	0.365	0.505	0.505
Flow depth (mm)	67.5	37.5	30.0
Ground water (%)	50.0	50.0	100.0
Loss of AC (av.)	0.66 mm/yr	0.72 mm/yr	1.0 mm/yr

In Table 6, no allowance has been made for the turbulent entry of ground water into the pipes (see Fig. 2) and the increased release of H_2S over that under normal conditions of flow. While the loss values in the table are speculative, the assumed flows are low. The predicted damage rates serve to illustrate that 5.8 mg/L of dissolved sulphide is a very substantial quantity. Data for pipe 3 have been included to show that, during periods of low flow of sewage from residences such as through much of the night, the damage from moderate amounts of ground water could have been considerable.

Other Observations

The outside surfaces of pipes were not attacked. The conversion of dissolved H_2S into gaseous H_2S and thence to sulphuric acid requires the evolution of the gas into an air space containing oxygen and aerobic bacteria. These conditions were not present outside the pipe. Sulphuric acid generation therefore occurred only once the ground water entered the pipe.

From Table 1, pipe condition tended to worsen with reducing R.L. This may relate to an accumulation of ground water downstream of pipe defects. It may also reflect dilution of the ground water closer to the water table, for example by rainwater.

It may also reflect that a fracture at depth suffers more head of water and a greater rate of ground water intrusion. Hydrogen sulphide is slightly heavier than air. The opportunity for migration of the gas in the upstream direction was therefore restricted.

It is not surprising therefore to find that condition generally worsened with depth and that significant changes occurred across some manholes.

CONCLUSIONS

Various observations and conclusions are made in the text. Of those, the principal conclusions are that the deterioration of the pipes was from sulphuric acid attack and that necessary sulphides came from abundant supply in the surrounding environment rather than from sewage. So severe were the intruding ground water properties, pipes of good quality decayed to the point of network failure within the early stages of their expected life.

The following points relate to management of asbestos cement and concrete pipe networks:

1. Unless the ground water conditions are known, it should not be assumed that ground water intrusion has the beneficial effect of flushing the system.

2. If doubt exists, a program of ground water sampling should be undertaken.

3. In the event of aggressive ground water, conductivity and rates of flow should be monitored at pumping stations to identify changes due to pipe fracture. Video cameras and flow meters can locate fractures.

4. Repair of fractures should be given high priority. Sulphide-bearing ground water is a problem only if it enters the pipes.

ACKNOWLEDGEMENTS

The author is indebted to Dr D K Brady for discussions concerning the origin of sulphides in ground water. The author also acknowledges the assistance of Dr J F Muller in conducting the field classification of pipes.

REFERENCES

1. COMITE EURO-INTERNATIONAL DU BETON, Durable Concrete Structures, Bulletin D'Information No 183, Thomas Telford, London, May 1992, 112 pp.

2. BOWKER, R P G AND SMITH, J M. Odour and Corrosion Control in Sanitary Sewerage Systems and Treatment Plants, Pollution Technology Review, No 165, Noyes Data Corporation, Park Ridge, New Jersey, 1989, 33 pp.

THEME FOUR:
REPAIR MATERIALS

Keynote Paper

LONG TERM PERFORMANCE CRITERIA FOR CONCRETE REPAIR MATERIALS

P S Mangat

F O'Flaherty

Sheffield Hallam University

United Kingdom

ABSTRACT. The paper presents a critical evaluation and discusses the limitations of the repair specifications adopted by agencies which manage reinforced concrete infrastructure. The key properties required of repair materials and their interrelationship with the substrate concrete are identified, based on wide ranging investigations both in the laboratory and on three reinforced concrete highway bridges which were repaired and instrumented insitu. The properties recommended for the primary specifications criteria are the elastic modulus, free shrinkage and creep of the repair material with the requirement that the elastic modulus of the repair material should be greater than that of the substrate.

Keywords: Repair, Long-term performance, Concrete, Hand applied materials, Spray applied materials, Flowing repair materials.

Professor P S Mangat is Head of the Built Environment Research Centre at Sheffield Hallam University. His current research interests are in the field of concrete deterioration and repair and the development of high performance materials based on industrial waste materials and mineral binders. He is currently engaged in a number of research projects funded by the EC and the Link Programmes.

Dr F O'Flaherty is a lecturer in the School of Construction at Sheffield Hallam University. His main research interests are concrete repair, rehabilitation and maintenance.

INTRODUCTION

The standards and specifications which control current practice of reinforced concrete repair, e.g. the Standard BD 27/86 of Highways Agency [1], pay insufficient attention to the long-term performance of repair patches and their structural interaction with the substrate. As a result, the key properties of repair materials and substrate concrete (and their required inter-relationships), which govern long-term performance (e.g. cracking, stress redistribution into the repair) are not systematically included in the design and decision making process of repair. Materials for repair are usually selected on a relatively ad hoc basis without an attempt to rationally optimise the selection process. Excessive reliance is placed on the information and advice supplied by manufacturers of repair materials. The data on properties of repair materials are frequently based on different test procedures and specimen size. It is well established that specimen size and test methods have a critical influence on the measured value of a property and manufacturers of repair materials are often inclined to select tests most favourable to their materials. Research on standardisation of test methods and specimen size for repair materials has tended to recommend smaller size of specimens compared with standard sizes used for testing concrete [2]. A result of this contradiction of different test specimen sizes is that any attempt to assess long-term structural interaction (stress redistribution) between repair and substrate on the basis of relative properties of materials becomes very difficult. In a rational design procedure for optimal repair, corrections to material properties are required to exclude the effect of specimen size mismatch.

Steel reinforcement corrosion is a major cause of deterioration in reinforced concrete structures, which disrupts the cover zone. Many types of repair materials are available on the market and selection of the most suitable material for reinstating the cover zone requires their performance data under service conditions. The principal requirements have traditionally been high dimensional stability and high early bond strength of the repair/substrate interface. Recent research, however, has raised concerns about renewed corrosion in the repair patch and incipient corrosion in the steel around the repair patch [3]. Consequently the emphasis of repair material selection has shifted to restoration of structural capacity of the corroded member, protection against renewed corrosion and maintaining long-term structural capacity if corrosion is re-initiated in the member.

The limitations to optimal repair materials selection, as outlined in the above discussion, are addressed in this paper on the basis of the results of major research projects which were concerned with determining

(a) the basic properties of generic repair materials [4]
(b) the long-term structural interactions between repair and substrate materials on three highway bridges which were repaired in situ and instrumented to monitor long-term performance of repair [5,6]
(c) the relationship between relative material properties (substrate and repair) and structural interactions and
(d) the relationships between the flexural strength (and deflection) and degree of reinforcement corrosion, before and after repair, using different generic repair materials [7, 8].

TEST PROGRAMME

Generic repair materials which are produced commercially were investigated, together with plain concrete mixes of similar grade.

Hand Applied Repair Materials

Material A is a blend of Portland cement, graded aggregate of 4mm nominal size and additives which control expansion and provide workability. The material complies with the standard BD27/86 (1). A water/powder of 0.13 is recommended.

Material B is a mineral binder based formulation with no coarse aggregate. It is very porous to allow leaching of salt from contaminated concrete. Recommended water/powder is 0.16.

Material C is a single component mortar containing microsilica, non-metallic fibres and styrene acrylic copolymer. Recommended water/powder is 0.16.

Material D is a heavy duty repair mortar containing a styrene acrylic copolymer, admixtures (including a water proofing agent), ordinary Portland cement, fibres, 6mm down graded aggregate and 10mm size granite aggregate.

Material E contains finely ground Portland blended cement blended with sulphoaluminate cement, microsilica, fibres, pozzolanic materials and spray dried styrene acrylic co-polymer. It is stated to be shrinkage compensated.

Concrete mix used as a control for comparison containing Portland cement, M grade sand, 10mm coarse aggregate, mix proportions 1: 2.24: 3.22, cement content 343 kg/m^3, w/c 0.56.

Spray Applied Repair Materials

Material F contains 5mm nominal size limestone aggregates, dust suppressants, rapid hardening PC, microsilica and a copolymer. It is designed for dry process spray application.

Material G is a blend of rapid hardening PC, microsilica, fibres, admixtures, spray dried styrene acrylic copolymer and sharp sand. It is designed for dry process spray application.

Material H is a blend of Portland cement, microsilica, limestone aggregate (maximum size 3mm) and admixtures. Recommended water/powder is 0.12.

Material J is based on Portland cement, silica sand of 5mm maximum size, admixtures, plastic fibres. Maximum water/cement is 0.35.

Concrete mix comprising Portland cement, M grade sand, w/c adjusted at the nozzle by an experienced operator.

Flowing Repair Materials

Material K contains 5mm maximum size aggregate, additives and shrinkage compensating agents. The cement content is 500 kg/m^3. Recommended water/powder is 0.37.

Material L is a rapid hardening material containing microsilica, shrinkage compensating admixtures, styrene acrylic copolymer and 6mm size aggregate.

Material M is a blend of Portland cement, graded aggregates and additives which produce a flowing, shrinkage compensated concrete. Recommended water/powder is 0.13.

Concrete (flowing) based on conventional materials. It consists of Portland cement, 10mm rounded aggregate, M grade sand, pfa, superplasticiser and polypropylene fibres, w/binder of 0.45.

Repair Methods

In the laboratory investigations, repair patches were applied by hand. In field repairs on highway bridges, repairs were applied by hand, by spraying and by pouring flowing repair material into water tight formwork (5,6). Hand applied repairs were carried out on soffits of bridge beams. Large areas of bridge abutments were repaired with spraying materials by means of the dry gunite process. Repairs on columns and beam-column joints of a bridge were applied by flowing materials.

Investigation of Corroded Beams

Tests were carried out to determine the structural performance of corroded beams before and following repair of the cover zone. Long-term structural performance of the repaired beams subjected to further corrosion after repair was also investigated. Details of the beam section are given in Figure 1. Each beam was singly reinforced with two deformed h.y. steel bars. The longitudinal reinforcing bars protruded out at one end of the beam as shown in Figure 1 in order to facilitate externally induced corrosion as described in a recent paper (7). Corrosion was induced by connecting an external power supply to the protruding rebars. The beams were placed in a $NaC\ell$ electrolyte and the current intensity of the external power supply was selected to achieve the desired degree of reinforcement corrosion in a fixed period of time. Each degree of corrosion (2RT/D)% was selected to represent a pre-defined percent reduction in rebar diameter.

R = corrosion rate (mm/year)
T = corrosion period since initiation of corrosion (years)
D = rebar diameter

The beams were tested under three-point bending by providing external shear reinforcement to ensure flexural failure.

RESULTS AND DISCUSSION

Compressive Strength and Elastic Modulus

The compressive strengths of the repair materials are listed in Table 1. These represent a range of data obtained from specimen sizes of 100x100x100mm cubes used in the laboratory (as for testing concrete to BS1881 part 116) to different sizes used by repair materials manufacturers. The strengths range between 30 and 70 N/mm^2. Specifying a repair material according to its compressive strength is rather meaningless unless, during its service life, the repair patch is expected to attract sufficient load from the substrate, resulting in high stress (equal to a significant proportion of its compressive strength). However, load redistribution from the substrate to the repair patch is entirely a function of the stiffness (elastic modulus) and creep property of the repair material instead of strength (6). Consequently, specification of a repair material should be based primarily on stiffness properties (elastic modulus, creep) with less emphasis on strength. A reliable co-relation between compressive strength and elastic modulus is expected of concrete materials (9). The data from Table 1 when plotted in Figure 2 show no such co-relation of these properties of the repair materials. Specifying high strength repair materials without the accompanying high stiffness will be counter productive for long-term efficiency of the repair patch since effective load sharing with the substrate will not be ensured.

Stress-Strain Relationship

Some examples of the stress-strain curves under compression of repair materials (A and C) together with similar grade concrete are given in Figure 3 (4). The falling branch of the longitudinal stress-strain curves is also shown in this figure. A careful study of the figure throws some useful light on the load-sharing capability of the repair materials with the substrate of a compression member.

Repair material C, which contains a styrene acrylic polymer, has a very high strain capacity at maximum load compared with concrete (approximately 5600 microstrain compared with 2400 microstrain). In any composite action with the substrate which would conform to strain compatibility, the very high strain capacity of the repair material C is of no benefit. In fact the high non-linearity of the stress-strain curve of material C indicates that the effective elastic modulus decreases rapidly with increasing stress so that continuous load re-distribution from the substrate in the long-term becomes impossible. The polymer modified repair material also has the highest creep strains under sustained loading (4) which, further aggravate its load sharing efficiency.

Repair material A (Figure 3), on the other hand, has a very high elastic modulus and its peak and falling branch strain characteristics are similar to concrete. It is, therefore, capable of effective load redistribution from the substrate of a compression member. However, the very high strength of material A is unlikely to be utilised.

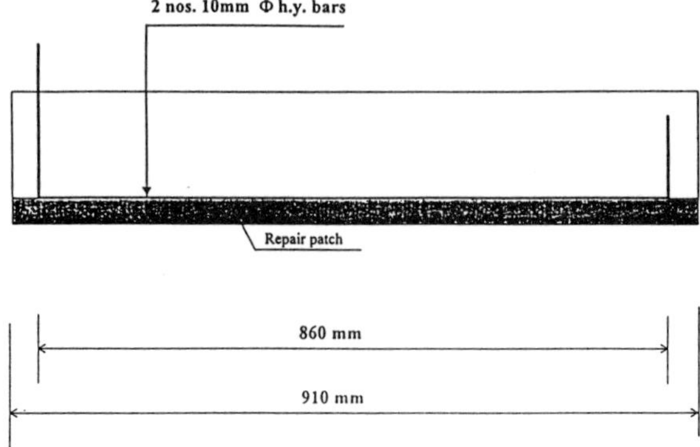

Figure 1 Corroded beam specimen

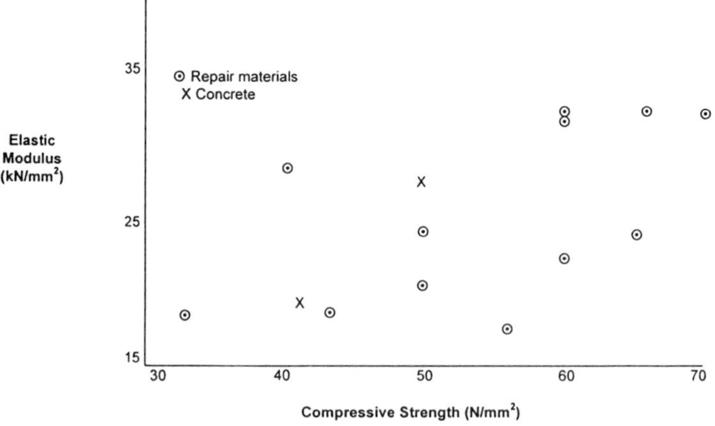

Figure 2 Compressive strength and elastic modulus of repair materials

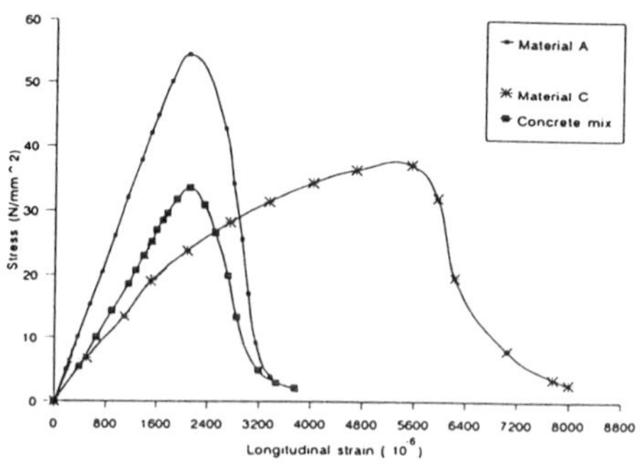

Figure 3 Stress-strain relationship of repair materials

Table 1 Properties of repair materials

REPAIR MATERIAL	APPLICATION	DENSITY (kg/m^3)	COMPRESSIVE STRENGTH (N/mm^2)	ELASTIC MODULUS (kN/mm^2)	FLEXURAL STRENGTH (N/mm^2)
A	Hand		64	32	7.7
B	Hand		33	19	4.2
C	Hand		44	18	3.7
D	Hand	2100	50	24	13.5**
E	Hand	1750	50	19.6	8.0**
Control	Hand	2100	41	19	4.4
F	Spray	2250	60	31	8*
G	Spray	2200	57	18	4.1*
H	Spray	2210	60	22.7	
J	Spray		60	29	
Control	Spray		46	24	
K	Flowing	2250	65	24	
L	Flowing	2250	70	32	
M	Flowing	2270	60	32	
Control	Flowing		45-50	27.4	

*Tested to BS 1881, Part 118 ** Tested to BS 4551

It is interesting to note from the creep data of repair materials (4), including the flowing repair materials (Figure 4), that creep of the control concrete mix (of similar grade as the repair materials) is much lower than that of the special repair formulations. The strain characteristics of the concrete mix (Figure 3) are also more compatible with a substrate concrete and, therefore, from the long-term load sharing point of view, repair materials based on standard concrete technology may offer some advantage compared with special formulations, especially polymer modified repair materials such as material C.

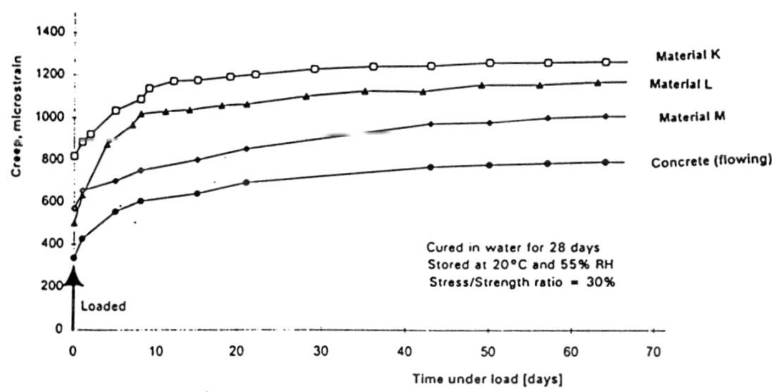

Figure 4 Creep of flowing repair materials

Shrinkage and Shrinkage Cracking

The typical drying shrinkage (free shrinkage) data of the repair materials A, B, C and concrete are given elsewhere (4). Shrinkage data of repair materials used in the bridge repairs (6) are listed in Table 2. These data show that all repair materials have significant free shrinkage despite sometimes misleading information given in manufacturers' catalogues. For example repair materials E, K, L, M (Table 2) are described as shrinkage compensated but display large drying shrinkage values. Amongst the lowest free shrinkage values for the hand applied, spray applied and flowing material categories are obtained from the control concrete mixes. This indicates that the special repair formulations generally shrink significantly more than designed concrete mixtures of similar grade. In general, repair materials incorporating higher proportions of coarse aggregate particles undergo lower levels of free shrinkage (4,6). This conforms with the behaviour of concrete in which coarse aggregate particles provide internal restraint to free shrinkage deformations (9).

Table 2 Drying shrinkage of repair materials used for bridge repairs

MATERIAL	APPLICATION	SHRINKAGE AT 100 DAYS (μSTRAIN)	AGGREGATE	ADMIXTURES
F	Spray	751	5mm graded	Copolymers
G	Spray	1311	Sand zone M	Styrene acrylic polymer
H	Spray	620	3mm graded	Admixtures
J	Spray	782	5mm graded	Admixtures
Control Concrete	Spray	717	Sand zone M	None
D	Hand	401	10mm graded	Styrene acrylic
E	Hand	1087	Fine	Styrene acrylic, shrinkage compensated
K	Flowing	740	5mm graded	Shrinkage compensating
L	Flowing	580	6mm graded	Copolymers Shrinkage compensated
M	Flowing	791		Shrinkage compensating
Control Concrete	Flowing	388	10mm rounded	Polypropylene fibres

Shrinkage cracking in repair patches is caused by the restraint to free shrinkage provided by the substrate concrete at the interface of the repair patch. This results in a tensile stress in the repair patch whose magnitude is proportional to the free shrinkage. Tensile cracking occurs when the tensile strain capacity of the repair material is exceeded by the restrained shrinkage. Tensile creep of the repair material, on the other hand, causes a relaxation of the tensile stress induced by restrained shrinkage. As a consequence, it can be envisaged that a repair material whose shrinkage and creep characteristics are both very high can perform as satisfactorily against restrained shrinkage cracking as another repair material which has both low shrinkage and low creep characteristics. Under long-term service, however, the latter repair material (low shrinkage and low creep) will be much more effective in sharing load with the substrate concrete. Many polymer modified repair materials fall in the category of high shrinkage and high creep properties. They may provide satisfactory resistance to restrained shrinkage cracking but their long-term structural interaction with the repaired member will be poor

Data obtained from the restrained shrinkage test method are often given in trade brochures of repair materials (10). The resistance to shrinkage cracking, as represented by crack size and spacing measured in the restrained shrinkage test, is a cumulative effect of free shrinkage, tensile creep, tensile strength and stiffness of the repair material. Consequently favourable restrained shrinkage characteristics can be obtained for high shrinkage, high creep (low stiffness) materials whereas in structural repair such materials will be ineffective for providing efficient long-term structural interaction with the substrate. For this reason, separate free shrinkage and creep data of repair materials is more useful for the design of optimal repair solutions than restrained shrinkage data.

Repair/Substrate Structural Interaction

Figure 5 shows the typical stress redistribution graphs in the different phases of a large repair patch applied by spraying to the abutment of Lawns Lane bridge (5). The (idealised) stress redistribution graphs have been derived from strains monitored in the substrate at the interface with repair, and in the steel reinforcement. The elastic modulus of the repair material (29.1 kN/mm^2) is considerably greater than that of the substrate concrete (23.8 kN/mm^2). Figure 5 shows that the stiffer repair material imparts significant compression to the substrate concrete during the shrinkage stage, which assists in reducing the restrained shrinkage tension in the repair material. In the long-term the repair material tension is neutralised by external load transfer from the substrate and a steady increase in compressive stress occurs. The repair performed satisfactorily, without cracking.

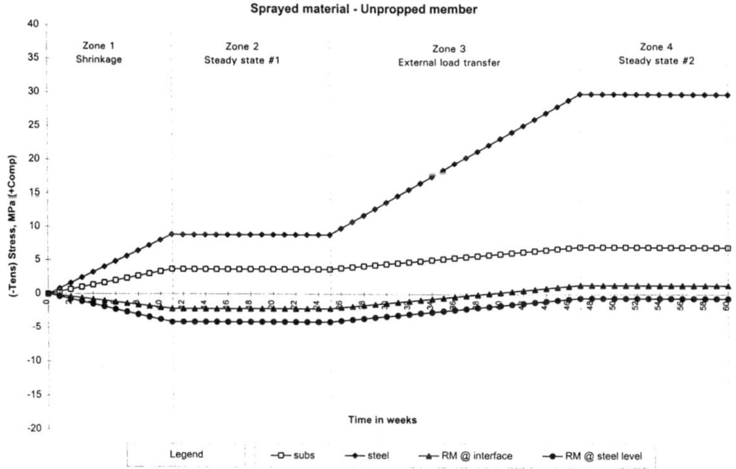

Figure 5 Stress Redistribution in the Repair Patch of a Bridge Abutment ($E_{repair} > E_{substrate}$)

An example of a repair patch applied with a repair material of lower elastic modulus (22.7 kN/mm^2) than the substrate (23.8 kN/mm^2) is shown in Figure 6. The stiffer substrate provides a more effective restraint to the repair patch during the shrinkage period compared with Figure. 5 where significant compressive strain was transferred into the substrate by the stiffer repair material. Extensive cracking in Figure 6 is due to excessive tension in the repair patch caused by the restraint provided by the stiffer substrate. It is interesting to note that repair material H in Figure 6 satisfies the Highways Agency specifications (1) whereas material J represented in Figure. 5 is an unapproved material.

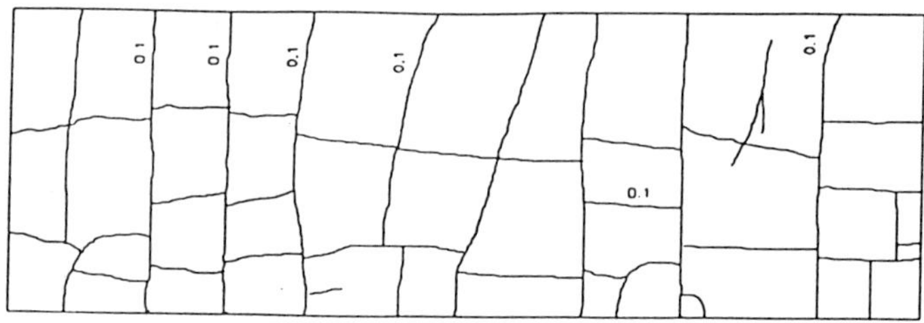

Figure 6 Cracking in a repair patch on a bridge abutment ($E_{repair}<E_{substrate}$)

Flexural Strength of Repaired Beams

Figure 7 shows the load-deflection graphs of repaired reinforced concrete beams which are not exposed to any further corrosion after repair. It represents beams which were subjected to an initial degree of corrosion of 2.5 percent and then repaired with hand applied materials A, B and the control concrete repair mix (Table 1). Figure 7 shows that the beams repaired with material A, which has the highest elastic modulus (E = 32 kN/mm^2), develop the highest stiffness. These are closely followed by beams repaired with repair material concrete (E = 19 kN/mm^2) which has much lower creep characteristics compared with material A (4). Repair material B, which has a low elastic modulus and relatively high creep, results in significantly reduced stiffness of the repaired beams. The flexural strengths, of the beams repaired with the different repair materials, however, were similar.

In field conditions, repaired structures can often be subjected to further corrosion. In order to simulate this, further corrosion was induced in the repaired beams to degrees ranging between 1.25 and 5 percent. The flexural strengths of these beams are represented in Figure 8. It is important to note that the performance of the beams repaired with the high stiffness, high strength repair material A has deteriorated significantly, resulting in similar residual flexural strengths as for beams repaired with repair material B. This effect is even more pronounced if the flexural load capacity at permissible deflections under serviceability conditions are considered (8). Beams repaired with the concrete mix perform considerably better. Repair material A allows less dissipation of corrosion products which, therefore, accumulate at the interface until longitudinal tensile cracking occurs along the steel reinforcement. Longitudinal cracking is less severe in the relatively porous and less stiff materials B and concrete and, therefore, flexural performance of the beams repaired with these materials suffers to a lesser extent due to further corrosion in the reinforcement.

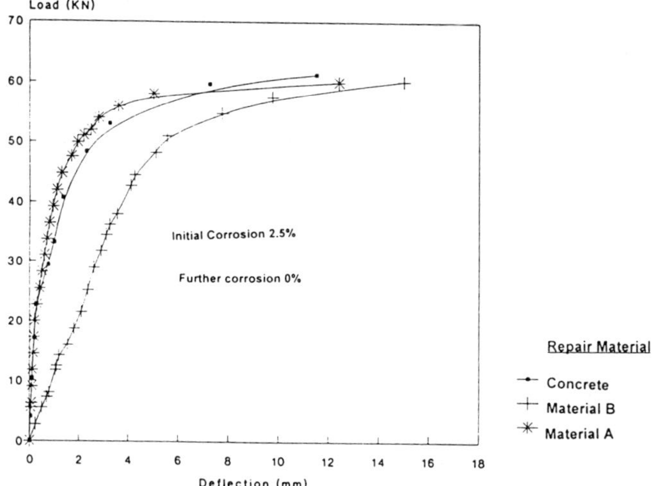

Figure 7 Load-deflection curves of repaired RC beams, initial corrosion 2.5 %, further corrosion, 0%

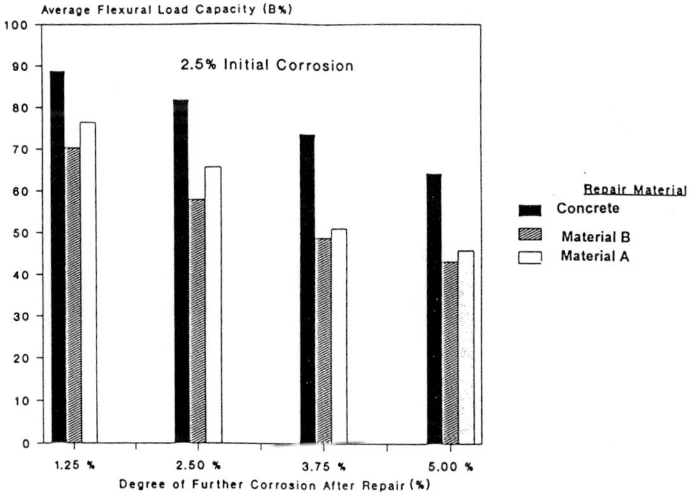

Figure 8 Flexural strength of repaired beams undergoing further corrosion

CONCLUSIONS

1. The elastic modulus, free shrinkage and creep properties of repair materials should be of prime concern in concrete repair specifications.

2. It is desirable to select repair materials of stiffness (elastic modulus) greater than the substrate concrete in order to prevent cracking during the shrinkage period (early age) and provide efficient structural interaction (load transfer from the substrate) in the long term.

3. In structures where continuing corrosion is likely after repair, the choice of relatively porous and lower stiffness repair materials is desirable to provide long-term structural efficiency. This requirement may be contradictory to the need for a more impermeable repair material to resist diffusion of aggressive agents. A compromise would, therefore, be required.

4. All repair materials reported in the paper undergo significant levels of free shrinkage under drying conditions (including those which are stated by the manufacturers to be shrinkage compensated).

5. Commercial repair materials generally undergo much higher levels of creep relative to concrete of similar grade.

ACKNOWLEDGEMENTS

The contribution of past research assistants M. S. Elgarf and M. C. Limbachiya is gratefully acknowledged. Funding from the Brite Euram Programme and the Link TIO programme has supported the research which forms the basis of this paper.

REFERENCES

1. DEPARTMENT OF TRANSPORT (1986), Materials for the Repair of Concrete Highway Structures, BD 27/86.

2. EMBERSON, N K, MAYS G C. "Significance of Property Mismatch in the Patch Repair of Structural Concrete, part 1: Properties of Repair System". Magazine of Concrete Research, Vol. 42, No. 152. September 1990, pp 147-160.

2. VASSIE, P R. "The Influence of Steel Condition on the Effectiveness of Repairs to Reinforced Concrete". UK Corrosion 88, pp 183-195, 3-5 October, Brighton, 1988.

4. MANGAT, P S, LIMBACHIYA, M C. "Repair Material Properties for Effective Structural Application". Cement and Concrete Research. Vol. 27, No. 4, pp 601-617, 1997. Elsevier Science Ltd.

5. MANGAT, P S, O'FLAHERTY, F J. "Long Term Performance of Sprayed Concrete Repair in Highway Structures". Proc. ACI/SCA International Conference on Sprayed Concrete/Shotcrete - Sprayed Concrete Technology for the 21st Century (Ed. S. Austin). 10-11 September 1996, E & F N Spon, London, pp 196-205.

6. MANGAT, P S, O'FLAHERTY, F J. "Long-term Performance of Bridge Repair Using High Stiffness Materials ($E_{rm}>E_{sub}$)", Magazine of Concrete Research, In Press.

7. MANGAT, P S, ELGARF, M S. "Flexural Strength of Concrete Beams with Corroding Reinforcement", ACI Structural Journal, January/February 1999.

8. MANGAT, P S, ELGARF M S. "Strength and Serviceability of Repaired Reinforced Concrete Beams Undergoing Reinforcement Corrosion", Magazine of Concrete Research, March/April 1999.

9. NEVILLE, A M. "Properties of Concrete", Fourth Edition, Longman.

10. DECTOR, M H, LAMBE, RW. "New Materials for Concrete Repair - Development and Testing", The Indian Concrete Journal, October 1993, 475-480.

EVALUATION AND REPAIR OF POST-TENSIONED CONCRETE STRUCTURES

J F Duntemann

P Plemic

Wiss Janney Elstner Associates Inc

United States of America

ABSTRACT. Prestressed concrete construction in the United States can trace its origin back to the early 1950's with the factory production of precast, prestressed concrete highway bridge girders. The use of unbonded post-tensioned tendons for slab systems began in the mid 1950's and has subsequently become relatively common. In the past several years, the incidence of corrosion and resultant tendon failure has also become more common in structures exposed to aggressive environments.

In a typical post-tensioned slab, the consequence of a local tendon failure may not be catastrophic, but can certainly be symptomatic of more serious problems. If an unbonded tendon fails at any point, the effective prestressing force of the entire tendon is lost. In most cases, due to the sudden release of energy, corrosion failures in unbonded construction produce dramatic evidence of the problem.

There are a variety of procedures used to repair existing post-tensioned structures where the post-tensioned tendons have failed. These procedures include tendon splicing, installing a new tendon in the existing sheathing, trenching the slab to install a new tendon, and external post-tensioning. There is also a variety of available hardware to facilitate these repairs, including some innovative center stressing devices. The most suitable method of repair may vary from one structure to the next and also depend on the condition of the existing post-tensioning system.

Keywords: Post-tensioned, Concrete, Unbonded, Tendons, Monostrand, Buttonhead, Corrosion

John F Duntemann is a Consultant with Wiss, Janney, Elstner Associates, Inc., Northbrook, Illinois, USA. He is a licensed structural engineer with over 20 years of experience in the assessment and rehabilitation of existing structures.

Petar Plemic is a Senior Engineer with Wiss, Janney, Elstner Associates, Inc., Northbrook, Illinois, USA. He has over 10 year of experience in the investigation and the design of repairs of post-tensioned parking structures.

INTRODUCTION

Post-tensioned concrete construction is classified as bonded or unbonded depending on whether the tendon ducts are filled with grout after stressing (bonded), or whether the tendons are greased and wrapped (unbonded). In the United States, unbonded post-tensioning systems have gradually become the more common type of post-tensioning system used in commercial concrete building construction. The four basic unbonded systems consist of monostrand tendons, single bar tendons, multi-strand tendons and multiwire tendons. Unbonded single strand or monostrand (12.7 mm diameter, seven wire) tendons are most common and, the principal subject of this paper.

Unbonded monostrand tendons were first used in building construction in the United States around the mid 1950's and consisted of greased and paper wrapped seven-wire strand. The spirally applied continuous paper strip served as the bond breaker between the strand and concrete and the grease coating provided corrosion protection. Plastic sheathing was developed in the mid-to-late 1960s and intended to serve as a bond breaker, protection from damage during handling, and as a barrier against the intrusion of moisture and harmful chlorides. The strand coating, or grease, reduces friction between the strand and sheathing and provides additional corrosion protection. Extruded polyethylene sheathing was first introduced in 1969, and is currently almost used exclusively.

Experience indicates that the life span of properly designed and constructed post-tensioned concrete structures is at least comparable to any other contemporary construction material or method. However, known incidents of corrosion of post-tensioned tendons and resultant durability problems emphasizes the need for corrosion protection and good detailing. In some cases, the cost to repair the existing post-tensioning damage can exceed the original post-tensioning system cost. Lessons learned from these problems enable us to improve current practice.

DURABILITY OF UNBONDED TENDONS

In comparison with mild-reinforced concrete structures, post-tensioned structures tend to have inherently enhanced durability due to the limitation of cracks that provide access of corrosive agents to the reinforcement. Since the axial force in the stressed tendon is transferred to the concrete by the tendon anchorages, however, the protection of the tendon and anchorages are paramount to the long-term integrity and performance of the structure.

The Post-Tensioning Institute (PTI), "Specification for Unbonded Single Strand Tendons," currently prescribes several requirements related to corrosion protection [1]. For applications in so-called "aggressive environments," PTI requires the sheathing to be connected to all stressing, intermediate and fixed anchorages in a watertight fashion, thus providing complete encapsulation of the prestressing steel from end-to-end. This level of encapsulation can be achieved by heat shrink tubing and proprietary encapsulation systems.

Nevertheless, there are many existing post-tensioned structures, particularly in geographical areas with relatively aggressive environments that have fallen short of the aforementioned durability expectations. Table 1 identifies several parking structures located in different geographical areas of the United States, which have all manifested corrosion related problems.

Many of these structures were simply constructed prior to the common use of polyethylene sheathing and epoxy-coated anchorages. However, experience with even more recent construction indicates a continuing need to improve on construction procedures and details to enhance the durability of these systems.

Table 1 Case histories of corrosion of post-tensioned parking structures

BLDG	TENDON DESCRIPTION	BUILT	TOTAL SUPPORTED AREA, SQ FT	CORROSION CONDITIONS	REPAIRED	REPAIR COST $ U.S.
A	Buttonhead wire	1964	660,000	Corrosion at intermediate anchorages.	1991-92	1.5 mil
B	Buttonhead wire	1964	960,000	Corrosion at intermediate anchorages.	1991-92	1.5 mil
C	Single bar tendon	1964	370,000	Failure at construction joint.	1992-93	0.9 mil
D	Strand, Paper wrapped	1965	400,000	Spalling and delamination of concrete along the construction joints.	1993-94	0.6 mil
E	Buttonhead wire	1969	125,000	Minimal concrete cover at tendon high points resulting in tendon corrosion.	1997-1998	0.6 mil
F	Buttonhead wire	1969	60,000	Corrosion stains on underside of slab.	1993-95	1.3 mil
G	Strand	1969-70	75,000	Corrosion of strand at delamination.	1990-91	0.5 mil
H	Strand	1972	80,000	Spalling of concrete on underside of slab.	1989-90	0.28 mil
I	Strand	1975	100,000	Damaged tendon sheathing during overlay installation.	Torn down In 1994	- -
J	Strand	1976	- -	Corrosion at intermediate construction joint.	Not yet Repaired	Estimated 0.5 mil
K	Strand	1984	86,000	Vertical displacement at intermediate construction joint.	1990	0.9 mil

EVALUATION OF POST-TENSIONED SYSTEMS

Prior to repairing a post-tensioned concrete structure, it is necessary to conduct a comprehensive engineering evaluation of the problem. Telltale signs of potential problems in post-tensioned slab systems include cracking, efflorescence or rust staining at construction joints or cracks, spalling adjacent to anchorage zones, and in extreme cases, relative displacement across construction joints. The methods for assessing an existing post-tensioned concrete structure are, for the most part, the same as for as for a concrete structure constructed with mild steel reinforcement. However, tendons and anchorages should be selectively exposed to examine their condition.

If an unbonded tendon fails at any point, the effective prestressing force of the entire tendon is lost. Section 18.18.4 of the American Concrete Institute Building Code Requirements for Structural Concrete (ACI 318-95) allows a two percent loss of total prestress due to unreplaced broken tendons [2]. Thus, for example, in the case of a post-tensioned slab structure that might contain a hundred tendons in a single floor, the loss of one strand without replacement would not be considered to have a significant impact on structural integrity.

While the consequence of a local tendon failure may not be catastrophic, it can certainly be symptomatic of more serious problems.

In most cases, due to the sudden release of energy, corrosion failures in unbonded construction produce dramatic evidence of the problem. The sudden exiting of an embedded tendon from a concrete slab or beam is generally referred to as a "blowout". Tendon blowouts are typically observed at the end anchorages or at the top of slab surface or soffit corresponding the high and low points of the tendon profile. An example of a tendon blowout at the top surface of the slab is shown in Figure 1.

The long-term performance of post-tensioned tendons requires that the sheathing and corrosion protection remains intact and undamaged. PTI currently prescribes a minimum thickness of 1 mm medium or high-density polyethylene or polypropylene sheathing. However, experience has shown that thin wall plastic sheathing is often damaged even before being placed in the forms. Tie wire between perpendicular reinforcement causes local indentations in the sheathing which tear the tendon when tensioned. Similarly, reinforcing bar indentations cause hard points that can also tear the tendon when tensioned.

The material standard specification for seven-wire prestressed concrete strand, ASTM A416, states that "slight rusting, provided it is not sufficient to cause pits visible to the unaided eye, shall not be cause for rejection" [3]. PTI prescribes five grades of strand, corresponding to Grades A through F, where Grade A exhibits no visible rust and Grade F exhibits heavy oxidation with strong flaking and pit formation. Strands used in new construction must be at least Grade C where pits or surface defects do not exceed 0.05 mm (can be felt with a fingernail) diameter or length. Reference 4 provides further discussion on this subject.

The principal vulnerability of sheathed single strand tendons generally corresponds to intermediate and end anchorages where the sheathing is removed to facilitate stressing. A polyethylene or polypropylene sheathed tendon, coated with a corrosion inhibiting grease, can provide ideal corrosion protection if the integrity of the sheath and grease coating is maintained. However, the common practice of removing the sheathing virtually eliminates the corrosion protection at the anchorage.

An accurate assessment of the condition of an embedded tendon is best achieved by exposing the tendon and examining it closely. This process generally consists of choosing a location in the structure where the tendon is more accessible, such as the high or low point of the strand profile.

The tendon can be readily located by use of an R-meter. Removal of the concrete around the tendon can be accomplished by using an electric or light pneumatic chipping hammer. Care must be exercised during the concrete removal to avoid damaging the tendon or undermining the tendon anchorage.

A view of an exposed monostrand tendon at an intermediate construction joint is shown in Figure 2. Note the absence of any corrosion protection at the anchorage. In order to examine the tendon, it is necessary to remove the wedges and shift the anchorage as shown in Figure 3. An example of the tendon condition after the wedges are removed is shown in Figure 4.

Figure 1 Vertical loop due to tendon blowout

Figure 2 Corrosion at intermediate anchorage

Figure 3 Removing wedges to examine tendon condition

Figure 4 Comparative examples of pitted and unpitted tendon

REPAIR METHODOLOGY AND CORROSION PROTECTION

There are a variety of procedures used to repair existing post-tensioned structures where the post-tensioned tendons have failed. These procedures include tendon splicing, installing a new tendon in the existing sheathing, trenching the slab to install a new tendon, and external post-tensioning. There is a also a variety of available hardware to facilitate these repairs, including some innovative center stressing devices. The most suitable method of repair may vary from one structure to the next and also depend on the condition of the existing post-tensioning system.

Monostrand Systems

Figure 5 illustrates two alternate methods of splicing a tendon that were performed on the same project. This particular structure had polyethylene sheathed monostrand tendons, and epoxy-coated anchorages and mild reinforcement. Corrosion damage occurred at the intermediate anchorages where the sheathing was removed to accommodate the original stressing operation. Examination of tendon locations away from the anchorage zones indicated that the remaining tendon was generally in excellent condition. Therefore, the repair solution consisted of splicing 4 m lengths of new tendon in the effected areas.

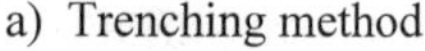

a) Trenching method

b) "Push-thru" method

Figure 5 Tendon splicing methods

The splices consisted of installing a standard coupler at one end of the splice and a center stressing device at the other end. As a general rule, the coupler and stressing device were installed near inflection points, providing sufficient room for coupler movement during retensioning. In most cases, the new tendon was installed in the existing sheathing by the "push-thru" method. In locations where the tendon could not be pushed through the existing sheathing, the slab was trenched to accommodate an entirely new tendon with sheathing. When multiple tendons were replaced, the repairs were performed in an alternating sequence to minimize in-plane eccentricities of the post-tensioning force. The slab was also temporarily shored during this process.

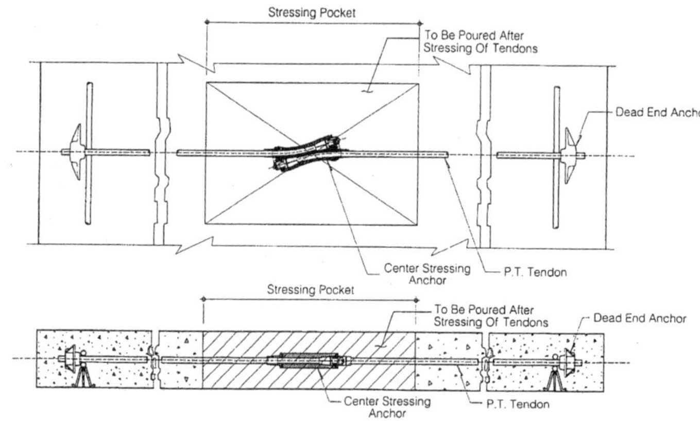

Figure 6 Proprietary center stressing anchor

A schematic view of a proprietary center stressing anchor is shown in Figure 6. This device develops 100 percent of the nominal ultimate strength of a 12.7 mm diameter, 270K (1861 mPa) strand, and is ideal for repairing broken strands in existing structures. Figure 7 illustrates the stressing operation. Figure 8 shows the completed strand repair with the epoxy-coated dowel reinforcement in place. The post-tensioning hardware is encapsulated by heat-shrink tubing and integrally attached to the polyethylene sheathing to prevent water intrusion. Note the size and positioning of the center stressing device relative to the 150 mm thick slab.

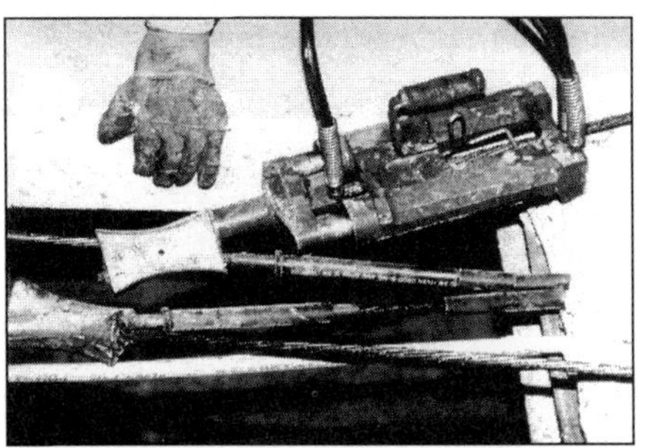

Figure 7 Center stressing operation

Figure 8 Completed strand repair with pvc wrap

Multiwire Systems

Many of the earlier post-tensioned structures were built using the paper-wrapped button-headed wire post-tensioning system. The tendons generally contain six or seven 6 mm diameter, 240K (1655 mPa) stress relieved wires. "Buttonheads" formed on the ends of the wires provide anchorage for the individual wires at the anchor plate. The tendons are stressed by pulling on the anchor plate until the desired elongation of the wires is achieved. Steel shim plates installed between the anchor plate and bearing plate maintain the tendon elongation and resultant prestress force.

In many instances, corrosion of these systems occur at high points in the tendon profile or at intermediate construction joints. Tendon splicing and restressing at these locations is accomplished by installing center stressing hardware. First, the tendon wires are detensioned. After the necessary concrete removal, the tendon wire terminations are cut and new end blocks are slid onto the wires. New buttonheads are formed on the ends of the wires by a cold-upsetting process. The tendon splice is completed by installing lengths of tendon wires through the end blocks and forming buttonheads. The connection of the end blocks serves as a center stressing device. The wires connect the left end block with the right jacking block and the right end block with the left jacking block. An illustration of the completed assembly, including the hydraulic jack is shown in Figure 9.

Stressing the tendon consists of placing a hydraulic jack between the two jacking blocks and forcing them apart until the required prestress forces is obtained. Shim plates are then placed between the jacking block to maintain the applied prestress force. After stressing the tendon, the wires are protected in a similar manner as for the monostrand repairs.

In some cases, the tendon wires are spliced at intermediate locations. The length of new tendon is connected to the existing tendon by a threaded coupler. An illustration of this coupler is shown in Figure10.

Figure 9 Multiwire center stressing assembly

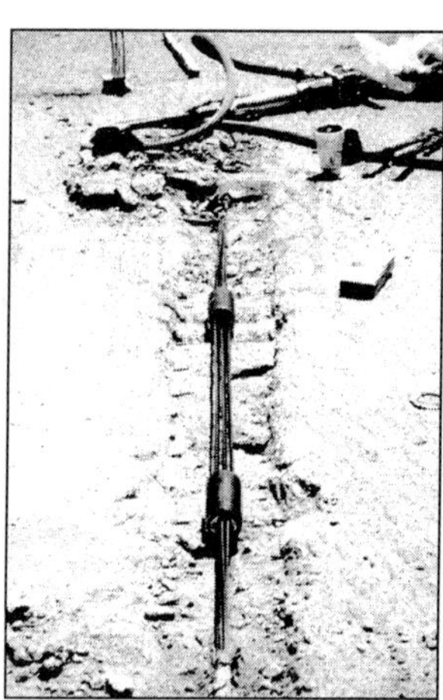

Figure 10 Multiwire tendon coupler

RECOMMENDATIONS REGARDING EVALUATION AND REPAIR OF POST-TENSIONED STRUCTURES

1. The methods for assessing an existing post-tensioned concrete structure are similar to the methods commonly used to assess mild reinforced concrete construction. However, tendons and anchorages should be selectively exposed to verify the existing conditions. In particular, unbonded tendons should be thoroughly examined at stressing locations and intermediate anchorages.

2. The long-term performance of post-tensioned tendons requires that the sheathing and corrosion protection remain intact and undamaged. Special inspection is recommended during construction to insure that this damage is avoided.

3. Existing tendons that exhibit pits or surface defects greater than 0.05 mm in diameter or length should generally be replaced. The material standards used in post-tensioned concrete rehabilitation should be consistent with the standards used in new construction.

4. There are a variety of procedures used to repair existing post-tensioned structures where the post-tensioned tendons have failed. The most suitable methods of repair will vary from one structure to the next and also depend on the condition of the existing post-tensioning system.

5. At the completion of repairs to a post-tensioned system, the tendons should be electrically isolated in a manner to minimize the possibility of recurring problems. There are a variety of proprietary systems on the market, including heat-shrink tubing and corrosion inhibiting tapes, that effectively encapsulate the post-tensioned components.

REFERENCES

1. POST-TENSIONING INSTITUTE, Specification for Unbonded Single Strand Tendons, Phoenix, 1993.

2. ACI COMMITTEE 318, "Building Code Requirements for Structural Concrete", American Concrete Institute, Detroit, 1995.

3. ASTM A416, "Standard Specification for Steel Strand, Uncoated Seven-Wire for Prestressed Concrete", American Society for Testing and Material, Philadelphia, PA 1991.

4. SASON, A "Evaluation of Degree of Rusting on Prestressed Concrete Strand"., PCI Journal, May/June, 1992 Vol. 37, No. 3, Prestressed Concrete Institute

NEW CREAM TECHNOLOGY FOR CONCRETE IMPREGNATION

R Hager

Wacker-Chemie

Germany

ABSTRACT. Silicones have been used very successfully over decades to protect concrete, especially reinforced concrete, against moisture, since water plays a key role in damaging processes such as reinforcement corrosion. Impregnating agents are normally based on silanes and siloxanes. Very often significant amounts of these low viscous fluids are lost due to evaporation of volatile ingredients and running off of excess material, particularly when working over head. To avoid this, recent developments have seen a new type of water repellent product which can be referred to as a cream. This material can be applied very easily in a single application step and at very high coverage rates, without loss of any active ingredients. It penetrates even high grade concrete well and leaves no visible residues on the surface. This paper provides an overview of developments and uses of this new technology for concrete impregnation.

Keywords: Silicone, Silane, Siloxane, Water repellency, Impregnation, Surface protection.

Dr R Hager is technical manager for silicone products in construction chemicals formulations at Wacker-Chemie GmbH, Burghausen, Germany. He is responsible for development and marketing of silicone based water repellent agents. One of his main subjects is the protection of concrete surfaces.

INTRODUCTION

Concrete offers attractive properties, particularly economy, durability and freedom of design, that have made it the most widely used construction material. Concrete is essential for modern industrial and public buildings, roads, tunnels and bridges.

Until a few years ago, concrete was thought to be resistant to all harmful effects. The many examples of damage demonstrate the contrary: concrete is vulnerable. Concrete structures are in peril [1].

Damage to concrete involves water. While water can cause purely physical damage, for example in the case of freeze/thaw stressing, it is also a medium for the transport of aggressive substances, such as chlorides from road salts, and forms a reaction medium for harmful chemical processes, particularly corrosion of the reinforcing steel [2].

The most efficient type of protection for concrete – and this naturally also applies to other construction materials – is thus protection against moisture [3]. In recent years, silicones have emerged as the class of materials most suitable for this purpose [4]. Silicones are pre-eminent among masonry protection agents, thanks mainly to their outstanding water repellency and durability. Practically no other substance is so inert to physical, chemical and microbiological attack. Assuming that the material has been chosen correctly, silicone impregnation can contribute significantly to the long-term preservation of a structure.

The impregnants used for concrete protection usually consist of low-molecular silanes or mixtures of silanes and siloxanes, applied either undiluted, diluted with organic solvents or as aqueous formulations. All of these preparations take the form of low-viscosity liquids. During application, this results in the loss of significant amounts of the active substances as the liquid runs off vertical surfaces. This applies particularly to overhead work. This drawback could be remedied by a new system: impregnation cream.

Before considering the properties and advantages of this new product grade, however, a brief look at the fundamental principles of concrete construction materials, at the concrete damage most commonly found, and at concrete protection shall be taken.

CONCRETE

Concrete As Construction Material

Concrete and reinforced concrete have drastically changed the course of building construction in the last few decades. When these materials were developed about 120 years ago, architects, civil engineers and builders were suddenly presented with materials that not only had outstanding mechanical and physical properties, such as compressive strength and tensile strength in bending, but which could be moulded like no other material before. It was possible to build either delicate and intricate structures or massive civil engineering works, such as bridges, towers and skyscrapers. Concrete's economy and durability made it indispensable.

Concrete and reinforced concrete consist essentially of Portland cement as binder, sand and gravel aggregates as well as, in the case of reinforced concrete, steel reinforcement to improve tensile strength. Water is also required, to harden the concrete and adjust its consistency for processing. One of the factors critical to the quality of concrete is its water/cement ratio: the ratio of the mass of mixing water to the mass of cement. Too high a water/cement ratio increases the formation of capillary pores in the cement matrix, leading to a loss in strength.

Assuming good workmanship and the correct ratio of ingredients, the resulting construction material has outstanding resistance to weathering and ageing.

Concrete Damage

Despite its high resistance to the effects mentioned above, serious damage to concrete occurs time and again, threatening the stability of building structures. The main cause of concrete damage is corrosion of the reinforcing steel resulting from environmental effects. While fresh concrete has a high alkalinity that passivates the steel, acidic gases in the atmosphere, particularly carbon dioxide, will over time neutralize the alkalinity of the surface. In the case of carbon dioxide, this process is called carbonation. Eventually, this non-alkaline carbonated zone reaches the reinforcing steel and destroys its passivating protective layer. Atmospheric oxygen and moisture can then begin to rust the steel. Since ferrous metals greatly increase in volume when they rust, the concrete layer above the reinforcement spalls, resulting in serious damage to the concrete. Similar patterns of damage are also caused by salts dissolved in water, particularly chlorides. Irrespective of the alkalinity of the concrete, chloride ions can cause catastrophic corrosion within an extremely short time. Since concrete chiefly absorbs chlorides via deicing salts, roads and bridges are particularly prone to this damage.

CONCRETE PROTECTION

Once the fundamental mechanisms of damage were known, researchers could work on achieving extremely durable and economical concrete protection. Investigations have concentrated on two main surface protection processes: water-repellent impregnation and film-forming coatings (Figure 1).

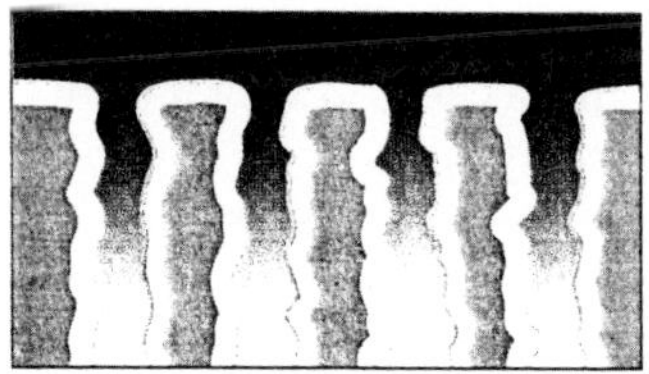

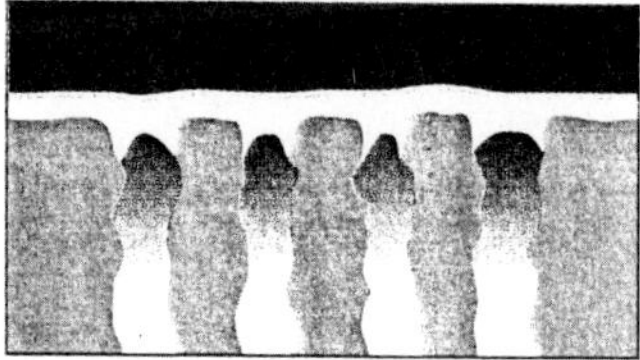

Figure 1 Schematic view of water-repellent and coated capillaries

Both processes focus on keeping out moisture, since water plays a key role in the absorption of harmful substances, eg, from deicing salts, as well as in corrosion processes.

In water-repellent impregnation, the pores of the concrete remain open. There is little or no effect on gas and water-vapour diffusion. Film-forming coatings, on the other hand, aim at preventing gas diffusion and thereby also the migration of carbon dioxide, the agent of carbonation. The two processes are often combined, ie, water-repellent impregnation is applied as primer for a subsequent coating.

Organosilicon compounds have a proven track record as impregnants. They are characterized by outstanding water-repellency, with no significant impairment of water-vapour permeability. They also show outstanding durability. The latter property results from the fact that silicones are extremely resistant to environmental effects (such as UV radiation, heat, chemically aggressive substances and microbes), and that silicone resin forms a stable, covalent bond with the construction material – concrete – on account of its chemical similarity (Figure 2).

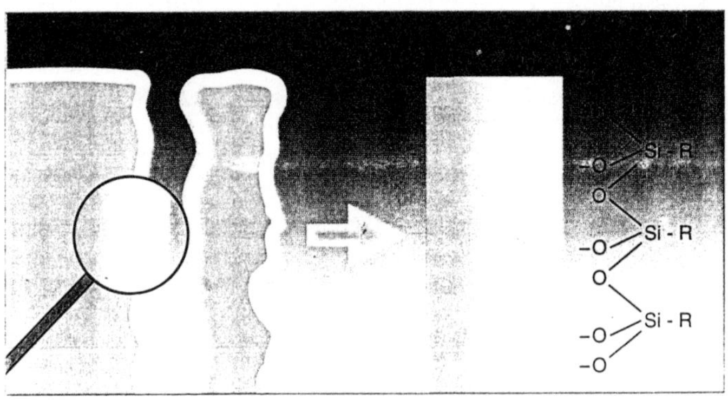

Figure 2 Physicochemical bonding of the silicone-resin network to the pore surface

Silicones must have two properties in particular for concrete impregnation. They must be able to penetrate the relatively dense concrete matrix, and they must not degrade under the highly alkaline conditions in fresh concrete.

Alkylalkoxysilanes, such as Octyltriethoxysilane, best meet these requirements. They are colourless, highly mobile liquids, which are usually applied undiluted to the concrete by flooding. There they react with moisture, eliminating alcohol, to form a three dimensionally crosslinked silicone resin chemically anchored to the concrete.

$$\mathit{Octyl}\text{ - }\overset{\displaystyle OEt}{\underset{\displaystyle OEt}{Si}}\text{ - }OEt \quad + \quad 3\,H_2O \longrightarrow \mathit{Octyl}\text{ - }Si \begin{cases} O\!- \\ O \\ O\!- \end{cases} \quad + \quad 3\,EtOH$$

Ocyltriethoxysilane   Water   Silicone resin network   Alcohol

Years of experience have clearly demonstrated that silanes are highly efficient and durable concrete waterproofing agents. However, they also have some disadvantages that should be eliminated where possible:

1. On very dry concrete surfaces (exposed to sun or wind), the silane lacks the necessary moisture for the crosslinking reaction, and considerable amounts of the active ingredient evaporate to the atmosphere

2. On vertical surfaces and particularly overhead surfaces, there is the risk of the material running off before it has penetrated into the concrete

3. Several application stages are usually necessary to apply effective amounts and obtain the required depth of penetration

To prevent evaporation of the active ingredient, there are two alternative counter-measures. Instead of pure long-chain alkylalkoxysilanes, either catalysed mixtures of silanes and oligomeric siloxanes are used, or aqueous products. Both methods have a good track record. To prevent loss of the active ingredient by uncontrolled flowing off of the impregnant, the aggregate state of the product must be changed. Low-viscosity liquids must be replaced by thixotropic, non-sag systems. This was the idea that formed the germ for the development of impregnation cream.

IMPREGNATION CREAM

Characteristics

Impregnation cream is doubtless one of the most revolutionary of recent innovations in the water-repellent treatment of concrete. The outstanding characteristics of this novel system can be summarized as follows:

- Outstanding penetration
- High content of active ingredient
- Optimum resistance to alkalis
- Drastic reduction in capillary water absorption
- No impairment of water-vapour permeability
- High protection against attack by frost/deicing salt
- Good adhesion of coating
- Solvent-free, aqueous preparation, which is therefore environmentally compatible
- Low volatility
- Thixotropic and can therefore be applied without loss

Active Ingredients

The impregnation cream contains essentially Octyltriethoxysilane as active ingredient in an amount of 80 %. The remaining 20 % is predominantly water and minor amounts of auxiliaries, such as emulsifiers. For an aqueous product, an active ingredient content of 80 % is unique. The emulsions used for concrete impregnation usually contain from 20 to 40 % active ingredient. The high content of active ingredient in the impregnation cream guarantees deep penetration, even when comparatively small amounts are applied.

The active ingredient and mechanism of the impregnation cream are the same as for conventional liquid silane impregnants. The silane reacts with water, eliminating alcohol, to form a polymeric silicone resin network, which forms a thin film on the surfaces of pores and capillaries. Its effect on the concrete's diffusion rate and water-vapour permeability is also no greater than that of liquid silanes.

Unlike conventional silane impregnants, the impregnation cream already contains the required amount of water for the crosslinking reaction. This prevents significant quantities of silane from evaporating when the preparation is applied to dry concrete surfaces. The new cream can be readily applied to surfaces fully exposed to sun and wind.

Processing

The biggest advantage of the impregnation cream compared to conventional low-viscosity impregnants is that even vertical surfaces and ceilings can be treated without the material trickling or dripping off in an uncontrolled way. This ensures that the entire surface is uniformly impregnated and protected.

The product is preferably sprayed by an airless process onto the concrete. It can also be readily applied by brush, lambskin roller or spatula to small areas.

A single application is usually sufficient. Depending on the absorbency of the substrate, up to 400 g/m^2 can usually be applied in a single operation – even to vertical surfaces and ceilings – with no waste. Only very high quality concrete, which has low absorption, should not be treated with more than about 200 g/m^2 in a single operation. When large amounts are applied, there is an increased risk of the cream layer being fluidized and starting to flow under the effect of the alkalinity of the concrete. A second application of cream may be used if required. However, 200 g/m^2 is usually adequate for high-quality concrete. With liquid products, this rate of impregnant application can usually only be achieved with three or more applications. Impregnation with cream thus saves significant amounts of time and money.

After application, the white cream coat on impregnated surfaces clearly distinguishes them from untreated concrete, giving a good visual indication of the uniformity of application. The thickness of the impregnant layer can be easily controlled.

According to the concrete quality and application rate, the active ingredient penetrates into the concrete within 30 minutes to some hours, and the milky white cream layer disappears completely.

Surfaces that have been treated with impregnation cream can subsequently be coated just as easily as those impregnated with conventional liquid impregnants.

Penetration Depth

The impregnation cream is formulated to provide the greatest penetration depth of the active ingredient into the concrete, and thus provide optimum protection against the absorption of water and harmful substances, as well as against frost/deicing salt.

The tremendous penetration depth is a function of the thixotropic consistency of the product, which ensures a long contact time of the silane active ingredient with the surface. It is also a function of the high concentration of active ingredient.
In addition to the active ingredient concentration, the penetration depth naturally also depends on the quality of the concrete. For different concrete grades (compression strength 15, 25, 35 and 45 MPa) the penetration depth is summarized in Table 1.

Table 1 Penetration depth depending on application rate and concrete quality

CONCRETE COMPRESSIVE STRENGTH MPa	PENETRATION DEPTH, mm	
	100 g cream / m^2	200 g cream / m^2
15	6	12
25	5	10
35	5	8
45	3	5

Water Repellency And Alkali Reresistance

The impregnation cream is tested for water repellency and alkali stability as part of the basic testing of surface protection systems according to Class OS-A of the Additional Technical Rules and Test Guidelines for Protecting and Renovating Concrete Structures (ZTV-SIB, published by the German Ministry of Transport [5]).

Mortar slabs (water/cement ratio 0.5) were treated by brushing with approximately 200 g/m^2 of the impregnation cream and aged for 14 days under standard climatic conditions (23 °C/50 % r.h.). After this treatment, the test specimens were immersed for 2 days in a 0.1 molar KOH solution, dried again and then immersed in water for 28 days. The water absorption characteristics shown in Figure 3 were obtained [6].

As it can be seen in Figure 3, the impregnation cream reduces water absorption by 64 % on average, and thus more than meets the requirements (water absorption of less than 50 % compared with untreated test specimens).

The penetration depth into the mortar slabs was up to 9.6 mm, a value that has never been achieved with liquid impregnants.

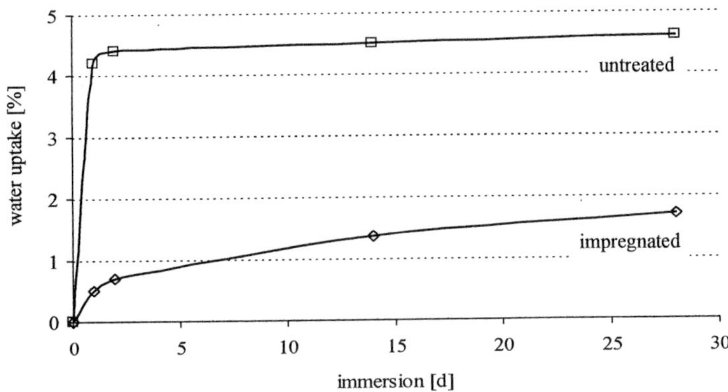

Figure 3 Water absorption of mortar slabs after alkaline exposure

Weight Loss After Frost/Deicing Salt Loading

This test was also carried out as part of the basic testing of impregnation cream as a surface protection system according to Class OS-A of the Additional Technical Rules and Test Guidelines for Protecting and Renovating Concrete Structures (ZTV-SIB) [5]. Concrete cubes (10 cm edge length, average compressive strength 38.7 N/mm^2, water/cement ratio 0.6) were treated with approx. 200 g/cm^2 impregnation cream and, after 14 days, immersed in a 3 % NaCl solution. Then the frost/thaw cycles were started (16 hours –15 °C, 8 hours +20 °C). Figure 4 shows the weight change of untreated and treated test specimens [6].

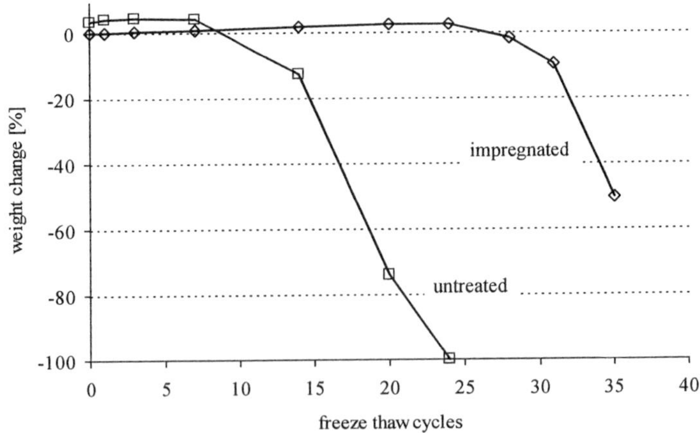

Figure 4 Weight change under frost/deicing salt cycles

As can be seen in the diagram, because of the protective effect of the impregnation cream, the treated samples last 20 cycles longer than the untreated cubes. The requirement is for only 15 cycles more than untreated cubes.

Impregnation Cream in Practice

One of the first examples of concrete repair using the new cream technology was the Fürstenland Bridge in St Gallen, Switzerland, built in the thirties. It is made of reinforced concrete that has suffered heavily from carbonation and the effect of road deicing salts.

After detailed preliminary tests by the laboratory of testing and material technology (LPM AG, Beinwill am See, Switzerland) [7], it was decided to use cream technology to render the concrete water repellent. This was not only because of the clear advantages for application, but also because cream technology obtains the best results for penetration depth and water repellency compared with liquid impregnants. At an application rate of 200 g/m^2 the active ingredient had penetrated to such a depth that the capillary water absorption was reduced by more than 80 %, even at a depth of 3 mm. Figure 5 shows the depth profile of capillary water absorption.

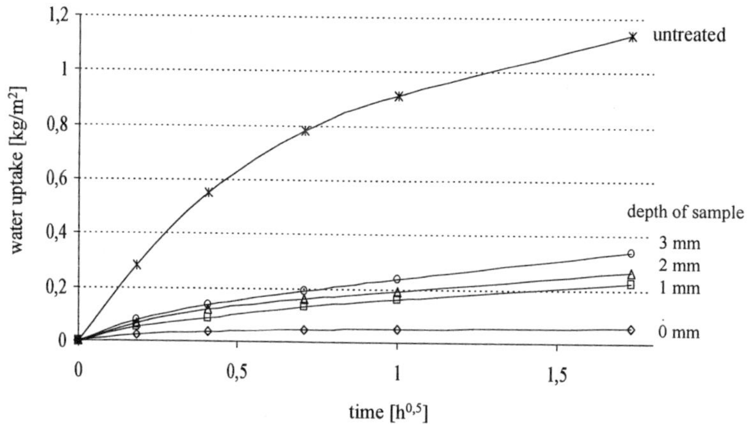

Figure 5 Capillary water absorption of drill cores from the Fürstenland Bridge in St Gallen, taken at different depths

CONCLUSIONS

Many experts are already agreed that cream technology, thanks to its intriguing properties, will change, even revolutionize, concrete technology, over the next few years. The results of extensive independent tests and practical experience demonstrate the outstanding properties of the impregnation cream. Its advantages include:

1. High active-ingredient content

2. Water-based preparation, and therefore solvent-free

3. Outstanding, easily controlled penetration behaviour

4. Easily processed by spraying, brushing or rolling

5. Time-saving, single-step application

6. Loss-free application

7. Problem-free "overhead" application

Impregnation cream is a good example of how, even after decades of successful use of silicone-based masonry protection agents, truly innovative developments can still be made, and indeed are necessary. In the field of concrete protection, impregnation cream will very likely become an essential part of any successful conservation and repair concept.

REFERENCES

1. WEBER, H. Reinforced concrete – Its afflictions and how to cure them. Bausubstanz, 1986, no. 1 and 2.

2. KLOPFER, H. Bautenschutz und Bausanierung, Vol. 1, no. 3, 1978, pp 86 – 97.

3. WEBER, H and WENDEROTH, G. Reinforced concrete – Damages and repair. Expert Publisher, 1987, Sindelfingen.

4. HAGER, R. Silicones for concrete protection. Proceedings of the International Conference, Concrete in the Service of Mankind, University of Dundee, Scotland, 1996, pp 361 – 367.

5. ZTV-SIB 90, published by the German Ministry of Transport. Guidlines for protection and repair of concrete surfaces. Verkehrsblatt-Verlag, Dortmund, document no. B 5230, 1990.

6. Test report no. A 3299 from ibac, Aachen. Basic test on a surface protection system of class OS-A according to ZTV-SIB 90, 30th April 1998.

BLOATING CONCRETE COATINGS TO IMPROVE FIRE-RESISTANCE OF BUILDING STRUCTURES

P V Krivenko

E K Pushkarjeva

M V Sukhanevich

Scientific Research Institute on Binders and Materials Kyiv

Ukraine

ABSTRACT. The maintenance of safety operation conditions of civil and industrial objects assumes taking measures increasing flame resistance of structures. The development of inorganic bloating fire protective coatings is reached at the expense of simulation of the processes of synthesis of substances, that are met in nature, and that are distinguished by the ability to increase its volume by 2-10 times at relatively low temperatures (to 500°C). The principles of synthesis of high flame resistant coatings are based on a long-term experience collected by the scientific school of the V.D.Glukhovskiy Scientific Research Institute on Binders and Materials, and consist of a dedicated creation of the zeolite- like new formations in the hydration products. The use of a proper tailored composition of the alkaline aluminosilicate binders enables to produce highly effective bloating coatings to concrete. The protective properties of this coating are adjusted by introduction of the additives of various application: strengthening the bloating layer; allocating gases, hindering to propagation of fire; regulating porous structure of the bloating heat-insulating layer. The use of alkaline aluminosilicate coating with thickness of about 4-5 mm enables to increase flame resistance of the concrete structure as long as to 1 hour.

Keywords: Bloating coating, Fire-resistance, Zeolites, Coefficient of bloating.

Professor P V Krivenko is Director of Scientific Research Institute of Binders and Materials (SRIBM), Kyiv University, Ukraine. He specialises in the field of alkaline binders and concretes. Professor Krivenko has many publications all over the world.

DrSc E K Pushkarjeva is Head of the Speciality Cements and Concretes Division, SRIBM. She specialises in the synthesis of heat-resistant, thermo-resistant and other special cements and concretes and the durability of such concretes when affected by high temperature environments.

Dr M V Sukhanevich is a Researcher in Speciality Cements, Scientific Research Institute on Binders and Materials, Kyiv, Ukraine. She specialises in the synthesis of fire-protecting bloating coatings to concrete and heat resistant heat insulating composites.

INTRODUCTION

The most widespread bloating coatings are the compositions based on organic components. In spite of high bloating ability at small initial thickness, all these materials have one defect – they contain organic components (carbomide, phenolformaldehyde etc.), which are known to be hazardous in terms of health of people as when these materials are drawing on the protective surface, as when effecting of temperature factor [1, 2].

It is known, that most ecological safe way of bloating is the way when materials bloat at the expence by allocation of chemical bound water (perlite, vermiculite, mica etc.) [3]. However, these compositions are bloating under relatively high temperatures (> 900°C) and ways which decrease temperature are not found [4].

At the same time some researchers noted that some kinds of natural zeolites have the ability to bloat under increased temperature. The name of zeolite as translated from Greek as a "bloating stone" [5].

With account statement the opportunity of reception of bloating coatings is realized at the expence of syntheses in structure of the zeolite-like new formation in the composition of the products hydration of alkaline cementitious systems [6, 7, 8]. Zeolite-like new formation are capable to dehydration with increasing volume under relative low temperatures (to 300°C).

EXPERIMENTAL DETAILS

Materials and Methods

Model systems based on synthetic zeolites and soluble glass were investigated for establishment of basic opportunity to obtain the bloating materials. Synthetic zeolites are received on experimental plant (Russia) and there compositions are listed in Table 1.

In the real alkaline cementitious systems are used reactive mixes based on aluminosilicate component, silicafume and soluble glass, in which components selected so that the compositions of these mixes are answered such formula $Na_2O \cdot Al_2O_3 \cdot 2mSiO_2 \cdot nH_2O$, where m=1-5. As aluminosilicate component applied metakaoline, which received under dehydration of kaoline at temperature 700°C. As a silicafume used were silica-containing wastes from production of the metalic silica.

Model mixes are prepared by mixing the soluble glass (Mc=2.8; ρ=1400 kg/m^3) and fine-growing zeolite. The solution to solid ratio is completed 3:1. The ratio oxides SiO_2/Al_2O_3 was limited from 2.4 to 115.58.

Mixes had been beared in closed vessel during 1 day and after that had been put on substrate by layer of 1-2 mm, optimum thickness of coating is three layers. Hardening coating happened in natural conditions. The technique of preparation reactive mixes and study degrees bloating of them is similar to technique, used under research model systems.

Table 1 Chemical composition of the synthetic zeolites

SYNTHETIC ZEOLITES	OXIDE CONTENT, MASS % :											RATIO SiO_2/Al_2O_3
	SiO_2	Al_2O_3	Fe_2O_3	TiO_2	CaO	MgO	K_2O	Na_2O	SO_3	loss on ignition	H_2O	
$Na_2O \cdot 0.9Al_2O_3 \cdot 2.2SiO_2 \cdot 1.7H_2O$	37.92	26.96	0.07	-	<0.1	0.06	0.12	17.85	0.04	16.54	8.92	2.40
$Na_2O \cdot Al_2O_3 \cdot 3.35SiO_2 \cdot 3.65H_2O$	46.48	23.88	0.12	0.02	<0.1	0.06	0.08	14.52	0.07	14.32	10.94	3.35
$Na_2O \cdot 3.3CaO \cdot 4.3Al_2O_3 \cdot 20.46SiO_2 \cdot 9.3H_2O$	53.07	18.99	0.18	0.06	8.04	0.14	0.25	2.68	0.06	16.12	7.24	4.70
$Na_2O \cdot 1.8K_2O \cdot 2.5Al_2O_3 \cdot 19.27SiO_2 \cdot H_2O$	6379	14.16	0.16	0.04	<0.1	0.10	9.76	3.47	0.06	7.97	6.88	7.73
$Na_2O \cdot 0.94Al_2O_3 \cdot 10.8SiO_2 \cdot 0.02H_2O$	73.01	10.82	0.31	0.04	<0.1	0.13	0.03	6.98	0.06	8.10	5.46	11.40
$Na_2O \cdot 0.65Al_2O_3 \cdot 27.6SiO_2 \cdot 16.2H_2O$	88.95	3.61	0.31	0.03	0.32	0.25	0.05	3.32	0.04	2.63	5.62	42.50
$Na_2O \cdot 0.34Al_2O_3 \cdot 39.3SiO_2 \cdot 0.005H_2O$	89.67	1.36	0.10	0.04	0.30	0.29	0.03	2.40	0.06	5.27	3.76	115.58

The composition of new formation was determined by complex phisico-chemical methods of analysis, including DTA, XRD, IRS and electromicroscopy.

The degrees bloating of the compositions evaluated with using of the coefficient which designed as the relation of thickness bloating layer to thickness of an initial coating.

Fire-resistance of bloating inorganic coating was investigated with the use of the laboratory installation designed by the Ukrainian Research Institute of Fire Safety (Kyiv). The heating of samples executed according to standard curve fire in account of the international standards ISO 834.

DISCUSSION OF RESULTS

For establishment of the basic opportunity of reception bloating materials based on alkaline cementitious systems and determination changes of degree bloating from ratio oxides in reactive mixes were investigated model system based on synthetic zeolites and soluble glass.

The analysis of received data shows, that heavies bloating of studied samples is marked at burning compositions under t=500°C which based on model mixes including synthetic zeolite with relation oxides SiO_2/Al_2O_3=4.7-11.4 in Figure1.

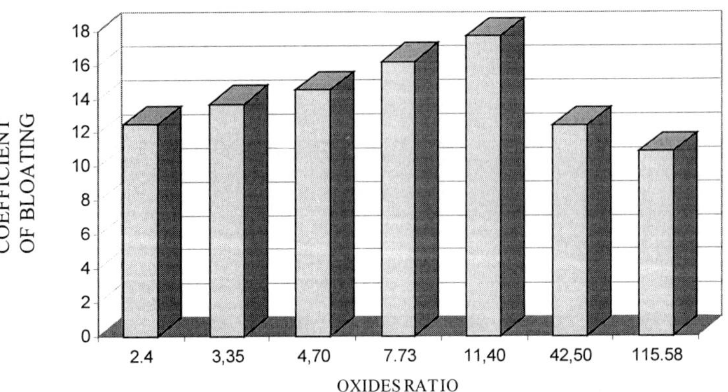

Figure 1 Change in coefficient bloating on ratio oxides SiO_2/Al_2O_3 in composition of synthetic zeolites

However it was noticed that the degree of bloating and the structure of the bloating material were influenc by other characteristics of initial components of model systems, namely: the ratio oxides Na_2O/Al_2O_3 in the synthetic zeolite and water contents (ratio H_2O/Al_2O_3).

Under researching of a reactive mixes were investigated the features of formation structure of material in dependences of ratio main oxides in system $Na_2O{\cdot}Al_2O_3{\cdot}(2\text{-}10)SiO_2{\cdot}nH_2O$ on the stages hydration (t=80°C, W=98%) and dehydration (t=500°C) .

Thus the choosing of ratio oxides was accepted in view of the results studing model systems. Quantity of water in studied compositions are chosen from condition of maintenance viscosity of mix 20-25 sm (on Suttard`s device) which ensuring necessary of covers ability of mixes at their drawing on a substrate.

The analysis of received data permits to notes, that the heaviest coefficient bloating is characteristed for reactive mixes which composition answered such formula $Na_2O{\cdot}Al_2O_3{\cdot}6SiO_2{\cdot}25H_2O$ in Figure 2.

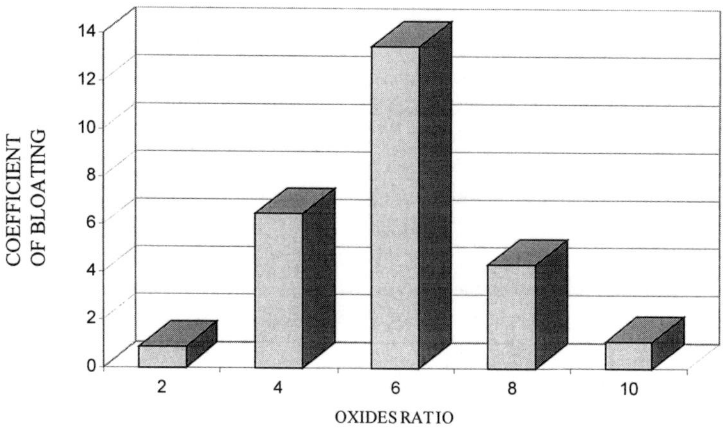

Figure 2 Change in coefficient bloating on ratio oxides SiO_2/Al_2O_3 in composition of reactive mixes

In the accordance of data XR analysis under hydration of composition $Na_2O{\cdot}Al_2O_3{\cdot}6SiO_2{\cdot}25H_2O$ new formation are submitted by heulandite $Na_2{\cdot}Al_2{\cdot}Si_7{\cdot}O_{18}{\cdot}6H_2O$ (d=0.489; 0.356; 0.307; 0.280; 0.266; 0.243; 0.227; 0.201; 0.166 nm) and remainders unhydrated metakaoline. After heat treatment is observed amorphisation of a structure, diffraction reflaction of heulandite are not fixed almost, it remain only reflactions of the metakaoline [9,10].

Thus, an ability to bloating of the alkaline aluminosilicate compositions is determined by the phase composition of the products of hydration, which shoud be submitted zeolite-like new formations of group heulandite $Na_2{\cdot}Al_2{\cdot}Si_7{\cdot}O_{18}{\cdot}6H_2O$ [11, 12].

The feature of minerals group heulandite including laumontite, phillipsite, desmin, stilbite, heulandite is low temperature dehydration of them (to 300°C) with partial or total amorphisation their structure. Zeolites this group differ also least heat resistant which is evaluated by temperature destruction of its srtucture. Also this zeolites characterized by irreversible changes of a structure after thermal processing in the temperature interval t=150-250°C [5].

As a result of the conducted researches the optimum ratio oxides in system Na_2O-Al_2O_3-SiO_2-H_2O can be determined namely: SiO_2/Al_2O_3=6-7; Na_2O/Al_2O_3=1-1.3; $H_2O/Al_2O_{3=}$ 22-24, which ensuring synthesis zeolite-like new formation compounds as heulandite, capable to dehydration with increasing of volume at t=150-250°C at the expence of availability in their cavities of desorientation molecules of water, connected by the weak hydrogenium bonds.

Proposed bloating compositions based on natural and technogenic raw materials were chosen as the binding for reception of fire-resistance bloating coating. With the purpose of regulation of physical-mechanical characteristics, as well as character formed of the porous structure in composition binding were entered of mineral fillings and aggregates.

The chosen fillings were executed on series of criterions: with accounts by their abilities to bloat at increase temperature, not to hinder bloated aluminosilicate binding, to improve the mechanical characteristics of received materials as well as to provide the allocation of gas phase, hindering to development of fire. At studied of influence fillings of various nature on process bloating composition in system Na_2O-Al_2O_3- SiO_2-H_2O is shown, that are the most effective (maximal coefficient bloating >20 and strength to 3.5 MPa) compositions including additives of perlite, SiC, $CaCO_3$ in quantity of 20%. Also its observed the allocation gas phasees CO_2 hindering to development of fire.

The choice of aggregates are dictated by requirements bore to fire-resistance bloating materials which providing the creation of heat-insolation porous layer on the surface constructions. The formation of such layer under bloating of a fire-resistance coating which describing closed of porous by structure is provided at the expense introduction of granulat-previously prepared from reactive mix on the technology of reception stekloporous [3].

Thus, the coating based on alkaline cementitious systems are received of the fire-resistance bloating which on the performances satisfy requirements of acting normative documents are competitive on the world, as relative to ecologically safety materials and preventing development of fire, allocating toxic l gases in Table 2.

CONCLUSIONS

1. The basic opportunites of reception bloating ecological safe inorganic materials in system Na_2O-Al_2O_3-SiO_2-H_2O being alternative of known analogous based on organic componenets are established. Is show that bloating of the compositions is stipulated by directed synthesis in structure of products hydration new formations zeolite-like structure capable to dehydration under low temperature with partial or total amophousation of the structure.

2. The structure and bases of technologies of receptions fire-resistance bloating coating based on reactive mixes are developed with ratio oxides SiO_2/Al_2O_3= 6-7; Na_2O/Al_2O_3=1-1.3; $H_2O/Al_2O_{3=}$ 22-24, ensuring receptions of materials with maximal coefficient bloating. The formation porous structure of bloating layer is achieved at the expence of introduction in the composition coating of granulat- previously prepared from reactive mix on technology of reception stekloporous, and also complex mineral fillings.

Table 2 Basic characteristics of the bloating fire-resistanct coating based on alkaline aluminosilicate cementitious materials and their analogs

NAME OF CHARACTERISTIC	UNIT OF MEASUREMENT	FIRE-RESISTANT BLOATING COATING		
		VPM-2	OVK-2	Propoused composition of bloating coating
Density: Coating before bloating Coating after bloating	 kg/m^3 kg/m^3	 - -	 1250 155-160	 1170-1190 145-152
Fire-resistance Limit	min	43-46	55-63	55-60
Adhesion to metal	MPa	0.83-0.86	0.81-0.95	0.78-0.84
Temperature of begining bloating	°C	185-190	170-185	200-220
Flexural strength	MPa	4.9-5.6	7.7-8.4	5.8-7.4
Compressive strength	MPa	8.7-9.5	14.5-16.0	11.2-13.7
Compressive strength of bloating layer	MPa	-	0.20-0.22	2.3-4.5
Coefficient of bloating	-	3.0-3.5	9.0-11.0	15.0-23.0
Water absorpsion for 5 days	%	-	11-13	8-10

3. The properties developed bloating fire-resistance coating are investigated and the influence of compounds and technological parameters on the coefficient bloating of coating is investigated. Is show, that at introduction granulat of optimum structure (size 2-4 mm) in quantity of the 65-75% over weights binding is formed the structure of bloating composite (Cb=20-23) with general porous 92-97%, heat-insulating λ=0.041-0.065 Vt/m·°C with thus the strength of bloating layer makes 4-6 MPa, adhesion to metal - (0.78-0.84 MPa), water absorption for 5 days does not exceed 10%.

REFERENCES

1. ROMANENKOV I G, ZIGEN-KORN V N. Fire-Resistance Building Structures from Effectiv Materials. Stroiizdat Publish., Moscow, USSR, 1984. 240 pp.

2. TOPF P. New Developments in Intumescent Materials. Proc. Conf. New Technol. Reduc. Fire Losses and Costs. Luxemburg, 2-3 Oct., 1986, London, New York. 1986, pp153- 169.

3. GORLOV U P. Technology Heat-Insolation and Acoustic Materials and Constructures. Visshaya shkola Publish., Moscow, USSR, 1989, 384 pp.

4. KRUPA A A. Phisico-Chemical Bases Resieving of Porous Materials from Vulkanik Glasses. Vischa shkola Publish., Kiev, USSR, 1978, 136 pp.

5. SENDEROV E E, HITAROV N I. Zeolites, There Synthesis and Conditions of Formation in Nature. Nauka Publish., Moscow, USSR, 1970, 123 pp.

6. GLUKHOVSKY V D, PETRENKO I U, SKURCHINSKAYA J V. On Synthesis of Crystalline Alkaline Aluminosilicates, Journal Dokl. Akad. Nauk UkrSSR, Nauka Publish, Kiev, USSR, 1968, pp 735-791.

7. KRIVENKO P V, PUSHKARYOVA E K, SUKHANEVICH M V. Development phisico-chemical bases directed synthesis inorganic binders in system Na_2O-Al_2O_3- SiO_2-H_2O for using bloating materials: Lournal Building Ukraine, Kiev, Ukraine , N 2, 1997, pp 46-49.

8. KRIVENKO P V, PUSHKARYOVA E K, SUKHANEVICH M V. Fire-Resistance bloating aluminosilicate coating: In book Problems of Fire–resistance Building Materials and Constructions, Lvov, Ukraine, 1994, pp 31-34.

9. BARRER R. Hydrothermal Chemistry of Zeolites. Mir Publish., Moscow, USSR, 1982, 420 pp.

10. BREK D. Zeolite Molecular Sieves. Mir Publish., Moscow, USSR, 1976, 776 pp.

11. ZIZISHVILY G V, ANDRONIKASHVILY T G, KIROV T N. Natural Zeolites. Chimiya Publish., Moscow , USSR, 1985, 224 pp.

12. ZIZISHVILY G V. Phisico-Chemical Properties and Filds Using of Natural Zeolites: in book Natural Zeolites, Mizniereba Publish., Tbilisy, USSR, 1979, pp 37.

EFFECT OF PASIVANT AND CATHODIC PROTECTION PRIMER SYSTEMS ON THE GALVANIC CURRENTS INDUCED BY LOCALIZED REPAIRS

E J Pazini

Federal University of Goiás

Brazil

P Castro

Centre of Research and Advanced Studies

Mexico

C Andrade **C Alonso**

Institute of Construction Science

Spain

ABSTRACT. The primers applied to the reinforcing steel as a repair method can present different protection mechanisms (repassivant, cathodic protection, barrier and inhibition). However, they can have influence on the non repaired areas as well as in the interfaces between the new and old material. Data of small beams that were repaired using repassivant and cathodic protection primer systems are presented and discussed in this paper. Several exposure conditions including curing, exposure to different relative humidities, partial immersion and drying were tested. Measurements of galvanic intensity, corrosion rates, corrosion potentials and ohmic resistance in the repaired, non-repaired and common interfaces were carried out. From the two studied systems, only the cathodic protection, at the early stages of the exposure, gave protection to the non-repaired zones. This effect disappeared after the repair curing after which its behavior behaved similar to that of the repassivant system. On the other hand , the interfaces of the repassivant system behaved anodic until the curing finishing only, while those of the cathodic protection system remained anodic through the most of the exposure period.

Keywords: Galvanic current, Corrosion rate, Cathodic protection, Inhibitors, Primers, Repassivation, Polarization resistance.

Dr Enio J Pazini is a Full Time Professor at the Federal University of Goiás in Brazil and researcher of the National Council of Scientific and Technological Developing (CNPq). His main interest is the concrete durability, repair materials and electrochemical techniques for corrosion monitoring in concrete structures.

Dr Pedro Castro is a Researcher at the Center of Research and Advanced Studies, Merida, Mexico. His main interests include the rehabilitation of corrosion-damaged infrastructure and the effects of aggressive marine environments on the durability of concrete structures.

Dr Carmen Andrade is a Research Professor of the Institute of Construction Science belonging to the Consejo Superior de Investigaciones Científicas (CSIC) from Spain. Her main interests are the service life prediction of concrete structures, measurement techniques in on-site condition and electrochemical repair techniques.

Dr Cruz Alonso is Researcher of the Institute of Construction Science belonging to the Consejo Superior de Investigaciones Científicas (CSIC) in Spain. Her main interests are the corrosion of steel in concrete, service life prediction and corrosion of prestressing steels.

INTRODUCTION

Localized repairs made by removing the damaged concrete are a common practice in deteriorated structures. These patches involve usually the application of a bond-agent between old concrete and the new material, a steel primer applied to the rebar, the repair material itself and oftenly a coating on the whole surface.

If the zones beside the repair material contain certain amount of chlorides, it has been stated [1] that this mode of operation may aim into an intensification of the corrosion in the rebar at both lateral sides of the repaired zone. This effect has been explained to be due to the new repassivating material that may enhance the anodic behavior of the neighboring non repaired zones still contaminated with chlorides.

Before repair, the zones adjacent to that to be repaired usually work as cathodes due to the strong anodic behavior of the corroding area. The situation after repair may progress in either: a) the repaired zone does not electrochemically influence the neighboring areas because the chloride concentration in the adjacent areas is above the threshold and therefore corrosion is based in micropiles, b) the repaired zone has galvanic influence on the neighboring areas because the chloride concentration in the adjacent areas is below the threshold and therefore there is a macropile corrosion contribution in the adjacent area or c) the chloride content in the adjacent area is below the threshold and therefore they depassivate only if additional chloride penetrates into the concrete.

The working mechanism of primers depends on their nature. Thus, simple alkaline repassivation, cathodic protection, barrier and inhibiting effects [2] are the usual mechanisms by which primers provide additional protection to the steel. It is expected that they will also influence in different manners the possible enhancing effect on the neighboring non repaired areas.

The aim of the present paper is to study the electrochemical behavior in the case that in the repair zone a zinc rich epoxy primer is used. The case where the repaired zone is repassivated by a standard mortar is also studied as reference.

EXPERIMENTAL DETAILS

Materials

Small beams as illustrated in Figure 1 were built. They had embedded a continuous and a segmented bar which enabled the electrochemical study of each zone of the beam separately. The beams were fabricated by means of mixing in the water a 0.7% of chloride ions (as $CaCl_2$) in relation to the cement weight.

The beams were fabricated in three steps: the first one was the bottom, later the side parts with 0.7% of chloride ions (as $CaCl_2$) in relation to the cement weight. This amount was previously studied [3] to be just above the threshold level able to induce depassivation of the steel. Finally, the central part was filled with the repair system in order to simulate a patch repair.

The concrete of the bottom and side parts was made of concrete with a density of 2,300 kg/m^3, using 297 kg/m^3 of an OPC and a w/c ratio of 0.65. The cement:sand:gravel ratio was of 1:2.9:3.2. As repair material in the central part was used a mortar having a density of

2,330 kg/m^3, made with 530 kg/m^3 of the same OPC, a w/c ratio of 0.45, sand/cement ratio of 1/3 and 1.5% by weight of cement of superplasticizer. The steel at the centre repair zone was never exposed to chloride contaminated concrete. The w/c ratio of 0.65 at the sides and bottom of the beams is typical of many structures that need repair in tropical environments.

The primer used to protect the steel in the central (repaired) zone was a paint consisting in a monocomponent of zinc rich epoxy. This primer contains zinc oxide (ZnO), silicium dioxide (SiO_2) and aluminium oxide (Al_2O_3).

As can be seen in Figure 1, the segmented bar was composed of 7 elements numbered from left to right in the Figure. The segments and the continuous rebar were made of corrugated steel bars of 0.6 cm in diameter and 60 cm of length (continuous bar) or 8 cm of exposed area in the segments. Electrical wires were welded to the extreme of the segments and the welding was isolated with epoxy resin and insulating tape in the extreme of the beam. These electrical wires were used to make the external electrical connections among the segments.

The repair was made 3 days after casting the beams. The curing period was then extended to 28 days from the casting of the beams. The environmental conditions used later are given in Table 1.

Table 1 Conditions at which the beams were exposed to and duration of each one

TIME	CURING	80-90% RH	PARTIAL IMMERSION	40% RH	85% RH	95% RH
Days	28	103	59	1106	90	110

This sequence was used in order to study the galvanic behaviour in very different moisture states of the beams.

Techniques

The variables that have been monitored are:

a) The galvanic current, Ig by means of a Zero Resistance Ammeter (ZRA) which is able to give the anodic or cathodic contribution of each segment [4]. While all the segments stayed connected, the ZRA was introduced between segment 1 and 2 being (by convention) the wire of 1 the cathode and the wire of 2 the anode. In this manner was measured the galvanic current between segment 1 and the rest of the segments. The cables between 1 and 2 were connected again and now the ZRA was positioned between segments 2 and 3 being 2 the cathode and 3 the anode. In this manner was measured the galvanic current between segments 1 and 2 and segments 3 to 7. The same operation was followed until finishing with the segment 7. The contribution of each segment to the galvanic current was obtained taking the first value as that of segment 1, the algebraic difference between this value and the next one was taken as the contribution of segment 2 and successively. This arrangement allows to measure the galvanic current for each element coupled to the system with an almost instantaneous nulling characteristics [5]. The sign of the galvanic current for each segment was assigned according to their active or passive character. The sum of all the Ig values with their respective signs must be zero since the anodic current must be equal to the cathodic one.

b) The corrosion current density, icorr, through the polarization resistance method, the corrosion potential, Ecorr, and the electric resistance, R, were measured with a commercial corrosimeter and with all the segments connected. The values of icorr, Ecorr and R were measured in the adjacent continuous bar in order to check for reproducibility and to detect, if any, some problem with the cables.

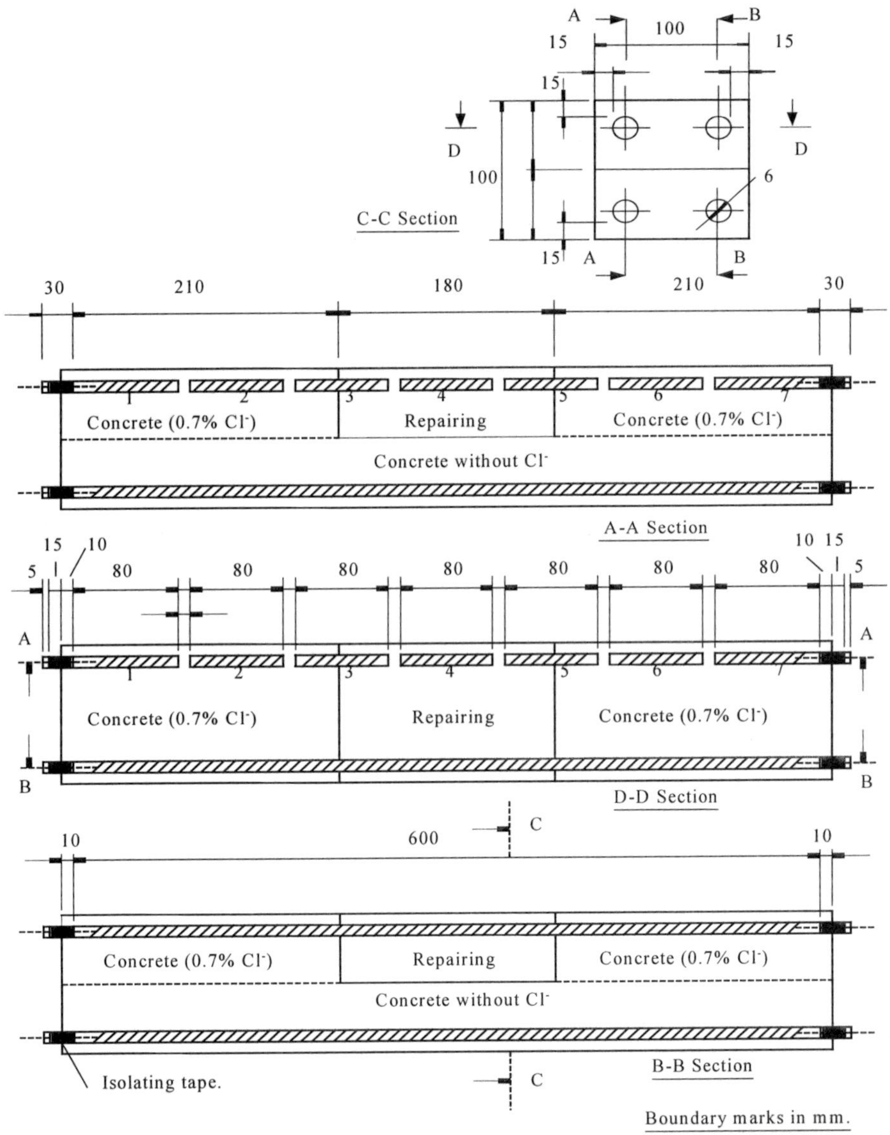

Figure 1 Details of the beam used in this work

RESULTS

Figures 2 and 3 show the evolution of the Ig since the repair (3 days) until the curing end (about 24 days) for the reference system and the cathodic protection system respectively. In general, Ig showed a symmetric behavior in both systems. It seems to be that the anodic behavior of the interfaces is controlling the cathodic one in the rest of the beam for the repassivant system. On the other hand, the anodic behavior of the repaired zone in the cathodic protection system controlled the cathodic one in the rest of the beam. During this stage, the use of the cathodic protection system reduced the anodic character of the interfaces.

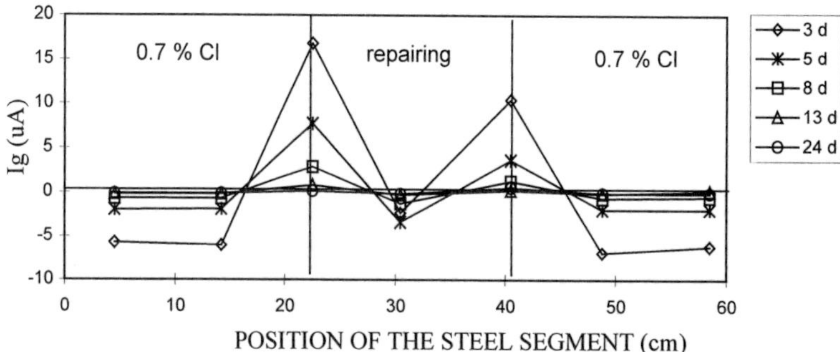

Figure 2 Ig evolution along the repair and curing for the repassivant (reference) system

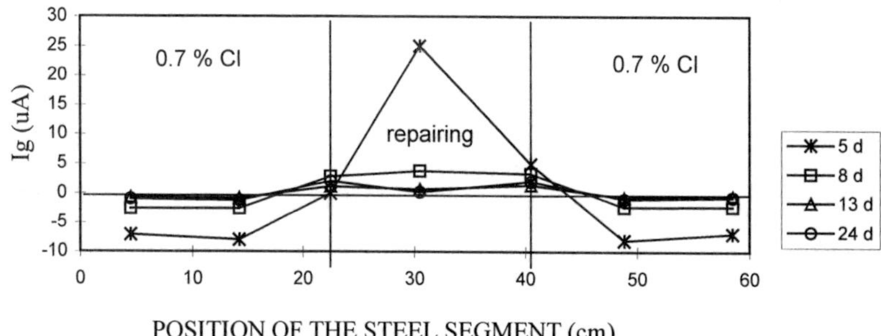

Figure 3 Ig evolution during repair and curing for primer with cathodic protection

The results symmetry observed during the repair and curing stages was reasonably maintained during the other exposure conditions. Therefore, and for simplicity, the results of all the exposure conditions were represented for three characteristic segments which were, from left to right in Figure 1, the number 2 in the non-repaired zone with 0.7% Cl^-, the number 3 in the interface between the repaired and non-repaired zones and the number 4 in the repaired zone.

Figures 4 and 5 show data of Ig (a), icorr (b), Ecorr (c) and R (d) for the reference and the cathodic protection system respectively. In general, the results of Ecorr, icorr and R for both systems behaved as expected during the exposure conditions.

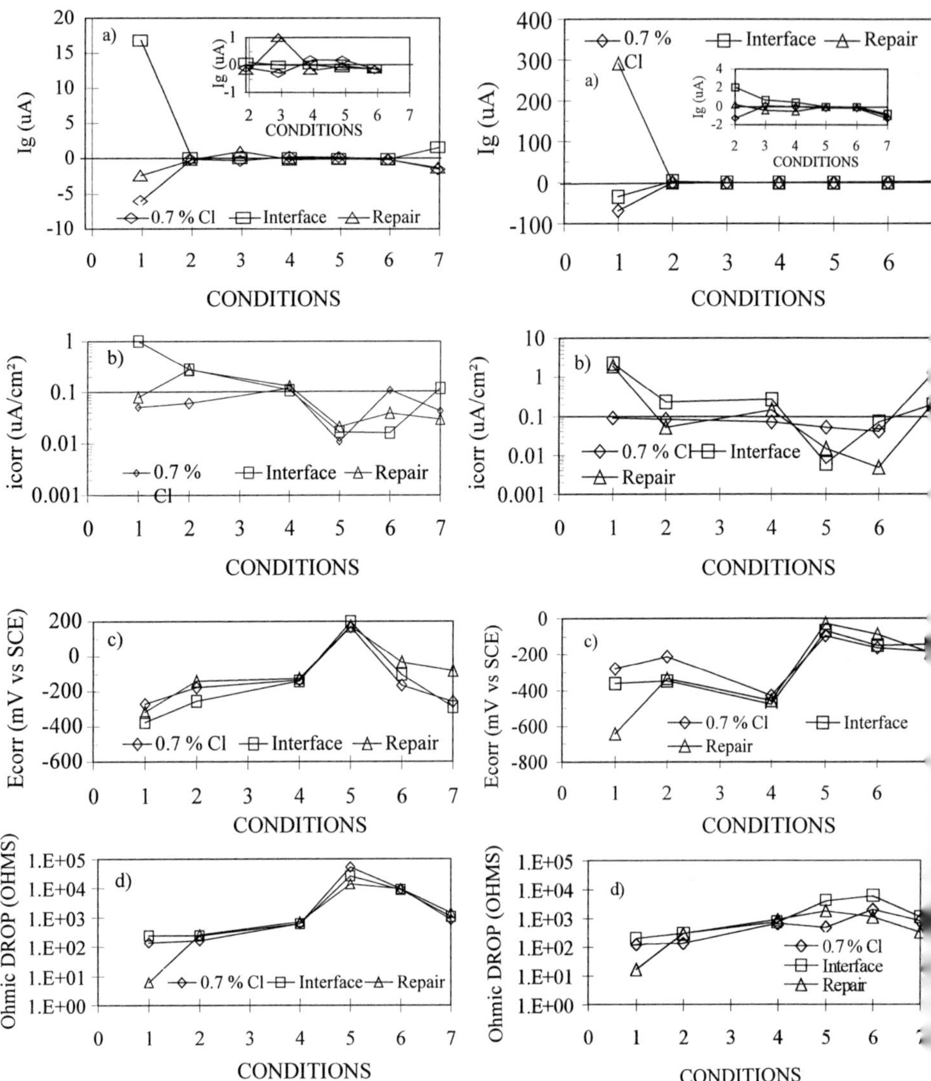

Figure 4 Ig evolution during all the exposure conditions for the reference system*

Figure 5 Ig evolution during all the exposure for the cathodic protection primer system*

*The conditions were:

1. Repair (3 days after casting)
2. Curing (21 days after repair)
3. 80%-90% of RH (103 days after curing)
4. Partial immersion (59 days after condition 3)
5. Drying (1106 days after 4)
6. 85% RH (90 days after 5)
7. 95% RH (110 days after 6)

This is to say that, in general, a high anodic behavior corresponded to a high icorr, a very negative Ecorr and a low R. From Figures 4 a) and 5 a) can be deduced that each segment behaves in different manner depending on the exposure conditions.

DISCUSSION

A chloride contaminated structural element can have several macro cells along its extension before being repaired. However, these macrocells can act in a different manner if a repair is made. In this work, the elements behaved in a different way depending on the type of repair system and exposure to the laboratory conditions. At the beginning, the reference system behaved as cathodic in the repaired and non-repaired zones and anodic in the interfaces. Later, the Ig decreases very much being the repaired zone first anodic (period 3) and later cathodic. On the other hand, the cathodic protection system behaved as cathodic in the interface and non-repaired zones and anodic in the repaired zone at the beginning. Later, the repaired zone becomes cathodic with respect to the interface.

During the period of casting (repair) the interface (segment 3) behaved more anodic in the reference system (Figure 4 a) than in the cathodic protection one (Figure 5 a). This was due to the zinc protection while dissolving in the repaired zone inducing the cathodic behavior in the other parts. After this period, the behavior in both systems was similar and tending to the cathodic site independently of the exposure condition. This is indicating that the electrochemical activity decreased presenting a general passivity. Although the icorr (Figures 4 b and 5 b) started to increase after the driest period, this was not clearly reflected in the galvanic behavior, maybe due to the small corrosion rate. The behavior of the interfaces is an indication of the short distance of the galvanic action.

The non-repaired zone (segment 2) shows a cathodic behavior in both systems during the most of the exposure conditions. This behavior was stronger in the beam with primer than in the reference just at the beginning and this was due to the protection given by the zinc dissolution in the repaired zone. Under these circumstances, it can be said that using the selected primer with zinc, the anodic character of the repaired zone does not decrease as observed in other inhibitor-based primer systems [6].

On the other hand, the repaired zone (segment 4) showed a higher anodic behavior during curing in the cathodic protection system than in the reference one. The galvanic behavior of both systems became random during the other exposure conditions but tending to be cathodic. It is important to note that the expected behavior of the sacrifying zinc in the repaired zone did not last as expected and that is the reason why the initial protecting power of the repaired zone over the interfaces and non-repaired zones disappears beyond the curing period.

An analysis of the primer with scanning electron microscope showed different contents of ZnO, SiO_2 and Al_2O_3 at the surface, centre and interface between steel and primer as observed in Table 2 [3]. It can be seen that the higher quantity of zinc was in the interface between the steel substrate and the primer (74.8%). This percentage is less than those suggested by BS 4652 and BRS DIGEST 109, which are 88% and 95% of zinc content in the dry film respectively. The presence of zinc and aluminum in the film was verified by SEM (Figure 6) through spherical and thin cylinder structures respectively [3]. To this lower Zn content is attributed the short duration of the non-protective power of the primer in the repaired zone over the interface and non-repaired zones beyond the curing.

Table 2 Percentage of zinc in the primer film

STUDIED AREA	ZnO (%)	SiO_2 (%)	Al_2O_3 (%)
Surface	56.77	31.81	11.42
Centre	65.03	25.00	9.97
Interface steel/primer	74.80	20.57	4.63

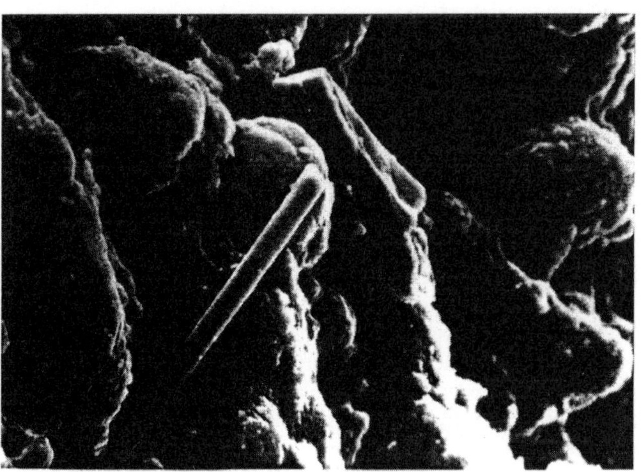

Figure 6 Zinc (bigger spheres) and aluminum (thin cylinders) present in the microstructure of the cathodic protection-based primer

From the above discussion it can be deduced that the primer based in zinc with a cathodic protection mechanism attenuates the enhancing of the anodic behavior in the neighboring of the repaired zones, although the length of this action is very short being irrelevant at long term due to the low corrosion rate registered, the zinc depletion or the corrosion products of zinc over the steel surface [7, 8].

The further activation of the beam maintains a similar difference between the beam with primer and the reference. For the evaluated systems, a practical implication seems to be that after the curing of the repair, the corrosion is governed by micro cell mechanisms instead of well defined macro cells.

CONCLUSIONS

The results presented here have been taken in chloride contaminated with chloride in quantities very close to the threshold for corrosion so that the conclusions found apply mainly to these conditions.

The conclusions reached can be summarized as follows:

1. The use of cathodic protection-based primer seems to attenuate the enhancing anodic behavior in the neighboring of the repaired zones, although due to the short duration of the corrosion activation, the presence of the primer can not be taken as relevant at long term.

2. At long term, it seems that the importance of the macro-cell between repaired and non-repaired zones disappears. Even the interfaces are subjected to a micro instead of a macro behaviour.

From the above points it can be deduced that after the curing of the repair the corrosion performance is governed by micro cell mechanisms instead of hypothetical macro cells.

ACKNOWLEDGEMENTS

The authors acknowledge the partial support from CNPq, CONACyT (2186PA), CSIC and CINVESTAV. One of the authors, PC thanks to CONACyT for its support through a posdoctoral grant. The opinions and findings are those of the authors and not necessarily of the supporting organizations.

REFERENCES

1. NAGATAKI, S, OTSUKI, N, MORIWAKE, A, MIYAZATO, S, SHIBATA, T, Macro-cell corrosion on embedded bars in concrete members with joints, Proceedings of the 7th International Conference on durability of building materials and components, Stockholm, Sweeden, 19-23 May, 1996, pp. 411-420.

2. PAZINI, E J, ANDRADE, C, Behavior of surface and barrier coatings on reinforcement in cement mortar containing chloride in Corrosion and corrosion protection of steel in concrete, Sheffield, U. K., Vol. 2, 1997: p. 1044.

3. PAZINI, E J, Evaluación del desempeño de revestimientos para protección de la armadura contra la corrosión a través de técnicas electroquímicas, PhD thesis, University of Sao Paulo, 1994.

4. ALONSO, C, ANDRADE, C, FARINA, J, LÓPEZ, F, MERINO, P, NOVOA, X R, Mats. Sci. Forum, 1995, 194, p. 899.

5. MANSFELD, F, KENKEL, J V, Laboratory studies of galvanic corrosion of aluminum alloys in Galvanic and Pitting corrosion-field and laboratory studies, ASTM STP 576, 1976, p. 20.

6. ANDRADE, C, ALONSO, C, PAZINI, E J, CASTRO P, Galvanic currents induced by a localised repair when using an inhibitor-based primer for reinforcing steel in: Rehabilitation of Corrosion Damaged Infrastructure, Eds. P. Castro, O. Troconis, C. Andrade, NACE International, Cancun, Mexico, 1998, pp. 126-133.

7. PAZINI, E J, CASTRO, P, ANDRADE, C, ALONSO, C, Corrientes galvanicas inducidas por una reparación localizada utilizando imprimaciones de la armadura (spanish), Proceedings of the IV Congreso Iberoamericano sobre Patología de las Construcciones, Puerto Alegre, Brazil, 1997, pp. 223-230.

8. CASTRO, P, PAZINI, E J, ANDRADE, C, ALONSO, C, Galvanic currents induced by a localised repair using steel primers (abstract), International Materials Research Congress, Cancún, México, 1997, p. 39

EPOXY INJECTION TO SALVAGE EXISTING RIVERGATE STRUCTURE

S Kulkarni

A Sabri

Kulkarni Consultants APC

United States of America

ABSTRACT. The $500 million Harrah's Casino in New Orleans, Louisiana, was designed utilizing existing Rivergate structure. The existing structure was built in 1963 and had undergone severe load variations due to its use as a Convention Center. The proposed new casino would demolish the post-tensioned shell type roof structure, but utilize the existing foundation to carry new column loads. The existing floor showed severe cracking around some of the columns. The existing structure was contaminated with asbestos and the likelihood of asbestos filling the cracks during its removal could not be ignored. Thus, the decision was made to inject epoxy to seal the cracks. Special devices were used to monitor crack progression. The final product was a stable structure that would house a brand new casino.

Keywords: Asbestos, Epoxy, Injection, Cracks, Foundation, Uplift, Existing structure, Rehabilitation.

Subhash Kulkarni is President and Principal Engineer of Kulkarni Consultants, APC, located in Metairie, Louisiana. He received a M.Tech from the Indian Institute of Technology, Bombay, India, and a MS in Structural Engineering from the School of Mines, Rapid City, South Dakota. He has served on ACI Committee 543, Concrete Piles, and 435, Deflection of Concrete Building Structures. Mr. Kulkarni has designed more than 500 structural engineering projects.

Aziz Sabri is Design Manager and Senior Structural Engineer at Kulkarni Consultants, APC. He received his MS and Ph.D. Degrees in structural engineering from Iowa State University, and has over 27 years of structural design and construction experience.

INTRODUCTION

The Rivergate Convention Center was built in New Orleans, Louisiana, in the early sixties to host small conventions. It was also used for Mardi Gras balls, which included several large floats. Over the years, the conventions became larger and larger and the existing Rivergate Center became too small to hold these conventions. The facility, thus, remained unused for a number of years, and a new Convention Center was built along the Mississippi River, which could accommodate more than 50,000 conventioneers.

In the early nineties, the State of Louisiana passed a law allowing gaming, thus opening an opportunity to have a land-based casino in downtown New Orleans. The Rivergate Center was chosen as the site of the new casino. The proposed Harrah's Casino had a construction budget of $840 million. Construction started in December 1994, and the budget was later scaled down to $500 million.

CONVENTION CENTER TO CASINO

The existing Rivergate was composed of a one story post tensioned structure extended over two spans 253 feet (77m) and 139 feet (42m) long, having 4'-0x4'-0 (1.2m x 1.2m) columns supported on 12 inch (300 mm) diameter steel pipe piles. The structure also had two levels of underground parking. The first floor, which was about 6 feet (1.83 m) above street level, was designed for 350 pounds per square foot (1700 Kg/m^2). The floor system was composed of conventionally reinforced concrete flat slab 8.25 inches (21 mm) thick with columns having drop panels. No expansion joints were provided in the slab. The lower basement slab was about 12 feet (3.7 m) below street level. Since New Orleans is located along the Mississippi River, and situated below mean water table, the 12-inch (305 mm) thick lower basement slab was designed for an uplift pressure of 710 pounds per square foot (3500 kg/m^2.)

The Rivergate facility contained asbestos that would have to be removed. Due to this and several other factors such as economics, time constraints, local and regional competition, particularly from our neighboring state, Mississippi, it was decided by the owners that the existing superstructure of the Rivergate would be torn down, whereas, the existing foundation would be utilized for the casino building.

PROBLEM DEFINITION

During preliminary site visits, the structural engineers from Kulkarni Consultants noticed several cracks in the first floor slab around the columns. The conical column heads did not show any distress, but the typical crack pattern in the slab followed the conical drop heads. The possible causes of these cracks might be one or more of the following:

1. Due to the lack of expansion joints in the structure, it was possible that the structure had developed cracks around the drop panels due to tensile stresses caused by expansion and contraction.
2. The floor was originally designed for a downward live load of 350 pounds per square foot (1700 Kg/m^2). During the years that the facility remained unoccupied, and thus with no live

load, only dead load resisted the upward buoyancy force. With a higher water table, the buoyancy might have pushed the structure upward causing the stress cracks around the column drops. Most of the cracks occurred on the top surface of the slab, thus making this argument plausible.

SOLUTION BY EPOXY INJECTION

The decision was made to seal the cracks through epoxy injection in order to salvage the first floor slab for the following reasons:

1. The demolition of the superstructure would trigger vibrations due to impact loading and thus, the cracks might become worse causing structural failure.
2. The debris during demolition would be piled up on the existing slab and might overstress the already cracked slab triggering failure. Figure 1 shows the Rivergate structure during demolition.
3. The asbestos removal may entrap asbestos particles into the cracks, thus exposing owners to future lawsuits.

Figure 1 Rivergate facility superstructure being demolished.

To evaluate the extent of cracking, an independent testing laboratory was employed to take core samples through the cracked areas. Several core samples were obtained from the first floor and upper basement level. It appeared that the fractured sections ran through the depth of the slab, but it was difficult to confirm whether some of that cracking might have been initiated due to vibrations during core drilling process. To further assess the extent of the cracking, epoxy ports were drilled around several cracks and epoxy injected. The Python Corporation of Slidell, Louisiana, was employed by the owners to perform the epoxy grouting. The opening date for the casino was mid-summer of 1996, thus putting the complete project on a fast track basis.

Considering that epoxy grouting of the slab was a critical item for the demolition work, it was placed on a high priority list. The epoxy grouting contract was more than $1 million, making it one of the largest in the country for work of that nature. The examination revealed that the crack sizes around the column heads varied in size from 0.004 to 0.080 of an inch (0.1 to 2 mm). Two different viscosity resins were recommended. The *Denepox 40 LV* manufactured by Deneef Construction Chemicals of Texas, was suggested for cracks between 0.005 to 0.015 inch (0.12 to 0.38 mm) width, and *Shul Inject LV* manufactured by Shul International Company of Georgia, was suggested for cracks wider than 0.015 inch (0.38 mm).

The following procedure was recommended by the Python Corporation:

1. Route the surface of the existing crack 0.25-in. wide by 0.25-in deep (6.4 mm by 6.4 mm) using a diamond saw router.

2. Using compressed air and a high-pressure air needle, flush out all impacted debris and sawdust by blasting at a 45-degree angle directly into the crack.

3. Install Lily injection ports spaced along the routed crack at specified intervals depending on crack width and slab thickness.

4. Seal routed void between cracks and around injection ports using *Shul Injection Gel* as manufactured by Shul International.

5. Inspect the bottom of the slab locations to be injected for any cracks that may protrude through the entire thickness of the slab.

6. Detection of these cracks will be by misting or spraying with low-pressure water and rapid evaporation of the surrounding surface to define cracks by water migration.

7. Any cracks that are visible that would allow an egress for the injected epoxy would then be sealed utilizing a 12 inch (30 mm) strip of *Shul Flex Seal* flexible adhesive as manufactured by Shul International.

8. Beginning at intervals of cracks at their widest location, the appropriate resin would be injected under pressure, using positive displacement plural component equipment capable of producing the necessary pressures to force the injected resin into the crack until it protrudes downline at the next port. A continuous crimping of ports and leap frogging of injection hoses is done until the crack is filled over its entire length.

9. After 24 hours of cure time, the surface areas of the cracks would be ground to a smooth condition removing any build up of sealing epoxy and ports that were installed for the injection process.

The operation of epoxy grouting was performed on a schedule of 16 hours a day, 6 days a week, and continued for about three weeks, with a crew of about 24 workers. Figures 2 and 3 show the epoxy grout application in progress.

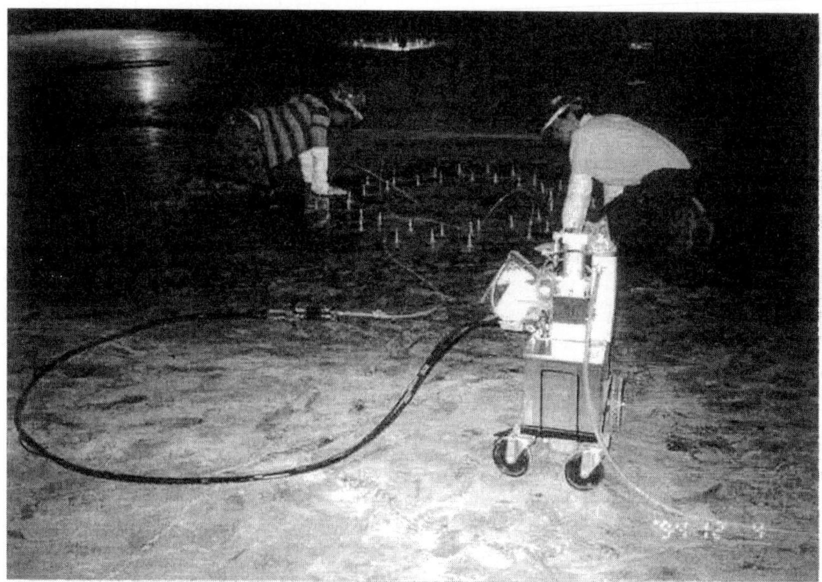

Figure 2 Epoxy grout application and crack monitoring devices

Figure 3 Sealed portholes around column heads, also seen are tertiary cracks

CONCLUSIONS

Assuming that most of the crack pattern was circular, the preliminary cost estimate was based on the perimeter of the column drophead. On an average, 35 to 40 portholes were drilled per each column head. During epoxy injection, it was noted that the epoxy would travel through the miniscule cracks connecting the circular cracks. To maintain the required pressure during grouting operation, these tertiary cracks had to be sealed. The contract for the epoxy treatment was on the basis of number of feet of sealed cracks. The cost to treat these tertiary cracks would be substantial. Also, it was noted that the tertiary cracks developed pressure during injection rather quickly indicating that they were thinner as well as shallower. To find the extent of cracking, it was decided to take concrete core samples in the areas around the columns, which had visibly noticeable cracks, as well as in the areas where tertiary cracks occurred. The information obtained from these core samples was valuable in determining which cracks might require epoxy injection and which might not. Because the epoxy held the core sample together, the core could be visually inspected (Figure 4) to study the extent and width of cracking. Although no additional tests were conducted to evaluate the strength of the concrete cores, the visual inspection showed that the crack thickness when sealed was less than 0.03 inches. The grouted areas in question would not accept any further epoxy grout indicating the cracks were completely sealed.

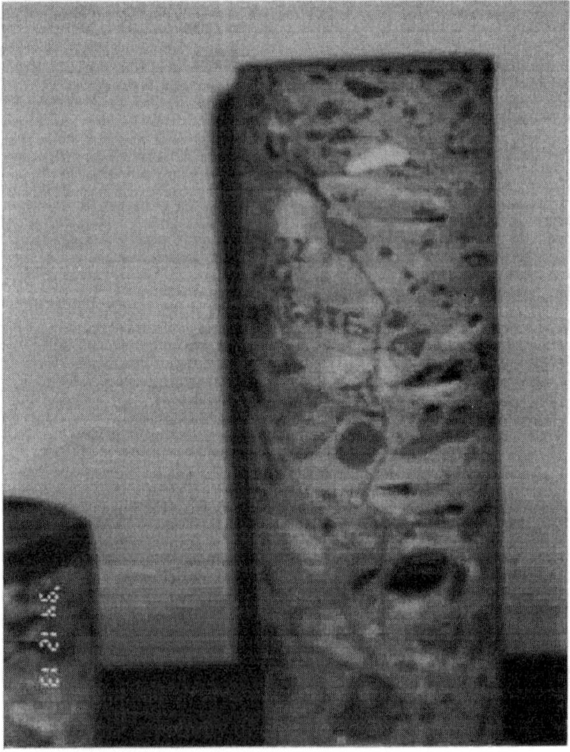

Figure 4 Core sample showing extent of cracks

The following was noted:

- Most of the cracks extended only through partial slab depths.
- The width of cracks as seen by the naked eye before epoxy injection were exaggerated.
- The assumption that the cracks were very thin was confirmed as the profile of the crack in the core sample could be visually inspected.

Therefore, it was decided that epoxy grouting was needed only in the areas where the worse cracking had been experienced. This exercise saved the owners hundreds of thousands of dollars in construction cost and time. The epoxy grouting was very effective in sealing the cracks, thus salvaging about 80,000 square feet (7,450 m^2) of existing first floor slab. With a good understanding of the structure, and observation of the core samples, a balance was struck between the use of such grouting and the economic savings for the owners. Figure 5 shows the Harrah's Casino under construction.

Figure 5 Harrah's Casino under construction

ACKNOWLEDGEMENTS

This project was the result of the efforts of several design and construction teams which included Perez Ernst Farnet/Modus, Inc., Architects and Planners, Kulkarni Consultants, APC, Structural Engineers, and Python Corporation, Concrete Restoration/Protection Specialists. Delta Testing and Inspection Company was responsible for the core drilling operation. The general contractor for the project was Centex Landis Construction Company.

REFERENCES

1. KULKARNI, S, P E, AND SABRI, A, P E, Ph.D. Designing for Transfer of Heavy Column Loads. Concrete International, Vol. 18, No 7, Durability, July 1996, pp. 52-56.

"ALUMPHS" – NEW SPECIAL COMPOSITE BINDERS FOR DESIGN

L B Svatovskaya **M N Latutova**

O U Makarova **A V Tarasov**

V L Shubaev **N P Chibisov**

M V Shershneva

Railway University St Peterburg

Russia

ABSTRACT. New coloured without roasting materials for decorative and artistic purposes are developed and investigated: stone paints, smalts and ferrosmalts, which are necessary in modern design and architecture. Developed materials hardens in natural conditions and possess increased strength, resistance to cold and waterproofness.

Keywords: Alumph, Aluminoconcrete, Resistance to cold, Waterproofness, Phosphoric acid, Smalt, Ferrosmalt.

Professor L B Svatovskaya is the Head of Department of Engineering Chemistry, Railway University, St.Petersburg, Russia. She specializes in the chemistry of the binders.

Dr M N Latutova, Department of Engineering Chemistry, Railway University, St.Petersburg, Russia. Area of scientific interests – silicate and phosphate binders.

Mrs O U Makarova, Department of Engineering Chemistry, Railway University, St.Petersburg, Russia. Area of scientific interests – phosphate materials.

Dr A V Tarasov, Department of Engineering Chemistry, Railway University, St.Petersburg, Russia. Area of scientific interests – materials technology.

Dr V L Shubaev, Department of Engineering Chemistry, Railway University, St.Petersburg, Russia. Area of scientific interests – materials technology.

Dr N P Chibisov, Department of Engineering Chemistry, Railway University, St.Petersburg, Russia. Area of scientific interests – materials technology.

Mrs M V Shershneva, Department of Engineering Chemistry, Railway University, St.Petersburg, Russia. Area of scientific interests – ecology.

INTRODUCTION

Work presents information about development of the special materials: stone paints, smalts and ferrosmalts, which obtained chemically by reactions between powdered aluminium hydroxide and phosphoric acid [1]. The problem of development of the hydraulic materials on the base of aluminium hydroxide and phosphoric acid was solved in accordance with modern concepts of thermodynamics of processes, role of surface nature in heterogeneous reactions and in heterogeneous catalysis. The hydraulic material were obtained in the case, when it is possible to obtain more base and less soluble aluminium phosphates. In the last case the increase of pH of the systems is a general feature of the process. From standpoints of thermodynamics, reactions in the given systems are characterised by the negative value of Gibbs free energy change ($\Delta G<0$).

Known binders of natural hardening from aluminium phosphates, obtained on the base of aluminium hydroxide, are not water-resistant, since products in the system are well soluble hydrophosphates:

$$Al(OH)_3 + 3H_3PO_4 \rightarrow Al(H_2PO_4)_3.3H_2O \ (\Delta H<0) \tag{1}$$

They gain resistance to water only after heating not below 200°C. It is known also that more base salts-phosphates are not soluble, for instance, variscite, according to R. Svenson and others, can be described by the formula $Al(OH)_2H_2PO_4$ and obtained in accordance with the scheme:

$$Al(OH)_3 + H_3PO_4 \rightarrow Al(OH)_2H_2PO_4 + H_2O \ (\Delta H<0) \tag{2}$$

Forming of the nonsoluble phosphates must give the material hydraulic characteristics in conditions of natural hardening. Realisation of scheme 2 is possible because of formation of difficultly soluble products and neutralisations of proton on acid-base mechanism that is reflected on pH, which value increases.

Materials

In the work was used $Al(OH)_3$ by different degrees of grinding in milling units: ball, porcelain, iron and vibroreducer, specific surface (5000 см^2/g). Redox reactions were investigated by the introduction into the system compounds of a ferric (II) and copper (I and II); as coloured catalysts were used pigments of inorganic nature: carbon; oxides, sulphides and selenides of d-metals. As aggregate was used aluminium hydroxide with the remainder on the screen № 008 60%, as mix liquid – phosphoric acid with concentration 55...65%.

RESULTS AND DISCUSSION

Materials were obtained on the base of aluminium hydroxide of certain grinding and phosphoric acid, liquid/solid<0,4.

Water-resistant hydraulic materials were named ALUMPHs. Composite materials, prepared from aluminium hydroxide, aggregate and phosphoric acid were named aluminoconcretes.Developed materials at the age of 28 days have compression strength after 2 days of saturation with water 10...18 N/mm^2. Through the time, strength of majority of hydraulic samples grows and in 3 year reaches 12...37 N/mm^2.

For obtaining of coloured hydraulic alumoconcretes – smalts for decorative purposes were used inorganic pigments (CdS; $CoO.Al_2O_3$; CdS.CdSe; Cr_2O_3; C (graphite); $Cr_2O_3.2H_2O$; FeO; Fe_2O_3). Characteristics of these materials are presented in the table 1. In the work were investigaed dependencies of concentrations of a pigment on strength of coloured alumoconcretes. Addition into the system of the pigments before 5% raises strength of samples after saturation with water to 20...50 N/mm^2. On the roentgenograms of hardened coloured alumoconcretes crystalline new growths are not discovered.

Table 1 The properties of alumoconcretes – smalts

PIGMENT	COLOUR OF ALUMOCONCRETE	COMPRESSIVE STRENGTH AFTER 2 DAYS SATURATION, N/mm^2 — Phosphoric Acid 55% 7 days	28 days	3 years
–	White	3,0	10,0	20,0
CdS·CdSe	Red	6,5	28,0	48,0
$CoO·Al_2O_3$	Blue	3,5	17,5	50,5
CdS	Yellow	7,5	36,0	36,0
Cr_2O_3	Green	7,0	29,0	37,5
$Cr_2O_3·2\text{-}3H_2O$	Emerald	14,5	25,0	27,5
CdS·CdSe	Claret	35,0	35,0	37,5
FeO; Fe_2O_3; SiO_2	Brown	9,0	18,0	28,5
C (graphite)	Black	8,0	12,0	30,5

The presence in the system of coloured materials influences upon the reactivity of hydroxide and phaseformation, that is prooved by the reduction of the effects at temperatures 250°C and 315°C on derivatograms, the decrease of the effect at the temperature 100°C in the case with CdS and disappearance of the effect in the case of with CdS.CdSe, Cr_2O_3 and graphite. The compression strength of developed coloured smalts increases on 30...140%, in contrast with the white composition. The change of reactivity of aluminium hydroxide is observed also on IR-spectrums – the shift of maximums of deformation fluctuations frequencies in the area of higher frequencies.

On the base of these studies it is possible to expect that coloured pigments are catalysts of hardening processes. Carried out experiments were used as the base for the technology of production of without roasting smalts. As strength characteristic was offered mark of the smalt – compression strength at the age 28 days of sample after 2 days of saturation with water. Developed coloured smalts has marks M 200, M 300, M 400 and M 500.

Table 2 The properties of alumoconcretes – ferrosmalts

ADDITIVES: OXIDE – 10% PIGMENT – 5%	COLOR OF ALUMO-CONCRETE	COMPRESSION STRENGTH, N/mm^2		
		After 2 days of saturation with water	After 30 cycles of freezing and defrosting	After 6 years of water keeping
–	White	2,0	–	10,0
FeO	Gray	7,0	12,0	17,5
FeO; Cr_2O_3	Dark-yellow	8,0	20,0	22,5
FeO; CdS; CdSe	Red	14,0	22,5	16,0
FeO; C	Black	28,0	40,0	34,0
FeO; $CoO{\cdot}Al_2O_3$	Dark-blue	8,0	16,0	21,0
FeO; pyrite cinders	Brown-lilac	23,0	23,5	36,0

The influence of oxidizers and reducers on the hardening processes in phosphate alumoconcretes and especially on the physicochemical conversions were studied by the Mossbauer spectroscopy. For this purpose ferric (II) oxides was used, added in aluminium hydroxide in the amount 5% and 10% mass. Mossbauer spectroscopy clearly shows the running of the process of oxidation of the ferric (II). With regard to these data it is possible to think of the following redox process:

$$4FeO + O_2 + 4H_3PO_4 \rightarrow 4FePO_4 + 6H_2O \qquad (3)$$

Process of oxidation is accompanied by the neutralization of proton, as a result increases pH that confirm results of studies. X-ray analysis of products have confirmed redox reaction. Forming $FePO_4.2H_2O$ (strengite), being crystal-chemical analogue of variscite, raises the share of nonsoluble hydrophosphates, that causes the growth of systems resistance to water. At the addition into the system of copper oxideses take place copper reduction to metallic and the raise of suspension pH. In the Table 2 are presented characteristics of strength and resistance to cold of coloured ferroaluminoconcretes – ferrosmalts, containing coloured pigments and ferric (II) oxides. Addition in the compositions of ferric (II) oxides in the combination with pigments raises strength of ferrosmalts to 7...28 N/mm^2. After 6 years of water keeping strength of ferrosmalts growth up to 17...36 N/mm^2. Ferrosmalts have marks M 200, M 300, M 350 [2,3,4].

PRACTICAL USES

On the base of developed coloured without roasting stone paints wall works are done – phosphate fresco, size 1.6x3.2 m^2 (Figure 1) and 2x3 m^2 (Figure 2). Smalts and ferrosmalts were used for making the mosaic paintings in the university chapel.

Figure 1 "Safo". Alumphs on concrete, painter Kuzmin V T

Figure 2 "Russia". Alumphs on concrete, painter Smirnova E V

CONCLUSIONS

1. New without roasting coloured materials for decorative and artistic purposes are developed: stone paints, smalts and ferrosmalts.

2. The most important physicomechanical and physicochemical characteristics of developed material are found.

REFERENCES

1. SVATOVSKAYA, L B, SYCHOV, M M, LATUTOVA, M H. Coloured cement. Patent № 759476, 1980.

2. SVATOVSKAYA, L B, LATUTOVA, M H, GOLOVINA, O A. Characteristics management of phosphate mixtures with regard to models of solids. Cement, № 5, 1990, pp 14...15.

3. SVATOVSKAYA, L B. Engineering Chemistry, Part 2. Press of SPb. Railway University, St.Petersburg, 1998, pp 92

4. LATUTOVA, M H. Abstract of doctor diploma. Press of SPb. Railway University, 1994, pp 24.

TECHNOLOGY OF ELECTIRC ACTIVATION OF CONCRETE MIXES DURING THE STAGE OF VIBRATION SEAL

V A Matviyenko

N M Zaichenko

Donbass State Academy of Civil Engineering

Ukraine

ABSTRACT. The article contains the results of theoretical and experimental investigations on electric-activation technology of concrete mixes at the stage of vibrocompacting with the purpose of improving of concrete mixes workability, increasing of reinforced concrete structures quality, and also economies material and power resources. Is shown, that at optimum parameters of processing of concrete mixes in a high-voltage electrical field their viscosity at vibration is reduced, the conditions for intensification of structure formation processes are created and the physical-mechanical characteristics of concrete products and structures are increased. On the other hand, specific power consumptions on electrical activation are minimum and make 1.5-2 Wt/m^3. The efficiency of electroprocessing is not identical to various kinds of concrete and rises with increasing in a blend composition the quantity of cement and small-sized fractions of aggregates (fillers) in the mix.

Keywords: Electric activation, Vibration, Rheological properties, Effective viscosity, Dispersed system, Double electrical layer (DEL).

Professor Vasiliy A Matviyenko is a Lecturer in Donbass State Academy of Civil Engineering and Architecture. His works has concentrated on electric activation forces in concrete technology. Professor Matviyenko has published widely and serves on many Technical Committees.

Nickolai M Zaichenko PhD is a Lecturer in Donbass State Academy of Civil Engineering and Architecture. He specialises in rheology and early concrete strength of concrete mixes.

INTRODUCTION

The vibromoulding of concrete mixes is a process, which combines breaking and recovery of bonds between the particles of dispersed phase, which are divided by water layers [1]. The shells of water on a surface of cement and fillers dispersed particles are structured within the limits of a double electrical layer (DEL). Incidentally the energy of interparticle interactions depends on a structure and properties of these shells and determines the rheological properties of concrete mixes.

In the technology of concrete the rheological properties (for example, increasing of mixes mobility) are regulated with the help of chemical additives of various spectrum operation. So, the additives of plasticizers (superplasticizers) are adsorbed on a surface of particles and shield the forces of attraction between them. As a result the altitude of a power barrier of repulsion between particles is increased, that affects on increasing of mixes mobility. However, in the future chemical additives can cause a number of negative phenomena: deceleration of the hydration process of cement (plasticizers), occurrence of salt-spots on a surface of products, corrosion of concrete reinforcement (electrolytes).

In order to regulate rheological properties of concrete, as well as promote the process of their structure formation, high-voltage electrical activation at the stage of vibrocompacting (i.e. at the vibromoulding of concrete products and structures) represents a novel approach. In that case the concrete mix is considered as a system "solid - liquid", in which the dispersed particles of cement and fillers with a surface charge are distributed in conductive dispersing environment [2].

EXPERIMENTAL DETAILS

Materials

The materials used in the research study were:

- Binders –Portland cement OPC-400; alumina cement “Secar-51”;
- Dispersed fillers – milled quartz sand (MQS) and limestone (ML) with specific surface 220-250 m^2/kg;
- Aggregates – mix of quartz sand and crushed granite rock in 10 mm and 14 mm single size.

Mix proportions

The mix proportions used are summarised in Table 1.

Table 1 Mix proportions

NO	MIX PROPORTIONS, kg/m^3					
	PC-400	Secar-51	MQS	ML	Aggregate	W/C
1	-	-	1400	-	-	0,22
2	-	-	-	1400	-	0,22
3	480	-	-	-	1400	0,35-0,45
4	-	480	-	-	1400	0,35-0,45

DETERMINATION OF EFFECTIVE VISCOSITY OF MIXES

The scheme of the installation for determination of effective viscosity of mixes at electroprocessing is shown in Figure 1. In viscosimeter 5 a ball 4 floats up under an operation of vibration from vibrating table 1 and the load of weight M with the certain speed. The electrodes from a high-voltage source of direct current 6 are located in such a way that one of them 3 has the contact with processing mix, another (with opposite polarity) is isolated with high-resistance insulation 2 for limiting of current density (less than 0,1 mA/m^2).

The viscosity at vibration was calculated using the formula [3]:

$$\eta_v = \frac{2}{6}\pi r(v_1 + v_2)\cdot[Mg + \frac{4}{3}\pi r^3 g(\rho_m - \rho_b) - \frac{m}{2l}(v_2^2 - v_1^2)],$$

where,
m = weight of a ball, kg
r = radius of a ball, m
v_1 and v_2 = float speed of a ball on the boundaries of mix sections, m/s
ρ_m and ρ_b = density of a mix and material of a ball, accordingly, kg/m^3

RESULTS AND DISCUSSION

It is known [4], that the balance of forces of molecular attraction (P_m) and electrostatic repulsion (P_e), determines the stability of coagulation of dispersed systems:

$$P_m = -2A/3r[1/S^2 - (S^2 - 8)/S(S^2 - 4)^2]$$
$$P_e = r\varphi_0^2 / 2(\varepsilon/(1 + \exp(\chi h))$$

where,
A = Gamaker' constant, Nm
S=H/r = (H = distance from a surface of a particle; r = radius of a particle, m)
φ_0 = potential of a surface, V
χ=1/δ = Debye' radius of screening (δ - thickness of DEL)

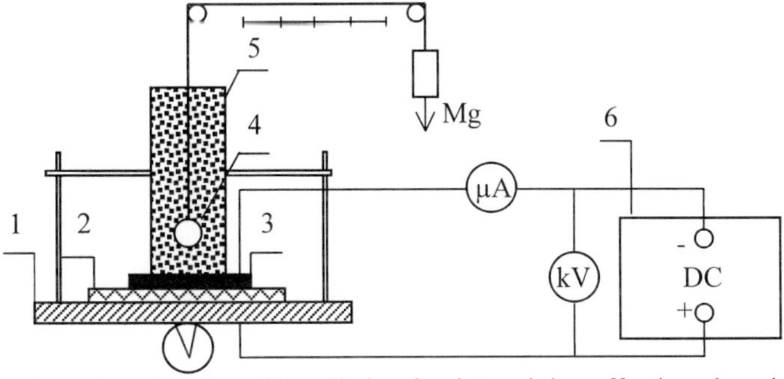

Figure 1 Schematic of installation for determining effective viscosity of mixes

When processing a dispersed system in a high-voltage electrical field, due to electro-adsorptive effects, the surface charge and structure of DEL will vary. Besides, in accordance with [5] if the polarity of the contact electrode from an external source of direct current coincides with the polarity of surface charge, the size of the latter is increased, and DEL extends. At different polarity - the charge of a surface is reduced, and the thickness of the diffusion area of the DEL decreases.

The calculations of pair load-carrying interaction (P_m and P_e) of cement particles, carried out by means of computer have shown, that in the first case the altitude of a power barrier of repulsion is considerably increased, and repulsion happens at a greater distance from a surface of particles. Therefore, this effect should influence the rheological properties of dispersed system (cement gel) as well as on the process of its structure formation. This hypothesis is confirmed by the outcomes of experimental researches of effective viscosity indexes of various mixes. Electroprocessing of mix consisting of milled quartz sand (has a negative charge of a surface) and water, the following is established. At negative polarity of the contact electrode the noticeable breaking of effective viscosity of mix is observed (Figure 2). In addition the indexes of viscosity are influenced by the size of the electrical potential. At the same time, electroprocessing of "milled limestone (has a positive surface charge) - water" a reduction in mix viscosity occurs with a positive contact electrode.

The particles of Portland cement, and also its hydration products (calcium hydrosilicates) have a negative surface charge. Thus, at electroprocessing of PC concrete reduction in the effective viscosity occur with a negative contact electrode (Figure 3). In contrast, alumina cement is characterized by a positive surface charge thus processing alumina concrete in an electrical field of positive polarity the reduction in the effective viscosity is reduced.

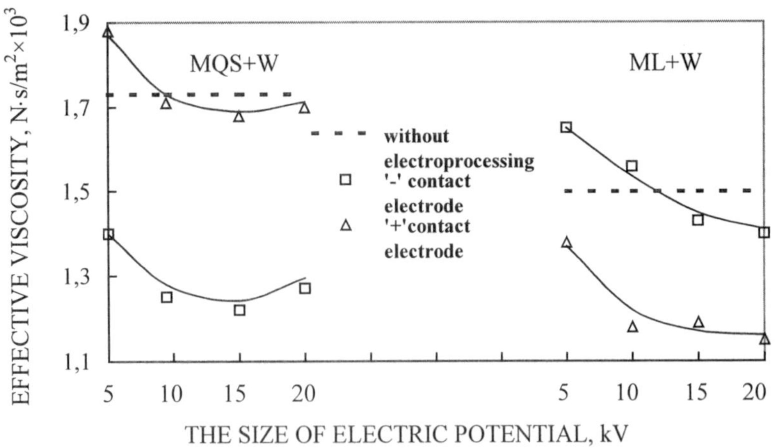

Figure 2 Dependence of effective viscosity of mixes on the size and polarity of electric potential

This is connected with the presence in the concrete mix structure of dispersed fillers (quartz sand) with a negative surface charge. In this case integral surface charge of all dispersed particles decreases, and between particles the electrostatic forces of attraction are exhibited. However, the efficiency of electroprocessing of concrete mixes depends on their water content. The maximum effect of viscosity reduction is reached in case of processing of moderately mobile fine-grained concrete mixes with w/c=0,375-0,425.

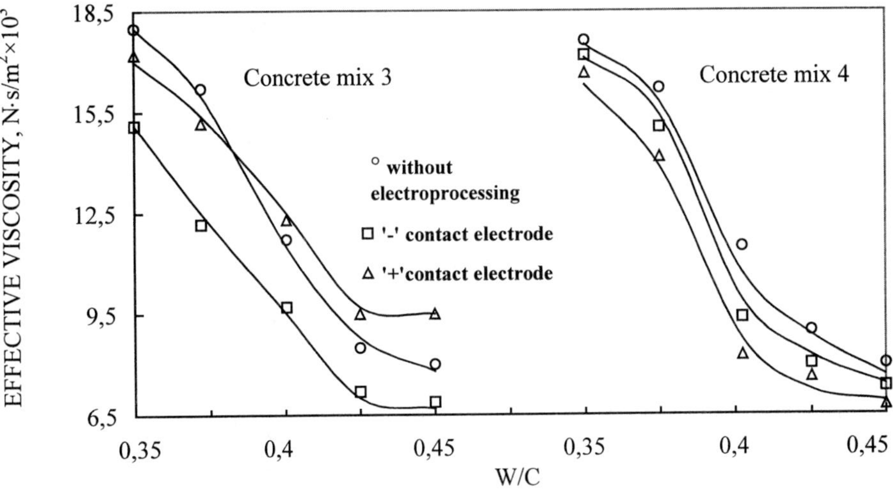

Figure 3 Dependence of effective viscosity of mixes on the base of OPC-400 (mix 3) and "Secar-51" (mix 4) on W/C

PRACTICAL APPLICATION OF RESULTS

In an industry of prefab reinforced concrete of Ukraine more than 90% of all products and structures are produced by vibration technology. In addition, as a rule, the factories are equipped with the vibration equipment, which is not adequate for the modern demands of industrial construction. In this respect, development of low power-intensive method of electrical processing of concrete mixes during vibrocompacting permitting to increase a production efficiency of reinforced concrete constructions, is neccessary.

Under factory conditions research which confirm efficiency of electrical activation of concrete mixes has been carried out. As an illustration of efficiency the compressive strength of fine grained concrete depends on the duration of vibration compacting (Figure 4). So, for mixes based on traditional technology, the dilution happens gradually and to reach the maximum concrete strength it is required not less than 120 seconds from working of the vibration equipment. The electrical activation of mixes during vibration promotes the faster their dilution - for 90 seconds. In addition the percentage increase in compressive strength of concrete is about 15-25%.

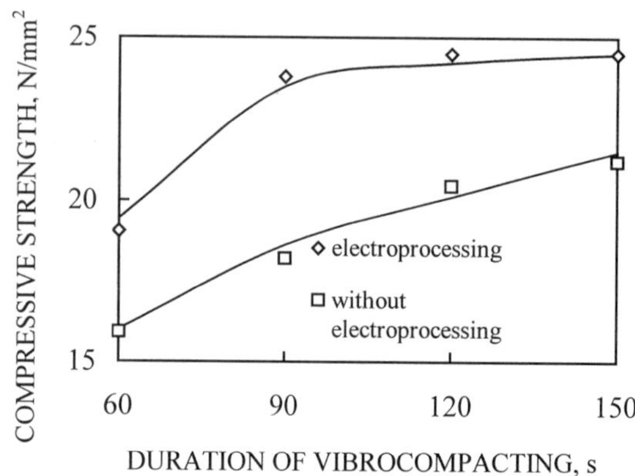

Figure 4 Dependence of concrete compressive strength on the duration of vibrocompacting

CONCLUSIONS

1. The effect of reducing the vibroviscosity of a concrete mix as a result of electrical processing at vibration compacting, gives increased concrete compressive strength of 20-40%.

2. The given effects can be reached under the following conditions:

- polarity of the contact electrode should correspond to an integral charge of the surface of dispersed particles of concrete mixes;
- A voltage of 5-10 kV; density of electric current - 0,5-1 μA/m^2 should be applied.

3. The method of electrical activation enables low power consumption in the processing of concrete mixes.

REFERENCES

1. UR'IEV, N B. Physical-chemical technologies of dispersed systems and materials [in Russian]: Moscow, Chemistry, 1988, 256 pp.

2. MATVIYENKO, V A. The influence of electric field voltage on the cement pastes strength. Journal of Applied Chemistry, No 9, 1991, pp 1857-1861.

3. GUSEV, B V AND ZAZIMKO, V G. The vibration technology of concrete [in Russian]: Kyiv, 1991, 60 pp.

4. BYBIK, E E. Rheology of dispersed systems [in Russian]: Leningrad State University, 1981, 146 pp.

5. MATVIYENKO, V A, GUBAR, V N AND ZAICHENKO, N M. Polarisation activation of cement past. "ConChem" – International conference proceedings, Les Pyramides, Brussels, 1995, pp. 529-538.

PROTECTIVE PROPERTIES OF SILICA FUME - Ca $(OH)_2$ MIXTURES AS REPAIR MATERIALS

N Kouloumbi

V Kasselouri

G Tsihlis   Th Tassios

National Technical University of Athens

Greece

ABSTRACT. As part of an extended research program, the aim of this experimental work is the study of Silica Fume-Ca$(OH)_2$ mixtures, with or without cement addition, as repair materials in reinforced concrete structures. The study of the hydration reactions' progress by XRD and SEM analysisof these mixtures show that Calcium Silicate Hydrates are formed, especially in pure mixtures of Silica Fume-Ca$(OH)_2$ and in those with low cement content. Damaged specimens repaired by the above mentioned materials also show low carbonation depth and lowest corrosion rate, close to that of undamaged specimens. This is attributed to the appropriate formation of CSH compounds at the interface between these mixtures and cement, resulting in good binding.

Keywords: Concrete, Silica fume, Repair materials, Corrosion of rebars.

Dr N Kouloumbi is an Associate Professor in the Chemical Engineering Department of the National Technical University of Athens (NTUA), Greece. His teaching and research fields include Electrochemistry, Corrosion and Protection of Metals.

Dr V Kasselouri is an Associate Professor at the Chemical Engineering Department of NTUA, Greece. Her teaching and research fields include, High Temperature Chemistry and Technology (Cement, Ceramics, Glass).

Mr G Tsihlis is a Postgraduate Student at the Chemical Engineering Department of NTUA, Greece.

Dr Th Tassios is an Emeritus Professor at the Civil Engineering Department of NTUA, Greece. His teaching and research fields include Concrete Technology.

INTRODUCTION

Damage of reinforced concrete structures occurs when the protective capacity [1,2] of the concrete is exhausted by carbonation or considerably reduced by contamination by aggressive ions e.g. by chloride ion penetration [3]. The steel rebars corrosion onset, the time interval before the attack starts and the rate at which it proceeds are dependent on factors influenced by the chemistry of cement, mineral additions, water binder ratio, curing and environmental conditions [4-9].

There is a wide range of cementerious repair mortars based on cement and components similar to those of concrete. The composition of repair mortars could sometime consist of more than one type of cement together with additions (i.e. Silica Fume, slag or fly ash), admixtures such as powder of plasticizers, air entraining and accelerators, polymer additives and fine polymer fibres [10-12].

In the present work, the efficiency of the use of Silica Fume (SF)-$Ca(OH)_2$ mixtures, with or without cement addition, as repair material of reinforced concrete structures, in stead of conventional cement-SF mixtures is examined.

EXPERIMENTAL DETAILS

The cement used in this work was an Ordinary Portland Cement, I45 (BS 12: 1996). The steel rebars used were S200 type with the following chemical composition (% wt): Fe, 99.23; Mn, 0.56; S, 0.07; C, 0.11 and P, 0.03. Calcareous sand of typical gradation of 0-0.2 mm (10% by wt), 0.2-1mm (30% by wt) and 1-5mm (60% by wt) and drinking water were used. The composition of mortar specimens is given in Table 1.

The repair mixtures consisted of densified Silica Fume of Norwegian origin (SF), (amorphous SiO_2>93% wt) and commercial calcium hydroxide. The calcium hydroxide was in slurry form containing 50% wt water.

The test specimens considered for the present study are shown in Figure 1. The steel bars were pretreated and weighted to 0.1mg before casting.

Table 1 Composition of specimens

COMPOSITION RATIO BY WT				
PC	Sand 0-0.2mm	Sand 0.2-1mm	Sand 1-5mm	Water
1	0.38	1.14	2.27	0.72

The steel surface in contact with mortar was equal to 2276.5mm^2. After the specimens being demolded they were stored in the curing room (RH=98±2%, T=19±1°C) for 28 days.

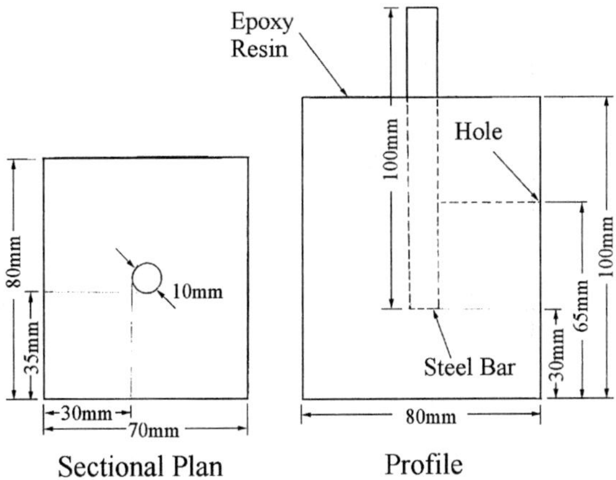

Figure 1 Schematic representation of reinforced mortar specimens

At the end of the curing a hole, 5mm in diameter, was drilled in each lateral side of the specimen at a height corresponding to the middle of the embedded steel bar and to a depth up to 7mm from the surface of the bar (Figure 1). Holes were filled by one of the mixtures given in Table 2. In order to improve the setting time of the SF-$Ca(OH)_2$ mixture, small amounts of cement have been added.

Table 2 Composition ratio by weight of repair mixtures

COMPONENT	CODE NAME OF REPAIRED SPECIMENS		
	C	D	E
Silica fume	1	1	1
$Ca(OH)_2$	3.7	3.7	3.7
Cement	0	0.15	0.3

Reference undamaged specimens (code name A), specimens with unfilled holes (code name B) and specimens with filled holes (code name C, D, E) were used. Finally, a part of all types of specimens was immersed in 3.5% NaCl solution and another part was stored at ambient conditions for carbonation depth measurements.

The study of the hydration products of the repair mixtures at predetermined ages was performed by X-Ray Diffractometer (XRD), by Differential Thermal Analysis (DTA-TG) and by Scanning Electron Microscopy (SEM).

The durability of the specimens was evaluated, by measuring the time dependence of the steel rebars corrosion potential, the carbonation depth and the corrosion rate of the rebars by the polarization resistance measurement. The results will be correlated with long term weight loss measurements.

RESULTS AND DISCUSSION

The XRD study deals with the examination of the hydration products of SF-$Ca(OH)_2$, as well as SF-$Ca(OH)_2$-cement mixtures at the age of 7 days, 28 days, 6 months and 1 year. In Figure 2 the XRD patterns after a 28 days hydration are presented. At all ages of hydration the main compounds observed are $Ca(OH)_2$ in the form of portlandite (2.63, 4.90, 1.93Å), small amount of $CaCO_3$ (3.015Å) due to the carbonation of $Ca(OH)_2$ and calcium silicate hydrates in the form of tobermorite (11Å). From the first ages the formation of calcium silicate hydrates in the form of tobermorite 11Å and 14Å, as well as in the form of $2CaSiO_3 \cdot 3H_2O$ (Riversideite: 14, 11.3, 12.5, 9.67, 3.07, 5.55, 2.93, 2.80Å) can be observed, either in the mixtures of SF-$Ca(OH)_2$, or in the SF-$Ca(OH)_2$-cement ones (13).

At the ages of 28 days, 6 months and 1 year the amount of calcium silicate hydrates is slightly higher in the mixtures of SF-$Ca(OH)_2$, than in the mixtures containing cement. This observation suggests that the addition of a small amount of cement (i.e. up to 6% wt) helps only the improvement of the setting time of the mixtures. The existence of a remarkable amount of unreacted $Ca(OH)_2$, even after a year of hydration, indicates the slow rate of the reaction between $Ca(OH)_2$ and the SiO_2 contained in the silica fume.

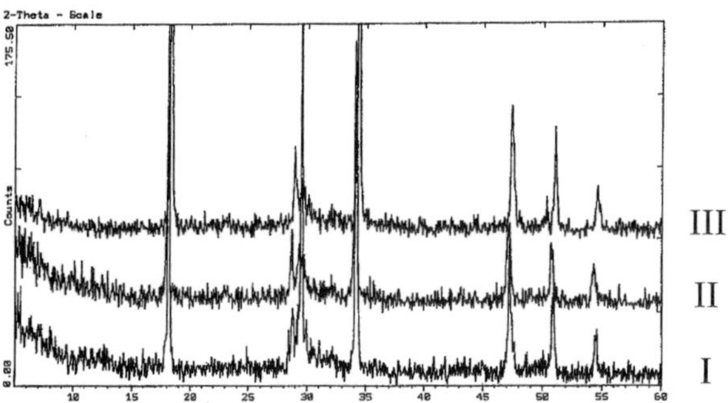

Figure 2 XRD patterns of hydrated repair mixtures after a 28 days of hydration, (I) Pure SF-$Ca(OH)_2$ mixture, (II) SF-$Ca(OH)_2$-3%wt OPC mixture, (III) SF-$Ca(OH)_2$-6%wt OPC mixture

The study by SEM aimed at the investigation of binding between the main mass of damaged specimen and the repair material. In Figure 3 a mortar part adjacent to the repaired hole after 56 days of hydration, is presented. A cluster of thin crystals of calcium silicate hydrate is observed. The microanalysis of these crystals showed that they contain 32.33% CaO and 31.55% SiO_2 ($CaO/SiO_2 \cong 1$). The crystals have slightly rounded edges, which indicates rapid crystallization. In Figures 4 and 5 the hydrated repair material (SF-$Ca(OH)_2$) is presented. In Figure 4 a grain of partly hydrated silica fume is shown. In the boundaries of the grain a calcium silicate hydrate has been formed. A focus on this boundaries area is presented in Figure 5. The microanalysis of Figure 4 gave as result 52.39% CaO and 46.54% SiO_2 ($CaO/SiO_2 \cong 1.26$). In the center of the grain's surface a cover of $Ca(OH)_2$ is observed (75.92% CaO and 22.04% SiO_2; $CaO/SiO_2 \cong 3.78$).

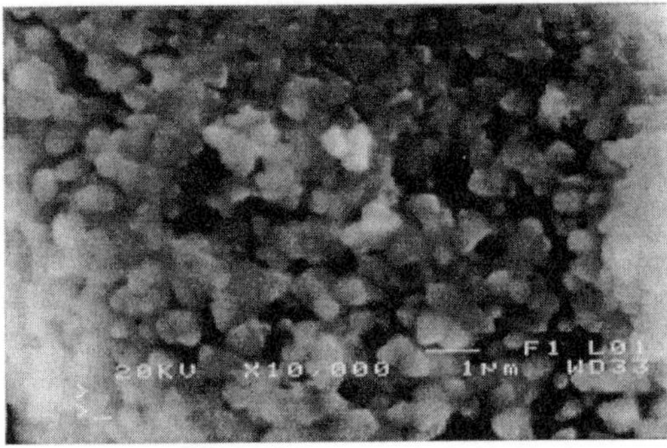

Figure 3 SEM of a 56 days hydrated cement mortar (M × 10,000)

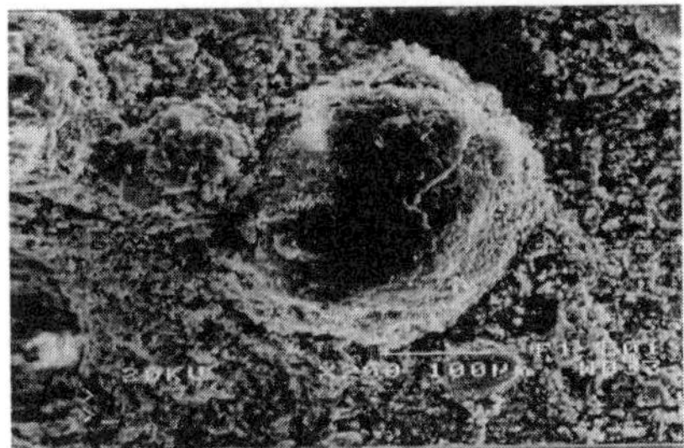

Figure 4 SEM of a 28 day hydrated repair material SF-Ca(OH)$_2$ (Mx200)

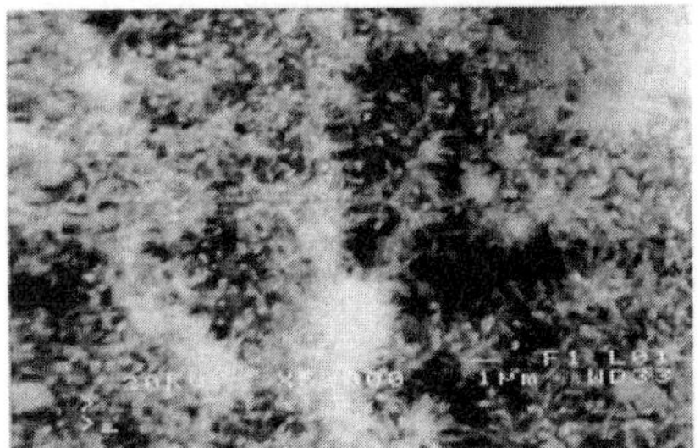

Figure 5 SEM of a 28 day hydrated repair material SF-Ca(OH)$_2$, details of grain boundaries (Mx5000)

The crystals shown in Figure 5 have a needle like form which indicates a better crystallization of the calcium silicate hydrates than this of the cement mortar. As it concerns the interface between the damaged mortar and the repair material, a formation of CSH can be observed (Figure 6). The form of the crystals is more similar to those of the hydrated cement mortar.

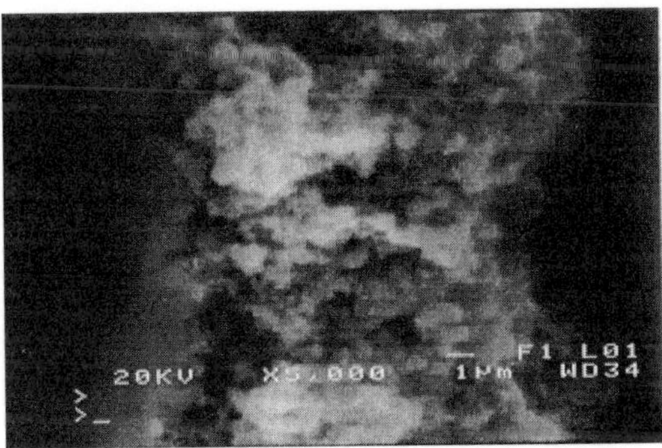

Figure 6 SEM of interface between mortar and SF-Ca(OH)$_2$ repair material (M × 5,000)

A DTA-TG measurement has been performed on a 3 months hydrated SF-$Ca(OH)_2$ mixture. The thermodiagram (Figure 7), shows three endothermic peaks at 125-130°C, 520°C and 780°C, corresponding to the dehydration of calcium silicate hydrates, the dehydration of $Ca(OH)_2$ and the decarbonation of $CaCO_3$ formed during the preparation of the sample. Respectively the exothermic peak at 975°C shows a crystallographic transformation of anhydrous tobermorites (13,14).

The corrosion activity or the reinforcing steel can be qualitatively estimated, regardless of the specimens' size or the depth of the mortar cover, by monitoring the change with time of the half cell potential of the steel bars. In Figure 8, the E_{corr} trend of the different types of samples is shown. All specimens, immediately after immersion in chloride environment showed free corrosion potential in the range of -330mV to -370mV. Thereafter a decay to more negative potential values was observed lasted from 4 to 7 days depending on the type of the specimens. Afterwards a more or less steady state situation is observed and the potential values remain almost stable around of -750mV. The potential decay to high negative values could be attributed to the experimental conditions, i.e. to the total immersion of the specimens into the NaCl solution, which leads to limited availability of oxygen and thus to cathodic polarization increase (15).

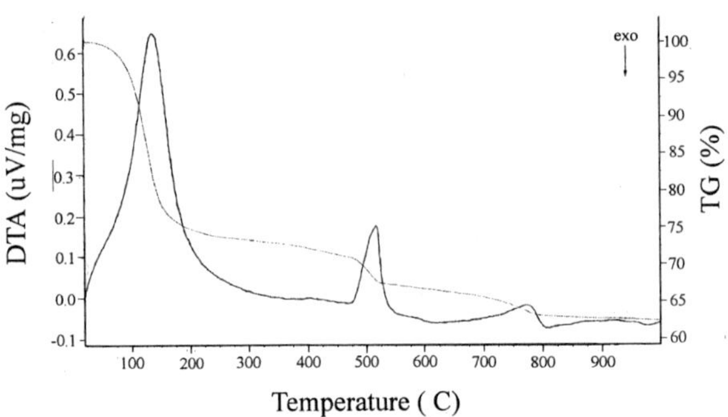

Figure 7 DTA-TG diagram of a 3 months hydrated SF-$Ca(OH)_2$ repair mixture

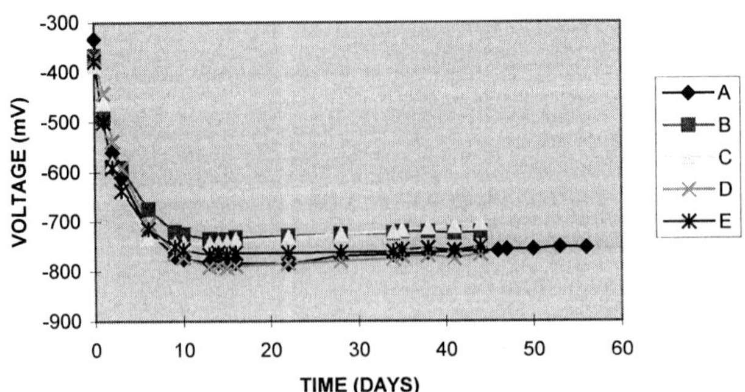

Figure 8 Development of half cell potential with immersion time

Regarding the significance of the numerical value of the potentials, it is well established (Test Method for Half-Cell Potentials of Uncoated Reinforcing Steel in Concrete, ASTM C876) that, if potentials are numerically greater than -275mV versus SCE, there is a greater than 90% probability that corrosion of reinforcing steel bars occurs.

From Figure 8 is clear that this condition is valid for all types of specimens. The small differences in plateau values do not permit any clear qualitative prediction of differences in the behavior of the difference types of specimens.

In Figure 9, the carbonation depth of all types of specimens 45 days after the repair of the hole is given. As it is shown, the undamaged mortar specimens present a very low carbonation depth, while in damaged but not repaired specimens the carbonation depth has reached the whole depth of the hole. On the contrary, in repaired specimens there is a significant decrease of the carbonation depth.

This could be attributed to the rehabilitation of the structure of the damaged specimen to an extent close to that of the undamaged one. This is a result of formation of calcium silicate hydrates of some importance in the case of pure SF-Ca(OH)$_2$ repair mixtures, as well as in mixtures containing cement.

Moreover these products are well crystallized on the surface of the hole and they exhibit good binding properties as it is proved by XRD and SEM measurements.

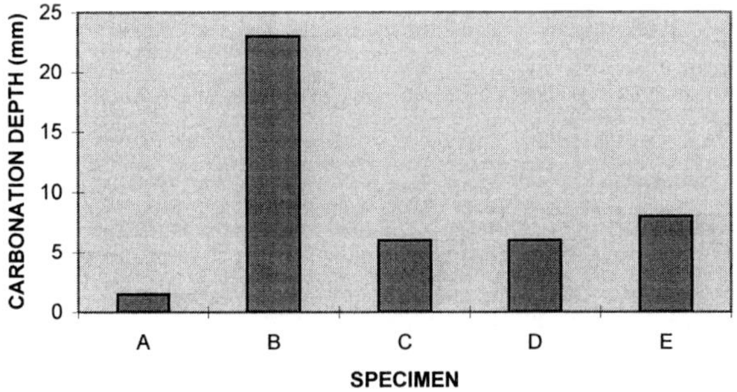

Figure 9 Carbonation depth after 45 days

The values of calculated instantaneous corrosion rates of steel rebars after 35 days of immersion in 3.5% NaCl solution are presented in Figure 10. These results show a similarity with those of carbonation depth. It is clear that the corrosion rate of the repaired specimens is very close to that of the undamaged ones and about half of that of the damaged specimens. This may be attributed to the above mentioned causes.

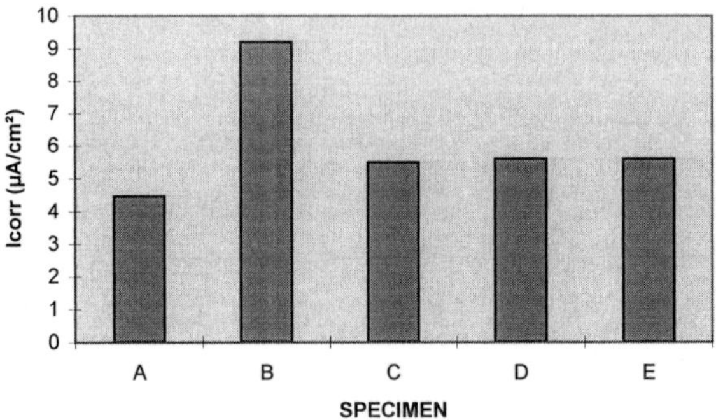

Figure 10 Corrosion intensity of steel bars after 35 days of immersion in 3.5% NaCl

CONCLUSIONS

The study of the use of Silica Fume-$Ca(OH)_2$ mixtures as repair materials instead of usual cementerious ones, leads to the following conclusions:

1. Calcium silicate hydrates are formed in appropriate extent, creating good binding properties to the repair mixtures.

2. The addition of cement in these mixtures does not ameliorate other properties, except that of the setting time.

3. Repaired specimens exhibit low carbonation depth.

4. In a corrosive environment of NaCl solution, corrosion rates of steel rebars in repaired specimens with pure Silica Fume-$Ca(OH)_2$ mixtures, as well as those containing cement, remain close to that of undamaged specimens.

REFERENCES

1. LEA F M, 'The Chemistry of Cement and Concrete.', 3rd Edition, Edward Arnold, Glasgow, Great Britain, 1970

2. WENGER, F, and GALLAND, I. 'EIS Study of the Protective Properties of Repair Products.', Eurocour'96, Nice, Session I, Corrosion in concrete, 1996, I OR 8-1.

3. BYFORS, K. 'Influence of Silica Fume and Fly Ash on Chloride Diffusion and pH Values in Cement Paste.', Cement and Concrete Research, 1987, Vol. 17, pp. 115-130.

4. KOULOUMBI, N, and BATIS, G. 'Chloride Corrosion of Steel Rebars in Mortars with Fly Ash Admixtures.', Cement and Concrete Composites, 1993, Vol. 14, pp. 199-207.

5. MIYAGAVA, T. 'Durability Design and Repair of Concrete Structure: Chloride Corrosion of Reinforcing Steel and Alkali-Aggregate Reaction.', Magazine of Concrete Research, 1991, Vol. 43, No 156, pp.155-170.

6. SORENSEN, B. and MAAHN, E., 'Penetration Rate of Chloride in Marine Concrete Structures.', Publication 1, Nordic Concrete Research, 1982, pp. 24.1-24.18.

7. MALAMI, CH, BATIS, G., KOULOUMBI, N, and KALOIDAS, V. 'Influence of Pozzolanicand Hydraulic Cement Additions on Carbonation and Corrosion of Reinforced Mortar Specimens.', Corrosion and Corrosion Protection of Steel in Concrete, 1994, Vol. II, International Conference held at the University of Sheffield, R.N.Swamy Ed., pp. 668-682.

8. KOULOUMBI, N, BATIS, G, and MALAMI, CH. 'The Anticorrosive Effect of Fly Ash, Slag and a Greek Pozzolan in Reinforced Concrete.', Cement and Concrete Composites, 1994, Vol.16, pp.253-260.

9. MALAMI, CH, KALOIDAS, V, BATIS, G, and KOULOUMBI, N. 'Carbonation and Porosity of Mortar Specimens with Pozzolanic and Hydraulic Cement Admixtures.', Cement and Concrete Research, 1994, Vol. 24, No. 8, pp. 1444-1454.

10. GUDMUDSSON, G, and OLAFSSON, H. 'Silica Fume in Concrete-16 Years of Experience in Iceland.', Alkali-Aggregate Reaction in Concrete (A. Shayan Ed.), Proceedings of the 10th International Conference, Melbourne, 1996, pp. 462-469.

11. LAGERBLAD, B, and UTKIN, P. 'Silica Granulates in Concrete-Dispersion and Durability Aspects.', Swedish Cement and Concrete Research Institute, CBI Report 3.93, 1993, pp. 44.

12. KASSELOURI, V, and PARISSAKIS, G. 'Stabilization of High Magnesia Cements by Adding Flying Ashes and Slag.', Silicate Industriels, Vol. 1, 1977, pp. 13-17.

13. FTIKOS, CH, KASSELOURI, V, TSIMAS, C, and PARISSAKIS, G. 'A Study on the Action of Seawater on Hydrated Cement Pastes.', 7e Congr. International de la Chimie des Ciments, 1980, Proced. Vol. IV.

14. KASSELOURI, V, FTIKOS, CH, and PARISSAKIS, G. 'DTA-TG Study on the $Ca(OH)_2$-Pozzolan Reaction on Cement Pastes Hydrated up to Three Years.', Cement and Concrete Research, Vol. 13, 1983, pp. 649-654.

15. BATIS, G, KOULOUMBI, N, and MALAMI, CH. 'Corrosion Resistance of Steel Rebars in Mortar Specimens with Santorin Earth.', OEBALIA, 1993a, Vol. XIX, pp. 53-60.

BEHAVIOUR OF REINFORCED CONCRETE DEEP BEAMS STRENGTHENED IN SHEAR

L L Jardim

S Melo

University of Brasilia

R B Gomes

Federal University of Goias

Brazil

ABSTRACT. The results of tests in nine reinforced concrete deep beams, simply supported and submitted to two top point loads (seven strengthened in shear) are shown. The seven strengthened beams were 800 x 150 x 1600 mm with f_c around 50 MPa, designed with insufficient shear reinforcement and loaded previously until service load. The main strengthened variables were the type, position and amount of the reinforcement positioned. The strengthening reinforcement were positioned with special mortar in ducts sawn on the surface of the beam. The results showed that the strengthening technique adopted worked properly without any problem of anchorage of the glued reinforcement. Besides this, the strengthened beams reached considerably higher ultimate loads and behaved very well in comparison with the two reference ones.

Keywords: Deep Beams, Strengthening, Repair, Reinforced Concrete, Shear

Ms Liana L Jardim is a former MSc student at the University of Brasilia. She finished her dissertation on the Behaviour of Reinforced Concrete Deep Beams Strengthened in Shear and was supervised by the two other authors.

Dr Ronaldo B Gomes is a Lecturer in Concrete Structures and the Coordinator of the Msc course in Civil Engineering at the Federal University of Goias, Brazil. His main research interests include punching in flat slabs and high strength / performance concrete.

Dr Guilherme S Melo is a Lecturer in Concrete Structures and Head of the Civil Engineering Department at the University of Brasília, Brazil. His main research interests include punching and post-punching in flat slabs and high strength / performance concrete.

INTRODUCTION

The useful life of concrete structures depends not only on the production and application of a durable concrete, but on appropriate design, detailing and construction methods, and appropriate levels of maintenance. Repair and strengthening are quite common nowadays, as many of the buildings are not built with the appropriate materials quality control, technique of execution, quality of workmanship, etc...

A lot of research has been done in repair and strengthening of concrete structures, that are damaged or subjected to unpredicted loads. However, research orientated to the investigation of the performance of deep beams that were previously loaded to service load and cracked, and then strengthened in shear are still insufficient. Another approach for this problem was adopted by Teng and others [1].

This paper presents some of the results found by the first author in her MSc Dissertation [2], under the supervision of the other two authors. The main objective of the research was to investigate the behaviour of reinforced concrete deep beams strengthened in shear, after being loaded to service load, in comparison to a monolithic beam with the full reinforcement. This happens when the reinforcement of the structure is not enough to resist the loading that is acting on the structure, or any extra load that is pretended to the structure. In a condition of safety, performance or durability being under problem, there is a need of strengthening. This work followed experimental investigations done at the University of Brasília, Brazil, in deep beams, by Bessa in 1994 [3], and in strengthened columns, by Vanderlei in 1996 [4], under two of the main research interests of the University of Brasília Structures Group, *"Experimental Analysis of Structures"* and *"Pathology and Recuperation of Structures"*.

EXPERIMENTAL PROGRAM

Table 1 shows the denomination and description of the beams tested.

The investigation was basically done in four steps: Beam 1 (reference with insufficient shear reinforcement) was initially tested until rupture (1st step). It was adopted after this first test that 600 kN (about 70% of the ultimate load) was the level that all the beams to be strengthened (2/S to 8/S) would be loaded in the first stage. This level was fixed based on the shear crack widths of beam 1 at this stage (0.3 mm). Beams 2/S to 8/S were then tested until 600 kN, when they were unloaded (2nd step), and then strengthened (3rd step). These beams and beam 2M (monolithic) were then tested up to failure (4th step). The strengthening reinforcement were positioned with high performance mortar in ducts sawn on the surface of the beam.

The behaviour of the beams tested were analysed through the strains of the flexural and shear reinforcement, and of the concrete, by the horizontal and vertical deflections, by the developing and widths of the cracks, and by the ultimate load and mechanism of rupture.

Table 1 Denomination and description of the deep beams tested

BEAM	DESCRIPTION
1	Reference beam with insufficient minimum web reinforcement
2/S	Strengthening stirrups close to the bottom edge and inside the vertical stirrups
3/S	Strengthening with orthogonal mesh, the horizontal stirrups positioned close the top edge
4/S	The same as 2/S, with the horizontal stirrups positioned outside the vertical stirrups
5/S	"V" type inclined strengthening reinforcement
6/S	The same as 2/S, inside the vertical stirrups without enveloping the beam
7/S	Inclined strengthening reinforcement distributed at the compressive concrete strut
8/S	Group of stirrups, close to the bottom edge, inside the vertical ones
2M	Reference monolithic beam, with minimum web reinforcement (same as 2/S)

The main reinforcement of all the beams were 8 deformed bars of 10,0 mm of diameter (f_y = 612.5 MPa), that were extended over the span of the beams, in four layers of two bars. The anchorage was done according to the recommendations of Leonhardt, with horizontals hooks at 180° [5]. The shear reinforcement of the reference beam (1) and for all the beams to be strengthened (2/S to 8/S) was similar, an orthogonal mesh with smooth bars (Horizontal – 3 x 5,0 mm; Vertical – 4 x 5.0 mm), with the horizontal bars anchored in the same way as the main reinforcement. The strengthening shear reinforcement were also of smooth bars of 5,0 mm (f_y = 720.0 MPa), as horizontal and vertical bars (2/S to 4/S, 6/S and 8/S), and of inclined reinforcement (5/S and 7/S). Figures 1 to 3 present the detailing of the beams 2/S, 5/S and 7/S respectively. The dotted lines in these figures are the initial reinforcement and the continuous line are the strengthening reinforcement positioned later.

Beams 2/S, 3/S, 4/S, 6/S and 8/S were strengthened with the same amount of reinforcement, but with different detailing. Beam 2/S was strengthened with vertical stirrups distributed at the shear spans and of horizontals stirrups situated between those already positioned, with one of those positioned near the bottom edge, Figure 1. The strengthening of beam 4/S was different only regarding the positioning of the horizontal bars, that were positioned outside the vertical stirrups previously positioned. Beam 3/S was strengthened with the same quantity as 2/S but with the positioning of one of the horizontal bars near the top edge of the beam. In beam 6/S the orthogonal strengthening reinforcement was positioned at the same position of beam 2/S but without enveloping the beam (the reinforcement of one face did not reach the other). For beam 8/S the three bars of the strengthening reinforcement of each face were positioned together close to the bottom edge of the beam. Beams 5/S and 7/S were strengthened with inclined reinforcement, as shown in Figures 2 and 3 respectively. The reinforcement of beam 5/S was of 6 inclined stirrups of 287 cm, where each group of three stirrups were positioned at one face of the beam, went through the thickness at the bottom of the beam, and then went up at the other side of the beam, with their ends anchored at the top part of the beam (Figure 2).

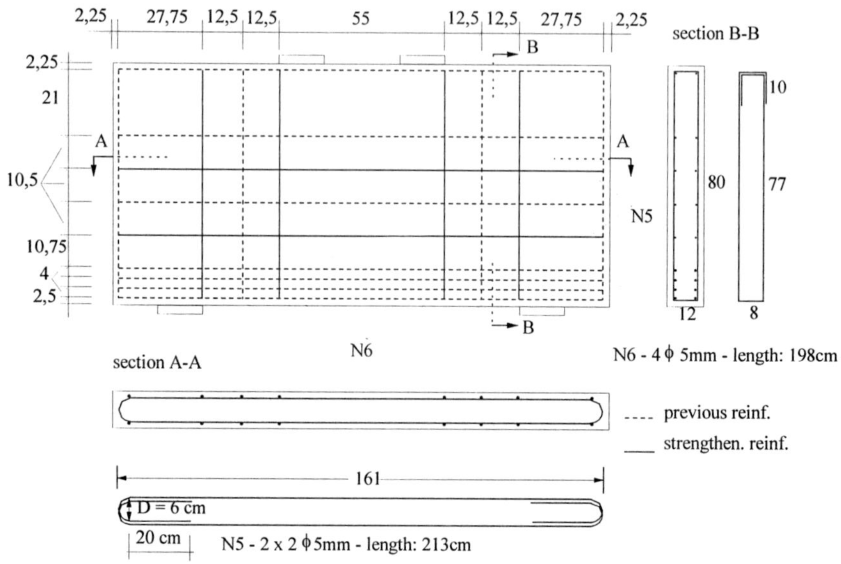

Figure 1 Beam 2/S strengthening reinforcement detail (cm)

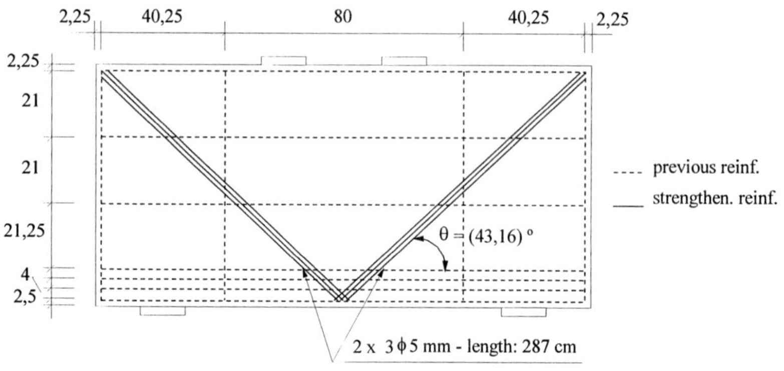

Figure 2 Beam 5/S strengthening reinforcement detail (cm)

The strengthening reinforcement of beam 7/S was five bars positioned through each of the "concrete struts" (Figure 3). Beam 2M was cast monolithically with the same amount and position of beam 2/S.

The strengthening reinforcement was positioned in ducts 5 cm thick, that were sawn at the concrete surface. The surface was then well cleaned with water pressure and compressed air, leaving the surface clean and dry. After that the beams were saturated in water and kept at the humid chamber for two days. The beams were cast in a vertical position, apart from beam 7/S that were cast in the horizontal position. Strains at the main, shear and strengthening reinforcement were monitored, together with horizontal and vertical deflections.

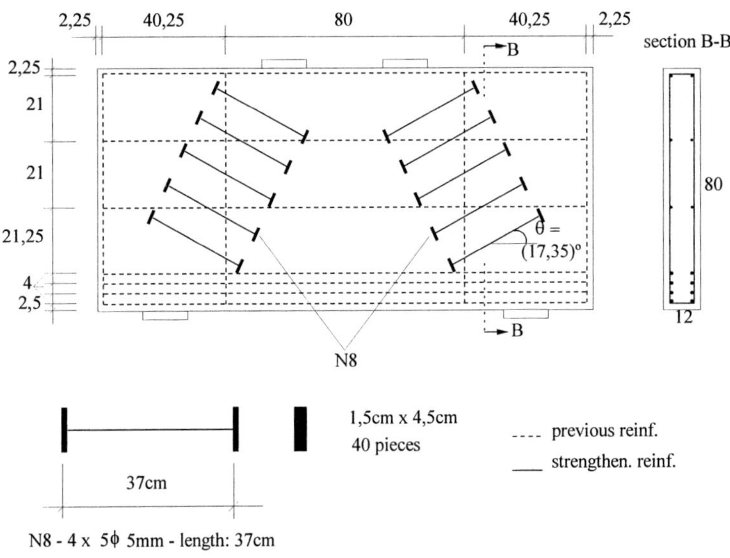

Figure 3 Beam 7/S strengthening reinforcement detail (cm)

RESULTS AND DISCUSSION

Figures 4 and 5 present total load x strain curves for the main flexural and the horizontal strengthening reinforcement, for beams 2/S, 4/S, 6/S and 2M, that have the same reinforcement pattern. They showed basically the same strains at the main bottom bar reinforcement until yielding. It was not possible to check the yielding of the main reinforcement of beam 4/S as the strain gauge did not work properly. The yield loading of beam 6/S was a little smaller than beams 2/S and 2M. This was possibly due to a partial contribution on the flexural resistance of the horizontal strengthening reinforcement of beam 6/S, positioned outside the vertical stirrups.

It is quite clear from Figure 5 that the horizontal stirrups positioned near the bottom edge were stressed in all beams before the stirrups at the top edge. It can be said that the stirrups contributed effectively for the strut resistance, and that they begun working for beams 6/S and 2M after the others, but were more stressed.

The flexural beam resistance is reached when the main reinforcement yields, presenting excessive plastic deformation, and collapse comes with crushing of the concrete at the top edge.

This mechanism happened for beam 7/S, before the ultimate shear resistance capacity was reached. Apart from beam 1 (monolithic), that failed by diagonal tension, without yielding the main reinforcement, and from beam 7/S, that failed by bending with yielding of the main reinforcement and crushing of the concrete, all the beams failed at the shear region, after the main bars had yielded. The rupture mechanism of these beams were defined as Diagonal Tension / Flexure (DT-F*).

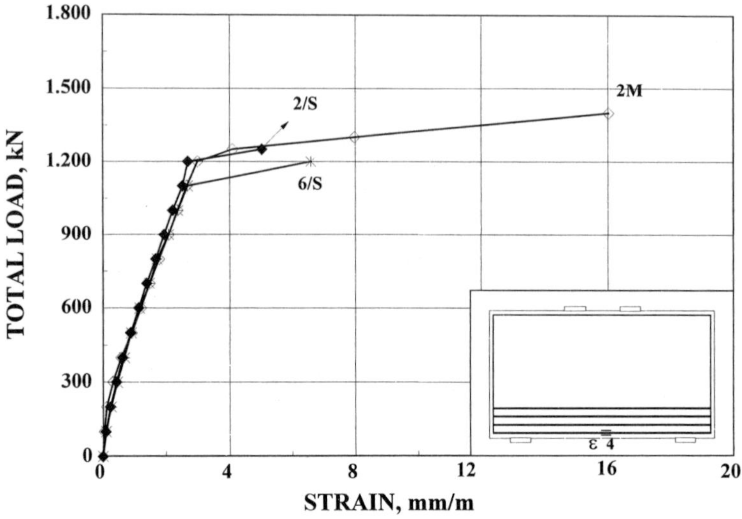

Figure 4 Total load v reinforcement strains for beams 2/S, 6/S and 2M

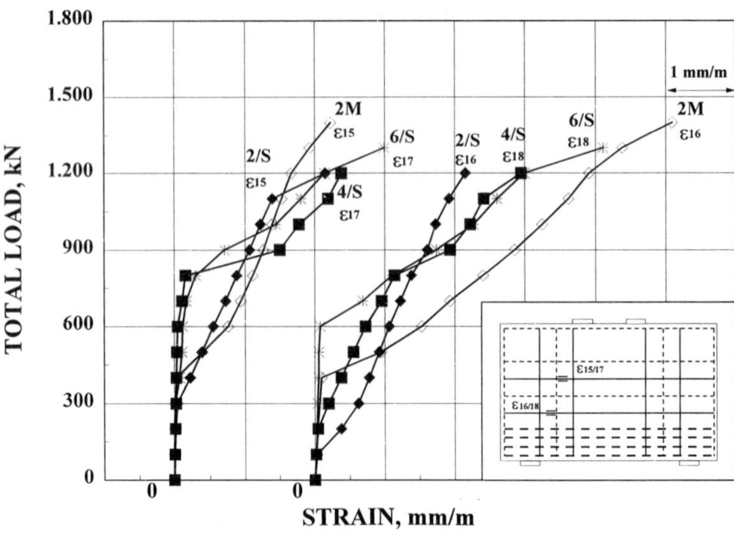

Figure 5 Total load v horizontal stirrups strains for beams 2/S, 4/S, 6/S and 2M

Nevertheless all the tested beams had the same main reinforcement ratio and disposition, it was observed in Figure 4 that beams 2/S, 6/S and 2M showed different yield loading. Table 2 presents, for each beam, the strength of the concrete and of the mortar used in the strengthening, the experimental ultimate load (P_u), and this load corrected by a “K” factor as a function of the resistance of beam 2M, defined as K = f_c (2M) / f_c (beam i), together with the type of rupture. It can be seen that positioning the reinforcement close to the top edge (3/S) increased 12% in the rupture load. Beams 2/S, 4/S and 6/S, with the same positioning for the strengthening horizontal reinforcement of beam 2/S, failed for loads corresponding respectively to 111%, 99% and 92% of the beam 2M (monolithic), showing that satisfactory levels were reached with the strengthening reinforcement. Beam 4/S, with the horizontal stirrups positioned outside the previous vertical ones, and beam 6/S, with the orthogonal mesh positioned without enveloping it, presented ultimate loads 14% smaller than beam 2/S. It can also be seen that the detailing of beam 4/S was not so effective as of the beam 2/S in ultimate load, despite that the previous was more effective in controlling crackwidths. The utilisation of a group of stirrups close to the bottom edge as web mesh (beam 8/S) allowed an increase of 18% in the shear rupture load, compared with beam 2M.

Table 2 Shear failure load, obtained experimentally, and estimated as function beam 2M

BEAM	$f_{c,a}$ (MPa)	f_c(MPa)	P_u(kN)	K	P_u(kN)	P_uK/P_uK (VP2M)	RUPTURE
1	-	33,3	850	1,72	1462,0	0,98	DT
2/S	49,2	46,4	1340	1,24	1661,6	1,11	DT-F*
3/S	57,4	49,9	1460	1,15	1679,0	1,12	DT-F*
4/S	57,1	50,4	1300	1,14	1482,0	0,99	DT-F*
5/S	66,6	57,5	1700	1,00	1700,0	1,14	DT-F*
6/S	53,9	55,0	1310	1,05	1375,5	0,92	DT-F*
7/S	65,7	57,5	*	*	*	*	Flexure
8/S	59,9	47,0	1450	1,22	1769,0	1,18	DT-F*
2M	-	57,5	1495	1,00	1495,0	1,00	DT-F*

$f_{c,a}$ – mortar comp. resist.; DT - Diagonal tension; DT-F* - Diag tension / yielding of the reinforcement;

Beam 5/S, with inclined strengthened reinforcement failed with a load 14% higher than the monolithic with orthogonal mesh (beam 2M). Nevertheless the shear critical crack appearing for a small load, the concentration and disposition of the inclined bars crossing the shear crack at mid level delayed its developing and controlled its crackwidths, delaying the formation of the second shear crack, that failed the beam in the end. Beam 7/S failed by flexure at 1760 kN, before reaching the ultimate shear load. The strengthening reinforcement increased the shear ultimate load in 17,7% in relation to beam 2M. Considering the types of strengthening reinforcement utilised in this study, beams 7/S, 5/S and 8/S presented, in this order, the best performance in rupture loads, compared with the monolithic beam 2M.

CONCLUSIONS

Reinforcement Strains and Deflections

1. Horizontal and group of horizontal stirrups, situated close to the bottom edge, in monolithic and in strengthened beams, enhanced the yielding load and consequently the flexural resistance. Vertical stirrups situated in the middle of the shear span were more efficient, compared with those situated close to the support or loading plate.

2. Inclined reinforcement is more efficient than orthogonal mesh in strengthening deep beams in shear, for the detailing studied here.

3. The presence of web reinforcement, as expected, reduced the vertical and horizontal deflections in deep beams. The strengthening orthogonal mesh, positioned closer to the bottom edge, inside the vertical stirrups, anchored as the main reinforcement and enveloping the beam, were more effective in reducing deflections.

Shear Resistance

1. Strengthened beams can reach ultimate loads even higher than monolithic beams, allowing also changing of rupture type to flexure.

2. Groups of horizontal stirrups close to the bottom edge and vertical stirrups in the middle of the shear span enhanced the shear ultimate loads and best predictions for these loads were found when the expression of Kong [6] was adopted. The ACI code [7] was conservative in estimating the shear failure load.

ACKNOWLEDGEMENTS

The authors would like to express their appreciation to CAPES (student scholarship), and to the Technical Control Department of FURNAS Centrais Elétricas, for the laboratory work.

REFERENCES

1. TENG,S, KONG,FK, POH,SP AND TAN, KH. Performance of strengthened concrete deep beams predamaged in shear. ACI Struct J, V.93, n°2, Mar-Apr 1996, p. 159-171.

2. JARDIM, LL. Experimental analysis of high resistance concrete deep beams strengthened in shear. M.Sc.Dissert, Univ Brasília, May, 1998, 217 p. (in portuguese).

3. BESSA, MAS. Experimental analysis of deep beams", M.Sc. Dissertation, University of Brasília, December 1994, 122 p. (in portuguese)

4. VANDERLEI, E. Reinforced concrete columns repaired with cross section recasting. M.Sc. Dissertation, University of Brasília, October 1996, 168 p. (in portuguese).

5. LEONHARDT, F. Poutres-Cloison: Structures planes chargées parallélement à leur plan moyen. Com Européen Béton, Bull d'inform, n°65, Paris, Fevrier 1968, pp. 1-113.

6. KONG, FK, ROBINS, PJ, SINGH, A, AND SHARP, GR. Shear-analysis and design of reinforced concrete deep beams. The Struct. Engineer, V.50, N 10, 1972, pp. 405-409.

7. AMERICAN CONCRETE INSTITUTE. Building code requirements for structural concrete (ACI 318-95) and commentary (ACI 318R-95). Detroit, 1995.

NON-DESTRUCTIVE INSPECTION METHOD FOR EVALUATING CONDITIONS OF REINFORCED CONCRETE BUILDINGS

G B Muravin **L M Lezvinsky**

Margan Physical Diagnostics

B Muravin

Tel-Aviv University

Israel

ABSTRACT. The paper describes a complexity of non-destructive test methods that the authors have designed, developed and used for examining reinforced concrete buildings. These include methods for the following:

- investigating damage development, detecting and pinpointing zones of visible and invisible defects, and discovering their dynamics;
- establishing whether the stress distribution in structural elements is uniform or not and what its absolute value is;
- determining the damage danger level in detected defects, the stability of the examined structural elements affected by hazardous influences and the stresses currently acting on them;
- rapidly measuring crack development dynamics and damage accumulation;
- detecting signs of concrete degradation in suspected zones.

Keywords: Reinforced concrete structures, Damage, Acoustic emission (AE) diagnosis, Evaluating condition, Stress relief, Stress measurements, Columns, Beams, Foundation.

Professor Gregory B Muravin is Chief Scientist of the Margan Physical Diagnostics, Israel. He specializes in acoustic emission and non-destructive testing of reinforced concrete, composite materials and metal structures, as well as in evaluating their condition by methods of Physics of Solids and Fracture Mechanics. He has published more than 190 articles and has patented more than 30 inventions.

Dr Ludmila M Lezvinsky is Senior Scientist of the Margan Physical Diagnostics, Israel. She specializes in the investigation of physical properties of concrete, composite materials and metal, and the prediction and physical modeling of their properties using AE data. She has authored more than 100 articles and inventions.

Mr Boris Muravin is a Postgraduate Student of Tel-Aviv University, Israel. He specializes in Applied Mathematics, which he uses in image recognition of defects in materials.

INTRODUCTION

According to the statistics of damage to civil and industrial buildings and their repair, the following are the main reasons for having to reconstruct buildings:

- instability of foundations, columns and supports connected to geological structures due to underground water activity, movement of soils, rocks and earth, and earthquakes;
- damage development such as the appearance and propagation of cracks due to local stress concentration zones, non-uniform distribution of strength in the material, cavities and inclusions, and non-uniform stress distribution in different structural elements;
- electrochemical and stress corrosion of steel reinforcement;
- fire, explosions, and natural cataclysms.

The authors have shown in several articles [1-8] the clear advantage of acoustic emission (AE) inspections in both standard and non-standard situations, for rapid and reliable diagnosis of different structures, including bridges, tunnels and refinery equipment. Let us examine the application of AE to diagnosing reinforced concrete buildings, excluding building damage due to natural cataclysms, fire, and explosions.

NON-DESTRUCTIVE TEST METHODOLOGY - PRINCIPAL REQUIREMENTS

The principal requirement of an effective non-destructive diagnostic system is the capacity:

- to achieve an integral evaluation of a structure's stability and to reveal defect zones in it;
- to select and measure those parameters that are necessary and sufficient to specify the type of damage and its danger levels.

In view of this, the following procedures in the suite of non-destructive test methods [1-8] that we have designed, developed and used for examining reinforced concrete buildings by the authors were included:

1. AE procedures:

   - to determine damage development, locate visible and invisible defective zones and also to discover the dynamics of the defects;
   - to identify damage types such as developing micro- and macro-cracks, local zones of stress concentration, decrease in concrete strength and electrochemical corrosion, and instability of columns, beams, supports and foundations;
   - to establish whether the stress distribution in structural elements is uniform or not and what its absolute value is;
   - to determine the damage danger level of detected defects, the current stress and how stable the examined structural elements are after hazardous influences;

2. AE and sclerometric tests of columns and mechanical tests of concrete specimens to determine the concrete strength and the uniformity of its distribution throughout the structure.

3. Laser investigation of the displacement of columns, beams, construction member joints and other elements from their initial positions.

4. The use of electron-fractography and X-ray tests to detect signs of concrete degradation in suspected zones.

5. The Image Recognition Method to determine the damage type and its danger level.

These were used these together with other fracture mechanics and physics of solids test methods to estimate the presence of damage, to evaluate the damage danger level and to assess the degree of operational risk. Acoustic emission systems for these tests were Physical Acoustic Corporation's Spartan and Mistras, and the Russian AWS-3 (Stress Wave Analyzer).

Characterisation of the structural state of the material according to the onset of micro-crack formation, the start of nonlinear creep or the development of main cracks and used the following statistical criteria of image recognition as the criteria of defect danger levels:

- micro-crack formation - requiring periodical monitoring - the third (lowest) level of risk;
- development of nonlinear creep and local fracture of adhesions between reinforcement and concrete - a critical process - the second level of risk;
- maximal danger of catastrophic failure occurring during the development of main power cracks in structural members, and/or mass fracture of adhesions between reinforcement and concrete - the first (highest) level of risk.

The ellipses of dispersion of the AE parameters were used to classify the damage development.

DIAGNOSIS OF BUILDING FOUNDATIONS

The building was 50 years old at the time of the AE examination. There were no faults in the structural elements nor were any cracks or other damage detected at its completion and when put into use. The first indications of damage to the building appeared two years before the AE test. At this stage, the construction department of the Municipality started to monitor crack openings and their lengths, periodically.

Analysis of the monitoring data confirmed the presence of seasonal crack openings as well as a tendency towards the development of cracks.

The principal aim of the work was to determine the state of the building. Initially the correlation between the calculated and the actual properties of the materials and the stress levels in structural elements and the stress levels in the structural elements was considered. Examination of the material's quality showed that the concrete strength in the structural elements was uniform and according to the design value.

Analysis revealed that the AE energy distribution around the building perimeter (Figure 1a) was not uniform. The maximal AE energy value was in zone "a". A main crack separating the building into two parts was located in the building wall above this zone (Figure 1c).

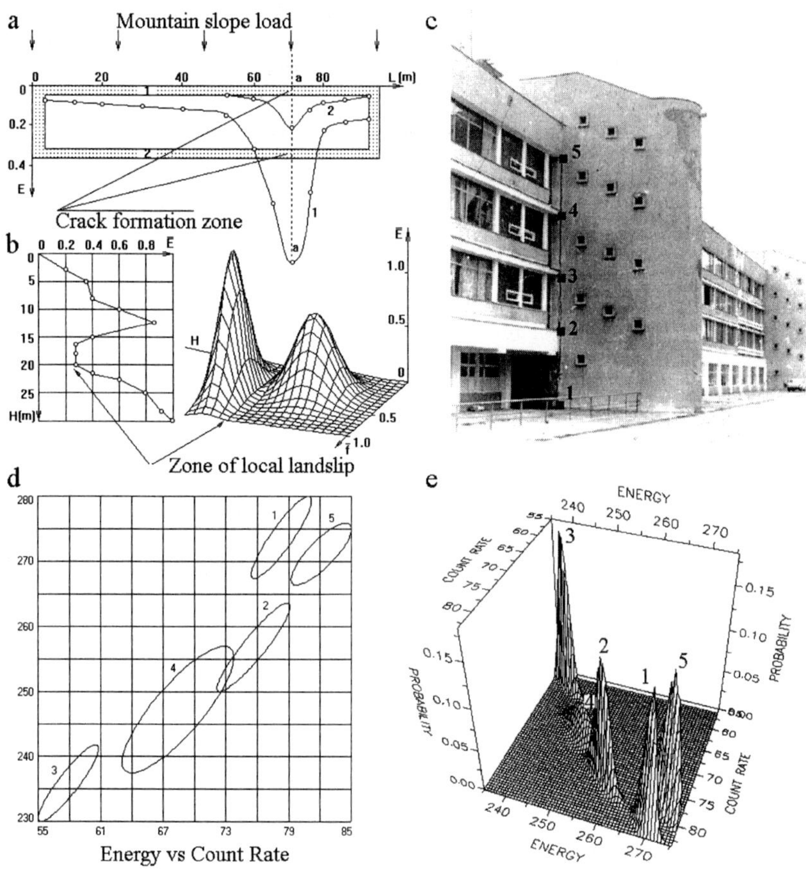

Figure 1 Results of acoustic emission tests

(a) Diagrams 1 & 2 - AE energy distribution along walls 1 & 2 respectively.
(b) AE energy and frequency at different drilling depths near point "a" on wall 1.
(c) The building examined.
(d) Ellipses of dispersion of AE pulses "average energy-average count rate" corresponding to the different locations of sensors on the building along the crack length (see "c").
(e) Probability density graphs of AE pulses "average energy-average count rate" corresponding to the different locations of sensors on the building along the crack (see c).

The distribution AE along the crack length (Figure 1c) was checked to determine the reason of its propagation, its dynamics and the direction of its elongation. It was established (Figure 1d) that maximal AE energy and AE count rate were at the crack's tips (Figure 1d, ellipses 1 and 5, and probability density graphs 1 and 5 respectively, Figure 1e). The minimal values of AE were recorded in zones 3 and 4 (Figure 1d, ellipses 3, 4, and probability density graphs 3 and 4 respectively, Figure 1e). The crack had maximal width in zones 3 and 4, indicating that the maximal intensity of crack development occurs near the building's foundation and the crack continues to elongate upwards, towards the top of the building. The crack almost stopped developing in zones 3 and 4, because the crack had achieved its maximal opening there.

It was supposed that the appearance of crack was connected with the underlying soil. To check this assumption, AE was measured at different depths under the foundations, with the help of special tubes driven into the ground (Figure 1b). This confirmed the asymmetry of the AE due to landslide at different ground depths around the foundation.

The suggestion was that the building foundations had been exposed to longitudinal shear caused by a local, hitherto unsuspected, landslide due to the construction of several buildings near the inspected building, without special reinforcement of the surrounding land. After defining the landslide plane in detail, it was reinforced and the building foundations were strengthened, after which the damage processes ceased.

DIAGNOSIS OF COLUMNS LOADING

Often at the time of inspection, data about the design load of the inspected building and its columns was not a variable. Nevertheless, it was reasonable to assume that the calculated design of the columns was for single axis compression and symmetrical loading. (the laser measurements and analysis of deformation did not establish any significant inclination of the columns from their initial position).

Having assumed single axis loading and load uniformity, it was expected that there would be no AE signals with energy higher than the level of concrete structural conversion energy. The AE inspection established that the distribution of AE signals recorded in the different supports was relatively uniform and did not exceed the level of micro-crack formation. This could indicate the absence of damage accumulation.

Several columns of the inspected building had significant AE indications of damage, such as local over-tension, and micro-crack development. Analysis of the AE signals' energy, amplitudes, count rate, frequency band and energy spectrum by the image recognition method indicated the developing propagation of individual non-correlated flaws (Figure 2a) [8].

The median frequencies of the signals and their relative energy levels were similar to those that we had observed experimentally in the case of developing micro-cracks, while testing reinforced concrete specimens in the laboratory and during field tests of other structures. Hence, it was concluded that the fault propagation in this case, too, was associated with micro-cracking in the concrete.

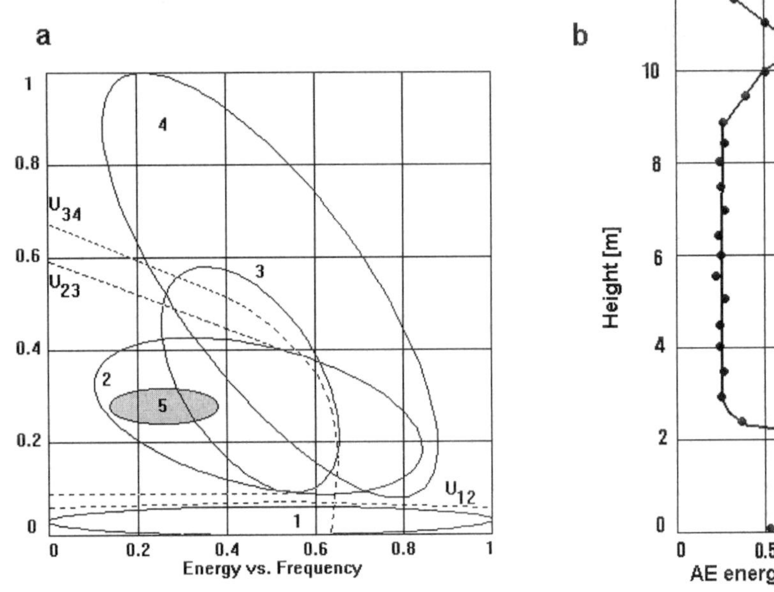

Figure 2

(a) The ellipses of dispersion of the AE pulses "average energy-average frequency" used to classify the damage development. 1- closing of primary micro cracks and pores; 2 - micro-crack formation; 3 - development of nonlinear creep; 4 - main cracks development; 5 - measurement data of unknown process.

(b) AE energy distribution along the column height.

Analysis of the location of zones with maximum stress concentration of columns(associated with maximal AE energy) is about 1.5 down from the ceiling and about 2 meters up from the floor (Figure 2b). The AE measurement data also show that the stresses are not uniformly distributed in the columns' cross-section. The maximal recorded values of AE energy were from the north side of the columns, where normal fracture micro-cracks were developing (Figure 2a, ellipse 5). The additional stresses from load redistribution here were tested by the AE dynamic method [1-2]. The measurements established that the stresses were about 105kg per cm^2, whereas in symmetrically loaded columns it was only 60kg per cm^2.

The inspection results and their evaluation lead us to affirm that the examined supports can be used on condition that preventative measures are taken against non-uniform displacement of supports.

ESTIMATING THE STATE OF REINFORCED CONCRETE BEAMS

Earlier [8] it was found that irreversible damage of bent beams and the danger of sudden, total failure of the system can be identified from the appearance of cracks in the compressive zone, the break of adhesive connections and the plastic yield of reinforcement. The AE indications of damage development due to electrochemical corrosion and the evaluation of failure danger level were described earlier in [5-7]. Here were shown that by the use of ellipses of dispersion of the AE signals "average energy-average frequency" (Figure 3a, and 3b), it is possible to distinguish between the three processes; corrosion product accumulation (ellipse 1), micro-crack development (ellipse 2) and macro-crack development (ellipse 3) Based on this information, 135 beams of industrial building were investigated under operational conditions. 20 beams were checked twice during "shutdown".

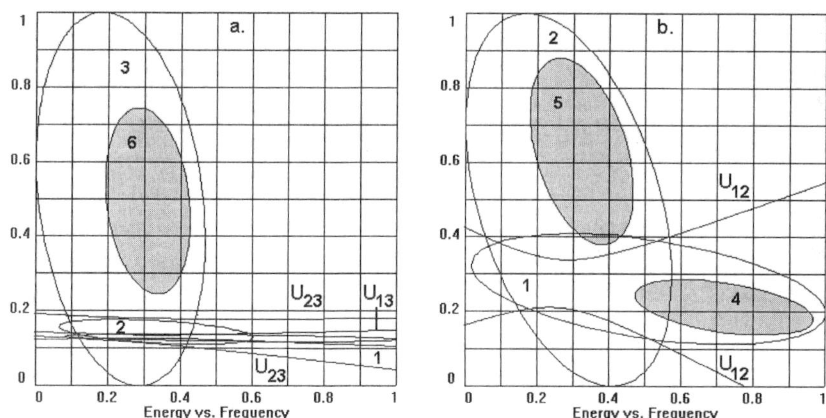

Figure 3 The example of the use of ellipses of dispersion of the AE signals "average energy-average frequency" to distinguish between the processes

The analysis of the AE data revealed differences in the energy, amplitudes, frequency, and other characteristics of the AE signals in the different beams of the structure. It was possible to arrange the results in three groups. In the first group we selected beams with low energy signals (up to 100 relative units) and probable frequency between 0.07-1.0 relative units. In this group, the AE flow was continuous with low amplitudes (not more than 40 dB). The authors have proved that this AE flow characteristics indicates local deformation and fracturing of the concrete in zones of weak adhesion

Our investigations of core samples from these reinforced concrete elements revealed the development of significant accumulations of corrosion products in them. Initially, these accumulations compressed the surrounding concrete. Finally, however, they shattered, producing many low-energy and low-amplitude AE pulses (Figure 3b, ellipse 4).

The pH measurement, X-ray analysis, scanning electron spectroscopy (SEM) observation and energy dispersive spectroscopy (EDýS) established that the pH values for concrete from suspected zones were more than 11, indicating that the condition of the concrete is normal. Slightly lower pH values found in the concrete from the damaged zone can be associated with early stages of $Ca(OH)_2$ decomposition. The ratio of the calcium phase to the silicon phase is much lower in "damaged" concrete than in "normal" concrete specimens. In the former, there is a higher density of micro-cracks. The micro-cracks occur mainly between the silicon-rich and the calcium-rich matrixes. These phenomena are the result of not applying the correct technology during the concrete casting.

Considering that the electrochemical corrosion process develops slowly, and individual micro-crack formation has a third - lowest level of risk, it is proposed to delay repair of the suspected beams to the last stage of reconstruction.

The existence of significant micro- and macro-crack formation and corrosion development was discovered in 10 reinforced concrete beams. Here the process of micro-crack development and the interaction of several micro-cracks together, is accompanied by signals with energy between 200 and 400 relative units and frequency of signal up to 0.6 relative units, and high correlation between AE average energy and AE average frequency. (Figure 3b, ellipse 5).

Since there were also indications that adhesive connections between reinforcement and concrete had fractured (a critical process - second level of risk), the repair of these elements in the second stage of reconstruction was proposed.

In the other reinforced concrete beams damage development was accompanied by the appearance of high energy AE signals, growing pulse amplitudes up 90 dB. The dispersion of the AE signals and correlation of AE characteristics increased substantially, too (Figure 3a, ellipse 6). All of this are associated with the fracture of bent elements: local crack development, broken adhesive connections between concrete and reinforcement, contraction of the compressive zone.

In the other reinforced concrete beams, tests revealed local crack development, broken adhesive connections between concrete and reinforcement, contraction of the compressive zone, all of which are associated with the fracture of bent elements.

Because methods for evaluating the durability and lifetime of reinforced concrete structures have not yet been adequately developed, and it is impossible to establish the operational risk level of structures with acceptable accuracy, we recommended the repair of those reinforced concrete beams was recommended as soon as possible.

CONCLUSION

The success in using the complex Acoustic Emission Non-Destructive Testing methods for examining reinforced concrete buildings was because they make it possible to establish reliably:

- the development of damage, the location of visible and invisible defective zones and also the dynamics of the defects;
- whether stress distribution in structural elements is uniform or not and what its absolute value is;
- the damage danger level of detected defects, and the present stress and the stability of the examined structural elements after hazardous influences.

REFERENCES

1. MURAVIN, G B, SIMKIN, YA V. ROZUMOVICH, E E, MERMAN, A I. Acoustic-Dynamic Inspection of the Stress State of Concrete.- Defektoskopiya, 1989, No. 12, pp 3-11.

2. MURAVIN, G B, ROSUMOVITCH, E E, MERMAN, A I, SIMKIN, Y V, LEZVINSKAYA, L M. Acoustic Emission Method of Determination of Stress Value in Reiforced Concrete Bridge Constructions. A. S. No 1632181. Declare No. 4732313 24.08.89 Registry date 01.11.90.

3. MURAVIN, G B, MERMAN, A I, LEZVINSKAYA, L M. Acoustic Emission Method of Estimation the Fracture Tougness of Concrete in Structures and Constructions. Defektoskopiya, 1991, No. 3, pp 10-16

4. MURAVIN, G B, LEZVINSKAYA, L M, MERMAN, A I, VOLKOV, S I, SNEZHNITSKY, YU S. Acoustic Emission Method of Determination Fracture Toughness of Materials in Reinforced Constructions. Positive decision on claim No. 4795584/33(024293). Date 27.02.90.

5. MURAVIN, G B, MAKAROVA, N O. Diagnostics of Process of Metal Corrosion in Concrete.- Defektoskopiya, 1991, No. 10, pp 22-29.

6. MURAVIN, G B, LEZVINSKAYA, L M, LEVITINA, I G, MAKAROVA, N O, VOLKOV, S I. Acoustic Emission Method of Determination Corrosion Damages in Materials of Constructions. Positive decision on claim No. 4806537 / 28 / 034808. Claim date 28.03.90.

7. MURAVIN, G B, LEZVINSKAYA, L M, MAKAROVA, N O, VOLKOV, S I. Acoustic Emission Method of Determination of Corrosion Damage Accumulation in Reinforced Concrete Constructions. Positive decision on claim 4832698 / 28 / 059917. Claim date 30.05.90.

8. MURAVIN, G B. State Estimation of Reinforced Concrete Beams by the Acoustic-Emission Method. - Progress in Acoustic Emission VII. Proceedings of the 12th International Acoustic Emission Symposium, Sapporo, Japan, October 17-20, pp 355-360, 1994, The Japanese Society for NDI.

MATERIALS FOR THE PROTECTION AND REPAIR OF CONCRETE: PROGRESS TOWARDS EUROPEAN STANDARDISATION

G C Mays

Cranfield University

United Kingdom

ABSTRACT. This paper provides an update on progress within Europe towards the standardisation of products and systems for the protection and repair of concrete. Included are surface protection products, concrete repair materials, structural bonding agents, crack injection systems and products for the anchoring of reinforcement. The general principles for the use of these products and systems were laid down in a prestandard (ENV 1504-09) in 1997. In future, specification standards will lay down the identification and performance requirements for a range of material properties and these, in turn, will be supplemented by a range of test method standards. The work is being co-ordinated within CEN/TC104/SC8. The paper identifies the significant progress made since the last Dundee International Congress in 1996.

Keywords: Concrete, Protection, Repair, Standardisation, Europe.

Professor Geoff Mays is Director of Civil Engineering and Dean of Faculty for Cranfield University at the Royal Military College of Science, Shrivenham, UK. He specialises in the repair and strengthening of concrete structures, with particular reference to the use of structural adhesives in bonded external reinforcement. He has been European Convenor of CEN/TC104/SC8 – Protection and Repair of Concrete – since 1996 and of its Working Group on Structural Bonding Products since 1990.

INTRODUCTION

For various reasons concrete structures may suffer damage or deterioration during their working lives and as a result may subsequently require repair. Alternatively, new or existing structures may be deemed to need protection from potential future damage or deterioration. The design of an appropriate repair or protection scheme is a complex process involving:

- assessment of the condition of the structure;
- identification of the causes of deterioration;
- deciding the objectives of protection and repair;
- selection of the appropriate principles for protection and repair;
- selection of methods;
- definition of properties of products and systems;
- specification of maintenance requirements following protection and repair.

The selection of products and systems for the protection and repair of concrete structures requires consideration of all of the above factors in order to ensure that the materials are appropriate for the intended use. New packages of European Standards are currently being developed to provide guidance in this selection process.

EUROPEAN CONCRETE REPAIR STANDARDS

European Standards for the protection and repair of concrete are being drafted by Sub-Committee 8 (SC8) of CEN Technical Committee 104, the secretariat being provided by the Association Française de Normalisation (AFNOR). The mirror organisation in the UK is BSI Committee B/517/8 and its members are playing a key role in developing the standards. The basic structure and responsibilities within SC8 for preparing the Standard pr EN 1504 – Products and systems for the protection and repair of concrete structures – is summarised in Table 1.

Each of the key product performance requirements of the relevant materials will be embodied within the "Specification Standards" pr EN 1504 Parts 2 to 6, respectively. The Working Groups are now developing "Test Method Standards" to evaluate material characteristics against those requirements. Over 60 new test method standards are currently within the work programme of SC8.

ENV 1504 Part 9, which defines the general principles for the use of products and systems for the protection of repair of concrete was published as a "Draft for Development" in 1997. Its sequel, pr EN 1504 Part 10 dealing with site application and quality control, is at an advanced stage of preparation and is also expected to be published as a "Draft for Development".

Table 1 Structure of CEN/TC104/SC8 and pr EN 1504

STANDARD	ACTIVITY	STATUS	GROUP	CHAIRMAN
EN 1504 Part 1[1]	General scope and definitions	Published 1998	SC8	Prof GC Mays[(1)] (UK)
pr EN 1504 Part 2	Surface protection	In preparation	WG1[(2)]	Dr R Stenner (Germany)
pr EN 1504 Part 3	Structural and non-structural repair	In preparation	WG2	Mr JDN Shaw (UK)
pr EN 1504 Part 4	Structural bonding	In preparation	WG3	Prof GC Mays (UK)
pr EN 1504 Part 5	Concrete injection	In preparation	WG4	Mr J Wiertz[(3)] (Belgium)
pr EN 1504 Part 6	Grouting to anchor reinforcement or to fill external voids	In preparation	TG[(4)] within WG2	-
pr EN 1504 Part 7	Reinforcement corrosion prevention	In preparation	WG7	Prof HR Sasse (Germany)
pr EN 1504 Part 8	Quality control and evaluation of conformity	In preparation	Ad-hoc Group	Dr H Davies (UK)
pr EN 1504 Part 9[2]	General principles for use of products and systems	Published 1997	WG7	Prof HR Sasse (Germany)
pr EN 1504 Part 10	Site application of products and systems and quality control of the works	In preparation	WG9	Mr F Dyton (UK)

Notes
(1) Formerly Mme A M Paillere (France)
(2) WG = Working Group
(3) Formerly Mme Y de Vinzelles (France)
(4) TG = Task Group

GENERAL PRINCIPLES FOR USE AND SITE APPLICATION

The scope of ENV 1504 Part 9 includes:

- the need for inspection, testing and assessment before, during and after repair;
- protection from and repair of defects caused by the influence of certain environments and chemical substances;
- the repair of defects from such causes as mechanical damage, differential settlement, loading, biological attack, inadequate construction or the use of unsuitable construction materials;
- protection and repair in order to decrease the progress of alkali-silica reaction;
- meeting the required structural capacity in repair by;
  - replacement or addition of embedded or external reinforcement;
  - filling of external voids between elements to ensure structural continuity;
- meeting the required structural capacity by replacement or addition of concrete;
- waterproofing as an integral part of protection and repair;
- protection and repair of pavements, runways, hard standings and floors, as an integral part of protection and repair;
- methods of protection and repair including:
  - treating cracks;
  - restoring passivity to reinforcement;
  - reducing the rate of corrosion of reinforcement by limiting moisture content;
  - reducing the rate of corrosion of reinforcement by electrochemical methods;
  - controlling corrosion of reinforcement with coatings.

The basis for the choice of products and systems is founded on 11 principles of protection and repair. These are based on the chemical and physical laws, which allow prevention or stabilisation of the chemical or physical deterioration processes in the concrete or the electrochemical corrosion processes on the steel surface.

These 11 principles, and associated methods of protection and repair covered by pr EN 1504, are summarised in Table 2 and 3 for defects in concrete and for reinforcement corrosion, respectively. The standard then tabulates those properties of products and systems which need to be considered in relation to each of the protection and repair methods that have been identified.

Whereas Part 9 is concerned with principles of use of products and systems, Part 10 deals with site application and quality control. As such its scope may be defined as:

- the preparation of the concrete or reinforcement before application of products and systems
- the minimum requirements as to environmental conditions for storage and application of products and systems
- controlling the quality of the repair work

The final draft of this Standard is expected to be submitted to CEN enquiry in 1999.

Table 2 Principles and Methods related to defects in concrete

	PRINCIPLE	DEFINITION	METHODS OF PROTECTION AND REPAIR
1	Protection against ingress	Reducing or preventing the ingress of adverse agents	Impregnation Surface coating Filling cracks
2	Moisture control	Adjusting and maintaining the moisture content in the concrete within a specified range of values	Hydrophobic impregnation Surface coating
3	Concrete restoration	Restoring the original concrete of an element of the structure to the originally specified shape and function	Applying mortar by hand Recasting with concrete Spraying concrete or mortar
4	Structural strengthening	Increasing or restoring the structural load bearing capacity of an element of the concrete structure	Installing bonded rebars Plate bonding Adding mortar or concrete Injecting cracks, voids or interstices Filling cracks, voids or interstices
5	Physical resistance	Increasing resistance to physical or mechanical attack	Overlays or coatings Impregnation
6	Resistance to chemicals	Increasing resistance of the concrete surface to deterioration by chemical attack	Overlays or coatings Impregnation

Table 3 Principles and Methods related to reinforcement corrosion

	PRINCIPLES	DEFINITION	METHODS OF PROTECTION AND REPAIR
7	Preserving or restoring passivity	Creating chemical conditions in which the surface of the reinforcement is maintained in or is returned to a passive condition	Increasing cover to reinforcement Replacing contaminated or carbonated concrete Realkalisation of carbonated concrete by diffusion
8	Increasing resistivity	Increasing the electrical resistivity of the concrete	Limiting moisture content by surface treatments, coating or sheltering
9	Cathodic control	Creating conditions in which potentially cathodic areas of reinforcement are unable to drive an anodic reaction	Limiting oxygen content by saturation or surface coating
10	Cathodic protection [(1)]		
11	Control of anodic areas	Creating conditions in which potentially anodic areas of reinforcement are unable to take part in the corrosion reaction	Painting reinforcement with coatings containing active pigments Painting reinforcement with barrier coatings Applying inhibitors to the concrete

Notes
(1) To be covered by pr EN 12696-1

SPECIFICATION STANDARDS

The Specification Standards pr EN 1504 Parts 2 to 6 will identify the performance characteristics of the particular family of products and systems, which are associated with the principles and methods of protection and repair described in Part 9. For example, Table 4, reproduced from pr EN 1504 Part 3, lists the performance characteristics for structural bonding agents associated with the principle "Structural Strengthening" (see Principle 4, Table 2) for each of the repair methods "Bonded plate reinforcement" and "Bonded mortar or concrete". Further tables in the Specification Standards then define the "Identification" and "Performance" requirements and the associated Test Method Standards.

An identification test is carried out on a component material, product or system to confirm to the user the identity, consistency and fundamental physical characteristics of the material under test. The requirements are usually stated in terms of the property falling within a specified percentage of the value provided by the manufacturer. A performance test is undertaken to demonstrate the adequacy and the durability of the product or system's performance to the user. The requirements are usually stated in terms of the property being less than or greater than a particular specified value. Examples of how the identification requirement for "pot life" and the performance requirement for "adhesion" of structural bonding agents are specified are given in Table 5.

TEST METHOD STANDARDS

pr EN 1504 currently makes reference to over 90 test method standards, of which over 60 are new methods being drafted by SC8. In most cases these are based upon proven existing techniques; in other cases further research has been, or will be, necessary before test methods for standardisation purposes can be recommended.

At the time of writing, 25 of the new test methods have been submitted to the CEN enquiry stage and a further 15 to the final CEN formal vote. However, it is not appropriate in a paper of this nature to enter into details of the proposed test methods.

EC MANDATE

The Standing Committee on Construction of the European Commission (EC) is currently preparing a "Mandate" for "Concrete, mortar, grout and related products", the latter including concrete protection and repair products. A mandate is a political request from the EC, as agreed upon by the Member States, addressed to CEN, in support of legislative work or an industrial policy action from the EC. Product mandates lead to the development of "Harmonised Standards" in support of "Essential Requirements" which allow the CE marking of the products. For concrete protection and repair products the harmonised standards will be the EN 1504 series. The essential requirements of the EC are laid down in the "Construction Products Directive" [3] and are summarised in Table 6.

The procedure for the preparation of a mandate involves a number of steps:

1) preparation of the first draft by the EC after consultation with experts (including CEN)

2) written consultation with the Members States

3) amended draft and proposal on "attestation of conformity" sent to CEN for information and comment

4) approval of the mandate and the proposal on attestation of conformity

5) issue of the mandate to CEN for preparation of the work programme

6) acceptance of the work programme by the EC and acceptance of the mandate by CEN

Table 4 Performance characteristics for principle and repair method

MAIN PERFORMANCE CHARACTERISTICS	REPAIR METHOD	
	STRUCTURAL STRENGTHENING	
	BONDED PLATE (NOTE 1)	BONDED MORTAR OR CONCRETE (NOTE 2)
1 Suitability for application:		
a) to vertical surfaces & soffits	■	■
b) to top horizontal surfaces	■	■
c) by injection	■	■
2 Suitability for application and curing under the following special environmental conditions:		
a) low or high temperature	□	□
b) wet substrate		□
3 Adhesion:		
a) plate to plate	■	
b) plate to concrete	■	
c) corrosion protected steel to corrosion protected steel	□	
d) corrosion protected steel to concrete	□	
e) hardened concrete to hardened concrete		■
f) fresh concrete to hardened concrete		■
4 Durability of composite system:		
a) thermal cycling	■	■
b) moisture cycling	■	■
c) freeze-thaw	□	□
5 Long term behaviour under mechanical load:		
(i) Fatigue under dynamic loading		
(a) during curing	□	
(b) in service	□	
(ii) Creep under sustained loading in service	□	■
6 Material characteristics for the designer:		
a) modulus of elasticity in flexure)	□	□
b) compressive strength) at min		□
c) shear strength) st'd,	■	■
d) open time) & max	■	■
e) workable life) service) temps	■	■
f) modulus of elasticity in compression)	■	■
g) glass transition temperature	□	□
h) coefficient of thermal expansion	□	□
i) shrinkage	□	□

■ = a material characteristic which shall be considered for all intended uses

□ = a material characteristic which shall be considered for certain intended uses

Table 5 Examples of a) identification and b) performance requirements

a) Identification Requirements

PROPERTY	TEST METHOD	REQUIREMENTS
Pot life	pr EN 29514	Within ± 20% of the value provided by the manufacturer

b) Performance Requirements

Principle: Structural Strengthening
Repair Method: Bonded Plate Reinforcement

PROPERTY	REFERENCE CONCRETE OR MORTAR	TEST METHOD	REQUIREMENTS
Adhesion	pr EN 1766 MC (0.40)	pr EN 12188	The tensile stress carried by the bonded joint in a pull off test shall not be less than 15 MPa. The slant shear strength of scarf-jointed prisms tested in compression at various angles θ shall not be less than the values x MPa tabulated below. θ — x(MPa) 50° — 50 60° — 60 70° — 70

At the time of writing (July 1998) the concrete mandate has reached step 3 but has not yet been approved by the Standing Committee on Construction. A number of performance requirements for concrete protection and repair products which are related to the essential requirements have been proposed. In addition, the current proposals for attestation of conformity comprise:

- for uses with low performance requirements in buildings and civil engineering works – System 4 (see Table 7)
- for other uses in buildings and civil engineering works – System 2+ (see Table 7)

Table 6 Essential requirements of Annex I of the Construction Products Directive (CPD)

1	Mechanical resistance and stability
2	Safety in case of fire
3	Hygiene, health and the environment
4	Safety in use
5	Protection against noise
6	Energy economy and heat retention

Table 7 Systems of Attestation of Conformity under the CPD

ANNEX III OF CPD	2(I)		2(II) FIRST OPTION		2(II) SECOND OPTION	2(II) THIRD OPTION
Conformity attestation: CEC numbering system	1+	1	2+	2	3	4
a) Tasks for the manufacturer						
1 Factory production control	✓	✓	✓	✓	✓	✓
2 Further testing of samples taken at factory according to prescribed test plan	✓	✓	✓	-	-	-
3 Initial type testing	-	-	✓	✓	-	✓
b) Tasks for the approved body						
4 Initial type testing	✓	✓	-	-	✓	-
5 Certification of FPC	✓	✓	✓	✓	-	-
6 Surveillance of FPC	✓	✓	✓	-	-	-
7 Audit testing of samples taken from the factory on the market or on construction sites	✓	-	-	-	-	-

2(i) = certification of conformity of the product by an approved certification body

2(ii) = declaration of conformity by the manufacturer

FPC = factory production control

CONCLUDING REMARKS

1. Significant progress has been made within Europe towards the standardisation of products and systems for the protection and repair of concrete since the last Dundee International Congress [4].

2. The development of a coherent set of specification and test method standards within the CEN framework and applicable across the European Union will potentially benefit all branches of the concrete repair industry. The whole industry will be working from a common set of tests and hence will be armed with comparable data. This should result in economic benefits because of the reduced need for multiple-testing and certification procedures.

3. Concrete protection and repair products are to be included within the EC mandate for concrete, mortar, grout and related products. This will lead to the development of harmonised standards in support of the essential requirements of the Construction Products Directive and allow the CE marking of these products.

REFERENCES

1. BRITISH STANDARDS INSTITUTION. BS EN 1504-1: 1998. Products and systems for the protection and repair of concrete structures – Definitions, requirements, quality control and evaluation of conformity – Part 1: Definitions.

2. BRITISH STANDARDS INSTITUTION. DD ENV 1504-9: 1997. Products and systems for the protection and repair of concrete structures – Definitions, requirements, quality control and evaluation of conformity – Part 9: General principles for the use of products and systems.

3. COUNCIL OF THE EUROPEAN COMMUNITY. The approximation of laws, regulations and administrative provisions of Member States relating to construction products. Council Directive 89/106 EEC, 21 December 1988.

4. MAYS, G C. European standardisation of materials for the protection and repair of concrete: Structural bonding agents. Proceedings of the International Congress on Concrete in the Service of Mankind, Dundee, 24-28 June 1996.

EVALUATION AND TESTING OF BOND STRENGTH BETWEEN ORDINARY AND EXPANSIVE CONCRETE

A Halicka

M Król

Technical University of Lublin

Poland

ABSTRACT. The adhesion between two contacting concretes is basic for repairs and strengthening of concrete structures. The parameter representing the adhesion is the bond strength. In the paper the theoretical models and samples usually used for testing of the bond strength are compiled.

The authors also present their own investigations of the bond strength carried out with the original "push off" type samples and the "patch test" type samples. The aim of the investigations was to estimate the bond strength in the contact zone between the expansive concrete and the ordinary one. The concrete based on the expansive cement produced in the Technical University of Lublin is used by authors for repairs and strengthening of concrete structures and shaping of the concrete composite structures.

The investigations proved that the bond between expansive concrete and ordinary old concrete is higher then the bond between two ordinary concretes both in the shear joints and in the "patch" repairs.

The obtained results recommend expansive concrete as an effective repair material. They also confirmed the usefulness of presented methods for concrete bond testing,

Keywords: Repairs of concrete structures, Bond strength, Methods of bond strength testing, Expansive concrete

Dr Anna Halicka is a Lecturer in Building Structures Department of Technical University of Lublin, Poland. She deals with methods of restoration of concrete structures and concrete composite structures.

Professor Mieczyslaw Król is Head of Building Structures Department of Technical University of Lublin, Poland. He is an expert in concrete structures and special concrete technology. He specialises in concrete based on expansive cement and their applications.

INTRODUCTION

Bond Strength in the Contact Zones of Two Concretes and the Models of Its Testing

The problem of the contact and co-operation of two different concretes appears most frequently in the following situations:

- Technological break while placing of the concrete,
- Repairs and strengthenings of structures (replacement of concrete in destroyed regions and casting new layer over existing concrete element),
- Interaction of the elements of the composite structures.

The physico - mechanical features of these two concretes should be similar (especially their compressive strength, deformability, durability). These concretes can be called: „basic concrete" (in the case of the repair it is repaired concrete, and it is a prefabricated element in the case of concrete composite structure) and „complementary concrete". The load capacity of the structural element made of two different concretes depends on their co-operation. The measure of this co-operation can be specified by bond strength. If the bond strength is too low, the failure of the composite or repaired element could occur in the joint surface instead of the failure in the most loaded cross - section. Moreover, if the bond strength is law, the tensile strength in the joint could be reached, and the separation of concretes could occur, as a reason of the shrinkage of complementary concrete.

The standardised or individually designed samples are used for bond testing. All types of samples can be divided into five groups according to the type of stress carried by the joint:

- the samples with tension in the joint,
- the samples with shear in the joint,
- the samples with shear and compression in the joint,
- the sample with torsion in the joint,
- others.

These samples are compiled in the Table 1.

The Essence and the Features of the Expansive Concrete

The main weakness of the Portland cement concrete is the shrinkage. The shrinkage is undesirable especially in the repairs of the structure, because it can cause the separation of the repaired zone and the original structure. Moreover, in the loaded repaired elements the shrinkage makes the redistribution of the internal forces impossible.

The concrete based on the expansive cement produced on the semi-technical scale in Technical University of Lublin is better in this respect. This expansive cement is type M cement (according to the specialistic terminology). It is produced of portland cement clinker, aluminous cement and gypsum.

The basic property of the expansive concrete, distinguishing it from the ordinary one is an increase of the volume and density. It is caused mainly by the formation of the ettryngite during the hydration. The increase of volume does not generate the density decrease (as in the case of gas entraining agents), but it causes the tightening of concrete. The shrinkage of the expansive concrete exists, but the shrinkage stress is about ten times lower then expansion one. During the curing time of such a concrete the full compensation of the shrinkage stress and unimportant decrease of the expansion stress can be noticed.

The maximum of the volume increase takes place, when the expansion is unconfined. It is so called free expansion. The free expansion strains can be ε_{CE} = 0,1% ÷ 1,0% and even more. The magnitude of strains depends on the power of the expansive cement. When the expansion is confined, the strains are smaller - the more rigid boundaries the smaller expansion strains. In this case the expansion causes the self-stress in the expansive concrete. The self-stress exerts pressure on the interface actively changing the initial distribution of the internal forces in the surroundings [3].

Table 1 The samples usually used for bond testing

STRESS IN THE JOINT	THE SHAPE OF THE SAMPLE
Tension	Direct Tension; Tension by Splitting; "Pull-off" Test; Tension in Flexure 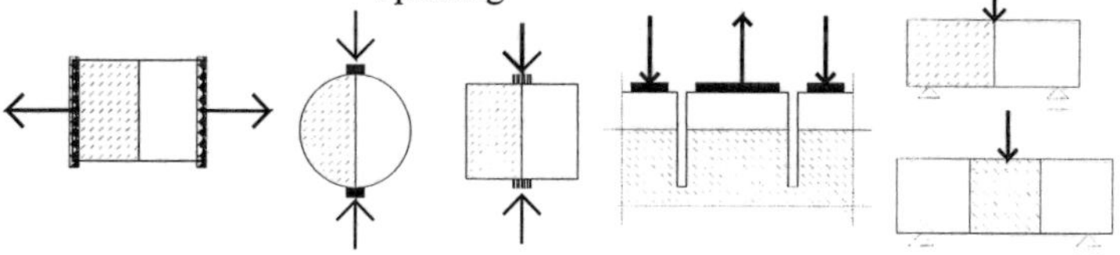
Shear or Cutting	Shear; Cutting 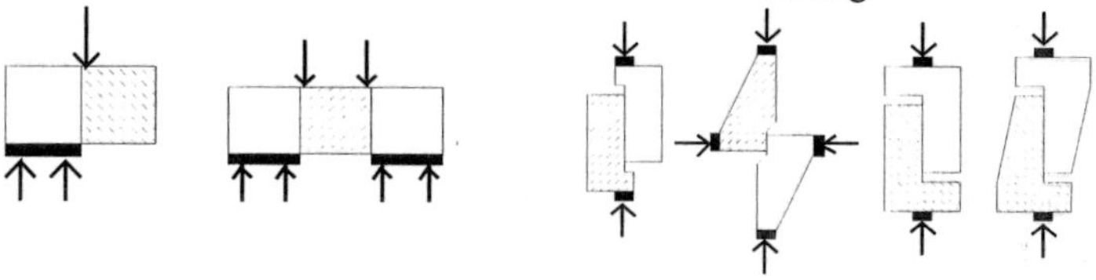
Shear and Compression	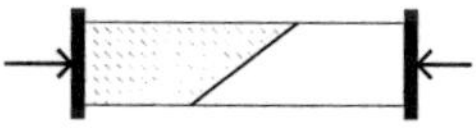
Torsion	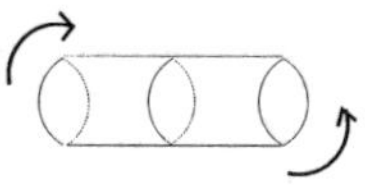
Indirectly Loaded Joint in "Patch Test" [2]	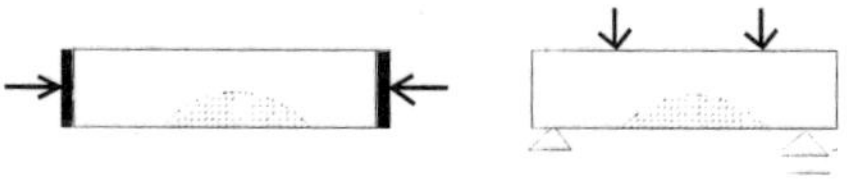

Factors Influencing the Bond Strength of the Expansive Concrete

According to the rules of estimation of the shear resistance of joints, as presented in the EC2 [1], the bond strength of not reinforced joints consists of two factors: adhesion (all physico-chemical phenomena in the contact zone) and friction:

$$\tau_{Rdj} = k_T \cdot \tau_{Rd} + \mu \cdot \sigma_N$$

The mechanical adhesion is recognised as the predominant physico-chemical phenomenon. It means that two concretes are interconnected by indentation (meshing). The adhesion factor is counted as the product of the coefficient of surface roughness k_T and shear strength τ_{Rd}. The friction factor is counted as the product of the coefficient of shear friction μ and the stress of external normal force σ_N..

The utilisation of the expansive concrete considerably improves both of these factors. The crystallised ettryngite penetrates into pores and irregularities of the surface of the basic concrete resulting in meshing them [4]. The second factor improving the bond strength is the self-stress. The basic concrete constitutes a boundary for the expansive concrete. It causes the self-stress exerting pressure on the interface and increase of frictional component.

The Aim of the Presented Research

The authors analysed a lot of results of the bond strength investigations made with different types of samples compiled in the Table 1. It was found that the results are scattered. They depend not only on the type of the sample, but also on the technique and the course of testing. Moreover it was found that those samples are not suitable for testing of the expansive concrete bond strength.

The aim of the taken out research was working out the all-purpose method of testing of the bond strength between two different concretes. In this paper this original „push off" type method is presented. The adaptation of the „patch test" for testing of expansive concrete bond strength is also proposed.

Below the results of the investigations of the bond strength between two different concretes are presented. The results can prove that the expansive concrete is good as a repair material and as a complementary concrete in composite structures.

THE INVESTIGATIONS OF THE SHEAR RESISTANCE IN THE JOINT OF TWO DIFFERENT CONCRETES

The Test Model

The test model designed by the authors is "push off" type model. The flat interface between two concretes was replaced here by the cylindrical one. The cubical 200 x 200 x 200 mm concrete sample was moulded with cylindrical hole 110 mm diameter. After 28 days of curing this hole was filled with complementary concrete (ordinary or expansive one). After curing of the complementary concrete (7, 14 or 28 days) the investigations were carried out. The test consisted in displacement of the complementary concrete in the relation to the basic concrete. The maximum force, causing the strip of the bond was a measure of the shear resistance. The scheme of test is presented on Figure 1.

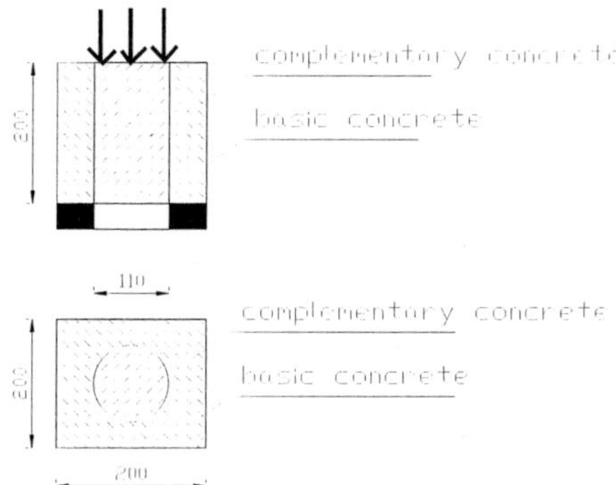

Figure 1 The scheme of loaded "push off" type sample of the shear resistance testing

Table 2 Mix proportions and compressive strength of concretes

TYPE OF TEST	CONCRETE	MIX PROPORTIONS, kg / m^3					AVERAGE COMPRESSIVE STRENGTH, MPa		
		Cement		Sand	Aggre-gate	Water	At 7 days	At 14 days	At 28 days
		PC	Exp.						
Push off	Basic Concrete	434	-	466	1283	200	-	-	41.83
	Ordinary Complementary Concrete	434	-	466	1283	200	26.37	35.00	38.49
	Expansive Complementary Concrete	-	600*	600	950	215	41.32	48.00	50.33
Patch Test	Basic Concrete	302	-	696	1148	200	-	-	36,8
	Ordinary Complementary Concrete	302	-	696	1148	200	-	-	36,3
	Expansive Complementary Concrete	-	600*	600	950	215	-	-	48,1

*) the amount of expansive cement is caused by the needed self-stress value

This test model is characterised by:

- complying with the requirement of the symmetrical position in the testing machine,
- increase of shear surface in comparison with the traditional samples and reduction of the possibility of local fluctuations of interface features,
- regular distribution of shrinkage and expansion stress on the interface,
- possibility of testing of the expansive concrete and estimation of the influence of the pressure caused by the expansion.

The Used Materials

Basic concrete was made with an ordinary portland cement. Complementary concrete was different in two series of test - ordinary concrete or expansive concrete. Mix proportions and compressive strength of 150 x 150 x 150 mm cubes are presented in the Table 2. Self-stress of the expansive concrete tested in the standardised dynamometer devices was 2,0 MPa.

The Results of the Investigations

The shear resistance of joint between complementary concrete and basic concrete was determined by the force causing the strip in the interface and displacement of the complementary concrete. The value of the bond strength τ_j was counted as quotient of the value of this force and the interface area. Then the results were compared to the compressive strength of the complementary concrete f_c and the index β was counted:

$$\beta = \frac{\tau_j}{f_c}.$$

The results were characterised by index of dispersion less ten 4%. Arithmetic average values of the results of two test series are presented in the Table 3 and on the Figure 2.

Table 3 The results of the "push off" type test

	FORCE , kN			BOND STRENGTH, MPA			INDEX β		
Age of Test	7 days	14 days	28 days	7 days	14 days	28 days	7 days	14 days	28 days
Ordinary complementary concrete	82,10	95,79	100,76	1,32	1,54	1,62	0,05	0,044	0,040
Expansive complementary concrete	227,65	430,42	440,38	3,66	6,92	7,08	0,08	0,14	0,14

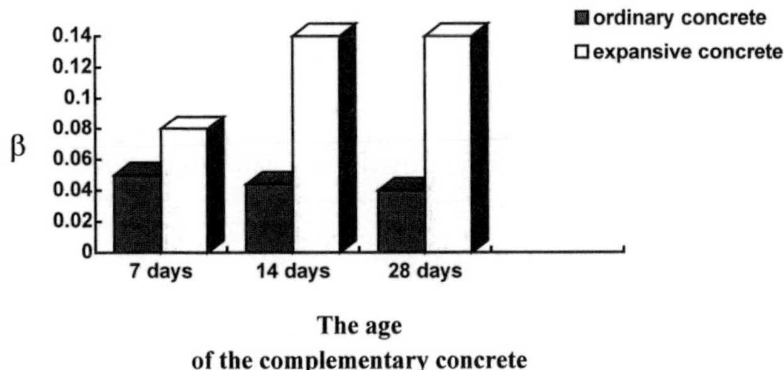

Figure 2 The value of index β characterising bond strength of ordinary and expansive complementary concrete during time of its curing

THE "PATCH" TYPE TEST OF LOAD CAPACITY OF THE JOINT

The Test Model

The samples modelled by "patch test" proposed in [2] were used. There were 100 x 100 x 400 mm concrete prismatic samples with trapezoidal incisions. After 28 days of curing these incisions were filled with ordinary or expansive complementary concrete. After 28 days of curing of complementary concrete these samples were tested by compression. The scheme of test sample is presented on the Figure 3. In order to estimate the effectiveness of repair, tests of comparative samples without incisions were carried out.

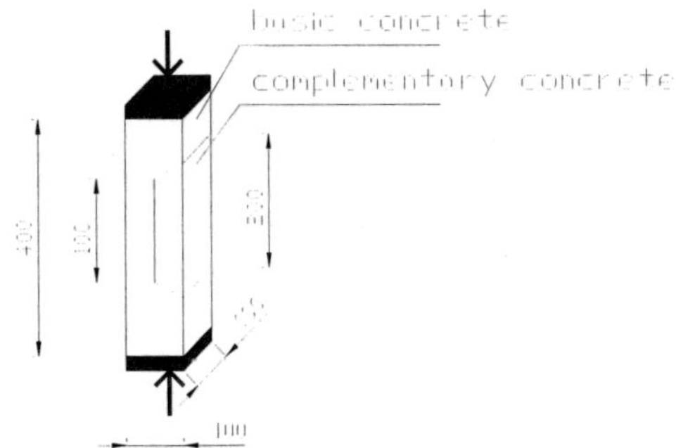

Figure 3 The scheme of loaded sample modelled by "patch test" [2]

The Used Materials

Basic concrete was made with ordinary portland cement. Complementary concrete was different in two series of test - ordinary concrete or expansive concrete.

Mix proportions and compressive strength of 150 x 150 x 150 mm cubes are presented in the Table 2. Self-stress of the expansive concrete tested in the standardised dynamometer devices was 1,95 MPa.

Table 4 The load capacity of "patch test" type samples

	THE AVERAGE VALUE OF LOAD CAPACITY, kN	EFFECTIVENESS OF REPAIR S
Comparative sample	353,3	100%
Ordinary complementary concrete	257,0	72%
Expansive complementary concrete	285,0	81%

The Results of the Investigations

The average values of results obtained in two series of testing are presented in the Table 4 and on the Figure 4. The effectiveness of the repair S was defined as the ratio of the load capacity of the repaired sample P_r and the load capacity of the comparative sample P_0:

$$S = \frac{P_r}{P_0}.$$

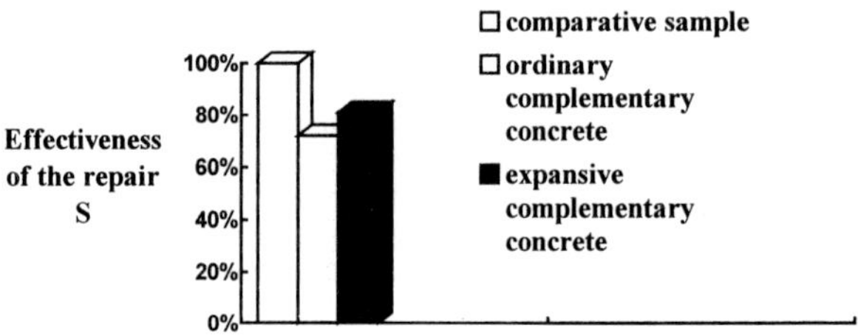

Figure 4 Effectiveness of "patch" repairs for different kinds of complementary concretes

CONCLUSIONS

Based on the results of the investigations it was found that the bond strength of the expansive concrete used as complementary concrete is higher than the bond strength of the ordinary one. It is especially visible in the original „push off" test. The ratio of shear resistance to compressive strength (index β) was three times higher in the case of expansive complementary concrete in comparison to ordinary one. Similarly in the „patch test" the effectiveness of the repair made of the expansive concrete was higher then repair made of the ordinary one, but the increase was only about 10%. It is probably the result of not rigid enough confines of strains during the curing and impossibility of arising of full self-stress.

The conclusion of the investigations is recommendation of the expansive concrete as the effective repair material and the complementary concrete in composite structures. The results confirm also that the proposed "push off" type method of testing is suitable for testing of bond strength of expansive concrete. The second "patch test" type samples should be improved by changing their dimensions.

REFERENCES

1. ENV 1992-1-3. Eurocode 2. Precast Concrete Elements and Structures - Basis of Design, pp.715-716

2. AUSTIN S A, ROBINS P.J. Development of patch test to study behaviour of shallow concrete patch repairs, Magazine of Concrete Research, Nr 164 - Sept.1993, pp. 221-229

3. KRÓL M, HALICKA A. Strategy of restoration of concrete structures with active compatible materials, Proceedings of International Congress „Concrete in the Service of Mankind", Dundee, Scotland June 1996, pp. 283-291

4. STARK J, CHARTCHENKO I.: Entwicklung der Quellzemente für die Baupraxis. Proceedings of the International Symposium „75 Jahre Quellzement" Weimar 11 - 13 Dezember 1995, Hochschule für Architektur und Bauwesen. Weimar, pp. 5-30

CONSERVATION OF STUCCOES USING NATURAL CEMENT

D C Hughes **M Parandian**

University of Bradford

S Swann

Swann and Associates

K Reeder

Halcrow Group Ltd

United Kingdom

ABSTRACT. The London Building Act of 1774 led to a demand for stucco as a means to produce a stone-like finish on a brick substrate. The great Regency developments in London and on the south coast, such as The Guildhall, London and The Royal Pavilion, Brighton employed Roman cement for this purpose. Roman cement, or more correctly Sheppey cement, was one of the most important natural cements, together with Harwich and Mulgrave cements, which pre-dated the development of Portland cement. The conservator of such works has two approaches to consider, i.e. the use of a modern equivalent cement or the production of small quantities of a "reproduction" cement using original septaria (cement stones). Three cements have been re-created, although not without difficulty, and some of their properties compared with Rapide, a modern natural cement, and normal Portland cement. The principal components are C_2S, C_2AS, a ferrite complex and CaO although there is large variation with source and calcination temperature. The properties of Mulgrave cement are similar to Rapide, which would be the preference of the conservator over PC. Solutions to the problems of calcination will extend the repertoire of the conservator.

Keywords: Natural cements, Normal Portland cement, Conservation, Stuccoes, Calcination, Strength, Permeability, Durability.

Dr D C Hughes is a Senior Lecturer in Materials Technology, University of Bradford, England, UK. His research interests include the influence of pozzolanic additions to cementitious binders such as Portland and Natural cements and Natural Hydraulic Limes.

Mr S Swann is a Conservation consultant interested in the restoration of lime and natural cement based historic fabric, with a particular interest in decay and failure processes in renders and plasters and appropriate conservation procedures.

Dr M Parandian is a former post-doctoral research assistant. He is currently a consultant and mathematics teacher.

Mr K Reeder, a former student, now works for Halcrow Group Ltd specialising in building and structures.

INTRODUCTION

Architectural applications were a major use of natural cements in the early 19th century. Prohibition of the use of external timbers by the London Building Act of 1774 and the increase in speculative building led to a demand for stucco as a means to produce a stone-like finish on a brick substrate. The latter was then considered to be a cheap substitute for stone, lacking its aesthetic appeal [1]. The great Regency developments in London and on the south coast, such as The Guildhall, London and The Royal Pavilion, Brighton employed Parker's Roman cement as a stucco. However, by the 1840s the place of natural cements for stucco work was being taken by other materials such as the Blue Lias Hydraulic Limes and the new Portland cement. An alternative form of stucco was used to create the artificial rock-work called Pulhamite. Pulham and Sons produced many grottoes, ferneries and rockeries, whose distinguishing feature was often the creation of realistic geological formations using various mortars. The cementitious component varied over the hundred years of the firm's existence but included natural cement, lime and Portland cement, individually or in combination.

Natural cement was produced by firing cement stones (septaria), sometimes found in the London Clay and Upper Lias beds, in a lime kiln. Pasley [2] recommended that, in the laboratory, they should be gradually heated to full red heat for 2 - 3 hours. However, in practice the full production process would have taken several days and was affected by factors such as the strength and direction of the wind. The determination of the precise temperature used by each manufacturer is made more uncertain as a result of a tendency to use lower temperatures than Parker in order to economise on grinding [3, 4]. The calcined septaria were then ground with millstones to an "impalpable powder" [5] with reputable manufacturers sieving the powder over a 34-mesh sieve (approx. 450 μm). The requirements in the USA were more stringent and a maximum residue of 8% on a No. 80 sieve was specified [6].

Unlike modern Portland cements, the natural cements showed a great variation in composition, engineering properties and colour. Pasley [5] stated that the use of the name "Roman cement" was a "nonsense" and that each cement should be referred to by name. Indeed, it would seem that some suppliers were hiding inferior product behind the general name. The superior natural cements were often considered to be Sheppey, Harwich and Yorkshire cements. The latter was also known by the names of Atkinson's in the south and Mulgrave in the north of England. All of the natural cements set very rapidly, albeit that Sheppey set quicker than Harwich. This had practical significance in the application of stucco. Donaldson [7] states that two layers of natural cement do not adhere unless the second is applied before the first coat has started to set. For large areas of stucco he recommended the use of a retarder.

The repairs to the Pulhamite “exposure” at Battersea Park provide an example of inappropriate conservation. The gunniting of the rocks has quite likely used a very different cement and resulted in a loss of the fine detail of the original. Successful conservation requires an understanding of the complete characteristics of that to be conserved to ensure comparability and compatibility. In selecting cements, the conservator has two approaches to consider, i.e. the use of a modern equivalent cement such as Rapide or the production of small quantities of a "reproduction" cement using original septaria. Whilst the latter approach should yield a high compatibility between the repair and the historic fabric, it requires an appreciation of both the manufacturing process and in-service performance of original and "reproduction" historic materials.

The work reported here was not conducted as part of a coherent research programme but rather as a series of individual small-scale projects over a period of some 10 years. This is an initial attempt to collate the data. Needless to say, over this period of time, insights and equipment have been gained which, with the benefit of hindsight, could have led to a more structured programme. These inadequacies are readily acknowledged but do not detract from the data to be presented; indeed, they strengthen the next phase of the research. The aim of this paper is to identify relevant factors in the reproduction of cements and begin a comparison with modern materials.

EXPERIMENTAL PROGRAMME

Samples of septaria were collected over a period of some 10 years from the old alum workings at Rock Holes Drift Mines, Mulgrave Woods (Mulgrave cement); from the same geological formation some 5 km to the east at Whitby East Cliff (Whitby cement) and from the Shotley peninsula near Harwich (Shotley cement). The Shotley septaria show more colour variation, both within and between stones, than do those from Mulgrave and Whitby. They range from grey to a dark brown with much evidence of iron staining. The grain size appears slightly larger than that of Mulgrave; fissures were prevalent with frequent calcite deposits. The Mulgrave septaria are light grey with some coarse crystalline regions.

The first series of sample preparations was conducted in 1988 with further series from 1996.

Series 1 Trials were conducted in 1988 using the Mulgrave septaria fired at 940°C (M 940/1) for 2.5 hours in an electric furnace. The resulting clinker was ground by hand and passed over a 150 μm sieve. The cement yielded setting characteristics and colour much as predicted by contemporary literature. Preliminary firings showed variability in the setting time of cements from different fragments fired under nominally identical conditions; this is an inherent problem with natural cements and hydraulic limes unless the source rock is unusually consistent. Consequently, it is not valid to claim that the cements produced during this programme are truly representative of the materials produced at least 150 years ago. However, the basis was provided for production of a larger batch for experimental purposes. Samples were prepared for compression testing on pastes of w/c = 0.65, this yielding the lowest water content that would allow fresh paste to be poured into the moulds before the initiation of the setting process.

Series 2 A lime kiln firing of Shotley septaria at 900°C and 1100°C was undertaken in 1996 [8]. The cements were crudely hand ground for qualitative analysis.

Series 3 Additional laboratory firings of Mulgrave at 1100°C (M 1100/3) and Shotley at 940°C and 1100°C (S 940/3 and S 1100/3 respectively) were made in 1998, the clinker being ground in a ball mill but without subsequent sieving. Samples for compression testing were produced as in Series 1. XRD analyses were performed.

Series 4 Septaria from both Whitby (W 940/4, W 1100/4) and Shotley (S 940/4, S 1100/4) were fired as in Series 3 but subsequently sieved, and reground as necessary, as in Series 1. Compression samples were produced at w/c = 0.55.

Compressive strength was measured on pastes using 22 x 22 mm diameter cylinders following water curing at 20°C. Such small samples and replicates (4 at each age) had to be used due to the scarcity of the available reproduction cements. Permeability of M 940/1 was assessed by a solvent exchange technique using methanol whilst permeameters with a pressure head of 0.69 N/mm^2 were used to assess Rapide. This figure is lower than that previously adopted for PC pastes, however, experience showed that Rapide was unable to sustain the higher pressure. Performance in sulphate solutions was measured on 160 x 40 x 10 mm prisms immersed in 0.35M Na_2SO_4 solution.

EVALUATION OF REPRODUCTION CEMENTS

Characterisation

The colour of the cements was markedly different with Mulgrave and Whitby cements being much the lighter. M 940 was a yellow/beige; M 1100 had salmon pink tones; both W 940 and W 1100 were yellow with the latter being slightly greyer; S 940 was chocolate brown with a purple tinge whilst S 1100 exhibited a slightly greyer shade of brown. These colours are not dissimilar from those cited in the contemporary literature [5, 7]. Particle size distributions were determined using a Malvern Instruments series 2600 analyser for the product from Series 1, 3 and 4. The median diameters were found to be 20.45 μm (M 940/1), 38.00 μm (M 1100/3), 22.93 μm (H 940/3) and 33.05μm (H 1100/3) which indicate slightly finer cements than Rapide (38.14 μm). Therefore, despite the lack of sieving in Series 3 the particle sizes are similar. It should be noted that the measurements on M 940/1 were obtained from a sample that had been stored for approximately 10 years in an airtight container. The additional sieving and grinding of Series 4 has reduced the variability in the particle size distribution between the cements. S 1100/4 is slightly coarser within the size range 5 - 70 μm whilst the remainder are very similar with a median diameter of 10 -12 μm. A comparison of the Shotley cements shows that whilst the maximum sizes are similar, the cements of Series 4 are finer than those of Series 3 by approximately 10 μm at many intermediate volume percentiles.

Qualitative analysis of the Series 3 cements shows the principal components to be C_2S, C_2AS, a ferrite complex and CaO, although the proportions of each vary. The ferrite is much more prevalent in Shotley cement with evidence of C_4AF, C_6A_2F, C_2F and CF so reflecting the higher iron content. S 1100 shows a trace of C_3A. Unreacted quartz is much more apparent in S 940 than M 940 whilst the latter shows the greater prominence of C_2S. The increase in calcining temperature to 1100°C yields consumption of quartz and a reduction in CaO with the C_2S being more prominent in Shotley cement; C_2AS appears to be relatively unaffected. In contrast, M 1100 seems to show more C_2AS and less C_2S than M 940. The identification of C_3S requires a refinement to the XRD procedure although its presence might not be expected at a calcination temperature of 1100°C. M 1100 indicates the presence of anhydrite to be more than the "trace" noted in M 940.

Setting

The Series 1 and 2 material behaved much as expected with very rapid setting times of 5 minutes for the lower calcination temperatures and approximately 20 minutes for the 1100°C material. M 940/1 proved difficult to remove from plastic cylinders, which were produced for a series of solvent exchange tests, and was attributed to expansion. This may have been due to the production of ettringite or a function of the 6% free CaO content.

However, Series 3 behaved quite differently despite being fired under similar conditions. Whilst both S 940/3 and S 1100/3 set rapidly, they expanded and disintegrated within 2 minutes when placed in the curing water. By way of contrast M 1100/3 did not set when mixed at w/c = 0.65; rather, it dried out following extensive bleeding to yield a highly friable sample. A second paste produced at w/c = 0.4 did set but with little strength and, like S 940/3 and S 1100/3, disintegrated when placed in water. Some samples of S 1100/3 were quickly removed from the curing solution and left in the laboratory air for 3 days. They were then re-submerged in water and most of them survived intact. When they were sawn for compression testing it was noted that whilst the outer few millimetres were hard, the core was soft and washed away by gentle rubbing and the action of the saw blade.

It is likely that the disintegration is related to the free CaO contents and associated unsoundness. After 24 hours exposure to the laboratory air, fracture surfaces of the disintegrated S 1100/3 samples were coated in a dense white layer of calcite. Swann has observed similar but thinner coats on renders made with the material of Series 2. One differentiating feature between Series 1 and 2 and Series 3 is the move from hand grinding to the use of a ball mill. The former is a very slow process during which it is likely that the cement will have reacted with atmospheric CO_2 and water. In contrast, the material from the ball mill was rapidly sealed in airtight containers. Thermal analysis of the 10-year old M 940 revealed a 6.0% CaO content, from 6.6% $Ca(OH)_2$ and 1.8 % $CaCO_3$, which conforms to the free CaO content measured on the fresh cement in 1988. This contrasts with a typical free CaO content of approximately 1% for a modern Portland cement. It is also recognised that the Series 2 cements were only crudely ground and relatively coarse so reducing the immediate exposure of free CaO to water but this is thought to be a second order effect.

Pasley [3] noted that fresh cement set too quickly and exposure to dry air for a few days yielded "the best state for general purposes" whilst Donaldson [7] considered that cements were "never better than when about a fortnight old". The C19th literature contains frequent references to the practice of maturing both Roman and Portland cements in air to overcome unsoundness. Knauss [9] has reported values of carbon dioxide and water in Harwich, Sheppey and Whitby cements of some 4.3%. This could imply either evidence of a post-production atmospheric reaction and /or an incomplete calcination. Consequently, samples of Series 3 cements were exposed to the laboratory air and the production of $Ca(OH)_2$ and $CaCO_3$ monitored by thermogravimetry (Table 1 shows the data for Shotley cement). It is apparent that exposure rapidly leads to the conversion of free CaO, with the production of $Ca(OH)_2$ being the more rapid and peaking at 2 days, after which the net effect is for $Ca(OH)_2$ and additional CaO to carbonate. The effect of the higher calcination temperature is seen in the reduction of CaO available for reaction. Thus, a period of aeration appeared to be a crude means of overcoming unsoundness.

Since the cements had already been calcined and ground the remaining samples of S 940/3, S 1100/3 and M 1100/3 were air matured for 4 days and samples produced for a limited number of compression tests. During the setting process a tendency to bleed was noticed. As before, the Mulgrave cement did not set whilst the Shotley cements set at a slightly slower rate than before. Upon immersion in the curing water the samples remained intact.

The slaking/carbonation processes are a function of surface area and since the Series 3 cements had not been sieved this element was included in Series 4. Samples were produced at the lower w/c of 0.55 to overcome the bleeding problems of Series 3. The rate of setting was temperature dependant with Shotley being the more rapid (W 940/4 - ~15 mins;

W 1100/4 - ~150 mins; S 940/4 - ~25 mins; S 1100/4 - ~35 mins). As noted in Series 3, the cements from the Whitby area are very sensitive to calcination temperature; such sensitivity with this class of cement has been previously noted [7, 9]. The samples remained intact when immersed for curing.

Table 1 Effect of exposure to air on Shotley cement

MATURATION, days	S 940/3			S 1100/3		
	$Ca(OH)_2$, %	$CaCO_3$, %	CaO, %	$Ca(OH)_2$, %	$CaCO_3$, %	CaO, %
0	0.6	0.5	0.7	0.6	0.2	0.6
1	1.0	1.7	1.8	0.4	0.6	0.6
2	1.0	2.2	2.2	0.6	0.6	0.9
3	-	-	-	0.6	0.8	0.9
4	-	-	-	0.4	0.7	0.7
5	1.0	3.4	2.7	-	-	-

The improved performance of Series 4 may be attributed to the sieving and regrinding both separating the agglomerated particles and also the producing higher surface area for subsequent reaction. TG analysis of S 940/4 reveals that 1.1% free CaO has been consumed to produce $Ca(OH)_2$ with no $CaCO_3$ being detected. The additional consumption in comparison with S 940/3 (Table 1) is small and the influence of other factors cannot be discounted. Although this procedure has overcome the short-term problem of unsoundness it should be acknowledged that Donaldson [7] considered that such a procedure "injures" cement. It is not clear exactly what he meant and it is possible that he was referring to a slight increase in setting time.

Strength

The early age strength is shown in Table 2 and the influences of the production process can be seen.

Table 2 Strength (MPa) of Mulgrave, Whitby and Shotley cements

TEST AGE, days	W/C RATIO 0.65			W/C RATIO 0.55			
	M 940/1	S 940/3	S 110/3	S 940/4	S 1100/4	W 940/4	W 1100/4
7	2.0	3.7	1.6	1.4	0.3	2.9	0.3
14	2.2	4.0	1.5	1.3	0.3	2.6	0.3
28	2.2	4.0	1.6	1.5	0.3	2.1	0.3
100	8.5	-	-	-	-	-	-

- The increase of kiln temperature from 940°C to 1100°C has reduced the 28 day strength of Shotley and Whitby cements. Gillmore [9] also found a dependence of degree of calcination on strength performance yet one which was of a complex function. When taken with the influence of temperature on setting the current data gives meaning to comments [7] that excessive heat may "entirely spoil" cements.

- Despite a reduction in w/c ratio the Shotley cements of Series 4 are weaker than their Series 3 counterparts. This may simply due to the variability between batches of cement produced from small and not necessarily representative samples of septaria or that sufficient free CaO remains which might either affect the resultant pore structure or the hydraulicity of certain compounds. Such results are not unique although Gillmore [9] attributed his differences to "imperfect manipulation".

- Series 4 data shows that at a kiln temperature of 940°C Whitby cement is stronger than that of Shotley at 28 days. The historical data on the relative merits of Yorkshire, Harwich and Sheppey cements is inconclusive. Whilst Yorkshire cement was available at a premium price at the time of the Great Exhibition, suggesting a superior product, Knauss [10] measured its soluble silica content to be approximately 50% that of Harwich and Sheppey cements. Since the dominant silicate is C_2S, a pronounced long-term strength development might be expected, as shown by M 940/1, which could yield a time dependant ranking. Additionally, the historical cements are likely to have been produced at optimal conditions for each source, although the key criterion was more probably that of setting time rather than strength.

- All pastes show essentially constant strength in the period 7 - 28 days with the exception of W 940/4, which yields a significant decrease. However, such a decrease was not unknown in C19th cements.

This study has revealed that setting and strength are affected by source of septaria, calcination conditions and post-production processing. In addition, the effect of residence time and atmospheric conditions in the kiln may also be factors to be considered. Until further optimisation studies have been undertaken the current observations remain to be confirmed. The formation and hydration of low temperature cementitious products within their own system parameters, rather than as an intermediate stage in the production of Portland cements, warrants investigation. Such data would also support the current resurgence of interest in hydraulic limes, which are close relatives of natural cements.

COMPARISON WITH MODERN CEMENTS

A limited comparison has been attempted between the Mulgrave cement of Series 1 and Rapide and PC. The initial strengths of Rapide and M 940/1 are similar but thereafter the strength development profiles diverge. After showing little strength gain in the first month, M 940/1 subsequently develops substantial strength to be the strongest cement at 12 months. This performance is symptomatic of a cement whose principal hydraulic compound is C_2S. Unfortunately, data is not available between 3 and 12 months and long-term tests are now underway with Series 3 and 4 cements which will yield a more comprehensive strength development profile. The profiles of Rapide and PC are similar although Rapide is consistently weaker. Skempton [4] cites a 28 day strength (ca 1850) of Atkinson's cement paste, at an unknown w/c ratio, as 1160 psi (~8 N/mm^2). Although the data was obtained for a cement paste several of the relevant factors are not known. It is known that Earle's produced cement using a combination of septaria from a number of local sources. This series of tests together with unpublished data indicate a wide variety of nodules within the Upper Lias of the Yorkshire coast. An important component of this variability within the Yorkshire septaria is the Fe_2S content which not only affects the colour of the cement but also its hydraulicity. Gillmore [9] reports US experience that the variations in cement stone precluded any assumption that the quality of cement from a given manufacturer could be maintained on a daily basis. Additionally, Series 3 and 4 data shows the substantial influence of kiln temperature. Until further investigations have been completed any comparison must be treated cautiously.

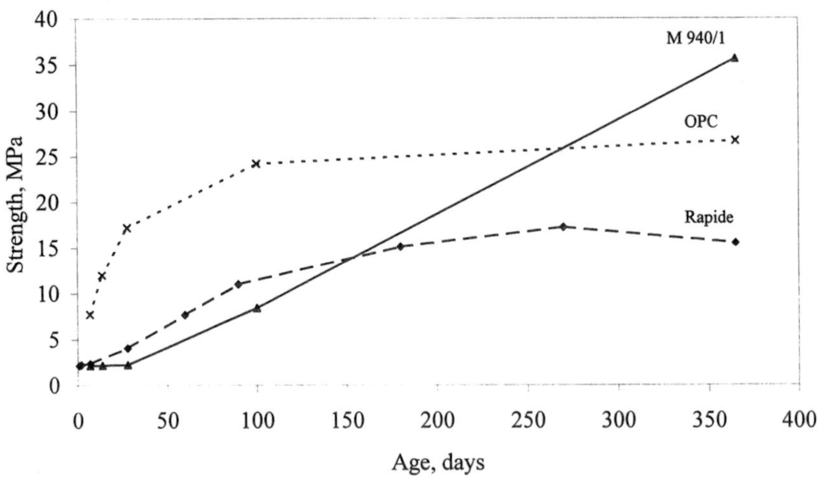

Figure 1 Strength development of various pastes at w/c = 0.65

Permeability

Methanol exchange was used to compare the 28-day "permeabilities" of M 940/1 and PC, where a higher rate of weight loss indicates a higher permeability. After 24 hours the rate of exchange was still linear and the respective values of weight loss were 0.166 mg/l and 0.095 mg/l; thus, M 940/1 was more permeable than PC. However, two factors preclude more detailed analysis. Following 15 days methanol immersion, M 940/1 had produced an orange discolouration to the methanol which became more pronounced with further immersion. In addition, as each disc was surface dried for weighing, small deposits of the cement paste were left on the paper towel which would have had a cumulative contribution to weight loss. The use of the technique was subsequently discontinued; this is unfortunate since it prevents a comparison with other data obtained in the laboratory on a wide range of materials.

Permeability data for Rapide is shown in Table 3. Unfortunately, it was not possible to obtain permeability data for Rapide at earlier ages than 28 days since the samples were unable to withstand the pressure in the permeaters. Tests on PC were not included in this particular programme. However, work by Nyame and Illston [11], conducted at w/c = 0.71 and using similar permeaters, suggests that the permeability of Rapide may be some 1 order of magnitude greater than that of PC at 28 days. However, Rapide shows a greater subsequent reduction in permeability than that found by Nyame such that the rank order may be reversed at later ages. Despite the previously stated problems with the use of methanol exchange with natural cements, data for both M 940/1 and Rapide suggest that they would exhibit similar permeability at an early age.

Table 3 Permeability (m/s) of pastes at w/c = 0.65

AGE, days	RAPIDE
28	227×10^{-13}
60	7.11×10^{-13}
90	1.44×10^{-13}
180	7.05×10^{-13}

Also of interest to the conservator is the water vapour permeability as result of a gradient caused by relative humidities, i.e. "breathability". A test procedure has been recently published as BS EN 1015-19 and work has begun to assess a range of natural cements and hydraulic limes.

Sulphate Resistance

A possible mechanism of degradation of stuccoes is the influence of salts that might migrate from the substrate or the ground, particularly in the form of sulphates. Immersion of M 940/1 and PC in 0.35M sodium sulphate showed that M 940/1 was the least resistant, with 1% expansion after only 5 days; PC attained this level of expansion after 22 days. The use of the Pozzuoli pozzolan with M 940/1 (30% replacement) yielded a material which was essentially immune to further sulphate attack after sustaining a small expansion of 0.13%

Whilst sulphate attack is not normally a problem associated with a new stucco on an old brick, all things being equal, it should be considered if part of the conservation involves replacement of an element of the substrate. Care should also be taken that neither the conservation nor subsequent deterioration alters the pattern of moisture movement such that any salts are mobilised within the substrate and diffuse into the repaired stucco, as might happen if gutters were to fail. Also of interest is the effect of seawater spray on restored stucco in coastal areas and this will be investigated in the next phase of the programme.

CONCLUSIONS

This work has shown that historic natural cements may be reproduced and confirmed that each has its own characteristics. There is substantial variability both within and between septaria from a single source which limits the generalisation of results. The principal components of the cements studied are C_2S, C_2AS, a ferrite complex and CaO although the proportions of each vary. The relatively high CaO contents lead to problems associated with soundness. This has been counteracted by either maturing the cements in the laboratory air for 4 days such as to produce $Ca(OH)_2$ and $CaCO_3$ or sieving the cement and re-grinding the coarsest fraction without resorting to the aeration cycle. The setting of the cements is rapid and is slowed by an increase in the calcination temperature from 940°C to 1100°C. Early age strengths are also reduced by the additional calcination.

Mulgrave and the modern natural cement, Rapide, in concrete have similar early age strength and permeability. Their permeabilities are higher and their strengths lower than normal PC. Whilst Rapide has a lower long-term strength than PC, the strength of M 940/1 at 1 year is higher. Thus, Rapide would appear to be a better source of restoration material than PC. Much more work is required to understand the historic natural cements if successful exploitation is to be made of any small local scale production of "reproduction" historic cement or universally successful use of modern materials in the conservation field.

REFERENCES

1. FRANCIS, A J. The Cement Industry 1796-1914: A History. David & Charles, Newton Abbot, 1977, p 319.

2. PASLEY, C W. Observations, deduced from experiment, upon the Natural Water Cements of England, and on the Artificial Cements, that may be used as substitutes for them. Establishment for Field Instruction, Chatham, 1830.

3. DANCASTER, E A. Limes and Cements. Crosby Lockwood and Son, London. 1920.

4. SKEMPTON, A W. Portland Cements, 1843-1887. Transactions of the Newcommen Society, Vol 35, 1964, pp 117 - 152.

5. PASLEY, C W. Observations on Limes, Calcareous Cements, Mortars, Stuccos and Concrete. John Weale, London. 1838

6. REID, H. A Practical Treatise on Concrete and How to Make It. Spon, London. 1869.

7. DONALDSON, T L. Encyclopaedia Metropolitana Vol XXV, London, 1845, pp 151 -183

8. SWANN, S. Roman Cement, 1796 - 1996. Lime News, Vol 5, 1997, pp 38 - 50

9. GILLMORE, Q A. Practical Treatise on Limes, Hydraulic Cements and Mortars. Van Nostrand, New York. 1890.

10. KNAUSS, C. Chemical Examination of Some English Hydraulic Limes. Dinglers Poytechnische Journal, Vol 135, 1855, pp 361 - 369.

11. NYAME, B, AND ILLSTON, J M. Capillary Pore Structure and Permeability of Hardened Cement Paste. 7th International Congress on the Chemistry of Cement, Paris, 1980. Vol VI, pp 181 - 185.

RETENTION OF REPLACEMENT MORTAR AND VARNISH

R Drochytka

Technical University of Brno

Czech Republic

ABSTRACT. The durability of concrete for cooling towers is strongly influenced by its surroundings, specifically dampness, concentration of gas (CO_2, SO_2), frost, etc. The durability of repair is heavily dependent on the retention of the mortar and varnish. This paper describes the causes of concrete corrosion in cooling towers and the problems of adhesion of new repair mortars on the original concrete.

Keywords: Carbonation, Sulphation, Durability, Concrete corrosion, Mortar of repair, Varnish, Coating, Cooling Towers, Bonding agent, Adhesive bridge

Associate Professor Rostislav Drochytka PhD is the head of the Institute of Technology of Building Materials and Components of the Civil Engineering Faculty at the Technical University in Brno. He specialises in the field of durability of various sorts of concrete, with an emphasis in the corrosion of concrete by carbonation, sulphation, etc. He also works in the field of maintenance and repair of concrete structures, with an emphasis on checking the quality of repair materials, ranging from mortars to coatings. He has made presentations at many vocational symposiums and conferences, mainly in the field of concrete repairs and repairs of cooling towers in the power industry. At present, he is also devoting himself to the problems of treating industrial waste for use as building materials.

INTRODUCTION

The lifetime of concrete claddings of cooling towers in power plants of any type is significantly influenced by the known phenomena of carbonation and sulphation, together with the effects of frost and moisture [1]. In the most severe cases even static failures due to increased deterioration of concrete may occur. In these cases, following a technically detailed building investigation in which the rate and the extent of the failure are determined, it is necessary to redevelop concrete and reinforcement. In technical papers, a number of concrete repair material producers - both of reprofiling mortars, spatulas or coatings - claim the excellent qualities of these materials such as high tensile strength, compressive strength as well as considerable adhesion to the infill concrete or even excellent diffusion parameters of the material varying in accordance with their applications, etc. The decisive criterion for the quality of the entire new structure is the adhesion of the newly produced concrete, mortar or coating on the base of the original concrete [2].

DETERIORATION MECHANISMS OF THE COOLING TOWER CONCRETE

Diffusion of Gases and Water Vapour in Capillary Porous Concrete

Water vapour and analogically also carbon dioxide and other gases from the atmosphere penetrate in a capillary manner through porous materials by diffusion. Water in the gaseous state is thus propagated into the construction by diffusion, liquid water by transmission. Both processes need not have the same direction; in cooling towers diffusion may very often take place in the wall from the inner surface to the outer one, i.e. in the direction of the pressure gradient of the water vapour, whereas water transmission can also go in the opposite direction, e.g. from the outer surface wetted by rain driven to the inside by the wind, thus in the direction of the concentration gradient of the water in the given material. Under practical conditions diffusion predominates [3].

Influence of Carbon and Sulphur Dioxide on the Cooling Tower Concrete

The action of atmospheric carbon as well as sulphur dioxide shows certain common features with the action of atmospheric moisture. Above all is common the circumstance that into the concrete which forms the cooling tower constructions, CO_2, SO_2 as well water vapour enter in the gaseous phase. Their penetration into the concrete, and the effects of this penetration can be substantially influenced, both positively and negatively, by the type of hydration product of the binders, by the porous capillary structure of the cement glue, and above all by the surface finishes on both the inner and outer surfaces of the concrete constructions.

Four Stages of Carbonation

The deterioration process of concrete on the basis of Portland cement can be divided into four adjoining stages:

In the first carbonation stage $Ca(OH)_2$ and its solution in the interglanular space is converted, onto insoluble $CaCO_3$. In the second stage the conversion of the remaining hydration products of the cement takes place through the action of gaseous CO_2. The forming $CaCO_3$ modifications, in conjunction with the amorphous silica acid gel, remain in the course of this process in pseudomorphoses after the hydration products of binder materials and as new very fine-grained crystalline formations of $CaCO_3$, respectively [3]. The third carbonation period is characterised by the re-crystallisation of previously created new carbonate formations, specifically from the intergranular solution. The fourth stage is characterised by an almost total degree of carbonation, in which the coarse aragonite crystals and particularly the calcite crystals penetrate the entire structure of the cement glue.

Course of the Frost Deterioration of Cooling Towers as Related to Concrete Quality

A significant defect ascertained during civil engineering technical surveys of these constructions is the appearance of "surface cavities" which appear in the vicinity of the two monitored surfaces of the tower jacket. These cavities are created in the jacket construction primarily through the action of frost in the presence of increased moisture inside the construction. Due to temperature changes of the surrounding environment, a gradual freezing and thawing of the moisture absorbed by the construction take place and a subsequent deterioration and delamination of the cohesion of the concrete [1].

STUDIES IN COHESION OF REPROFILING MORTARS WITH THE GROUND

Achieving an adequate bond between repair materials, other materials and existing concrete is a critical requirement for durable surface repairs. Various techniques are available to achieve the required bond (Figure 1) [4].

The bond at the interface between the repair material and concrete substrate is likely to be subject to considerable stress from volume changes, freeze, thaw, gravity, and sometimes impact and vibration. The stress states that occur at the bond lines will vary considerably, depending on the type and use of the structure. For example, the bond on a bridge deck overlay may be subject to shear stress in conjunction with tensile or compressive stress induced by shrinkage or thermal effects, and to compression and shear from service loads.

It is essential that the repair material achieves a strong bond to the substrate and that subsequent stresses are not so great as to cause debonding. Repairs which have bond lines in direct tension have the greatest dependency on bonding.

Repairs that are subject to shear stresses at the bond line are able to resist stress not only by bonding mechanisms, but also by aggregate interlocking mechanisms, which add greatly to shear bond capacity. It should be remembered that high initial bond strength is generally not as important as bond durability. Concrete repair materials with drying shrinkage will contract in volume if unrestrained. Drying shrinkage is a long-lasting process (Figure 2).

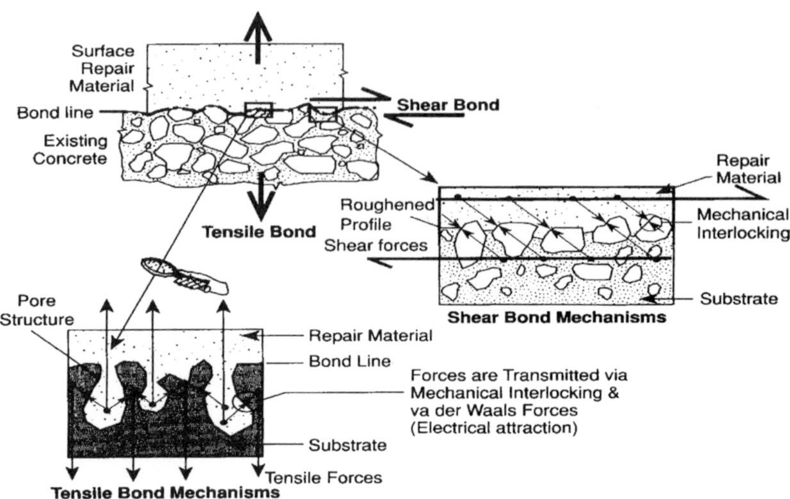

Figure 1 Schematic view of bonding repair materials to existing concrete

Since shrinkage (strain) is restrained by the substrate from occurring, the repair material will accumulate internal tensile stress [4].

The minimal cohesion of the whole reprofiling system of concrete repair mortars with the ground is a largely discussed problem. Evidently, the decisive factor is the quality of the redeveloped concrete. A calculation was made of loading conditions of surface layers, modelled on the method of finite elements, which paid particular attention to the consequential influence of shrinkage of the reprofiling mortar [5].

Two calculation models were selected for the solution of the problem. The first option was carried out by means of a plenary element, which enables the treatment of problems in the condition of plain deformation. In this case a rotation symmetrical condition was chosen. A spatial problem with an application of eight-node isoparametric element was chosen as the other option.

With respect to each problem, the state of stress of a 15-mm thick layer of concrete repair mortar applied to an existing iron concrete structure was solved. A 7.5-mm layer of thickness was included in the calculation for this structure. The regions were always divided into 1.5-mm layers of thickness. This means that five layers of old ground concrete and ten layers of concrete repair mortar were included in the calculation.

The assumption that the concrete ground structure is as a whole stiff and the state of stress is produced by the effects caused by the shrinkage of the layer of mortar was introduced by means of the border conditions. The values of shrinkage in this case were in the range of 0.3 to 1.5 per mill.

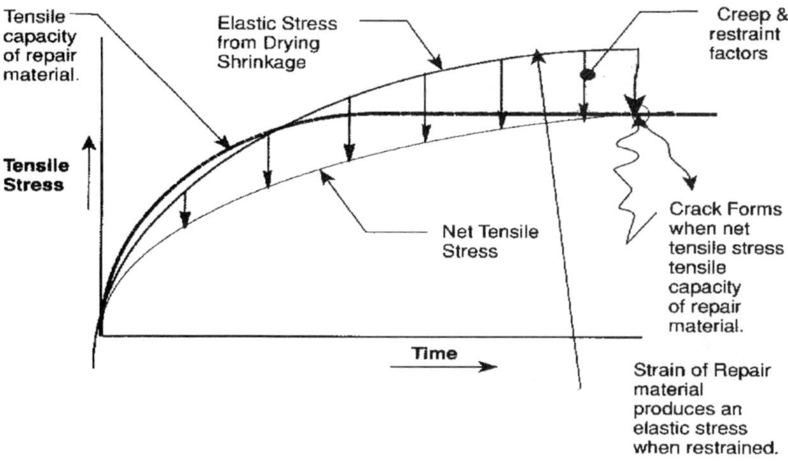

Figure 2 Volume change effects - dry shrinkage process by repair

In the first option, the assumption was introduced that there would arise entire circular regions without cracks with a diameter ranging from 12.5 mm (minimum) to 200 mm (maximum). In the spatial problem, square regions with an edge length ranging from 12.5 mm to 200 mm were introduced. For the old concrete the modulus of elasticity E_b=30.000 N/mm^2, for mortar E_b= 27.000 N/mm^2 was considered. In the layer of mortar problem, the modulus of elasticity was corrected by the influence of the plastic flow by the acknowledged reduction coefficient 0.2573 to the resulting value:

$$E_b = 0.2573 \times 27.000 = 6,946 \text{ N/mm}^2.$$

The maximum tensile stresses then move in the span 16.2 N/mm^2 in the region with a diameter of 200 mm and 3.83 N/mm^2 in the region with a diameter of 12.5 mm. In the spatial problem, at the shrinkage 0.3 per mill, the maximal tensile stresses ranged from 1.51 N/mm^2 to 5.75 N/mm^2 [6].

THE INFLUENCE OF ADHESIVE BRIDGE ON THE ADHESION OF MORTARS

Adequate bonding can be achieved by placing repair material directly against properly prepared substrate. There are special conditions when bonding agents are used. Three main types of bonding agents are frequently used: cement-based slurries, epoxies, and latex emulsions (Figure 3).

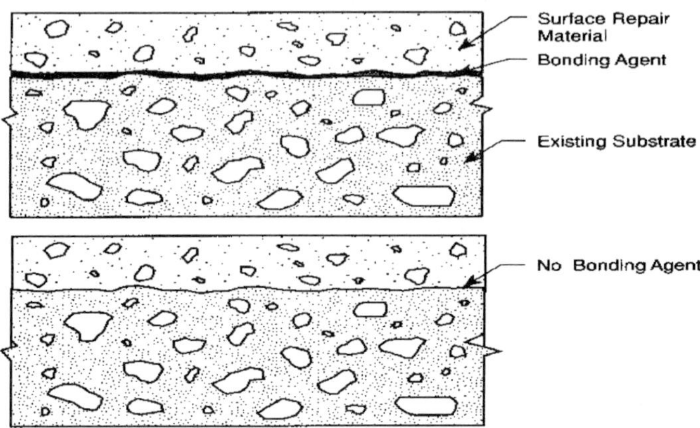

Figure 3 Schematic view of concrete repair with bonding agent

For Portland cement based repairs and overlays, cement or sand-cement slurry is used. After the substrate has been prepared, and immediately before applying the repair material, a thin coating of "creamy" grout must be vigorously and thoroughly brushed into the prepared surface.

In the case of latex-modified or microsilica-modified repair materials or overlays, the bonding grout can be brushed in directly from the mix.
Latex bonding agents are also used in the industry. The following latex product are used as bonding agents:

- Styrene Butadiene (SBR)
- Acrylic
- Polyvinylacetate (PVA)

Re-emulsifiable Polyvinylacetate (PVA) bonding agents should not be used in structural applications. This agent can re-emulsify after being subjected to wet-dry cycles, resulting in eventual loss of bond. A variety of epoxy products are available for use as bonding compounds. Use of an epoxy-bonding agent may produce a vapour barrier, resulting in the failure of the bond.

CONCLUSIONS

1. The extent of concrete corrosion in cooling towers influences the technology of the repair of the concrete structure and the selection of concrete repair materials.

2. The durability of concrete repair depends mainly on the adhesion (bonding) of concrete repair mortars and coatings.

3. As a result of shrinkage, high stresses which almost always exceed the limit of strength occur in concrete repair mortar or in the contacting joint between it and the ground

concrete and result in cracks which divide the area in the regions into "acceptable values" of stress. The calculated width of these cracks is approximately 0.001 mm.

4. The decisive factor which in principle influences the state of stress of the surface layers is the modulus of elasticity of the concrete repair mortar, in particular its working diagram in the region of tensile stresses.

5. It is possible to state that the requirements for the cohesion of reprofiling layers with the ground with a minimal value 1.5 N/mm^2 are fully justified; however, their absolute values must be established with respect to the modulus of elasticity of concrete repair mortars. The values of resulting stresses in the surface layers will decrease in proportion to the lower values of the modulus of elasticity of reprofiling layers.

6. To increase the adhesion of the concrete repair mortar it is suitable to use an adhesive bridge either in the form of penetration or spatula which will considerably increase the values of cohesion of the new mortar to the original ground concrete.

REFERENCE

1. DROCHYTKA R., HELA R. Defects of cooling tower concretes in the Czech Republic, Natural Draught Cooling Towers Kauserlauten, pp. 445-449, Balkema, Rotterdam, 1996.

2. DROCHYTKA R, AND HELA R. Draft of the methodology of tests of the quality of concretes and repair mortars during the maintenance of cooling towers of the STAVEXIS bureau Ltd., Brno, April 1993

2. MATOUŠEK M, AND DROCHYTKA R. Influence of carbon and sulphur dioxide on the life of concretes. STAVIVO 63, 1985, pp. 394 - 398.

4. EMMONS, P H. Concrete Repair and Maintenance Illustrated, RS Means Company, Kingston, MA, 1993, pp. 155-163.

5. DOHNÁLEK J, PUMPR V, AND MYNÁŘ, M. Cohesion of reprofiling mortars with the ground on the level 1.5 MPa - necessity or luxury?, V. Symposium REPAIR OF CONCRETE STRUCTURES, Brno, 1995, pp. 196-199.

6. DROCHYTKA, R. Repair materials on the silicate basis, Magazine REPAIR OF CONCRETE STRUCTURES, No. 3, 1995, pp. 15-20.

REPAIR OF DETERIORATED BRIDGE DECK SLAB BY D-RAP METHOD

W Koyanagi
Gifu University

T Aoki
Aichi Institute of Technology

M Yasui
Nogoya Road Maintenance

T Watanabe
Japan Highway Public Corporation

H Matsushima
Dainichi Consultant Inc
Japan

ABSTRACT. Various repairing methods of damaged deck slab have been developed and adopted in site. Recently, D-RAP method has been proposed to rehabilitate damaged slabs. It belongs to the top face reinforcing method, where after removal of asphalt pavement, deteriorated concrete slab surface is cut out and then pre-fabricated panels are glued over the deck slab in two layers with epoxy resin mortar. Fundamental test results and actual application are presented. These test results showed that the D-RAP method is an effective reinforcing method of restoring deteriorated slabs.

Keywords: Repair, Bridge deck slab, D-RAP method, Epoxy resin mortar, Prefabricated panels, Beam tests, Fatigue tests.

Professor Wataru Koyanagi is Director of the Concrete Technology, Gifu University, Japan. His research interest covers fracture of concrete, failure of concrete structural members, especially slabs, and maintenance of concrete structures. He has published many books and serves on many Technical Committees.

Professor Tetsuhiko Aoki is Director of the Concrete and Steel Technology, Aichi Institute of Technology, Japan. His main research interests include the fracture mechanism of concrete and steel structure. He has also published many books and serves on many Technical Committees.

Mr Masayuki Yasui is a Director Engineering Department in Maintenance of Concrete Slabs, Nagoya Road Maintenance, Japan. He proposed D-RAP method.

Mr Takaharu Watanabe is a Manager Engineering Department in Maintenance of Highway, Japan Highway Public corporation.

Mr Hideo Matsushima is a Deputy Director Engineering Department in Maintenance of Concrete Slabs, Dainichi Consultant Inc, Japan.

INTRODUCTION

Reinforced concrete slabs on the steel girders of old Highway Bridge get damaged occasionally. Many cracks appear at the bottom of the slabs and in some cases, partial-punching damage can be noticed. These are caused as a result of fatigue mainly due to heavy traffic load. Seepage of rainwater into concrete slabs makes the damage more serious. Various repairing methods of damaged deck slab have been developed and adopted in site. The repairing methods are classified into two types. One is so-called the bottom surface reinforcing method such as epoxy bonded steel plate method and epoxy bonded carbon fibre sheet method. The other one is so-called the top surface reinforcing method such as slab depth increase method by in-situ concrete placing where steel fibre concrete is always used.

Recently, D-RAP method has newly been proposed. It belongs to the top surface reinforcing method, where after removal of asphalt pavement, deteriorated concrete slab surface is cut out and then pre-fabricated panels are glued over the deck slab in two layers with epoxy resin mortar. Gluing epoxy resin mortar layer plays also a role of water proofing layer on the slab and it gives the slabs a long fatigue life. "D-RAP" means Deck Restoration by Adhesive Panels.

Various tests were made on the mechanical behaviour of beams and slabs reinforced by this method [1-4]. Small and large size model beam tests were made and the effects of the panel arrangement and those of ambient temperature up to 60 were examined. The effects of the imperfect gluing and the change of loading point were tested by the small-scale slab tests. Fatigue tests of full-scale slab on two alternative points were also made. The property for panels and gluing mortar was also examined.

These test results showed that the D-RAP method is an effective reinforcing method of deteriorated slabs. It increases load –carrying capacity and fatigue strength as high as newly constructed slabs. It can also minimise the removal of damaged slab portion. The method has been applied in several deteriorated bridge deck restorations in site of Japan Highway. The fundamental test results and actual field application will be presented in this paper.

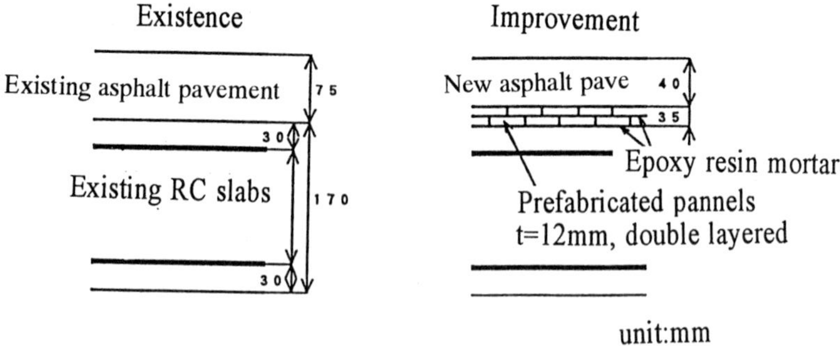

Figure 1 D-RAP Method

THE DEVELOPMENT OF D-RAP METHOD

The concept of the top surface strengthening method of RC slabs is based on the fact that reinforcement of bridge deck slab did not yield and remained in elastic range even when punching damage occurred by over loading heavy traffic. D-RAP method is a kind of the top face reinforcing method. It is important that the repaired materials have to uniform the surface of existing concrete slabs. Epoxy resin mortar was used for unification between the repaired materials and slab surfaces. We also used the pre-fabricated panels as a repaired material.

The size of pre-fabricated panels was decided by handling facility under construction. At the joint of panels discontinuity arises in strengthening effect. We arranged the panels alternatively in two layers so that the joint did not meet in a same line and the effect was expected all over in the slab.

PREVIOUS PROJECTS

The first D-RAP method was applied on the off ramp of the Meishin Highway across the Aoki River in Japan, 1992. The bridge named Shimohira Bridge consists of three continuous supported 20.5m long spans. PC box girders spaced at 1.7m support the 180mm thick concrete deck slab. The size of prefabricated panels was 0.3m×0.45m×0.012m. The mix proportion of epoxy resin mortar was 1:1 (resin, sand) by weight. The repaired area was 430m^2.

The D-RAP method was further applied on the Meishin Highway across the Ibi River, 1993. It is a five span continuous steel plate girders bridge spanning each 69.6m length named Ibi-gawa Bridge (Figure 2).

The repaired area was 590m^2 in the middle span. After that, the D-RAP method was applied in 9 bridges. We used the waterproof lining on the D-RAP method since 1996. Experimental results of fatigue test clarified the D-RAP method can fulfil the function to waterproof for existing concrete slab. However, there were some potholes of pavement occurred after the first field application. We confirmed from actual application that the waterproofing agent of prefabricated panels was not perfectly effective for inner fracture of panel. But it was effective for waterproof of existing concrete slabs.

MATERIALS

Prefabricated Panels

Applied prefabricated panels were non-asbestos fibre-cement flat sheets with non-autoclave curing. The size was 1.2cm×30cm in cross section and 45cm in length. The mix proportion was 69:4:1:1:12.5:12.5 (cement, wood pulp, fibreglass, vinyl resin, $CaSO_3$, $CaCO_3$) by weight. Table 2 shows the characteristics of non-asbestos board.

Figure 2 Application of D-RAP Method in Ibi-gawa River Bridge

Table 1 Characteristics of fibre-cement flat sheets

	DENSITY (kg/m^3)	BONDING STRENGTH (MPa)	MODULUS OF ELASTICITY (GPa)	ELONGATION (%)	COMPRESSIVE STRENGTH (MPa)	PEELING STRENGTH (MPa)	WATER ABSORPTION RATE (%)
Dry	1.71	30.4 [18.8]	32.9 [24.9]	0.162	107.9	2.2	16.2
Wet	1.71	19.7 [18.8]	15.8 [13.9]	———	72.9	1.0	———

Epoxy resin

Adhesive used was epoxy resin mortar and the mix proportion was 3:1:8 (resin, hardener, sand) by weight. Table 3 Shows the characteristics of epoxy resin mortar.

Table 2 Characteristics of epoxy resin

DENSITY (kg/m^3)	COMPRESSIVE STRENGTH (MPa)	TENSILE STRENGTH (MPa)	MODULUS OF ELASTICITY (MPa)	BONDING STRENGTH (MPa)
1.75	81.6	23.5	4440	Min. 3.1

Table 3 Dimensions for specimens of full scale model

SPECIMEN	KIND	DIMENSION UNIT : cm				NUMBER OF SPECIMEN
		Width	Depth	Effective depth	Length	
1-1,2	Normal concrete	330	17	14	180	2
2-1,2	Normal concrete	330	20	17	180	2
3-1	Normal concrete	330	22	19	180	1
4-1,2,3,4,5	by D-RAP method (t=3cm)	330	17 (20)*	14	180	5

*: Total depth in D-RAP method

THE PURPOSE OF EXPERIMENT

Concerning the D-RAP method, various tests were made on materials of panels and gluing, panel arrangement and strengthening effect under various conditions. Static flexural loading tests on small RC beam specimens were made with various items, such as ambient temperature, imperfection of gluing, content of water absorption, repetition of wet and dry cycles, etc.

The flexural loading test result of small beams indicated as follows:

(1) The D-RAP method increased load-carrying capacity as high as newly constructed beams of same depth.
(2) Ambient temperature over 60°C decreased load-carrying capacity.
(3) Imperfection of gluing area over 50% decreased load-carrying capacity.
(4) It is better to avoid the application of D-RAP method during rainy day. The wetting on contact area between panels and concrete decreased load-carrying capacity in proportion to wet ratio.
(5) Joint spacing of panels was the week point compared to the non-joint panels. Shifting of panels in double layers arrangement was effective.

STATIC BENDING TESTS

Flexural loading test of large beams were made. The purpose of this experiment is as follows:

1. The effects of reinforcing by D-RAP method.
2. The effects of wet ratio on contact area between panels and concrete.
3. The failure mode of deteriorated beam.
4. The effect of repetition of loading.

Figure 3 and Table 3 give the dimensions of large scale beam specimens. The span of these simply supported beams is 1.5m span length and subjected to two point loading system at L=0.5m and 1m respectively. The specimen 2-1, 2 were reinforced by D-RAP method. The specimen 3-1, 2 were reinforced by carbon fibre sheet at bottom surface of concrete beams.

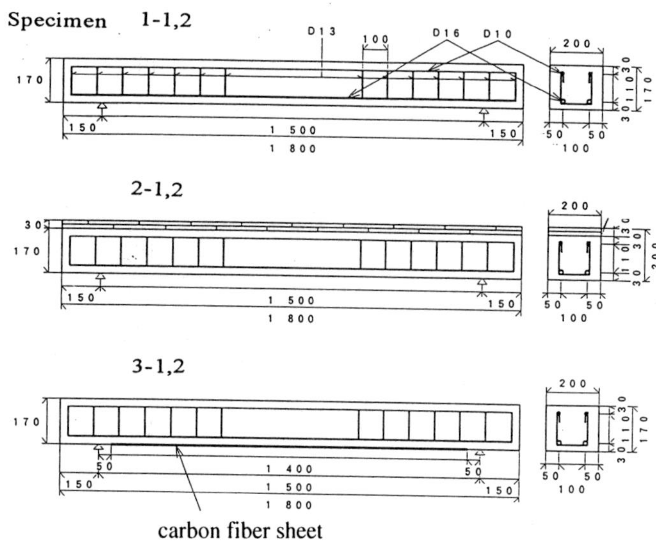

Figure 3 The specimen of static bending test

Figure 4 shows the load deflection curves. It shows that D-RAP method increased load-carrying capacity. It indicates the D-RAP method had greater deflection before the beam failure compared with ordinary RC beams and carbon fibre sheet glued beams.

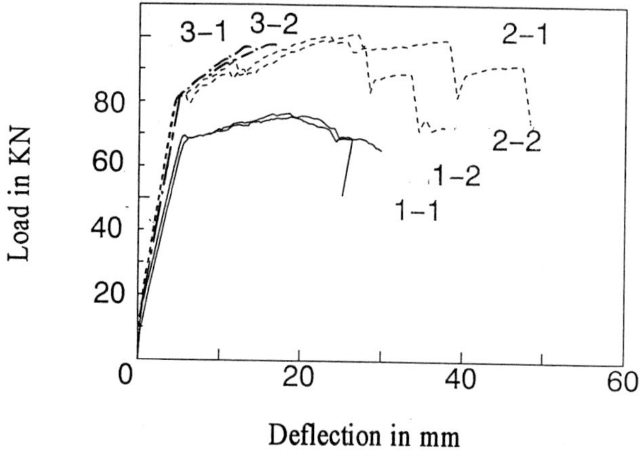

Figure 4 Load deflection curves

FATIGUE TEST

Fatigue test was carried out, for the effect of reinforcing by D-RAP method and confirming of fatigue failure.

The dimensions of full scale test specimens were 0.17m×1.8m×3.3m on which was reinforced by D-RAP method of 0.03m depth. Strength of concrete was 32MPa. Yield stress of reinforcing bars was 361MPa and tensile strength was 529MPa. The specimen was simply supported and the longitudinal

Span length was 1.5m. Transverse edges were unsupported. Figure 5 shows the layout of the fatigue test. We applied sinusoidal cyclic loading with a frequency of 4Hz by a servo-controlled fatigue-testing machine. The vertical load was on two alternative points, at span centre in the direction of simple support and 70cm apart from each other for simulating the actual traffic loads. Water was retained over the slabs for the confirmation of waterproofing effect.

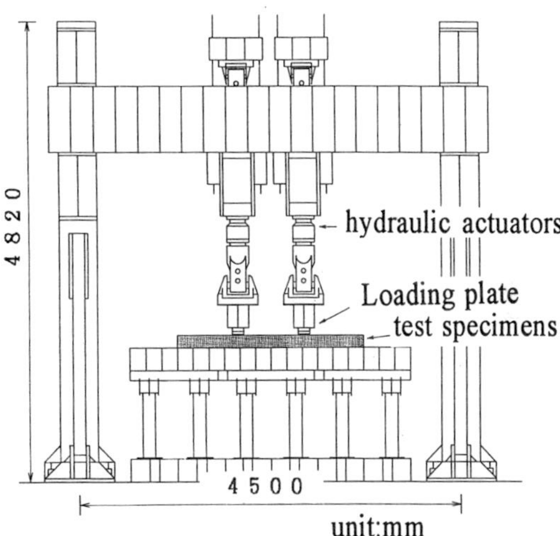

Figure 5 Layout of fatigue test

Table 4 shows the test results. The D-RAP method is more effective for extending fatigue life.

CONCLUSIONS

The test results showed that the D-RAP method is an effective reinforcing method for deteriorated slabs. It increases load-carrying capacity and fatigue strength as high as newly constructed slabs. It can also minimise the removal proportion of damaged slab. The method has been applied in several deteriorated bridge deck restorations in Japan Highway.

Table 4 Summary of fatigue test results

	LOAD METHOD	ULTIMATE STRENGTH IN kN	NUMBER OF CYCLES	FAILURE MODE
1-1	Static	584	————————	punching shear
4-1	Static	663	————————	punching shear
1-2	Fatigue	Upper load 304	180,000	punching shear
2-1	Fatigue	363 431 490	2,000,000 1,500,000 250,000 sum=3,750,000	punching shear
2-2	Fatigue	431 490	2,000,000 337,000 sum=2,337,000	punching shear
3-1	Fatigue	363 431 490	2,000,000 2,000,000 3,000,000 sum=7,000,000	not failure
4-2	Fatigue	235 304 363	2,000,000 250,000 1150,000 sum=2.3650,000	punching shear
4-3	Fatigue	304 363 431	4,000,000 1,000,000 58,000 sum=5,058,000	punching shear
4-4	Fatigue	363 431	2,000,000 115,000 sum=2,115,000	punching shear
4-5	Fatigue	431 490	2,000,000 18,000 sum=2,018,000	punching shear

ACKNOWLEDGEMENTS

This research was supported by the grant from Japan Highway Public Corporation.

REFERENCES

1. AOKI, T. et al. The fatigue test of RC slabs by D-RAP method. Proceedings of the 52nd annual conference of the Japan Society of Civil Engineers, Japan, 1997, V-561, pp348-349. (In Japanese)

2. KOYANAGI, W. et al. The flexural loading test of RC beams by D-RAP method. Proceedings of the Japan Concrete Institute, Japan, 1995, 17.2, pp 923-928. (In Japanese)

3. YASUI, M. et al. The fatigue test of RC slabs by D-RAP method. Proceeding of the 49th annual conference of the Japan Society of Civil Engineers, Japan, 1994, V-331, pp662-663. (In Japanese)

4. MATSUSHIMA, H. et al. Fracture and fatigue strength of slabs repaired with D-RAP method. Proceedings FRAMCOS-3, Fracture Mechanics of Concrete Structures, Vol. 3, 1998, pp1873-1882.

USE OF STRESS WAVES IN THE EVALUATION OF STRUCTURES AFFECTED BY ASR

M Thomas

S Tesfamariam

University of Toronto

A Sadri

ANDEC Manufacturing Ltd

Canada

ABSTRACT. This paper discusses the potential application of ultrasound techniques for evaluating concrete structures affected by alkali-silica reaction (ASR). Concretes examined include a hydraulic structure, 1-metre-cube blocks on an outdoor exposure site in the U.K. and concrete beams exposed in Kingston, Ontario. Ultrasound data was collected using traditional through-transmission, pulse-echo and impact-echo techniques. The collected data were analyzed in both the time and frequency domain. In addition to velocity, various "damage indices" including the attenuation (α-) coefficient and quality (Q-) factor were determined. These indices are more sensitive to ASR-damage than simple velocity measurements, but they also suffer from sensitivity to coupling effects. The pulse velocity correlates well with expansion and provides a means of monitoring in-place concrete properties as ASR progresses. Also, velocity data serves as a suitable input for determining the distribution of concrete quality within a given structure (e.g. tomographic surveys). The good correlation between ASR expansion and certain damage indices calculated from impact-echo is encouraging as such techniques are particularly applicable to the evaluation of pavements (and similar structures) that are not totally suited to through-transmission measurements.

Keywords: Alkali-silica reaction, Concrete, Non-destructive testing, Ultrasound.

Dr Michael Thomas is an Associate Professor in the Department of Civil Engineering at the University of Toronto.

Solomon Tesfamariam is a graduate student in the Department of Civil Engineering at the University of Toronto. He is currently working on his M.A.Sc. Thesis which in concerned with the use of ultrasound techniques for evaluating concrete.

Dr Afshin Sadri is a geophysicist working with ANDEC Manufacturing Limited of Toronto.

INTRODUCTION

There are several non-destructive, non-invasive techniques for determining various in-situ properties of concrete. Most of these only measure properties of the concrete at, or close to, the surface. However, ultrasonic pulse velocity measurements may be made across large concrete structures (up to 50m) to provide an indication of the quality of concrete between the points of transmission and reception; generally, higher velocities are indicative of good quality concrete. Such measurements have been widely used to assess concrete condition in the laboratory and field since the 1950's and a comprehensive review of the technique, including details of instrumentation, effects of various parameters, standardization and applications is given by Naik and Malhotra [1].

Pulse velocity has been used as an indirect measurement of many physical properties of concrete including, elastic modulus, Poisson's ratio, strength, setting time, homogeneity, depth of cracks, layer thickness, and the degree of deterioration due to physical or chemical attack on the concrete. However, the relationships between pulse velocity and these properties is affected by many variables and no unique relationships exist for predicting concrete properties from ultrasound measurements. Relationships between pulse velocity and concrete properties are further complicated by the presence of steel reinforcement (longitudinal pulses may travel up to three time faster in steel) and corrections have to be made to the calculated velocity.

In spite of these limitations, the pulse velocity provides a relative measure of concrete quality. Generally, low velocity indicates the presence of low quality concrete and the velocity has been shown to be sensitive to a number of factors including:

- poor workmanship, e.g. inadequate compaction [2,3]
- low strength, e.g. conversion of high alumina cement [2,4]
- elevated temperature/fire damage [5-7]
- frost damage [8-18]
- alkali-silica reaction (see below)
- sulfate attack

Alkali-silica reaction may result in expansion and cracking of concrete, which in turn leads to reduced mechanical properties (i.e. strength and stiffness). Ultrasonic pulse velocity has been used in field studies of ASR to determine the extent of the reaction and associated damage within a structure [11-15]. Significant reductions in pulse velocity are observed in concretes that have suffered ASR. For example in studies of ASR on the piers of the Hanshin Expressway in Japan, Imai et al. [14] observed the velocity to decrease with increasing cracking down to values below 2000 m/s compared with velocities of more than 4400 m/s in unaffected piers. Swamy and Wan [16] reported that changes in pulse velocity were observed prior to the onset of visible cracking. On the other hand, Hobbs [17] suggested that pulse velocity was an "insensitive indicator of ASR damage". Although, this statement is hard to reconcile with his data, which indicated a decrease in velocity with cracking.

In-situ testing has a specific advantage over testing cores removed from concrete structures affected by ASR. The expansion due ASR may be partially or wholly restrained in concrete structures due to the presence of embedded steel reinforcement or other external constraints.

This prevents the accumulation of damage in-situ. However, when cores are removed from the structure, they are free to expand and subsequent testing will indicate inferior mechanical properties to the in-place concrete. This phenomenon was recently illustrated in a study by Jones et al. [18]. Non-destructive techniques such as ultrasound provide the opportunity to evaluate the concrete under the prevailing stress conditions in the structure.

Although the time of arrival of the first wave (and hence the velocity of the leading wave) is the most commonly used quantity from ultrasound measurements, the velocity may not be sufficiently sensitive to many anomalies in concrete. The pulse velocity has been shown to be significantly reduced in badly damaged and extensively cracked concrete. However, individual cracks are often "invisible" to velocity surveys. When a pulse passes through an air-filled crack or void the velocity will be reduced, however, the overall increase in transit time of the pulse across a large section of concrete with a narrow crack will not be noticeably increased. Furthermore, if the dimensions of the crack are small in the direction normal to wave propagation, some waves will travel around the crack and the resulting increase in distance (and hence travel time) will be small.

Attenuation is a far more sensitive indicator of cracks in concrete as the amplitude of the wave is reduced considerably due to reflection and refraction at the air-concrete or water-concrete interface. However, attenuation measurements are complicated by extreme sensitivity to transducer-concrete coupling [2,19], but this may be overcome by selection of a suitable couplant [20]. The sensitivity of amplitude to cracks was demonstrated by Whitcomb et al. [21] in their analysis of signals received through concrete blocks with and without flaws (0.1 mm crack). Although the transit time was hardly affected by the presence of the flaw, the amplitude of the first "burst" was reduced by 68%. Successive bursts were apparently unaffected by the flaw and these may indicate the arrival of waves that circumnavigated the flaw. Attenuation measurements have been made in the field to evaluate the condition of concrete structures [e.g. ref. 22].

Attenuation is frequency-dependent, with high frequency waves being attenuated more easily. Recently, there has been increased interest in analyzing the "frequency content" of received signals in concrete [21,23-27]. Whitcomb et al. [21] showed the frequency spectra, obtained by Fast Fourier Transform, was significantly altered by the presence of small cracks.

Ultrasonic waves are physically dispersed as they propagate through concrete due to differences in acoustic impedance between cement paste and aggregates, and between concrete and various anomalies (e.g. air filled or water-filled cracks and voids). Consequently, the amplitude of the wave decreases (attenuation) with distance from the source [28, 29]. The extent of attenuation depends on the size of the aggregate particle or anomaly in relation to the wavelength (λ) of the wave and, generally, anomalies with dimensions smaller than λ have little effect on wave propagation. Short wavelength (high frequency) pulses are more sensitive to small defects, and a higher spatial resolution may be achieved using them. However, if the wavelength is smaller than the maximum aggregate size, the wave is scattered at the paste-aggregate interface and rapidly dispersed. This effect necessitates the use of low frequency (long λ) for most concrete applications.

This paper presents ultrasound data from a number of preliminary studies of ASR-affected concrete including a 40-year-old mass concrete dam, field-exposed concrete blocks in the

U.K. and Canada, and concrete prisms stored under laboratory conditions. Ultrasound data were collected using pulse-echo and impact-echo in addition to standard through transmission techniques. The data collected were analyzed in both the time and frequency domain to establish a variety of damage coefficients and test their sensitivity to ASR-damage.

EXPERIMENTAL STUDIES

Concrete Dam

The dam examined was constructed in Northern Ontario between 1949 and 1951. During construction a field trial was carried out examine the potential for using fly ash to control temperature rise in mass concrete. This trial involved the instrumentation and testing of concrete in the top lifts of two similar sections of the dam. The dimensions of the sections were 4.42 x 10.98 x 3.66 m and details of the concrete mixes used are given in Table 1.

Table 1 Details of concrete mixes used in dam

	PC CONCRETE	PFA CONCRETE
Portland cement (kg/m^3)	305	215
Fly Ash (kg/m^3)	-	90
Water (kg/m^3)	188	177
W/CM	0.62	0.58
28-day strength (MPa)	25.3	29.2

The monitoring of these sections included pulse velocity readings, which were taken periodically from the downstream to upstream face (4.42 m) from the age of 24 hours to 6 months, using an instrument (the Soniscope®) developed at Ontario Hydro [19]. As part of a wider study on the long-term performance of fly ash concrete, pulse velocity readings were made at the same locations approximately 40 years after construction. The Soniscope® was also used at this time, although the instrument had evolved somewhat since the 1950's version, the principal change being the adoption of a digital system (using a PC) rather than the original analogue system (oscilloscope). In both cases 20-kHz transducers were used with there being little significant modification between the two testing periods.

Figure 1 shows the change in pulse velocity readings between 24 hours and 40 years for the two concrete sections. The velocity increased from around 3150 m/s to 4100 m/s between 1 and 28 days, with there being little significant difference between the data for the two different concrete mixes. However, between 28 days and 6 months the velocity of the fly ash concrete increased slightly more than the PC concrete, although differences were still more (3% higher velocity for the fly ash concrete). The relative changes in velocity observed for the two concretes were consistent with the changes in compressive strength through the same period.

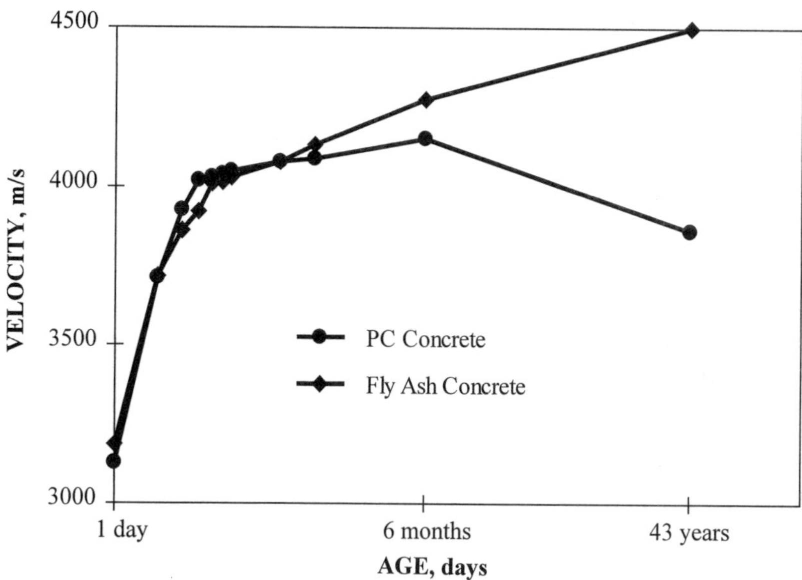

Figure 1 Pulse velocity results for dam

After 40 years the difference between the measured velocities for the two concretes is substantial, the velocity of the fly ash concrete being nearly 17% higher than the PC concrete. Although the early-age and 40-year readings are not directly comparable because of changes in the equipment, the data indicate that whilst the quality of the fly ash concrete had continued to improve there is evidence of some deterioration in the properties of the PC concrete. These results were unexpected and prompted a more detailed investigation of cores taken from the structure.

A petrographic examination of the concrete samples using polished and thin sections revealed the presence of alkali-silica reaction in the PC concrete. Reaction was found in association with granitic gneiss and, to a much lesser extent, rhyollite in both the coarse and fine aggregate. Expansion and micro-cracking was observed in association with these phases, however, the extent of cracking was not sufficiently advanced to be immediately apparent on the exposed vertical faces of the section. Samples of the coarse and fine aggregate, retrieved from material archives, when tested in accordance with ASTM C 1260 produced 14-day expansions of 0.069% and 0.109%, respectively. Minor problems with turbine alignment and clearances consistent with an overall growth of the concrete have been experienced at this structure. No evidence of ASR or internal micro-cracking was observed in the samples taken from the fly ash concrete block.

Concrete Blocks at BRE

A series of large concrete blocks (approximately 1 metre cubes) containing an alkali-silica reactive flint sand and different contents of high-alkali cement and fly ash were cast and placed on an outdoor exposure site at the Building Research Establishment in the U.K. during the period 1989 to 1991. These blocks have been continuously monitored for linear expansion (using surface-mounted reference points) and pulse velocity using a commercially-available PUNDIT for determining the travel time across the nominally 1-metre wide block. Details of the materials used and results for laboratory-stored specimens can be found elsewhere [30].

Figure 2 shows the expansion and velocity data for a concrete block with 550 kg/m^3 of high-alkali cement (1.15% Na_2O_e) and no fly ash. No cracking was observed during the first 4 years exposure at the end of which period an average expansion of 0.037% was recorded. However, between 2 and 4 years the pulse velocity was observed to reduce by approximately 14%. Over the next year significant expansion (average = 0.280%) occurred and this was accompanied by extensive cracking of the block and a further reduction in velocity (28% lower at 5 years compared with the 2-year result). The expansion and pulse velocity data are in very good agreement and it is interesting that a noticeable reduction in velocity occurred before the significant cracking was observed at the surface.

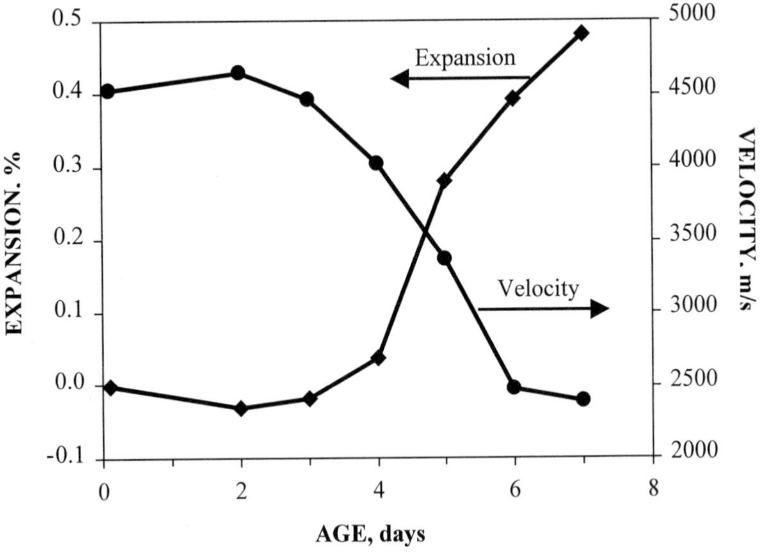

Figure 2 Results for concrete block at BRE

Concrete Blocks in Kingston, Ontario

The materials tested in this field study were four large (0.6 x 0.6 x 2.0m) concrete beams which were cast using a reactive siliceous limestone and various combinations of low- and high-alkali cement, fly ash, slag and silica fume. The total cementitious content and water to cementitious materials ratio was nominally the same for all mixes. Details of the materials used, the preparation of specimens and the exposure conditions have been given elsewhere [31]. Ultrasound measurements were made on these beams after 8 years exposure to the elements. Reported linear expansion results and the visual condition of the beams at the time of examination are given in Table 2. Only two of the four beams showed any visual sign of distress. However, a third beam showed a map-pattern of discolored lines on the surface typical of minor exudation activity frequently observed prior to the onset of cracking in laboratory specimens.

Table 2 Linear expansion data and visual observations for the beams examined

BEAM #	EXP., %	VISUAL OBSERVATIONS
4	0.01	No visible cracking or surface discoloration
3	0.02	No visible cracking, some map-pattern discoloration (exudation)
5	0.05	Faint map-cracking, cracks generally tight (< 0.3 mm) at the surface
6	0.12	Visible map-cracking, cracks up to 1 mm wide at the surface

The equipment used for this study was an AndecScope® capable of performing measurements in the three modes of through-transmission, impact-echo and pulse-echo. For the through-transmission technique, 50-kHz piezoelectric transducers were used for both transmission and reception. In the case of the pulse-echo testing, a 50-kHz piezoelectric transmitting transducer and a 150-kHz broadband, flat-response piezoelectric receiving transducer was used. For the impact-echo testing, an automatic electronic impactor and a 150-kHz piezoelectric broadband receiver were used.

In addition to determining the velocity of stress waves, the data were analyzed in the time and frequency domain to determine various indices reportedly sensitive to the internal condition of the concrete. The first of these is the attenuation coefficient, α, which is derived from the following exponential function used to describe the decay in stress wave intensity with time [32]:

$$I_t = I_0 e^{-\alpha t} \quad (1)$$

where I_0 and I_t are the intensity at time zero and time t, respectively. The second coefficient used was the "quality factor", Q, which is a measure of the spectral width for the peak frequency and may be calculated using the following equation:

$$Q = \frac{f_r}{f_2 - f_1} \qquad (2)$$

where f_r is the peak frequency and f_2 and f_1 are the frequencies either side of the peak at 70.7% of the spectral peak amplitude [32].

Figures 3 shows the measured values of the α-coefficient and Q-factor plotted against the reported expansion results for the four beams. Unfortunately, it is not possible to ascribe the differences in the ultrasound data solely to the ASR-damage due to differences in the cementitious materials used for the four beams. The use of mineral admixtures such as silica fume, fly ash and slag are likely to affect the mechanical properties of the mature concrete and consequently the transmission of stress waves. Despite this fact, the generally good correlation between the indices from ultrasound measurements and the linear expansion indicates that both α and Q are sensitive to the damage due to ASR. Beams 5 and 6 only differ in composition in terms of the chemistry of the Portland cement used. Both are CSA Type 10 cements but Beam 5 was cast with low-alkali cement and Beam 6 with high-alkali cement. Thus the main factor influencing the values of α and Q for these beams is the extent of ASR damage. Significant reductions in both coefficients were observed due to the increase in expansion from 0.05 to 0.12%.

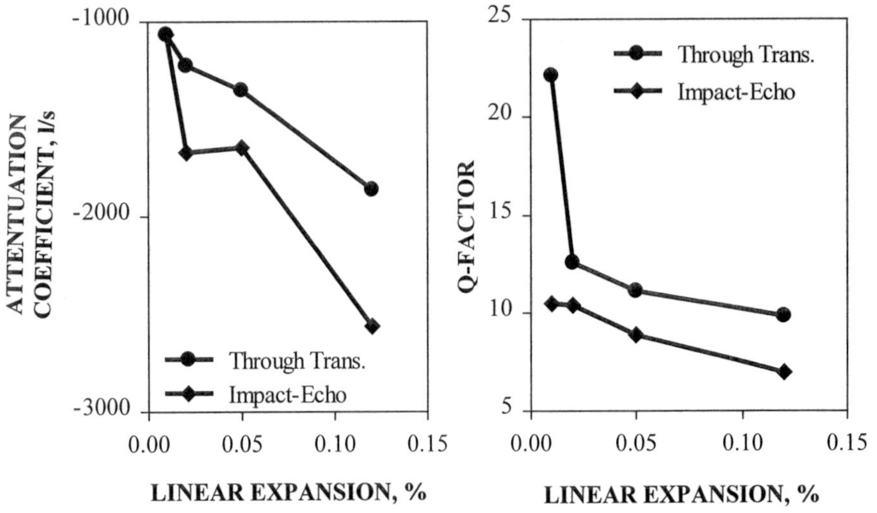

Figure 3 Ultrasound damage coefficients for concrete beams at Kingston

Both the α-coefficient and Q-factor appear to be more sensitive than velocity, as there was little significant difference between the pulse velocity measured for the four beams.

Other indicators of internal damage such as peak frequency shift were also examined in this study as was the use of pulse-echo techniques. A detailed description of this field study is presented elsewhere (33).

DISCUSSION

The internal micro-fracturing that results from alkali-silica reaction in concrete clearly influences the transmission of stress waves impacting both the velocity and the rate of attenuation. As such ultrasound testing techniques have considerable potential for evaluating concrete structures affected by ASR. However, ultrasound data obtained from concrete structures are sensitive to a number of parameters besides the condition of the concrete itself; these may include specimen geometry, transducer geometry and frequency, nature of the pulse generated, coupling between transducer and concrete, and data processing (e.g. filtering). Consequently it has proven difficult to calibrate the technique to the extent that an instrument can be taken out into the field and used to collect definitive data regarding the extent of ASR within a structure. Thus some level of calibration may be required on a case-by-case basis. Alternatively, ultrasound measurements can provide valuable information on a comparative basis. This might involve monitoring the evolution of damage with time for a given concrete, comparing performance of different concretes, or determining the distribution of damage (or other parameters) within a structure. Taking this last point to the extreme, tomographic techniques have been employed to determine the 2-dimensional (or even 3-dimensional) distribution of concrete quality within large concrete elements (34,35).

Although velocity measurements appear to be sensitive to ASR as it progresses in a particular specimen, it may not be sufficiently sensitive to detect subtle differences between different concretes. Other factors based on signal attenuation or frequency content appear to offer more promise. However, these factors are strongly influenced by nature of the coupling between the transducer and the concrete, and efforts need to be made to eliminate these effects.

The use of impact-echo (or pulse-echo) for evaluating concrete suffering from ASR, either through velocity measurements or other damage indices, provides a useful adjunct to through-transmission techniques. The latter requires access to opposite sides of an element which limits its applicability to certain types of structures. The authors are currently collaborating with other investigators to develop techniques based on impact- and pulse-echo for monitoring the performance of pavements suffering from ASR and the evaluating the effect of remedial treatments (e.g. impregnation with lithium-based compounds).

CONCLUSIONS

1. Cracking due to ASR causes a reduction in the velocity of stress waves transmitted through the concrete. In some cases significant reduction in the velocity was observed before visible cracking manifested itself at the surface of the concrete.
2. Damage indices based on attenuation (α-coefficient) and frequency (Q-coefficient) from both through-transmission and impact-echo techniques appear to be more sensitive to ASR damage than velocity.

REFERENCES

1. NAIK, T.R. and MALHOTRA, V.M. "The ultrasonic pulse velocity method." CRC Handbook on Nondestructive Testing of Concrete (Ed. V.M. Malhotra and N.J. Carino), CRC Press, Boca Raton, 1991, pp. 169-188.

2. JONES, R. "Testing of concrete by ultrasonic-pulse technique." Proceedings of the Highway Research Board, 32, 1953, pp. 258-275.

3. KAPLAN, M.F. "Effects of incomplete consolidation on compressive and flexural strength, ultrasonic pulse velocity, and dynamic modulus of elasticity of concrete." ACI Journal, 56 (47), 1966, pp. 853.

4. NEVILLE, A.M. "A study of deterioration of structural concrete made with high alumina cement concrete." Proceedings of the Institution of Civil Engineers, 25, 1963, pp. 287.

5. ZOLDNERS, N.G., MALHOTRA, V.M. and WILSON, H.S. "High-temperature behaviour of aluminous cement concretes containing different aggregates." Proceedings ASTM, 63, 1963, pp. 966.

6. CHUNG, H.W. and LAW, K.S. "Diagnosing in situ concrete by ultrasonic pulse technique." Concrete International, October 1983, pp. 42.

7. RAMAKRISHNAN, V., SHAFAI, H.F. and WU, G. "Cyclic heating and cooling effects on concrete durability." Proceedings of the Second International Conference on Durability of Concrete (Ed. V.M. Malhotra), ACI SP-126, Vol. 1, American Concrete Institute, Detroit, 1991.

8. COHEN, M.D., YIXIA, Z. and DOLCH, W.L. "Non-air-entrained high-strength concrete - Is it frost resistant?" ACI Materials Journal, 89 (4), 1992, pp. 406-415.

9. KUKKO, H. and MATALA, S. "Effect of composition and aging on the frost resistance of high-strength concrete." Proceedings of the Second International Conference on Durability of Concrete (Ed. V.M. Malhotra), ACI SP-126, Vol. 1, American Concrete Institute, Detroit, 1991, pp. 229-248.

10. STURRUP, V.R., VECCHIO, F.J. and CARATIN, H. "Pulse velocity as a measure of concrete compressive strength." In Situ Non-Destructive Testing of Concrete (Ed. V.M. Malhotra), ACI SP-82, American Concrete Institute, Detroit, pp. 201-227.

11. BLIGHT, G.E., MCIVER, J.R., SCHUTTE, W.K. and RIMMER, R. "The effects of alkali-aggregate reaction on reinforced concrete strutures made with Witwatersrand quartzite aggregate." Proceedings of the 5th International Conference on Alkali-Aggregate Reaction in Concrete, Cape Town, 1981, pp. 13.

12. BLIGHT, R.E., ALEXANDER, M.G., SCHUTTE, W.K. and RALPH, T.K. "The effects of alkali-aggregate reaction on the strength and deformation of a reinforced concrete structure." Proceedings of the 6th Intenational Conference on Alkalis in Concrete, (Ed. G.M. Idorn and Steen Rostam), Danish Concrete Association (Dansk Betonforening, DBF), Copenhagen, 1983, pp. 410-410.

13. BLIGHT, G.E. and ALEXANDER, M.G. "Assessment of AAR Damage to Concrete Structures." Proceedings of the 7th International Conference on Concrete Alkali-Aggregate Reactions (Ed. P.E. Grattan-Bellew), Noyes Publications, New Jersey, 1986, pp. 121-125.

14. IMAI, H., YAMASAKI, T., MAEHARA, H. and MIYAGAWA, T. "The deterioration by alkali-silica reaction of the Hanshin Expressway concrete structures - investigation and repair." Proceedings of the 7th International Conference on Concrete Alkali-Aggregate Reactions (Ed. P.E. Grattan-Bellew), Noyes Publications, New Jersey, 1986, pp. 131-135.

15. ISHIZUKA, M. UHTO, S. and KUZUME, K. "Characteristics of road structures damaged by AAR on the Hanshin Expressway due to continuous observation." Proceedings of the 8th International Conference on Alkali-Aggregate Reaction, (Ed. K.Okada, S. Nishibayashi and M. Kawamura), Kyoto, 1989, pp. 771-778.

16. SWAMY, R.N. and WAN, W.M.R. "Use of dynamic nondestructive testing methods to monitor concrete deterioration due to alkali-silica reaction." Cement, Concrete and Aggregates, 15 (1), 1993, pp. 39-49.

17. HOBBS, D.W. "Some tests on fourteen year old concrete affected by the alkali-silica reaction." Proceedings of the 7th International Conference on Concrete Alkali-Aggregate Reactions (Ed. P.E. Grattan-Bellew), Noyes Publications, New Jersey, 1986, pp.342-346.

18. JONES, A.E.K., CLARK, L.A. and AMASAKI, S. “The suitability of cores in predicting the behaviour of structural members suffering from ASR.” Magazine of Concrete Research, Vol. 46, No. 167, 1994, pp. 145-150.

19. LESLIE, J.R. and CHEESMAN, W.J. "An ultrasonic method of studying deterioration and cracking in concrete structures." ACI Journal, 21, p. 17-36. Proceedings ACI, 46 (1), 1949.

20. THARMARATNAM, K. and TAN, B.S. "Attenuation of ultrasonic pulse in cement and mortar." Cement and Concrete Research, 20, 1990, pp. 335-345.

21. WHITCOMB, R.W., JACOBS, L. and AREF, L. "Quantitative ultrasonic evaluation of concrete." The International Conference on Nondestructive Testing of Concrete in the Infrastructure, Society for Experimental Mechanics, 1993, pp. 238-255.

22. OLSON, L.D. "Sonic NDE of structural concrete." Proceedings of Nondestructive Testing of Concrete Elements and Structures, ASCE, 1992, pp. 70-81.

23. AKASHI, T., AMASAKI, S. and TAKAGI, N. "The estimate for deterioration due to alkali-aggregate reaction by ultrasonic methods." Proceedings of the 7th International Conference on Concrete Alkali-Aggregate Reactions (Ed. P.E. Grattan-Bellew), Noyes Publications, New Jersey, 1986, pp. 183-187.

24. WEI-DU, L. "Frequency spectrum analysis of ultrasonic testing signal in concrete." Proceedings of Nondestructive Testing of Concrete Elements and Structures (Ed. F. Anscri and S. Sture), ASCE, 1992, pp. 104-114.

25. BOCCA, P. and ROSA, G. "Frequency spectra and ultrasonic pulse attenuation analyses for the assessment of damages in concrete." The International Conference on Nondestructive Testing of Concrete in the Infrastructure, Society for Experimental Mechanics, 1993, pp. 218-227.

26. BOCCA, P., BOSCO, C. CARPINTERI, A., INDELICATO, F., IORI, I. and VALENTE, S. "Nondestructive characterization of concrete and damage/fracture diagnosis of civil structures and infrastructures." The International Conference on Nondestructive Testing of Concrete in the Infrastructure, Society for Experimental Mechanics, 1993, pp. 1-20.

27. POPOVICS, J.S. and ROSE, J.L. "Survey of developments in ultrasonic NDE of concrete." IEEE Transactions on Ultrasonics, Ferroelectrics, and Frequency Control, 41 (1), 1994, pp. 140-143.

28. GALAN, G. Combined Ultrasound Methods of Concrete Testing, Elsevier, Oxford, 1990.

29. SANSALONE, M. and CARINO, N.J. CRC Handbook on Nondestructive Testing of Concrete (Ed. V.M. Malhotra and N.J. Carino), CRC Press, Boca Raton, 1991, pp. 275-304.

30. THOMAS, M.D.A., BLACKWELL, B.Q. and NIXON, P.J. "Estimating the alkali contribution from fly ash to expansion due to the alkali-aggregate reaction in concrete." Magazine of Concrete Research, Vol 48 (177), 1996, pp 251-264.

31. AFRANI, I. and ROGERS, C. "The effects of different cementing materials and curing on concrete scaling." Cement, Concrete, and Aggregates, Vol. 16 (2), 1994, pp. 132-139.

32. KRAUTKRAMER, J., and KRAUTKRAMER, H. Ultrasonic Testing of Materials. 4th Ed., Springer-Verlag, Berlin, 1990.

33. TESFAMARIAM, S., THOMAS, M.D.A., SADRI, A. and WIESE, D. "The use of stress-waves to evaluate concrete affected by alkali-silica reaction." International Conference on Structural Faults and Repair - 99, Kensington, England, July 1999.

34. SMITH, J. and DYER, B.C. 1990. "Seismic tomography as a tool for monitoring the condition of a concrete dam." Proceedings of the Institution of Civil Engineers, Part 2, 89, pp. 289-293.

35. THOMAS, M.D.A., CARATIN, H. and WIESE, D. "Three-dimensional visualization of concrete structures." In Innovations in Non-Destructive Testing of Concrete (Ed. S. Pessiki and L. Olson), ACI-SP 168, American Concrete Institution, Detroit, 1997, pp. 233-243.

REHABILITATION AND RETROFITTING OF AN ARCH DAM

A C Singhal

Arizona State University

United States of America

ABSTRACT. The Stewart Mountain Dam is a 64.6 m high double-curvature thin arch dam. The structure is located near Phoenix, Arizona, USA and was completed in 1930. It has experienced alkali-silica reactions within the concrete and exhibited no bond across horizontal construction lift surfaces. The dam could be subjected to upgraded maximum credible earthquake (MCE) of magnitude up to 6.75. Alkali-silica reactions and expansions have caused visible surface cracking. This structure was analyzed for gravity-, reservoir-, temperature-, and earthquake-induced loads. Results indicated an unsafe structure for earthquake conditions. Several measures for prevention of further deterioration and strengthening such as complete dam replacement, replacement of top layer of concrete, post-tensioning and epoxy-coated post-tensioning were considered. Epoxy coated post-tensioning cables were finally selected to provide seismic strengthening. Cable design uses the stiffness and acceleration response spectra methods. Vertical post-tensioned cables were installed during the 1990-92 construction phase. Post-tensioning cables are found to be a viable solution for the dynamic stability of a thin-arch dam. This paper summarizes various laboratory, field, construction and performance studies.

Keywords: Concrete, Thin-arch dam, Alkali-silica reactions, Post-tensioning, Earthquake design, Epoxy coating, Retrofitting, Rehabilitation.

Professor Avinash C Singhal is Professor of Civil Engineering, ECE 5306, Arizona State University, United States of America.

INTRODUCTION

The Stewart Mountain Dam located 66 km east of Pheonix, Ariz., on the Salt River, was completed in March 1930. The structure contains an arch dam, two thrust blocks for simulating abutments for the arch dam, three gravity dams, and two spillways. The arch dam measures 64.6 m high at the maximum section, 2.44 m thick across the crest, 10.36 m thick across the base, and 177.7 m in length along the crest. Four-keyed vertical contraction joints with copper water stops separate the arch into distinct concrete sections called cantilevers. Dams over 15.25 m in height are considered potentially susceptible to damage from seismic loading. Most of these high dams, built before computers became available for static analyses, do not meet current seismic standards. The concrete structure of the dam under consideration has experienced alkali-silica reactions and has exhibited no bond across horizontal construction lift surfaces. In addition, the dam could be subjected to upgraded maximum credible earthquake (MCE) of magnitude 6.75.

The arch dam had basically three problems: (1) the dam was not bonded together. Construction practices in 1930 did not recognize the importance of cleaning horizontal construction surfaces before placing subsequent lifts. Core drilling vertically through the dam in 1984 indicated 13 unbonded joints out of the 16 joints. These unbonded joints showed a thin layer of laitance along the surface. Having a high percentage of the unbonded lift surfaces at every 1.5 m height was a significant concern. The dam cantilevers would not perform like a monolithic structure as designed. (2) Concrete practices at the time did not recognize that certain aggregates react with the alkali in cement, causing a chemical reaction and producing a gel that causes voids and swelling of concrete. Measurements showed significant permanent displacement of 150 mm upstream and 75 mm vertical of the crest. (3) The postulated maximum credible earthquake at the site was a magnitude of 6.75 at 14 km producing an estimated peak ground acceleration of 0.34g. This amount of shaking was predicted to cause enough inertia force in the upper arch to cause instabilities. AS a result, many engineering questions that arose during the investigations and inspections: (A) What caused the poor lift surface bond and what was its extent?; (B) What is the serviceability of the existing or deteriorating concrete?; (C) At what rates are the alkali-silica reactions deteriorating the concrete?; (D) Is the concrete more brittle due to the micro-fracturing from the reactions?; (E) Why does the upper arch appear more susceptible to the alkali-silica reactions than the other areas of the dam?; and (F) Why do deflection measurements of the crest indicate a slowing or stopping of the rate of permanent drift towards the upstream direction?

Many dams built before 1945 and located in the southwestern United States, such as the Coolidge, Stewart Mountain, and Parker dams in Arizona and the Riant and Matilija dams in California, have shown signs of alkali-silica reactivity in the concrete. The Matilija dam [1] showed permanent displacement upstream at the crest [2] with concrete cores indicating alkali-silica reactions and deterioration in the upper 7.6 m. Modifications made to the Matilija dam included notching and enlarging the spillway. The Railroad Canyon Dam [2] in southern California has similar horizontal lift surface problems [1]. The dam, completed in 1928, consists of an arch dam portion with supporting thrust blocks. The dam was stabilized by placing additional concrete on the thrust blocks and installing six 890 kN post-tensioned cables in each abutment.

Current analyses overestimate seismic stresses in dams. Concrete dams are analyzed with state-of-the-art, finite element, computer programs for various combinations of thermal, hydrodynamic and seismic loads in order to simulate the conditions under which a dam must operate [3,4]. These analyses are based on the assumption of a linear, elastic continuous dam structure. It is these elastic analyses of existing older dams, which indicate that they do not meet current seismic standards and are potentially susceptible to severe seismic damage and attendant losses due to flooding of populated areas. Extensive seismic studies of existing concrete dams using computer codes based on the assumption of a linear, elastic, continuous structure predict unacceptably high dynamic, tensile stress [5-9].

A recent seismic analysis of Morrow Point Dam predicts seismically induced, tensile stress of over 131.1 N/mm^2 [8]. Analyses accounting for the additional flexibility provided by nonlinear, line discontinuities present in real dams would increase the accuracy of predicted, seismic stress. All dams, by their very nature, violate assumptions of their being continuous linear structures. Horizontal lift lines and vertical construction joints violate the assumption of continuity of the dam mass and cannot be adequately modeled linearly [8,10]. Recent computer codes include the effects of nonlinear discontinuities, e.g. ADINA and ABACUS [11,12,], but linear, continuous-structure codes are still the only ones extensively used by various governmental agencies [6,8] and many of these are outdated, some being over 40 years old. Many new non-linear finite element codes are under development [13,14].

FIELD DATA

Alkali aggregate reaction in the dam concrete was especially noticeable in the upper elevations. A core drilling and laboratory program was initiated in 1984 to determine the vertical extent of alkali aggregate reaction in the dam, to determine engineering properties and to determine the remaining potential for further alkali aggregate reaction. This effort showed: (1) Unbonded lift surfaces were caused by construction practices. (2) The interior concrete was relatively strong with a compressive strength of 37.21 N/mm^2. (3) Deflection measurements at the crest had slowed significantly since 1964. (4) Alkali aggregate reaction had run its course and further expansions of the dam from alkali aggregate reaction were not expected.

The foundation is pre-cambrian intrusive rocks, mainly quartz diorite cut by irregular dikes of granite and smaller diabase and silicic dikes. The rock becomes increasingly more fractured from the right to left abutment. Major shear exist in the original river channel beneath the arch dam (tailrace shear) and beneath the spillway (spillway shear).

MODIFICATION CONSIDERED

Seismic analysis of the dam showed that the arch portion of the dam is potentially unstable during a MCE seismic event. Justification of the decision to modify the structure and the chosen method of modification is based upon the following findings.

1. Inertia forces at the crest of the arch are large judging from the resulting peak accelerations of 2.32g at the crest.

2. A linear finite element analysis of calculated tensions indicates that the arch dam pulls apart horizontally with a duration of up to 0.1 seconds, long enough for concrete blocks to slide.

3. Horizontal lift surfaces are laitance filled and exhibit little or no cohesion.

4. Vertical contraction joints are keyed but provide little resistance against the sliding of the massive concrete blocks.

5. Uniaxial compression tests on 152 mm cores extracted from the dam interior indicate very strong concrete of about 37.21 N/mm^2 compressive strength. Alkali-aggregate reaction has not deteriorated the dam to the point requiring its total replacement.

Seismic analysis revealed that the dam will not perform dynamically as a monolithic unit, because of the unbonded horizontal lift surfaces.

POST-TENSIONING

Alternatives to improve the dynamic stability of the dam included complete dam replacement, replacing the top 12 m of concrete, epoxy injection and post-tensioning cables. Post-tensioned cable were by far the least expensive and practical alternative because of the strong interior of the existing concrete. Post-tensioned tendons increase the normal force on the unbonded horizontal arch lift line surfaces and consequently the frictional component of sliding. Cables also produce three-dimensional stresses throughout the arch section depending on orientation and eccentricities. Post-tensioning induces two equal and opposite loads at the ends of the free length. The load at the top transfers through the bearing plate into the concrete. The load can be considered a concentrated force. Load at the bottom develops through bond along the embedment length of the cable.

Dynamic finite element analyses were done to calculate the inertia forces of the concrete blocks in the dam during the MCE. To inhibit the concrete from sliding during this seismic event, 62 cables consisting of 22 seven wire strands (15.24 mm diameter strands) were required to produce enough normal force across the horizontal lift surfaces at design load.

LABORATORY TESTING ON THE TENDONS

Full Scale Grout Test

Full scale laboratory tests were made for designing the grout mix and observing how grout flows through the cable assembly. A clear 15 m high 254 mm diameter plexiglass pipe was vertically supported and equipped with tendons, grout pipes, and spacers inside the pipe. This provided a viewable tendon hole. As grout flowed from the bottom of the pipe, the flow patterns were observed. The amount of bleed water and settlement of the grout column was also determined.

Strand Tension Tests

Tension tests were performed on the individual epoxy-coated tendon strands to observe the interaction between the steel strand and the epoxy coating. The strand and epoxy stayed together at all stages.

Strand Pullout Tests

Pullout tests were performed on bare steel strand and epoxy coated strand grouted into concrete cylinders. One end of the strand protruded from the cylinder to be tensioned. The other end was flush with the end of the cylinder to observe the initiation of slippage. Once slippage initiated, the bare strand unraveled (i.e. the wires separated) and pulled out of the cylinder; the epoxy-coated strand did not readily pull out.

PRE-INSTALLATION TESTING ON THE TENDONS

Full Scale Load Tests

Full scale pullout tests with 3 m and 6 m bond lengths were made in the good right abutment, poor tailrace shear zone, and fractured left abutment. Even with these relatively short bond lengths, only the 3 m bond length in the tailrace shear zone failed.

Survey Evaluation Test

Down hole surveys were made with an Eastman-Christensen Seeker-1 rate gyro survey equipment to test the accuracy of the equipment. A metal pipe was hung from the from the spillway at a slight angle and surveyed by conventional equipment. The results showed the acceptability of the equipment.

Drill Evaluation Test

Drilling of the 254 mm diameter cable holes was performed with a Casagrande Model C-12 diesel hydraulic drill, equipped with a down-the-hole hammer. The first hole drilled in the arch dam was an evaluation test. This was one of the more difficult holes at 70 m in depth, slanted at 7° 10' off the vertical, and oriented at a bearing of 344°. The hole was within 140 mm of the desired location at 70 m. Four geophones on the dam surface measured a peak particle vector velocity sum of 3.75 mm/sec. This is well below the allowable criterion of 51 mm/sec for blasting at Reclamation structures.

CABLE CAPACITY AND CONSIDERATIONS

Designing and constructing the cable system for the dam was especially challenging as it is curved in plan and in section. The aforementioned inertia forces were computed at fifteen locations along the crest. A design cable load is 3114 kN per 3.05 m spacing along the crest. The cables were positioned within the arch dam as close to the centerline of the vertical section as possible.

Finite element studies showed a beneficial stress distribution within the arch dam created by the cable during normal operating conditions. Special design considerations and requirements were developed for drilling methods, drilling accuracy and tolerances, tensioning sequence, placement within the arch, corrosion protection, grouting, monitoring, and pre-stressing.

For drilling two tolerances were established. First, to position the cables within the dam as close as possible to the centerline of the vertical sections and second, to minimize friction losses, cable damage and corrosion damage. For corrosion resistance the preferred alternative was epoxy coating.

The design intent of the tendon was to provide a uniform normal force on the unbonded horizontal lift surfaces for dynamic stability. As the tendons were not able to be placed at equal spacing due to contraction joints, it required each tendon to have a different load. Hence, to load the arch uniformly the tendons were stressed in a specific sequence. Once the free length of the tendon is grouted, the load on the dam and within the cables cannot be removed. Hence, a prudent measure was to evaluate the structural performance of the tendons before grouting itself. This was done through laboratory tests such as: full scale grout test, strand tension tests and strand pullout tests, and through pre-installation field tests such as: full scale load test, survey evaluation test and drill evaluation test.

Special care was also taken at the time of cable installation and during construction. The performance of the dam was checked after construction of the cable system.

CONCLUSIONS

The Stewart Mountain dam has been deteriorated by alkali-silica reactions and exhibits no bond across horizontal lift surfaces. In addition, it is now required to be subjected to an upgraded maximum credible earthquake. Trends from historic deflection measurements, concrete coring programs, and laboratory tests indicate that the deterioration from the alkali silica reactions is contained. A system of post-tensioning for arch stabilization was chosen. Ease of design and cable load control were among the factors in this selection. The continuous monitoring of the performance at various stages have shown satisfactory results. Post-tensioned cables are a viable solution for the dynamic stability of a thin arch dam.

ACKNOWLEDGEMENTS

Financial support for this research work was received through the U.S Bureau of Reclamation. Mr. L. Nuss of the USBR worked on this project. Mr. K. Ramanathan, a graduate student at the Arizona State University, participated in this research.

REFERENCES

1. INTERNATIONAL ENGINEERING CO., "Matilija dam- Stress Investigations," Report for the Department of Public Works, County of Ventura, International Engineering Co., Inc., Ventura, Calif, 1972,

2. WOODWARD CLYDE CONSULTANTS, "Railroad Canyon dam safety evaluation," Final report for TEMESEAL Water Company, Woodward Clyde Consultants, San Francisco, Calif., 1984.

3. SINGHAL, A. C. , "Nonlinear dynamic response of bonded and unbonded rings under high intensity blasts," J. Indian Inst. Eng., 1969, **50**, pp. 1-11.

4. SINGHAL, A. C. , "Comparison of computer codes for seismic analysis of dams," Comput. Struct., 1991, **38**(1), pp. 107-112

5. NUSS, L. K. , AND SINGHAL, A. C. , "Static and dynamic stability at Stewart Mountain Dam" Technical Memorandum, SM-220-01-86, US Bureau of reclamation, Denver, CO, 1987.

6. SINGHAL, A. C. , "Structural engineering issues-dynamic analysis of concrete structures of Kortes Dam," Technical memorandum, KD-221-1, Dam and waterway design, US Bureau of Reclamation, Denver, CO, 1986.

7. SINGHAL, A. C. , "Evaluation of the effects of water compressibility on seismic behavior of dams," Technical memorandum, SM-221-1, Dam and waterway design, US Bureau of Reclamation, Denver, CO, 1986.

8. SINGHAL, A. C. AND NUSS, L. K. , "Dynamic earthquake analysis of Morrow point dam structures," Technical memorandum, MP-221-1, Dam and waterway design, US Bureau of Reclamation, Denver, CO, 1986.

9. SINGHAL, A. C. AND NUSS, L. K. , "Cable anchoring of deteriorated arch dam", J.Perform. Constructed Facilities, ASCE, 1991, **5**(1), pp. 19-36.

10. BOGGS, H. L. AND MAYS, J. R. , "Effects of vertical construction joints on the dynamic response of the arch dams," Joint meeting, NSBIR 87-3540, US Department of Commerce, National Bureau of Standards, Washington, DC, 1987.

11. ADINA ENGINEERING, "Automatic dynamic incremental non-linear analysis," Rep AE84-1, ADINA Engineering, Watertown, MA, 1984.

12. ROW, D. AND SCHRICKER, V., "Seismic analysis of structures with localized nonlinearities," Proc. 8th World Conf. On Earthquake Engineering, San Francisco, CA, 1984.

13. SINGHAL, A. C. and ZUROFF, M .S., "Dynamic analysis of dams with nonlinear slip-joints," J. of Soil Dynamics and Earthquake Engineering, Vol. 17, 1998, pp. 185-196

14. SINGHAL, A. C. AND NUSS, L. K. , "Performance of a retrofitted arch dam," Engineering Mechanics: A Force for the 21st Century, Proceedings of the 12th Engineering Mechanics Conference, American Society of Civil Engineers, New York, 1998, pp. 845-848.

THEME FIVE:

DEVELOPMENT AND DIAGNOSTIC TECHNOLOGY

Keynote Paper

OPERATION SUCCESSFUL, PATIENT DIED – THE ASSESSMENT OF STRUCTURES AS AN ENGINEERING (OR A MEDICAL) PROBLEM

T Vogel

Swiss Federal Institute of Technology

Switzerland

ABSTRACT. The starting point is parallels and differences between medical diagnosis and structural assessment. The development of radiography and other medical diagnostic procedures is described. Some similar tasks of medicine and the preservation of structures are listed. The state of the art of non-destructive testing is illustrated with the problem of not properly grouted prestressing tendons in multispan bridges. Different procedures for the preservation of structures are treated, showing the level of consideration. Probable and desirable trends are extrapolated, comparing with known developments in medicine. A possible solution for the above-mentioned example is sketched.

Keywords: Ducts, Grouting, Imaging methods, Monitoring, Non-destructive testing (NDT), Prestressing tendons, Screening, Voids.

Professor Thomas Vogel is staff member of the Institute of Structural Engineering of the Swiss Federal Institute of Technology (ETH), Zurich. His main interests in research are in the field of preservation of structures in general and in the evaluation of concrete structures in particular. He is vice-chairman of Working Commission 1 of IABSE and till March 1999 chairman of the Zurich section of the Swiss Society of Engineers and Architects.

INTRODUCTION

The assessment of structures is an engineering problem of growing importance, because the stock of structures is getting older and is not being replaced before reaching its service life as has been the case in phases of expanding infrastructure.

Medical doctors have always been faced with a similar problem, i. e. to determine the state of the health of a patient or in the case of bad health to diagnose the disease.

This paper is written by an engineer with some experience in structural engineering but no special knowledge in medicine, being just an interested consumer of newspapers and other sources of information. It does not summarise solutions recently found but attempts to show in what direction development may be taken to have a fair chance of success.

MEDICAL COMPARISON (DIAGNOSIS)

Parallels

Structural engineers design and construct bridges and buildings and hand them over to the owners for commissioning. Users do not care much about structures as long as they perform well, but as soon as they doubt structural safety or fault serviceability they remember the structural engineers. In the last few decades the idea arose that maintenance should be provided by engineers in order not to wait until problems arise.

Medical doctors have a similar task regarding human beings. Until recently they left the genesis of individual life to nature but afterwards we need them during our whole lifetime as soon as mind or body do not function properly anymore. They also would prefer to be asked preventatively, instead of acting only when the damage has already occurred. Their working method is similar to that of engineers. The medical history of the individual is recorded, actual findings are added, and with all this information and the knowledge of the doctor a diagnosis is made, so that the problem can be defined. Subsequently one tries to solve the problem, using all available methods.

Differences

There are, however, some differences between patients and structures: To fight for the life of a patient is a purely medical task; extending the lifespan of a structure is urgent only within economical constraints. Regarding the rising costs of our health system, however, economical considerations on what treatment is justified for what patient are not out of the question anymore. On the other hand, some structures have to be preserved regardless of costs, being part of the cultural heritage or are irreplaceable. Patients can be interviewed, structures cannot and show signs of damage only occasionally.

So the better comparison would be with veterinarian medicine, because most animals have primarily an economical value and cannot answer the questions of the doctor. With structures economical constraints, however, do not work in all cases and the answers of patients are not always reliable and have to be interpreted as well. So let us stick to our comparison between human medical diagnosis and the examination of structures.

SOME SUCCESS STORIES IN MEDICINE

Let us start with some success stories of medicine. As patients do not want to be harmed by a routine health examination, the diagnosis of a disease or just for the sake of medical progress itself, medicine developed a large range of diagnostic methods that do not harm the patient. As engineers we would call them non-destructive testing methods.

X-Rays

Starting with the discovery by Wilhelm C. Röntgen in 1895 X-rays have become an indisputable tool of any diagnosis, Figure 1. As the first applications were by no means harmless, doses could be reduced subsequently and methods were developed to amplify the rays after penetration of the body and to enlarge contrasts.

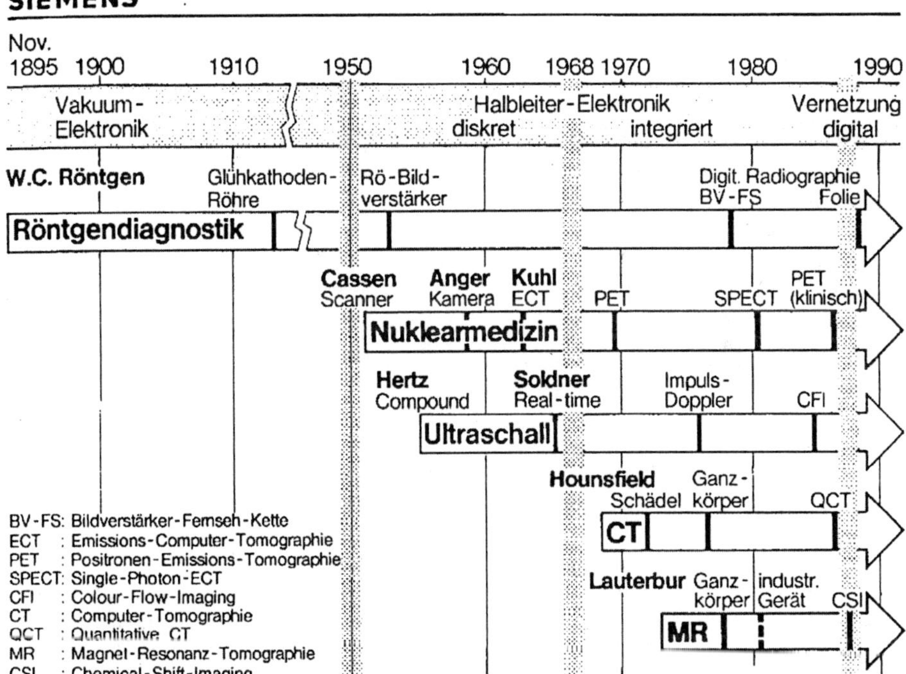

Figure 1 Innovation steps in imaging diagnostics [1]

X-ray plates are always projections of a spatial body on a plane and therefore involve a loss of information. The growing speed of data processing and the mathematical algorithms developed allowed the construction of computed tomography that cuts virtual slices out of a spatial object.

Since imaging techniques were further improved, these pictures can be recognised or even understood by a lay person, for instance by the patient himself.

Other Imaging Methods

Other physical phenomena like acoustic wave propagation, radioactivity of artificial isotopes and magnetic resonance were used as diagnostic tools as soon as the corresponding equipment was available.

Figure 1 shows the development of radiology, nuclear medicine, ultrasonics, computed tomography and magnetic resonance tomography over the last hundred years. Medicine played a predominant role in the development of these methods and often devices were ordered before they existed and purchased before it was defined what they were needed for [1].

One reason for their success is that the results can be displayed on a screen, printed or saved as a picture in another way. That means that the most powerful interface between machinery and man, the visual sense, is used.

Combination of Methods

Every time a new method appeared it was doubtful if the older ones would survive because of their known disadvantages. They always did because only the skilled combination of different methods using their respective advantages gives the best results.

SOME PROBLEMS IN MEDICINE

Health Costs

The increase of life expectancy and the low reproduction rate lead to an ageing of the population. Diseases that are caused by wear and defective genetic codes increase.

Technical methods to detect these diseases are available and therefore used, even if they cannot prolong lifetime. It is doubtful if that patient who refuses to fight an early detected cancer lives longer than that one who does not know anything until it is too late. On the other hand, using all the technical medicine available, the last year of a patient costs as much as that of the rest of his life before.

Economics feed back and market laws are not working, because the one who causes the costs is not the one who has to pay them. A competition may work for standards asked but not for the associated costs.

Education

Physicians belong in all countries to the professions with the highest reputation and in spite of the increasing health costs society honours them with a high income.

In most universities the faculties of medicine are crowded, restrictions of funds do not allow increasing the educational staff and the quality of education is endangered. Other restraints like the number of patients, the willingness to participate or even the number of corpses for autopsies limit the number of students as well.

Medical faculties react by increasing the requirements for applicants and in doing so raise the reputations of those passing even further.

SOME PROBLEMS IN ENGINEERING

Let us face the corresponding problems in engineering.

Maintenance Costs

In developed countries the built infrastructure reaches a high level. In Switzerland for instance concrete used amounts to 80 m^3 per capita, a volume that would correspond to 20 mm if it were spread over the whole country.

A considerable part of that concrete is used in structures. It is commonly agreed that one to two percent of the capital invested in the infrastructure should be spent per year for maintenance and repair to maintain the level of its performance. All industrial countries are faced with the problem of raising the required funds, to spend the money wisely and to set priorities in cases where the necessary budgets cannot be reached.

Education

Civil engineering faculties have problems keeping up the number of students because the field is considered to be less demanding than other disciplines and without a future. Further easing of access to studies are discussed, despite the demanding questions we are facing.

DETERIORATION OF CONCRETE STRUCTURES

Although reinforced concrete structures are in most cases durable, configurations exist where this is not the case. Let us concentrate on the nastiest ones, namely hidden defects due to bad detailing and workmanship and damages caused thereby.

The Harmfulness of Destructive Testing

The first idea is to search for such defects and damage wherever they are expected. In contrast with humans, structures do not heal. Holes from core drilling and joints in waterproofing accumulate and are often the initiation of further deterioration.

Non-Destructive Testing

Reinforced concrete is a complicated material for all non-destructive testing methods, because concrete itself is a composite with constituents of great variety and reinforcement normally is placed along the surfaces and shields the core in every respect. Nevertheless, many methods of non-destructive testing are available and have been developed considerably. They work under laboratory conditions and for modest size and scale. As far as real structures are concerned, success depends on the experience of the assessor and the questions asked.

Example

One of the difficult questions for post-tensioned concrete bridges is, whether all ducts of the prestressing tendons have been grouted properly. Many countries have found examples within their stock of bridges where this has not been the case.

In Germany a comparative test on concrete specimens with an only partly grouted prestressing tendon (Figure 2) has been made [2]. The test methods applied were radar, impact echo and various ultrasonic pulse echo methods. The thickness of the specimen and the location of the duct could be found by all methods more or less accurately. Void and filled parts of the duct, however, could only be distinguished by impact echo and two ultrasonic pulse echo methods using image reconstruction and for each of these three methods from one side only. Table 1 shows an overview on the methods used and their success in distinguishing between void and filled ducts.

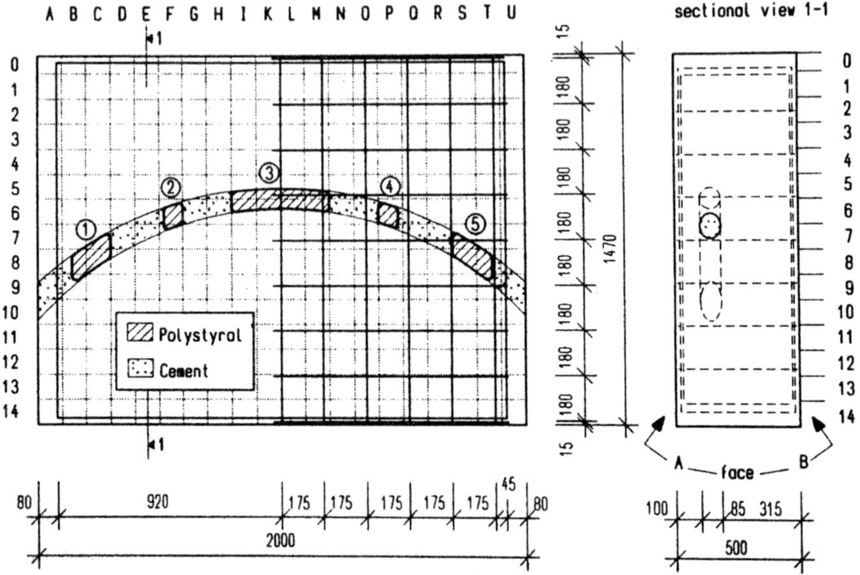

Figure 2 Plan of concrete specimen of comparative test [2]

Table 1 Results of comparative tests [2]

NO	METHOD		DATA REPRESENTATION	DISTINCTION VOID/FILLED DUCT
1	Radar		B-scan	Impossible[a)]
2	Impact echo		Frequency domain	From surface A
3	Ultrasonic pulse echo	Single element	A-scan	No results
		Separate transmitter ,receiver	A-scan	No results
4		Array	A-scan	No results
5		Single element	B-scan	No results
6		Single element	B-scan, LSAFT[b)]	From surface A
7		2D-Aperture (scanning laser vibrometer)	Phase shifted superposition	No results
8			3D-SAFT[c)]	From surface B

a) Due to electromagnetic shielding

b) Linear synthetic aperture focusing technique

c) Three dimensional synthetic aperture focusing technique

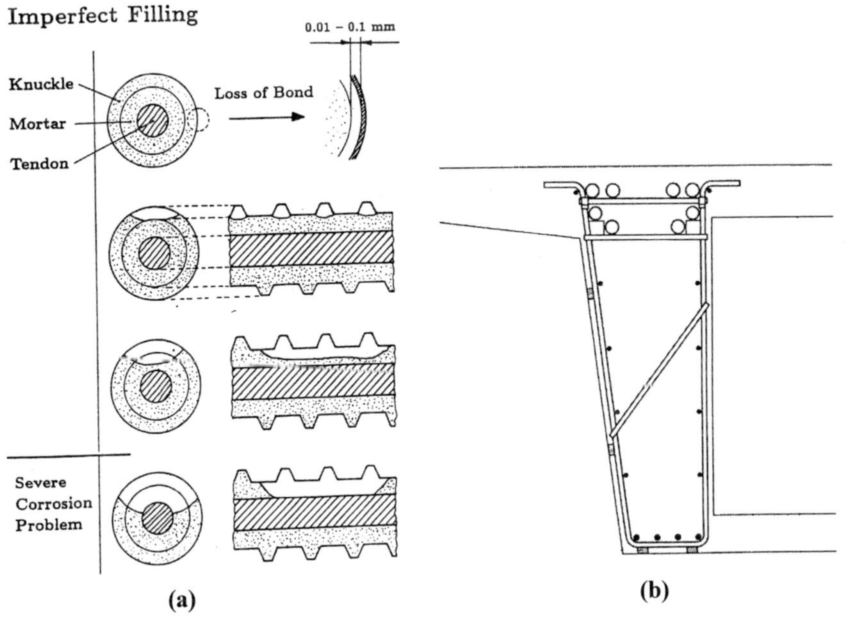

Figure 3 Grouting of post-tensioning tendons in a bridge section
(a) Different stages of imperfect filling of a duct
(b) Example of a pier section of a bridge

The principal difficulty is that the size of the voids to be detected is close to that of the maximum size aggregates that are of no interest. Furthermore, small delaminations, that occur by shrinking of the grout but do not cause harm, give similar signals to those of empty ducts. Even voids detected properly are not necessarily defects that need to be repaired as long as the prestressing steel is still covered by grout (Figure 3 (a) from [3]).

Although the specimen used for the comparative tests looks quite realistic, real multispan bridges are normally more complicated. Bundles of four or even more tendons are concentrated and the lateral access to the upper region is impeded by the deck slab (Figure 3 (b)). The deck slab is normally covered by a waterproofing membrane and an asphalt layer that should not be removed for testing. This configuration causes problems for other non-destructive testing methods that could be applied to detect the voids or the subsequent corrosion as well.

THE ENGINEERING TASK

Questions Asked

Neglecting all aspects of serviceability and concentrating on safety, the engineering problems mentioned above can lead to the following questions:

- Are our structures safe?
- How much can rehabilitation funds be lowered without affecting safety?
- Is that very special structure safe and how long will it remain safe?

As always the simpler the questions the more complicated the answers.

Different Levels of Considerations Required

As already implied by the questions the measures to answer them will depend on the level of consideration and decisions [4]. Let us distinguish the following tasks depending on the questions asked.

Screening

We know that critical configurations and details of a whole population of structures exist because

- unsuitable materials have been used (concrete showing alkali-silica reaction, sigma-steel etc.)
- actions have been underestimated in frequency and magnitude (increasing axle loads, impact, earthquake etc.)
- bad workmanship was tolerated and control was not effective enough (honeycombs, poorly grouted ducts etc.)

Although such structures could be detected by having a look at every single one, faster and cheaper methods are needed to concentrate on those few, where a more detailed investigation is justified. That is why screening of data bases like construction plans or facility and bridge management systems is required.

Such data and procedures are helpful in the case of a major structural failure as well. After such an incident the press, the public and therefore also the politicians want to know within hours what similar structures are endangered, to be able to show resolution in forbidding something. The more precise we can answer this question, the fewer overreactions must be feared.

Inspections

Inspections are part of normal preservation procedures and should be made regularly on all major structures. The first and most important method is visual inspection, because it is simple and not limited to expected phenomena. If all critical situations and details have to be detected, additional methods have to be used, which should not increase the costs considerably, however. That is why only such methods can be taken into account that are simple or have at least the potential to reduce personnel costs. Results should be recorded in an objective way to be compared over a longer time period.

Examinations

An examination consists of a condition survey (assessment in a narrow sense) and a review (also called evaluation) that leads to a recommendation to the owner and is executed on special occasions only, i. e. when significant defects or damage have been observed, a modification of the construction is considered or a change of use or accidental loads have occurred. This sentence, taken from the Swiss recommendation on the preservation of concrete structures [5], sounds clear and assumes, that a potential defect or a potential critical stage of a deterioration process is detected already by prior screenings or inspections.

An examination is always an extraordinary procedure, that is not undertaken periodically. That is why more sophisticated methods are justified, because the option to do nothing at all normally does not exist, which means that costs will incur anyway. The investigation has to proceed stepwise, from the general to the details, always keeping in mind the aim. To decide whether the money should be spent either for examination and testing in order to know more or for strengthening or replacement even if not absolutely necessary, is a typical engineering problem. Even with a detailed and comprehensive examination not all questions will be answered. Where questions remain it has to be judged if undetected defects and damage may lead to a sudden or brittle failure or if there is a chance that they will be detected in time by users, inspectors or monitoring systems.

Monitoring

Structures other than machines should function without permanent control of key values. Observant users, however, provide a certain amount of control also for structures. In special cases a permanent monitoring can be appropriate, for instance to ensure safety during the planning of a rehabilitation or to extend the residual service life of a damaged structure before replacement. It is essential to monitor parameters of a structure that give a warning early enough to take appropriate measures. Such parameters will be either global ones that give information on the whole structure or local ones if the most critical parts are identified. It is also possible to record relevant incidents such as heavy loads or breaking wires and strands. Such recording systems, however, need a high reliability over long periods, because missed incidents distort the evidence.

LEARNING FROM MEDICINE

Application of Non-destructive Testing Methods

At first let us try to extrapolate the state of the art of non-destructive testing methods following the success story of medical application.

Imaging methods and interpretation of data

Whenever possible data should be provided online and in a visual form. In that way the amount of data is already condensed, it fits best to visual inspection and can be combined with what the assessor sees, hears, feels and smells. In that case the assessor has to be an educated person that understands the physical principles the testing method is based on. As with radiologists that are doctors and not nurses, such condition survey should be done by specialised engineers.

In cases where this procedure becomes too expensive, measures have to be taken so that the valuable information of the site can be transferred to the office. In tunnel maintenance powerful tools have been developed to scan surfaces automatically, using visual and infrared scanner and optical distance measurements. Suitable is almost every method that does not need to touch the concrete surface but can be applied from a distance of some metres (Figure 4 (a) from [6]). Like this all relevant data can be stored and transferred to the office, where the specialist interprets the graphs more efficiently.

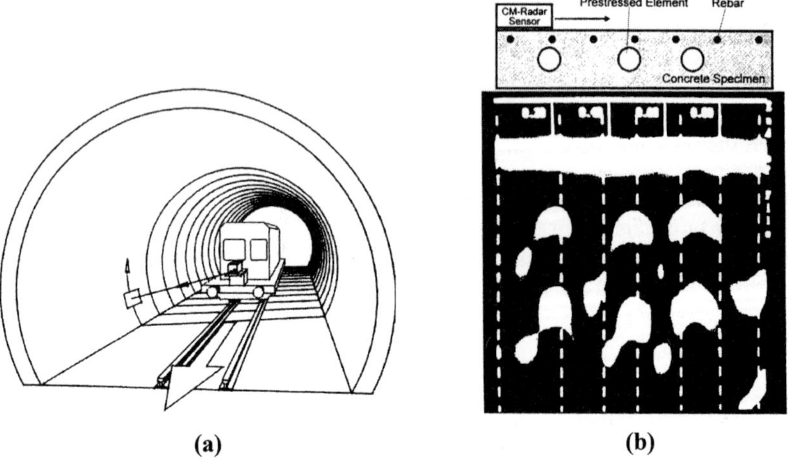

Figure 4 Application and combination of imaging methods (a) Tunnel scanner on rails (b) Combination of covermeter and radar

Data processing

Probably the procedure of interpreting can be partly automated in the long term, using neuronal networks and other pattern recognition procedures. At least those algorithms already available to enlarge small differences of response signals may be used and tested for relevance.

Combination of methods

To combine different methods a common orientation system is needed. Like this the results can be used as different layers of the same picture. A covermeter for instance has been combined with radar allowing to judge whether a radar signal is caused by a mild steel bar at the surface or a prestressing strand underneath (Figure 4 (b) from [7]).

Assessment of Structures

Let us now try to transfer some procedures of medical diagnosis to the examination of structures.

Screening

In medicine risk profiles exist for almost every disease. In information technology tools like data mining and knowledge discovery have been developed to draw relevant information out of a growing amount of data that cannot be handled anymore. Until now these tools were primarily used in marketing to construct profiles of potential customers, but application in other fields will follow [8], [9], [10], [11].

Inspections

Pulse and blood pressure are measured at every consultation, because it is so simple and deviations from the normal have to be explained. For structures global parameters like deflections and eigenfrequencies give an information on the overall behaviour. The actual condition of critical parts of a structure should be recorded, like the biannual X-ray picture of the teeth by the dentist. Imaging methods are suitable, in the simplest case a photograph taken from a well defined position under well defined conditions. Even if the filed phenomena cannot be explained their development in time can be observed.

The crucial question is whether all critical parts have been detected. The patient goes to the doctor when he experiences pain. We should consider a similar measure for structures. Acoustic emission has the potential to give early warnings when former stress levels are exceeded due to increasing actions or declining resistance.

Examinations

As mentioned before the danger is evident that methods are used just because they are available without ascertaining whether the eventual results will influence the subsequent actions. In that field doctors could learn from engineers, i. e. to solve a problem with a certain amount of money within a certain period of time.

Examinations should concentrate on safety and durability problems. Serviceability criteria may play a predominant role in design, but they need not be the central point of an examination. A structure in use is judged by the users and they decide whether it is satisfactory or not. To objectify these subjective opinions, however, measurements are justified.

As far as safety is concerned emphasis should be concentrated on sudden and brittle failures because in these cases no time is left to take other measures.

Monitoring

Monitoring can be compared to an intensive care unit. That means that it is suitable for a restricted time only and is connected with substantial costs. It cannot be an aim of civil engineering to turn structures into machines because we cannot afford such structures over decades or centuries.

Investigations during Demolition

Due to change of use and other reasons some structures are decommissioned and demolished before they have reached the end of their service life. Like in pathology these cases are a valuable source of information that otherwise could not be gained. They could be better integrated into education to provide a feed-back from real behaviour over time to improve design.

CONCLUSIONS

Consequences for the Example

To conclude let us come back to the example with the improperly grouted ducts and propose a strategy, how safety can be guaranteed as long as non-destructive detection of the deficiency does not operate.

To prevent prestressing steel from corroding even if it is not covered by grout, humidity has to be kept out by providing and maintaining a reliable waterproofing membrane. The long-term performance of that membrane can be checked by thermography and ground penetrating radar [12].

Of course this procedure is not as reliable as a direct investigation of the prestressing strands. That is why a worst case scenario should be calculated, assuming that one or all prestressing strands are corroded and the flexural resistance of a pier section is decidedly reduced. With simple methods of the theory of plasticity (Figure 5) it can be shown that continuous beams in most cases could still carry at least the dead load with a reduced safety margin that is also used for hazard scenarios with accidental loads [13].

To be sure that this strategy works, prestressed concrete bridges that are demolished for other reasons should be investigated systematically concerning the state of the prestressing steel and the quality of grouting.

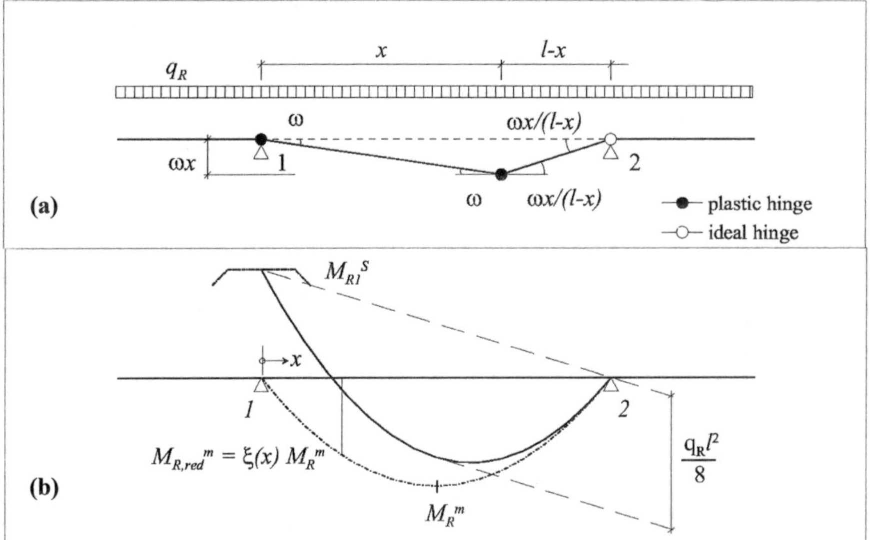

Figure 5 Estimation of the structural capacity of a damaged intermediate span
(a) Mechanism of upper bound method
(b) Bending moments of lower bound methods

REFERENCES

1 GUGERLI, D. Die Automatisierung des ärztlichen Blicks (The automation of medical observation). Inaugural lecture ETH Zurich March 18,1998, http://www.tg.ethz.ch/, pp. 1-12.

2 KRAUSE, M et al. Comparison of pulse-echo methods for testing concrete. NDT&E International, 1997, Vol. 30, No. 4, pp. 195-204.

3 JANSOHN, R, KROGGEL, O and RATMANN, M. Detection of Thickness, Voids, Honeycombs and Tendon Ducts Utilising Ultrasonic Impulse-Echo-Technique. Proc. Internat. Symposium 'Non-Destructive Testing in Civil Engineering' (NDT-CE), Berlin September 26-28, 1995, DGZfP, Berlin 1995, pp. 419-427.

4 VOGEL, T. From Structural Assessment to Appropriate Preservation Strategies, Report IABSE Symposium 'Extending the Lifetime of Structures', San Francisco August 23-25, 1995, Vol. 73/1, pp. 681-686.

5 SIA 162/5. Erhaltung von Betontragwerken (Preservation of concrete structures). Recommendation, Edition 1997, Swiss Society of Engineers and Architects, Zurich 1997, 44 pp.

6 RÖSHOFF, K, MEIXNER, J, WIESLER, A. Laser Scanning for Tunnel State Assessment. Proc. IABSE Colloquium 'Tunnel Structures', Stockholm June 4-6, 1998, International Association of Bridge and Structural Engineering, Vol. 78, pp. 457-468.

7 PÖPEL, M, FLOHRER, C. Combination of a Covermeter with a Radar System - an Improvement of Radar Application in Civil Engineering. Proc. Internat. Symposium 'Non-Destructive Testing in Civil Engineering' (NDT-CE), Berlin September 26-28, 1995, DGZfP, Berlin 1995, pp. 737-743.

8 DILLY, R. Data Mining, An Introduction. Queen's University of Belfast, Parallel Computer Centre (PCC), Online Version 2.0, Feb 1996, http://www-pcc.qub.ac.uk/tec/courses/datamining/ohp/dm-OHP-final_1.html

9 DEJESUS, E X, Data Mining. Byte online, October 1995, http://www.byte.com/art/9510/sec8/art1.htm

10 WATTERSON, K. A Data Miner's Tools, Byte online, October 1995, http://www.byte.com/art/9510/sec8/art8.htm

11 TECHNISCHES KOMMITTEE DER DAGM. Positionspapier zu Datamining und Knowledge-Discovery: Was für Mustererkennungsprobleme sind zu lösen? (State-of-art report on data-mining and knowledge-discovery: What pattern recognition problems are to be solved?), Deutsche Arbeitsgemeinschaft für Mustererkennung, http://www.ipb.uni-bonn.de/DAGM/positionspapier.html

12 de BOSSET, C, ROBERT, A. Non-destructive testing methods applied to bridge decks in order to localize the defects associated with the waterproof membrane, Proc. Internat. Symposium 'Non-Destructive Testing in Civil Engineering' (NDT-CE), Berlin September 26-28, 1995, DGZfP, Berlin 1995, pp. 713-720.

13 VOGEL, T. Safety for Progressive Failure of Concrete Bridges with Damaged Prestressing Strands, Proc. Symposium 'Advanced Design of Concrete Structures', Chalmers University of Technology, Gothenburg, June 12-14, 1997, International Center for Numerical Methods in Engineering (CIMNE), Barcelona 1997, pp. 207-214.

DATA ANALYSIS OF STRAIN GAUGES DETECTING STEEL REBARS CORROSION

G Batis

Th T Routoulas

National Technical University of Athens

Greece

ABSTRACT. A brief presentation of a laboratory technique of detection of steel rebars corrosion in mortar specimens with Strain Gauges (SG) is given. The analysis of SG measurement data is discussed in this paper. The present technique is based on the phenomena of tension state, in which the mortar mass comes near the rebar area, during the formation of corrosion products. SG embedded in reinforcing mortar specimens during casting, monitor the internal stresses, after the compensation of disturbing parameters. Several types of mortar specimens with admixtures and corrosion inhibitors tested under impressed potential, immersed in a 3,5% wt NaCl solution to accelerate corrosion conditions. The applied potential, the resulting current and the SG signal converted by a bridge in mV were plotted versus time. Corrosion data measured through SG was correlated with that obtained from charge resistant transfer,(calculated from current plots). The dynamic response of SG sensors in variations of corrosion potential applied, was also investigated. The reliability of the technique was also evaluated by the gravimetric weight loss determination of the rebars. The test results obtained, indicate that the SG technique, directly related to corrosion products, is suitable for the laboratory study of the corrosion factors and the influence of concrete admixtures in corrosion protection.

Keywords: Strain gauge, Corrosion, Reinforcing rebars, Electrical charge transfer, Volume changes compensation, Weight loss.

George Batis Dr Chem Eng Asst Professor, National Technical University, Chemical Engineering Department, Materials Science & Engineering Section, Athens, Greece.

Th T Routoulas Chem Eng MSc, Technological Educational Institute of Piraeus, Physics, Chemical and Material Technology Department, Egaleo, Athens, Greece.

INTRODUCTION

It is widely accepted that the alkaline environment of the cement paste around steel reinforcement protects the steel from corrosion due to the formation of a Fe_3O_4 or γ-Fe_2O_3 passivating very thin film.[1,2]. Penetration of the porous concrete matrix by chloride ions, water and oxygen in a corrosion environment can destroy this passivation and lead to corrosion of the steel and forms corrosion products. Having significantly greater volume (2 to 4 times) than the original iron the corrosion products may develop stresses in the steel rebar/concrete interface. This results in cracking and spalling of the concrete cover and potentially failure of the reinforced structure.

Therefore, the early detection and the checking of the condition of reinforcing bars by nondestructive techniques are very important.

Of the laboratory techniques studied to check the corrosion of steel reinforcement, we could point out the electrochemical methods such as polarization curves have been widely used, mainly to compare the effect of concrete admixtures [3].

The polarization resistance technique Rp developed by Stern and Geary, was applied to steel bars embedded in concrete or mortar specimens, to study the corrosion rate, utilizing a number of variables related to the concrete and cement by Andrade and Gonzalez.[4,5] Recently John et al [6]and Venger et al applied the AC impedance technique after study and proposal of electrical models simulating the steel in concrete system.The above techniques were generally accompanied by the gravimetric weight loss method, in order to confirm their accuracy.

The method of electrical resistance, based on the weakening of the cross-section of the reinforcement and the subsequent increase in its electrical resistance, as result of the corrosion has also been used [7]. Recently this method has been improved using the so called corrosion sensor by V.Zivica [8] . This improvement overcomes some disadvantages of the former method eg the influence of ambient temperature changes.

The presented technique is based on the mentioned stresses caused by steel corrosion products, monitored by Strain Gage extensivmeters embedded in mortar specimens. Disturbing effects during the measurements such as, ambient temperature and specimen volume changes are compensated. The reliability of the SG technique is determined by directly relating the corrosion products formation, and further evaluated by other independent methods of corrosion determination, simultaneously executed.

EXPERIMENTAL DETAILS

Materials

The chemical composition of PC (Portland Cement), CKD (Cement Kiln Dust), P35 (PC with pozzolans as fly-ash or Santorini Earth) and steel used in casting the reinforced specimens are shown in Tables 1 and 2. Sand (BS 4550: Part 6). Drinking water and corrosion inhibitors are also used.

CKD is a by-product of low-alkali cement manufacturing process, mainly containing calcium carbonate and is classified as calcareous additive.

Three types of corrosion inhibitors of steel in concrete are used:

Inhibitor-A is a corrosion inhibitor based in aminoalcohol action.
Inhibitor-D is a corrosion inhibitor based in $Ca(NO_2)_2$ action.
Inhibitor-M is a corrosion inhibitor based in alcanolamines action

Specimens Casting

The mortar test specimens were in the form of 80x80x100 mm prisms. The steel bars (diameter 12mm) were machined on a lathe to a final diameter of 10 mm and prepared according to ISO/DIS 8407.3 .

The bars were embedded in the mould up to a depth of 15 mm.Thus the area of bars active to corrosion equals to 26,7 cm^2. The mortar moulds were stored in the curing room. The specimens, after being demoulded were cured in water at 20° for 7 days and then left dry for 24 h. A copper wire cable is connected to each steel bar and the specimens were covered with epoxy resin Araldite to protect the connection of steel with copper cable against corrosion, as shown in Figure 1.Then they were immersed in a 3,5% NaCl solution.

Table 1 Chemical composition of materials (%)

OXIDE	PC	CKD	P35
SiO_2	20,67	13,68	27,38
Al_2O_3	4,99	4,36	9,10
Fe_2O_3	3,18	2,30	5,65
CaO	63,60	42,59	45,39
MgO	2,73	1,23	2,73
K_2O	0,37	0,79	0,94
Na_2O	0,29	0,28	0,56
SO_3	2,44	0,10	2,71
LOI	1,52	--	5,04
IR (ISO)	0,21	--	--
CaO (f)	2,41	--	2,67

Table 2 Chemical composition of steel (% wt)

Fe	Mn	S	C	P
99,23	0,56	0,07	0,11	0,03

Table 3 Categories of specimens - Composition Ratio (wt)

CODE NAME	P35	PC	SAND	WATER	INHIB.-M	INHIB.-D	INHIB.-A	CKD
SGH	1	-	3	0,5	-	-	-	-
SGM	-	1	3	0,49	0,01	-	-	-
SGO	-	1	3	0,5	-	-	-	-
SGC	-	1	2,94	0,5	-	-	-	0,06
SGD	-	1	3	0,47	-	0,03	-	-
SGA	-	1	3	0,47	-	-	0,03	-

Mix Proportions

Six categories of specimens were cast. The proportion of materials used and their code names are shown in Table 3.

Methods

The sensors used for measurement of corrosion rate of reinforcing steel are strain gauge (SG) of type KM-30-120 , of KYOWA .

Two SGs were embedded in the mortar specimen during casting. The deposition of SG into the mortar was made by hands and their alignment was aided by the lead wire cable. Distance and directions between SG are shown in Figure 1.

One of the specimens (horizontally mounted) was used for corrosion result measurements and the other (vertically mounted) for compensating the temperature and other parameters (creep, wetting) influencing the specimen volume. A third SG was also used free in teh working area for all experiments for temperature compensation of the electrical circuit. The test setup, including a potentiostat for applying the corrosion potential, a Strain gauge bridge-amplifier circuit and a multimeter for SG resistance is shown in the schematic diagram Figure 2. The application of the corrosion potential to the specimen by the potentiostat is made by means of working bar electrode, 80x100x20 mm graphite auxiliary electrode and reference saturated calomel electrode(SCE).

Impressed Potential Test

The corrosion process in specimens immersed in the 3,5% NaCl solution was accelerated by impressing 1000 mV of fixed anodic potential for some days. In order to modify the acceleration two other values of fixed potential were applied. Another procedure of impressed potential application was by impressing 1000mV anodic potential with increment of 100 mV per day.

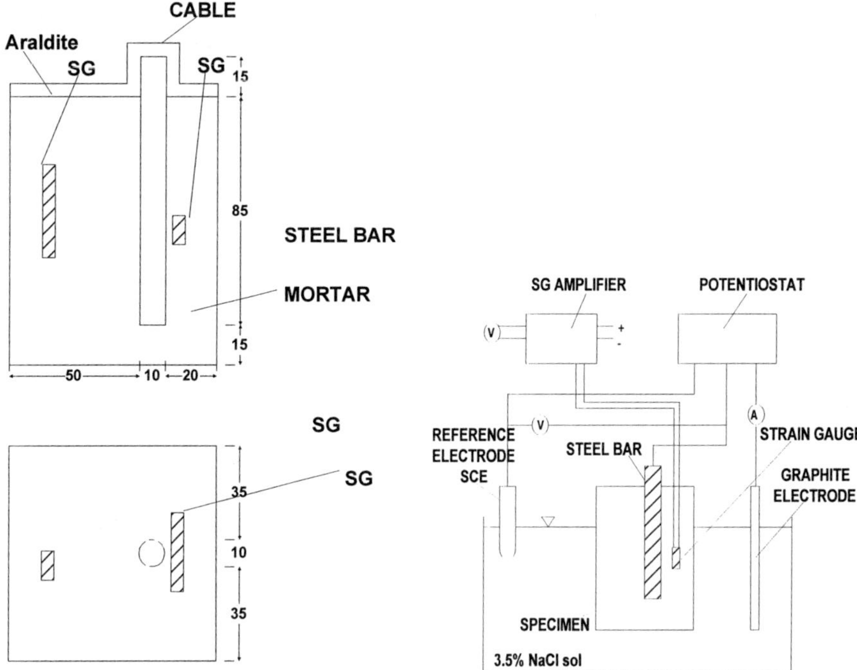

Figure 1 Dimensions of Specimens

Figure 2 Test Setup Diagram

SG Measurements

Before the applying anodic potential to the specimen immersed in 3,5% wt NaCl solution, the output voltage of each SG is measured in order to have the initial values as SGE_0, SGT_0 and SG^*_0 [common (E), volume changes compensation (T) and corrosion (*)]. After the application of the corrosion potential to the specimens in the given time (t), the output voltage of SG amplifier for each SG (SGE_t, SGT_t, SG^*_t) measured at daily intervals, plotted against time, makes it possible to indicate the corrosion of the steel bars.
The following differences:

$SGE_t - SGE_0$,corresponds to circuit temperature changes.
$SGT_t - SGT_0$,corresponds to specimen volume changes and circuit temperature changes.
$SG^*_t - SG^*_0$,corresponds to corrosion plus volume changes plus circuit temperature changes.
$(SGT_t - SG^*_t) - (SGT_0 - SG^*_0)$, corresponds to the compensated corrosion status.

For each SG with initial electrical resistance R, included in a Wheastone bridge it is known [10]that:

$$SG = \frac{V_i}{4} \cdot \frac{dR}{R} \text{ and } \frac{dR}{R} = G \cdot \frac{dL}{L}$$

where: Vi = input Voltage
R = Electrical Resistance of Strain Gage dR= Resistance change
L = Resistance length of Strain Gage dL= Resistance length change
G = Gage factor SG =Output voltage

From the values SG measured in the test it is possible to calculate the change of the length of SG resistance which corresponds to the stresses developed by corrosion products formation into the specimen.

Corrosion Evaluation Through Current Measurements

The anodic current of specimens was measured and plotted versus time. From the data analysis of these plots we can obtain the actual charge flow through specimens and the related corrosion.

Assuming that for the procedure of impressed potential application with linear increment the corrosion rate of the system was controlled by activation polarization, the resistance Rp=dE/dI is related to Icor through the relationship [5] $Icor = B/Rp$ and thus B.dI/dE is analog to corrosion rate dI/dt and consequently the plot I against time is correlated to the total corrosion against time.

For the procedure of impressing fixed potential E, for time t, the current –time plots were analyzed by using the relationship I=E/Rw+Rp, where Rw the ohmic resistance, Rp values produced and the $\int(1/Rp)dt$ can be correlated to the actual charge flow.

Mass Loss Determination

At the end of the test, specimens were broken, pieces of mortar and the Araldite were removed. The steel bars were immersed in inhibited hydrochloric acid for 15 min, washed with water, alcohol and aceton, were weighted, and their weight loss was found. The comparison of weight loss and SG* values for different specimens enables to confirm the degree of corrosion.

RESULTS AND DISCUSSION

The results obtained for the SGO, SGH, SGC, SGA and SGD specimens tested by the SG technique and the related electrochemical measurements are illustrated in Figure 3 to Figure 7 as a function of time. The initial time is the moment of application of anodic potential to the specimen. In all cases there is an increase in SG values, related to the corrosion. The electrical charge plot shows a relatively good agreement with SG plot and could be considered for a first confirmation of reliability of the SG measurements. We also observe a good dynamic response of SG in the potential increases was also observed.

This view, supports the fact that SG measurements are in agreement with the electrical charge flow through specimens, e.g. the corrosion caused, until the moment of mortar destruction.

The comparative analysis of the above plots SGO,SGC,SGD and SGA, tested under similar conditions, with expected different corrosion behavior due to their composition, shown in Figure 8, confirms that the SG measurements curves, are in agreement with the corrosion behavior. The admixture CKD acts protectively against corrosion but with lower efficiency in comparison with Inhibitor-D and Inhibitor-A. The corrosion behavior of CKD as admixture in concrete has been studied in previous paper [9].

Test results obtained for specimens SGO and SGM by impressing potential of analog increment are illustrated in Figure 9 and Figure 10. The current-time plots related to the total corrosion are comparable to the SG curves.

For a further explanation of the results, the relationship between SG measurements before cracking and corrosion of steel bars, was investigated via gravimetric weight loss determination of each specimen bar ,as shown in Table 4 and Figure 11.

In cases of specimens broken during testing the weight loss was estimated ,based on the final determination and the electrical charge transferred through the specimen. The weight loss values and the related SG measurements for each specimen (shown in Figure 11) must be divided by the duration of test in order to give the corrosion rate.

A constant relationship between SG and weight loss values is observed and it is of the order of 200mV per gr. weight loss of steel.

According to the above relationship it is proven that the elongation of SG sensor is related to the formation of the corrosion products and consequently the SG measurements correspond to the real corrosion caused in steel bars.

The porosity measurement (with mercury porosimeter Carlo Erba 2000)of the mortar of above specimens with similar values (12,26 to 12,79) confirm that the different SG values plots of various specimens of the same porosity were caused by other factors influencing corrosion (chloride ions, passivating film).

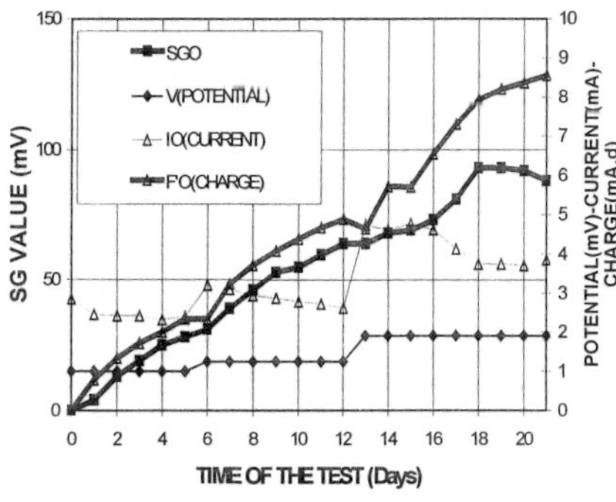

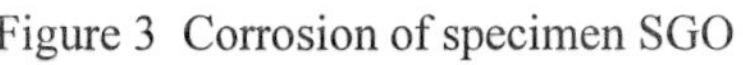
Figure 3 Corrosion of specimen SGO

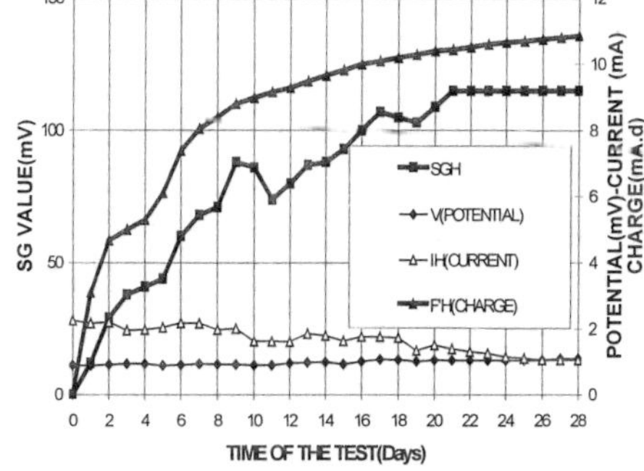

Figure 4 Corrosion of specimen SGH

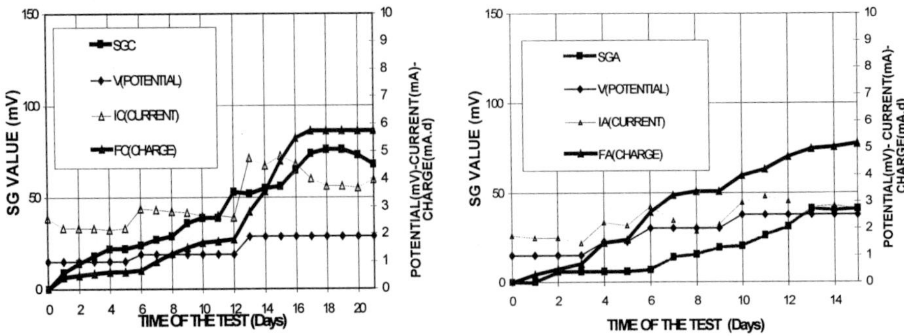

Figure 5 Corrosion of specimen SGC

Figure 6 Corrosion of specimen SGA

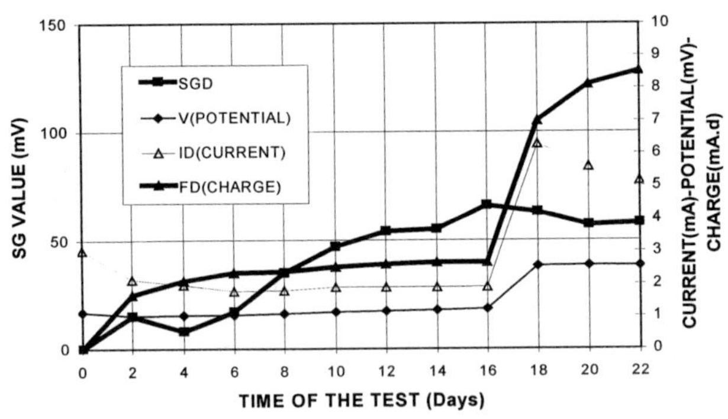

Figure 7 Corrosion of specimen SGD

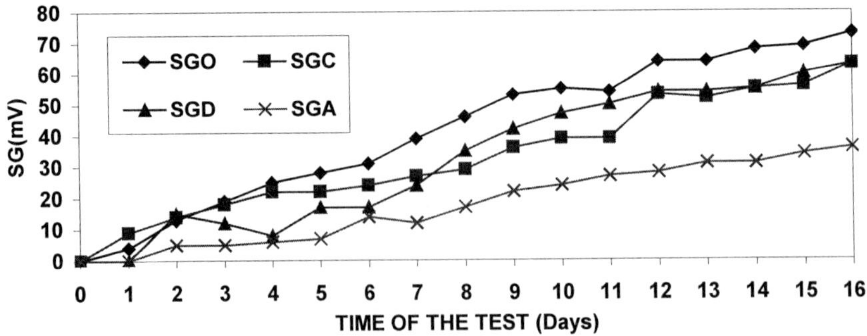

Figure 8 Comparative SG curves for different specimens

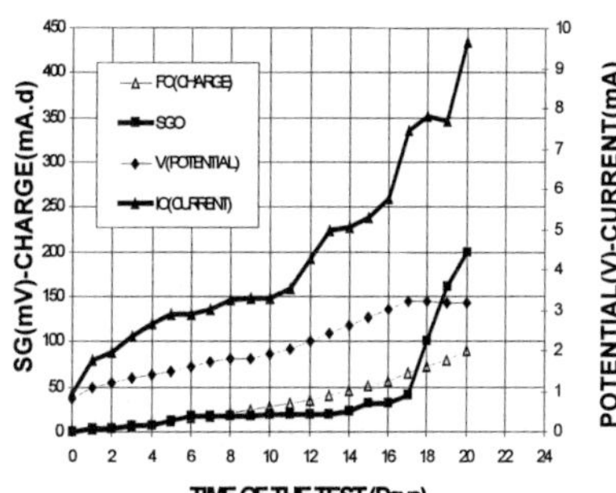

Figure 9 Corrosion of specimen SGO

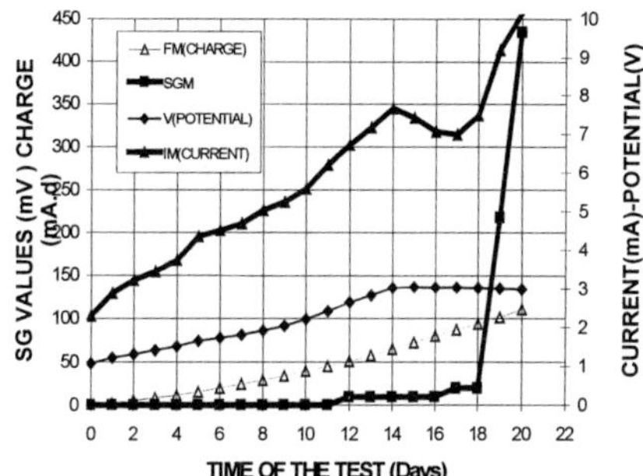

Figure 10 Corrosion of specimen SGM

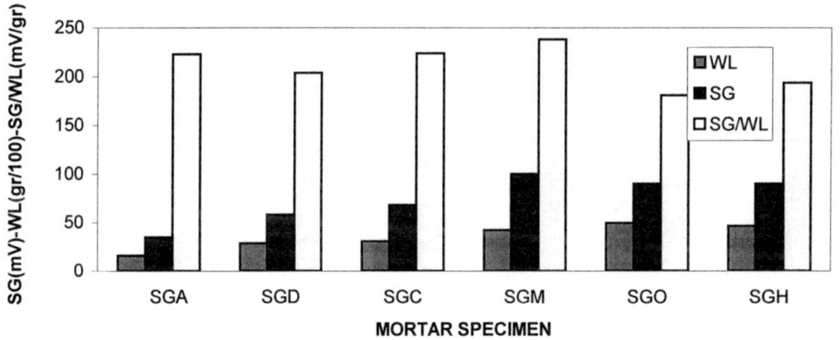

Figure 11 SG Values- weight loss comparison for different specimens

Table 4 Weight loss-SG values comparison for different specimens

SPECIMEN CODE	WEIGHT LOSS $(gr*10^{-2})$	SG VALUE (mV)	SG PER WEIGHT LOSS (mV/gr)	CORROSION TIME (days)	CORROSION RATE (mg/cm^2.day)
SGA	15,67	35	223	16	0,39
SGD	28,43	58	204	17	0,67
SGC	30,27	68	224	17	0,71
SGO	49,74	90	181	20	0,99
SGH	46,30	90	194	9	2,05
SGM	42,00	100	238	21	0,80

CONCLUSIONS

1. The correlation between SG and electrochemical data measurements for all categories of specimens provides comparable results of corrosion evaluation, confirmed by mass loss.
2. The comparison of plots obtained by SG technique, for different specimens, confirms the expected corrosion behavior of specimens.
3. The test results obtained, suggest that this method can be used for an efficient inspection of the state of steel reinforcement and for the corrosion rate measurements.
4. This nondestructive technique is suitable for the laboratory study of corrosion factors, including the evaluation e.g. of various admixtures.

REFERENCES

1. MEHTA, P K. Durability of Concrete Exposed to Marine Environment - A Fresh Look. ACI SP 109-1 ,1988, pp.1-29
2. NEVILLE, A. Chloride attack of reinforced concrete: an overview.Materials and Structures,28 1995,pp. 63-70.
3. POURBAIX , M. CEBELCOR, Rapport Technique No 205,1972.
4. ANDRADE, C and GONZALEZ, J A. Quantitative measurements of corrosion rate of reinforcement steel embedded in concrete using polarisation resistance measurements. Werkstoff Korrosion, V 29 1978 pp 515-519.
5. GONZALEZ, J A and ANDRADE, C. British Corrosion Journal,Vol 17 No 1, 1982, pp.21-28.
6. JOHN, D G, COOTE, A T, TRENDAWAY, K W J and DAWSON, J L. Corrosion of Reinforcement in Concrete Construction. Society of Chemical Industry, London, June 1989, pp.263-286.
7. SCHIPPA, G. Influence du $CaCl_2$ sur les caracteristiques du mortieres de ciment .in Proceedings,Colloque International sur les Adjuvants de Mortiers et Betons, Bruxelles, August 1967, Rapport IV/10, pp. 127-138.
8. ZIVICA, V. Improved method of electrical resistance- suitable technique for checking the state of concrete reinforcement. Materials and Structures, V 26.1993 pp. 328-332.
9. BATIS, G, KATSIAMBOULAS, A, MELETIOU, C A and CHANIOTAKIS, E. Durability of reinforced concrete made with composite cement containing kiln dust. Concrete for Environmen Enhancement and Protection. Edited by K.Dhir and T.D.Dyer.(1996) pp.67-72
10. COLOMBO, G. Automazione Industriale. Vol.4. Dott. Giorgio Torino.1986.

EVALUATION OF STRUCTURAL DAMAGE IN REINFORCED CONCRETE BY DYNAMIC SYSTEM IDENTIFICATION

J Vantomme **J M Ndambi**

Royal Military Academy, Brussels

J de Visscher

Vrije Universiteit Brussel

B Peeters

Katholieke Universiteit Leuven

Belgium

ABSTRACT. The paper presents experimental results from vibration analysis of reinforced concrete beams subjected to progressive cracking. Experiments are made on three reinforced concrete beams of 6 metre length. Cracks are artificially induced in these beams by static loading tests in a certain number of load-steps. After each static-load step, a complete experimental modal analysis is performed in order to determine the dynamic characteristics of the beams. Their evolutions allow conclusions to be drawn about the suitability of dynamic system identification techniques. This investigation is a first exploratory step in the development of a dynamic system characteristics (DSC's) based technique for monitoring damage in large concrete structures. Progressive cracking of concrete leads to a measurable decrease of eigenfrequencies, and to an increase in modal damping ratios, which is far more difficult to explain. The non-linear behaviour of the beams is investigated; the effect on eigenfrequencies is limited, but its influence on damping remains to be examined.

Keywords: Structural monitoring, Concrete durability, Non destructive testing, Dynamic system identification, Modal analysis.

Dr J Vantomme, is a Lecturer at the Civil Engineering Department of the Royal Military Academy in Brussels, Belgium. His research interests include the use of dynamic methods for material and structural characterization in the area of concrete and composite materials.

Mr J M Ndambi, is a Research Student at the Royal Military Academy. He is currently completing his PhD on the subject of material damping and damage in concrete structures.

Dr J de Visscher, is a Researcher at the Mechanics of Materials and Structures Department of the Vrije Universiteit Brussel, Belgium. Her work focuses on the evaluation of damage in civil engineering structures by means of dynamic system identification methods.

Mr B Peeters, is an Assistant at the Katholieke Universiteit Leuven, Belgium.

INTRODUCTION

Damage in concrete structures is associated with structural modifications, which can be observed through changes in modal parameters: eigenfrequencies, modal damping ratios, mode shapes and their derivatives (curvatures, local stiffness, etc.). Damage detection by changes in the dynamic properties of structures, is a subject that has received considerable attention during the last few years. Several researchers have investigated changes in these parameters with simulated damage, using real and model structures [1]. The idea of using the modal parameters to detect damage in structures is already accepted, but the practical exploitation is difficult [2,3]. One of the main difficulties is that modal parameter estimates are affected by data processing techniques and the methods of excitation used. If the influence of these items on modal parameters is not well understood, wrong conclusions could be drawn as to the accuracy of the detection results and the suitability of the method.

Promising results have been obtained in the domain of fibre-reinforced plastics (FRP), where damage under the form of matrix failure, loss of adhesion between fibres and matrix, etc. can be related to the decrease of modal frequencies and the increase of damping [4]. The aim of the present research project is to look at the applicability of the results for FRP to the case of concrete structures. Questions have to be answered such as: which modal parameter or derivative is most appropriate for use as a damage detection parameter and how is the transposition to be made from global measurements towards local defects?

In this paper attention is focused on the evolution of eigenfrequencies and modal damping ratios of reinforced concrete beams submitted to progressive cracking. Two excitation techniques are used: impact hammer testing and two electromagnetic shaker signals: pseudo-random and swept-sine. The Frequency Domain direct Parameter Identification (FDPI) technique is used to determine all the modal parameters.

DESCRIPTION OF THE TEST BEAMS

The test specimens are three reinforced concrete beams of 6 meters length. These beams have been manufactured with the same design parameters. They are designed in order to achieve the following:

- the first eigenfrequencies have to be comparable with the lowest eigenfrequency generally encountered in civil engineering structures (e.g. bridges); a compromise is needed because of the preponderate effect of the difference in dimensions between a laboratory beam and a bridge;
- the cross section is chosen to be rectangular in order to avoid the coupling effect between horizontal and vertical vibration bending modes;
- the reinforcement ratio is chosen to have a sufficient difference between the first crack moment and the failure moment of the beams;
- transverse reinforcement ensures that failure occurs in bending.

In meeting these requirements this gives reinforced concrete beams of 6 metres length (L), 250 mm width (l) and 200 mm height (h). The beams are reinforced with longitudinal steel (S500) bars with 16 mm diameter, equally distributed over tension and compression side, corresponding to a reinforcement ratio of 1.4 %; the transverse reinforcement consists of stirrups with 8 mm diameter each 200 mm. The eigenfrequencies are proportional to h/L^2: a height of 0.2 m and a length of 6 m result in a first eigenfrequency of about 20 Hz. For each of the three beams, a total mass (m) of 750 kg result in a density of reinforced concrete of ρ = 2500 kg/m^3 [5]. The cross section of the beams is presented in Figure 1.

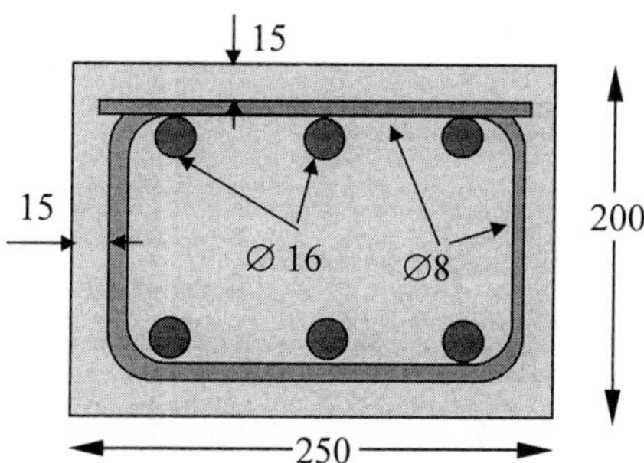

Figure 1 Cross section of the beam

MEASUREMENT SYSTEM

The tests are prepared and worked out in the laboratory of the Civil Department of the K.U.Leuven (BE). Two types of experiments are combined: static loading test and dynamic measurements.

Static Test

The static loading test is performed to introduce cracks in the beams. The three-points (beams 1 & 2) and four-points (beam 3) bending test configurations are used (Figures 2 and 3).

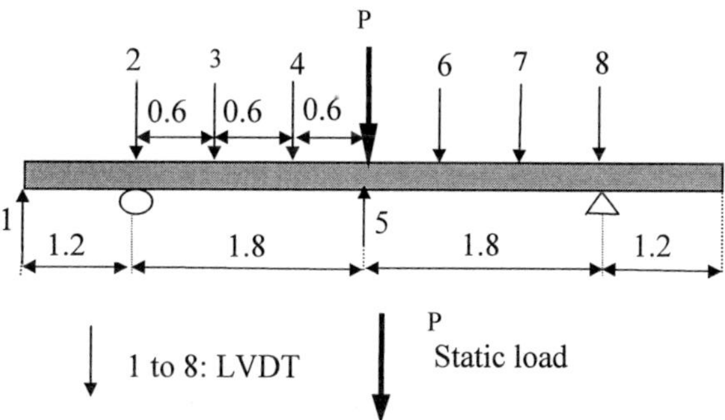

Figure 2 Static test configuration 3-points bending test:
Dimensions in (m)

The beams are simply supported. The positions of the supports are calculated in order to minimise the bending moment due to the own weight of the beams.

Damage is induced in different steps during which accumulation of cracks is observed. Load is applied by means of a hydraulic system. At each intermediate load step, the load forces are measured by means of load cells, vertical displacements are measured by means of Linear Variation Displacement Transducers (LVDT) and strains are measured by means of extensiometers.

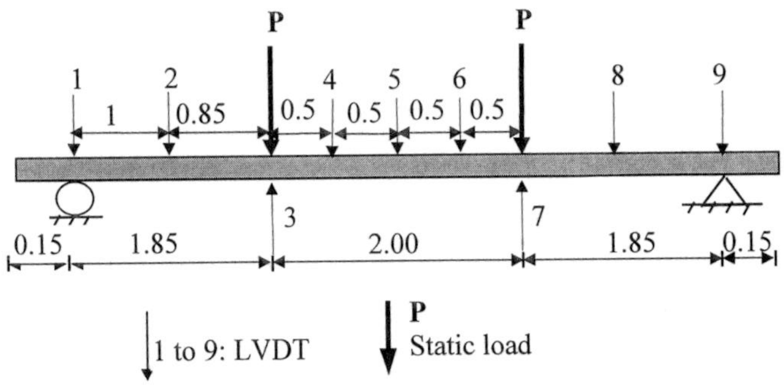

Figure 3 Static test configuration 4-points bending test: Dimensions in (m)

The Tables 1 and 2 show the static loading test programmes for Beams 2 and 3. The last load steps correspond to the failure of the beams.

Table 1 Load steps beam 2

DATE (TEST)	LOAD STEP	LOAD (kN)
96-30-10	ref	0
96-13-11	1	8
96-19-11	2	15
96-21-11	3	24
96-02-12	4	32
96-05-12	5	40
96-12-12	6	50
96-19-12	7	56

Table 2 Load steps beam 3

DATE (TEST)	LOAD STEP	LOAD (kN)
97-16-10	ref	0
97-16-11	1	4
97-16-11	2	6
97-16-11	3	12
97-17-12	4	18
97-17-12	5	24
97-18-12	6	25.3

Dynamic Measurements

Dynamic measurements are performed after each static load-step in order to determine all modal parameters of the beams. Free boundary conditions are simulated by suspending the beams by means of four elastic springs attached at the theoretical nodal points of the fundamental (first) bending mode of the beams: Figure 4.

Acceleration measurements are made in 62 points divided in two rows of 31 points on each side of the top surface in order to detect both torsion and bending vibration modes of the beams.

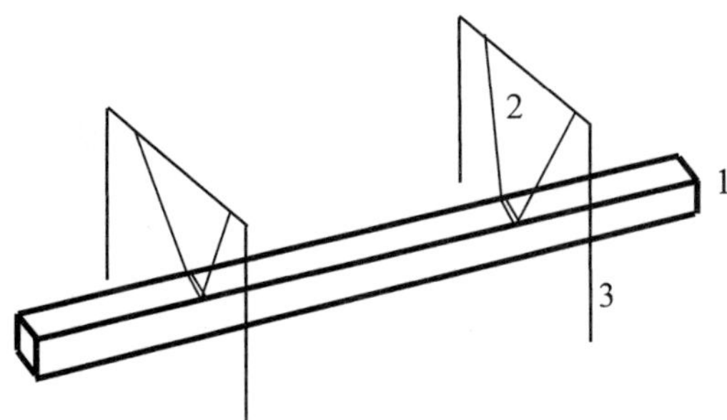

Figure 4 Dynamic test configuration; 1: Beam; 2: Springs; 3: Frame

Modal parameters are determined by using modal testing techniques. Modal tests consist of measuring the dynamic response of the beams to the applied loading. This loading is performed by hitting the beams with an impact hammer or by excitation by means of an electrodynamic shaker: Figure 5.

A hammer type PCB SN4503 with sensitivity ± 0.7 mV/lb is used; by a proper choice of the hammer tip, it is possible to generate a reasonable response containing frequency components in the frequency range of interest (0-1000 Hz). An electrodynamic shaker type MB Dynamics with force sensor type PCB SN12737 with sensitivity ± 52 mV/lb is also used. The shaker is driven by signals generated by the FFT Analyzer ONO-SOKKI CF-350 signal-out channel and by the DIFA signal generator of the DIFA SCADAS II frontend of the CADA-X system.

Twelve accelerometers type PCB 338A35 and 338B35 with sensitivity ± 100 mV/g are used to register the dynamic response of the beams. Force and acceleration signals are sent trough an amplifier and an analogue filter and stored on Digital Audio Tape (TEAC DAT-Recorder). The digitally stored signals are replayed by the TEAC DAT-recorder, producing analogue signals that can be acquired by DIFA SCADAS II frontend of the CADA-X system. The CDA-X system was used to extract the modal parameters[6].

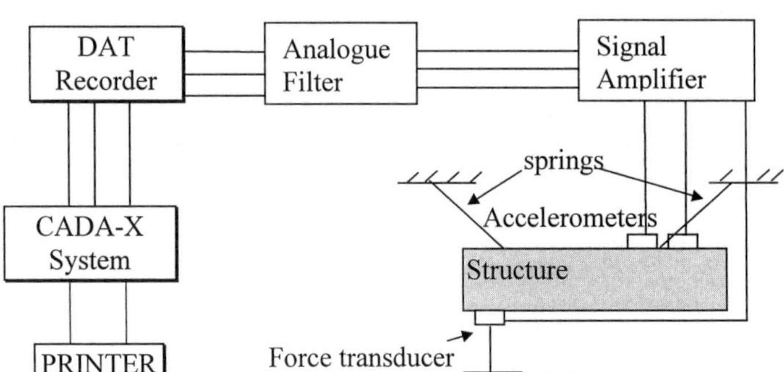

Figure 5 Test set-up:Dynamic measurements

EXPERIMENTAL RESULTS

Static Loading Test

Only the results for Beam 2 are presented here (three-point bending test): Figure 6 shows the force versus displacement curve of the seven static loading steps, measured at the mid-section of the beam.

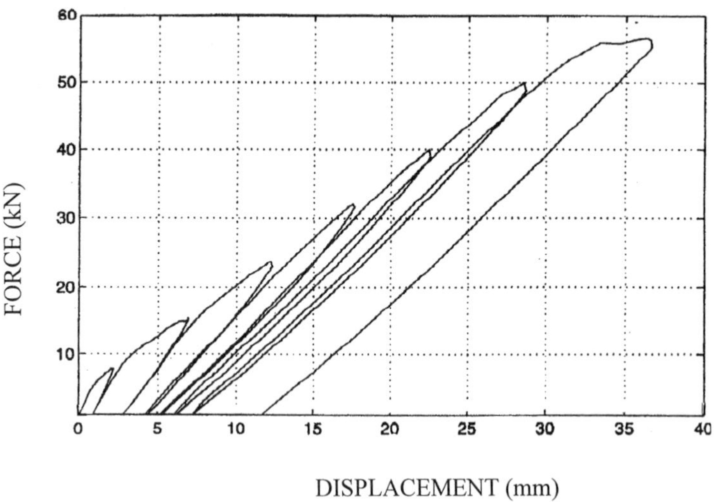

Figure 6 force-displacement curve at the mid-section of the beam - sensor 5

The static load is applied in seven steps. In the first step the maximum force is limited to 8 kN (before the occurrence of the first crack). The failure of the beam occurs at 56 kN.

The first visible crack is observed in the vicinity of the mid-section of the beam. Failure occurs when the crack at the mid-section of the beam of the beam has grown over the entire section

Dynamic Test

For Beam 3, three different excitation techniques are used: impact hammer and two electromagnetic shaker signals. The eigenfrequencies and the modal damping ratios are presented in Tables 3 and 4 for the three different excitation methods.

Table 3 Eigenfrequencies and modal damping ratios (Pseudo-random and swept-sine methods) for flexural (F) and torsional modes (T)

	PSEUDO-RANDOM METHOD		SWEPT-SINE METHOD	
Modes	Freq [Hz]	Damp [%]	Freq [Hz]	Damp [%]
1F	20.3878	0.3336	20.442	0.3618
2F	57.1242	0.3432	56.600	0.3112
3F	112.0886	0.3390	112.525	0.4312
1T	169.2678	0.4435	169.670	0.2648
4F	183.8405	0.2984	184.133	0.2690
5F	273.9919	0.3653	274.718	0.3018
2T	363.6219	0.2775	364.099	0.3112
6F	373.6741	0.3152	374.585	0.3394

Table 4 Eigenfrequencies and modal damping ratios (Impact testing method)

	IMPACT TESTING METHOD	
Modes	Freq [Hz]	Damp [%]
1F	19.962	0.943
2F	56.093	0.899
3F	110.287	0.709
1T	168.498	0.619
4F	181.283	0.788
5F	270.716	0.722
2T	362.740	0.423
6F	369.751	0.634
7F	475.289	0.649

The difference between the resonance frequencies for the first two methods (pseudo-random and swept-sine) is negligible (max. 0.5 %). On the other hand, between these methods and the impact testing method, the difference in resonance frequencies reaches 3%. This difference may be due to the influence of the shaker on the dynamic behaviour of the beam; the shaker makes it to behave in a more rigid way, resulting in apparently higher resonance frequencies.

Eigenfrequencies as Function of Load Steps

Figure 7 shows the relative variation of the eigenfrequencies as function of load steps, obtained with the pseudo-random excitation method (beam 3).

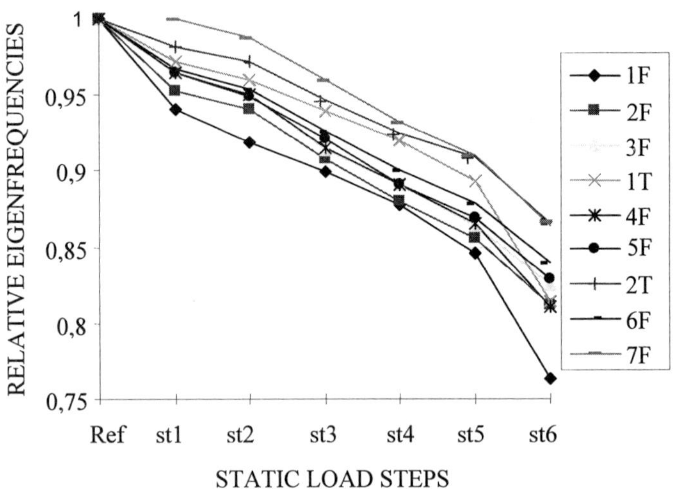

Figure 7 Eigenfrequencies as function of load step;
for bending modes (F) and torsion modes (T) – pseudo-random

It appears from this figure that:

- all the eigenfrequencies are affected by the accumulation of cracks: they decrease. This phenomenon can be explained by the fact that when cracks occur in a concrete structure, its EI stiffness decreases and the resonance frequencies vary in the same way [2]. The first cracks are visible at about a static load of 4 kN. The drop in resonance frequency after this step is already 6 % for the first bending mode;
- for the beams 1 and 2, which are subjected to the 3-points bending static test, the first bending mode is clearly more sensitive than the other modes, because damage is concentrated exactly in the middle of the beams, where the first mode has the highest amplitude. For the beam 3 on the other hand, the damaged area is larger and affects the symmetric, asymmetric and the torsion modes in an equivalent way.

These results confirm the results obtained from the first beam investigated [5].

Modal Damping Ratios as Function of Load Steps

Figure 8 shows the evolution of the modal damping ratios as function of static load steps for the pseudo-random excitation method.

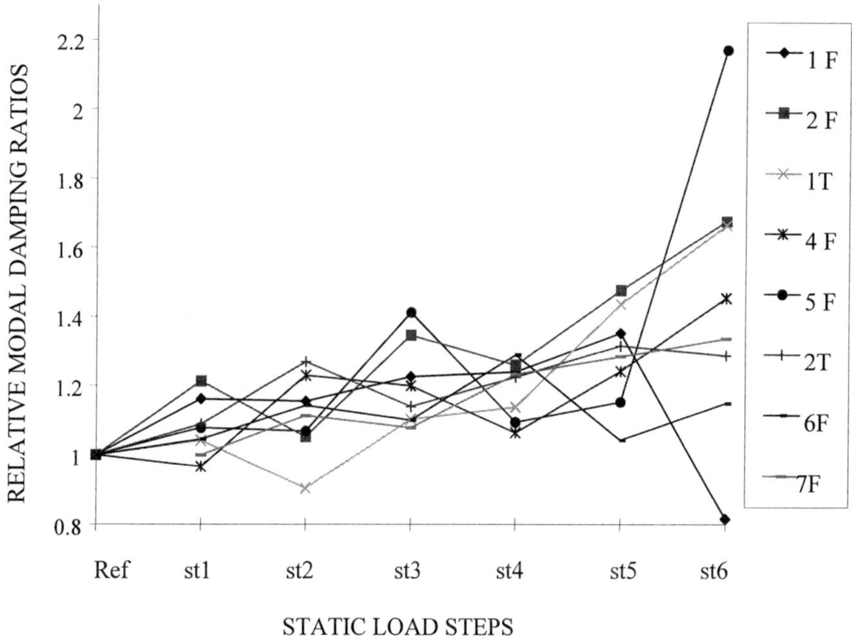

Figure 8 Relative Modal Damping ratios as function of load step, for bending modes (F) and torsion modes (T) – pseudo-random

The variation of the modal damping ratio with the induced damage is not monotonical: according to these experiments, the damping increases and decreases again at the highest level of damage. This could be explained by the interaction of crack surface friction and the redistribution of stresses when damage occurs. The last argument requires further experimental work and theoretical modeling, which is focused on stress-strain energy balance calculations, based on available micro-mechanical models for cement matrix based composites.

It should be noted that modal damping estimates are affected by data processing techniques, which are used to average time-series or spectra data, to smooth spectrum values and to separate closely spaced modes [7]. All these areas need systematic investigations. As a conclusion up to now, it is clear that the modal damping ratio is a parameter that is difficult to estimate; it is at the moment not a very suitable parameter for detecting damage in concrete structures.

Eigenfrequencies and Modal Damping Ratios as Function of Excitation Amplitude

Experimental modal analysis assumes the structure to behave in a linear way. In order to investigate whether this assumption is acceptable for the test beam, complete experimental modal analysis were performed at different amplitudes of the excitation signal, for the case of the swept-sine method.

Figures 9 and 10 show the evolution of the relative resonance frequencies and the relative modal damping ratios with the increase of the excitation amplitude.

These figures show that:

- the resonance frequencies decrease with increasing excitation amplitude;
- the decrease of resonance frequencies is observed as well before as after the induction of damage;
- in the variations of the modal damping ratios, no particular trend can be observed;
- non-linear behaviour is observed, even at the very low amplitude at which the modal analysis is performed. However, the variations for the eigenfrequencies appear to be relatively small (and even negligible) (±0.01% - ±0.5%) compared with the variations of eigenfrequencies due to the accumulation of cracks (± 6%- ±24%).

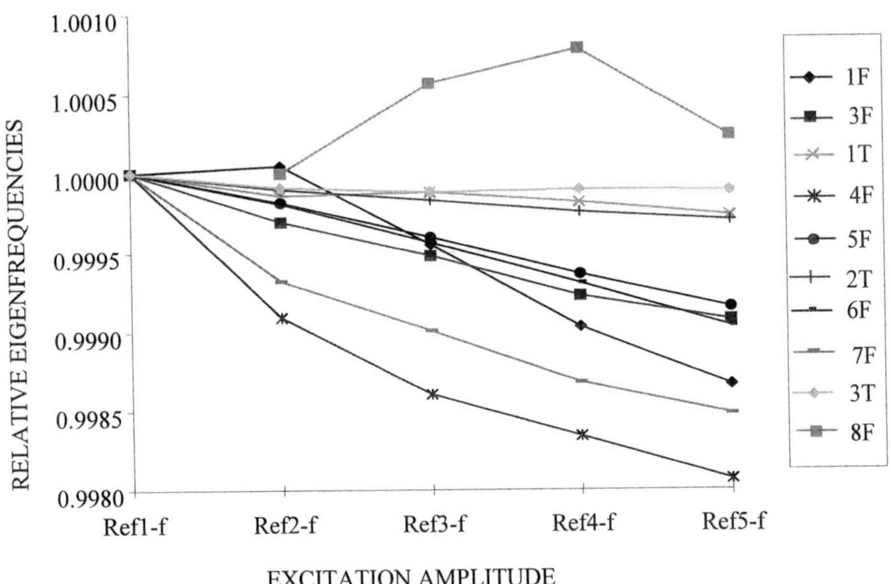

Figure 9 Relative eigenfrequencies as function of excitation amplitude – reference measurements

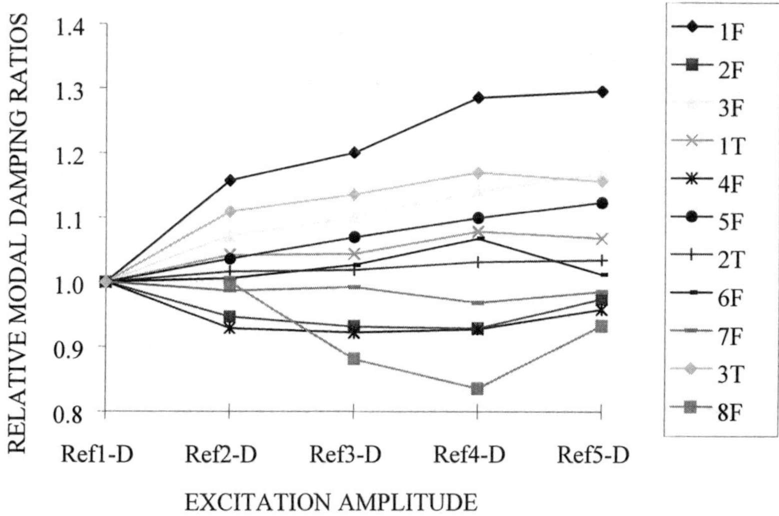

Figure 10 Relative Modal Damping ratios as function of excitation amplitude – reference measurements.

CONCLUSIONS

The paper describes the static and dynamic testing of reinforced concrete beams that are subjected to progressive cracking. At each intermediate step, a complete experimental modal analysis is performed in order to determine the changing modal parameters of the beams. The non-linear behaviour of the third beam is also investigated. It appears from these experiments that:

- the progressing of cracks is reflected in a measurable frequency shift (6 - 24%);
- the variation of the modal damping ratio with the induced damage is not monotonical. It increases and decreases again at the highest level of damage. Further analysis is planned on this item;
- non linear behaviour is observed, but at first sight, this does not seem to cause a main problem for the evolution of eigenfrequencies. On the other hand, non linear effects may cause erroneous conclusions as to the damping values.

ACKNOWLEDGEMENTS

This work is a part of a research project sponsored by the National Fund for Scientific Research. Its financial support is gratefully acknowledged.

REFERENCES

1. DOEBLING, S W, FARRAR, C F, PRIME, M B, SHEVITZ, D W. Damage identification and health monitoring of structural and mechanical systems from changes in their vibration characteristics: a literature review, Research report LA-13070-MS, ESA-EA, Los Alamos National Laboratory, Los Alamos, New Mexico, 1996.

2. ZAKIC, B D. Vibration in diagnosis of damages in concrete bridge structures, Proceedings of 2nd RILEM International Conference, Diagnosis of concrete structures, Strbske Pleso, Slovakia, 1996, pp 320-323.

3. MARTIN, J, HARDY, M S, USMANI, A S, FORDE, M C. Accuracy of NDE in bridge assessment, Engineering Structures, Vol. 20, No. 11, 1998, pp 979-984.

4. VANTOMME, J, A. Parametric study of material damping in fibre-reinforced plastics, Composites 26, 1995, pp 147-153.

5. PEETERS, B, WAHAB, M A, DE ROECK, G, DE VISSCHER, J, DE WILDE, W P, NDAMBI, J M, VANTOMME, J. Evaluation of structural damage by dynamic system identification, 21st Int. seminar on Modal Analysis (ISMA 21), K.U.Leuven, Belgium, 1996.

6. CADA-X; Modal Analysis user manual; LMS international, Rev. 3.4; Belgium, 1994.

7. BENCAT, J. Statical and Dynamical Testing of Structure; Proceedings of 2nd RILEM International Conference, Diagnosis of concrete structures, Strbske Pleso, Slovakia, 1996, pp 259-266.

COMBINED IN-SITU DETERMINATION OF THE DEPTH OF CARBONATION AND OF THE CHLORIDE CONTENT OF CONCRETE

P J Guisa

H-P Gatz

Federal Highway Research Institute

Germany

ABSTRACT. The transportable drilling equipment, which can be used in-situ, permits the changes in concentration of the boring dust produced in the drilling and as transported in the drilling fluid to be determined directly. The concentrations as determined can be assigned in a precise manner to the depth of drilling. At the present time the parameters of pH-value and chloride content, which are important for assessing the condition of the concrete near the surface, can be determined directly for the particular depth of drilling. The determination of both values is made by examining the boring dust obtained from the same hole. In the test a hole with a diameter of approx. 19 mm and a depth of around 50 mm is drilled, so that it's allowed to call it of a low-destructive test.

Keywords: In-situ, Drilling process, pH-Value-measurements, Carbonation, Chloride content

Peter J Gusia Dipl.-Ing., Dipl.-Ing. (FH), studied civil engeneering in Hagen and Bochum, Germany, worked with builders and constructors, since 1991 with the Federal Highway Research Institut (Bundesanstalt für Straßenwesen, BASt), Germany, in the bridge department, with concrete constructions. Questions of new concrete technologies, constructions and maintenance methods of bridge constructions, further quality and workmanship of concrete bridge constructions.

Hans-Peter Gatz Dipl.-Ing. (FH), studied civil engeneering in Wuppertal, Germany, precedently traffic and road construtions, worked for several years in a road material laboratory with the german construction industry, since 1974 with the Federal Highway Research Institut, in the bridge department, with concrete constructions. Activities in several national and international working groups.

INTRODUCTION

Experience gained in the course of repairs to concrete structures has shown that the early recognition and elimination of damage to concrete can contribute substantially to reducing costs. The idea of developing the drilling process was stimulated by the realisation that up to the present time there has been no process, which is suitable for use in-situ and which can be considered to be low-destructive, for determining at the same time the depth of carbonation and the chloride content. The objective of the project was therefore to develop such a process with the aid of which the substance of different structures could be compared and evaluated and then drawn up in order of priority. This in turn would enable the funds available for maintenance measures to be spent in a more specifically aimed and effective manner.

THE DRILLING PROCESS

The drilling device consists of individual elements which permit continuous and impact-free drilling at a depth of about 50 mm into concrete components of any desired alignment. In the process a measuring solution transports the bore dust in a closed circuit. With the help of special electrodes, the alkalinity and the chloride content are measured independently of one another. The measured values are electrically converted, digitalized and stored and processed by a DP program so that the pH value and the chloride content can be represented in graphical form as a function of the drilling depth immediately after the completion of the drilling. Figure 1 shows an outline diagram of the drilling process.The drilling device itself consists of

1. a suction base to fix the drilling device to the surface of the concrete (special seals are for sealing the described liquid circuit system to the concrete. In this way it is guaranteed that the measuring solution circuit remains closed),
2. a drill guidance device in the housing to guide the drill, to center it and to seal it against the housing,
3. a hollow shaft with exchangeable hollow and core drills with a fitted diamond bit,
4. a path measurement device with which the penetration depth of the drill into the concrete can be continuously recorded,
5. a drill drive with a battery-operated drill unit,
6. a circuit board for potential-free power supply of the plug-in units and for the separation of the potential of the electrodes and the
7. analog signal conversion, which again is composed of the plug-in units for the determination of the pH values, Cl-values, drilling depth and flow quantity.

During the drilling process, the measuring fluid is circulated in a closed system with the help of a pump. It is circulated to the following stations

1. The *measuring fluid* is directed in a hose pipe from the *pump* via a *flowmeter* to the hollow drill and flows through it to the *drill bit*; the bore dust generated during drilling is transported in the measuring fluid via a return hose pipe to the

2. *deaerator box.* The measuring fluid is there deaerated and flows into the *cyclone.*

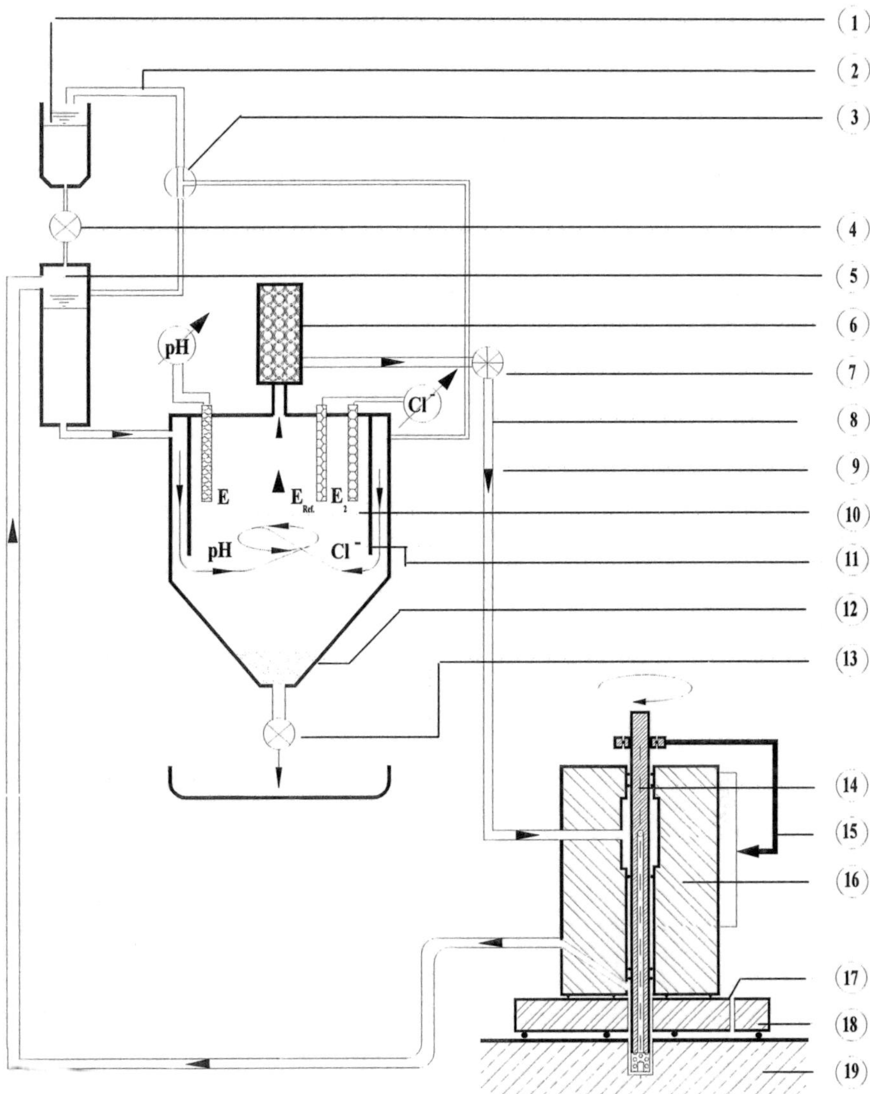

Notes in Figure 1

1. Supplies Container
2. Vent Line
3. Three-Way Tap
4. Spherical Valve
5. Deaerator Box
6. Pump
7. Flowmeter
8. Measuring fluid
9. Hose Pipes
10. Cyclone
11. Guidance Device
12. Settling Cone for the Drill Dust
13. Drain Cock
14. Hollow Shaft with Added Thread for the Hollow Drill
15. Path Transmitter
16. Drill Mounting
17. Vacuum
18. Vacuum Base
19. Concrete

Figure 1 Outline Diagram of the Drilling process

3. In the area between the outer *cyclone wall* and a concentrically arranged *guidance device* the incoming and circulating measuring fluid is set in rotation. Due to the centrifugal forces, the drill dust particles move towards the wall and fall there into the settling cone of the cyclone. During the time between the very fine fragmentation of the concrete by the diamond drill and the depositing of the drill dust in the settling cone of the cyclone, the Cl^- in the concrete dissolves.

4. The measuring fluid freed of the coarse bore dust, which can badly damage the pump, rinses, driven by the centrifugal forces of the cyclone, the *electrodes* E_1 for the determination of the pH value and E_2 together with E_{Ref} for the determination of the content of Cl^- ions.

5. After flowing through the cyclone, the measuring fluid is again drawn in by the pump and directed via the flowmeter to the hollow drill. Thus the liquid circulation flow is completed.

The transportable units for the innovative drilling process, which can be used in-situ, enable the depth-dependent concentration changes in the measuring fluid to be determined. One hole of about 19 mm diameter is drilled so that it can therefore said to be a low-destruction method. A diamond drill is used to open the concrete.

The measuring fluid is pumped through the hollow drill and is separated from the 'coarse' drill dust in a cyclone. It circulates in a closed circuit and sensors measure continuously and simultaneously their pH value and Cl^- content in the cyclone.

At the present time the two parameters, pH value and chloride content, which are important for assessing the condition of the concrete near the surface, can be determined for the respective drilling depth. The determination of the two values is made by examining the bore dust from the same hole. The concentration can be directly allocated to the drilling depth.

With the help of a minicomputer the measurement results are prepared and presented in diagram form directly at the place where the measurements were taken (cf. Figures 2 and 3).

Using the measurement protocols printed directly after the measuring has taken place, an immediate decision can be taken on whether further drilling is necessary for statistical substantiation of the sampling. The drill unit, including the necessary measurement computer and printer, is battery-operated.

Other ion-selective sensors can also be used in the cyclones with which, for instance, the loads resulting from salts used near ground-level which damage buildings can be determined as part of measures to protect historical monuments.

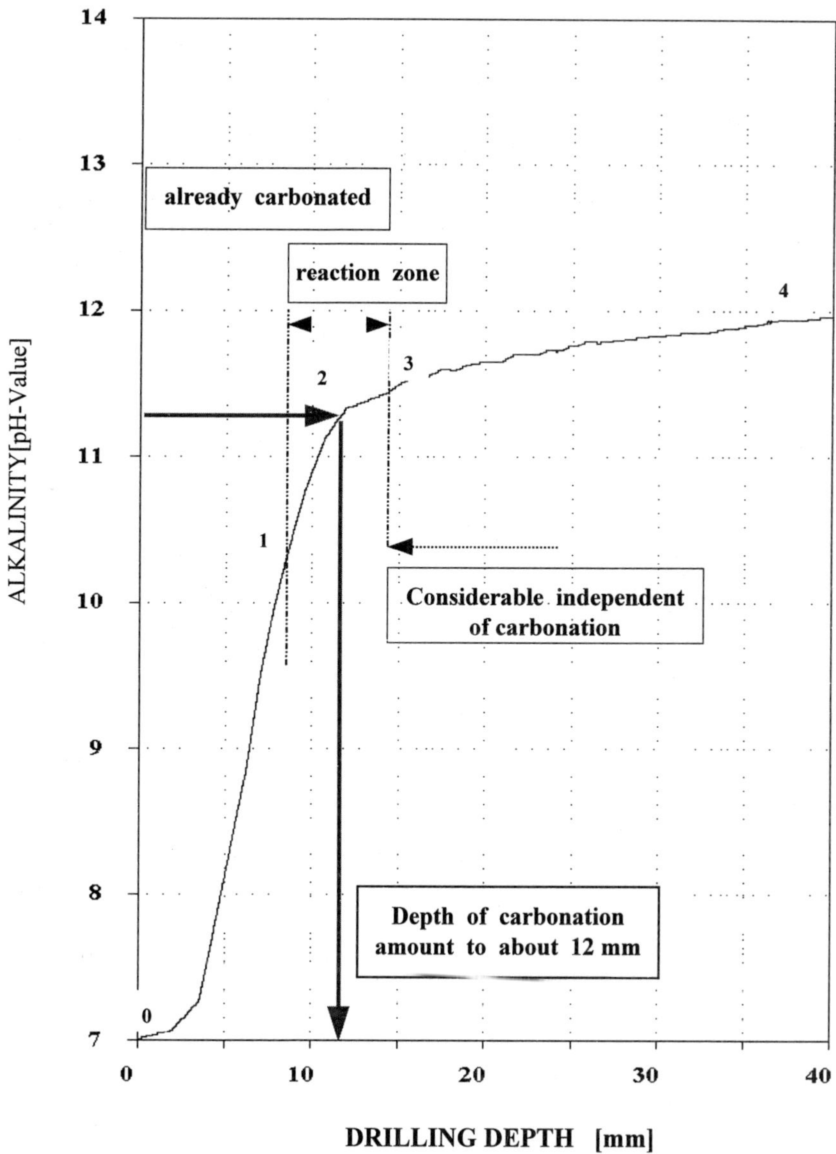

Figure 2 Representation of a pH-Value-Measurements in-situ, reached drill depth > 40 mm

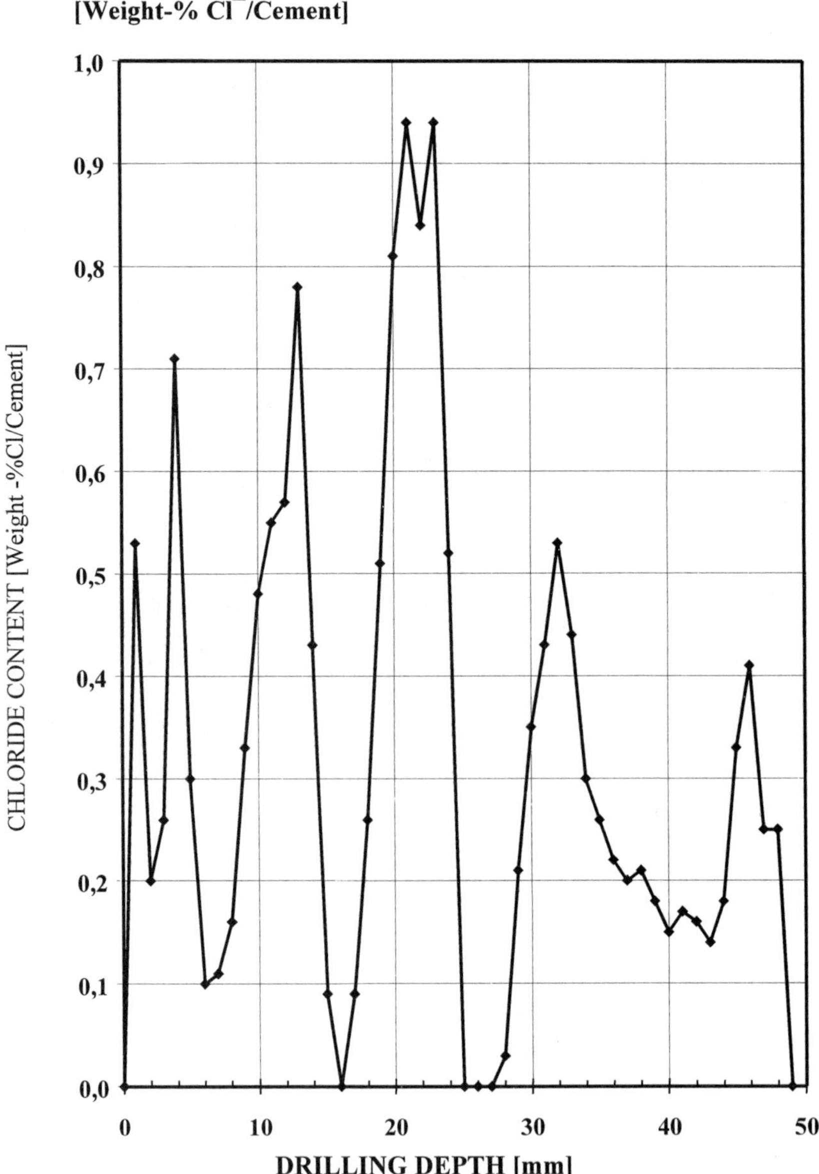

Figure 3 Representation of Chloride Content Measurements in situ, reached drill depth 49 mm

PRACTICAL IMPLICATIONS

The results established as part of the laboratory investigations to date prove that the drilling process is suitable for a rapid and reproducible near-surface concrete investigation, in particular:

1. for the estimating of the risk of corrosion forming in the concrete steel and thus for estimating the current state of a concrete component and

2. for supporting in-situ the decisions to be taken in connection with the maintenance measures which are in progress.

If the depth of the upper reinforcement, i.e. the thickness of the concrete protective covering, is known, then it can be estimated with sufficient accuracy when damage will occur and whether or when appropriate maintenance measures appear meaningful. Under certain circumstances, it is even possible to make a corresponding statement during the guarantee period.

A comparison of the determination procedures currently in use with the drilling process show that the conventional methods supply average Cl^- levels or the end point of reaction of the phenolphthalein; however, a differentiated assessment of events, processes and developments dependent on time and place is not possible.

Test methods used to date to determine the carbonation depth and the chloride content in concrete buildings are inadequate. With the newly developed drilling process, the corrosion risk can be estimated before damage caused by corrosion becomes visible.

PERFORMANCE TESTS ON THE IN-SITU BUILDING AT CARDINGTON

P Chana

C Judge

R Moss

Building Research Establishment

United Kingdom

ABSTRACT. Early in 1998 a seven storey in-situ concrete building was constructed at BRE's Cardington Laboratory under the auspices of the European Concrete Building project. It is designed as a commercial office building with constraints imposed as if it were built on a real site in Bedford. The building incorporates a number of innovative features which include:

- Flat slab construction with no upstands or downstands;
- Use of high strength concrete at the lower column levels to permit the same column size to be used over the height of the building;
- Innovative approaches to the provision of punching shear reinforcement;
- Steel cross-bracing used to provide lateral stability obviating the need for shear walls or cores.

The completed frame is now available for interested parties to carry out performance testing and demonstration projects. This paper outlines the projects which are currently underway, those which are planned and the broad research themes for which the building provides a unique opportunity.

Keywords: In-situ concrete, European Concrete Building Project, Flat slab construction, High strength concrete, Punching shear reinforcement, Performance testing, Fire, Dynamics, Settlement, Structural monitoring.

Dr Pal Chana is Technical Director – Concrete Structures at the BRE. Previously, he was Director of the Concrete Research and Innovation Centre and Senior Lecturer at the Department of Civil Engineering at Imperial College, London. He is working with BRE on proposals for performance testing on the in-situ concrete building.

Mr Chris Judge is the project manager responsible for co-ordinating the performance testing on the in-situ concrete building at Cardington.

Dr Richard Moss is a Senior Research Engineer within the Centre for Concrete Construction at BRE. He has extensive experience in the fields of load testing and structural monitoring.

INTRODUCTION

Design of the In-situ Frame

Although the construction is within the protected environment of the Cardington hangar, the design brief has been made as realistic as possible. A real site was identified in central Bedford which posed a realistic scenario for the construction of an office building.

For realistic research into the whole building behaviour of a completed in-situ frame building it is generally considered necessary to have a minimum of 3 bays in plan in each direction. For this building it was considered that there would be additional benefits if the plans were increased to 3 bays by 4 bays, so that there was only one axis of symmetry. The bay size is 7.5 m.

The minimum number of storeys required was considered to be five. However when consideration was given to the potential benefits for the research, it was decided that there would be a significant advantage in increasing the number of storeys to seven.

For design purposes the external walls are assumed to be a precast concrete cladding system. The floors are assumed to carry a propriety medium duty raised floor system. Ceilings are assumed to comprise demountable metal tiles. The floor to ceiling soffit height is 3.5 m.

To maximise the benefits from the performance research, the structure has been designed to minimise any redundancy, with the frame structure being as structurally efficient as possible whilst still complying with EC2. In particular a live load of 2.5kN/m^2 has been assumed.

The flat slabs are 250 mm thick and incorporate a range of reinforcement designs over the height of the building. Different arrangements of loose bars and prefabricated mats have been used as part of a parallel project being run by the Reinforced Concrete Council on rationalised reinforcement. A summary of the reinforcement configuration for all seven floors, together with other information on the construction is given in Table 1.

Innovative Features

The in-situ concrete frame incorporates a number of innovative features which have helped optimise the construction process.

High strength concrete columns with normal strength slab connections

To keep the column size uniform throughout the height of the building, high strength concrete (C85) has been specified for the lower three floors. One process constraint with this solution is the requirement in codes to use high strength concrete in the slab element between the two columns. This is felt to be unnecessary and hence a normal strength slab connection (C37) was provided to demonstrate adequate performance.

Table 1 Data summary for each floor

	DESIGN METHOD	TOTAL REBAR MASS, tonnes	MAIN REBAR TYPE	PUNCHING SHEAR REINFORCEMENT		MIX TYPE	CUBE STRENGTH*, N/mm²			STRIKE AGE, hours	MAX DEFLECTION DUE TO STRIKING, mm
				Gridlines A and B	Gridlines C and D		1 day	7 day	28 day		
1	conventional sub-frame analysis	17.66	traditional loose	traditional links	traditional links	C37N	13	48	67	49	11
2	conventional sub-frame analysis	17.9	traditional loose	traditional links	traditional links	C37N	24	44	51	25	8
3	deflection unlikely to impair performance	16.7	rationalised loose	traditional links	traditional links	C37P	12	42	67	43	8
4	yield line gridlines1-3 conventional gridlines 4-5	19.85	blanket cover loose	ROM shear ladders	ACI stirrup	C37F	7	53	60	49	11
5	conventional sub-frame analysis	19.88	one-way mats	Square Grip shear hoops	Square Grip/ DEHA stud rails	C37N	14	40	47	45	9
6	finite element analysis	25.52	blanket cover one-way mats	n/a	Anco-PLUS shear studs	C37PG	7	n/a	39	72	10
7	Birchwood Omnia	23.42	traditional loose	square hollow section	square hollow section	C37N	25	46	53	44	9

N = normal workability P = plasticiser added F = flowing mix PG = slag/cement mix
based on Temperature Matched Cured cubes

Novel permanent formwork system used for roof construction

Half of the roof of the in-situ concrete building has been constructed using permanent formwork panels joined together with a special material known as DENSIT. DENSIT is a very high strength mortar incorporating a large percentage of steel fibres which gives it excellent anchorage properties.

The panels include the bottom reinforcement which is lapped within the DENSIT joints allowing the construction of two-way spanning floors using units which normally span in one direction only.

Prefabricated stairs

Precast concrete stairs have been adopted for use with the building and DENSIT has again been employed to join the two halves of the concrete stair together. The monolithic unit formed has no intermediate support between floor levels resulting in a very elegant method of construction.

Composite bracing systems

Bracing is generally provided by monolithic concrete walls which are complicated to detail and lengthen the construction cycle. The system utilised in the in-situ concrete building comprises steel flats acting compositely with the concrete frame.

PROCESS RESEARCH

A series of projects looking at the process of constructing in-situ concrete frame buildings have been going on whilst the building has been constructed. These have covered the overall construction process together with more detailed projects looking at all aspects of the construction operations and are not considered in this paper. Further details are given elsewhere [1].

PERFORMANCE TESTING

Current Research Projects

Brick cladding

Leeds Metropolitan University are leading a project looking at the loads and strains induced in brickwork panels used in combination with a concrete frame. Conventional wisdom says that concrete shrinks whilst brickwork tends to expand usually requiring the provision of soft joints between the two materials.

This project will assess the scope for effectively prestressing the brickwork as a result of this phenomenon. Different types of brick and mortar combination will be investigated with brick skins both 'trapped'' between the floor slabs and running the full height of the building before being “trapped” by purpose-made steel restraints.

As a precursor to this project Leeds Metropolitan University have installed DEMEC gauges on selected columns at different heights within the building which they have monitored.

They have also cast comparison creep and shrinkage specimens which are being stored and monitored alongside the structure to gain a fuller understanding of the mechanisms involved.

Slab/column connection behaviour

Queen's University Belfast are working closely with Durham University to study the behaviour of slab/column connections. The in-situ concrete building at Cardington is providing the opportunity for full-scale validation of earlier laboratory-scale tests.

Special strain-gauged bars termed "Durham bars" produced by the University of Durham have been installed in one area of the 6th floor. The bars are highly specialised, comprising the installation of a large number of strain gauges along the centre-line of the bar. Each instrumented bar is sawn in half longitudinally to enable the strain gauges to be attached before being glued back together. Wiring for all the gauges is run along the bar and taken out to a remote measurement point.

Instrumenting the bars in this way enables very detailed information to be gained on the strain and hence stress distribution within the bar without the problems associated with measuring surface strains.

Bars of different shapes and sizes have been installed within the 6th floor slab as replacement for the bars specified on the drawings. Checks on the bars post-construction have indicated that they have been installed successfully without being damaged during the construction process. A series of tests is being carried out looking at the behaviour of slab/column connections in the area in which the instrumentation has been installed.

Dynamic characteristics of thin flat slab floors

As slab depths are reduced, it is vital to check dynamic characteristics to ensure that problems will not arise. It is planned to carry out induced vibration tests aimed at determining the dynamic characteristics of thin/slender flat slab floors with varying boundary conditions provided by the building. The data on the natural frequency and mass which will be provided will be extremely useful in the serviceability design of such floors, both under normal loads, eg people walking, car park loading etc., and accidental dynamic overload situations, eg dance floors.

Demonstration Projects

Precast concrete cladding

Concrete Connections, part of BRC-Square Grip are keen to organise the erection of some concrete cladding panels on the in-situ frame. Initially this is seen very much as a demonstration project for the use of proprietary Deha panel fixings which can be used in new build or with retrofit cladding.

The current proposals are for the erection of 6No self-supporting sandwich panels at ground and first floor level and 4 (possibly 6No) panels hung on the end elevation. Once the panels are erected it is hoped that possibilities for research on the panel system and fixings will be pursued.

FUTURE RESEARCH THEMES

It will be important to validate the innovative design features within the building and to this end a Performance Tasks Committee has been established comprising representatives from universities, industry, BRE and the principal funding bodies EPSRC and DETR. Some of the aspects which are being considered are described below. The benefits of full-scale testing have been previously identified by Armer and Moore [2].

The EPSRC has already made a considerable investment in terms of providing funding for the installation of a considerable amount of built-in instrumentation and control specimens.

Serviceability Performance - Design of Thinner Flat Slabs

Minimising slab depths has knock-on benefits in terms of reducing overall construction self weight and hence costs, but the scope for doing this is currently limited by the perceived potential for deflection problems and the need to increase the amount of reinforcement to help overcome them.

The in-situ concrete building at Cardington provides ideal conditions for measuring the deflections of flat slabs during construction and under known service loads over a number of years.

As part of the research which has gone on during construction, the deflections of the slabs due to striking and subsequently have been measured by precise levelling. This has been supplemented by short-term deflection measurements of the formwork movements during the concreting process. Displacement transducers have been left in position which we are now measuring the shortening of the columns and the deflection of an internal panel at each floor level. Precise level measurements can be taken at any point in the future to determine further changes.

The slabs are currently unloaded apart from their self-weight but it is hoped shortly to apply simulated service loading to them using sand bags.

Dr Vollum from Imperial College has been developing a non-linear finite element programme to calculate the long-term deflections of reinforced concrete slabs. Their work has highlighted the shortage of comprehensive test data available. Data collected from Cardington is being used to assess and improve existing methods for the design of flat slabs at the serviceability limit state [3].

Dynamic Performance of Structures

Tests are planned to measure the overall dynamic characteristics of the in-situ concrete building with and without the lateral bracing being effective. The dynamic performance of the frame has been assessed at various times during the construction phase by monitoring response to natural ambient vibration using a remote laser interferometer.

This will be supplemented by forced vibration tests on the completed building using large eccentric-mass vibrator systems installed on the top and on different levels of the structure. Data will be obtained on the natural frequencies of the different modes of vibration, mode shapes, coupling of modes, overall stiffness and damping characteristics.

The information from this series of tests will be used to verify theoretical models that predict the overall behaviour of the building. These tests will be repeated when cladding, finishes, etc are installed.

The dynamic characteristics of the building will also be measured after the introduction of localised structural damage to see if such damage alters the characteristics measured. Such damage will be induced as a result of local static overload tests, after fire tests, after subsidence tests and after gas explosion testing.

It is also planned to investigate the response of the building to ground borne vibrations introduced remotely from the building. Both continuous and intermittent (impact) vibrations will be investigated.

In-Service Monitoring and Smart Structures Technology

The in-situ concrete building at Cardington provides an opportunity for validating novel instrumentation techniques eg use of optical fibre sensors for strain measurement.

One particular project planned for the building has the intention of artificially inducing the effects of subsidence and heave at one corner and at one end of the building. The scope for use of Artificial Intelligence techniques as a means of evaluating performance using data collected from strategically located instrumentation as the input will be investigated.

The particular advantage of using the concrete building for the development of this technology is that, since it will be a test building, it will be possible to introduce gross deformations and try out techniques which would not be possible in a conventional structure.

Ultimate Load Performance

Owing to the requirements to maintain serviceability loads for a considerable period and the need to secure funding, it is unlikely that ultimate load testing will proceed before the year 2001 at the earliest. The following topics are of interest.

Punching shear and redistribution of moments

Few full-scale tests have been carried out on representative flat slab structures. Most codes are based on the results of full-scale tests on isolated components (eg edge columns and internal columns) or scale models of one or two panels.

To facilitate the tests which will be carried out at Cardington a large number (400) strain-gauged bars have been incorporated within representative slab areas on two floor levels in the building as part of the built-in instrumentation referred to earlier.

A variety of different punching shear reinforcement systems have been used at different locations within the building including the use of large structural steel shearheads which

enable large openings to be formed in the slab adjacent to the columns. Coupled with the different reinforcement arrangements used at the different floor levels and the different grades of reinforcement (in terms of ductility) which have been used, there is plenty to be learnt from localised structural tests at these different locations.

The results of these tests can be used to help assess:

(a) The effect of reinforcement ductility on the redistribution of moments;
(b) Comparative performance before and after fire damage;
(c) The effectiveness of different innovative solutions to the provision of punching shear reinforcement;
(d) The reserves of strength available from continuity and other effects not currently considered in design.

It is envisaged that tests on the full-scale structure will proceed in parallel with tests on smaller scale sub-assemblies and analytical work.

Loading on concrete floors

The floor slabs have been designed to receive a live loading of 2.5kN/m^2 as recommended by the British Council of Offices. In practice many buildings are designed to receive higher loadings than these and in any event it is sometimes necessary to apply a higher patch load of 4kN/m^2 say in a corridor area. By testing localised parts of the structure through to failure it is hoped to demonstrate the huge reserves of strength available from mechanisms not normally assumed in design (eg membrane action). This will demonstrate that design, according to current procedures using low assumed values for imposed loading is perfectly adequate, even when it is suspected that the actual loadings might be slightly higher.

Demonstrations of the validity of assuming lower values for imposed loads and the development of new more sophisticated and accurate design approaches will allow savings to be made in construction costs thereby making concrete framed buildings more competitive.

Fire Performance

It is proposed to carry out an extensive test programme involving fully developed fires within compartments in the building. The main objectives of this programme are:

(a) To provide accurate data on the structural response of complete concrete structures to fully developed fires;

(b) To study the thermal response of a compartment bounded by concrete construction to fully developed fire;
(c) To use the information gathered to contribute to the development of a robust approach to the structural fire engineering design of concrete structures.

Fire compartments will be established at a number of different levels in the building. In particular a large compartment fire encompassing half the floor area will take place on the ground floor.

As mentioned earlier, high strength concrete has been used for the construction of the columns up to 3rd floor level. The high strength concretes used (C70/C85 grade) have included both microsilica and metakaolin at different levels and positions on plan. A major issue with the use of high strength concrete is its susceptibility to spalling in fire. Polypropylene fibres incorporated within the concrete can help reduce this problem and such fibres have been included within some of the high strength concrete columns within the building. Fire tests will be carried out on columns with and without fibres included to compare performance.

This work will provide an extension to laboratory test work on the performance of individual high strength concrete columns which is currently underway.

To facilitate the large fire test, box-outs have been provided within the first floor slab to allow for the inclusion of instrumentation.

Cladding Research

There is clearly plenty of scope for research into different cladding systems and their performance on the in-situ concrete building with large areas of elevation available. Some of the cladding projects already underway or planned are mentioned earlier, being of both a research and demonstration nature. Ideas and suggestions for other projects in a similar vein are very welcome. The possibilities are broad and could include:

(a) Weathertightness including rain and air penetration and condensation;
(b) Innovative systems and products;
(c) Thermal insulation;
(d) Fit and tolerances;
(e) Fixings and anchorage;
(f) Accuracy of erection;
(g) Blast resistance;
(h) Toughening and laminating of curtain walling systems.

Response to Gas Explosions

The in-situ concrete building offers great potential for research into this important area. The objectives of studies planned are:

(a) To evaluate the effectiveness of building products designed for improved performance under blast loading relative to conventional practice;
(b) To improve understanding of the way in which real buildings (ie clad, glazed and otherwise equipped) respond to blast loading, how this behaviour can be predicted and how to design for improved serviceability.

Studies will include investigations into:

(a) Blast wave propagation around and inside the structure, in 3 dimensions;
(b) Response of individual and grouped glazing and cladding panels, including post-failure behaviour such as debris dispersal;
(c) Transfer of load through fixings to the structural frame;
(d) Response of the frame itself.

It is also intended to develop improved analytical and design models of the various aspects of building response to blast listed above. These models will be used in the development of improved building products and design philosophies.

Accidental explosions have been shown to cause both considerable hazard to public safety and extensive damage to buildings. The safety of the public within and around such buildings is influenced by the way in which the structure and its individual components perform during and subsequent to, this extreme form of loading.

Once the immediate chaos relating to the explosion itself has passed and public safety has been assured, the building will require refurbishment. If the cost of refurbishment is predicted to exceed or approach that of demolition and rebuilding then the latter option is likely to be selected.

The cost of replacing the cladding and services will far exceed the cost of the structure. Thus, improved design of these elements to ameliorate the effects of accidental explosions will result in significant financial benefits while at the same time improving safety.

It is clearly desirable to improve the structural performance of cladding and glazing systems under such extreme loads. To do so requires an improved understanding of the way in which an explosive load acts upon and interacts with a real building and its neighbours, as well as the way in which the various components interact with each other when loaded in a manner for which they may not have been designed.

Repair and Strengthening

Repair and strengthening of concrete structures is a growing field with many new products and techniques being put into the market place. Clearly these need to perform adequately from the point of view of both strength and long term durability.

The in-situ concrete building offers an opportunity for assessing the strength of these repair systems with reference to concrete frame structures. Some of the topics worthy of investigation include:

(a) Retrofitting and strengthening of concrete slabs with large openings using steel or composite fibre plates;
(b) Strengthening of column/slab connections (sometimes necessary in car park structures);
(c) Repair and strengthening of damaged columns.

Demolition and Recycling

The in-situ concrete building represents an ideal opportunity for different demolition techniques to be investigated and compared. The methods of demolition to be employed need to be considered in conjunction with the desire to recover and re-use materials.

Government policy is aimed at making construction more sustainable by minimising waste production and improving management practices; there is considerable potential for using recycled demolition waste for example as a substitute for primary aggregate.

The demolition industry is committed to increasing significantly the use of recycled and recovered materials resulting from the demolition process. Nevertheless this is an area in which the UK currently lags some way behind some other European countries.

The TMR Scheme

The Training and Mobility of Researchers (TMR) programme run by the European Commission encourages and finances the participation of, preferably young, researchers in unique and innovative research projects across the European Community and Associated States.

The Large Scale Facilities Programme allows access to unique facilities within the Community, which are not normally available in individual member countries.

The BRE Cardington Large Building Test Facility (LBTF) has been designated as a Large Scale Facility and has been awarded a 2 year contract running until May 2000 to allow researchers to be involved in ground breaking whole building projects not just on the in-situ concrete building but on the steel and timber frame buildings also.

Expressions of interest are welcome from any eligible parties and further details about the scheme can be found on the BRE web site http://www.bre.co.uk.

CONCLUDING REMARKS

There are clearly many people and organisations who can derive mutual benefit from collaborating in research, demonstration and development projects on the in-situ concrete building.

This paper has identified projects which are underway and has attempted to identify research themes under which future proposals can be developed. However this list is not exhaustive and the authors would welcome suggestions for activities in other areas.

For any particular collaboration the aim is to enable the partners involved to invest a relatively small amount of effort and finance and derive a highly geared return. A viable collaboration is achieved by a number of organisations each with their own independent and non-conflicting objectives and funded from different sources having a common interest.

The advantage of this form of collaboration is firstly that the cost is spread between a number of organisations and secondly if a single element fails to receive the necessary funding, the project as a whole can still probably proceed.

Those with an interest in any of the topics described in this paper or other related studies are invited to contact the authors and assist in the planning of the research. By proceeding in this manner, the opportunity offered by this unique and exciting project will be maximised.

ACKNOWLEDGEMENTS

The authors wish to acknowledge the contribution made by DETR in funding the preparation of this paper, and all those who have contributed to the project to date.

REFERENCES

1. CHANA, P S AND JUDGE, C J. Radical re-design of the in-situ concrete frame process. Proceedings of the Concrete Communication Conference, July 1998, published by British Cement Association.

2. ARMER, G S T AND MOORE, D B. Full-scale testing on complete multi storey structures. The Structural Engineer, Vol. 72, No. 2, 1994 P.30-31.

3. VOLLUM, R L, HOSSAIN, T R AND PAVLOVIC, M N. Slab deflectons in the Cardington in-situ concrete building: A comparison of theory and practice. Proceedings of the Concrete Communication Conference, July 1998, published by the British Cement Association.

CALCULATION OF CONCRETE STRUCTURES RESIDUAL SERVICE LIFE

S N Leonovich

Belarussian Polytechnical Academy

Republic of Belarus

ABSTRACT. As a theoretical base for concrete durability prediction consideration was given to voids and cracks in the development of a model. This had a two-level structure: 1) matrix of hardened cement paste with including (particles of fine and coarse aggregate); 2) voids of different form (cracks) as the result of external actions (influences). The principal of the method was the development of a parameter of defining crack resistance, i.e. critical stress intensity factor

Stress intensity factors in tip of a defect (crack) in a material, can include in their values the influence of micro-crack and macro propagation. It is possible to take of these defects at all levels of the hierarchic composite structure. Every pore or crack in concrete creates some field of stresses around itself. If these pores and cracks are graded in terms of volume in concrete periodically, the fields of stresses and deformations can be superimposed enabling complicated stress-deformation conditions to be considered.

Keywords: Durability, Cracks, Model of concrete, Fracture mechanics, Stress intensity factor.

Dr Siarhci Leonovich is Associate Professor of Department of Reinforced and Stone Constructions, Civil Engineering Faculty, Belarussian Polytechnical Academy, Minsk, Republic of Belarus. He specialises in the use of fracture mechanic parameters for calculation of concrete durability.

INTRODUCTION

The following parameters are accepted for calculation of concrete durability when using approaches of fracture mechanics (generalising criteria).

1. Concrete is elastic, quasi homogenize media, which consists of: a) matrix - hardened cement paste and elements of sand and crushed stone; b) voids – capillaries - cracks and pores with initial cracks.

2. The voids in matrix are present in a five-level system with dependence on form and dimensions. Under external influences the voids, reaching critical dimensions, transfer from this level to next level in accordance with following processes: stabilisation of dimensions, delocalisation, accumulation, critical concentration in elementary single volume transfer to the next level.

3. The process of formation and growth of cracks is considered as the result of external influences on the basis of fracture mechanic principles. This stress-deformative conditions are evaluated by the following parameters: stress intensity factors, fracture toughness, or energy for creation of a unit crack area.

4. The general constant of concrete crack resistance, its resistance of emergence, accumulation in volumes and formation of leading main cracks of critical (τ) sizes is $K_{cij}(\tau)$. $K_{cij}(\tau)$ is the algebraic sum of critical values K_{ij} in all system at all levels of cracks-voids, which fill the canonical volume until critical concentration.

$$K_{ij} = \sqrt{G_{ij} E_{ij}} \tag{2}$$

5. External temperature, moisture and corrosive actions create the field of stresses in tip of a crack. This is considered by using theory of concrete growing old.

6. The processes of concrete destruction by cracks is considered as a general condition in some canonical volume. This canonical volume poses physical peculiarities as a composite with elastic and rheological properties. Phenomenological peculiarities of physical processes are plausible and reasonable by experimental data K_I and K_{II} (samples cubes and prisms with sizes of cross section 100 x 100 mm.)

$$K_{ic}(\tau) = K_{ic}(\tau_o) D \tag{3}$$

All defects of the concrete structure are divided into 5 types: 1. Round pores with sub-microcracks on the boundary (around). 2. Elliptical pores with microcracks on the boundary. 3. Mezocracks in concrete body. 4. Cracks on the border. 5. Cracks near surface of fine and coarse aggregate.

Deformation and strength properties in a unit concrete volume are provided by a system of active and reactive forces in the structure.

$$\sum N_{act} - \sum N_{react} = R_i \tag{4}$$

If the external conditions (temperature, humidity, pressure) are changed, the strains arise in the defects of the structures (pores, capillaries, cracks) filling by liquid, steam, ice.

The sizes of defects, quantity and properties of ties (connections) are changed and this influences the level of R_i, E_j.

THEORETICAL BASIS EXPLANATION AND ANALYTICAL SOLUTIONS

Some elementary volume of cement stone includes some level of capillaries. Depending on the external conditions these capillaries contain some free water which under the action of environment's temperature can change to gaseous or solid (hard) states. The main parameters to model the capillaries of a system (concentrator of stresses, initiator of cracks formation) are l_c - capillary's length; a_c - capillary's diameter; b_c - depends on humidity of cement stone; W - humidity; t - temperature of cement stone.

Condition 1

T = const, W ≠ const, P ≠ const.
Stress intensity in the tip of a capillary is given by

$$K_{I,t}^{I,1,c} = \frac{4\sqrt{\pi}}{\sqrt{2}} \cdot \frac{a_c \cdot \cos\theta \cdot \alpha_0 \cdot (1 - \frac{t}{t_k})}{g_c \sqrt{l_c \left[1 - (1 - \frac{W}{100})^2 \right]}} \tag{5}$$

where α - surface tension of water; θ - angle of wetting; g_c - distance between two adjoining capillaries; t, t_k - critical temperature.

If $a_{c1} = 10^{-7}$ m, $K_I = 0{,}087$ MPa m$^{1/2}$. The value K_{ic}^{cs} is changed from 0,1 - 0,32 MPa m$^{1/2}$ depending on W/C ratio, and determined from equation

$$K_{ic}^{cs} = -0{,}0148 + 0{,}1330 C/W + 0{,}0058 R_c - 0{,}0082 W \tag{6}$$

The result of these stresses are deformations of cross dislocation at the capillary's tip. This process is described by a stress intensity factor K_{II}. The value of stress intensity factor is calculated from the expression:

$$K_{II} = \tau \sqrt{\pi l_c} \tag{7}$$

If $K_{II} = K_{II}^{cs}$, the mechanism of the growth of microcrack longitudinally is the mechanism of cross dislocation.

Condition 2

T = const, W = const, P ≠ const.

It is useful to examine the cement stone at the macro-level. The cement stone consists of non hydrated grains (particles) and the hydrated mass. The hydrated mass consists of pores (capillaries) and chrystaline system (micro-level).

In the hydrated mass, micro-defects of two-types are observed: I – capillaries, II – radial cracks.

Radial cracks are created as a result of differences in modulus of elasticity and linear expansion coefficients of non-hydrated particles and hydrated mass.

For instance, for ice the main equation for calculation of stress intensity factors are:

$$K_{I,t}^{I,2,i} = \alpha_{t,i} \cdot \Delta t \cdot E_i \cdot \sqrt{\frac{\pi l_c}{2}} \left[1 - \left(\frac{2}{\pi}\right) \arcsin\left(\frac{2b_c}{l_c}\right) \right] \tag{8}$$

$$K_{I,t}^{I,2,cs} = \alpha_{t,cs} \cdot \Delta t \cdot E_{cs} \cdot \sqrt{\frac{\pi l_c}{2}} \tag{9}$$

$$K_{I,t}^{II,2,i} = 2\alpha_{t,i} \cdot \Delta t \cdot E_i \cdot \sqrt{\frac{l_c}{\pi}}\, arccos\left(\frac{b_c}{l_c}\right) \left[1 + f\left(\frac{b_c}{l_c}\right) \right] \tag{10}$$

$$K_{I,t}^{II,2,cs} = 3{,}523 \cdot \alpha_{t,cs} \cdot \Delta t \cdot E_{cs} \cdot \sqrt{\frac{l_c}{\pi}} \tag{11}$$

where E_i, E_{cs} are modulii of elasticity of ice and cement stone respectively; $\alpha_{t,i}$, $\alpha_{t,cs}$ are coefficients of linear temperature expansion.

It is possible to use this model on a macro-level for cement-sand mortar and for concrete, i.e. this model is a hierarchic system, which is useful for describing destructive processes at any level of observation in a structure. Capillary on the macro-level are observed as micro- and macro-cracks, the non-hydrated mass as particles of fine cement-sand mortar, or coarse (concrete) aggregates, and the hydrated part - as a cement stone on macro-level (cement-sand mortar)or mortar (concrete).

Using appropriate characteristics for materials of aggregate and cement-sand mortar it is possible to get deformative, strength parameters, parameters of propagative cracks for concrete by change its humidity and temperature. It is very significant to know the difference in destruction of light-weight aggregate concrete and heavy-weight aggregate concrete.

The contact cracks on the boundary matrix-aggregatre don't arise and that's why values K_{Ic} and K_{IIc} are not determined by expressions. The conformity between experimental and theoretical curves for ordinary heavy-weight concrete is very good. It is confirmation about possibility of using this model on all levels of concrete.

DETERMINATION OF CONCRETE CRACK–RESISTANCE

The quantity of pores and cracks of types I–IV depends on porosity of concrete and of type V depends on volume concentration of particles of fine and coarse aggregate.

Moreover, these cracks on the grain boundary characterise mezo- and macro-level of concrete structure. The volume concentration of every kind of pores and cracks on concrete depends on porosity and volume concentration of aggregate. The its stress intensity factor corresponds to every kind of damages and depends on quantity of given kind of pores and cracks in volume of concrete.

Using the principle of super-position and considering, that pores and cracks are distributed proportionally within the concrete volume, it is possible to write - for submicro, micro and mezo-levels, the following equation,

$$K_1 = n_1 K_I^I + n_2 K_I^{II} + n_3 K_I^{III} + (1 - n_1 - n_2 - n_3) K_I^{IV} \tag{12}$$

where $K_I^I, K_I^{II}, K_I^{III}, K_I^{IV}$ - stress intensity factor accordingly in tips of micro-defects of types I, II, III and IV;

n_1, n_2, n_3 - the relative content of types I, II, III and at typical levels of micro-damages; for mezo- and macro- levels.

$$K_1 = n_1 K_I^I + n_2 K_I^{II} + n_3 K_I^{III} + n_4 K_I^{IV} + (1 - n_1 - n_2 - n_3 - n_4) K_I^V \tag{13}$$

where K_I^V - stress intensity factor at the tips of contact cracks at the boundary of particles of fine and coarse aggregates; n_4 - the relative content of cracks on the border.

Thus, finally

- for mezo- and macro-levels

$$K_1 = p\sqrt{\pi}\left\{ \begin{aligned} & n_1 A\sqrt{l_1\left(1+\frac{d}{L_1}\right)} + n_2 B\sqrt{l_2\left(1+\frac{d}{L_2}\right)} + n_3 C\sqrt{L_3} + \\ & + n_4 D\sqrt{L_4} + (1 - n_1 - n_2 - n_3 - n_4) E\sqrt{\frac{LS}{2}} \end{aligned} \right\} \tag{14}$$

where R = $5 \cdot 10^{-7}$ m; $L_1 = 1 \cdot 10^{-8}$ m; $L_2 = 1 \cdot 10^{-8}$ m; $L_3 = 5 \cdot 10^{-3}$ m; $L_4 = 5 \cdot 10^{-3}$ m; $L_5 = D_{max}^{ag}$; a = $5 \cdot 10^{-6}$ m; D_{max}^{ag} - the maximum size of particles of fine and coarse aggregate in matrix or concrete; L_5 - the length of radial crack, E - coefficient.

The Calculation of Concrete Durability

Considering, that the critical length of crack of normal alienation is the identical as by short duration action as by long duration action of force, the relaxation of value of stress intensity factor in time is written

$$K_{Ic}^{b}(t) / K_{ic}^{b} = \frac{1}{\sqrt{1 + 2E(1-\nu^{2})c_{b}(t,\tau)}} \tag{15}$$

anal the measure of creep of concrete is expressed, using the theory of growing old

$$C_{b}(t,\tau) = C_{b}(\infty,\tau)\left[1 - e^{-\gamma(t-\tau)}\right] \tag{16}$$

where $C_b(\infty, \tau)$ - the limit value of measure of concrete creep; γ - the speed of growth of deformations of creep.

The durability of concrete are calculated from following equations and estimated as minimal val:

$$t = \frac{2(K_{IC}^{CT} - K_{I}^{N})}{(K_{IW}^{TW} - K_{IC}^{TW})\psi}(years) \tag{17}$$

$$lg\,t = \frac{K_{I}^{CT}}{K_{IC}} lg\,28\sqrt{1 + 2E_{b} \cdot c(\infty,28)(1 - e^{-\gamma(t-28)})}(days) \tag{18}$$

CONCLUSIONS

Calculation of concrete structures residual service life is possible using parameters of concrete porosity and fracture mechanics.

REFERENCES

1. LEONOVICH, S N AND GUZEEV E A. Prediction of Concrete Structures Durability; Another Look. Proceedings of the 13th FIP CONGRESS ON CHALLENGES FOR CONCRETE IN THE NEXT MILLENNIUM, Vol, Amsterdam, 1998. pp 983-986.

PREDICTION OF SERVICE LIFE AND CHOICE OF REPAIR STRATEGY

C Henriksen

C Michaux

In-Situ SA

Luxembourg

ABSTRACT. The increasing damage to concrete structures during the last few years has led to an increasing need for repair works and is not likely to change in the future, until all old concrete structures have been repaired or secured against repair through preventive maintenance. As the extent of repair works exceeds available funding there is a need to optimise spending of the available funds. Service life calculations have become a necessary part of this process. No established well documented service models are however available. The paper consider that there is an extensive need for the developing of more reliable models, but that it is possible to carry out reliable service life evaluations based on existing knowledge. Furthermore it is highlighted that the sampling of the data to be used in the models is far more important for reliability of repair assessment needs than the model itself. A simple model, therefore, is preferable and the paper sets up a guideline for such model and finally how to prepare realistic repair strategies.

Keywords: Deterioration, Corrosion, Frost, Alkali-aggregate reactions, Service life modelling, Economic calculations, Repair strategies.

Carsten Henriksen is Manager of In-Situ SA., Luxembourg. He specialises in testing of concrete structures, concrete durability, corrosion and evaluation of repair strategies for reinforced concrete structures.

Christophe Michaux is Fifield Engineer. He specialises in testing of concrete structures, concrete durability, corrosion and evaluation of repair strategies for reinforced concrete structures.

SERVICE LIFE MODELS - PROBLEMATICS

Service life modelling has become a necessary part of the evaluation of existing structures.

- Data sampled from the structure are put into a theoretical model and the time of initiation of damage and the remaining service life are calculated ref. Figure 1.

Most models available are based on the second law of Ficks[1,2,3]. It is wellknown that this diffusion model is not all time valid e. g:

- The model is based on the diffusion of the aggressive agents (CO_2 and chlorides primarily). Chloride ingress in not submerged structures however will take place through capillary suction as well. This process is a far more rapid process than that of diffusion.

- The model requires a permanent water filled pore system. The moisture content in not submerged structures however is varying with the surrounding climate. Saturation takes place in wet periods and drying out in dry periods. In most climates the concrete cover layer over years will tend to be in an equilibrium with 85-90 % RH (the annual moisture average the most places). The requirement to saturation rarely can be fulfilled in practise.

Another critical parameter is the threshold value for the chloride content. Chlorides are the main reason of the most durability problems to reinforced concrete structures [3]. Consequently the most modelling concerns chloride initiated corrosion. The initiation depends on the size of the threshold value. However we have no all-time valid threshold value for chloride.

On this bases it is hard to claim that reliable service life models are available. Better or theoretically more correct models are being developed but if they will improve the decision basis is difficult to know.

Most concrete durability problems are related to hidden execution faults[3]. Even the most advanced theoretical model can not foresee where execution faults are located. The quality of the sampling in practise (to detect where these faults are located) therefore is more decisive for the quality of the estimate than the model itself. A simple service life model therefore should be preferred despite the uncertainty.

Finally it must be stressed that the evaluation of the load carrying capacity is a crucial element in the service life calculation. This is often forgotten discussing durability and service life.

SAMPLING OF DATA

Based on the sampled data the times t1, t2, t3 and t4, ref. Figure 1, are estimated. The sampling of data itself is crucial for the estimation. To ensure the reliability of the sampled data a certain procedure has to be followed from time to time. In the following a test procedure is suggested. The procedure is valid in cases where durability problems are caused by chloride induced corrosion.

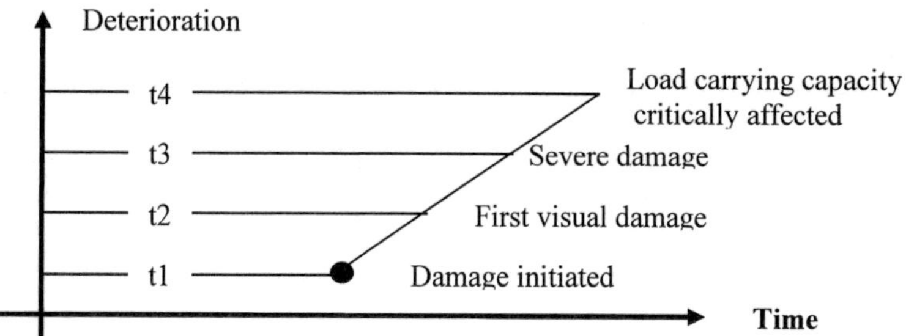

Figure 1 Service life curve for a concrete structure

The suggested procedure for the data sampling is:

Phase 1: The Overview

- structural evaluation by a structural engineer(detection of areas critical to the load carrying capacity, expected crack patterns etc.)
- visual inspection of the structure to detect cracks and other obvious visible concrete damage.
- hammer tapping and eventually Impact Echo testing or similar to detect areas with internal flaws.
- potential measurements to detect areas with corrosion(corrosion will not be initiated without effecting the electrochemical potential).

Phase 2: The Confirmation

- Based on the potential measurements the structure is <u>preliminary</u> divided into e.g. high risk, medium risk and low risk corrosion zones ref figure 2. This can be done automatically by equipment available on the market. Or following e.g. the ASTM-standards(even it is wellknown that they are not all time valid).
- chloride profile and carbonation measurements are carried out in each corrosion risk zone[4]. It is important that chloride profiles are sampled to a depth behind the reinforcement layer in relevant intervals e.g. 0-15mm, 15-30mm, 30-50mm and 50-70mm.
- covermeter measurements are carried out in each risk zone
- break-ups to visually inspect the actual state of corrosion in each corrosion risk zone. Even the rebars in the low risk zone have to be visually inspected.
- cores are cut for the purpose of detecting areas with delamination, evaluating the concrete quality or the risk of frost/alkali aggregate reactions[5]. The locations of the cores are chosen based on the results of the visual inspection, the hammer tapping and the Impact-Echo testing.

Phase 3: Re-evaluation

This phases is initiated if Phase 1 and 2 have revealed results not being logic or not fully confirmed.

Typically problem: No correlation between potential measurements and the visual inspection of the exposed rebars e.g.:

- *the measurements indicate corrosion and no corrosion can be visually observed.* Visual corrosion products develop over time. Not immediately after the initiation of the corrosion process. If the concrete is carbonated or chloride contaminated it is obvious to conclude that a corrosion process actually is developing. Repair/maintenance means will be needed. If the concrete is not carbonated or chloride contaminated other parameters than active corrosion are affecting the potentials(shall not be detailed in this paper). Consequently there are no need to initiate repair/maintenance means. A future survey might be preferable.
- *the measurements indicate no corrosion, the concrete however is chloride contaminated/carbonated and corrosion has developed.* The potential measurements and the ranking in low risk, medium risk and high risk zones has to be re-evaluated. The evaluation of the potential measurements carried out during Phase 1 is preliminary only. The correct correlation between the visual observations and the potential measurements has to be set-up during phase 3. This is done changing the potential limits preliminary used (e.g. the ASTM-standard) in a more positive direction (from e.g. -300 to -250mV, from -250 to -200mV etc) until the correlation is correct. More break-ups might be needed.

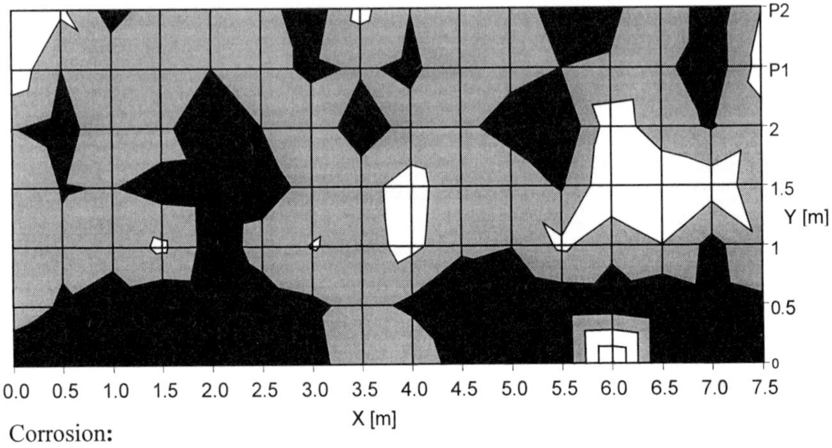

Figure 2 Results of the potential mapping of a concrete structure (a wall of a swimmingpool) and related estimates on the time of initiation, ref Figure 3

SERVICE LIFE CALCULATION

General

In the following a simple model of service life calculation is described. This model requires uncracked conditions. If the concrete is cracked the model is not valid.

In case of cracking conclusions have to be based on a structural evaluation/calculation.

If the concrete appears with delaminated or spallen areas the model can not be used neither. Such areas anyway have to be repaired. Which mean to be initiated has to be decided upon based on a structural evaluation/calculation. Finally it must be stressed that the definition of the service life is an individually choice. In this paper the time of initiation is defined as the service life. The time until the damage has affected the structure (t2, t3 or t4) however might be defined as the service life.

Calculation

In the following a calculation model based on a simplified version of Fick's 2. law is used[1,2]. The model requires that any transport takes place as diffusion and that the concrete is saturated.

The theoretical time of initiation is calculated according to the formula:

$X = K\sqrt{t}$ where
X = concrete cover layer thickness (mm)
K = constant
t = age (years)

Example 1: Calculation

The threshold chloride level (e.g. 0,05% mass of dry concrete) is located in the depth of 20 mm(X) in a 25 years (t) old structure ref. Figure 3.

The constant K then is:
$K = 20/\sqrt{25} = 4$

With K=4 the time(t1) until the chlorides penetrate the concrete cover layer (X1) can be calculated:

$t1 = (X1/K)^2 = (30/4)^2 = 56$ years

The structure is 25 years(t) old. The remaining service life until corrosion theoretically will be initiated is:

Remaining Service Life = t1-t = 56-25 = 31 years

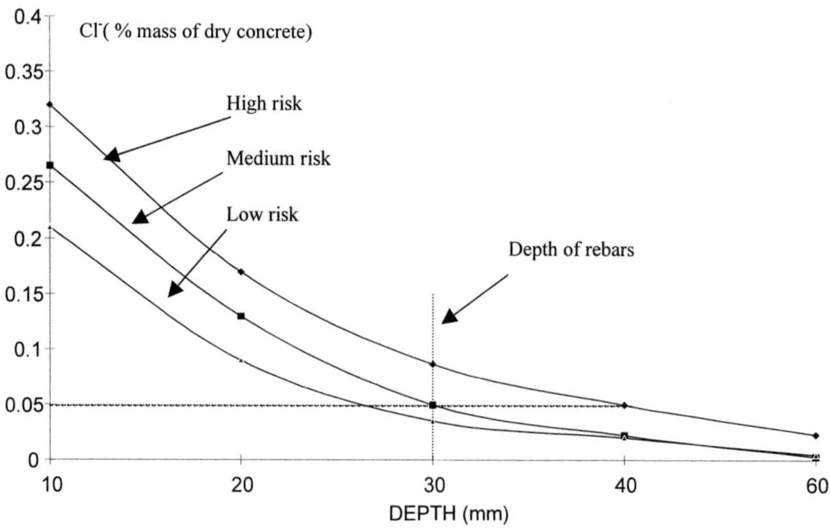

Figure 3 Chloride profiles measured in each of the corrosion risk zones ref Figure 2

Figure 3 illustrates 3 chloride profiles sampled in each of the 3 corrosion risk zones defined in Figure 2.

Based on the formula described in example 1 the time of initiation (defined as the remaining service life) of each risk zone can be calculated(threshold chloride value: 0.05% mass of dry concrete):

- High risk: -11 years
- Medium risk : -3 years
- Low risk: + 8 years

In the high risk zone corrosion already has been initiated and corrosion damage extensively developed. In the medium risk zone the corrosion too has been initiated but the reinforcement and the concrete normally still will be without extensive damage. In low risk zones the corrosion has not yet been initiated.

CHOICE OF REPAIR STRATEGY/ ECONOMIC CALCULATIONS

For each corrosion risk zone the remaining service life has been estimated ref. Figures 2 and 3. Based on the potential measurements the area (m^2) of each zone can be estimated.

Normally the number of repair strategies available is limited. Furthermore each of them require that the structure has a certain remaining service life to prevent from early age repair damage, ref. Figure 4.For each risk zone the available strategies and costs are assessed based on Figure 4.

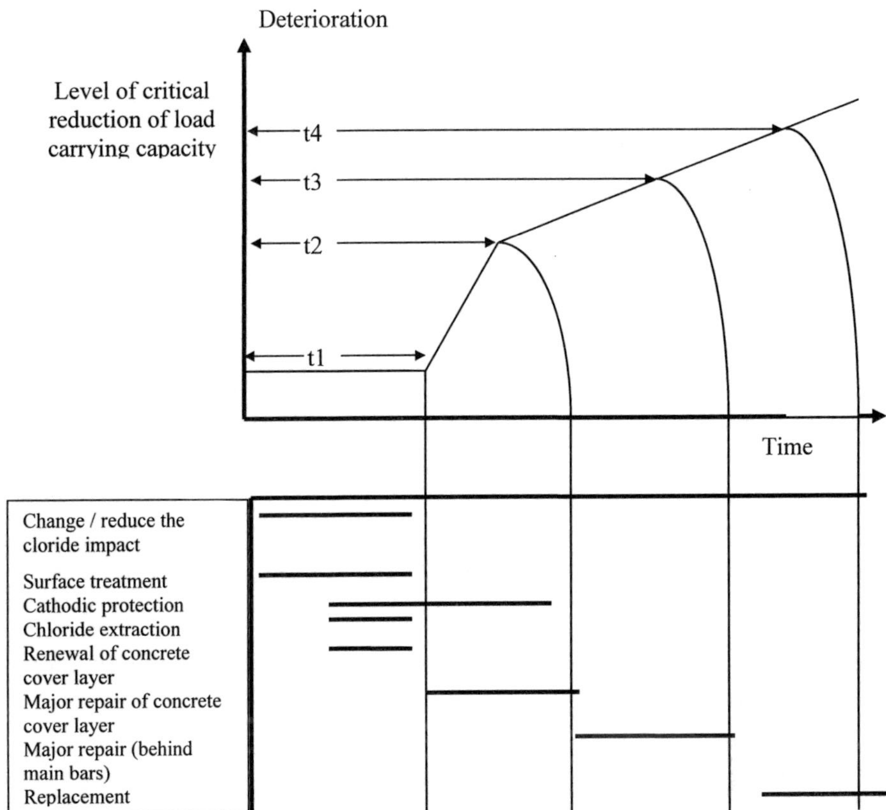

Figure 4 Relevant methods depending on the level of deterioration. The service life curve is calculated based on a special investigation

The final choice of relevant strategies has to be a mix suiting the best the actual situation. It is obvious that repairs initiated step by step to the different risk zones at different times are not realistic. An overall repair strategy has to be initiated for the entire structure. Either initiating immediate actions or choosing a different strategy to be executed later. In the following 3 different strategies for the wall are described.

Strategy 1: Local major repair and surface treatment
Major repair of all high and medium risk zones and finally surface coating of all areas to prevent from further damage. The work must be carried out immediately. The strategy includes regular repair of the coating every 10 years.

Strategy 2: Overall major repair
The load-carrying capacity will not be affected the first 10 years. The repair is postponed 10 years. There is extensive damage after 10 years. Major repairs have to be carried out to the entire structure(all high, medium and low risk zones). A surface treatment is included to prevent future damage(to be repaired every 10 years)

Strategy 3: Local minor repair and cathodic protection
Only the most necessary minor repairs are carried out to high risk zones. Further damage to medium and low risk zones is prevented using cathodic protection(to be repaired every 15 years).The strategy has to be initiated as soon as possible to limit the extent of concrete repairs. Cost for the surveillance of the cathodic installation has to be included.

The financial requirements over a 20-year-period are shown in Table 1.

Table 1 Costs of 3 different strategies

YEAR	STRATEGY		
	1	2	3
0	175,000		145,000
5			10,000
10	40,000	475,000	10,000
15			50,000
20	40,000	40,000	10,000
Total costs	255,000	515,000	225,000
Net present value *	206,000	250,000	180,000

* Discount rate 7%

Choice of Strategy/risk Analyses

Strategy 1 and 3 are technically and economical equal, strategy 3 however having the lowest cost over time(20 years).

If money is available now, strategy 3 should be chosen. If no funds are available it is apparent that the technical and economic consequences will be extensive. If the repair works are postponed an extensive development of damage has to be foreseen and finally call for another repair strategy(strategy 2).

Strategy 2 is however not attractive in any respect and stress that a postponement- policy in most cases is not beneficial.

FINAL REMARKS

Service life modelling based on in-place testing will be a necessary tool when trying to optimise investments in concrete repairs and maintenance in the future. With today's knowledge, evaluation of service life is somewhat uncertain. But this is no obstacle to an evaluation as long as the technical and economic consequences are also evaluated. In most cases the effect of uncertainties is minor. Only in cases where long time predictions are made (> 20-25 years) care should be taken to confirm the predictions through regularl controls over the period.

The model described in this paper is of a very simple nature that does not pretend to be very accurate. However as a tool in the hands of an experienced engineer it will be of a sufficient accuracy to prepare repair strategies within the time perspective of 10-25 years. This time perspective normally covers the most cases.

The experienced engineer normally will know how to handled deterioration problems even without any model. The model in fact does only support his evaluation and helps him to detail the time of execution, helps him to evaluate if preventive measures are more beneficial than repair means and finally helps him to describe correctly the extent of areas in the need of repair to prevent early age repair damage.

Future developments however have to be implemented in order to improve our basis for decision-making. This benefits both the individual engineer and society in general.

REFERENCES

1. POULSEN, E. 13 betonsygdomme, Beton 4, SBI 1985. (E. Poulsen, " 13 Concrete Diseases " Concrete 4, SBI 1985)

2. POULSEN, E. Klorider og 100 ars levetid, DBF-Publikation nr 36, (E. Poulsen, " Chlorides and 100 years service life ", DBF Publication n° 36)

3. Chloride-induced Corrosion, The Danish Road Directorate, October 1991

4. GERMANN, C, PETERSEN, ANKER HANSEN, O G. Dansk Beton Nr 1 1991, RCT-Metoden, (Danish Concrete Nr 1, 1991, The RCT Method by C. Germann Petersen and Anker, Hansen)

5. HENRIKSEN, C. Consequences of alkali-aggregate reaction in Denmark, Presented at the 6th international DBMC, 1993, Japan)

6. HENRIKSEN, C. Service Life Prediction; Choice of Repair Strategy for RC Structures, Presented at the Materials Research Society Fall Meeting, December 1997 in Boston.

DURABILITY CONTROL OF CONCRETE SEWER PIPES

S Saegrov

SINTEF Civil and Environmental Engineering

Norway

ABSTRACT. Water and sewer pipelines are built to last for a period of more than 100 years. There are, however, no methods for specifying or achieving acceptable durability. A system may be based on a number of independent test methods, which can be combined into a "durability number". Such a system has been developed for the analysis of a number of old Norwegian sewer pipes. Comprehensive laboratory analyses were undertaken, including methods for strength, porosity and chemical analysis. This durability number was developed as a measure of the total result of the strength and porosity tests. For each test, an index was calculated as the proportion between the test result and reference number representing perfect conditions. The indices of each test method were combined to form one unique durability number. Durability numbers calculated for pipes from 60 different networks are compared with the general knowledge of the pipelines, and with the result of a visual inspection of the pipe surface. There was a strong relationship between the results from the test system and the pipe condition. Based on this concept, the Norwegian concrete industry association and the SINTEF research foundation have developed a system for durability assurance of concrete pipe production. This system has been tested for pipe samples received from various concrete pipe industries. The system has proved to be appropriate for the durability control of concrete pipe production, and is now to be implemented in the control schemes of the concrete pipe industry.

Keywords: Sanitary technology, Concrete, Sewer pipe, Durability, Control system.

Dr Sveinung Saegrov is a Senior Research Engineer at SINTEF Civil and Environmental Engineering, Department of Water and Waste water, Trondheim, Norway. His main research interests include water and water pipe materials, as well as methods for analyses of water and waste water system performance.

INTRODUCTION

Serious leakage and structural problems are found in a large part of the Norwegian sewer systems. Pipe leakage is a major reason that 2/3 of the treated waste water is unpolluted water infiltrated into the sewer pipelines, and 1/4 of the sanitary sewage never reaches the treatment plants. Structural problems may also lead to pipeline collapses with the risk of basement flooding.

Pipelines more than 25 years old do not satisfy modern performance requirements with regard to strength and imperviousness. They are exposed to mechanical and chemical deterioration. It will cost $ 6 billion to renew the 13,000 km of pipeline, which were laid before 1970. This is far beyond the economical resources of the Norwegian municipalities. The cost may be reduced by a better knowledge of the factors, which govern the state and breakdown of sewer pipes, and by methods for effective durability control of existing and new concrete sewer pipes. Based on this knowledge, preventive maintenance and renewal plans may secure a cost-effective investment in the sewer system.

A selection of laboratory methods can be applied for the study of the deterioration due to mechanical and chemical influences on concrete sewer pipes. Associated with this, it is an important task to find criteria for pipeline evaluation and appropriate test methods, which may form the basis for durability control.

It is a matter of fact that deterioration affects the strength and the porosity of the concrete pipes. The test methods should therefore measure parameters relating to these properties

RESEARCH PROGRAMME

General

In a Norwegian research project, a total of 80 pipes from 60 different networks were undertaken in a comprehensive test program. For the final discussion, the results were presented as "quality numbers", the number containing results from each of the tests. This gives a basis for the development of a system for quality assurance of concrete sewer pipes.

Selection of Pipes to be Tested

Digging and transportation of utilized concrete sewer pipes are expensive, and it has therefore been necessary to restrict the pipe selection to pipes under reconstruction or repair. Importance have been emphasized, for the inclusion of pipes with different life histories and problems, such as cracked pipes, sags, concrete corrosion and leakage. It has also been of great importance that the selection should represent various ages.

About one third of the pipes which are investigated are reconstructed due to pipe material defects. The selection is equally distributed with regard to age for the period 1920 - 1980. Pipes with diameter over 300 mm are underrepresented in the selection. This is due to lack of available test items. The test items were collected from all parts of Norway.

Laboratory Programme

The laboratory program should include relevant parameters for the strength and durability. The resistance against chemical corrosion is closely linked with the concrete porosity. The microstructure tests can give important information about the rate of deterioration. Table 1 gives a view of test methods included in the program. The visual registration is concentrated on surface roughness and visible cracks.

Table 1 Laboratory program for analyses of concrete sewer pipes [3]

METHOD	MEASURE
a) Test preparation	
1. Cleaning, dimensions, photo	Preparation and identification
2. Visual evaluation	Condition
b) Strength measurement	
3. Fracture strength	Pipe strength
4. Compressive strength	Compression Strength
5. Flexural strength	Strain strength
c) Porosity measurement	
6. Water absorption	Porosity, pore volume
7. Capillary action	Porosity, pore volume distribution
8. Carbonization indicator	Carbonization depth
d) Microstructure	
9. Electron-microscopy of thin sections	Microstructure, minor cracks
10. Electron probe micron analysis	Chemical composition

Strength measurements are a major way to evaluate the concrete quality. The program therefore includes measurement on fracture strength, compressive strength and flexural strength.

The porosity is measured in three ways: Water absorption, capillary action and carbonization. The thin section method is based on very thin items prepared by a special grinding technique. The items are studied in the microscope, fitted with several filters for various investigation purposes.

The electron probe micron analyses are also based on special grinded and polished test items. These are analyzed in a microscope equipped with a X-ray analyser, and give as result the chemical composition on chosen spots on the item.

A Joint Method For Result Evaluation

A simple evaluation system has been developed for the analyses of the connection between laboratory results and the general pipe conditions, i.e., observed cracks, surface corrosion and age. A non-dimensional "Quality number" is calculated for each pipe. This number is a common parameter for the strength and porosity test methods. The system is based on the following assumptions:

1. A reference value (R_i) is defined for each test method. This value should be representative for acceptable test results. The measured value (M_i) is divided with the reference value.

2. The quotients of each test method are added. It is possible to use different weights (W_i) for the various tests. The quality number is thus defined as

$$Q = \sum_{i=1}^{n} W_i \times \frac{M_i}{R_i} \tag{1}$$

where i represents the test methods.

ANALYSES OF TEST RESULTS

Reproduction of Test Results

The evaluation system is utilized and quality numbers calculated for each of the test pipes. The laboratory program is based on one pipe for each pipeline. To check if the results are representative, in some cases two parallel pipes from the same line have been investigated. This test shows that the uncertainty is larger for the strength measurements than the porosity measurements. It is also shown that the test result reproduction between parallel test items from the same pipe is better than the results from different pipes from the same line. Pipes from the same line will naturally give far better reproduction values than the combination of all the results of the program.

Quality Number and Pipe Age

The test material is divided into four age categories. Figure 1 shows mean values and standard deviation for quality numbers for the categories. The results confirm that the older pipelines are likely to be in a poorer condition, however with larger quality differences. Standard hypothesis testing shows that there is a probability of 62 % that pipes from the period 1945-59 have higher strength and durability qualities than pipes from the period 1920-44. The corresponding probability for the periods 1960-69 versus 1945-59 is 88 % and 1970-79 versus 1960-60 is 95 %.

This confirms that pipes produced in the 1970's have a higher and more stable quality compared with pipes from the 1960's, which in turn are better than pipes produced earlier.

The technical development towards better process equipment, better process routines and better control procedures that has taken place after 1960 has given a significant improvement of results in terms of strength and durability.

The rather large quality variations show that age should not be taken as the main parameter for pipe reconstruction decisions.

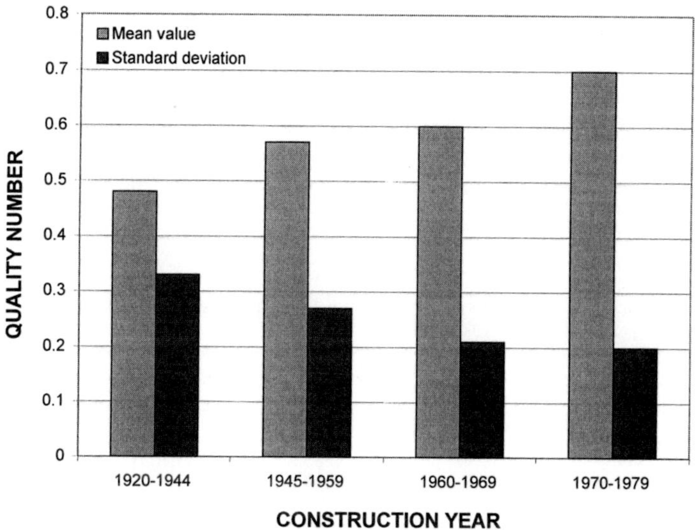

Figure 1 Quality numbers and age. Mean values and standard deviation [3].

Other results

Based on the laboratory results, the following additional conclusions can be drawn:

1. There is a strong connection between the visual inspection and laboratory results. It is, however, shown that visual inspection alone may give misleading conclusions.

2. There is a connection between bad material quality and pipe damage history. Bad material quality is a major reason for pipe damages.

3. Thin section analyses and electron probe micron analyses reveals detailed information on pipe deterioration, and confirm the results from strength and porosity test methods.

APPLICATION

The quality number concept has shown to be an appropriate system for evaluating concrete pipes and give a documentation of the durability of new concrete pipe products.

The evaluation system has been tested for unused pipes, received from concrete pipe manufacturers short time after the production. These tests have revealed some practical problems of the test performances, which includes the design of bored cylinders for compressive strength testing and pre-treatment of samples for porosity measurements. However, these problems have now been overcome. The system obviously also has a large potential for the evaluation of existing pipelines, prior to the decision of renewal.

An evaluation of actual test methods has concluded that fracture strength, compressive strength, water absorption and capillary porosity should be included in the quality number system. These methods have shown to be among the most stable for a fixed concrete pipe quality, and are also sensitive to quality variations. The parameters "water absorption" and "capillary porosity" are directly related to the pipe durability, whereas the "fracture strength" and the "compressive strength" indirectly measure the probable technical life length. The system is related to common test procedures, which are performed in the industry as a part of internal control. The limits of acceptance for the various test methods are presented in Table 2.

Table 2 Durability control system parameters and limits of acceptance [2,4]

PARAMETER	REFERENCE LEVEL
Fracture strength	Individual values for every diameter and wall thickness (method by Hillerborg, [1])
Compressive strength	50.0 MPa
Water absorption	4% (weight %) plus water absorbed in aggregates (max 1.2%)
Capillary porosity	10% (volume %)

The limits of acceptance have been estimated from the research program, and reflect the pipe condition after service periods from 20 to 80 years. To be accepted, the test samples should fulfil demands on the quality number for the combined system, and demands on each system parameter. A pipe is accepted if:

1. The quality number (eq. 1) is more than 1.
2. Each sub-number (relation M_i/R_i) is more than 0.9.

Example

A pipe manufacturer sent samples taken from a 600 mm non-destructive pipe (class FAVA). The pipe samples were tested at SINTEF. The results (mean values) were:

Fracture strength: 94.2 kN/m

Compressive strength: 59.8 MPa

Water absorption: 3.0%

Capillary porosity: 10%

The corresponding quality number (mean values of samples) are:

$$K = \frac{1}{4}(\frac{94{,}2}{80} + \frac{59{,}8}{50} + \frac{4}{3{,}0} + \frac{10}{8{,}6}) = \frac{1}{4}(1{,}18 + 1{,}20 + 1{,}33 + 1{,}16) = 1{,}22$$

The corresponding sub-numbers for each parameter (mean value minus standard deviation for each parameter) are:

$$K = \frac{1}{4}(\frac{88.5}{80} + \frac{52.9}{50} + \frac{4}{3.2} + \frac{10}{9.98}) = \frac{1}{4}(1.11 + 1.06 + 1.25 + 1.00)$$

With a quality number of 1.22 and no sub-numbers (mean value minus standard deviation) below 0.9, this particular production will pass the acceptance criteria with a safe margin.

CONCLUSIONS AND FURTHER WORK

The results from a number of laboratory tests on concrete strength and porosity have been combined into a common number, a quality number. The connection between this number and the pipe condition measured by damage history, visual inspection and advanced investigation methods has been studied, and it is shown that a combined quality number based on simple laboratory test methods can give representative values for the concrete pipe condition and deterioration.

The idea has been further developed to a system for specifying and achieving concrete pipe durability. Based on a limited number of strength and porosity tests, routine measurements will be carried out by Norwegian pipe manufacturers.

Based on these tests, a common "durability number" is calculated for each production series and compared with standard values. This is meant to secure a stable production quality. The system will be calibrated against advanced methods based on micro-structure measurements.

Additionally, the concept might have a capacity for the evaluation of concrete sewer network conditions. By taking samples of representative pipelines, a pipe condition overview could be established for the network, as well as an accurate description of the state of the pipeline where the sample(s) have been taken.

ACKNOWLEDGEMENTS

The author wishes to thank Prof Odd E. Gjorv, Norwegian University of Science and Technology, research engineer colleagues Ola Skjoelsvold, Kaare Johansen and Inger Meland at SINTEF and Mr Jan G Eckhoff and Mr. Terje Reiersen of the Norwegian Concrete Pipe Industry for valuable support during this work.

REFERENCES

1. GUSTAFSSON, P J, AND HILLERBORG, A. Improvements in concrete design achieved through the application of fracture mechanics. Application of fracture mechanics to cementitious composites, NATO-ARW, Northwestern University, USA

2. JOHANSEN, K, SKJOELSVOLD, O, AND SAEGROV, S. Durability of concrete pipes. Specifications of methods for quality numbers calculations. SINTEF reports STF70 A95089, STF22 A96810 and STF22 A97324 (in Norwegian).

3. SAEGROV, S. Durability of concrete sewer pipes (in Norwegian). Dr. Thesis 1992:21 Norwegian University of Science and Technology, Trondheim.

4. SAEGROV, S, JOHANSEN, K, MELAND, I, AND SKJOELSVOLD, O. Concrete pipe durability. A documentation system for pipe manufacturers. SINTEF report STF60 F95015 (in Norwegian)

AN EXPERIMENTAL STUDY OF DAMPING IN REINFORCED CONCRETE

J De Visscher

Free University of Brussels

J M Ndambi

Royal Military Academy Brussels

B Peeters

Catholic University of Leuven

Belgium

ABSTRACT. This paper discusses the experimental identification of the modal damping of reinforced concrete structures. Because of the nonlinear constitutive behaviour of reinforced concrete, the damping is sensitive to the vibration amplitude. The experimental technique used for the present experiments takes account of this amplitude dependence. It is shown that the damping properties depend not only on the vibration amplitude, but also on the loading history of the test structure. This is explained by the fact that loading the structure induces internal damage, under the form of microcracks. A better understanding of the phenomenon of damping in reinforced concrete, including its evolution with damage, could in time lead to the use of damping measurements for assessing internal damage in reinforced concrete members in a nondestructive way.

Keywords: Damping, Modal analysis, Reinforced concrete beams, Non destructive evaluation, Damage

J De Visscher is a post-doctoral researcher at the Free University of Brussels (Vrije Universiteit Brussel), Belgium. Her research interests are in the field of dynamics and the use of mixed numerical experimental techniques for material and structural identification.

J M Ndambi is a doctoral researcher at the Royal Military Academy, Brussels, Belgium. He specializes in the use of dynamic system identification for the characterization of reinforced concrete structures and the influence of damage on the dynamic properties.

B Peeters is a research/teaching fellow at the Catholic University of Leuven (Katholieke Universiteit Leuven), Leuven, Belgium. His research activities are in the field of dynamic system identification of civil engineering structures.

INTRODUCTION

The constitutive behaviour of reinforced concrete, subjected to cyclic loading, is known to be highly nonlinear. The vibrational damping, characterized by the surface of the cyclic load-deformation loop, is therefore dependent on the vibration amplitude.

Dynamic system identification techniques that assume the system to be linear, such as classical experimental modal analysis, are acceptable for the identification of resonant frequencies and mode shapes, provided the excitation amplitudes are sufficiently small. For modal damping ratios (MDRs) on the other hand, results from different experimental modal analyses show a large experimental scatter, mainly due to the alnplitude dependence. This can be observed when tests are repeated at varying amplitudes of excitation.

This is one of the conclusions from an experimental program, that was set up to investigate the use of dynamic system characteristics for the evaluation of damage in reinforced concrete structures [1]. A further conclusion could be that MDRs are unsuitable parameters for the purpose of damage evaluation, since conclusions regarding structural health can not be based on uncertain parameters. Yet, it is very often suggested that damping in reinforced concrete should be very sensitive to damage [2,3]. The idea is that the accumulation of internal cracks leads to an increase of internal friction, associated with increasing energy dissipation. Therefore, it was decided to gain a better insight in the phenomenon of damping in reinforced concrete, its amplitude dependence and its evolution with internal crack damage.

EXPERIMENTAL SET UP AND TECHNIQUES

Test Beams and Set Up

The test beams are reinforced concrete beams of 6 m length, with normal and shear reinforcement bars as shown in Figure 1. Three identical test beams were produced and tested in the programme.

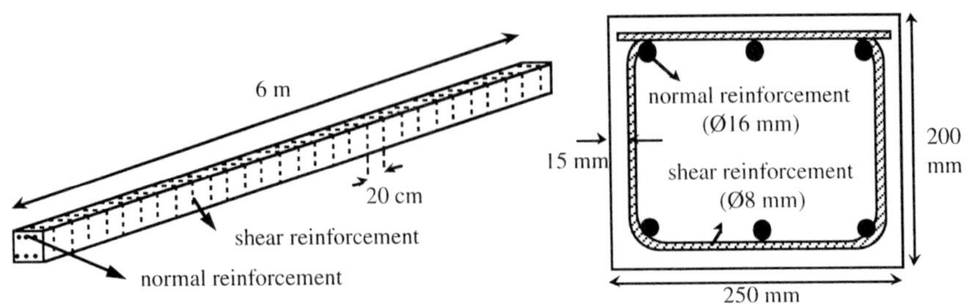

Figure 1 Geometry of the test beams

The dynamic measurements are performed while the test beam is suspended by means of elastic springs, attached at the modal lines of the first flexural eigenmodc (see Figure 2). A grid of 62 measurement points (2 rows of 31 points) is marked on the top surface of the beam, which allows to detect the vertical bending modes and the torsion nodes.

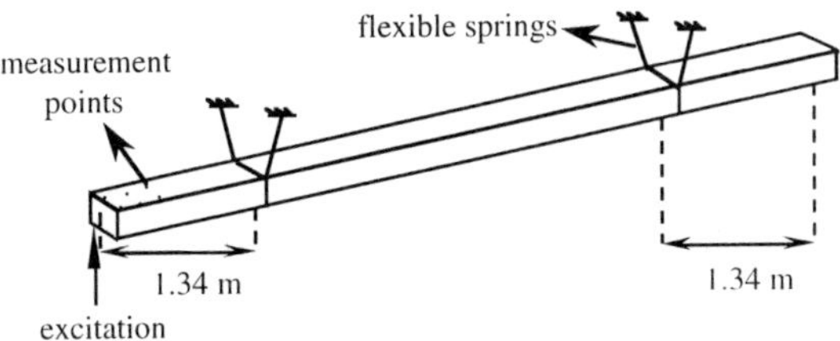

Figure 2 Set up for dynamic measurements

Modal Damping Identification Techniques

In the present paper, the results from two identification techniques are compared: classical experimental modal analysis and a single mode time domain method, which allows to control the vibration amplitude for every particular eigenmode.

Experimental modal analysis

An excitation is applied at one corner of the beam in the vertical direction. The frequency response functions (FRFs) are measured in the 62 measurement points. Curve fitting the measured FRFs yields the dynamic system matrices, which allow for the calculation of the complex eigenvalues (resonant frequencies and MDRs) and the eigenmodcs. This technique assumes the structure to be linear. The amplitude of every eigenmode is different for every measured FRF, dependent on the position of the measurement point with respect to the nodal lines of the eigennnode. Consequently, it is not possible to associate vibration amplitude to the identified MDRs.

Single mode time domain method

The beam is excited by a sine signal, with a frequency tuned to the resonant frequency of the considered eigenmode. This signal is transmitted acoustically to the test structure by means of a loudspeaker. The free damped modal vibrations are recorded fi-on1 the moment the excitation is stopped. The recorded time signal is then fitted by an exponentially decaying sine function:

$$x(t) = X\sin((2\pi f)t-\phi)e^{-\xi(2\pi f)t}$$

where ξ is the MDR, X the amplitude, f the eigenfrequency and ϕ the phase.

The measurement is repeated for every individual eigenmode. The expression for the free vibrations given above is also based on the assumption of linear behaviour. Yet, the amplitude of the modal vibrations can be controlled, allowing for repeatable measurements. Another advantage is that loudspeaker excitation is non-contacting, so that no artificial damping is added to the test structure by the excitation device.

EXPERIMENTAL RESULTS

Comparison of Different Experimental Techniques

Experimental modal analysis was used to identify all the modal parameters of the test beams. Two different excitation techniques were used: impact excitation (applied by a hammer), and pseudo-random excitation (applied by means of an electrodynamic shaker). The resonant frequencies and mode shapes obtained with the different techniques are well in agreement. The MDRs however, show a large experimental scatter. Table 1 lists the identified MDRs for test beam n°2, including the results from the single mode method.

Table 1 Comparison of the MDRs obtained with different techniques

		EXPERIMENTAL MODAL ANALYSIS		SINGLE MODE METHOD
		Pseudo-random	Impact	Sine
Mode	Frequency (Hz)		MDR (%)	
1 (B)	22.5	0.319	0.441	0.305
2 (B)	62.7	0.287	0.422	0.263
3 (B)	122.9	0.301	0.508	*
4 (T)	180.9	0.609	0.659	*
5 (B)	198.9	0.303	0.354	0.264
6 (B)	295.1	0.325	0.411	0.288
7 (T)	377.7	0.331	0.335	0.294
8 (B)	401.7	0.338	0.349	0.265

(B): Bending mode (T): Torsion mode

The following observations can be made:

The MDRs obtained with the impact excitation are systematically higher than for the other techniques. This can be explained by the amplitudes involved in the measurements: for the hammer excitation, accelerations were larger than for the other excitation techniques. Further evidence of this will be given in the next section.

The single mode method always yields the lowest values. There are 2 explanations for this observation. As was already mentioned, the method has the advantage that it is uses a non-contacting excitation device (loudspeaker), in contrast to the electrodynamic shaker. To estimate the influence of the shaker on the MDRs, a few measurements were performed with the single mode method, while the shaker was attached to the test beam (the excitation still being provided by the loudspeaker). This resulted in MDRs that were +10% higher than the results obtained without the shaker. Another explanation is the influence of vibration amplitude as the single mode measurements were taken at the smallest possible amplitudes.

Influence of vibration amplitude

To investigate the influence of vibration amplitude in further detail, the MDRs of beam n°3 were measured as function of increasing amplitude. For the experimental modal analysis, a swept sine excitation was used for which the amplitude was amplified from one analysis to the other. In this way, the amplitudes of the modal contributions vary, proportionally to the excitation. Figure 3 shows the results for mode 2 (asymmetric bending) and mode 3 (symmetric bending) as function of the maximum modal acceleration, measured at the end of the beam.

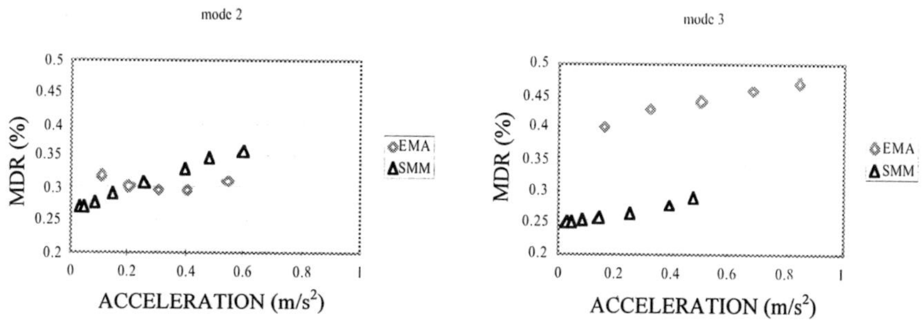

EMA: Experimental Modal Analysis SMM: Single Mode Method

Figure 3 Influence of vibration amplitude on the identified MDRs

For mode 3, both methods show a similar amplitude dependence. The MDRs increase with amplitude, which explains why the MDRs obtained with the hammer excitation, as shown in the previous section, are systematically higher. Also, the results from the swept sine method are higher than for the single mode method, which is explained by the influence of the shaker.

For mode 2, the single mode method shows the same evolution: an increase with increasing acceleration amplitude. The results from the experimental modal analysis deviate from the general trend. Probably, the results of the MDR identification are also affected by the data processing parameters used in the modal analysis, as a consequence of the nonlinearity of the structure.

Influence of crack damage

The test beams were subjected to static loading tests with gradually increasing load levels, in order to induce controlled damage. After every static loading test, the beams were subjected to series of nondestructive dynamic tests, allowing for the identification of the MDRs. The evolution of the MDRs can then be plotted as function of the static load level, which is a measure for the crack damage in the beam. Figure 4 shows the results for beam n°3, which was subjected to static 4-point bending tests, as obtained by the single mode method.

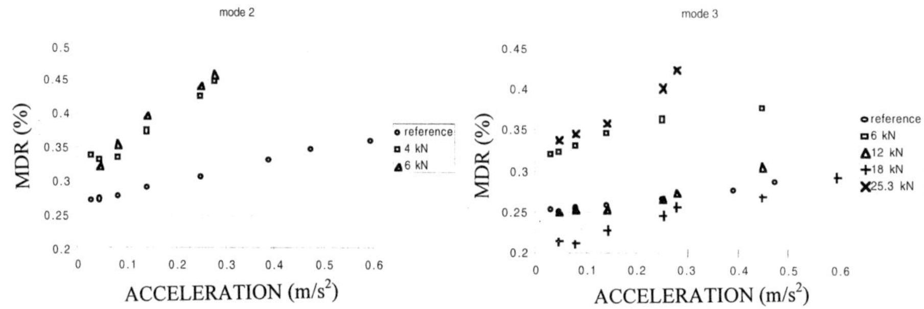

Figure 4 Evolution of the MDRs with load level of the static test (as function of vibration amplitude)

The variations of the MDRs with damage are indeed large. Variations of the order of 30 to 50% are observed, while for the resonant frequencies, the maximum variations were 20%. Yet, the variations of the resonant frequencies with damage are easier to understand. Resonant frequencies decrease with increasing damage, which is explained by the loss of bending stiffness due to crack damage. The evolution of the MDRs with damage is more complicated. For small load levels, the damping increases with increasing damage. This is indeed what is intuitively expected: since the integral of crack surfaces over the structure increases, more vibration energy is dissipated by internal friction. For higher load levels however, the damping, decreases and for the fully cracked beam, a large increase is observed.

To explain this evolution requires a detailed study of all the damping mechanisms that take place at the microscale of reinforced concrete. The friction taking place between contacting crack surfaces is just one of these mechanisms. Other mechanisms that could possibly play a role are: the compression and decompression of air or water in the pores of the concrete, stress induced flow of moisture and internal stress redistributions due to damage propagation.

A decrease of MDRs with increasing damage may be difficult to understand at first sight. Therefore, one has to consider the reinforced concrete beam as a composite structure consisting of a concrete matrix and steel reinforcement. The MDR of an eigenmode can be written in terms of the energy dissipated in one oscillation cycle (D) and the maximum strain energy attained in that cycle, known as the modal strain energy (W), [4]:

$$\xi = \frac{D}{4\pi W}$$

The energy dissipated in the reinforced concrete structure can then be written as the sum the energy dissipated in the concrete phase and the energy dissipated in the steel phase:

$$\xi = \frac{D_C + D_S}{4\pi W} = \frac{W_C}{W}\xi_C + \frac{W_S}{W}\xi_S$$

Where W is the total modal strain energy
D_C and D_S the energies dissipated in the concrete and steel phases
W_C and W_S the modal strain energies stored in the concrete and steel phases
ξ_C and ξ_S the material damping of concrete and steel

When damage occurs under the form of cracks in the concrete phase, a redistribution of stresses takes place from the cracked concrete to the steel bars. Consequently, the modal strain energy of the steel increases while that of the concrete decreases. According to the equation shown above, the contribution of steel damping to the reinforced concrete's damping thus increases. As the material damping of steel is small compared to the material damping of plain concrete, this results in an overall decrease of the modal damping of the reinforced concrete structure. A decrease of the MDR due to crack damage, as explained by this simple theory of stress redistributions, is also frequently encountered in fibre reinforced polymeric composites, when the damping of reinforcing fibres is less than that of the matrix material.

CONCLUSIONS

MDRs of reinforced concrete structures are difficult to identify correctly.

1. They are very sensitive to the experimental conditions. In this practical application, the use of an electrodynamic shaker increases the MDRs by +10%, while its influence on the resonant frequencies is negligible.

2. The MDRs are dependent on the vibration amplitude. MDRs resulting from dynamic analyses carried out at different vibration levels can thus not be compared.

3. The evolution of the modal damping with damage is not straightforward: it can either increase or decrease. Consequently, a direct correlation between the variation of MDRs and damage can not he found.

The 2 former conclusions both lead to the final conclusion that, with the present technology for identifying MDRs and with the present understanding of damping mechanisms in reinforced concrete, MDRs can not yet be safely used for the purpose of damage evaluation.

ACKNOWLEDGEMENTS

This research was performed in the framework of a major research project (FKFO G.0243.96), sponsored by the Fund for Scientific Research of Flanders. The authors therefore wish to express their gratitude for the financial support.

REFERENCES

1. VANTOMME, J ET AL. Evaluation of structural damage in reinforced concrete by dynamic system identification. Proceedings Int. Congress Creating with Concrete, 1999.

2. AGARDH, L. Modal analyses of two concrete bridges in Sweden. Structural Engineering International, April 1991, pp 35-39.

3. SALANE, H J, BALDWIN, J W. Identification of modal properties of bridges. J. of Structural Engineering, ASCE. July 1990. Vol. 116, No.7, pp 2008-2021.

4. UNGAR, E E, KERWIN, E M. Loss factors of viscoelastic systems in terms of energy concepts. The Journal of the Acoustical Society of America. 1962. Vol. 34, 7, pp 954

CHARACTERISATION OF CONCRETE SURFACES USING DIGITAL IMAGING PROCESSING

S Misra **O Dikshit**

Indian Institute of Technology Kanpur

A J Patil

Government Polytechnic Jalgaon

India

ABSTRACT. Deterioration of concrete often results in changes to the surface characteristics, appearance of cracks and discoloration. Rapid and scientific quantification of these characteristics would be extremely useful in monitoring the performance of concrete structures and developing a methodology for their maintenance and repair. An effort has been made here to study possible use of concepts in digital image processing to estimate the extent of cracking, and develop a better understanding of the surface characteristics (e.g. roughness) of concrete.

Digital data from photographs of a cracked RC column in a building and mortar specimens subjected to accelerated deterioration in the laboratory has been used in the study. Edge detection algorithms have been used to obtain information regarding the width and length of cracks. Statistical textural features based on the gray level difference histogram method have been used to characterize the mortar surfaces.

Results obtained clearly illustrate the possibility of using digital image processing (DIP) techniques for nondestructive evaluation and monitoring of concrete structures.

Keywords: Concrete, Deterioration, NDT, Digital image processing, Roughness, Cracks

Sudhir Misra is Associate Professor at the Indian Institute of Technology Kanpur, India. His main research interests are deterioration and durability of concrete, construction materials and nondestructive testing and evaluation of concrete structures.

Onkar Dikshit is Assistant Professor at the Indian Institute of Technology Kanpur, India. He specialises in digital image processing and geoinformatics, interpretation of satellite data, and application of remote sensing techniques in civil engineering.

Anil J Patil is Lecturer at the Government Polytechnic at Jalgaon, India, where he teaches analysis and design of RC structures, nondestructive testing, and construction practices.

INTRODUCTION

Results from the development in digital image processing (DIP) [1] have been used in diverse areas such as medicine, metallurgy, robotics and land use studies. In civil engineering, the principles have been used in town planning, water resource management and environmental sciences. It is sought to make use of the same principles and study the possible application of DIP for the nondestructive monitoring and evaluation of concrete structures. Deterioration of concrete often results in changes in surface roughness, appearance of cracks, discoloration, etc. Quantification of these characteristics could be useful for, monitoring the performance of concrete structures, developing a methodology for their maintenance and repair, and improving understanding of the overall integrity of the structure. Results presented here form part of an ongoing research effort to utilize DIP in nondestructive evaluation of concrete structures. Further work for relating changes in surface characteristics to structural integrity and mechanical properties, such as strength, is presently in progress. The results reported here largely focus on developing a better understanding of the extent of cracking in concrete structures and roughness of concrete surfaces.

RESEARCH SIGNIFICANCE

The need to develop nondestructive testing and evaluation methods that can be used without having to physically approach the structure cannot be over emphasized. The method suggested in this study could be useful in, regular monitoring, identifying areas that require closer attention and maintenance and repair of concrete structures. The fact that photographs used were taken using an ordinary camera is also significant in this context.

METHODOLOGY

A basic flow chart showing the different steps followed in the study is given in Figure 1. Digital data from photographs taken using a commercially available SLR camera has been used in the study. Commercially available software (IDRISI) has been used for image analysis and new programs have been developed wherever required.

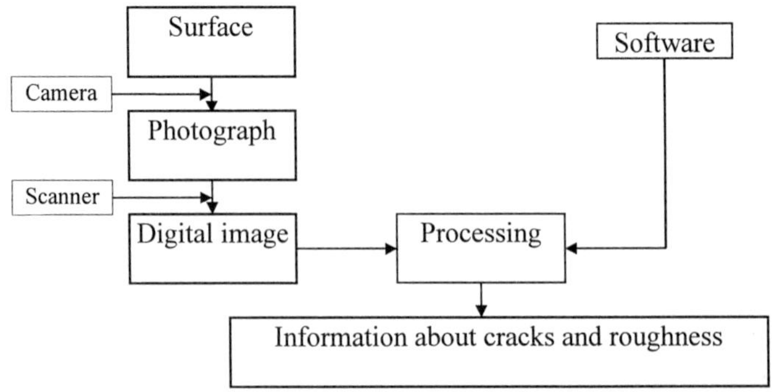

Figure 1 Methodology

To study the cracking in concrete structures, a black and white photograph of the cracked portion in a RC column at one of the buildings at IIT Kanpur was taken from the distance of 4.5m with a camera aperture of 5.6 (Figure 2). This photograph was scan-digitized using a UMAX flatbed scanner at different dpi values (300, 600, and 1200).

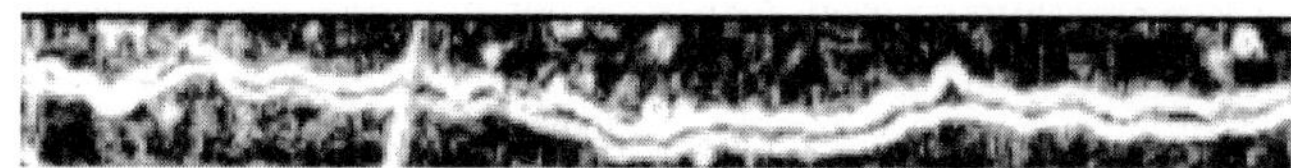

Figure 2 Rotated output edge image of a crack in RC column with Sobel operator

For studying the changes in roughness in concrete surfaces, a study using mortar specimens was carried out. Mortar cubes using OPC were cast using locally available sand and tap water. The specimens were water cured for 28 days, and then immersed in 0.5N and 1.0N sulfuric acid to induce rapid changes in surface roughness. Black and white photographs of the cubes were taken at the beginning of the exposure tests, and then after 1 and 3 months of immersion in acid. A representative set of these is given in Figure 3 (immersed in 0.5N sulfuric acid). Photographs were taken indoors so that, illumination, distance from the camera, aperture of the camera, etc. could be controlled and maintained at the same level throughout the study.

(a) Initial (b) After 1 month (c) After 3 months

Figure 3 Appearance of cube surface

DETECTION OF CRACKS

Model

As shown in Figure 4, the crack has been assumed to be made of two edges taken to be piece-wise linear between scan lines. The coordinates (x_i, y_i) were measured at intersections of the edge and scan lines drawn at 30 mm intervals. The extent of cracking was quantified in terms of length, area and the width of the crack. The length of the crack has been measured along the centre line, and the crack area computed assuming the area to be a trapezium bounded by succeeding scan lines and the two edges. Crack width was calculated using the model shown in Figure 4 and explained below:

(a) C-method: distance between the two edges along the perpendicular to the centre line [2]. This method gives two values (w_1 ,w_2), depending upon the segment of the centre line used to draw the perpendiculars. Min{w_1,w_2} has been taken to be the crack width.

(b) P-method: distance between the two edges along the perpendicular to the edges. This gives values (w_3,w_4,w_5,w_6), depending upon the point from which the perpendicular is dropped and the edge segment used. Min{w_3,w_4,w_5,w_6}, has been taken to be the crack width.

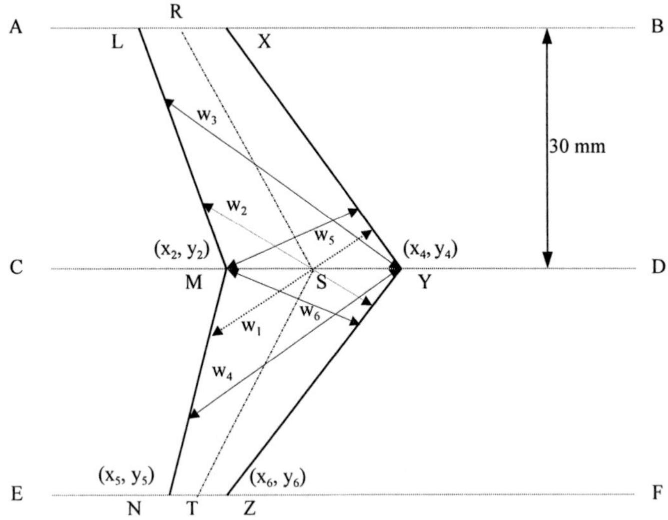

Figure 4 Model of crack adopted for the study

Edges in an image have been detected using sudden and substantial change in grey level values in a neighborhood. After initial trials with different edge detectors or operators [3], 1992), the Sobel operator was used. The output edge is thresholded to convert it into a binary image on which edge measurements were carried out. The scanning was carried out at different dpi values to study the effect of resolution on the accuracy of the results.

Results

The crack has been assumed to a single long crack (approx. 800 mm long, Figure 2). Measurements were carried out at 28 locations over the length. The crack width was computed using the two models described above and physically measured using a Vernier caliper (Table 1). Though only the average values are given in Table 1, a detailed study was carried out by comparing the observed and estimated crack widths at the various locations. It was found using linear regression that the crack width estimated using the P-method was closest to the observed values and the accuracy also improved at higher resolutions. Table 1 also gives details concerning the area and length of the crack. Though the importance of these parameters in the present case is limited, the possibility of calculating these parameters, if required (e.g. AAR), is clearly established. It may be noted that though a 'very wide crack' has been used in this study, appropriate scanning resolution (dots per inch; dpi value) can be used when applying the method for cracks with smaller width. The only condition is that the crack should be discernible in a photograph.

Table 1 Measurements for various crack parameters

DPI VALUE	300		600		1200	
Method of estimation	C-method	P-method	C-method	P-method	C-method	P-method
Average width (mm) (*)	13.264	12.883	12.892	12.521	12.822	12.448
Length (L) (mm) (*)	832.775		833.228		830.591	
Area (A) (mm^2)	11160.000		10856.250		10693.130	
Average width (A/L) (mm)	13.400		13.029		12.874	

* Observed average width and length are 11.671 and 916.621 mm, respectively

CHANGES IN ROUGHNESS

Method

The roughness of mortar surfaces has been characterized in terms of image textural properties that describe the spatial variation of grey levels. After initial experiments and considering the memory space and processing time, all photographs were scanned at 150 dpi, where each pixel represents a square of 0.329 mm side. Though efforts were made to take all photographs under similar conditions, all the images were "histogram equalized" before further processing [4] to eliminate possible effect due to non-uniformity in illumination.

Several methods are available to calculate image texture by using window-based texture features. In these methods, the spatial arrangement of grey level values is contained in the intermediate co-occurence matrices or grey level difference histograms (GLDH). From these, several statistical textural features, such as, mean, entropy, energy, etc. can be computed [4, 5]. In this study, only two texture features - energy and entropy, have been used. These features are a measure of the homogeneity and heterogeneity in a surface, and are therefore negatively correlated. In the context of the work reported here, for a rough surface, the computed energy values should be lower and the entropy values should be higher, than a smooth surface.

The energy and entropy features were generated by using the GLDH algorithm for window sizes varying from 5 to 15 pixels with a distance metric of one pixel in all four principal directions, i.e. horizontal, vertical and right and left diagonal [2]. The features were scaled on a 0-255 scale to obtain a textural image. To account for possible changes in the size of the cubes due to leaching, erosion, etc., it was decided to follow the changes in roughness of a central area on a face of the cube. Therefore, a 50x50 pixels area was chosen from the centre of the textural image to extract relevant information.

Though the discussion in this paper is limited to average feature values, a detailed study using the minimum, maximum, standard deviation and coefficient of variation of these features was also carried out [2].

Results

As can be seen from Figure 3, continued exposure to the acid, leads to progressive increase in the roughness of the mortar surface. The effort, as outlined above, is focused on quantification of this change. Representative values of the average energy and entropy (for the 50 x 50 pixel area) just before immersing in the acid and at later stages, for one of the faces of the mortar cube are given in Table 2 for a window size of 7 pixels.

Table 2 Textural parameters for one of the faces in a cube

	ENERGY		ENTROPY	
TIME	0.5N	1.0N	0.5N	1.0N
0	10.76	19.72	211.65	190.04
1 month	5.68	3.76	225.53	234.68
3 month	3.47	2.61	235.20	239.33

Table 2 clearly indicates the expected trend in the changes in the energy (decrease) and entropy (increase) values with time and change in the roughness of the surface. It was however, found that little significance could be attached to numerical (absolute) values of these parameters. For example, the initial average energy and entropy values for the four faces of the cube were (10.47, 12.39, 10.76, 17.24) and (213.68, 205.88, 211.65, 193.56), respectively. Also, though it appears from the raw data (Table 2) that the changes in values are more rapid in the initial one month and become slower in the latter part of the experiment, more experimental data is required to make a definite observation in that regard.

The data also suggests that higher acid strength leads to a more rapid change in the entropy and energy values in the initial one month, signifying a more rapid deterioration of the surface.

CONCLUDING REMARKS

It was found that the width of a crack could be successfully estimated from its photograph using suitable DIP algorithms. The possibility of using these methods for estimating parameters such as the total crack length and the area of the crack was also established.

It was also found that statistical textural features, such as energy and entropy, could be successfully used to characterise changes in surface roughness of mortar cubes.

The results obtained complement visual changes in the roughness of cubes upon immersion in sulfuric acid.

The study establishes the possibility of using DIP in nondestructive monitoring and evaluation of concrete structures, quality control during construction, and identifying critical areas requiring closer observation, on the basis of changes in the surface characteristics.

ACKNOWLEDGEMENT

The support by the Department of Science and Technology, Government of India through a grant (No. III 5 (83)/94 ET, of 11.12.96) for part of this study is gratefully acknowledged.

REFERENCES

1. GOOL, I V, DEWAELE, P AND QOSTERLINCK, A. Computer Vision, Graphics, and Image Processing, Texture analysis, anno 1983, 1985, Vol. 29, pp 336-357.

2. PATIL, A K. Characterization of Concrete Surfaces Using Digital Image Processing, M.Tech Thesis submitted to the Dept. of CE, Indian Institute of Technology, Kanpur, 1997.

3. PRATT, W K. Digital Image Processing, John Wiley, New York, 199278.

4. HARALICK, R M, SHANMUGAM, K AND DINSTEIN, I.. Textural features for Image Classification, IEEE Trans. on Systems, Man, and Cybernetics, SMC-3, 1973, No. 6 PP 610-621.

5. CONNERS, R W AND HARLOW, C A.. A theoretical comparison of texture algorithms, IEEE Trans. on Pattern Analysis & Machine Intelligence, PAMI-2, 1980, No. 3, PP 204-222.

TOWARDS AN EXPERT SYSTEM FOR THE EVALUATION OF DAMAGE IN CONCRETE STRUCTURES IN SAUDI ARABIA

H M Z Al Abideen

M M Khalifa

Ministry of Public Works and Housing

A S Eldin

King Saud University

Saudi Arabia

ABSTRACT. The paper is a report on an ongoing research project and the results achieved so far. The objective of the project is to develop a Knowledge-based system for the evaluation of concrete cracks & damages of buildings in the Kingdom of Saudi Arabia. The system integrates two major technologies databases and expert systems in order to mimic experts' actions for the considered problems. The system uses facts and rules of thumb in addition to the data stored in the database. It has to identify the probable causes of cracks in concrete based on their shape, pattern, density, etc. Environmental conditions of Saudi Arabia and Gulf countries are taken into consideration. This is considered at three different levels; namely, macro-, meso- and micro-environment. This project will solve problems, which exist not only in Saudi Arabia but also in other similar countries. It incorporates state-of-the-art computer technology. While existing systems rely on text-only description of defects; this research attempts to use descriptions of defects by pictures and illustrations. The major objectives of the research may be summarized are acquisition and formalization of the relevant knowledge of causes of cracks amd building a flexible expert system which can be used for the diagnosis of defects in concrete buildings.

Keywords: Expert system, Concrete cracking, Causes of cracking, Assessment of structures, Environmental conditions, Knowledge based systems.

Dr Habib M Zain Al Abideen is the Deputy Minister - Ministry of Public Works and Housing –Riyadh – Saudi Arabia. He has authored several books and numerous papers in the area of concrete design and technology.

Dr Ahmed Sharaf Eldin is Assoc. Professor – College of Computer & Information Sciences, King Saud University - Riyadh – Saudi Arabia. His research interests are in artificial intelligence applications, databases and mathematical software. Authored more than 100 papers and two books.

Dr Magdi M Khalifa is Structural Engineering Advisor – Ministry of Public Works and Housing - Riyadh Saudi - Arabia. He has wide experience in Structural Engineering and design including structural dynamics. He specializes in concrete durability, assessment of existing structures and methods of repair.

INTRODUCTION

Due to the weak tensile strength of concrete, it is expected to crack under tension, flexure and shear. Therefore, steel is used with concrete to resist tensile stresses. Reinforced concrete as well as plain concrete may exhibit deterioration in both normal and severe environments. Assessment of concrete structures that undergo defects i.e. cracking and /or deterioration have assumed great importance in the industry in the past 20-25 years especially in regions of severe environment and in particular the Arabian Gulf area where urban and infrastructure developments have proceeded in a very fast pace and in volumes unprecedented before. The subject of assessment needs a special type of knowledge; that is not normally taught to engineering students in under-graduate programs. It also needs considerable practical experience. Awareness of this fact is very important and the task of assessment should only be assigned to the right person who has the expertise in the field [1]; otherwise, many problems and delays can happen in making the decision.

Creating expert systems to overcome some of these problems was the subject of several research papers. Which also tried to collect experiences in order to form a knowledge base which may help in training and qualifying engineers. These attempts to our knowledge were not so successful in achieving the objectives to solve the problems and help the expert to reach the right decision [2].

The objective of the research is to establish a comprehensive system, which combines data bases with expert systems. The system attempts to simulate the thinking of the expert and his logic in inspecting the subject building. His observation of the defects, the facts he knows are translated by his experience into rules about the probabilities of causes and reasons behind defects. On one hand it is assumed that the engineer who can utilize the system is educated and trained in this area [3, 4]. On the other hand the system educational features may be very useful not only for teaching purposes but also for assessment of trainees.

EXPERT SYSTEMS FOR CONCRETE

Computer expert systems are considered as one of the most important applications of artificial intelligence, which found considerable importance in practical problems. They attempt to simulate the human thinking (logic) in solving a particular problem [5,6]. They consist of the following components:

1. Knowledge base where various knowledge related to the topic are stored. There are several methods to represent knowledge; the simplest and perhaps the best is storing knowledge in the form of IF-THEN rules.
2. Inference Engine which compares stored knowledge rules against given data in order to eventually reach a decision.
3. Working – storage working where temporary results are stored during inference.

The applications of expert system technology in the field of concrete are mainly in structural analysis and design, concrete mix design, construction management, and Diagnosis of problems and suggestion of repairs.

Prior to starting this research a literature review was carried out that revealed the following about available Expert Systems (ESs):

1. Existing ESs are not very practical; they rely mainly on text only in describing defects; therefore, they are limited in use and did not utilize multi – media features.
2. Most of these ESs were developed for and in countries and environments having conditions that differ from those prevailing in KSA, Gulf and many other developing countries. E.g. hot climate, technologies, work conditions, unskilled labor just to mention a few.
3. Expert systems reviewed do not have training programs to help beginners to assess a case.
4. They assume that the user is an expert in the field, therefore it is designed in a way that it cannot be used as an educational tool.
5. Most of those proposed systems were not tested on actual practical diagnosis cases.

The current research attempts to overcome the shortcomings in the systems reviewed .The literature review included some forty systems. Examples are given in Ref. [7-10].

The objectives of this research are five – fold:

1. Developing an expert system which takes into consideration the practical and environmental conditions as well as materials and methods of construction.
2. Utilizing multi-media in improving the ES especially in the user-machine. Interaction.
3. Obtaining a system that is as comprehensive as possible such that it is practical and be used by engineers who have experience in this topic after they are trained on utilizing the ES.
4. Developing an educational system in parallel and connected to the ES, which can be used to help beginners in this field.
5. Testing new ideas such as the use of knowledge dictionary.

SYSTEM DEVELOPMENT

General Framework for the Developed System

Figure 1 depicts the general framework of the developed system and its components. In order to recognize how the system works we have to mimic the procedures that an expert follows to determine the cause of defects in a structure and select the appropriate repair methods. We cannot over emphasize the necessity of following a clear procedure and the importance of avoiding unnecessary studies and testing. Figure 2 depicts a flow chart of the recommended procedures which is very useful as a guideline for the investigator.

Development Environment

The system was developed for use on personal computers compatible with IBM using Windows 95. In order to develop the system the GURU development environment is being used. It is a collection of comprehensive tools, which may be used to build the expert system. The environment used for the educational training system is “Knowledge Pro” which has the necessary features such as hypertext and Multi-media.

User
Data
Consulting
User Interface
Knowledge Management
Knowledge Rules
Entrance Engine

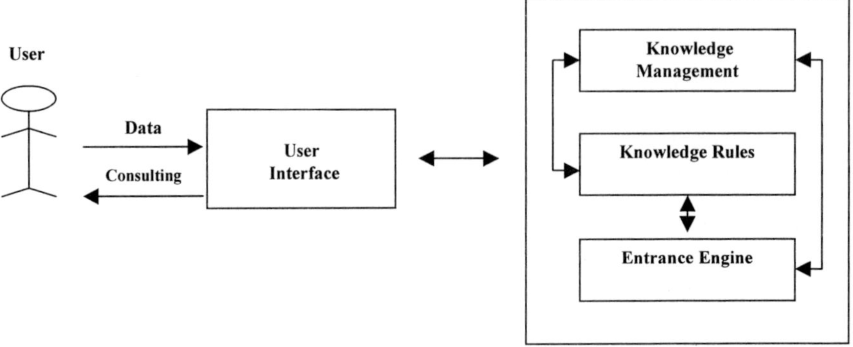

Figure 1 General framework of the expert system

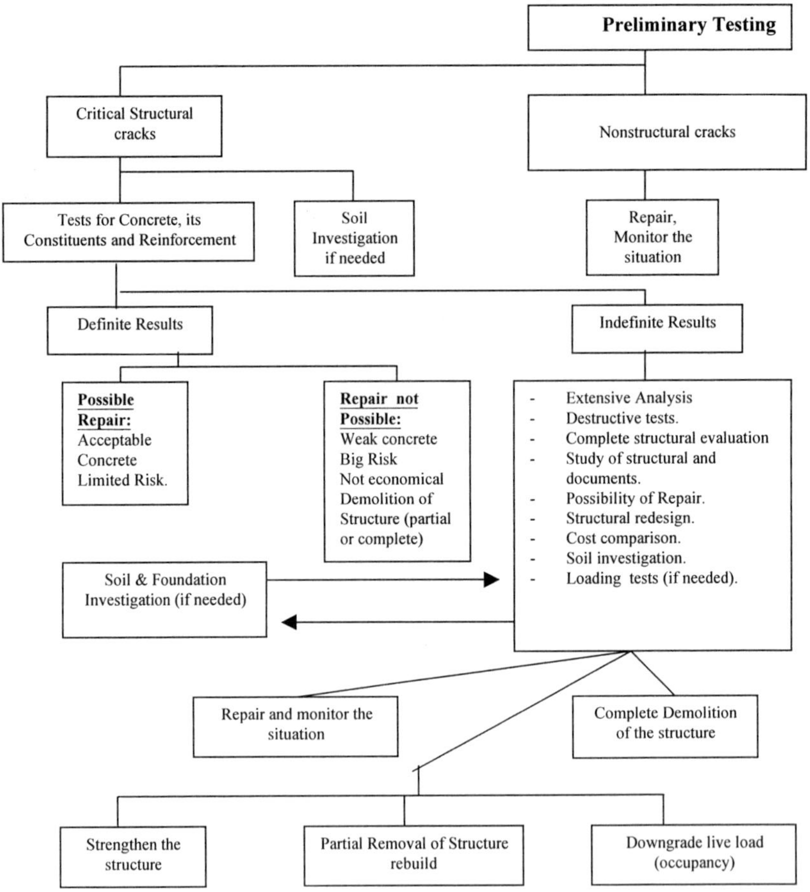

Preliminary Testing
Critical Structural cracks
Nonstructural cracks
Tests for Concrete, its Constituents and Reinforcement
Soil Investigation if needed
Repair, Monitor the situation
Definite Results
Indefinite Results
Possible Repair: Acceptable Concrete Limited Risk.
Repair not Possible: Weak concrete Big Risk Not economical Demolition of Structure (partial or complete)
- Extensive Analysis
- Destructive tests.
- Complete structural evaluation
- Study of structural and documents.
- Possibility of Repair.
- Structural redesign.
- Cost comparison.
- Soil investigation.
- Loading tests (if needed).
Soil & Foundation Investigation (if needed)
Repair and monitor the situation
Complete Demolition of the structure
Strengthen the structure
Partial Removal of Structure rebuild
Downgrade live load (occupancy)

Figure 2 Procedure for safety assessment of RC structures

Knowledge Acquisition

In general, knowledge acquisition from experts is a formidable task. This is understandable, because the expert can solve a problem directly without realizing how he invokes the knowledge, information and experience that he developed over the years in his subject. Information accumulation has generally complicated knowledge. Methods of knowledge acquisition may be divided into two major type [11]:

a. Knowledge acquisition from the source directly (expert, books, references … etc.)
b. Knowledge acquisition through "Machine Learning" where numerous practical cases are given to the system with anticipated results and the system extracts rules automatically.

Type (a) above is the method that is usually used especially in difficult cases whereas type (b) is still in the development phase and has been tested only on simple cases. In this research, method (a) was used through sessions with experts and specialist engineers together with obtaining knowledge from available references (books, research, periodicals …etc.) in addition to two-thousand some cases investigated by Ministry of Public Works and Housing in Saudi Arabia,

Knowledge Dictionary

Having common ground in terminology among the experts and engineers in this field is very important in order to make sure that using the ES efficiently is not hampered by inaccurate definitions or different interpretations. It was, therefore, essential to establish a knowledge dictionary as a tool to collect definitions and concepts in a common reference.

This dictionary is not only useful for the ES but also for the educational features. As a matter of fact it is considered the backbone of the educational part.

The dictionary includes the following:

- Basic definitions of all types of defects, (cracking, deterioration and textural damages).
- Types of tests, destructive and non-destructive.
- Statistical measures used in assessment of concrete strength.

The idea of establishing knowledge dictionary is one of the advantages and is a new development of the ES. The concept of a knowledge dictionary relevant to knowledge bases is similar to data dictionary pertinent to databases. The dictionary is stored in the system as a database related to the ES.

Knowledge Acquisition Form

A form was designed in order to collect expert knowledge or actual cases in a formal and unified way. Factors related to diagnosis of damages (defects) in concrete structures were also classified in the following categories:

1. Environmental factors.
2. -Type of building and structural system.
3. Part or member of the building.
4. Type of damages and defects and surrounding conditions.

Details are given in Reference [12].

Handling of Imprecision Knowledge

In this area of engineering there are some knowledge which are not absolutely certain; and the human expert often works with such information. For example, there are two types of pattern cracks which follow the steel reinforcement in one-or two- directions; however, the cause is different one is due to concrete plastic settlement while the other is due to steel corrosion. It is possible to assign a level of confidence to model this system. This confidence level indicates how certain we are that the data is true. It was agreed between the experts consulted on this work to express the level of confidence qualitatively as (definite, likely and probable). A certainty factor (CF) on a scale of (1-100) is used; one shows very low certainty while 100 indicates complete certainty. Confidence of complex rules which depend on previous rules are calculated by super position [13] as follows:

$$CF\ (A \text{ and } B) = \text{Min}\ [CF\ (A),\ CF\ (B)]$$
$$CF\ (A \text{ or } B) = \text{Max}\ [CF\ (A),\ CF\ (B)]$$

Selection of Development Environment

The selection of development environment for expert systems must satisfy certain criteria in order to save time and effort in fulfilling its objectives. Selection must satisfy the following basic conditions:

1. Development tools must be running on IBM personal computers and compatibles under Windows 95.
2. It must allow Multi-Media features directly or indirectly.
3. It is preferable that the development tools can provide extracting files that may work independently of the tool itself (EXE Files).
4. Possibility of writing programs in different computer languages such as C or C++.

The search for development tools in the international markets was based on the available literature where papers on this topic mentioned the tools the authors used, Other sources were the Internet, universities and experts in the field. The comparison between the tools that were found was conducted scientifically using the Multi Attribute Utility Theory [14].

After the initial survey to cross out tools that did not satisfy the minimum requirements the comparison was conducted among the following systems: Exsys, Guru, Vp Expert, Rule Master, Personal consultant, Advisor, Kappa-pc, and Knowledge Pro.

The study proved that GURU is the best tool for developing the thought expert system whereas Knowledge Pro was the best for the educational training system.
It is worthy to mention that Knowledge Pro has capabilities to construct an ES; however, it is not capable to represent imprecise data and rules.

APPLICATIONS TO CONCRETE MEMBERS

The investigators and the group of experts started with a knowledge acquisition form. The form was very helpful in collecting information on different buildings and numerous types of cracks. This was very useful in its own merit and may be used to establish a data base, which can be recalled later by the computer ES for reasons of comparison in order to improve the confidence with which a decision is made. This effort was also very useful as a means to focus the expert's attention on the objectives of the work and helping them to identify requirements in order to translate their knowledge in computer form.

Several sessions were held to develop an efficient means of this conversion process of knowledge in the expert's mind into rules that the machine (computer) can handle. Investigators, consultants and experts went through true brain storming. It was obvious that the experts' thoughts were convergent and became coherent which reflected the fact that they have been working together for many years through which they developed a unified approach and similar procedures.

It was suggested to establish the system on the basis of following the logic of a systematic investigator (not all investigators are systematic) when he enters the subject building. The first question (fact) that comes to his mind is the type of member whether it is a slab, a beam, a column, a wall ... etc. Then he sees a crack or defect. The second step in describing the crack will be the direction (orientation) of the crack he observed, vertical, horizontal, inclined ... etc. The computer asks some questions analogous to the intentional or unintentional questions that the expert asks himself, each answer would confirm or negate the thought that comes to the investigator's mind. The process continues until the computer decides on the type of crack (cause type) with some degree of confidence (probable, likely, definite) then it continues to attempt to identify the reasons because of which the crack or defect had occurred.
In the following sections, example of slabs which exhibited defects are considered. A flow-chart is given which explains the process in a logical sequence as the expert or engineer sees it, while invoking the knowledge and experience stored as memory in his mind. The chart is open-ended. This is the preliminary attempt to start a dialogue in computer language provided by the software. They have been tested and found to be working; however, they need to be refined and completed. They are going to be pursued further to more testing and provided with a quantitative measure of confidence by assigning degrees of certainty at each step.

Example – Reinforced Concrete Slabs

Slabs may be divided generally into one and two-way slabs. In this example we consider two way solid slabs with drop beams.

Figure 3a depicts the logical basic steps (procedure) to describe damages (or defects) of the slabs. Figure 3b shows the different defects that may be observed from top, while Figure 3c shows the same from the bottom.

The following lines describe one of the rules coded in the expert system in knowledge GURU language (KGL):

```
RULE: RULE13
        PRIORITY: 50
        IF:     X1= "Slab' and  X2S = "y" AND X3S = "Bottom" AND X10S = "Yes"
        THEN : CRACK= "may lead to corrosion"
        REASON:     In cases where there is spelling and cracks from bottom, it may lead to
                    corrosion cracks.
RULE: RULE15
        PRIORITY: 50
        IF      X1 = " slab" AND X2S = "Y" AND X3S = "Bottom' AND X12S = "Yes"
        THEN : CRACKS = "Formwork failure during cast"
        REASION :   In cases where there is large sag of slab seen from the bottom , it is
                    formwork failure during cast.
```

The following example explains how to use the expert system:

Concrete Defects Expert System – CEDS

KACST Project AR-15-14

PI: Dr. Habib Zein al-Abideen – CI: Dr. A. Sharaf Eldin

Press Enter to continue

What is the member type (Beam/Slab/Column/Wall)?Slab

Does the crack continue past plastering (Y/N)?Y

Are Cracks from (Top/Bottom)?

Top

Cracks Pattern (Rectangular/Random Map/Parallel to beams/Perpendicular on beams)?

Random Map

Are Cracks accompanied by spalling (Yes/No)?

No

Cracks Spacing (Very closely spaced/closely spaced/Widely spaced)?

Closely spaced

It seems that it is Plastic shrinkage cracks

.....Press ANY key to continue

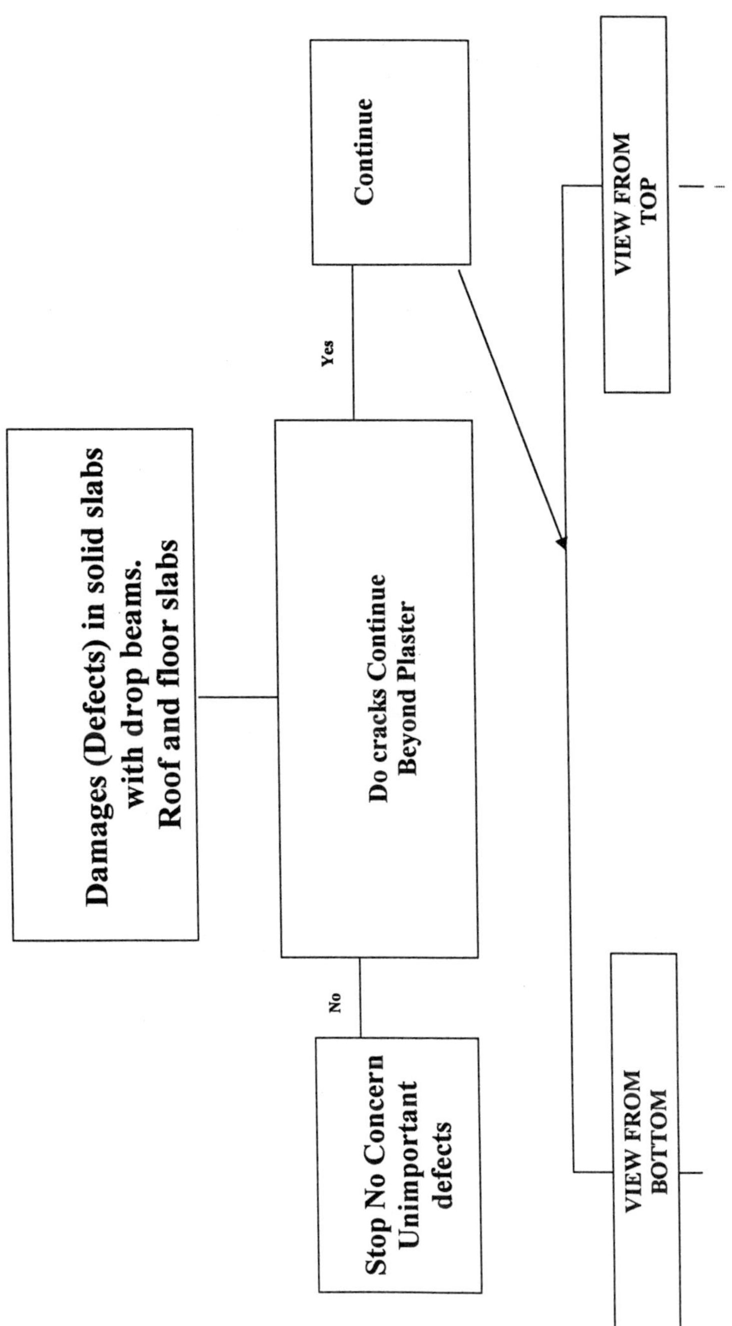

Figure 3a Cracks or defects in solid slabs with drop beams

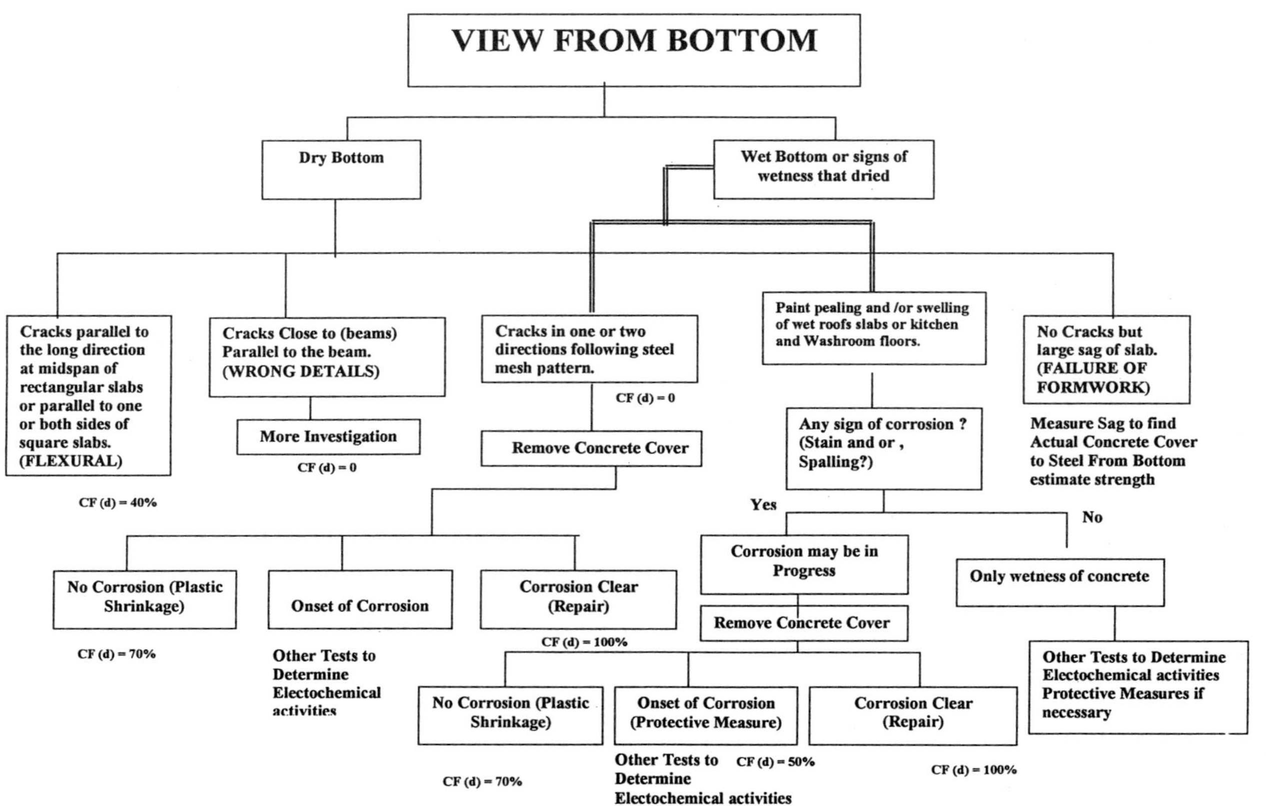

Figure 3b Cracks or defects in solid slabs with drop beams (contd)

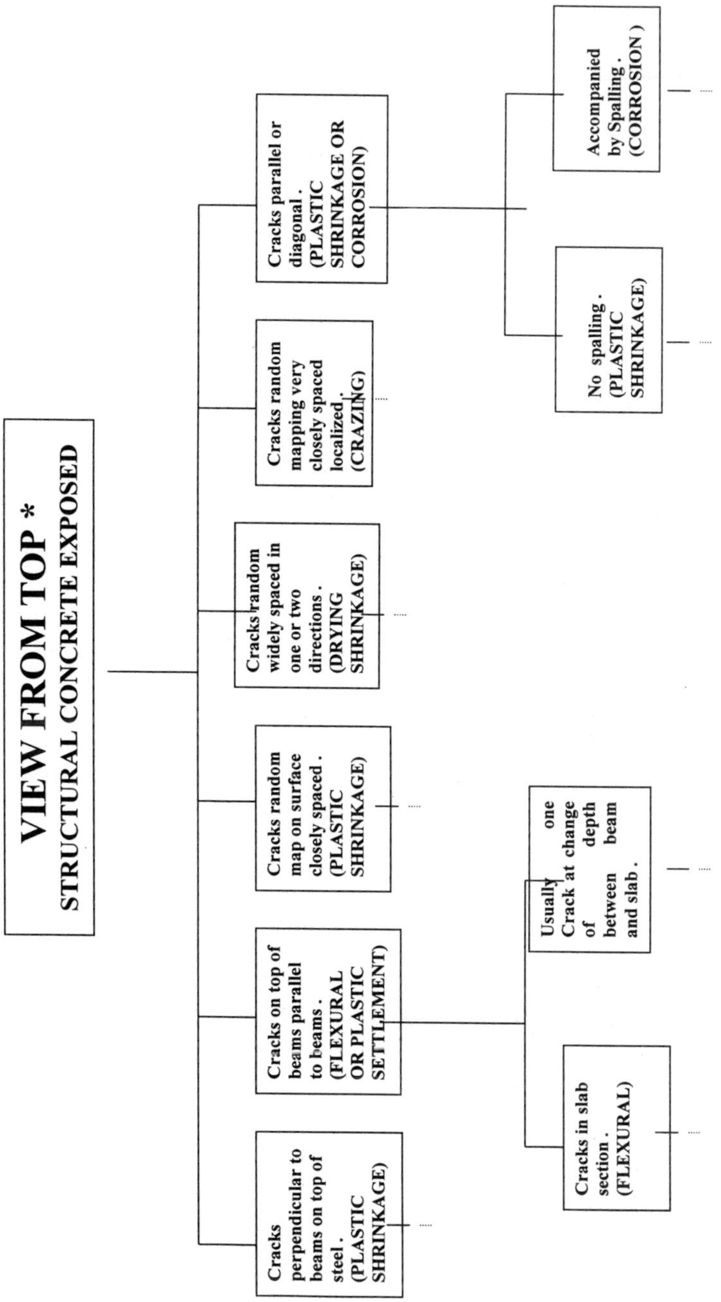

VIEW FROM TOP *
STRUCTURAL CONCRETE EXPOSED

- Cracks perpendicular to beams on top of steel. (PLASTIC SHRINKAGE) ......
- Cracks on top of beams parallel to beams. (FLEXURAL OR PLASTIC SETTLEMENT)
  - Cracks in slab section. (FLEXURAL) ......
  - Usually one Crack at change of depth between beam and slab. ......
- Cracks random map on surface closely spaced. (PLASTIC SHRINKAGE) ......
- Cracks random widely spaced in one or two directions. (DRYING SHRINKAGE) ......
- Cracks random mapping very closely spaced localized. (CRAZING) ......
- Cracks parallel or diagonal. (PLASTIC SHRINKAGE OR CORROSION)
  - No spalling. (PLASTIC SHRINKAGE) ......
  - Accompanied by Spalling. (CORROSION) ......

Figure 3c Cracks or defects in solid slabs with drop beams (contd)

CONCLUSIONS AND RECOMMENDATIONS

Artificial intelligence technologies especially expert systems are very useful in developing computer programs which accumulate the experience of groups of experts. One of the important applications in this field is the assessment of distressed and deteriorated concrete buildings. This area has a lack of experts not only at the national or regional levels but also internationally. The assessment procedure is complex and involve many factors which include environmental parameters, materials used, construction details and procedures as well as the function and use of buildings …etc [14-15]. In this research we attempt to combine computer technologies; namely, ESs, multimedia, Data Base systems with the experience which was developed in qualified organization during twenty five years in the area of assessment of concrete buildings and the repairs they require.

The proposed system has the following characteristics:

- Using knowledge dictionary.
- Automatically evaluating the knowledge obtained.
- Handling imprecision knowledge and data.
- Knowledge acquisition, codifying and storing.
- Allowing the input of new knowledge.

Some important recommendations are:

1. Training a new generation of engineers in the field through utilization of the educational system.
2. Allowing a great number of users and collecting their feedback and criticism to improve the system. This can be achieved through a location on the Inter-Net.
3. Introducing the subject of concrete building assessment to civil engineering curricula at universities.
4. Encouraging the utilization of computer technologies in this field.

ACKNOWLEDGMENTS

The authors are grateful to King Abdul-Aziz City for science and Technology (KACST) for the financial support of research number AT-15-14. Thanks are due to experts of the Ministry of Public works & Housing in the Kingdom of Saudi Arabia. H. Izindal , H. Gockel , M. Foss, M. AboulelUlla, S. Althikry and B. Eltoony.

REFERENCES

1. ZAIN AL ABIDEEN, H M. “Who Assesses the Safety of Buildings” A lecture given to a symposium on “Deterioration of Concrete Buildings and Methods of Repair”,Engineering Committee 1-4 jumada Ula 1409 Riyadh, 7-10 Rajab 1409 Dammam (In Arabic).

2. ZAIN AL ABIDEEN, H AND SHARAF ELDIN, A. An Expert System for Evaluation of Concrete Deterioration in Saudi Arabia, first progress report, KACST, 1997.

3. ZAIN AL - ABIDEEN , H M " Expensive Errors in Buildings Assessment and Repair " 5th International Conference " Deterioration and Repair of Reinforced Concrete in the Arabian Gulf "Bahrain pp.595-662 October ,1997

4. KHALIFA, MAGDI M. "Concrete, Durability in the Arabian Gulf Region", International Association for Bridge and structural Engineering Symposium on Extending the Lifespan of Structures, San Francisco, August 23-24, 1995.

5. RICH, E. Artificial Intelligence. McGraw-Hill, 1985

6. HARMON, P AND KING, D. Expert Systems, A Wiley Press Book, 1985.

7. SADEK, A W; ATTIA, T M; SAMAN, S; MORSY, E H. *Prototype knowledge based system to diagnose building defects*, Modelling, Measurement & Control B: Solid & Fluid Mechanics & Thermics, Mechanical Systems 55 4 1994. p 53-64

8. FLORA, K S. *DR. LEO: An Expert System for the Diagnosis and Repair of Reinforced Concrete,* Final rept., California State Dept. of Transportation, Sacramento. Office of Materials Engineering and Testing Services, 1994.

9. SHIRAISHI, N; FURUTA, H; UMANO, M AND KAWAKAMI, K, *Expert System for damage assessment of a reinforced concrete bridge deck*, Fuzzy sets and systems, v 44, no. 3, Dec. 1991, pp. 449-457.

10. MELCHOR-LUCERO, O; FERREGUT, C. *Toward an expert system for damage assessment of structural concrete elements*, Artificial Intelligence for Engineering Design, Analysis and Manufacturing: AIEDAM v 9 n 5 Nov 1995. p 401-418, 1995.

11. PRERAU, D. "Knowledge Acquisition in the Development of Large Expert Systems", AI Magazine, vol. 8, no. 2, 1987, pp. 43-52.

12. ZAIN AL ABIDEEN, H. AND SHARAF ELDIN, A. An Expert System for Evaluation of Concrete Deterioration in Saudi Arabia, second progress report, KACST, 1997.

13. GRABOT, B, AND CAILIAUD, E. "Imprecise Knowledge in Expert System: A simple Shell", Expert Systems with Applications, vol. 10, no.1, pp. 99-112, 1996.

14. KEENEY, R AND RAIFFA, H. Decision with Multiple Objectives: Preferences and value tradeoffs. John Wiley & Sons, 1976.

15. ZAIN AL ABIDEEN, H M."Safety Assessment of Reinforced Concrete Structures" Ministry of Public Works and Housing, Riyadh . Saudi Arabia 2nd edition 1992.

16. ZAIN AL ABIDEEN, H, AND SHARAF ELDIN, A, An Expert System for Evaluation of Concrete Deterioration in Saudi Arabia, third progress report, KACST, 1998.

DIAGNOSING MATERIAL PROBLEMS IN CONCRETE CONSTRUCTION

A M Dunster

K Hollinshead

Building Research Establishment

United Kingdom

ABSTRACT. This paper describes investigations of durability problems in two structures, both of which contained Calcium Aluminate (also known as High Alumina), cement

The first investigation concerns an assessment of polymer modified patch repairs to a reinforced concrete structure in saline conditions. A few months after application, many repairs were found to be cracked and de-laminating, often with a white deposit on the surface. The causes of the phenomena observed and their significance for the future performance of the repairs are discussed.

The second investigation was an assessment of pre-cast concrete components in a building. The durability of the components was assessed and this showed that reinforcement corrosion, chemical attack and carbonation were beginning to affect the performance of the concrete at some locations.

Keywords: Calcium Aluminate cements, Pre-cast concrete, Patch repair materials, Durability

Dr Andrew M Dunster is a Project Manager with the Materials Diagnostic Unit, Building Research Establishment, UK. He specialises in the assessment of the performance of Calcium Aluminate based materials in structures. Other research interests include water reducing admixtures and moisture conditions within concrete.

Dr Kate Hollinshead is a Consultant to the Construction Best Practice Programme and was until recently, Team Leader of Corrosion Prevention and Repair in the Centre for Concrete Construction at the Building Research Establishment. Research interests include monitoring, prevention and remediation of reinforcement corrosion.

INTRODUCTION

Through its consultancy work, BRE's Materials Diagnostic Unit investigates the causes of a range of materials durability problems. Diagnosing these problems and their implications often involves a detailed analysis of materials to assess their condition and identify any deterioration. Success in this complex field demands expertise in a range of materials problems and the use of a combination of analytical techniques. This paper is based on two of BRE's investigative case studies. Its purpose is to ensure that those involved at the sharp end of designing, building and maintaining concrete structures are made more aware of potential durability problems in cement based materials.

Case Study A concerns the diagnosis of the condition of patch repairs to a reinforced concrete structure in saline conditions. Case Study B describes an assessment of pre-cast concrete components in a commercial building. Both structures contained Calcium Aluminate cements (CAC). CAC is also known as High Alumina cement (HAC).

CASE STUDY A

Background

In 1997, the authors were asked to investigate the causes of white efflorescence, cracking and de-lamination that had developed with concrete patch repairs. Advice on the implications of this deterioration for the design life of the structure was also provided.

More than seven hundred patch repairs were applied to a 60-year-old concrete structure between 1994 and 1997. The structure comprised seven Portland cement concrete dolphins (or moorings) in a dock. Each dolphin was approx. 160 metres x 5 metres. The upper part of the original structure had been demolished, the reinforcing steel of the demolished columns was cut back and the area repaired with proprietary concrete patch repair materials (concrete and mortar). It was the intention of the owners to build a new superstructure over the repaired areas and to protect the structure using a cathodic protection system with a view to achieving a maintenance-free life of a further 60 years. The patch-repaired areas were below the normal impounded water level in the dock and would be fully submerged in brackish water under normal circumstances. The original structure had been shown to be heavily chloride contaminated.

Dolphins 1 and 2 had been repaired in 1994-95 without significant problems with the repairs. During the winter of 1996-97, repairs to dolphins 3 to 7 were carried out. Although many of the repairs were in an acceptable condition, a proportion had developed cracking and a white deposit on the surface and this was reported to have occurred within a few weeks of placing. De-lamination between layers within the repairs was also found at some locations. In late April 1997, work on the site was suspended on the grounds of de-bonding failure, poor compaction, cracking, low pull-off strength results, and the presence of white deposit on the surface of some repairs on dolphins 3 to 7. It is now known that Dolphins 1 and 2 had been repaired with a Portland cement based material and the remainder with a material based on a CAC. Both materials contained SBR (styrene butadiene rubber).

BRE were asked to visually inspect the repairs on site, to examine samples of the raw materials, cores from the site, and physical test data (pull-off tests and strength tests) provided by the resident engineers. The main objective of the work was to advise on assessment criteria for acceptable repairs and on the potential durability of repairs in deteriorated areas. A series of mixing trials were also undertaken and the specimens cast were immersed in sea water under laboratory conditions in an attempt to replicate the formation of the white deposit.

Site Inspection

Figure 1 shows the structure of a typical dolphin. The patch repairs had been carried out at locations on the crossbeams (locations 2 and 3), and at their ends (locations 1 and 4). Repaired areas on the crossbeams were typically approx. 300 x 300 mm and 150 mm deep. The shear walls of the new superstructure had already been constructed on top of the repairs at a number of locations at the time of the visual inspection.

The exposed top surfaces of the repairs were examined at over three hundred locations. There were numerous micro cracks and larger cracks on the top surfaces and the top layer of repair material had de-bonded at several locations. Cracks were also observed in the parent concrete adjacent to the repair material, indicating that the parent concrete had been damaged during repair. A white deposit, both at repair/substrate interfaces and within horizontal voided zones within the repair material, was present on the sides of some of the repairs.

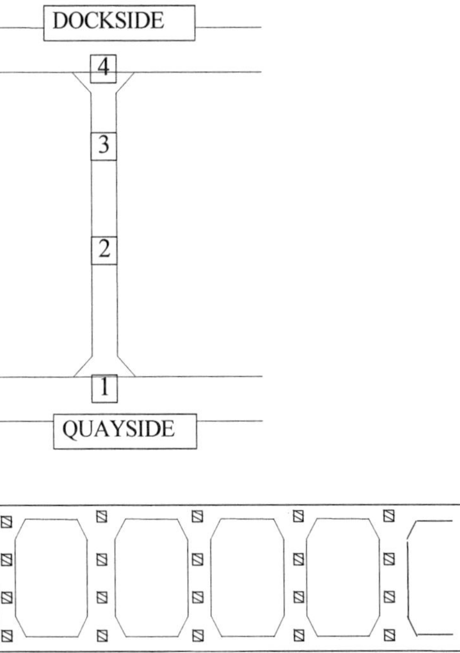

Figure 1 Case Study A - numbering of repair locations

The concrete patch repair on dolphins 6 and 7 (which had been repaired by a different contractor to the other dolphins) showed a significantly better finish than the repairs to dolphins 3 to 5. This was reflected in the number of patch repairs without cracks in the two groups of dolphins: 40 % for dolphins 3 to 5 and 55% for dolphins 6 and 7. It is understood that better care was taken with curing and with securing the stability of the formwork on Dolphins 6 and 7, which may explain the lower incidence of cracking.

All the cracks observed in the patch repairs appeared to have been generated during the early life of the repair. The most likely cause was a response to poor curing (principally plastic shrinkage) although plastic settlement and surface crazing caused by overworking of the surface probably also contributed to the extent of cracking. The de-lamination between the layers within the repairs probably indicated incorrect preparation and application of the repair products as these were placed in layers with a bonding agent applied at interfaces. Polymer-modified bonding agents have a relatively short drying time and bonding is very sensitive to the conditions at the interface. If used incorrectly, bonding agents can significantly reduce bond strength.

The cause of cracking was not the main consideration in judging whether the patch repairs were fit for purpose. The main criterion was whether the cracks in the patch repair would affect the durability of the patches. There is no clear guidance on acceptability criteria for cracks in patch repairs. It is generally accepted, however, that cracks occur in concrete without the concrete necessarily being defective. Cracks with widths greater than 0.5 mm were present in the patch repairs on the dolphins (this is greater than the upper limit of 0.15 mm for "exterior members exposed to aggressive environment" given in reference [1]. There was a strong possibility that these cracks would extend into the bulk of the repair material, severely reducing its durability.

Analytical Methods

Optical microscopy was used to examine cores taken from the repairs and from the substrate concrete in repaired and un-repaired areas. Sub-samples from the cores were sliced and made into thin-sections 30 microns thick. The sections were examined using a petrological (optical) microscope. The trapped air voidage of a number of thin sections of the repair material was determined using a point counting method [2]. Selected thin sections, a polished sample, and a sample of the white surface deposit were examined using a scanning electron microscope (SEM). Differential Scanning Calorimetry (DSC) and X-ray diffraction (XRD) were used to identify the constituents of the hydrated and unhydrated repair materials and the white surface deposit.

Pull-Off Tests

Acceptance criteria for pull-off tests

Reference [3] outlines a scheme for classifying acceptance using pull-off test data based on the nature of the failure and the failure stress. Where pull-off values greater than the target strength are achieved, the test has clearly been passed.

However, where the target value is not achieved, the assessment should then consider the frequency and exact nature of such failures. Frequent low pull-off values with failure in the substrate may indicate a weak or poorly prepared substrate. There is currently no standard criterion for interpretation of pull-off data. However, the general view within the industry is that pull-off values of at least 1 MPa should be achieved, whilst the Concrete Society Model Specification gives a target of 0.8 MPa[4]. In this case it was decided that the latter would be adopted.

Results of pull-off tests

Figures 2 to 3 summarise BRE's assessment of the pull-off results obtained from Dolphins 5 and 6 in terms of achievement of the target strength of 0.8 MPa and the mode of failure.

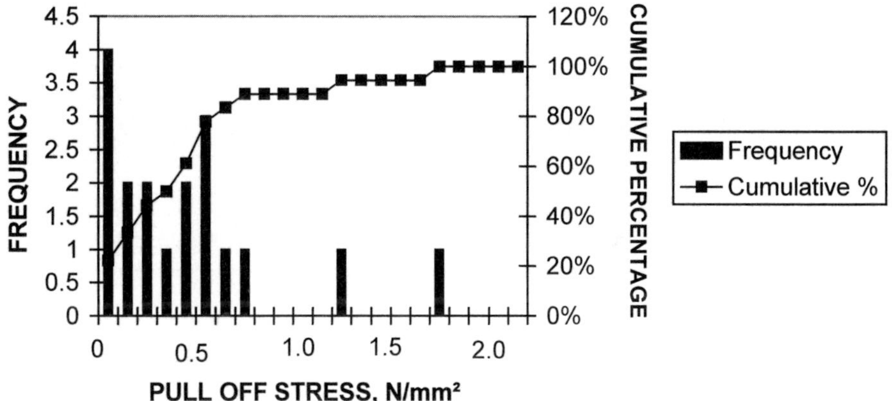

Figure 2 Distribution of pull-off data from Dolphin 5

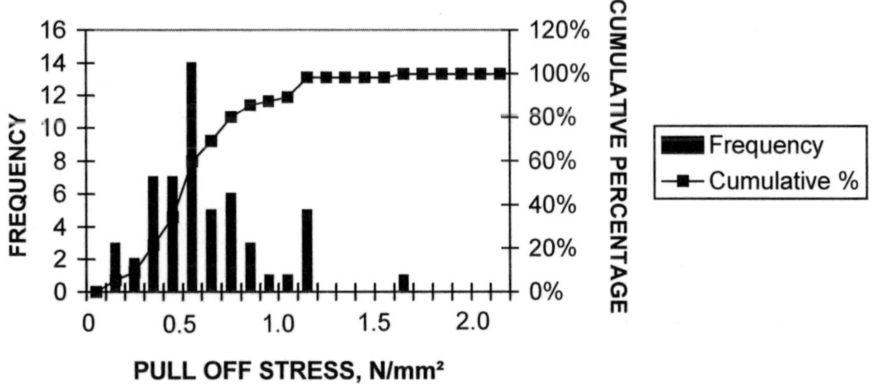

Figure 3 Distribution of pull-off data from Dolphin 6

This target strength was often not achieved with these repairs, with failure tending to occur within the repair material, at repair-substrate interfaces, or within the substrate near the interface (Table 1). A high proportion of the failures at strengths below 0.8 MPa on all three dolphins involved at least some failure of the substrate thus potentially indicating a weak or poorly prepared substrate. From BRE's assessment; 12 out of 16 (Dolphin 5), 22 out of 47 (Dolphin 6) and 4 out of 6 (Dolphin 7) failed in this way.

Table 1 Analysis of pull-off results from Dolphins 5, 6 and 7

FAILURE MODE	DOLPHIN NUMBER					
	Pull-off > 0.8 MPa			Pull-off < 0.8 MPa		
	5	6	7	5	6	7
Adhesive	0	0	0	1	7	0
Failure in repair	1	1	0	1	10	1
Repair/substrate interface	0	1	0	2	8	1
Mixed interface/substrate/ repair material	0	4	0	8	17	1
Substrate	1	2	0	4	5	3
Sub-totals	2	8	0	16	47	6

Laboratory Investigation of Patch Repairs

Petrography and electron microscopy of repairs

Four cores from dolphins 3, 4 and 5 were examined. The cement paste was identified as a CAC and, although its matrix was of low capillary porosity, there were a large number of entrapped air voids, often within lateral zones within the material. There was also a high proportion of un-hydrated cement clinker. Some of the air voids contained a surface deposit of fine white or colourless fibrous crystals or bulk crystalline material. There was no evidence to indicate that the deposits were the source of failure of the repair. In some areas, entrapped air voidage (determined by point counting) constituted more than 30% of the cross sectional area examined. By comparison, levels of voidage achieved by mixing the materials in BRE's laboratories were all less than 10 % by volume. There was no evidence under the optical microscope of any form of chemical attack.

Examination of white deposit from site

The white deposits present in the larger voids and on the surface of some repairs were associated with poor compaction at the interfaces between the repair material layers.

XRD and DSC showed that this material contains bayerite (a form of AH_3), non-crystalline AH_3, calcite, quartz and sodium chloride. The deposit may also contain hydrotalcite (M_4AH_{10}). It seems likely that secondary AH_3 formed in the large voids from the hydration product calcium aluminate hydrate gel and from alpha alumina (which was a minor component of the cement). SEM analysis indicated that the deposit was quite variable in composition on a sub micron scale. The deposit is caused by leaching of hydration products from the repair material (AH_3 is a normal hydration product of CAC's), and interaction with the brackish water in the dock. It was, therefore, a symptom rather than a cause of the problems with the repair material.

Formation of the white deposit under laboratory conditions

Specimens cast from the repair materials in the laboratory were immersed in seawater or deionised water at 20°C. Porosity and the presence of saline water were clearly factors in the formation of the white surface deposit as it only developed on the most porous specimens stored in seawater. Where present, the deposit appeared within 48 hours of exposure. DSC analysis of material removed after 7 days showed the presence of AH_3.

Laboratory Assessment of Substrate Concrete

Physical tests

Compressive and tensile strength tests were performed by Technotrade Structural Services on pairs of cores (100 mm diameter) taken from the substrate concrete on Dolphins 2, 3, 5 and 7. Compressive strength tests were undertaken as recommended in Reference [5]. The tensile tests were undertaken by bonding the cores between two steel plates and applying a load in tension until failure. The results are shown in Table 2. The mean estimated in situ compressive strength was 69.0 MPa and the mean strength in tension was 2.5 MPa. In all cases the tensile failure mechanism was partial bond failure. The tensile strength results can be considered as a lower bound for strength with the actual tensile strength of the cores in excess of the values determined. The results of these tests, and results from pull-off tests on unrepaired areas of substrate concrete on site (data not included here), did not indicate that the substrate concrete in its initial condition was generally sufficiently weak to account for the low pull-off strengths obtained for the repairs. Consequently, BRE undertook petrographic analyses of concrete adhering to four cores taken during pull-off testing of repaired areas and of substrate concrete in two cores taken from un-repaired areas.

Petrographic analysis of substrate concrete

In all the samples from repaired areas examined, extensive micro cracking was observed in the original substrate concrete near the interface with the repair material. It is probable that the majority of this cracking was the result of mechanical damage to the substrate during preparation prior to repair.

In the larger cracks and defects at the repair/concrete interface, the repair concrete had "flowed" some distance into the underlying material indicating that these defects were "open" at the time of repair.

Table 2 Strength test results from cores taken from Dolphins 2, 3, 5 and 7. (after Technotrade Structural Services)

DOLPHIN	ESTIMATED IN SITU COMPRESSIVE STRENGTH, MPa	STRENGTH IN TENSION, MPa
2	68.5	2.2
3	61.0	2.9
5	77.0	2.0
7	69.5	2.5

Examination of the concrete cores from the substrate concrete taken away from any repair or breaking out showed no evidence of any significant micro cracking. A few very minor micro-cracks and voids containing ettringite were the only features of any distinction in an otherwise sound material. This provides evidence that the original substrate concrete was sound at the un-repaired locations examined.

Durability of CAC Repair Materials

The authors did not identify any published data on the durability of cements and mortars made with the relatively pure form of CAC used in the repair materials as this type of cement is used mainly in refractories. However, long term data for the Ciment Fondu type of CAC (which is less pure than the type used here) are available and there is no reason to suppose that the durability of the refractory cement would differ significantly from that of Ciment Fondu. It is clear, however, that the use of calcium aluminate cement was not the cause of the problems in this case.

CAC's were originally developed for their resistance to sulphate bearing environments and seawater [6]. There is evidence from the literature that repair mortars made with Ciment Fondu can perform well in marine conditions and have shown good chemical resistance and durability [7,8]. However, there appears to be little available data on the durability of SBR-modified CAC systems.

Conclusions

The preparation of the substrate concrete and placement of the repair materials on the dolphins did not meet current good practice recommendations for substrate preparation compaction and curing.

There was therefore a risk that the repairs would not fulfil the intended maintenance-free life requirement for the structure. BRE advised that the top layers of patch repairs exhibiting the most serious cracks, as well as the de-laminated areas, should be removed. The exposed surface of the remaining layer should then be inspected for cracking and de-lamination. If extensive cracking or de-lamination was found this layer should also be removed.

The final decision on the course of remedial action depended largely upon the perceived impact of the repairs, in particular the preparation of the substrate, on the maintenance-free life and stability of the structure. Unfortunately, there was insufficient long-term information available on the performance of patch repairs to be able to accurately assess the increase in risk of the maintenance-free life not being achieved.

It was demonstrated that the substrate concrete on dolphins 3 to 7 was damaged during preparation. However, it was not always practical to take the action necessary to remedy this, as it would involve the complete removal of many repairs (some of which had already been built upon), followed by removal of the damaged layer of substrate concrete.

This would have involved considerable effort and could have caused more damage to the concrete. Durability in terms of corrosion protection to the reinforcement will probably depend largely upon the successful operation of the upgraded cathodic protection system.

Recommendations

When considering options for protection or repair, clients should consider carefully the performance required from their repaired or protected structure. They must also consider the maintenance strategy they wish to adopt; for example, do they want a "once and for all" treatment or is periodic patching an option.

A consideration of the level of maintenance that they are prepared to give the structure and the intended residual service life of the structure is also important.

There is little published information on the main factors causing failure of repairs in practice, particularly over a period such as sixty years. However, evidence collected over several decades by the United States Department of the Interior Bureau of Reclamation [9] has indicated that, in their experience, the most common cause of repair failure is improper or inadequate preparation of the old concrete prior to application of the repair material. UK experience may differ from that of America but this emphasises the importance of substrate preparation.

It is generally acknowledged within the industry that water blasting is the best method of surface preparation due to the nature of the surface produced. If water blasting is not considered appropriate, damage can be minimised by the use of properly sharpened tools and avoiding the use of large equipment [4,10].

A lack of proper substrate preparation, proper curing (by application of a curing membrane), and adequate formwork were obviously key factors in this case.

CASE STUDY B

Background

This case study describes an investigation of a commercial building containing CAC concrete floor beams. During a routine inspection of the building in 1997, consultants to the building owner noticed severe cracking of the soffits of two of the beams in a boiler room. It was also discovered that the records of the structural assessment (or design check) of the floors undertaken in the 1970's had been mislaid. BRE was asked by the building owners to advise on any action or remedial works required in the deteriorated areas and to provide a prognosis for the future durability of the building. It was agreed that this initial investigation would be followed by a design check of the structural capacity of the floors.

The building was built in the early 1970's. It consists of a two-storey block over a basement (which contains a swimming pool, boiler room and service rooms), with a four-storey block attached. The floor slabs comprise pre-cast pre-stressed CAC concrete "I" and "X" beams below a screed topping with lightweight concrete blocks between the beams.

Site Inspection and Sampling

The authors inspected the building in late 1997. The boiler room ceiling provided support for a number of water and waste pipes, several of which were leaking at the time of the site visit. One of the CAC concrete beams in this area, although not currently wet, showed evidence of cracking due to reinforcement corrosion. A further beam in this area of the boiler room (which was, unfortunately, inaccessible for sampling), showed severe cracking of the soffit associated with corrosion of several reinforcing wires. The soffit was visibly very wet at the time of the site visit. Except for these isolated locations, there was no evidence of water ingress to the CAC concrete in other areas of the building.

Fifteen concrete lump samples were taken from the soffits of the beams and the condition of the exposed steel, where encountered, was noted. Three samples were taken from areas of the boiler room near the water and waste pipes and twelve from areas throughout the building which were unaffected by leakage. Except where the concrete was cracked due to reinforcement corrosion, all the lump samples were difficult to remove and the concrete was hard and dense.

Chemical and Materials Analysis

Optical microscopy (petrography) was used to establish the depth of carbonation [11,12] and to check for chemical attack. Selected concrete lump samples were analysed in the laboratory using the following techniques: acid soluble chloride by ion selective electrode; cement content by Al_2O_3 determination (Atomic Absorption Spectroscopy); Na_2O and K_2O determinations (Atomic Emission Spectroscopy), to check for alkaline hydrolysis with carbonation in accordance with reference [11].

Results of Site Inspection

Condition of reinforcement

Table 3 summarises the results of the general inspection throughout the structure. From a total of sixteen locations eight wires were uncorroded, four wires were slightly corroded, and two beams (in the boiler room) showed evidence of significant reinforcement corrosion which had led to cracking of the concrete cover. With the exception of areas of the boiler room where there was clear evidence of water ingress, there were no differences in the severity of reinforcement corrosion between the different parts (and environments) within the building.

Where corrosion had occurred, this was found to be associated with the presence of chlorides and/or carbonation (see below).

Table 3 Condition of pre-stressing wires (16 beams)

AREA OF BUILDING/ ENVIRONMENT	NUMBER OF BEAMS		
	Uncorroded	Slightly Corroded	Heavily Corroded
Boiler room and maintenance room (warm, dry, locally wet in boiler room)	3	3	2
Kitchen (high humidity)	1	0	0
Swimming pool (high humidity)	4	1	0
Total	8	4	2

Results of Laboratory Investigation

Table 4 shows the results of the laboratory tests for selected beams in the boiler room and their relationship to the site observations. Results of chemical analysis of a sample from the four-storey block (H4, taken from a nominally dry area), are also given as a control. One of the deteriorated beams in the boiler room (H2) showed significant reinforcement corrosion due to carbonation and chlorides.

One beam (H8) in the boiler room that was not carbonated to the steel showed evidence of slight reinforcement corrosion. This was associated with the presence of chlorides and the low carbonation depth was probably due to the effect of water keeping the soffit of the beam damp. The chloride content (2.82 % by weight of cement) in the deteriorated beam H2 is more than forty times greater than the values of approx. 0.05 % identified in nominally dry areas (H4 and H6). A high level of chloride (3.08 % by weight of cement) was also found in H8 where there was only limited corrosion. However, this beam was further from the source of leakage and would therefore not have remained as continuously wet as H2 in the hot conditions encountered in the boiler room.

Table 4 Results of site inspection and laboratory analysis, (boiler room)

REF NO	EVIDENCE OF WATER INGRESS?	CONDIT-ION OF STEEL*	CARBON-ATED TO STEEL ?	% BY WEIGHT OF CEMENT			ALUMINA/ EQIV ALKALIES
				Cl^-	Na_2O	K_2O	
H2	Yes	3	Yes	2.82	2.95	0.37	12
H7	Yes	2	No	0.07	0.24	0.27	91
H8	Yes	2	No	3.08	2.78	0.33	13
H6	No	1	Yes	0.07	0.31	0.25	89
H4 (control)	No	**	Fully carbonated	0.05	0.26	0.24	94

* 1 = reinforcing wire uncorroded; 2 = reinforcing wire slightly corroded; 3 = reinforcing wire heavily corroded.
** steel set deep in web of beam and therefore not examined.

The ratio of Al_2O_3 to alkalies (total equivalent Na_2O) are shown in Table 4. The values for control samples H4 and H6 (from dry locations) were above the lower limit of 70 stipulated in the BRE Digest 392[11].

Samples H2 and H8 both contained high levels of sodium and low levels of potassium (Table 4) and the low values of the alumina/alkali ratio (12 and 13) indicate that there would be the potential for alkaline hydrolysis with carbonation to occur if the beam were to remain wet.

Petrography showed the CAC concrete to have good compaction, low excess voidage and low capillary porosity. None showed any signs of sulfate attack, alkaline hydrolysis with carbonation, or other forms of chemical attack.

Results of the BRAC Design Check Assessment

The results of the materials investigations throughout the building showed that, with the exception of a few isolated areas in the boiler room, the durability of the pre-cast units was adequate for continued use of the building in the medium term.

A desk exercise to check the beam/floor capacity against the DoE guidance issued in 1976 [13] was therefore carried out. As a part of this, it was necessary to expose the ends of the beams to ascertain the types of components as recommended in BRE Digest 392.

The BRAC Design check showed that the strength of the floors was generally satisfactory although provision of additional support was considered necessary in the critical areas of the boiler room.

Discussion of Results

The severely disrupted beams in the boiler room (which have resulted from water penetration and damp conditions), were not typical of the conditions that were likely to develop elsewhere in the building. Enquiries have shown that the chemicals disposed of via these pipes contained sodium hydroxide and chlorine compounds. The presence of high levels of chlorides and sodium (with low potassium) in the concrete samples from these areas provided evidence that a bleach type cleaning solution may have been responsible. Carbonation under the hot dry conditions prior to water penetration is probably also a factor in the deterioration observed.

Although Sample H2 had carbonated to the depth of the steel, the presence of high levels of chloride (from leaking pipework) in the sample meant that chlorides had probably contributed to the severity of corrosion at this location. There was evidence of slight chloride-induced corrosion in a chloride contaminated sample (H8) where the concrete was uncarbonated. The relatively small amount of corrosion in H8 (which had a similar chloride content to H2), highlighted the importance of both chlorides and persistent moisture to promote corrosion.

Remedial Measures

The provision of permanent strengthening measures was recommended for the deteriorated areas of the boiler room which were propped as a temporary measure. Trials are in progress using an epoxy bonded carbon fibre plates with the objective of enhancing strength in the critical areas and corrosion inhibitors in other areas where there may be a risk of corrosion.

CONCLUSIONS

1. The high levels of chloride and sodium in the deteriorated areas were associated with leaking pipework used for disposal of cleaning fluids.

2. All samples from nominally "dry" areas showed advanced or patchy carbonation to the depth of the steel. This means that the steel was no longer protected from corrosion at these locations and corrosion could occur in the presence of moisture and oxygen.

3. This case study confirms the view presented in BRE Digest 392 that ageing CAC concrete units are at risk from reinforcement corrosion in the presence of persistent moisture. Building owners should ensure, through regular maintenance, that such units are not subjected to such conditions. As is the case with PC concrete, reinforced CAC concrete can suffer chloride-induced corrosion so prolonged contact with chloride solutions should also be avoided.

REFERENCES

1. CONCRETE SOCIETY. Non-structural cracks in concrete, Technical Report No. 22. 3rd Edn, 1992. 48 pp.

2. AMERICAN SOCIETY OF TESTING MATERIALS. Designation C457-82a, Standard practice for microscopical determination of air void content and parameters of the air void system in hardened concrete. 1982, pp 222-232.

3. LAMBERT, P AND ECOB, C R. Evaluation and testing of sprayed concrete, Proc. ACI/Sprayed Concrete Assoc. Int. Conf. on Sprayed Concrete/ Shotcrete, E&FN Spon, London. 1996, 5pp.

4. CONCRETE SOCIETY. Patch repair of reinforced concrete: Model Specification and Method of Measurement, Technical Report 38, The Concrete Society, London. 1991, 88 pp.

5. BRITISH STANDARDS INSTITUTION. BS 6089; Guide to assessment of concrete strength in existing structures. BSI, London, 1981.

6. CONCRETE SOCIETY. Calcium Aluminate Cements in Construction, Technical Report No. 46. The Concrete Society, London, 1997 63 pp.

7. EL JAZIERI, B AND BANFILL, P F G. The properties of High Alumina Cement based materials for rapid repair of concrete. Proc. Structural faults and repair 89, Vol 2, 1989, pp 307-314.

8. BAKER, N C AND BANFILL, P F G. Durability of High Alumina Cement mortars for marine repair works. Proc. Structural faults and repair 93, Vol 3, 1993, pp 133-138.

9. UNITED STATES DEPARTMENT OF THE INTERIOR BUREAU OF RECLAMATION, Standard Specifications for Repair of Concrete, Denver, 1996, 63 pp.

10. PULLAR STRECKER, P. Corrosion damaged concrete assessment and repair. CIRIA, London, 1987, 99 pp.

11. BUILDING RESEARCH ESTABLISHMENT. Digest 392. Assessment of existing High Alumina Cement concrete in the UK. March 1994, 12 pp.

12. DUNSTER, A M. Assessing carbonation depth in ageing HAC concrete. BRE Information Paper IP11/98, June 1998, 6 pp.

13. DEPARTMENT OF THE ENVIRONMENT. Building Regulations Advisory Committee. Report by Sub-Committee P (High Alumina Cement concrete). BRAC (75) P40, 1975.

ACCURATE MEASUREMENT OF ELECTROMAGNETIC PROPERTIES OF CONCRETE FOR NON-DESTRUCTIVE EVALUATION AT MICROWAVE FREQUENCIES

D K Ghodgaonkar

W M B W A Majid

R B A Majid

Mara Institute of Technology

Malaysia

ABSTRACT. Complex permittivities (ε^*) of concrete are a function of moisture content, frequency, temperature and concrete mix constituents. By using effective medium theory, it is possible to determine the moisture content from dielectric measurements of dry and wet specimens. A free-space microwave measurement system is used for reflection coefficient measurements of metal-backed specimens in the frequency range of 8.0-12.5 GHz. Complex permittivities of concrete specimens are determined from reflection coefficient values by using an algorithm which finds zeroes of the error function. The key components of the measurement system are a pair of spot-focusing horn lens antennas, mode transitions, coaxial cables and network analyzer. The results include complex permittivies and moisture contents of specimens with specified compressive strengths of 30, 40 and 50 N/mm^2. It is found that the real as well as imaginary part of ε^* is significantly higher for wet specimens as compared with dry specimens. Also, moisture content values obtained using the dielectric method are close to the actual values.

Keywords: Microwave nondestructive testing, Complex permittivity, Dielectric constant, Loss tangent, Concrete and moisture content.

Associate Professor Dr Deepak Kumar Ghodgaonkar is a member of the Faculty of Electrical Engineering at the Mara Institute of Technology, Shah Alam, Selangor, Malaysia. His research interest is in the area of microwave nondestructive testing of materials He is a senior member of IEEE, a Fellow of IETE, India and a member of IEM.

Professor Dr Wan Mahmood B W A Majid is a member of the Faculty of Civil Engineering at the Mara Institute of Technology, Shah Alam, Selangor, Malaysia. He has published more than 50 papers in leading journals and conferences. His main research interests include characterization of construction materials such as concrete and timber. He is a Fellow of the Institution of Engineers, Malaysia (IEM) and a member of Institution of Engineers, Australia. Currently, he is a vice president of IEM.

Rosnoizam Binti Abdul Majid has completed a B. Eng. (Hon.) in Electrical Engineering in May, 1998 from the Faculty of Electrical Engineering, Mara Institute of Technology, Shah Alam, Malaysia.

INTRODUCTION

Microwave nondestructive testing techniques (such as ground probing radar, free-space microwave techniques) are increasingly being used for quality control and condition assessment of concrete structures [1-3]. It is known that complex permittivies of concrete are a function of moisture content, frequency, temperature and concrete mix constituents [1]. So, the measurement of complex permittivities can be used for nondestructive evaluation of concrete moisture content. Also, knowledge of complex permittivities of construction materials such as concrete is also important in radio propagation related applications such as cellular mobile systems and wireless local area networks [4].

In this paper, complex permittivities (ε^*) of concrete specimens were measured for dry and wet specimens using a free-space microwave measurement (FMM) setup [5,6]. Then, the moisture content is estimated by applying effective medium theory to measured complex permittivities [3]. The FMM setup consists of a pair of spot-focusing horn lens antennas, mode transitions, coaxial cables, microwave vector network analyzer (VNA) and a computer. Complex reflection coefficients of metal-backed specimens were measured using FMM setup after performing LRL (line, reflect and line) calibration. An algorithm, which finds zeros of the error function is used for calculation of complex permittivities [5].

In the FMM setup, the inaccuracies in measurements of reflection and transmission coefficients is due to 1) differaction effects at the edges of the specimen and 2) multiple reflections between horn lens antenna via the surface of the specimen. The spot-focusing antennas are used for minimizing differaction effects due to the edges of the specimen. The free-space LRL calibration technique is used along with the smoothing function of VNA for eliminating errors due to multiple reflections. For each grade of 30, 40 and 50 concrete, two concrete specimens were prepared for complex permittivity measurements. These correspond to concrete specimens prepared for specified compressive strengths of 30, 40 and 50 N/mm^2. Moisture contents for these specimens calculated using effective medium theory (from complex permittivities of dry and wet specimens) were compared with moisture content values obtained from weights of dry and wet specimens.

EXPERIMENTAL DETAILS

Free-space Measurement System [5,6]

Figure 1 shows a block diagram of the free-space microwave measurement system for complex reflection coefficient and complex transmission coefficient measurements of concrete. The measurement system consist of a pair of spot focusing horn lens (transmit and receive) antennas mounted on a large table (1.83 m × 1.83 m). The spot-focusing horn lens antenna (Model 857 from Alpha Industries, Woburn, MA, U. S. A) consists of two equal plano-convex dielectric lenses mounted back to back in a conical horn antenna. The combination of two lenses produces an electromagnetic wave at the focus. The ratio of focal distance to diameter of the lens (F/D) of these antennas is equal to 1 and D is approximately 305 mm. At 10 GHz, the 3-dB beamwidths in the H and E planes of the spot-focusing antennas are 43.7 mm and 32 mm respectively. The 10-dB beamwidths in the H and E planes of the spot-focusing antennas are 78.5 mm and 54.9 mm at 10 GHz. These values are calculated from the radiation patterns of the antenna measured at the focus by the manufacturer. The depth of focus is 8.4 wavelengths (252 mm) at 10 GHz.

Because of the spot focusing of the lens antennas used in this measurement system, the diffraction effects at the edges of the specimen are negligible if the minimum transverse dimension of the specimen is greater than three times the E-plane 3-dB beamwidth of the antenna at its focus.

A specially fabricated sample holder is mounted at the common focal plane for holding planar specimens. The transmit and received horns have been mounted on a carriage and the distance between them can be changed with an accuracy of 25.4 μm. The antennas are connected to the two ports of Wiltron 37269B Vector Network Analyzer by using coaxial cables, rectangular to circular waveguide adapters and coaxial to rectangular waveguide transitions. The network analyzer is used to make precision measurement of S-Parameters of the specimen after performing free-space LRL calibration.

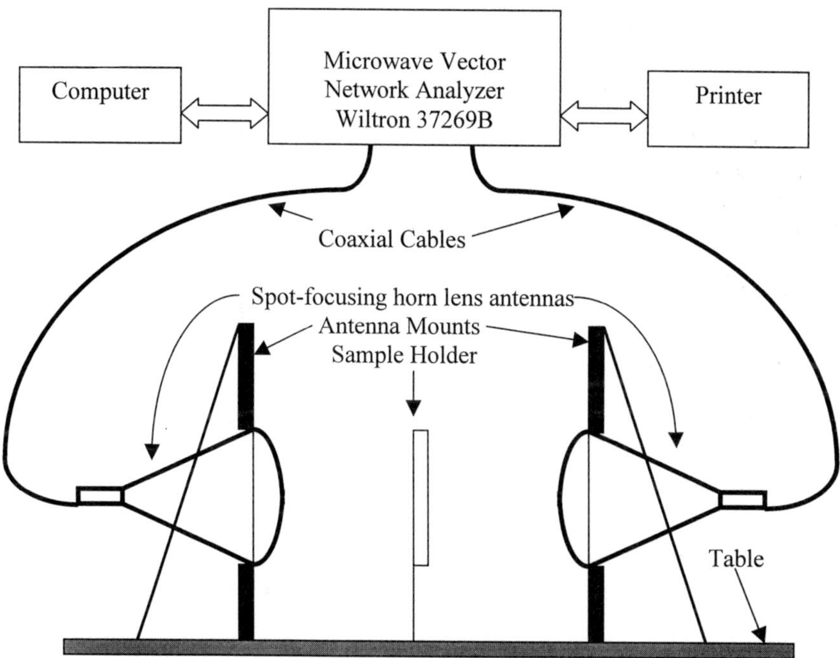

Figure 1 Free-space microwave measurement system for measuring complex permittivity

In the measurement set up shown in Figure 1, there are multiple reflections between spot-focusing horn lens antennas and mode transitions via the surface of the sample. We have modified LRL calibration kit for coaxial medium of the network analyzer for free-space measurements of reflection and transmission coefficients. Two line standards for LRL calibration are of zero length and quarter wavelength in free-space at the center of the band. These standards are realized in free-space by having a common focal plane for the first line standard and by quarter wavelength separation between focal planes of two antennas for the second line standard.

The reflect standards for port 1 and port 2 are obtained by placing a metal plate at the focal planes of transmit and receive antennas, respectively. After calibration, complex reflection coefficient (S_{11}) and complex transmission coefficient (S_{21}) of a planar sample can be measured. The reference planes corresponding to the transmit and receive antenna are located at the front and back face of the specimen, respectively.

Specimen Preparation [7]

Two specimens were prepared for each grade of concrete which were 15.24 cm × 15.24 cm in width and height, respectively. Thickness of these specimens is given in Table 1. Grades 30, 40 and 50 have the following cement (C) , sand (S), aggregate (A) and water-to-cement ratio (W/C).

Grade 30: C : S : A = 1 : 2 : 4 with W/C =0.5
Grade 40: C : S : A = 1 : 1 : 2 with W/C =0.6
Grade 50: C : S : A = 1 : 1 : 2 with W/C =0.5

Due to the small quantity of concrete constituents used for specimens to satisfy the size requirements, the mixture is hand-mixed. The mould was made of plywood and the diesel was applied on the inner surface of the mould to prevent water from seeping out. Diesel also acts as mould releasing agent. Mixture is placed in the mould and vibrated using vibration machine. Then, the concrete mixture is compressed and leveled with a trowel. Table 1 gives the thickness and weight of specimens.

Table 1 Thickness and weight (dry) of concrete specimens

SPECIMEN	GRADE OF CONCRETE	THICKNESS cm	WEIGHT IN GRAMS
30 A	30	3.05	1540
30B	30	3.00	1512
40 A	40	2.78	1394
40 B	40	2.72	1386
50 A	50	3.06	1538
50 B	50	3.00	1479

EXPERIMENTAL RESULTS

Complex reflection coefficients S_{11} were measured for metal-backed concrete specimens after calibrating the measurement system in free-space by using LRL calibration. The frequency range was 8-12.5 GHz. The measured data was modified by the smoothing function of the network analyzer to remove post-calibration errors. Because errors due to smoothing are large at the band edges, the data from 8.5-12 GHz was used for calculation of complex permittivities. Formulation for calculation of ε^* from S_{11} is given in the references 3 and 5 which involve finding zeros of the error function. Because concrete specimen is more than half-wavelength (in the specimen material), there will be multiple zeros of the error function which will result in multiple values of ε^*. To find a unique value of ε^*, it is necessary to have a good initial estimate of ε^*. In case of dry concrete specimens which have low-losses, there are multiple minimums in the plot of $|S_{11}|$ versus frequency which are used for initial estimates ε' (real part of ε^*) and ε'' (imaginary part of ε^*) as per equations 8 and 11 in reference 3. Then, we get ε^* values for dry concrete specimens by using the zero finding algorithm because of good initial guesses. For wet specimens, concrete is soaked in water for one day. Then, S_{11} is measured for the metal-backed specimens. There is an absence of distinct minimums for wet specimens because of higher losses due to the presence of water. So, a guess for ε' for wet specimens is assumed to be equal to ε' of the dry specimens. This is a good guess for wet specimens because of low values of moisture contents. Then, we get ε^* values of wet specimen from the zero finding algorithm. Table 2 gives complex permittivities for dry and wet specimens at frequencies close to $|S_{11}|$ minimum of dry specimens (for highest accuracy in measurements).

Table 2 Measured complex permittivities for dry and wet specimens

SPECIMEN	FREQUENCY IN GHZ	ε^* DRY SPECIMEN	ε^* WET SPECIMEN
30 A	9.77	5.11- j 0.35	6.03 – j 1.07
30 B	9.92	5.10 – j 0.4	6.07 – j 1.00
40 A	10.34	5.55 – j 0.4	6.40 – j0.75
40 B	10.27	5.81 – j 0.4	6.57 – j 0.78
50 A	8.85	6.30 – j 0.64	7.11 – j 0.92
50 B	9.28	5.96 – j 0.58	6.73 – j 0.69

By weighing dry and wet specimens, the moisture content (MC) can be determined by the weighing method. Following formula is used.

MC = (weight of wet specimen - weight of dry specimen) / weight of dry specimen

MC values computed using weighing method are given in table 3. In the dielectric method, moisture content is calculated from measured dielectric constants of dry and wet specimens by using the following formula [3].

$$MC\left[\frac{\varepsilon_{wa}-\varepsilon_{w}}{\varepsilon_{wa}+2\,\varepsilon_{w}}\right]+(1-MC)\left[\frac{\varepsilon_{d}-\varepsilon_{w}}{\varepsilon_{d}+2\,\varepsilon_{w}}\right]=0$$

ε_d and ε_w are the measured dielectric constants (real part of ε^*) of dry and wet concrete specimen. ε_{wa} is the dielectric constant of water at midband (10 GHz) which is equal to 55.0 [8]. Table 4 compares MC values obtained using dielectric method with the weighing method.

DISCUSSION AND CONCLUSIONS

From table 2, it is observed that the real as well as imaginary part of complex permittivity is significantly higher for wet specimens as compared with dry specimens because of the presence of water in wet specimens. Also, in Table 3, the moisture content values calculated using dielectric method (employing effective medium theory) are close to the actual values of moisture content obtained by the weighing method.

Table 3 Moisture content values of concrete specimens by weighing and dielectric method

SPECIMENS	MC (%) WEIGHING METHOD	MC (%) DIELECTRIC METHOD
30 A	2.47	6.77
30 B	2.91	7.17
40 A	3.3	6.06
40 B	3.4	5.34
50 A	4.49	5.40
50 B	4.46	5.32

Discrepancies in MC values in the two methods is due to inaccuracies in dielectric measurements, inhomogeneity of concrete specimens and inadequacies of effective medium theory [1]. Also, there is an increase in real as well as imaginary part of complex permittivity as the specified compressive strength of concrete specimens increases from 30 N/mm^2 for grade 30 to 50 N/mm^2 for grade 50.

REFERENCES

1. HALABE, U B, SOTOODEHNIA, A, MASER, K R AND KAUSEL, E A. Modeling the electromagnetic properties of concrete. ACI Materials Journal Vol. 90, November-December 1993, pp 552-563.

2. ACI COMMITTEE 437. Strength evaluation of existing concrete buildings. American Concrete Institute, Detroit, ACI 437R-91, pp. 437R-1 to 437R-24.

3. AL-QADI, I L, GHODGAONKAR, D K, VARADAN, V K AND VARADAN, V V. Effect of moisture on asphalt concrete at microwave frequencies. IEEE Transactions on Geoscience and Remote sensing, Vol. 29, 1991, pp. 710-717.

4. SATO, S, MANABE, T, IHARA, T, KASASHIMA, Y AND KATSUNORI, K. Measurement of the complex refractive index of concrete at 57.5 GHz. IEEE Transactions on Antennas and Propagation, Vol. 44, 1996, pp 35-40.

5. GHODGAONKAR, D K, VARADAN, V V and VARADAN V K. A free-space method for measurement of dielectric constants and loss tangents at microwave frequencies. IEEE Transactions on Instrumentation and Measurements, Vol. 37, 1989, pp 780-793.

6. GHODGAONKAR, D K, VARADAN, V V and VARADAN V K. Free-space measurement of complex permittivity and complex permittivity of magnetic materials at microwave frequencies. IEEE Transactions on Instrumentation and Measurements, Vol. 39, 1990, pp 387-394.

7. MAZNI BINTI MUNIR. Microwave non-destructive testing of cement concrete using free-space techniques. B. Eng. (Hons) Electrical Engineering Thesis, MARA Institute of Technology, Shah Alam, Malaysia, 1997.

8. VON HIPPEL, A. Dielectric Materials and Applications. Artech House, p. 361, 1954.

MONITORING FRACTURE PROCESS IN CONCRETE STRUCTURES USING ACOUSTIC EMISSION ANALYSIS

M Shigeishi

M Ohtsu

Kumamoto University

Japan

M C Forde

University of Edinburgh

United Kingdom

ABSTRACT. The phenomenon of acoustic emission (AE) is the propagation of elastic waves generated from a source, known as a micro-crack in an elastic material. The research reported herein is for the application of AE techniques to damage assessment of reinforced concrete (RC) structures. Repeated-static or cyclic-dynamic bending tests of RC beams were carried out and AE signals were detected during the tests. In the repeated-static bending tests of RC beams, the difference between AE activity in the damaged beam and the undamaged beam was by detected AE accumulation. Furthermore, the results of AE moment tensor analysis showed the fracture process of the RC beam in cyclic bending.

Keywords: Acoustic emission (AE), Beam, Bending, Corrosion, Crack, Fatigue, Fracture, Moment tensor, Reinforced concrete (RC)

Dr Mitsuhiro Shigeishi is an Associate Professor in the Department of Civil Engineering and Architecture, Kumamoto University, Japan. His main research interest is the application of acoustic emission techniques to the construction field. Dr Shigeishi has developed the computer software for AE source inversion named SiGMA under Professor Ohtsu. He is continuing the study of non-destructive testing for infrastructure using acoustic methods and other techniques.

Professor Masayasu Ohtsu is Head of the Composite Construction Material Group in the Department of Civil Engineering and Architecture, Kumamoto University, Japan. He specialises in the elastic behaviour and the fracture mechanics of rock and concrete materials as well as the non-destructive investigation of concrete infrastructure. He is a chairman of the Structural Engineering Committee for Damage Assessment of Structures of the Japan Society of Civil Engineers.

Professor Michael C Forde is Head of the Institute for Research in Engineering and holds the Tarmac Chair of Civil Engineering Construction, University of Edinburgh, Scotland, UK. He specialises in the investigation of existing structures using radar and other non-destructive testing techniques. Professor Forde has published widely, chairs the Institution of Electrical Engineers PG S6 Committee on non-destructive testing and serves on many international technical committees. He is the Editor-in-Chief of the Journal: Construction and Building Materials, published by Elsevier Science.

INTRODUCTION

The phenomenon of acoustic emission (AE) is the propagation of elastic waves generated from a source, which is a micro motion caused by rapid dislocation of mass in an elastic material. The most typical AE source is the release of internal energy due to propagation of an existing crack. In addition, contact of the faces and friction at the existing crack surfaces are regarded as sources of AE. Thus, AE behaviour, activity and waveforms contain data on source mechanisms and the conditions of the medium during propagation. Accordingly, it is expected that the existence and degree of damage can be evaluated using AE measurement and analysis.

The AE technique enables real-time and comprehensive monitoring of structures without interfering with their service. In addition, the behaviour of structures can be observed under dynamic loading. In view of these advantages, the AE technique has been actively developed to diagnose condition and behaviour in the field of civil engineering [1].

In this study, AE activity in reinforced concrete (RC) beams, some undamaged and some damaged by electrolytic corrosion, were subjected to repeated-static bending and monitored. The experimental application of moment tensor analysis was attempted on these real AE waveforms to identify and classify the micro cracks occurring in the cyclic-dynamic bending tests of the RC beams. The result showed the onset of damage with the progress of fatigue quite clearly.

RC BEAM BENDING TEST DETAILS

RC Beam Specimens

The configuration of test specimens is shown in Figure 1. These were designed to bear 44.1 kN static bending in the tests. The specimen had a square section of 100 mm and a length of 400 mm. Type A contained a single deformed steel bar, type D10, which had a diameter of 10 mm approximately. A normal steel bar of 10 mm diameter was embedded in the Type B specimen. Each bar was arranged at a covering depth of 20 mm. Air entrained concrete was used and the compressive strength of concrete aged 28 days was 31.4 MPa.

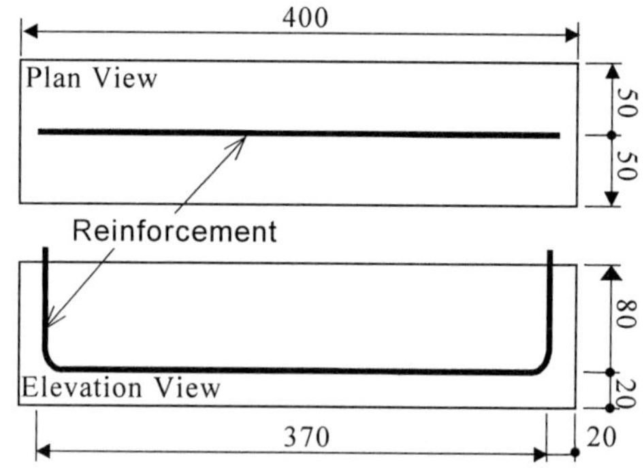

Figure 1 Configuration of RC beam specimen

A damaged RC beam specimen (Type A') was created using a galvanic corrosion by introducing a 30 mV direct current for 6 days. This electrolytic corrosion of the reinforcement in the Type A' beam caused a crack at the center of the bottom surface in parallel with the reinforcement direction.

Bending Test and AE Measurement

The experimental set-up of the four-pointed bending test is shown in Figure 2. For comparison the AE activities between the non-damaged (Type A) and damaged (Type A'), loading and unloading was statically repeated five times whilst increasing the peak load to 10, 20, 30, 40 kN and then yield. For the observation of the fatigue process under bending, cyclic-dynamic loading from 0.98 kN to 22.05 kN was automatically applied one hundred times to another Type A and Type B beam. The loading speed of the ram was set to 10 mm/s.

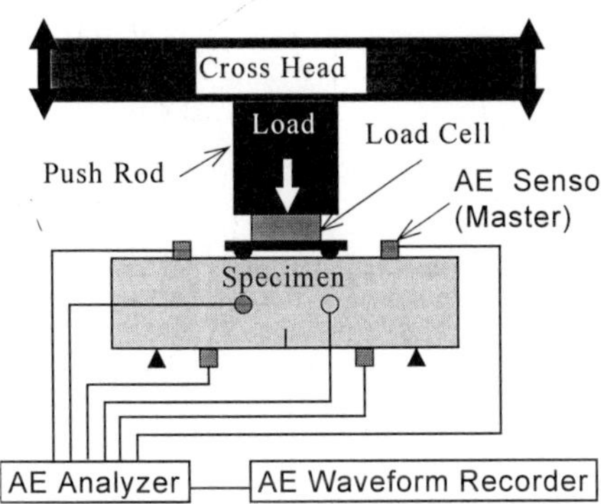

Figure 2 Experimental set-up of bending test

During the test, AE waveforms, which were detected at the six points, were continuously recorded by a six-channel AE measurement system. When a signal arrived to the master sensor, the system triggered all channels to start recording the AE signals. The signals were digitised into 2048 words length at 2 MHz sampling rate and recorded on the hard disk.

AE BEHAVIOUR OF RC BEAM UNDER BENDING

Figure 3 shows the relationship between cumulative number of AE events and bending load in a non-damaged RC beam (Type A), together with the load-displacement curve. Figure 4 like Figure 3 shows the AE and displacement behaviour under bending in a damaged RC beam (Type A'), which has corroded.

There is no significant difference in yield load and bending stiffness except for early tangent modulus until second time loading between Type A and Type A'. The difference in displacement between these two emerged in early AE behaviour. In the case of Type A, even if first loading was applied up to 10 kN and removed, the AE event was hardly recognized. The occurrence of the AE was noted over 10 kN load at the second bending. Otherwise, in the case of Type A', AE events increased as the load increased.

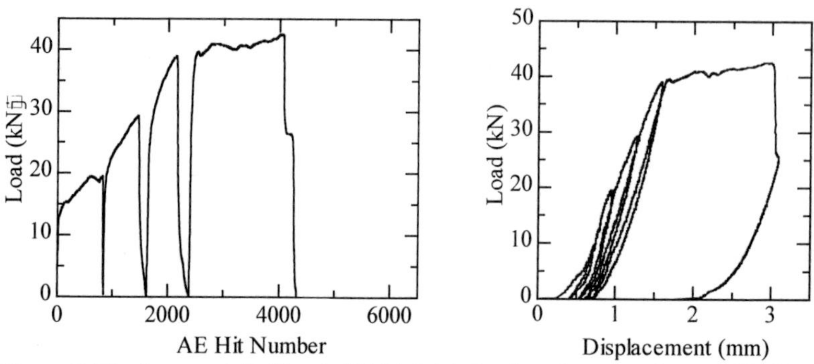

Figure 3 Displacement (left) and AE (right) behaviour in a non-damaged RC beam

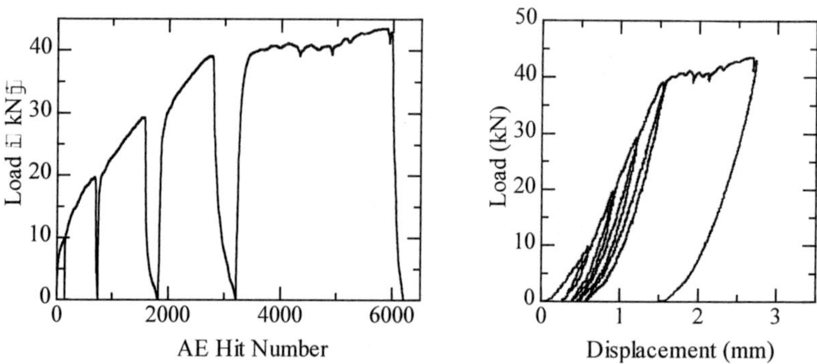

Figure 4 Displacement (left) and AE (right) behaviour in a damaged RC beam

FATIGUE PROCESS OBSERVATION USING AE SOURCE INVERSION

The AE wave is emitted in a variety of radiation patterns [2] depending upon micro cracking at its location. To identify the AE source kinematics [3], that is location and crack-type, an AE inversion method of quantitative AE waveform analysis was proposed [4]. Figure 5 shows the implications of AE source locations and crack types in a concrete beam reinforced by a single deformed steel bar (Type A) under cyclic loading. Figure 6 shows the results in a RC beam reinforced by a normal bar (Type B). In these figures, the arrow symbols (↔) indicate the micro-cracks resulting from tensile motion and its tensile directions. Otherwise, the cross symbols (×) indicate the micro shear cracking. Propagation of the visible cracks on the surface are also showed by the dotted lines in these figures.

In the case of Type A, micro-cracks are generated by compressive stress at the upper part at the 15th loading. Some micro-cracks created near the reinforcement indicate either the loss of bond or the slipping off. Most micro-cracks are generated along the visible cracks after the 15th loading. In the case of Type B, before the 16th loading, the propagation of the visible cracks is hardly observed. However, some of micro-cracks are scattered across a wide area from the visible cracks. This means that the fracture process zone is spreading out gradually. At the final stage, the area, where the micro-cracks are generated, spreads out further. Thus, it was found that the shape of the deformed bar contributed significantly in preventing damage.

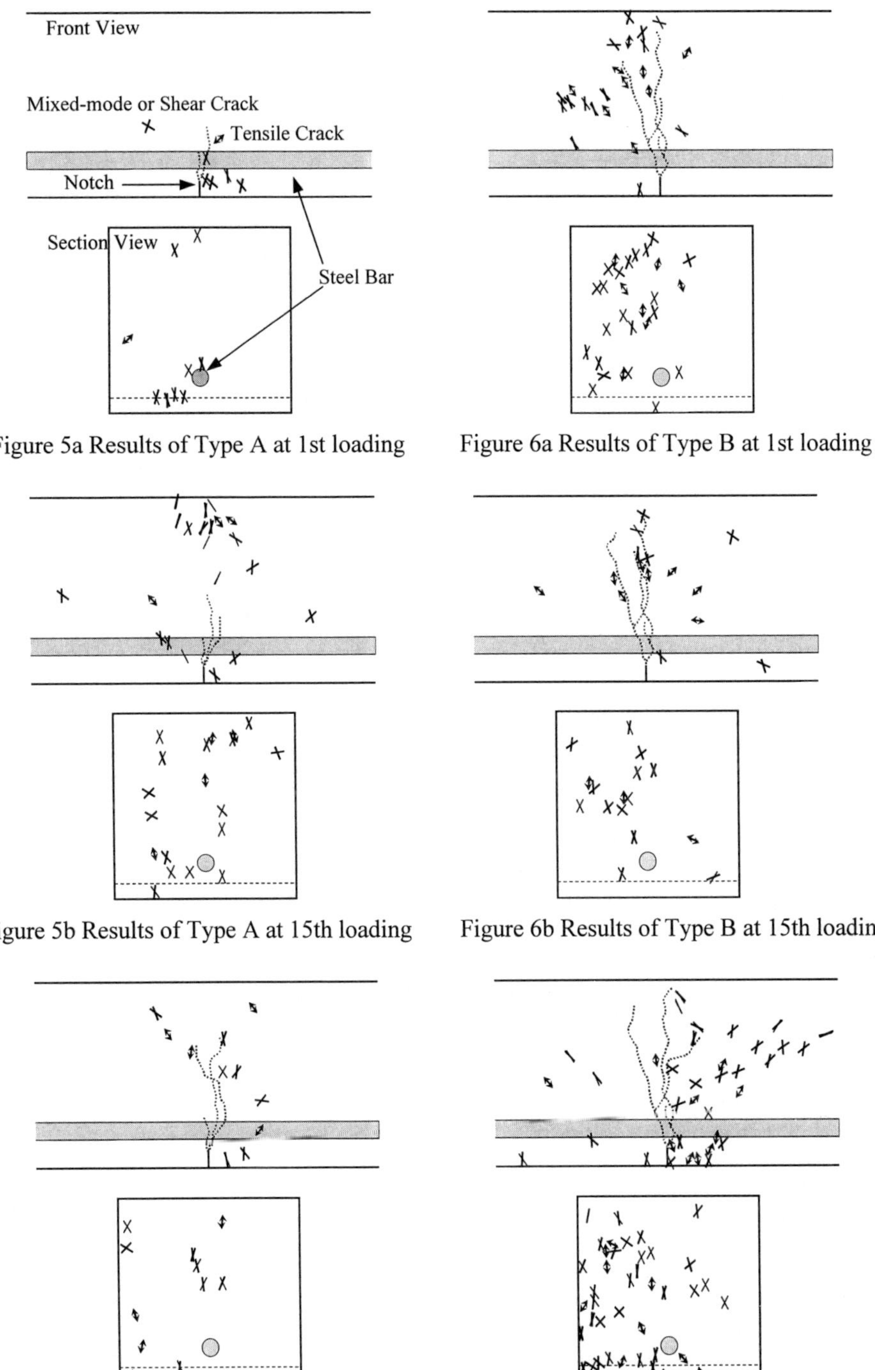

Figure 5a Results of Type A at 1st loading

Figure 6a Results of Type B at 1st loading

Figure 5b Results of Type A at 15th loading

Figure 6b Results of Type B at 15th loading

Figure 5c Results of Type A at 100th loading

Figure 6c Results of Type B at 100th loading

From the above results, the difference in the damage processes resulting from the shape of reinforcement is clearly observed. In Beam type A, the visible cracks propagate only a little and a few micro-cracks are produced at the tip of the notch. In contrast, the visible cracks in Type B grow and almost reach the top of the specimen. The micro-cracks have a wide distribution around the visible cracks. Thus it was confirmed that Type B was already severely damaged at this early stage.

CONCLUSIONS

AE measurement experiments and analysis were performed under controlled laboratory conditions as part of the development of AE technique applied to damage assessment of existing concrete structures.

AE activity in undamaged and electrolytic corrosion damaged RC beams was observed under repeated bending. Different AE behaviour was noted depending upon the cumulative repeated load mechanism to failure. Different AE behaviour was observed during the early loading of damaged and undamaged RC beams.

In order to monitor the fatigue process, moment tensor analysis was attempted on experimental AE waveforms in the RC beams under the cyclic-dynamic bending. The result clearly showed the clustering and distribution of micro cracks depending upon the progress of fatigue. The influence of deformed bar reinforcement in the fatigue process was confirmed using AE.

ACKNOWLEDGEMENTS

The authors would like to express their appreciation for the research grants made by The WESCO Foundation of Japan to carry out the reported work, which forms part of the major research programme on the application to diagnosis of existing civil structures. We thank Mr. Nobuyuki Tsuji, Mr. Daisuke Yasuoka and Mr. Koji Suetake for the various analyses whilst carrying out their graduate research. In addition, the research guidance made by Mr. Daisuke Yukawa and Mr. Yuichi Tomoda contributed greatly to the progress of this research.

REFERENCES

1. OHTSU, M. The History and Development of Acoustic Emission in Concrete Engineering, *Concrete Library of the Japan Society for Civil Engineers*, Vol 25, 1995.

2. OHTSU, M. and ONO, K. A Generalized Theory of Acoustic Emission and Green's Function in a Half Space. *J. of AE*, Eds. K Ono, AEWG, Vol 3, No 1, 1984, pp 124-133.

3. OHTSU, M. Determination of Crack Orientation by Acoustic Emission. Material Evaluation, *ASNT*, Vol 45, No 9, 1987, pp 1070-1075.

4. OHTSU, M and SHIGEISHI, M. Determination of Crack Location, Type and Orientation in Concrete Structures by Acoustic Emission, *Magazine of Concrete Research*, American Concrete Institute, Vol 43, No 155, 1991, pp 127-134.

PERFORMANCE OF EPOXY REPAIRED AND PLATE BONDED SHEAR WALL COUPLING BEAM JOINT

A I Abu-Tair

A Nadjai

M Gross

W Cousins

University of Ulster

United Kingdom

ABSTRACT.. The performance of different designs and constructions affected by earthquakes has been dramatically illustrated in a number of major earthquakes in recent years. Typical concrete frame buildings performed very poorly, with many buildings collapsing. Buildings built post-1981, performed better than older ones. Some were extensively damaged, but most had only slight damage. The buildings that fared best, and those without significant damage, had extensive concrete shear walls. More often than not, shear walls are pierced by openings. Because of this, shear walls are coupled by beams, which transmit shear forces from one wall to another. This paper is concerned with the behaviour of the joints between such beams and shear walls under static loading. A test programme was conducted to evaluate the performance of joints, before and after repair. Two techniques were used, one where only resin injection was used and the other where resin injection was used in combination with plate bonding. The test programme demonstrated the ability of such repair technique to restore all the strength of such joints and also most of the original stiffness.

Keywords: Coupled shear walls, Seismic damage, Resin injection, Plate bonding, Static loading.

Dr A I Abu-Tair is a lecturer of Structural Engineering at the University of Ulster, He is director of the Sustainable Materials Research Group. He is a member of the CIB W8. His main research interests include the durability, repair and strengthening of concrete structures.

Dr A Nadjai is a lecturer of Structural Engineering. His main research interests are, computer modelling and simulation of structural problems using fine elements. He is a joint editor of the 7th International Conference on Civil and Structural Engeering Computing, Oxford 1999.

Mr M Gross is a graduate Civil Engineer of University of Ulster, now working as a design engineer in Germany.

Dr W Cousins is a lecturer of Material Science and Structures at the University of Ulster, his research interests centre on concrete technology, he is a former committee member of the NI branch of the Concrete Society.

INTRODUCTION

Multi-story buildings can be subjected to severe lateral forces caused by wind and seismic effects. The results of such lateral forces are illustrated in Figure 1. The performance of different designs and constructions affected by earthquakes has been dramatically illustrated (e.g. in Kobe; Japan 1995, Los Angeles; USA 1994, etc.). Typical concrete frame buildings performed very poorly, with many buildings collapsing (see Figure 2). Buildings built post-1981, performed better than older ones, some were extensively damaged, but most had only slight damage. The buildings that fared best, and those without significant damage, had extensive concrete shear walls [1] (see Figure 3).

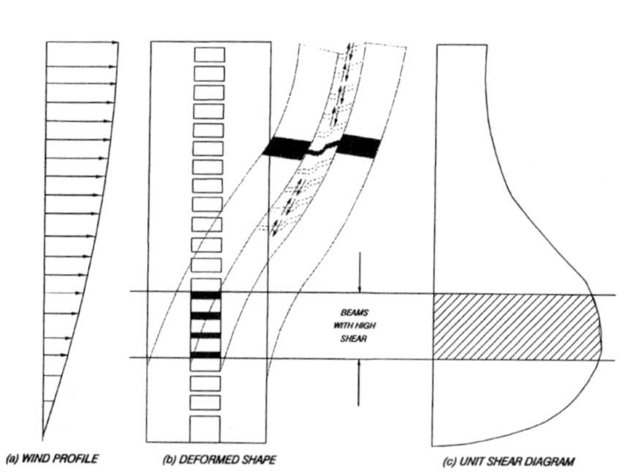

Figure 1 Behavior of coupled shear walls

Figure 2 Collapsed concrete building in Kobe

Figure 3 Modern parking garage in central Kobe, the building was undamaged

More often than not, shear walls are pierced by large openings. Because of this, shear walls are coupled by beams, which transmit shear forces from one wall to another. This paper is concerned with the behaviour of the joints between such beams and shear walls under static loading. The paper includes a review of building behaviour under lateral forces, the different methods of designing the coupling beams and common repair methods. A test programme was conducted to evaluate the performance of joints, before and after repair. For reparation of the damage two techniques were used, one where only resin injection was used and the other, where resin injection was used in combination with plate bonding.

The main aims of the research programme were two fold. Firstly, to develop a better understanding of the behaviour of connecting beams with coupled shear walls and secondly, to examine the use of resin injection techniques and plate bonding as means of effectively repairing damaged joints in coupled shear walls.

BACKGROUND TO THE RESEARCH

Strength and behaviour of CSW beams

Observations of earthquake damage have repeatedly indicated the failure by diagonal tension of coupling beams containing insufficient web reinforcement. Clearly such failures which result in a high rate of strength degradation under cyclic loading, must be suppressed if satisfactory seismic resistance is to be provided. Irrespective of the design loads, the shear strength of a coupling beam must be equal to or larger than its flexural capacity. This requirement may impose an upper limit on the flexural steel content in such beams. The action of the shear force acting through the point of contraflexure produces maximum bending moments at the end supports, with the development of flexural cracks. When the applied shear force is increased, the flexural cracks progress towards the compression corners. Eventually, crushing will take place in the compression corners, marking the ultimate failure of the beam [2].

The pure shear deformation requires both top and bottom surfaces of the beam, all along its length, to be in tension. There is compression and tension along the diagonals. An element of the beam near the midspan is subjected to a biaxial compression-tension state of stress. When the tensile stress in the concrete along the diagonal reaches the limiting tensile strength of concrete, it cracks in a diagonal splitting mode [2,3].

The actual deformation of the coupling beams is a combination of the flexural and shear deformations. In beams where flexure governs, the overall deformation is still accurately represented by the flexural type deformation.

In beams where shear governs, the overall deformation of the beam is much more complex. The flexural deformation, which causes the beam to bend in double curvature, conflicts with the shear deformation. In the first the double curvature results in changes to the stress from tension to compression on the same face of the beam, with opposite stresses on the other, in the second, the beam goes into tension on both surfaces along the whole length.

Repair and Strengthening of Concrete Structures

Reinforced concrete structures are capable of achieving their design lives, which can be 120 years or more, provided that they are designed, detailed, constructed and maintained correctly. Deterioration can be caused by a number of reasons, including corrosion of the reinforcement, fire damage, chemical attack, impact. Structural strengthening may also be required due to changes in loading, use or regulations.

Each deterioration condition requires a clear understanding of what is expected of the repair. Three general performance requirements are protection,, appearance, and load carrying. The process of repair design and specification consists of determining the exact function of the repair so that the correct repair materials can be specified [4,5,6].

Various materials and methods are available for strengthening purposes. One way to repair or strengthen concrete structures is by adding reinforcement either surface embedded or in drilled holes. Another kind of external reinforcement is plate bonding. Therefore different materials like steel, fibre reinforced plastics, and carbon fibre reinforced polymers can be used [7,8,9]. For repair of small cracks, crack width between 0.02mm and 6mm, epoxy resin systems are generally used [10,11].

EXPERIMENTAL PROGRAM

Specimens

The test specimens were designed as part of a coupled shear wall, which was connected to a beam with the latter being cut at midspan. The specimens are part of a shear wall, which has dimensions of 400 mm in width, 300 mm in depth, and 950 mm in height. The wall section of all six specimens was reinforced with 16 mm diameter high tensile steel bars in the longitudinal direction, and with 8mm diameter links. The concrete was designed in accordance to DoE "Design of Normal Concrete Mixes" (1975) for $f_{cu} = 50$ N/mm^2.

Two types of beams were used. In the first three specimens the beams were made of reinforced concrete, the other three were I-section steel beams. Figure 4 shows the dimensions and different types of specimens.

The reinforced concrete beams had a cross-section of 250 mm deep, 100 mm wide and 800 mm long. Load was applied to the beam at a distance from the wall of 700 mm. To protect the end of the beam against failure where the load was applied, the dimensions were increased to 350 mm deep, 200 mm wide and 200 mm length.

The beams were reinforced with six No. 16 mm high tensile bars in the longitudinal direction, and with 6mm links every 120 mm c/c. A summary of material properties can be found in Table 1.

Table 1 Summary of material properties (design values)

	COMPRESSIVE STRENGTH N/mm^2	TENSILE STRENGTH N/mm^2	MODULUS OF ELASTICITY kN/mm^2	VISCOSITY MP
Concrete	50	3.5	35	
16 mm bar		460	210	
8 mm bar		460	210	
6 mm bar		250	210	
3 mm bar		250	210	
Epoxy resin	> 80	> 35	2.7	95

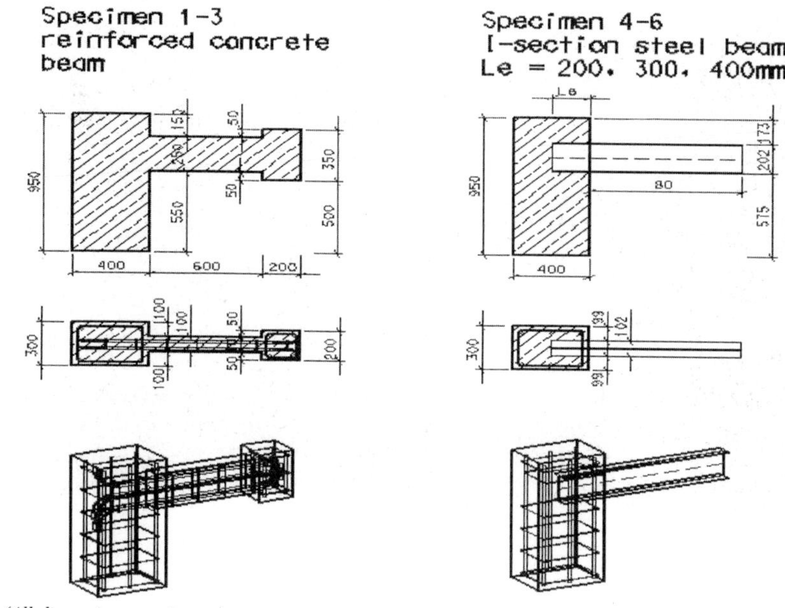

Figure 4 Test specimens

The second type of specimen had steel I-beam sections only, measuring 102 mm x 202 mm. For this kind of beam the embedded lengths had different arrangements. This was to show the effect of the embedded length on the failure mode and also on the effectiveness of the repair. The embedded lengths were 200 mm, 300 mm, and 400 mm.

Test Set-up and Instrumentation

The test specimen was anchored to the structural floor by means of threaded anchors. In addition to this, in the bottom of each specimen two screws were anchored to provide fixed end restraints. A support was provided to minimise the horizontal movement against the wall section of the specimen. The quasi-static load was applied by using a hydraulic jack with 700 mm distance from the wall face. The pressure was controlled and monitored using a 200 kN load cell which was placed on top of the hydraulic jack.

To monitor the movement and deflection a dial gauge was installed on the top surface of the enhanced section, at a distance of 700 mm from the face of the wall. Additional to the readings taken manually, instrumentation was used to plot load versus deflection during the test. The test set-up is shown in Figure 5.

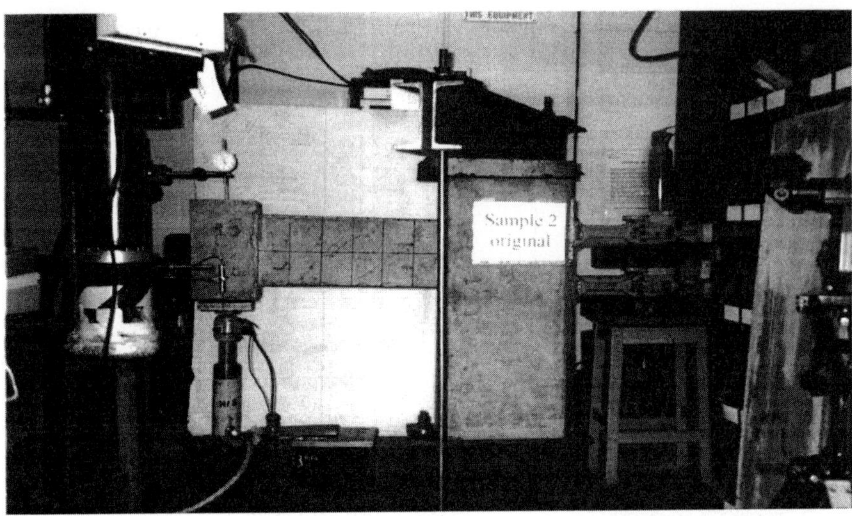

Figure 5 Test set-up

Repair to the Damaged Specimens

After the first test series on the original specimens, the specimens were prepared for the repair stage. The main cracks were measured, using a crack microscope. Specimens 1, 2, and 3 had cracks measuring 0.1mm, and for Specimens 5 and 6, the cracks measured 1.0 mm. Specimen 4, with an embedded length of 400 mm failed on the base and could not be used for any further testing. The resin injection method was used with Specimen 1, Specimen 3 was repaired using the plate bonding system and Specimen 2, 5 and 6 were repaired using resin injection and plate bonding.

Resin injection

Concretin® IHS 93, an epoxy resin system was used as repair material. The Concretin® IHS 93 is a two component epoxy resin with very low-viscosity of around 95 MP, with a compressive strength of 80 N/mm^2, flexural strength of 45 N/mm^2 and a tensile strength of 35 N/mm^2 after two days curing. The bond strength is usually so high that any failure will occur in the concrete rather than in the repair material.

To inject the epoxy resin in the crack a compressor with a pressure regulator was used. This was necessary to control the injection pressure, which is limited by the seal to the crack. To seal the crack at the surface a Polyurethane material, Concretin® PUH, was used. The material cures to a hard but tough-elastic material and was ready after one day for injection pressure up to 60 bar. Before the cracks were sealed it was necessary to clean the surface area around the cracks to provide adequate bonding. Compressed air was forced into the cracks to remove any remaining loose particles

Specimens 5 and 6 were injected with a pressure of up to 10 bar, this was not possible for Specimens 1 and 2 where the crack width was too small, at around 0. 1 mm. For the latter Specimens the cone system combined with a small pressure was tried, because at higher pressures the cone kept separating from the injection point rendering the injection impossible. Even at the lower pressure, it was not possible to finish the injection as the viscosity of the resin decreased significantly due to the length of time involved.

Plate bonding

To strengthen the damaged specimens plate bonding was used as a second repair method. After failing to resin inject Specimens 1, 2, and 3 it was decided to repair them all with external reinforcement bonded to the surface by means of epoxy resin. The crack pattern after the first test showed two different types of failure. Firstly shear failure in the beam in Specimens 1, 2, and 3 and secondly tension cracks in the wall section in all specimens. The plates used for external reinforcement were glued to the concrete surface by means of an epoxy resin system Concretin® IHS 93. The resin was applied on the concrete surface using a brush. The cleaned mild steel plates were then placed in the resin.

To strengthen the beam against shear failure 3 mm thick mild steel strips were used. The dimensions were 250 mm in length and 19 mm in width. The strips were placed in line with the links and were located using a cover meter. All five specimens were strengthened in the wall section with 3 mm mild steel plates against tension failure. The dimensions of these plates were 600 mm in length and 75 mm in width. The plates were placed on each side of the specimens, 20 mm from the front edge.

EXPERIMENTAL RESULTS

Specimen 1 was tested first, during this test it became clear that the horizontal support provided, was not adequate to prevent large horizontal movement. Therefore, a more stable horizontal support was used for the other specimens. This change brought a decrease in the vertical deflection at 62 kN for Specimens 2 and 3 of 7.9 mm and 7.6 mm respectively, compared to the maximum deflection of Specimen 1. Therefore, the test results of repaired Specimen 1 were compared with the original test results of Specimen 3, this can be seen in Figure 6. The behaviour of Specimens 2 and 3 was nearly identical.

The test results achieved are shown graphically in Figures 6 to 9 and all the results of the test are summarised in Table 2 and 3. The graphs plotted during testing were used to monitor the behaviour of the specimens and to stop the test before serious damage occurred, making repair impossible. Typical crack patterns caused are shown in Figures 10 and 11. After the repair was carried out, Specimens 1, 2 and 3 performed similarly in the after repair test. The cracks in the beams during the original test re-opened at loadings of 35 kN, 40 kN, and 30 kN for Specimens 1, 2, and 3, respectively When the load was increased further, these cracks developed only between the mild steel strips. Furthermore, the shear crack sizes were less than in the original test. This showed that in all specimens, the external reinforcement provided was adequate to strengthen the beams against shear failure. It was observed that no crack crossed the external reinforcement and no detachment of the strips occurred.

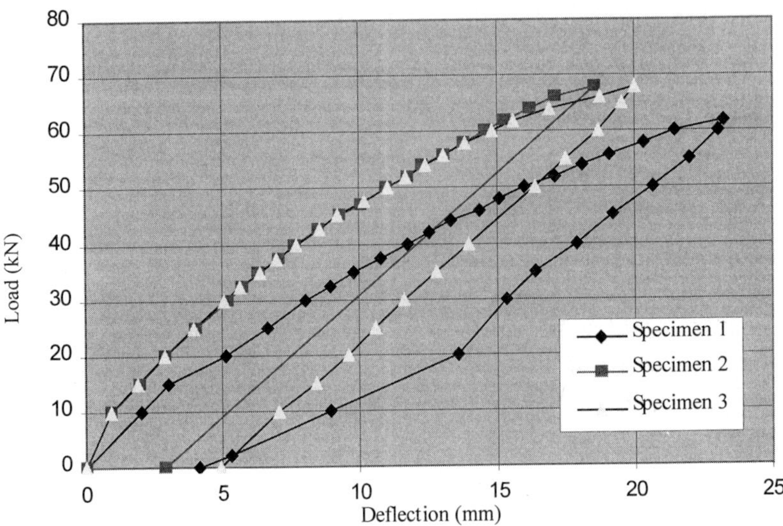

Figure 6 Comparison of original test results for specimens 1,2 and 3

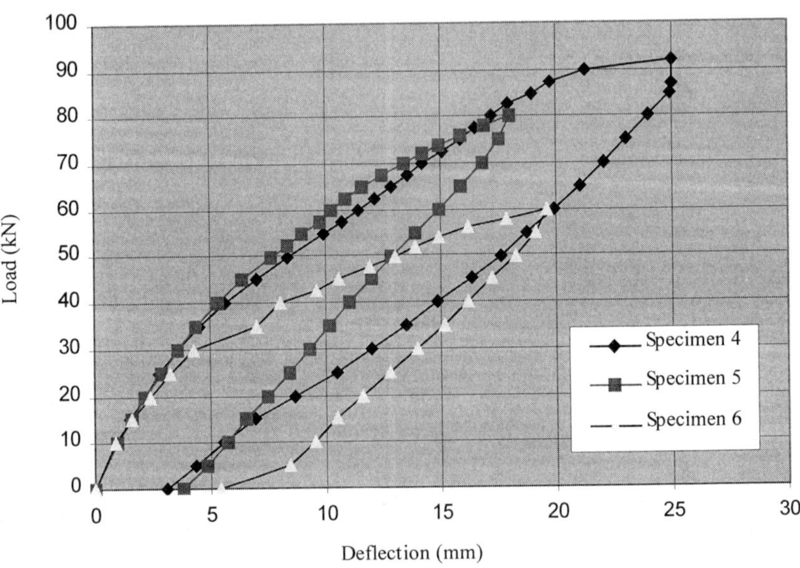

Figure 7 Comparison of original test results for specimens 4,5 and 6

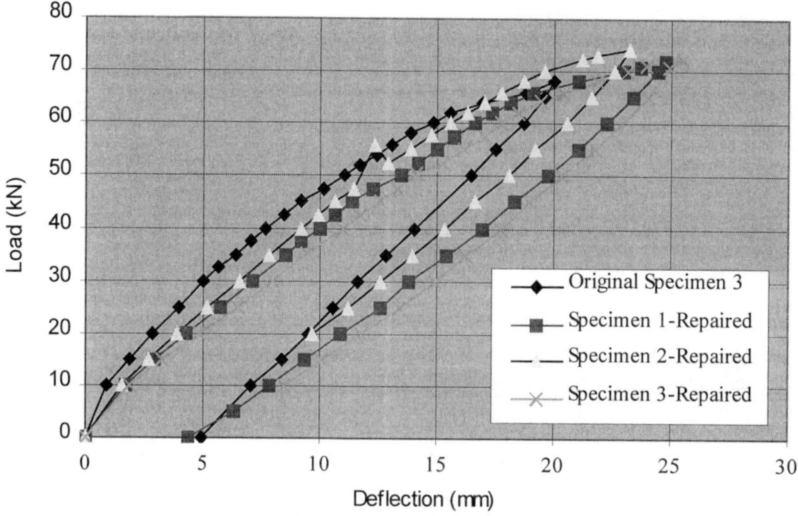

Figure 8 Comparison of test results after repair of specimens 1,2 and 3 with the original results for specimen 3

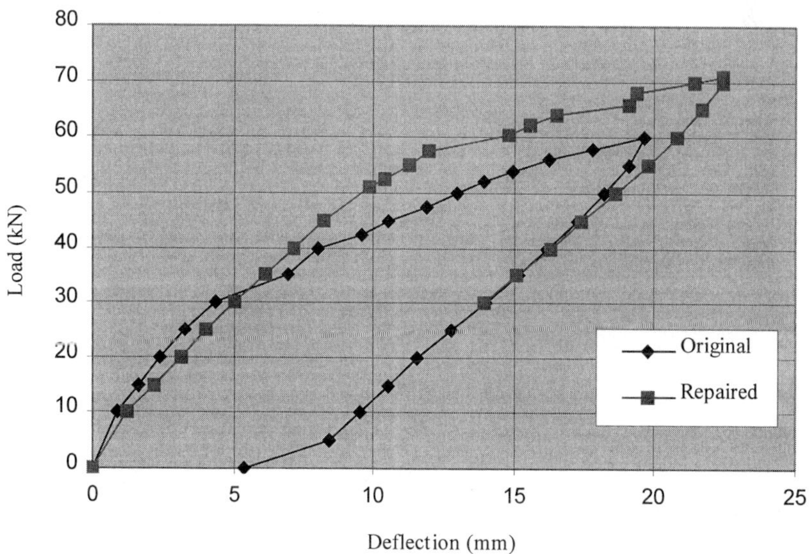

Figure 9 Comparison of the test results for specimen 6

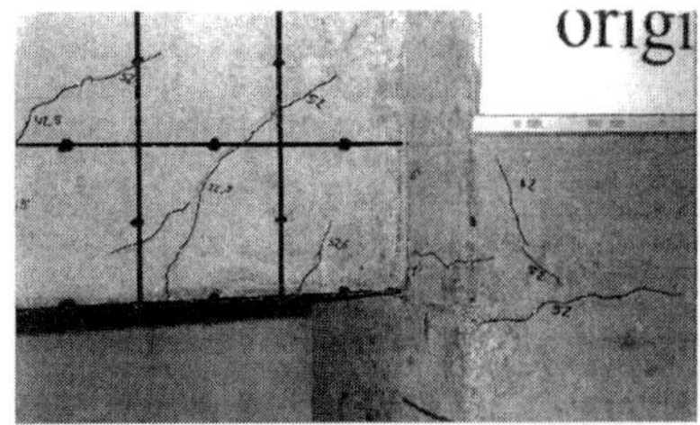

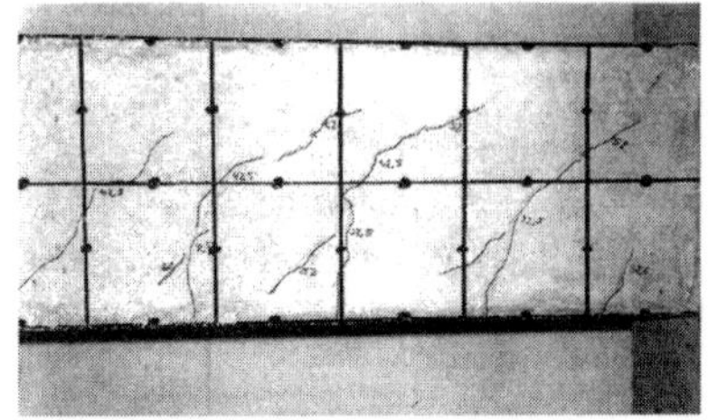

Figure 10 Crack Patterns for Specimen 3

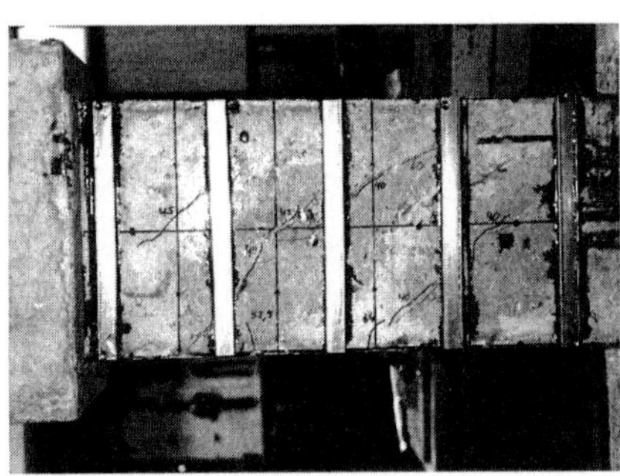

Figure 11 Crack patterns for specimen 1 after re-testing

Specimen 5, with an embedded length of 300 mm, showed after repair a significant loss in its stiffness. With Specimens 1, 2, and 3 the small loss of stiffness could be explained by the previous cracks which were not injected with epoxy resin. However, these re-opened again quite easily. The reason for this behaviour could not be ascertained. During the re-testing it was observed on Specimens 5 and 6 that the crack patterns were similar to the original test but the cracks injected with epoxy resin have not opened again. Furthermore, the important influence of the embedded length to the stiffness and load capacity of the beam wall structure was very clear.

DISCUSSION

The stiffness of the beam wall system was determined from initial slopes of the specimen's load deflection curves. The main test results and the calculated initial and final stiffnesses are presented in Table 2. In the original test of Specimen 1, a different horizontal support was used, the initial stiffness was more than 50 % lower than the other specimens, the movement of the support caused this. Specimens 2, 3, 4, and 5 had similar initial stiffness. Specimen 6, with the shortest embedded length of 200 mm, had an initial stiffness of 11.24 kN/mm which was about 9 % higher than Specimens 4 and 5.

Table 2 Summary of test results and properties

TEST PARAMETER	SPECIMEN NUMBER.					
	1	2	3	4	5	6
Beam material and section	Concrete 100mm widex250mm deep			Steel I Section 102mmx202mm		
a) Original test						
Maximum load (kN)	62	68	68	92	80	60
Maximum deflection (mm)	23.2	18.6	20.0	25.0	18.0	19.6
Final stiffness (kN/mm)	2.67	3.66	3.40	3.68	4.44	3.06
Deflection at 10 kN (mm)	2.05	0.94	0.91	0.97	0.97	0.89
Initial stiffness (kN/mm)	4.88	10.64	10.99	10.31	10.31	11.24
Crack width-unloaded (mm)	0.1	0.1	0.1	1.0	1.0	1.0
Load to first crack (kN)	30	30	32.5	55	50	40
Failure mode		Shear in beam		Base failure	Tension in wall	
b) Repair method						
Resin injection					X	
Plate bonding-tension plate	X	X	X		X	X
Plate bonding-shear strips	X	X	X			X
c) Repaired specimens						
Maximum load (kN)	72	74	72		54	71
Maximum deflection (mm)	24.7	23.3	25.5		20.6	22.4
Final stiffness (kN/mm)	2.91	3.19	2.84		2.62	3.17
Deflection at 10 kN (mm)	1.70	1.55	1.82		01.79	1.25
Initial stiffness (kN/mm)	5.88	6.45	5.49		5.59	8.00
Load to first crack (kN)	35	40	30		45	55
Failure mode	Tension in wall	Beam-wall joint (BWJ)			Base	BWJ

The final load reached was similar to the design ultimate strength of the coupling beam wall system. The principle failure in Specimens 1, 2, and 3 could not be clearly identified. The theoretical ultimate shear capacity was estimated to be 60 kN. It was noted that the repair of the specimens strengthened the specimens. The mild steel strips used in the repair, did not increase the stiffness but they reduced, quite remarkably, the crack width and increased the shear capacity of the beam.

The steel plates provided on the wall section increased the tension capacity as well as the stiffness. The latter can be seen for repaired Specimen 1 shown in Figure 8. The slope of the load-deflection graph decreased, when the first plate detached at a load of 50 kN and at 66 kN when the second plate pulled off. The ultimate load increased by 6 % to 72 kN. The bonded steel plates on Specimens 2 and 3 performed better, the failure occurred in the beam wall joint and the ultimate load was also increased by 6 %.

The epoxy resin injection was successful for Specimens 5 and 6 as they had wider cracks of approximately 1.0-mm in width. This technique worked well in restoring the bond, strength stiffness and energy dissipation capacity of the specimens. In general, the epoxy repaired cracks did not reopen in the tests on the repaired specimens. New cracks tended to develop in the concrete adjacent to the repaired cracks.

CONCLUSION

- In general, the repaired specimens were capable of resisting the same and even higher static loading than the originals.

- The epoxy resin was effective in the repair of structural cracks. The epoxy repaired cracks did not reopen in tests on the repaired specimens. New cracks tended to develop in the concrete adjacent to the repaired cracks

- The plate bonding was used successfully in this test programme. The tension plates glued to the wall section and the shear strips placed on the coupling beams increased the ultimate load capacity of the specimen. The tension plates increased the stiffness, whereas the shear strips principally controlled the developing and re-opening cracks.

- To further investigate the effectiveness of coupled shear wall structures under seismic loading, it would be more relevant to consider using cyclic loading.

REFERENCES

1. MAHIN, S A et al. Non-linear Seismic Response of Coupled Shear Wall Systems. Journal of the ASCE Structural Division Dec. 1976, pp 1759-1780.

2. PAULEY, T. Coupling Beams of Reinforced Concrete Shear Walls. J. Struct. Div. ASCE, No. 3, Vol. 97, 1971, pp 843-861.

3. SUBEDI, N K. RC-Coupled Shear Wall Structures, I: Analysis of Coupling Beams. J. of Struct. Div. ASCE, Vol. 2, March 1991, pp 667-679.

4. KAY, E A. The European Standard on Concrete Repair Principles. Construction Repairs 6 Vol. 11, No. 2, March/April 1997, pp 52-55.

5. KING, E S & ECOB C L. Review and Specification of Concrete Repair Materials, Structural Faults and Repairs 93- London, Vol. 2, pp 211-215.

6. ROBERY, P. Maintenance and Repair Strategies. Construction Repair: Concrete Repairs 6, Vol. 11 No. 2, March/April 1997, pp 33-38.

7. OEHLERS, D J et al. Upgrading Continuos Reinforced Concrete Beams by Gluing Steel Plates to their Tension Faces, J. of Structural Engineering, March 1998.

8. GHAZI, J. et al. Shear Repair of Reinforced Concrete Beams by Fibreglass Plate Bonding, ACI Structural Journal, July-August 1994.

9. SUBEDI, N K. Reinforced Concrete Beams with Plate Reinforcement for Shear, Proceedings Inst. Civil Eng. Part 2 Sep 1987, pp 377-399.

10. ABU-TAIR, A I et al. The Effectiveness of Resin Injection Repair Methods for Cracked RC Beams. The Structural Engineer/ Vol. 69, No. 19, Oct 1991, pp 335-341.

11. ABU-TAIR, A I et al. The Use of Resin Injection to Repair and Impact Damaged Motorway Bridge, The Structural Engineer, Vol 73, No. 12, June 1995, pp 200-201.

EXPERIMENTAL INVESTIGATION OF CONCRETE SUBJECTED TO ELEVATED TEMPERATURES

M R Resheidat

M S Ghanma

Al-Balqa' Applied University

Jordan

ABSTRACT. This paper presents the test results of concrete cylinders subjected to elevated temperatures of 300, 500, and 700 °C. The compressive strengths of these were compared with reference concrete stored at room temperature to establish the effect of strength loss due to heating. This study is part of a research project that will consider other parameters such as the inclusion of fibers, bond strength, tensile strength and cyclic temperature changes. The loss of concrete strength was significant and increased with increase of temperature and heating time. A prediction model was developed to estimate the ultimate strength of concrete as influenced by both heating temperature and heating time.

Keywords: Concrete, Fire, Colour, Cracking, Durability, Structural assessment.

Professor Musa R Resheidat is a Professor and Dean of Engineering. His research interests are concerned with structural design of reinforced concrete structures and assessment and development of materials. Professor Resheidat has published some forty papers in his field. He is an ASCE fellow, an ACI member, and a registered Expert with Jordan Engineers Association.

Mwafag S Ghanma is an Instructor at the Department of Surveying and Geomatics Engineering. His research interests include computer-aided analysis and design of concrete structures, engineering related MIS and web project management. He teaches many engineering and programming courses. He is a member of the Jordan Engineers Association.

INTRODUCTION

Concrete structures could be exposed to elevated temperature conditions. Examples of such conditions are concrete foundations for launching rockets, concrete structures in nuclear power stations, or those exposed to fire.

When concrete is subjected to high temperatures, there is deterioration in its properties such as compressive strength, elastic modulus, tensile strength, bond with reinforcement, etc.

Interest in high temperature behavior of concrete structures starts at lower bound temperatures of 100°C. Immediately above that, free water starts to be driven off. The engineering properties of concrete vary slightly from those measured at room temperature. However, above 150°C, the progressive continuum of cement dehydration reactions, thermal incompatibilities between paste and aggregates, and physiochemical deterioration of aggregates lead to high thermal stresses, micro cracking, and rapid worsening in most mechanical properties of engineering. On heating above 300°C, the colour of concrete can change [1, 2] from normal to pink (300-600°C). The change in colour is associated with a loss in mechanical properties. Such loss is influenced by several interrelated factors. These include; type of aggregate and cement, free moisture in concrete, concrete mix properties and the targeted nominal design compressive strength.

Several research studies have been conducted to assess the concrete properties due to fire in terms of colour change, cracking and spalling [3, 4].

When the Safeway International building was subjected to fire [5], the only source of available information about concrete behavior was the international literature. No experimental studies relevant to the local cements and aggregates were available. Accordingly, this paper presents a pilot study aimed at studying the loss of concrete compressive strength when it is subjected to elevated temperatures ranging from 360°C to 900°C.

EXPERIMENTAL PROGRAM

Materials

The Portland pozzolanic cement manufactured by the Jordanian cement factory in accordance with the Jordanian standard specification No. 118 was used. Coarse and medium aggregates were prepared from crushed limestone. Wadi sand was used as fine aggregate. The water was from the main water supply. A retarding agent with a highly plasticizing effect was also added to the mix.

Concrete Mix Design

The method followed in design of the mix is the one outlined in the British Research Establishment-BRE Publication, "Design of normal concrete mixes" The international conference: concrete mixes". All relevant ASTM standards such as ASTM:C 136-93, C 117-90, C127-93, C 128-93, and C 182-98.

A summary of trial mixes is shown in Table 1.

Table 1 Recommended properties of concrete constituents.

CONCRETE CLASS	COARSE AGGREGATE (Kg/m^3)	MEDIUM AGGREGATE (Kg/m^3)	WADI SAND (Kg/m^3)	CEMENT (Kg/m^3)	WATER ($litre/m^3$)	PLASTICIZER ($litre/m^3$)
C 20	526	564	790	300	220	1.20
C 25	526	564	790	320	200	1.28
C 30	513	550	770	360	207	1.44
C 35	502	538	753	400	206	1.60

The concrete, as designed, was prepared by a ready mix concrete plant that was the supplier for concrete used for Al-Balqa' Applied University projects. Samples were taken from the delivery trucks during concreting. The standard 6x12 cylinders were cast and cured according to the standards.

Heating to Equilibrium Temperatures

A Webcot kiln model 3090, as shown in Figure 1, was used to heat the concrete cylinders. Its maximum operating temperature is 1300°C. The approximate capacity of firing chamber is 285 liters.

Figure 1 The Webcot kiln model 3090 used for heating

The Concrete specimens were cured for 28 days and subjected to heat at the age of 90 days. Nine concrete cylinders were heated to the equilibrium temperature of 300, 500 and 700°C, 3 specimens were soaked for 3 hours, 3 specimens for 6 hours, and 3 specimens for 9 hours. The same procedure was followed for the equilibrium temperatures 500°C and 700°C.

Each batch has also 3 control specimens that were normally cured to form a reference of the nominally design ultimate compressive strength. All specimens were subjected to compressive test.

Test Results

The results of ultimate compressive strength are given in Table 2.

Table 2 Test results

TEMPERATURE (°C)	HEATING TIME (HOURS)	COMPRESSIVE STRENGTH (MPa)	AVERAGE STRENGTH (MPa)	STANDARD DEVIATION (σ)	CONTROL STRENGTH (MPa)
300	3	23.5	22.67	0.918	30.0
		22.0			31.5
		21.3			30.4
	6	15.5	15.8	0.216	
		16.0			μ= 30.63
		15.9			σ = 0.634
	9	12.1	12.7	0.535	
		13.4			
		12.6			
500	3	19.8	20.53	0.573	32.5
		21.2			33.1
		20.6			31.7
	6	13.8	14.4	0.245	
		14.1			μ = 32.43
		14.4			σ = 0.573
	9	6.2	9.23	2.191	
		10.2			
		11.3			
700	3	8.4	8.27	0.419	26.3
		8.7			25.2
		7.7			24.3
	6	5.8	5.7	0.455	
		6.2			μ = 25.27
		5.1			σ = 0.828
	9	Damaged			
		Damaged			
		Damaged			

The average concrete ultimate strength results are also presented in Figures 2, 3, and 4.

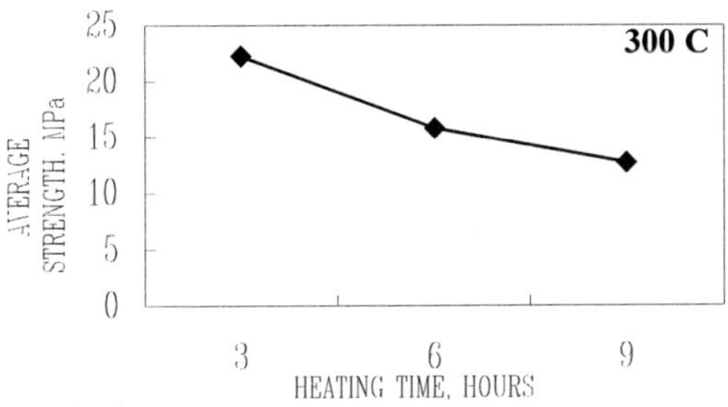

Figure 2 Concrete strength vs. heating time at 300°C

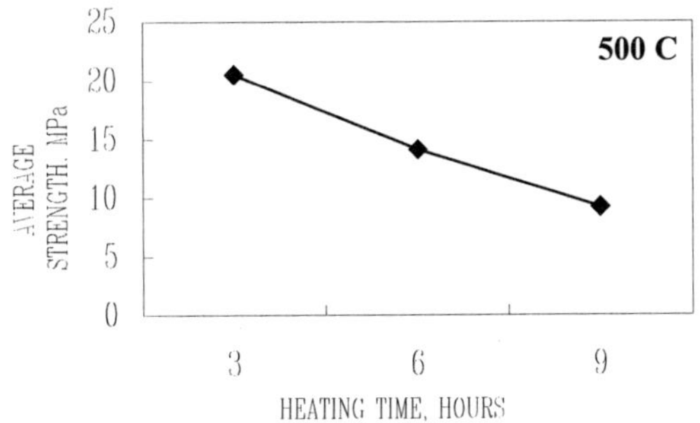

Figure 3 Concrete strength vs. heating time at 500°C

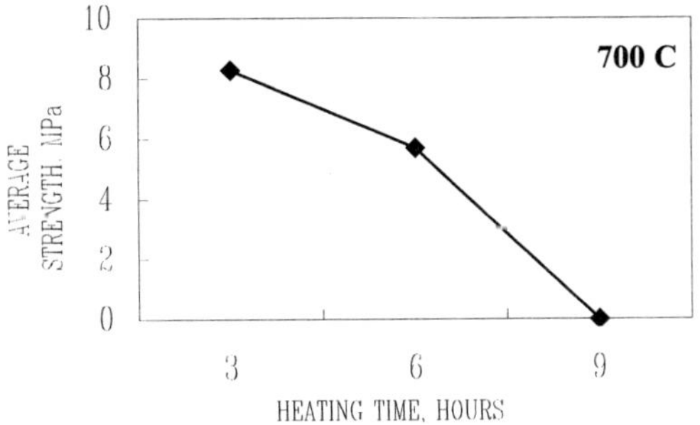

Figure 4 Concrete strength vs. heating time at 700°C

DISCUSSION OF RESULTS

It can be easily observed from Figures 2, 3, and 4 that the ultimate compressive strength of concrete exhibits loss in values. This loss is influenced by both variables: the elevated temperature and the heating time.

At 300°C, the loss starts with 27% and ends with 60%. At 500°C, the loss starts with 37% and ends with 72%. At 700°C, the loss starts with 67% and ends with complete failure.

Similar observation could be outlined if the heating time is considered. An attempt to quantify the prediction of strength, a multiple regression is carried out to present the following formula:

$$Y = 1.24 - 0.041\,T - 1.167\times10^{-3}\,C \qquad (1)$$

Where,

Y = Normalized predicted value of ultimate strength.

T = Heating time.

C = Heating temperature.

From Equation 1, and considering the fact that the actual loss of strength starts at heating temperature of 100°C, the model shows that at such temperature concrete heating time should be 5.44 hours. After that the strength will decrease with time.

CONCLUSION AND RECOMMENDATIONS

The following conclusions could be drawn from this study:

1. The compressive strength of concrete is highly influenced when it is subjected to elevated temperatures. Loss in concrete strength is proportional with the increase of temperatures values.
2. Duration of heat increases also the loss of concrete strength.

The study presented herein is limited to the prediction of concrete strength as presented in Equation 1. However, more work is needed to predict other parameters such as cracking, colour change, and other mechanical properties of concrete such as the tensile strength and the elastic modulus.

ACKNOWLEDGEMENTS

The authors would like to acknowledge all the assistance and help provided by the Engineering Office, the staff of the Metallurgical Laboratory at BAU. Thanks are extended to Dr. Amjad Barghouthi for his support, assistance in carrying out the tests, and the valuable data he provided.

REFERENCES

1. HUNT, R W G. The specification of colour appearance, Colour Research and Application, Vol. 2, 1977, pp 55-68, and pp 109-120.

2. HUNT, R W G. Colour technology, Colour Research and Application, Vol 3., 1978, pp 79-87.

3. GUISE, S E, SHORT N R, AND PURKINS, J A. Colour Analysis for assessment of fire damaged concrete, Proceedings of The International Conference, Concrete in the Service of Mankind- Concrete Repair, Rehabilitation and Production, University of Dundee, Scotland, UK, 1996, pp 53-63.

4. MALHOTRA, H L. Spalling of concrete in fires, Technical Note 118, Construction Industry Research and Information Association, London, 1984.

5. ARABIC CENTER FOR ENGINEERING STUDIES. Assessment effect of fire on concrete structures for Safeway International Existing Building, Amman, Jordan, SP 93041, 1994, pp 53.

EVALUATING THE CONDITION OF REINFORCED CONCRETE STRUCTURES AFTER FIRE

G Muravin

L Lezvinsky

Margan Physical Diagnostics

B Muravin

Tel-Aviv University

Israel

ABSTRACT. The paper examines the complex methodology of acoustic emission (AE) and traditional techniques and their application to the inspection of reinforced concrete structures that have been affected by extreme conditions, particularly, fire. The proposed procedure presents a reliable method for detecting the presence and location of weaknesses caused by fire, damage zone limits, structural elements that are unsuitable for further use. It includes an investigation of the criteria for evaluating damage danger levels as well as methods for estimating the stability of construction elements and structures, when in continued use.

Keywords: Reinforced concrete structures, Fire, Acoustic emission (AE), Evaluating condition, Stress relief, Chemical bond water, Stress measurements, Hydrolysis, Hydration.

Professor Gregory Muravin is Chief Scientist of the Margan Physical Diagnostics, Israel. He specializes in acoustic emission and non-destructive testing of reinforced concrete, composite materials and metal structures, as well as in evaluating their condition by methods of the Physics of Solids and of Fracture Mechanics. He has published more than 190 articles and has patented more than 30 inventions.

Dr Ludmila Lezvinsky is Senior Scientist of the Margan Physical Diagnostics, Israel. She specializes in the investigation of the physical properties of concrete, composite materials and metal, and the prediction and physical modeling of their properties using AE data. She has authored more than 100 articles and inventions.

Mr Boris Muravin is a postgraduate student of Tel-Aviv University, Israel. He specializes in Applied Mathematics, which he makes use of for the image recognition of defects in materials.

INTRODUCTION

Even though reinforced concrete or metal structural elements show no outward signs of significant damage after a fire, they may well represent future hazards. From an analysis of data available in literature including patent and invention certificates in 1982, it was clear that there were no conventional methods of inspection suitable for determining whether such structures have suffered dangerous damage by fire.

This was specifically true regarding:

- industrial measurement of damage to such structural elements or of stress distribution in them;
- diagnostic methods for determining weakness caused by fire or for detecting and pinpointing of stress concentration zones, micro- and macro-cracks, and zones of concrete creep, in them;
- criteria for determining damage zone limits;
- methods for evaluating structural load limits.

Prior to our work in 1982, there were no methods available for estimating, even approximately, the degree of damage that the evaporation of chemically bonded water and the breakdown of hydration may cause to concrete [1,2]. Data had yet to be published about the normal and critical recovery rate of concrete cracks after a fire or about acceptable deviations from initial, pre-fire rates of recovery. In the absence of such methods and data it was impossible to assess the structural integrity of composite materials in constructions accurately after fires and to decide whether it was possible to repair them.

While searching for answers to the above problems, it was found possible to create, patent and apply in practice a necessary and sufficient set of data for use in the examination and testing of reinforced concrete and composite constructions that have been damaged by fire. This article presents our findings on the subject.

NON-DESTRUCTIVE TEST METHODOLOGY

Before examining a structure for fire damage, one must first define what types of defects might appear and then formulate the criteria necessary for their identification. Classification of defect types is usually a straightforward matter. They include: change in the structural integrity of concrete and metal structure, micro- and macro-crack development, isolated sites of excess overstresses and deformations, deviation of structural elements from their initial position, unstable structural loading and rapid ageing of structural material. We, therefore, proposed the following criteria as aids in determining the existence of defect types and distinguishing between them:

- the absence of defect development;
  - the non-appearance and non-reproducibility of AE under repeated loading (Kaiser effect),

- a linear relation between the mechanical deformation energy of an object and its AE energy;

- defect type and its danger potential;
  - statistical image recognition criteria, namely average frequency $\overline{f}$, average energy $\overline{E}$, dispersion of $\overline{E}$, $\overline{f}$ and the correlation coefficient between E and f , which together combine to form dispersion ellipses,.

- concrete and steel condition;
  - uniformity of structural integrity and local internal stress distribution,
  - absence of critical micro- and macro-cracks,

- specific requirement for concrete condition;
  - restoration of structural integrity after fire,

- condition of structure;
  - additional stresses from load redistribution,
  - deviation from initial position,
  - isolated excess stresses.

The fracture toughness criteria of eight concrete compositions was investigated having different structural and mechanical characteristics obtained by varying the contents of binder, inert filler, air-entraining additives, cement, and other factors. To do this we prepared twelve prisms measuring 7 cm x 7 cm x 28 cm from each composition, half with a 0,1 x 1,75 notch. The test was by means of a three-point bending system, recording loads, bending deflections, and AE parameters.

Strain-gauge studies were carried out by means of an SIIT-2 system, and an AWN-3 instrument was used for the AE measurements. The fracture rates in each test were estimated by the rupture of wire gauges and reflective amalgam strips. A computer processed measurement data on the instruments and also performed a spectral analysis of the shapes of the individual AE pulses. The load was changed in fixed steps at fixed intervals.

The critical stress intensity factor (K1C) and the resistance to crack propagation (G), were calculated up to the instant of the main crack fissured. These characterize the elastic and elastoplastic strains, respectively, The criteria K1C, and G were calculated on the basis of the measurement data.

The structure of the concrete and the metal was examined, comparing samples of the material both in the original state and after undergoing heating by fire (400° C). For the latter we made use of material from a construction that had had a fire lasting for about one hour before being extinguished with water. The purpose of the tests was to define AE parameters for samples of material with different types of defects.

The examination combined a number of scientific procedures, including AE and mechanical tests, optical and electron microscopy, and X-ray diffraction analysis with the help of the following appliances:

- a "Neofot 32" microscope, to examine the structure of the concrete and the metal;
- a "URS-10" with an RKD camera, for the X-ray studies - grain size and X-ray patterns calculated by conventional methods;
- a YEBM-100K microscope under an accelerating voltage of 100 kV using the "foils" method and the disk technique, for the electron microscope inspection.

A "stress-strain" diagram in the course of the mechanical tests was prepared. The principal procedure in the laboratory and field tests was the AE emission test. Sounds associated with damage (acoustic emission signals) were detected, amplified, and filtered with SPARTAN, IMPULS and AWS-3m acoustic emission devices, and with a digital storage oscilloscope. Recorded data included AE signal amplitude, energy, RMS, counts, duration, rise time, counts to peak and average frequency as well as the time of the test. The data was then analyzed, using special computer programs.

The indirect strength of the concrete was determined by Schmidt hammer tests on different parts of the construction, both inside and outside the fire zone. Samples were compared by means of the homogeneity criteria of distribution (t- Student criterion); dispersion (F- Fisher criterion of 5%) and coincidence of mean quantities (V- criterion), using the AE data. A significance level of 5% was established when verifying the homogeneity hypothesis.

INVESTIGATION OF UNDAMAGED CONCRETE EXPOSED TO FIRE

Compression Test of Concrete Specimens Under Ordinary Conditions

Elevated temperatures due for instance to fire, can cause the chemical products of hydration and hydrolysis to breakdown in the cement matrix of concrete. This can have an adverse effect on its structural and mechanical properties, and can change its fracture toughness and its ability to regain its structural integrity. To make a quantitative evaluation of these failures, and thus to be able to judge the condition of concrete after a fire, we selected certain phenomena as the specific criteria for assessing concrete condition. These included the onset of micro-crack formation; the start of non-linear creep, the development of main cracks, resistance to crack propagation and crack “healing” (restoration of structural integrity after fire) [3,4].

In experiments, various grades of concrete of different composition were examined. AE method as the principal procedure for testing the condition of the concrete was used. However, we also used ultrasound (U.S.) and strain measurement methods, and made electron-microscope observations. Furthermore, we analyzed the spectral composition of the AE signals.

When subjected to uniaxial pressure with loads ranging from 0.1 up to 0.55 σ/R, prisms made from various types of concrete exhibited phenomena of crystal-boundary displacement and the intrusion of crystalline hydrates into pores (Figures 1,2). Compacting the concrete produced an increase in ultrasound velocity. Furthermore, there was an increase in AE energy as a result of the fracture of discrete crystal aggregations under slip and the destruction of micro-cells due to the displacement of free water.

When loads increased above 0.55σ/R, micro-cracks began to appear in the micro-pore walls, splice planes and the large pores between crystal planes. The onset of micro-crack growth corresponds to the break in the curve E/E_{max} and the point R_c^0 on the ultra-sound graph Figures 1,2) The Ultrasound velocity decreased as the decompression of the material. AE energy increased considerably with the onset and growth of micro-cracks but also when their propagation slowed.

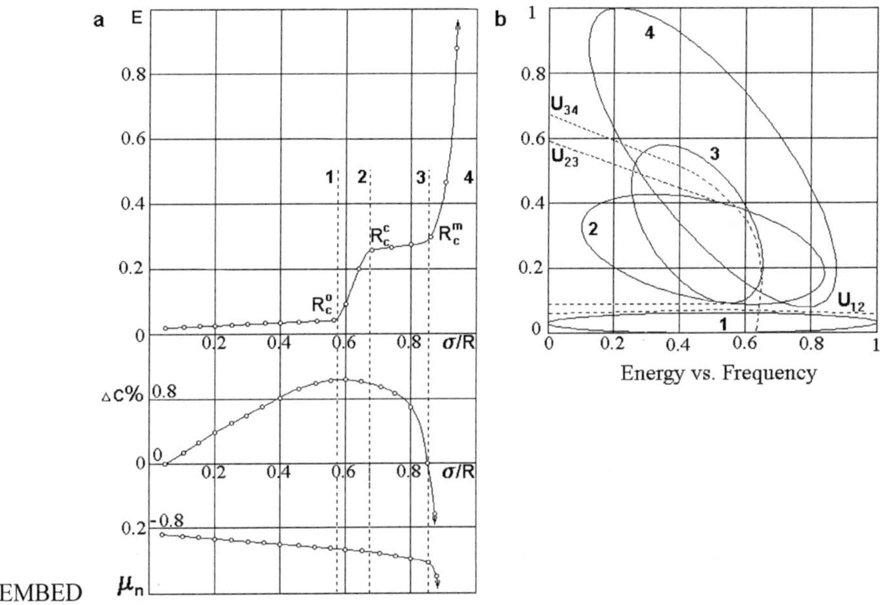

EMBED

Figure 1 Sample testing data

a) Change of acoustic emission energy (E), ultrasonic velocity (C%), total transverse deformation coefficient (μ_n)-in concrete under load. 1- high limit of material compacting zone and low limit of micro crack formation zone; 2 - high limit of micro crack formation zone and low limit of creep commencement; 3 - high limit of creep and commencement of main crack propagation zone. R_c^o, R_c^c, R_c^m, stress of micro crack formation, creep and main crack propagation commencement, respectively.

b) Dispersion ellipses of AE signals corresponding to: 1 - concrete compacting; 2 - micro crack formation; 3 - creep; 4 - main crack propagation. U_{ij} - discriminant function between processes.

Under a load of 0.65σ/R, individual cracks developed from separate pores and then joined together. Unstable fractures appeared near the boundaries of main breaks. Simultaneously, the AE intensity decreased slightly, concurrent with the reduction in the volume of the deformed material and with the increase in its porosity. Loads above 0.85σ/R led to an abrupt increase in the number of main cracks, as shown in the fractograms (Figure 2). This was accompanied by a considerable (up to twofold) increase in AE intensity and amplitude, phenomena which are characteristic of unstable crack development. Simultaneously the ultrasound velocity decreased by 0.8%. A sharp change in direction appears on the coefficient of change curve of the total transverse deformation, evidence of the nonlinear character of this increase (Figure 1). Under increased loads the specimens soon collapsed.

Examination of the spectral composition of the AE signals at various stages of deformation, showed that during the development of discrete faults, there is, as a rule, no correlation between individual AE signals (Figure 1). The ellipse of dispersion of average AE energy and median impulse frequency is relatively small. During micro-crack formation, the correlation coefficient increases but not exceeds 0.6.

The emission correlation grows substantially (up to 0.9), during non-linear creep and the development of unstable cracks in the concrete. The average energy values and median frequencies of the AE signals grow considerably due to the high speed of fault formation, the breakdown of very hard inclusions, the redistribution of released energy around crack tips and the unstable fracturing.

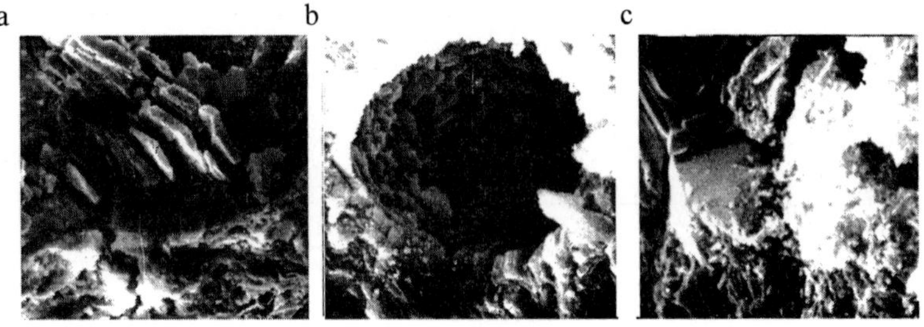

Figure 2

a-Large crystallites in pore walls are stress concentrators: that is where the first cracks develop when $\sigma < 0.55R$

b-Cracks developed on the pore contour in crystal junction planes, begin to grow when $\sigma > 0.55R$

c-Spall steps on grains boundary and new cracks, developing at 30-60° angle when $\sigma \geq 0.85R$

At each stage, the AE can be characterized appropriately by average energy/frequency dispersion ellipses. These significant differences in signal characteristics in the given range simplify the precise determination of the real state of the concrete under examination. It also makes it easy to detect the stage of the deformation within an accuracy of a few percentage points, when plotting the discriminant function (Figure 1).

Fracture Toughness Test of Concrete Under Ordinary Conditions

Previous attempts to estimate the state of the material in concrete and reinforced concrete structures by the fracture toughness criteria K1C and resistance to crack propagation (G),were reported in other papers [5,6]. In the present study we explored the possibilities of determining K1C and G as factors for estimating the resistance of concrete to fracture, before and after fire. Table 1 presents the results of the tests.

Table 1 Results of fire tests

MIX	H (MPa)		R (MPa)		K_{var}		R_c^o/R		R_c^c/R		R_c^υ/R		J_{1c}		G	
	1*	2*	1*	2*	1*	2*	1*	2*	1*	2*	1*	2*	1*	2*	1*	2*
1	35	40	35	30	0.09	20	0.59	0.32	0.78	0.86	0.87	0.87	37	20	130	82
2	36	39	36	29	0.08	22	0.64	0.45	0.80	0.89	0.85	0.91	39	24	120	90
3	36	42	36	31	0.09	17	0.62	0.40	0.82	0.90	0.87	0.92	38	23	115	60
4	38	45	38	32	0.08	20	0.65	0.38	0.83	0.92	0.91	0.92	40	19	102	60
5	45	50	45	35	0.09	18	0.67	0.41	0.82	0.92	0.9	0.93	42	18	96	70
6	52	50	52	40	0.07	23	0.70	0.43	0.85	0.94	0.92	0.95	46	21	89	50
7	55	49	55	42	0.08	21	0.70	0.42	0.87	0.93	0.94	0.94	48	25	90	36
8	62	50	62	49	0.07	22	0.75	0.45	0.90	0.95	0.96	0.96	49	28	96	40

1*- before a fire, 2* - after a fire, H -concrete hardness, R - concrete compressive strength, K_{var} -coefficient variation of concrete strength, R_c^o/R - lower boundary of microcrack formation, R_c^c/R - boundary of creep development, R_c^υ/R - boundary of development of unstable cracks, J_{1c} - critical value of resistance to crack propagation, G- energy of crack propagation.

Concrete Subjected to Elevated Temperatures

For this we tested concrete samples from the site of a fire that had damaged a concrete structure. We took eight batches each of twelve samples of concrete from the zone of the fire's origin, after determining precise sampling points by acoustic emission tests that revealed sources of elevated AE. All the samples had the same composition and had endured a high temperature of 400° C for one hour during the fire and half of them had been rapidly cooled with water during its extinguishing. We examined all the concrete samples for structural and mechanical characteristics, fracture toughness criteria and resistance to crack propagation.

Six samples with a notch and six without were tested in each batch by the same system of measurement used for concrete under ordinary conditions. The object was to learn the extent to which the heat of fire (400^0 C) and the rapid cooling had altered the concrete's structural and mechanical characteristics and to judge the physical state of the construction after the fire was extinguished. Table 1 presents the results of the tests.

The tests revealed that the effect of "fire" and of "fire and rapid cooling" was to:

- decrease the load at which micro-and macro-crack formation commenced;
- expand the limits of creep stresses;
- increase the coefficient of concrete strength variation;
- decrease the resistance to crack propagation, G.

The hardness of concrete after a fire can sometimes increase. This effect could be connected with evaporation of water from the surface of concrete samples or to peculiarities of Schmidt hammer test. The increased hardness can sometimes lead to a strengthening of fracture toughness. But the coefficient of variation of both these characteristics (hardness and fracture toughness) is usually significantly higher in samples of concrete damaged by fire than in those that had not been subjected to fire temperatures (400^0 C).

Evaluation of Concrete's Ability to Restore Its Structural Integrity

As a result of the irreversibility of plastic deformation, the majority of engineering materials retain an historical memory. They do not emit stress waves under repeated loading until the effective forces exceed the initial maxim load that caused the plastic deformation. This property, known as the Kaiser effect, is widely used when inspecting the condition of structures.

As we reported elsewhere [3,4], there is a close relation between the Kaiser effect in concrete and the structural state of the material. This is observed when loads are less than the cracking limit. However the phenomenon is temporary, owing to the recovery of the structural integrity of the cement cells. At the stage of non-linear creep, the Kaiser effect begins to be violated as a consequence of general expansion of the concrete and the occurrence of a considerable number of cracks in it. It can no longer be observed during the stage of unstable crack growth.

Complete "healing" of the cracks and complete restoration of AE occurs after a month and a half or more in different compositions of concrete under loads lower than the micro-crack formation level. In our experiments on eight different compositions of concrete that had been affected by excessive heat (three-years old samples loaded within the limits of their elastic deformation) the total restoration of AE was observed only after 4 months. Nevertheless, a partial regeneration of AE during reloading (Figure 3) occurred after 3 to 7 days. This is connected with the restoration of structural integrity and healing of defects due to the continuous process of hydrolysis and hydration of the cement.

The difference in the rate of regeneration of AE after loading suggested the possibility of using AE data to evaluate the capacity of concrete to recover its structural integrity .

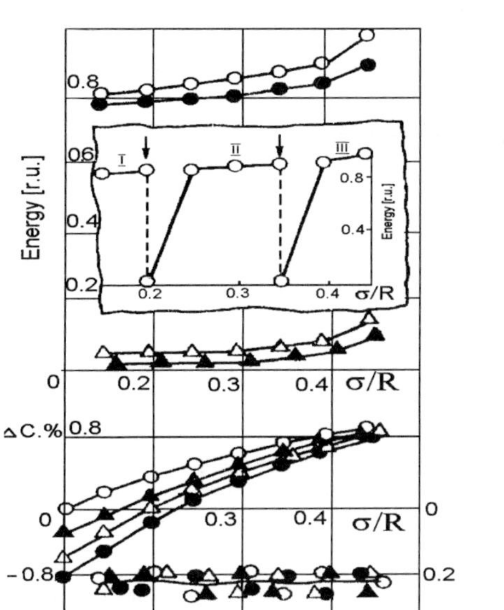

Figure 3 Change AE energy, Ultrasonic velocity (ΔC), complete transversal deformation coefficient (μn) in concrete under loading up to σ=0.45R,
□- 1-stloading: ▲-2-nd loading after 4 days of "rest"; Δ-3-d loading after 8 days of "rest";
• - 4-th loading after 36 days of

Analysis of experimental results for the eight batches of samples revealed that the restoration of AE after fire differed according to the different composition of the concrete and the time of the "fire's" influence (Figure 4).

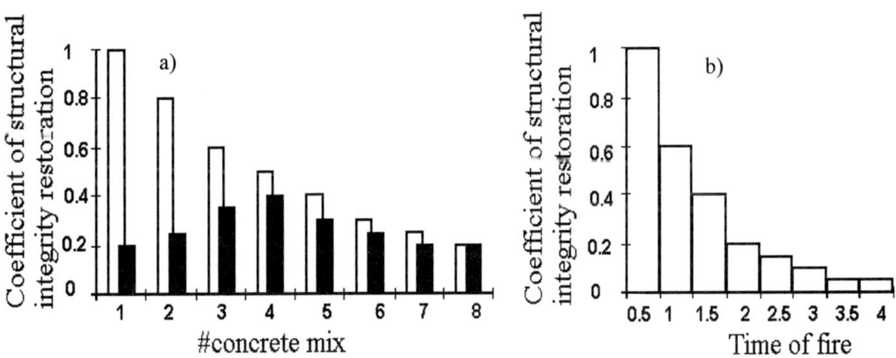

Figure 4 a) AE energy restoration after 36 days of "rest" for concrete mixes
□ - normal condition; ■ - after one hour of a fire
b) AE energy restoration after different time of a fire. Mix # 4. 56 day of a "rest".

The concrete that originally displayed a low activity of structural integrity restoration, exhibits almost no 'healing' capacity after the fire. Conversely, the concrete that originally exhibited a greater capacity for structural integrity restoration, displays a significantly better 'healing' capacity. Furthermore, the longer the concrete was subjected to the effect of the 'fire', the lower the possibility of recovering structural integrity (Figure 4).

EXAMINATION OF A CONCRETE STRUCTURE AFFECTED BY FIRE AND VERIFICATION OF DAMAGED AREA LIMITS

Visual inspection revealed many beams and columns with localized damage, and much wall displacement. There were also zones where visual inspection had not revealed damage but showed signs of fire or other indications of possible damage. We, therefore, applied a combination of acoustic emission test methods and conventional methods to discover the damage zone limits and to estimate damage danger levels. Specifically, acoustic emission images of the fracturing processes were compared with corresponding images from normal concrete samples.

It was taken into account that AE will always occur during the continuous evolution of structural integrity in concrete, and the formation of hydrates and crushing of crystalline cells in it [7,8]. In each of these processes, the spectral and energy composition of the AE and, the ellipses of dispersion of the AE flow display distinctively different characteristics, making it possible to ascertain the concrete condition (Figure 5).

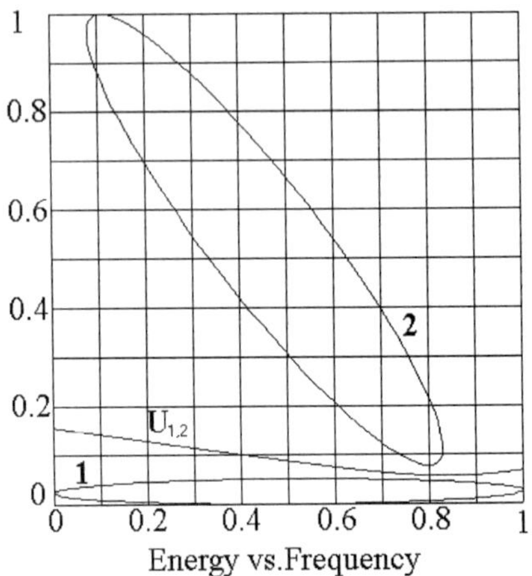

Figure 5 Dispersion ellipses of AE signals corresponding to:
1 - formation of concrete structure;
2 - crack propagation.
U_{12} - discriminant function between processes.

The characteristics of the AE from the formation of crystalline structures and cell-fracturing are invariable and dependent of factors such as temperatures or the applied stress level. Nevertheless, these specific AE signals disappear when fire has stopped hydration in the concrete. This makes it possible to detect where the processes of concrete cell strengthening and crystalline structure formation have ceased and also where micro- and macro-cracks are developing.

One must measure AE at a number of points along the structural elements and compare the resulting data with previously stored indicative parameters in a data bank (Figure 5). The presence of those AE signals that are specific to the formation of structural integrity would show that fire has not interfered with the hydration process. Their absence would establish irreversible damage to the concrete. AE signals specific to the fracturing process also help to pinpoint where the micro- and macro -cracks are developing.

This information was used to evaluate the condition of the beams and columns of the structure at the fire zone and define its limit. The structure was divided provisionally separated conditionally into zones with radii 2.5 to 5 m and AE signals were measured in all of them.

The object was to find whether there was any fire damage and how dangerous such damage was, and to estimate which of the structural elements did not require replacing.

Inspection revealed that the AE characteristics in most of the elements were in accordance with the transformation processes normal during concrete hardening. The acoustic emission measurements did not reveal damage accumulation or crack development. The AE parameters did not exceed the standard reference values that we had established in laboratory tests on undamaged concrete specimens.

There was very intense AE in some structural elements. This indicates crack development and continuing redistribution of stresses in the structure. The image identification method with dispersion ellipses and data banks helped to pinpoint the origin of these signals, making it possible to discover in which elements there were localized over-stresses, growing cracks and extreme deformation.

The indirect strength of the concrete in all the suspected zones of damage was evaluated by Schmidt hammer tests. This characteristic was also checked statistically in 12 parts of the structure where AE test had not revealed damage development and where the analysis of the test results had shown that the strength of the concrete and the coefficient of variation of concrete strength were according to design criteria. The same parameters had, however, significantly different values in the suspected zones (Table 2). There, the coefficient of variation of concrete strength was significantly bigger (about 23% as opposed to of 9%).

Three cores were taken from each zone that was undamaged by fire (type #1 samples) and two from each suspected zone (type #2 samples).Examination of the cores and analysis of the AE data (Figure 6) established that micro-cracks did not start to develop in samples of type #1 before the load was 0.55R, while in samples of type #2 a load of 0.25R was sufficient for this.

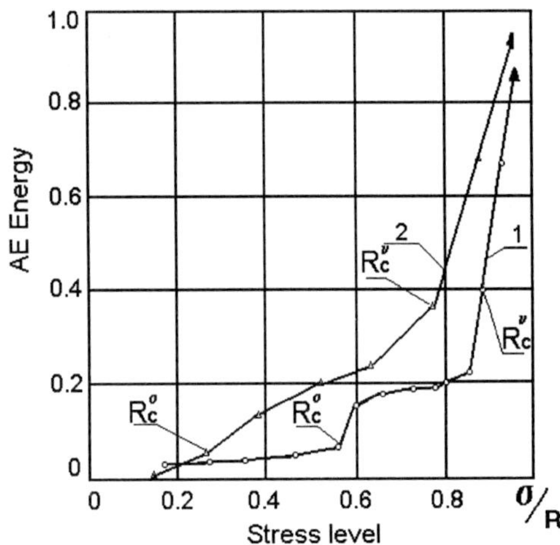

Figure 6 Cores testing. Energy vs. stress. R -stress of fracture. 1- concrete in normal condition; 2- concrete after fire

Tests showed that non-linear creep appeared in type #1 concrete at loads of 0.6R, compared with 0.4R in type #2. Increasing the load to above 0.88R in the case of undamaged concrete and above 0.75R in the case of concrete that had been subjected to fire, led to a significant (up to double) increase in AE activity, which is typical of unstable crack-propagation. The samples shattered rapidly when the loads increased beyond these values.

The fracture toughness tests of cores type #1 and #2 established that the characteristics of the undamaged concrete corresponded to material in a normal condition. The fracture toughness of the samples from constructions damaged by fire (core type #2) was significantly (from 40% to 50%) lower.

The process of structural integrity restoration was examined in samples of both types, #1 and #2. The undamaged concrete (samples type #1) did not radiate AE signals during repeated loading, owing to the historical memory referred to above.

After an interval of 56 days, however, all AE activity renewed. In concrete that had been subjected to fire (sample type #2), AE signals appeared immediately after the first loading.

However, the energy of these signals was about 30% of the value recorded during the first loading of undamaged samples. Furthermore, complete crack "healing" and restoration of AE did not occur in the samples of type #2 after 56 days from the first loading, unlike sample type #1.

CONCLUSIONS

It was demonstrated that the Complex Acoustic Emission Non-Destructive Testing procedure is a powerful and successful technique for use in investigating suspected damage to concrete in large-scale reinforced concrete structures, where high temperatures from fire (400° C) may have caused chemically bonded water to evaporate and hydration products to break down. The proposed procedure presents a reliable method for detecting :

- the presence and location of;
  - weaknesses caused by fire,
  - damage zone limits,
  - structural elements that are unsuitable for further use;
- where stresses sufficient to cause micro- and macro-crack formation are lower than usual;
- where the coefficient of concrete strength variation is greater than usual;
- decreased resistance (G) to crack propagation;
- differences in the rate of regeneration of AE after loading, as an indication of the concrete's continued capacity to recover structural integrity.
- the presence and location of weakness caused by fire in large-scale reinforced concrete structures.

REFERENCES

1. MURAVIN G.B. Method of Determination Unfitness to Operation of Construction Subjected to Fire. USSR patent (A.S.) # 4780370 19.12.89 Registry date 15.01.91.

2. MURAVIN G. B., Palei Yu. M., Sneznitskii Yu. S., Volkov S. I. Diagnostic of Condition the Constructions Subjected to Fire. - BBT, 1991, N 6, pp. 19-23.

3. MURAVIN G.B., Simkin Ya.V. Gur'ev V.V. Method of Determination of Stress Value in Constructions. USSR patent (A.S.) # 1523994. Declare N 4343914 15.12 87. Registry date 22.07.89.

4. MURAVIN G.B., Gur'ev V.V. Kaiser Effect and Structural State of Concrete.- Defektoskopiya, 1986, N 10, pp.22-27.

5. MURAVIN G. B., MERMAN A.I., LEZVINSKY L.M. Acoustic Emission Method of Estimation the Fracture Toughness of Concrete in Structures and Constructions. Defektoskopiya, 1991, N 3, pp. 10-16.

6. METHODOLOGICAL GUIDELINES: Crack-Resistance Characteristics of Concrete in Short-Term Static Loading /in Russian/. Standartov, Moscow (1989).

7. MURAVIN G. B., PAVLOVSKAYA G.S., SHCHUROV A.F. Study of Acoustic Emission by Hardening Concrete. - Defectoskopiya, 1984, N 10, pp. 77-81.

8. MURAVIN G.B. SNEZNITSSKII YU.S. PAVLOVSKAYA G.S. Investigation of the Process of Hardening of Concrete at Low Temperature by Acoustic Emission Method.- Defektoskopiya, 1989, N 10, pp. 9-15.

THERMOPHYSICAL PROPERTIES OF CONCRETE EXPOSED TO A FIRE

R Černŷ

J Toman

T Klečka

P Bouška

Czech Technical University

Czech Republic

ABSTRACT. Thermal conductivity, specific heat, and linear thermal expansion coefficient of concrete are measured in a wide temperature range of 201C to 1000'C. The measured results indicate that both thermal diffusivity and thermal conductivity increase significantly in the high-temperature region after an initial decrease up to 4001C which is a negative information concerning the fire-protecting abilities of the material because the heat transfer is accelerated. On the other hand, the thermal expansion coefficient increases in much less significant way, and therefore the resulting thermal stress is not much higher than as calculated from the room temperature data.

Keywords: Thermal conductivity, Specific heat, Density, Linear thermal expansion coefficient, High temperatures

Dr R Černŷ is an Associate Professor at the Department of Structural Mechanics, FCE CTU Prague. He works in the field of development of measuring methods for determination of thermal and hygric properties of building materials.

Professor J Toman is Professor of Physics at the Department of Physics, FCE CTU Prague. His main research interest is in measuring thermal properties of building materials under nonstandard conditions.

Dr T Klečka is Director of Klokner Institute, CTU Prague. He specializes in measuring thermal and mechanical properties and analyzing porous structure of building materials.

Dr P Bouška is Senior Research Worker at the Klokner Institute, CTU Prague. He specializes in measuring mechanical and thermal properties of building materials.

INTRODUCTION

Concrete is a material that can survive severe thermal conditions. There are examples of concrete structures that were exposed to a big fire and after reconstruction they were able to serve again (for instance Great Exhibition Palace in Prague). Nevertheless, not only mechanical but also thermophysical properties of concrete exposed to a fire can be changed significantly due to the temperature increase and to the chemical processes at elevated temperatures, and it is useful to know in what extent.

In the determination of fire resistance of building structures, the time period when the construction is capable of performing its heat-insulating function and protecting the other parts of the building from a fast temperature increase is one of the most important parameters. Duration of this period depends primarily on the external conditions such as the temperature of the fire. However, also the variations of thermal material parameters such as thermal conductivity or specific heat with temperature can play an important role because they can be so significant that the calculations performed with room temperature data are not of any use.

A specific role in evaluating the fire-protection capabilities of a structure plays the linear thermal expansion coefficient. In the conditions of a fire, concrete structures undergo significant thermal stress that can result in a damage of the structure. However, thermal expansion of concrete is commonly measured at room temperature only, and hightemperature region is usually not very interesting for the engineers and designers. This may lead to bad mistakes in the evaluation of the structure response to a fire because for most materials, thermal expansion coefficient increases significantly with temperature and the resulting thermal stress is higher than expected from the room temperature data.

In this paper, the main thermophysicaJ properties, namely thermal conductivity, specific heat, and linear thermal expansion are measured in the temperature range from WC to 10000C.

METHODS FOR MEASURING THE THERMOPHYSICAL PARAMETERS

Theoretical analysis

In the classical theory of linear irreversible thermodynamics (see, e.g., [1-3], the Fourier's Law can express the conduction of heat in absence of cross effects

$$q = -\lambda \text{ grad } T \tag{1}$$

where q is the heat flux, λ is the thermal conductivity and T the temperature.

In the case of materials containing moisture, the heat transfer processes have to be combined with the moisture transfer that leads to the replacement of the classical thermal conductivity λ for "pure" heat conduction by the "common" thermal conductivity λ+

$$\lambda^+ = \lambda + \rho\delta \left[\Theta \frac{\partial \mu}{\partial u} \frac{\rho}{\rho_s} - T\left(\frac{\partial \mu}{\partial T} + \frac{\partial \mu}{\partial u}\frac{\partial u}{\partial T}\right)\right], \tag{2}$$

where u is the moisture content, $u = (m_m - m_d)/m_d$, m_m is the mass of the moisten sample, m_d is the mass of the dried sample, ρ_s is the volume mass of the porous skeleton, μ the chemical potential, δ the Soret's thermodiffusion coefficient, Θ the thermodiffusion ratio, $\Theta = \delta T/D$, D is the diffusion coefficient. In addition it is necessary to include one generalized thermodynamic force more, the moisture gradient, in the relation for the heat flux,

$$q = -\lambda^{+} \text{ grad } T - \beta \text{ grad } u, \tag{3}$$

where β is the Dufour's coefficient.

In the case that also phase change processes and chemical reactions occur in the material, the corresponding volume source/sink terms S_F and S_R, respectively, have to be included into the heat conduction equation. These terms can be replaced using the relations $S_F = \text{div}\vec{I}_F$, $S_R = \text{div}\vec{I}_R$, where $\vec{I}_F$, $\vec{I}_R$ have the meaning of heat fluxes due to phase changes and chemical reactions. Thus, $\vec{I}_F$ and $\vec{I}_R$ can be formally included into the value of the generalized thermal conductivity λ^* which leads to the heat balance equation formally identical with the Fourier's heat conduction equation. In the case of one-dimensional heat transfer we obtain:

$$\rho c_p \frac{\partial T}{\partial t} = \frac{\partial}{\partial x}\left(\lambda^* \frac{\partial T}{\partial x}\right) \tag{4}$$

with

$$\lambda^* = \lambda + \rho\delta\left[\Theta \frac{\partial \mu}{\partial u}\frac{\rho}{\rho_s} - T\left(\frac{\partial \mu}{\partial T} + \frac{\partial \mu}{\partial u}\frac{\partial u}{\partial T}\right)\right] + \beta\frac{\partial u}{\partial T} + \frac{I_F(T,t)}{\frac{\partial T}{\partial x}} + \frac{I_R(T,t)}{\frac{\partial T}{\partial x}}. \tag{5}$$

Determination of the Generalized Thermal Conductivity λ^*

The relation (5) for λ^* is rather complicated and its direct calculation would be probably not very easy for example due to the difficulties in the determination of the chemical potential and the Soret's and Dufour's coefficients. Therefore, the most convenient way to determine λ^* is an experiment. Taking the advantage of the formal agreement of the new formulated heat conduction equation with the classical Fourier's Law, we can employ some of the dynamic methods based on the analysis of the temperature field. The basic principle of these methods consists in measuring temperature fields in the sample at one-sided heating and the subsequent solution of the inverse heat conduction problem (see, e.g., [4] for details). All the mentioned methods require, however, the knowledge of the temperature dependence of the density and the specific heat. The density can be measured directly from the mass and the volume of the sample, the specific heat at high temperatures can be determined for example by the nonadiabatic method from [5].

Linear thermal expansion

In measuring the high-temperature linear thermal expansion we employed the experimental device currently developed in our laboratory (see [6]).

The device consists of a cylindrical, vertically placed electric furnace with two bar samples located in the furnace. The first sample is the measured material, the second sample is a reference material with the known dependence of the thermal expansion coefficient on temperature. The length changes of the samples are measured mechanically outside the furnace by thin ceramic rods that pass through the furnace cover and axe fixed on the top side of the measured sample. These ceramic rods pass by an indefinite temperature field, therefore their elongation is not possible to be determined mathematically and a comparative method of determining the elongation of the rod is used.

MATERIAL SAMPLES

In the experimental measurements of thermophysical parameters, we studied the samples of cement mortar that was chosen instead of real concrete mainly for its better homogeneity, considering the dimensions of samples necessary for the experiments.

The composition of the mixture for one charge was the following: Portland cement ENV 197 - 1 CEM I 42.5 R (Kraluv Dvur, CZ) - 450 g, natural quartz sand with continuous granulometry I, II, III (the total screen residue on 1.6 mm 2%, on 1.0 mm 35%, on 0.50 mm 66%, on 0.16 mm 85%, on 0.08 mm 99.3%) - 1350 g, water - 225 g.

The mortar was prepared by mixing and compacting using a mixing machine and vibrator. The dimensions of the samples were 40 x 40 x 120mm for the thermal expansion, and 70 x 70 x 70mm for the specific heat and for the thermal conductivity. The samples were left in moulds for the first 24 hours in a high relative humidity environment under wetted cloth. After mould removal, the time remaining to 28 days spent the samples in 20°C water and then they were put in protected external environment (a metal-sheet shed) with the relative humidity approximately 65%.

EXPERIMENTAL RESULTS

Figures 1 to 4 summarize the main experimental results of measuring hightemperature material properties of cement mortar. Figure 1 shows that the linear thermal expansion coefficient a was in the expected range up to approximately 40WC, then it began to increase and achieved a maximum at ~ 500°C, and finally at 1000°C it decreased to a value approximately equal to that at 0°C. The peak at ~ 500°C is most probably a consequence of structural changes (removal of crystallically bonded water, etc.) in this temperature range. The measured data for specific heat c in Figure 2 correspond well to these findings. The values of c were approximately constant up to 400°C and then c(T) began to increase.

The behavior of the thermal conductivity vs temperature relation λ(T) was the most interesting in our measurements. The λ(T) function decreased as expected only up to ~ 400°C, and the differences between the Fourier thermal conductivity (preheated sample to 1000°C) and generalized thermal conductivity were only ~30% in maximum which means that the material was relatively well stable even after preheating. However, we have observed a quite unusual behavior for temperatures higher than 400°C where the λ(T) function began to increase fast, and at 8001C it was already higher than at room temperature.

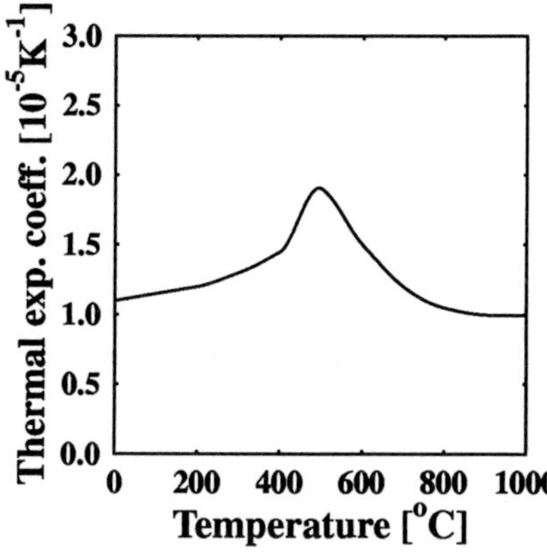

Figure 1 Linear thermal expansion coefficient of cement mortar

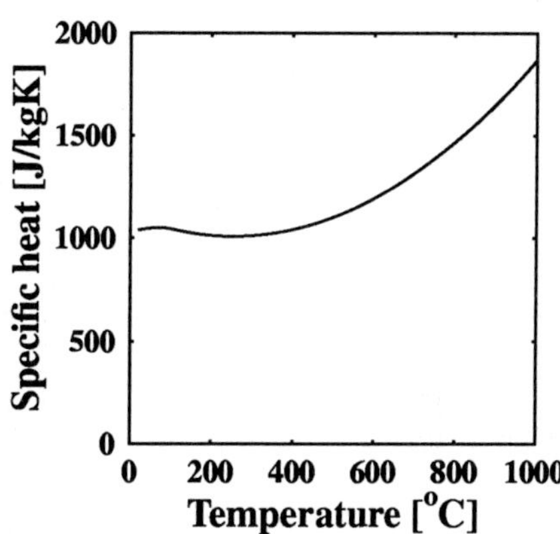

Figure 2 Specific heat of cement mortar

For the sake of comparison, we have done also the measurements of thermal conductivity by classical hot wire method at room temperature. Figure 3 shows that our results axe in a good agreement with these independent measurements and the extrapolation of our measured values to 25°C would give a very similar value as is the hot-wire measured thermal conductivity. The results obtained for the thermal diffusivity (see Figure 4) axe qualitatively very similar to those for thermal conductivity, therefore, the changes of thermal conductivity in the high-temperature region dominated those of specific heat and density.

CONCLUSIONS

The basic thermophysical parameters of cement mortar were determined in the high-temperature region up to 1000°C. A common feature was observed in the measurements of all parameters, the appearance of structural/chemical changes at about 400°C, resulting in significant variations in the material's behavior for temperatures higher than ~ 400°C.

For the fire protection of building structures, the measured results indicate that the heat transfer in concrete is accelerated relatively fast once temperature achieves 400-500°C which is very dangerous because the fire-protecting ability of such a structure decreases compared to the values calculated with data measured at lower temperatures. Using of the high-temperature data in the evaluation of fire-protecting abilities of concrete structures is therefore strongly recommended.

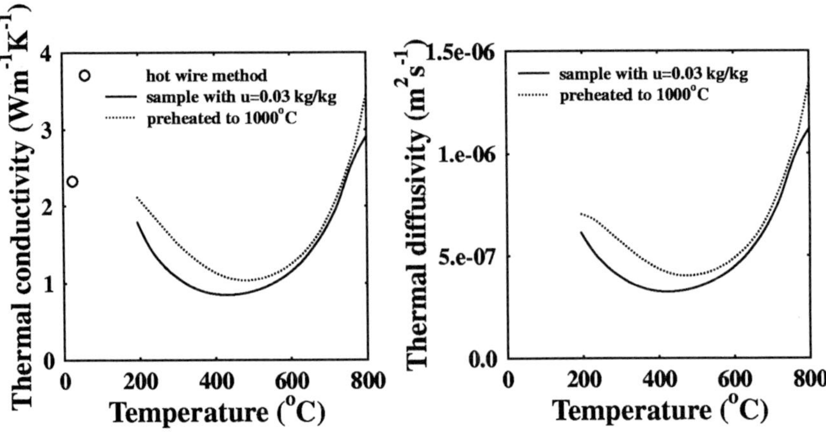

Figure 3 Thermal conductivity of cement mortar

Figure 4 Thermal diffusivity of cement mortar

On the other hand, the thermal expansion in the high-temperature region did not increase very much, and therefore the real thermal stress would not be much higher compared to the calculations with room temperature data that is a positive information.

ACKNOWLEDGEMENTS

This research has been supported by the Grant Agency of the Czech Republic, under grants # 103/97/0094 and # 103/97/K003.

REFERENCES

1. DE GROOT, S R, MAZUR, P. Non-equilibrium Thermodynamics. North-Holland, Amsterdam 1962.

2. PRIGOGINE, I. Introduction to Thermodynamics of Irreversible Processes. Charles C. Thomas, Springfield 1955.

3. KYASNICA, J. Thermodymanics (in Czech). SNTL, Prague 1965.

4. ČERNÝ, R, TOMAN, J. Determination of Temperature- and Moisture-Dependent Thermal Conductivity by Solving the Inverse Problem of Heat Conduction. Proc. of International Symposium on Moisture Problems in Building Walls, V.P. de Freitas, V. Abrantes (eds.), Univ. of Porto, 1995, p. 299.

5. TOMAN, J, ČERNÝ, R. High-Temperature Measurement of the Specific Heat of Building Materials. High Temp, High. Press., Vol. 643, 1993, p. 643.

6. TOMAN, J, ČERNÝ, R. Measuring the Thermal Expansion of Building Materials at High Temperatures. Proceedings of the Seminar: Reseaxch Activities of Departments of Physics in Czech and Slovak Republic, p. 51, STU Bratislava 1997 (in Czech).

NUMERICALLY ASSESSED BEHAVIOUR OF CONCRETE WALLS EXPOSED TO FIRE

J Šelih

National Building and Civil Engineering Institute

Slovenia

A C M Sousa

University of New Brunswick

Canada

ABSTRACT. An analysis which combines heat and mass transport with the calculation of thermal stresses due to non-uniform temperature distribution is presented for a concrete wall exposed to a short duration high intensity fire scenario. The mathematical formulation of simultaneous heat and mass transfer is based on the averaging procedure over a representative elementary volume of the porous medium, and the calculation of the thermal stresses assumes elastic strains only. Particular attention is given to the effect upon the wall response of the modulus of elasticity changes induced by temperature, which, as the results indicate, have a pronounced influence upon the stress distribution in a fire-exposed wall. The results also indicate that thermal stresses higher than those resulting from vapour trapped within the material may develop in concrete walls exposed to fire conditions.

Keywords: Fire, Concrete, Numerical simulation, Thermal stresses.

Dr Jana Šelih is the Head of the Laboratory for Concrete, National Building and Civil Engineering Institute, Ljubljana, Slovenia, and an Adjunct Professor at University of Ljubljana. Her current research interests are durability of concrete exposed to varying deterioration processes and serviceability of concrete structures.

Dr Antonio C M Sousa is a Professor of Mechanical Engineering at University of New Brunswick (Canada), and an acknowledged expert on the numerical simulation of single- and two-phase industrial flows. He has published extensively, with over 150 papers and 1 text to his credit, and his experience and knowledge are much in demand.

INTRODUCTION

Fire is one of the main causes for the failure of a building, as the occurring elevated temperatures may impose high thermal loads to the structural elements. Concrete structures exposed to fire deserve special attention as concrete is weak in tension, and the stresses which develop as a result of the non-uniform temperature distribution may cause serious material failure [1].

To prevent damage, which may cause the eventual collapse of the structure and even the loss of human lives, safety codes, particularly for buildings, have enforced measures designed to guarantee their integrity and survivability in the occurrence of a fire. Usually, the codes require an assessment of the building material by standard experimental tests [2], however, there is a considerable uncertainty related to fire endurance tests and the fundamental phenomena involved in a fire [3].

Combustion in a closed compartment, as it occurs during a fire in a building, involves highly complex mechanisms. As a first approximation, the heat transfer from the fire to the adjacent structure can be expressed by a time-dependent fire temperature. Several time-dependent relations to describe this temperature have been proposed in the literature, e.g. [2,3]. In the present work, numerical predictions are obtained for a fire temperature model obtained by using an overall energy balance equation for the fire compartment [3], the so-called *short duration high intensity fire* (SDHI) temperature model. In this model the temperature is described by a curve that exhibits an early peak followed by a decaying period.

The objective of the present study is to enhance the capability of the existing numerical tool for fire safety assessment of concrete structures [4] by implementing the calculation of the thermal stresses within a concrete wall, as proposed in [1]. Special emphasis is placed to the softening of the concrete exposed to elevated temperatures described by a decreasing modulus of elasticity.

Concrete is a porous material, with its pores filled with water. Elevated temperatures occurring in a concrete structure during fire affect moisture migration within the porous material. Heat supplied from the fire is being partly absorbed by the evaporation of the liquid water, and when the evaporation rate is higher than that of the vapour migration, high pore pressures may develop. Due to the pore structure of concrete, a physically realistic mathematical formulation for combined mass and heat transfer in concrete has to describe free water flow, adsorbed water movement, water vapour and air migration, in addition to the energy changes associated with water phase change.

MATHEMATICAL FORMULATION AND NUMERICAL MODEL

The details of the mathematical formulation developed to describe heat and mass transfer in concrete are fully described in [4], [5], and will not be discussed here. The governing equations of heat and mass transfer are based on Whitaker's formulation [6], and control volume formulation [7] is employed to discretize in space the governing differential equations. Results of the code are generated in terms of spatial distribution of temperature, saturation and total pressures of the gas phase at a pre-specified time.

To calculate the thermal stresses in the concrete wall, the cross-section is divided into n layers as shown in Figure 1. It is assumed that the vertical movement of each layer is prevented. The layer i has a thickness Δx_i, and its stress is determined as:

$$\sigma_{1,i} = -\alpha_{T,i}\, E_i\,(T_i - T_i^0) \tag{1}$$

where $\alpha_{T,i}$ is the thermal expansion coefficient, E_i modulus of elasticity, and T_i and T_i^0 current and initial temperature of layer i, respectively. Vertical movement is prevented, therefore the force acting on the entire cross-section is:

$$F = -\Sigma\, \sigma_i\, A_i' = \sigma_2 A' \tag{2}$$

where A_i' is the reduced area of the layer i, $A_i' = b\,\Delta x_i \xi_i$, and A' the total area of the cross-section, $A' = \Sigma A_i'$. Externally applied stress $\sigma_2 = F/A'$ is acting on each layer, and ξ_i represents the reduction factor that accounts for the changes of modulus of elasticity due to the temperature changes, $\xi_i = E(T_i)/E_{ref}$. E_{ref} denotes the reference modulus of elasticity at 20°C. The moment of inertia of the entire reduced cross-section is:

$$I = \sum_i (A_i' y_i^2 + \Delta x_i^{\,2} \xi_i b/12) \tag{3}$$

where y_i is the distance from the layer i to the centre of gravity, $y_i = |x_{CG} - x_i|$.The distance to the centre of gravity, x_{CG}, is determined with $x_{CG} = (\Sigma A_i' x_i)/A'$, where x_i is the distance to the layer i. The additional bending moment, *M*, defined with:

$$M = \sum_i (\sigma_{1,i} + \sigma_2)\, y_i A_i' \tag{4}$$

yields the stress condition $\sigma_{3,i} = \dfrac{M y_i}{I}$. The total stress due to the non-uniform temperature distribution in the layer i can be defined as $\sigma_i = \sigma_{1,i} + \sigma_2 + \sigma_{3,i}$.

Deflection of the wall, δ, is calculated from the expression δ = (M L^2)/(8EI), where L represents the height of the wall.

RESULTS AND DISCUSSION

To demonstrate the capability of the numerical model, the analysis was performed for a 0.2 m thick wall made of concrete exposed to SDHI fire as depicted in Figure 1 (a). The initial ambient temperature and relative humidity are taken as 20°C and 30%, respectively, on both sides of the wall. Ambient conditions adjacent to the wall not exposed to fire are assumed constant, and at the fire-exposed side, the ambient temperature increases according to the selected fire exposure curve. The mass transfer coefficient is taken as 0.01 m/s, and the heat transfer coefficient and emissivity on both sides of the wall have the value of 20 W/(mK) and 0.5, respectively. The initial saturation and temperature in the wall are 90% and 25°C. The details on the selection of the constitutive relations, mass transfer coefficient, permeability, porosity and irreducible saturation levels are given in [5]. The dependence of the modulus of elasticity, *E(T),* and thermal expansion coefficient, α_T , upon temperature, is shown in Figure 1 (b) and (c) as reported in [8].

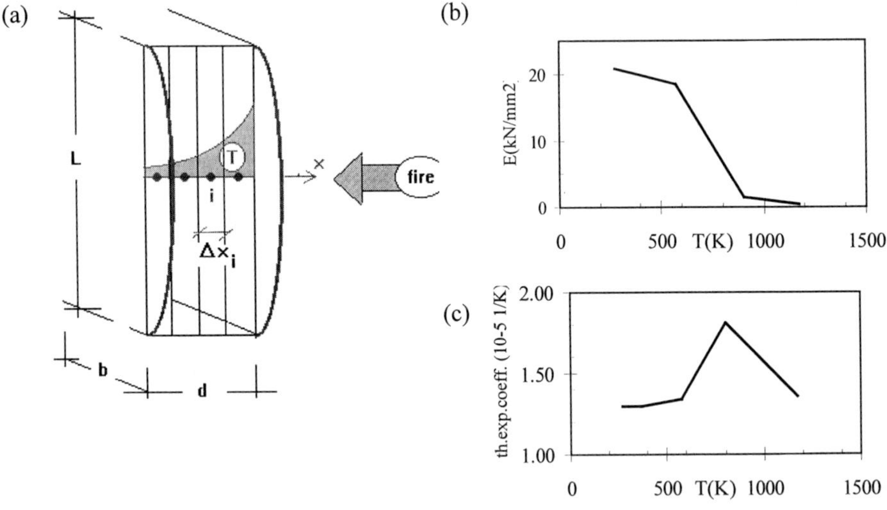

Figure 1 Schematic depiction of the wall exposed to fire: (a), dependence of modulus of elasticity; (b) and thermal expansion coefficient (c) upon temperature [8]

Figure 2 presents temperature and thermal stress distribution in the wall exposed to SDHI fire for exposure times of 25 and 45 minutes; the SDHI fire temperature [3], Figure 3 (a), exhibits the peak value after approximately 20 minutes of fire exposure. The temperature distribution in the wall, Figure 2 (a), shows that while the wall surface exposed to the fire starts cooling down after the peak value is reached, temperatures throughout the wall continue to increase. Figure 2 (b) clearly shows the importance of taking into account the reduction of the modulus of elasticity at high temperatures. At peak temperatures, a thin layer of the wall next to the exposed surface is in tension due to the reduced modulus of elasticity; however, when a constant modulus of elasticity was assumed in the calculation of stresses, the above mentioned layer was under compression. The central part of the wall is in tension while both sides of the wall are in compression due the large increase of the temperature combined with the assumption of elastic strains in the thermal stresses calculation. The assumption of a constant modulus of elasticity yields higher stresses in both tensile and compressive region at equal exposure times.

Temperature, thermal stress and deflection development with time is presented in Figure 3. Several typical sections of the wall are analysed. Deflection is calculated for a wall height of L=3 m. It can be seen that for the time domain analysed, the temperatures at the unexposed side of the wall (x=0.0475m) and at mid-plane (x=0.0975m) increase slowly. Close to the fire-exposed surface (x=0.1925m), significant temperature increase can be observed with peak temperature value of 867 K achieved after 25 minutes of exposure. A temperature-dependent modulus of elasticity has a pronounced influence upon the development of both stresses and deflection with time. Thermal stresses are larger when constant modulus of elasticity is taken into the account throughout the time domain analysed. Deflection calculated with constant E starts decreasing once temperatures and temperature gradients within the wall start decreasing, which is not the case when temperature-dependent E is taken into the account.

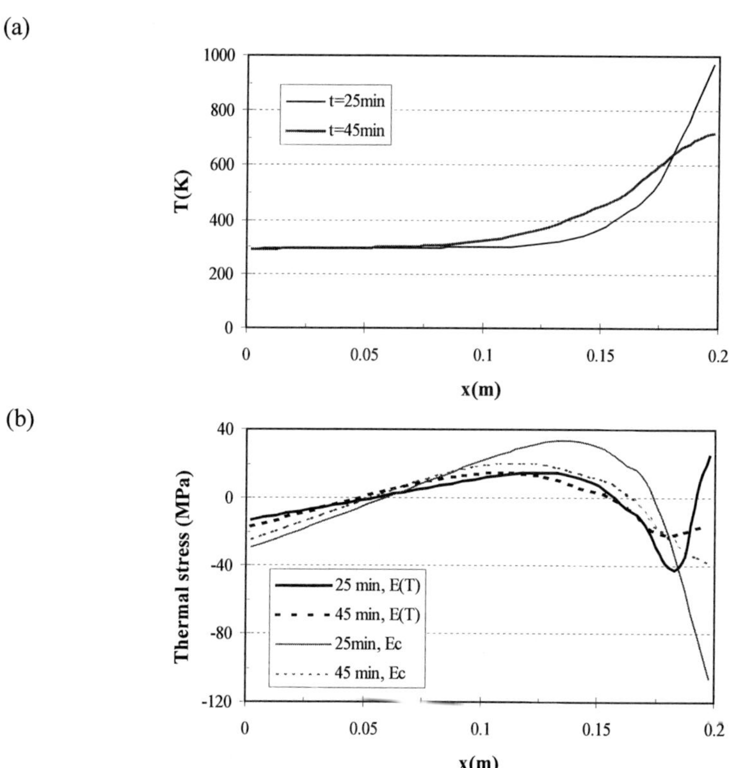

Figure 2 Temperature (a) and thermal stress (b) distribution in the cross-section of the wall after 25 and 45 minutes of SDHI fire exposure; stresses are calculated for constant (Ec) and temperature-dependent modulus of elasticity (E(T))

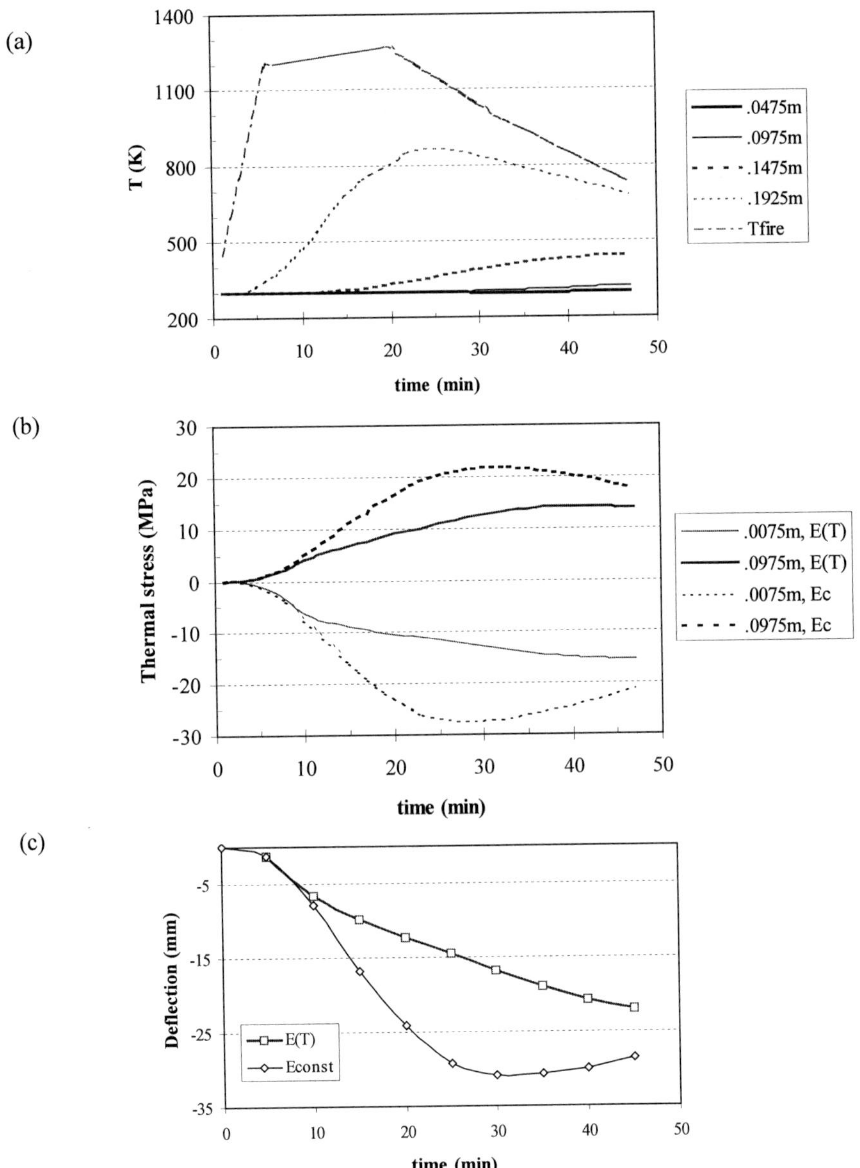

Figure 3 Effect of temperature dependent modulus of elasticity, E(T), upon time development: a) temperature development in characteristic points (x=.0475, .0975, .1475, .1975 m) and time development of SDHI fire temperature [5], b) thermal stress development in time for 2 characteristic points, x=.075 and .0975 m, c) increase of deflection in midplane (x=.10m) with time for a wall height L=3 m

Temperature at the unexposed side of a fire-resistant wall should not increase more than a specified amount [2]. Therefore, response in a section located 0.0475m from the un-exposed wall (section A, Figure 4) is analysed for walls of varying thicknesses. The values of the thickness, d, range from 0.1 to 0.3 m. For wall thicknesses larger of equal to 0.2 m, section A is in compression. Tensile stresses are present in this section only when wall thickness equals to 0.1 m. This agrees with the observations related to Figure 2, as for this thickness, section A almost coincides with the mid-plane of the wall. For the time domain analysed, significant temperature increase and saturation drop in section A is predicted only for the 0.1 m thick wall.

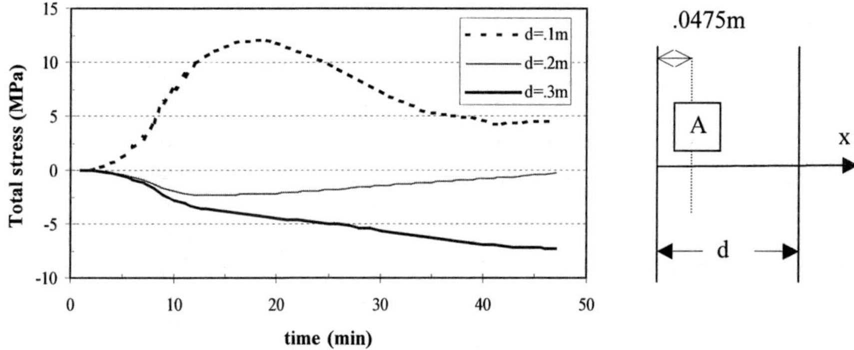

Figure 4 Influence of wall thickness, d, upon thermal stress evolution with time at a distance of x=0.0475 m (section A) from the unexposed side of the wall

CONCLUDING REMARKS

The numerical model presented has the capability of providing a time-dependent analysis of heat and mass transfer in concrete combined with the computation of thermal stresses. Simulations were conducted for concrete walls with different thicknesses subjected to short duration high intensity (SDHI) type of fire.

The results show that the occurring thermal stresses in the wall are large and they may affect seriously its structural integrity, i.e. the predicted values may be larger than typical concrete compressive or tensile strengths. The simplification of the model yielding a linear stress-strain relationship leads to stress distributions far more conservative than those obtained with temperature-dependent Youngs' modulus.

Typically, the central part of the wall can be expected to be in tension, while the sides of the wall are in compression. Based on the predictions obtained, it can be concluded that the magnitude of thermal stresses can be the most important single factor in the risk assessment of a concrete structure exposed to fire.

REFERENCES

1. LENCZNER, D. Movements in buildings, Pergamon Press, New York, 1981.
2. ASTM Standard E-119. Standard test methods for fire test of building construction and materials, ASTM, Philadelphia, Pa., 1989.
3. ELLINGWOOD, B R. Impact of fire exposure on heat transmission in concrete slabs, J. Struct. Engrg., ASCE, Vol. 117, No. 6, 1991, pp 1870-1875.
4. ŠELIH, J, SOUSA, A C M BREMNER, T W. Moisture and heat flow in concrete walls exposed to fire, J. Engrg. Mech., ASCE, Vol. 120, No. 19, 1994, pp 2028-2043.
5. ŠELIH, J. Movement of water during drying of saturated concrete, Ph.D. Thesis, University of New Brunswick, Fredericton, NB, Canada, 1994.
6. WHITAKER, S. Simultaneous heat, mass, and momentum transfer in porous media: a theory of drying, Adv. in Heat Transfer, Vol.13, 1977, pp119-200.
7. PATANKAR, S V. Numerical heat transfer and fluid flow, Hemisphere Publ. Corp., New York, NY, 1980.
8. KEATS, C. Effect of elevated temperature on concrete, M.Sc. Thesis, University of New Brunswick, Fredericton, NB, Canada, 1994.

EVALUATING STRUCTURAL CONCRETES USING SUBSURFACE RADAR

A Goodier

S L Matthews

S Massey

Building Research Establishment Ltd.

United Kingdom

ABSTRACT. This paper reports work undertaken by the BRE on the use of subsurface radar to evaluate the properties of concrete. This work forms part of a large Brite Euram research project which examined the use of subsurface radar as a means of investigating a range of building and construction industry materials and structures. The investigations reported in this paper are as follows: A study investigating the influence of various concrete mix and conditioning parameters upon the dielectric properties of structural concrete at frequencies generallly used for structural investigations; The development of analytical dielectric models of plain structural concretes; and the use of a large reinforced concrete laboratory model to demonstrate the effectiveness of the technique in realistic conditions.

Keywords: Subsurface radar, Dielectric properties, Analytical dielectric models, Laboratory model studies, Finite difference time domain modelling.

Mr A Goodier is a Senior Research Engineer in the Centre for Construction Repair and Refurbishment at the Building Research Establishment Ltd. His main research interests include the application of advanced computing techniques to construction industry problems, developments in subsurface radar and structural monitoring.

Dr S L Matthews is the Director of the Centre for Concrete Construction at the Building Research Establishment Ltd. He has gained a wide variety of experience in the investigation, appraisal and monitoring of existing buildings, including the assessment of deteriorating structures and the specification and design of remedial measures for repair and restoration.

Dr S Massey is a Senior Research Scientist in the Centre for Heritage, Archaeology, Stonework and Masonry at the Building Research Establishment Ltd. His main research interests include computer modelling of pore systems and dielectric responses, weathering and deterioration of materials.

INTRODUCTION

This paper describes several studies relating to the non-destructive testing of concrete using subsurface radar. The breadth of the studies indicates the range of research being undertaken in this field by the BRE, and illustrates some of the capabilities of subsurface radar to evaluate concrete. The work reported is part of a wider programme of research to improve the effectiveness of sub surface radar for the investigation of concrete structures. The studies reviewed in this paper include:

- Dielectric measurements on plain concrete slab specimens. This involved measurements on a range of concrete specimens, environmentally conditioned in a number of ways using a commercial subsurface radar.
- Development of an analytical dielectric model calibrated using experimentally derived data. If concrete dielectric properties have been estimated via subsurface radar the next step is to use these to give an indication of the physical properties of concrete, such as moisture content. This requires the use of a suitable analytical dielectric model. The BRE developed a new modelling approach.
- Investigations of large scale models. Assessing the dielectric properties and hence concrete properties of plain concrete is relatively straightforward. However, for more complex structural forms this becomes considerably more difficult. A step towards assessing the dielectric properties on field structures is the use of laboratory physical models. An example study is presented.

Subsurface Radar

Most commercial radar systems operate by radiating pulses of radio frequency electromagnetic energy from a transmit antenna into the body under investigation. A receive antenna detects the energy reflected from discrete changes in the conductivity or permittivity of the medium concerned. Further understanding can be gained from the Figures 1a. to d.

- Figure 1a shows a simplified representation of a reinforced concrete wall with a void behind it.
- Figure 1b shows the transmit antenna emitting a pulse of electromagnetic energy.
- Figure 1c shows the pulse travelling through the wall, changing as it does until it encounters the wall / air interface of the void.
- Figure 1d shows some of the energy being reflected back, whilst some continues to travel forward from the various interfaces within the model. The energy reflected from the interfaces is recorded at the receive antenna.

If the collations of the reflections are plotted against time a plot similar to that shown in Figure 1e is created. This form of presentation is known an A-scan. If a set of A-scan plots from a number transmit / receive positions are collated together this creates a plot like that shown in Figure 1f. This is generally known as a radargram. Further introductory information upon subsurface radar can be found in a number of publications [1,2].

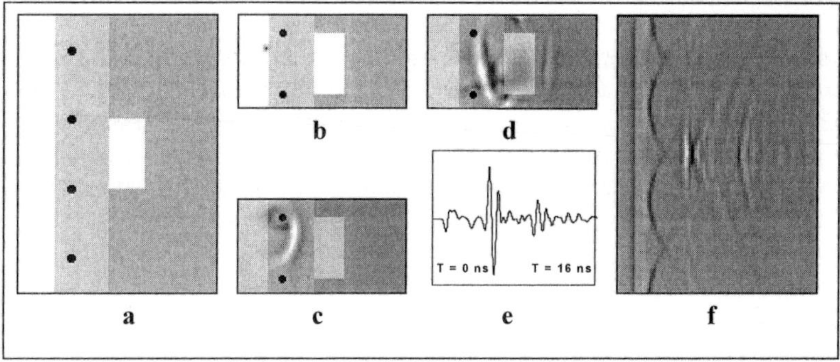

Figure 1 Subsurface radar operation and presentation of data

DIELECTRIC MEASUREMENTS UPON CONCRETE SLAB SPECIMENS

Ten concrete mixes were examined typical of those used in the construction industry. The concrete mix designs incorporated commonly used constituent materials including Portland Cement (PC) and cement substitutes Pulverized Fuel Ash (PFA) and Ground Granulated Blastfurnace Slag (GGBS). The specimens had a compressive strength within the range of 20N/mm^2 to 60N/mm^2. Other potential influences studied included poor compaction (honeycombing), air-entrainment, coarse aggregate size, the addition of steel fibres, salt contamination, temperature variation, curing and resaturation. The specimens were subjected to various forms of environmental conditioning involving wetting / drying, different types of curing regime and the introduction of chlorides (de-icing salts). Moisture contents varied from saturated to air dry (for the final drying conditioning the specimens were heated to 60°C).

A total of 24No slab specimens measuring 0.6m by 0.6m and either 150mm or 300mm thick were cast using 8No types of concrete (EC1 - EC8). These were sealed on 5 faces with an epoxy resin to promote uniaxial drying via the remaining (0.6m by 0.6m) unsealed face.

Chloride contamination was achieved by introducing dried specimens into a sodium chloride (NaCl) solution. Final chloride contents / profiles in the concrete slabs were evaluated from incrementally drilled dust samples by chemical (potentiometric titration) techniques.

The electromagnetic measurements were performed using two commercially available subsurface radar systems, these being a Pulse Ekko PE1000A system with (nominal) 1.2GHz antennas and a GSSI SIR 3 system with a (nominal) 1GHz antenna. A range of measurements was made in both transmission and reflection modes to obtain:

- wave velocity, determined by transmission measurements
- measurement of reflections from the front and rear surfaces

The measurements reported in this paper were made with the antennas located on the surface of the concrete specimens. Measurements were performed from both the sealed (wetter face during drying conditions; drier face during wetting conditions) and the unsealed faces. In addition, data was obtained with and without a metal plate being present on the rear face of the specimen.

For transmission and reflection measurements involving the determination of travel time through the specimen, the real part of the relative dielectric permittivity was estimated from the relative wave velocity in the conditioned concrete using the following equation for low loss dielectrics:

$$\varepsilon' = \left(\frac{c\,.t}{z} \right)^2 \quad (1)$$

ε' = real part of relative permittivity (also commonly referred to as the *dielectric constant*)
c = velocity of light in vacuum
t = travel time of electromagnetic wave (one-way or two-way travel as appropriate)
z = distance travelled by electromagnetic wave (one-way or two-way travel as appropriate)

Results of Dielectric Measurements

Influence of moisture content: The dominant influence was found to be the moisture content of the concrete. Both the real part of the relative permittivity and the loss (attenuation) values were found to increase with moisture content. However the real part of the relative permittivity - moisture dependency exhibited a number of interesting characteristics.

- Drying after curing: This resulted in reductions in both the real part of the permittivity and the loss component as the concrete became drier. The PC concretes of different strength grades exhibited an almost linear relationship between the real part of the permittivity and volumetric moisture content over this part of the data set.
- Wetting after curing: Slabs submerged in water after standard moist curing exhibited reductions in both the real part of the permittivity and the loss component as the concrete became wetter. This effect is thought to relate to the ongoing hydration taking place in the maturing concrete specimens.
- Resaturation after drying: Slabs immersed in water after drying took up water very quickly, leading to a close grouping of measurement points at relatively high moisture contents. However, it appears that the path followed during resaturation is likely to be similar to that followed during drying after curing.
- Re-drying after saturation: Although the very slow rate of moisture loss from the concrete slabs stored under air-drying conditions restricted the range over which data was obtained, the slope of the trend lines for the PC concretes of different strength grades (20, 40 and 60N/mm^2) appear to be comparable with those obtained during initial drying.

Influence of curing of concrete: There was no apparent change in the relationship between the real part of the complex permittivity and volumetric moisture associated with the curing of the concrete.

Compressive strength of concrete: PC concretes of different strength grades (20, 40 and 60N/mm^2) had different values for the real part of the complex permittivity and loss.

Size of coarse aggregate : The change from 10mm to 20mm coarse gravel aggregate made no noticeable difference to the dielectric properties of the concrete.

Use of cement replacement materials : The addition of GGBS and PFA to the concretes investigated altered the relationship between the real part of the complex permittivity and volumetric moisture content. The addition of GGBS had the most significant effect, noticeably increasing the value of the real part of the complex permittivity relative to that of the comparable 40N/mm^2 PC mix.

Influence of voidage and honeycombing: In the case of the honeycombed concrete slabs, the replacement of a large proportion of the fine concrete solids (matrix) by air produced a significant drop in the real part of the complex permittivity relative to that of the dense concrete mixes examined.

The dielectric properties of the concrete mixes containing entrained air (nominally 5% and 9%) were similar to those of the 20 N/mm^2 PC mix, suggesting that entrained air in the form of the small diameter spherical inclusions has little influence upon either the :

- relationship between the real part of the complex permittivity and volumetric moisture content, or
- the real part of the complex permittivity.

Influence of salinity: Chloride contents of up to some 0.9% by mass of sample caused an appreciable increase in the apparent real part of the relative complex permittivity, by a value of about 2 units, and a dramatic increase in the loss (attenuation) value.

ANALYTICAL DIELECTRIC MODELLING OF PLAIN CONCRETE

An essential part in many of the analytical dielectric models suggested to date [3,4,5] is the assignment of dielectric values to the dry components and to the water phase. In previous work these values have mostly been determined from laboratory specimen measurements, curve fitting or assumed as *a priori* values gathered from the literature.

Laboratory measurements usually make the assumption that simple extrapolation to zero moisture will give the correct value. However, the nominal zero value derived from such experimental data may not correspond to the state where there is absolutely no water in the specimen. This then raises the question - *What is the true value for a 'waterless' concrete ?*

The BRE has addressed this by developing a model referred to as the Molar Refractive Model (MRM). The MRM also addresses questions about the composition of the concrete derived from the cement composition, the aggregate content and the void fraction with respect to the dielectric properties of concrete.

Basis of the MRM

Within organic chemistry it is know that the optical properties of various compounds are related to the molecular mass and density by means of what is defined as the molar refraction R [6]:

$$R = \left[\frac{n^2 - 1}{n^2 + 2}\right] * \left[\frac{M}{\rho}\right] \qquad (2)$$

n = index of refraction, M = molecular weight, and ρ = density.

For organic compounds the molar refraction is additive with respect to mixtures of compounds and also with individual elements and their associative bond structures. This additive property is based on the assumption that there is no interaction between the various components in the mixture. The basic physics derives from assuming that an individual molecule is located roughly within a spherical hole in a liquid (or solid) and the electric field within the hole is enhanced by a factor of 1/3 giving the Clausius-Mossotti equation:

$$3 * \frac{\varepsilon - 1}{2 + \varepsilon} = K\beta \qquad (3)$$

The numbers 2 and 3 appear due to the factor of a 1/3, which would be different had the hole, in which the molecule is situated, been other than spherical. K is the number of molecules per unit volume and β is the mean atomic polarisability of the molecule. Taking β/3 as constant at a particular frequency and using $\varepsilon = n^2$ then the origin of the molar refraction can readily be seen. The form of the Clausius-Mossotti equation also governs the form of the volumetric models suggested by others [7].

Given that the index of refraction is merely the high frequency value of the permittivity, the method enables calculation of the permittivity at optical frequencies. The assumption being made is that it is possible to estimate the permittivity of a mineral at 1 GHz from the optical index of refraction by means of a linear equation.

In order for the technique to work there needs to be a consistent value of the molar refraction for each molecular unit or element across the cement chemistry and natural rock forming mineralogical groups. It can be shown that a mineralogical suite comprising the basic cement chemistry for both the hydrate and unhydrated states plus the rock forming minerals quartz, calcium carbonate and feldspar form a consistent group of molar refraction values for either:

- The main cement groups C, A, F, S (e.g. the individual components forming C_3A, tricalcium silicate), the mineral groups (e.g. CO_3, carbonate in calcium carbonate), and water (H_2O), or
- The individual elements H, C, Si, O, Mg, Al.

The precise values for each of the components depends on the particular groups and sub-sets of minerals chosen. To apply the approach to concrete some form of model is required to represent the various mineralogical fractions present within the concrete matrix.

The simplest of these is the one initially developed by Powers for PC [8]. The permittivity values predicted by the MRM for 'waterless' concrete mixes are shown in Table 1.

The values without air voids are required for the three phase mixture models. The inclusion of the air voids allows comparison with nominally dry concrete values. These permittivity values have a range of 0.4, which is barely detectable with current radar systems.

It should be noted that the only unknown in the 'waterless' concrete model is the degree of hydration once the initial unhydrated cement chemistry and the water/ cement ratio has been determined. Thus concretes of the same composition should exhibit the same moisture response - given that they have reached the same state of hydration.

Table 1 Results obtained from the MRM

MIX	'WATERLESS' DIELECTRIC PERMITTIVITY (AIR VOIDS NOT INCLUDED)	'WATERLESS' DIELECTRIC PERMITTIVITY (AIR VOIDS INCLUDED)
EC1	6.70	5.6
EC2	6.66	5.7
EC3	6.74	5.9
EC4	6.72	6.0
EC5	6.70	5.9

EC1 = 20MPa PC, EC2 = 40MPa PC, EC3 = 60MPa PC,

EC2 = 40MPa PC + PFA, EC2 = 40MPa PC + GGBS

Assuming that the cement paste is 50 to 60 % hydrated, the values for the total porosity (i.e. saturated moisture content) of the mixes agree to ± 2% of the model values. Within the model is possible to differentiate between the gel moisture content and the total moisture content. The calculated gel moisture content, which is assumed to be trapped within the gel, is again in good agreement with the measured moisture content after six months drying. These facts suggest that the concrete model is reasonable not only for estimating the composition of the concrete but also the void space.

Whilst it was not the intention to use the model to estimate the effects of moisture content, it is relatively easy to estimate a figure for R_{H2O} and appropriate ranges of moisture content. Using the experimental data a best estimate value for R_{H2O} is 95 for the water interaction. This is the value which most nearly fits all the results for the mixes EC1 - EC5. Better fits can be achieved for the individual mixes, but it is assumed that there is only one true value.

The inclusion of moisture in the model generally produces values for the permittivity to within 0.5 units of the measured values at 1GHz. Given that the observed (and estimated) measurement error is of the order of 0.3 - 0.4 units, the results are acceptably close. This is for a model which is based on PC, with no account being made for additions to the concrete mix such as PFA and GGBS. Thus it might be possible to conclude that such additions result in fairly minor changes to the mix.

Comparison of Selected Dielectric Models

Taking the 'waterless' concrete values as the initial starting point for each of the selected dielectric models (Linear, Complex Refractive Index Model (CRIM), De Loor (using inclusion or host material as the matrix) and *Continuous grain size distribution model (CGSDM)* and solving for the dielectric permittivity, either explicitly or by numerical means, shows that the Linear and CRIM methods overestimate at high moisture, while De Loor (either) and CGSDM are reasonable estimators. The values used in the simulation were $\varepsilon_{wat} = 80 - 7i$ [9], $\varepsilon_{air} = 1$, ε_{conc} using 'waterless' values without air as tabulated earlier.

EXPERIMENTAL MEASUREMENT OF LARGE SPECIMEN CONTAINING A INDUCED HONEYCOMB DEFECT

Aims of the Model Study

- An aim of the study was to establish the dielectric properties of the concrete comprising the model. This splits down into a number of sub-aims, which included:
  - Investigating how difficult it was to detect honeycombing within concrete elements using subsurface radar. That is, can the different types of concrete be differentiated?
  - Investigating whether the form of the interface between the sound and honeycomb concrete effected the detection of the different types of concrete.
  - The position and size of the different types of concrete could be determined.
  - Investigating what factors can make it difficult to detect the different types of concrete.

The Subsurface Radar Surveys Undertaken

A number of surveys of a honeycombed beam (HCB) specimen were undertaken. The data was collected using commercial radar with 1.2GHz nominal centre frequency antennas. Subsurface radar data was collected using two main methods: (i) simple traverses, (ii) on a 5cm grid. For both methods the data was collected in reflection and transmission modes.

The data collected using the first method was examined either as a radargram or as an A-scan. The second method of data collection enabled a datacube to be produced. This was treated in a manner similar to that suggested by Annan *et al* [10].

Results of the Subsurface Radar Surveys

How difficult is it to differentiate between the different types of concrete? Figure 3 shows a radargram from a traverse along the centre line of the face of the HCB (the rough side of the honeycombing is nearest the antenna). The effect of honeycombing is visible as a group of weak reflections and by a reduction in the travel time for the reflection obtained from the back face of the HBC. From this the depth to the honeycombing can be estimated.

What factors can make it difficult to detect the different types of concrete? Figure 4 shows a radargram for the same traverse as that shown in Figure 3, but with the antenna orientated to maximise the response from the reinforcing bars. In this case identification of the honeycombed volume is much more difficult. This is due to the disruption introduced by the combination of the long pulse radiated by the antenna and multiple reflections from the reinforcing bars. This causes a 'ringing' which repeats down the radargram. It is difficult to establish a change in the travel time for the back face reflection. However it can be seen in Figure 4 that the appearance of the response is different in the zone containing the honeycomb insert.

How does the form of interface between sound and honeycomb concrete effect the detection of the different types of concrete? Figure 5 shows the radargram taken from a traverse across the vertical face of the beam nearest the smooth side of the honeycombed volume. The response from this side is noticeably different to that obtained from the rough side - refer Figure 3.

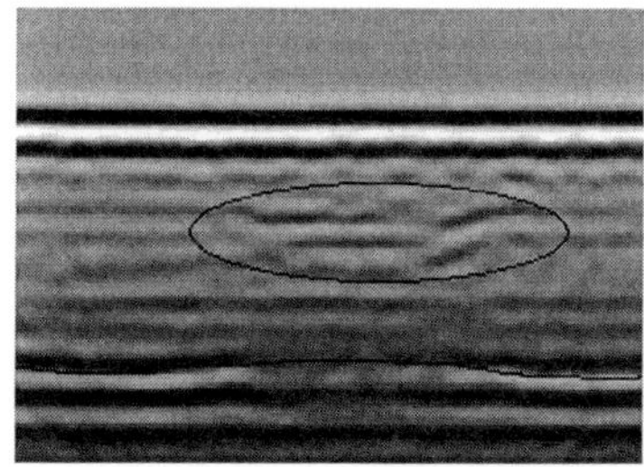

Figure 3 Radargram from face nearest the rough side of the honeycombed volume

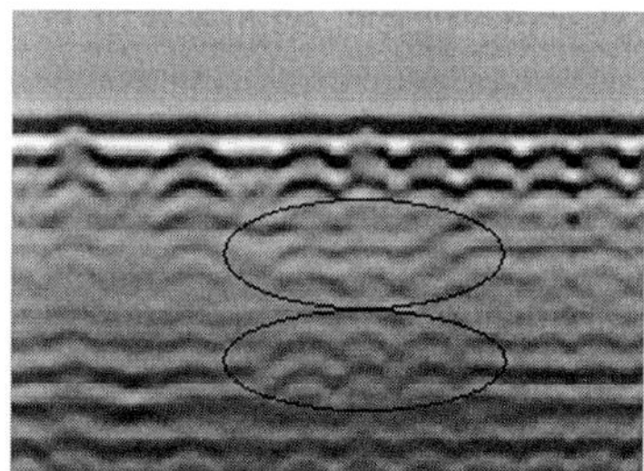

Figure 4 Radargram with antenna orientated to maximise reinforcing bar response

Can the position and size of the different types of concrete be identified? An estimate of the length of the honeycombed volume can be made from the radargrams presented as Figures 3, 4 and 5. However more information can be obtained from the results of the datacube study - refer Figure 6. In this case not only can the honeycombed volume be easily identified, but its approximate length and width dimensions can also be established.

Deriving the Dielectric Properties of the Concrete Laboratory Model

Establishing the dimensions for this model is easy. The depth of the honeycomb volume is can be estimated and the dimensions of the beam are measurable. From this the permittivity of the sound concrete was estimated as 9. The dielectric measurements on plain concrete specimens for this type of concrete indicate that the permittivity is associated with a moisture concrete of approximately 12%. If the sound concrete used in the model had not been the same as one of those measured, estimates of the moisture content could have been made using the analytical modelling approach. The permittivity of the honeycombed concrete was 4, from which the moisture content was estimated to be approximately 6%.

Figure 5 Radargram from HCB face nearest the smooth side of the honeycombed volume

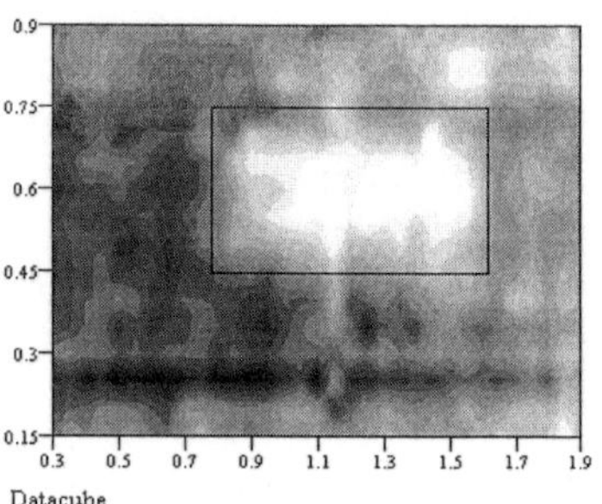

Figure 6 Time slice from the datacube study

CONCLUSIONS

This paper has demonstrated a number of research studies concerned with the evaluation of structural concrete using subsurface radar. The paper has shown how linked studies can be used in conjunction to improve interpretation of radar measurements and the estimation of materials properties such as moisture. The next stage is the evaluation of field structures where even less is known about the nature and composition of the components forming the structure. Studies have been undertaken on this including the use of numerical modelling and neural networks to assist the evaluation and interpretation of the data. These will be reported in other publications.

ACKNOWLEDGEMENTS

Financial support for this work was received from the Department of the Environment, Transport and the Regions, UK and the European Commission. The work forms part of the BRITE EURAM III project; *Subsurface Radar As A Tool For Non-Destructive Testing And Assessment In The Construction And Building Industries*; Project No: BE 95-2109.

REFERENCES

1. MATTHEWS, S L. Application of Subsurface Radar as an Investigative Technique. BRE Report No 340, CRC Ltd, Watford, 1997.

2. CONYERS, L B AND GOODMAN, D. Ground Penetrating Radar, an Introduction for Archaeologists, Altamira Press, California, 1997.

3. VAN BEEK, L K H. Dielectric Behaviour of Heterogeneous Mixtures, Progress in Dielectrics, Vol. 7, 1967, pp 69-114.

4. DE LOOR, G P. Dielectric Properties of Heterogeneous Mixtures, Applied Science Research, 1964, B11, pp 310-320.

5. TSUI, F AND MATTHEWS, S L. Analytical Modelling of the Dielectric Properties of Concrete for Subsurface Radar Applications, Construction and Building Materials, Vol. 11, No. 3, 1997, pp149-161.

6. WEAST, R C, ASTLE, M J AND BEYER, W H (Eds.). CRC Handbook of Chemistry and Physics, 66th Edition, 1985.

7. SAROKA, I. Portland Cement Paste and Concrete, Macmillan Press Ltd., 1979.

8. HASTED, J B. Aqueous Dielectrics, Chapman and Hall, London, 1973.

9. MATTHEWS, S L, GOODIER, A, MASSEY, S AND VENESS, K. Permittivity Measurements and Analytical Dielectric Modelling of Plain Structural Concretes, GPR '98 : 7th International Conference on Ground Penetrating Radar, USA, May 1998, Vol. I, pp 363-368.

10. ANNAN, A P, DAVIS, J L AND JOHNSTON, G B. Maximising 3D GPR Image Resolution - A Simple Approach, Sensors and Software, 1997.

THEME SIX:

CONTRACT CONDITIONS/ MANAGEMENT

ARBRITRATION COSTS - IN REAL TERMS

C G Thompson

Con-Tech Associates Ltd

United Kingdom

ABSTRACT: The construction industry is by definition an adversarial industry. Almost all construction forms of tender include for the provision of some means of settling disputes that may arise during the construction period. Relatively few construction contracts of reasonable size reach completion without the inclusion of claims and counterclaims being made, many of which result in arbitration. Numerous claims consultants advocate that arbitration is the shortest and most cost-effective means of settling construction claims, since it is headed by a chosen construction professional (the arbitrator) who is often an expert in the area involving the dispute and whose appointment must be agreed by both parties. However, case studies will confirm that, arbitration is not always a commercially realistic option. This paper is written in an effort to emphasise the pitfalls and the question to be addressed before embarking on a course of arbitration.

Keywords: Construction disputes, Construction claims, Claims consultants (C.C.), Bills of quantities (B.O.Q.), Forms of tender, Client's role, Consultant's role, Contract administrator (C.A.), Contractual interpretation, Arbitration, Adjudication, Arbitrators ruling, Alternative dispute resolution (A.D.R.).

C G Thompson is the Projects Director of a construction company specialising in concrete repair, cathodic protection, grouting, ground stabilisation, shotcreting, protective coatings and natural stone repairs and replacement. He has worked throughout the United and has extensive experience in the Middle East and Africa both in contracting and consultancy. During the past several years he has represented his company in a lengthy dispute involving arbitration. As a result of his experience he has complied this paper, hopefully to enlighten those who are currently considering arbitration as a means of settling their dispute.

INTRODUCTION

Arbitration (definition): The hearing and settlement of a dispute between two parties; as between two countries; two persons; a company and a union; between a contractor and a client; by the decision of a third party or court to which the matter is referred by the contestants as a means of avoiding war, a strike, a lawsuit or other extreme action.

These definitions have been taken from an international dictionary and are to a large extent reasonably accurate. However, many construction professionals would argue that in construction terms, the above definitions are academic and do not warn the parties of the often protracted, painstaking and time consuming input in order that the ensuing claim stands a reasonable chance of success.

It does not warn the parties of the huge costs which may be incurred especially where the respondent decides to engage the services of a consultant who at any early stage in the proceeding, decides to justify their existence by introducing a procedure known as 'Security of Costs' followed by a counterclaim usually based on clever contractual or legal points raised in a bid to camouflage the issue.

Again the definition does not prepare the parties that where lengthy periods of claim preparation have resulted, the entire emphasis of the case will shift from fighting the 'Claim' to 'Protection of Costs'.

CONTRACTUAL DISPUTES

The Costs

It is estimated that disputes, culminating in litigation and arbitration are costing the industry upwards of £300 million a year. It is generally felt that this is a conservative estimate since it does not account for cases settled out of court.

The Issues

Various articles state that personality clashes between consultant and contractor and the ensuing polarisation is the reason for the majority of contractual disputes.

However, publications of case studies would contradict this theory listing the following as amongst the most prominent:

A. Additional works leading to prolongation of contract; the contracts administrator (C.A.) refuses to issue a certificate for extension of time; refutes contractor's claim for loss and or expense.

B. Unfair instructions such as partial possession, resulting in the contractor's access or egress being reduced, having to make alternative arrangements for storage, additional or off-site storage, accelerated progress refusing to reimburse the contractor.

C. Inadequate preparation of contract documents; inadequate description of the works or sections of the works; misleading or ambiguous instructions; inaccuracies in dimensions; absence of dimensions.

D. Other contractual problems leading to additional works; additional management inclusion resulting in additional costs, out of sequence operations, contractual frustration.

THE INTERPRETATION

It is generally accepted that a contractor faced with the above problems in any combination, must consider the possibility of legal action. However, before embarking on this route it would be advisable to consider a diplomatic approach, in an effort to reach an accommodation whereby a reasonable settlement would follow. Legal action should be a *last resort.*

Nonetheless, construction professionals have never been noted for their diplomacy or their diplomatic application and as a result, what should have been a reasonable objection followed by a reasonable response has developed into a contractual dog-fight with threats, and counter threats claims and counter claims replacing common sense and the desire to construct.

The resulting stand-off irrespective of who is to blame leaves the contractor no option but to pursue a legal course of action since failure to do so could result in liquidated and ascertained damages being applied, compounding the loss.

At this stage, the contractor will inform the client through his agent that legal action appears to be the only means of settling the dispute stating the relevant contractual clause.

The serving of such notice does not mean that legal action is obligatory, the contractor may withdraw the notice in the event of both parties reaching an accommodation.

THE DECISION

To arbitrate or not to arbitrate is a difficult decision and one, which necessitates a great deal of thought. However, in many instances the decision is dictated by legal obligations and finances, in otherwords *can we afford not to arbitrate?*

In the event of arbitration being considered necessary, the next step to be deliberated is either to engage the services of an external consultant or to attempt the handling of the case in-house.

Since few contractors' staff have sufficient training and expertise in construction law and since they are often too close to the case to make a valued judgement, it would probably be prudent to employ an external consultant with a background in dispute resolution and the fluency and articulation to achieve maximum results.

THE PROCEEDINGS

Pre-Arbitration

On realising that time is of the essence, the average contractor will probably select their legal consultants by perusing the various construction publications usually making their selection on company profile. This often means that the large multi-discipline practices of national acclaim are at an advantage since these practices can afford to advertise their services and boast their successes (few consultants advertise their failures). As a rule, the fees charged by these large consultants are as proliferious as their advertising campaigns.

Prior to the appointment, the legal consultant's representative will meet the contractor. This meeting will be referred to as an initial or familiarisation meeting during which problems, reasons, events and indeed all other relevant issues will be discussed and cogitated.

History informs us that during these meetings, legal consultants vying for business have a tendency to inform the contractor of the watertightness of the case. This can have the effect of luring the over-zealous contractor into a false sense of security and in doing so can obscure the difficulties that lie ahead.

The claim's consultant may be duly appointed given express instructions to represent the contractor from the date of appointment forth.

The client's response to legal representation may be to agree to dialogue or conversely to diametrically oppose the claim.

In the event of the latter being the order, the legal consultant will inform the client that it is the intention of the contractor to refer the dispute to arbitration in accordance with the relevant contractual clause and also requesting concurrence in the appointment of an arbitrator.

The client's response, agreeing to the appointment of the arbitrator signals the official commencement of the arbitration process following which the contractor will be referred to as the claimant and the client thereafter called the respondent.

The Process

The claims consultant will allocate a representative (possibly a partner) to be responsible for the case. Should the case be of a complex nature and most construction cases are, the partner will probably have a team of understudies or junior colleagues to assist with the inevitable mountains of paperwork.

The claimant will furnish the legal team with all claim details, correspondence, diary entry dates, times etc. The claimant will probably be of the opinion that since all relevant information has been collated in-house, culminating in the form of a draft claim document, the main purpose of employing a legal consultant is to have the claim scrutinised and modified where necessary in order that all legal requirements are met. In otherwords the legal consultancy will merely act as a legal support team.

However, this situation changes dramatically when the arbitrator calls for a preliminary hearing to discuss a programme and during the preliminary hearing, the respondent, may make representation to the arbitrator for security of costs alleging that the claimant may be forced into receivership in the event of the ruling be found in favour of the respondent.

Should the arbitrator award security of costs in favour of the respondent, the arbitrator will order the claimant to provide such security in the form of banker's draft, bank guarantee or whatever means the arbitrator finds acceptable.

The introduction of such an award can leave a claimant with hefty problems thus:

i) The immediate payment of the entire sum of money as directed by the arbitrator.
ii) The effect this may have on a nervous bank manager, especially if the contractor's recent profit margins were relatively low in relation to turn over.
iii) Budget provisions had not been made for such payments.

The claimant on realising the dilemma will undoubtedly question if:

a) The claims consultant should have known and warned of such an event.
b) The case is turning in favour of the respondent.
c) The costs can be capped or will they soar out of control.

The claimant's problems may be further augmented by the respondent who on realising the first hurdle has been won decides on frustration as a tactic and a means of providing a defence and counterclaim.

These tactics have a protracted effect on the case and make it difficult for the claimant to provide answers to questions that have little relevance to the case, the net result of which leads to the deployment of additional staff thus additional costs.

As a further consequence, the claimant realises that action must be taken in an effort to gain control by applying financial constraints.

However, the legal team will probably emphasise that whilst they appreciate the contractor's concern, they must be given financial autonomy to provide the necessary defence, otherwise success can not be guaranteed and since costs follow the event, it is imperative that all efforts are made and provided to enable the case to be drawn to a conclusion in favour of the claimant.

This news can leave the claimant reeling. Claimants confronting difficulties of this proportion often find themselves (armed with a letter of comfort explaining the order of events and merits) needing to approach their board of directors or indeed their bank to obtain additional funding.

In order that additional funding may be obtained the claimant is often asked to produce a set of management accounts so that full financial appraisal of the company may be assessed. This accountancy exercise adds to the claimant's costs.

On realising that the proceedings have some considerable time left to run, a further frightening aspect has been realised. The legal costs, which are still ongoing, are expected to dwarf the claim figure. Therefore, the emphasis of the case will substantially alter. The fight will now focus on protection of costs and not the contractual claim, which was the reason for pursuing a course of arbitration in the first instance.

The reality of the above situation coupled with the prospect of further costs, can leave the claimant facing dire financial problems, problems which often force the claimant into making the commercial decision of drawing the proceeding to a close, the nett result of which pressurises the claimant into accepting a reduced settlement.

To add insult to injury, the arbitrator when ruling on costs decides on a reduction of the same (commonly referred to as an abatement) stating that the cost analysis presented by the legal consultant is excessive and can not be met in full.
In essence, the arbitration case has left the contractor out-of-pocket, on balance between claim reduction and cost abatement.

The contractor feels dejected and that justice has not been administered. This point is often expressed by the contractor who openly states that the only beneficiary from the entire proceeding was the legal profession. It is worth mentioning that small to medium contractors facing financial problems of this magnitude may quite easily find themselves in financial ruin.

THE REMEDY

In the event of a dispute arising – a dispute that has the potential of costing the contractor a considerable sum of money in extreme cases financial ruin – is there a course of action available to a contractor the cost of which can be controlled even though the complexities of the case may be far reaching?

The answer to the above question is fraught with difficulties and one, which can not be answered in general terms or textbook manner.

However, one answer may be given with a measured degree of accuracy and that is, should a contractor be sufficiently far sighted to embark on a training course that leaves the contractor reasonably conversant with the mechanisms of arbitration, the contractor in question will find arbitration more accessible and less frightening.

The trained contractor will find it easier to approach and if necessary interview several legal consultants before deciding to make an appointment.

Armed with a list of relevant questions, the contractor will be in a stronger position to propose the terms and conditions of appointment encapsulating the specific needs of the company with respect to the case. Since the decision to arbitrate is always of a commercial nature, the questions posed by the case must broadly cover the following salient points:

1. What are the chances of success and conversely what are the chances of failure? What factors effect these points?

2. What discipline or profession should the arbitrator be sought from i.e. Engineer, Quantity Surveyor, Architect etc?

3. What is the likelihood of security of costs being introduced. If so, how may this be opposed and if unsuccessful in opposition, can a reduction of security be pleaded? It would be prudent to emphasise that preparation of this item should be treated with the same urgency and detail as any other section of the claim.

4. What percentage of the claim can be defined as quantum, can this be settled as part of a pre-arbitration settlement in an effort to cut costs?

5. What percentage of the claim can be defined as contractual issues and what is the anticipated timetable?

6. Provide quotation of scale of fees. Utilising data from previous questions, what are the anticipated legal costs and how accurate can the forecast be during the pre-arbitration stage?

7. In the event of an over-run can a reasonable reduction of legal fees be expected? If so, can a fee reduction chart be prepared prior to appointment?

8. Can an addendum be included in the terms and conditions of appointment requesting all parties to use their best endeavours and ensure the proceedings are kept to a minimum in terms of time, thus costs? In the event of prolongation being unavoidable can the contractor's staff/employees be responsible for the preparation of the claim in an effort to keep the legal cost to a minimum.

9. Can the contractor make a request to the arbitrator to have costs capped?

10. What is the likelihood of the arbitrator deciding on fee abatement? If so can an estimate of abatement be considered at this point, quoting maximum and minimum percentages?

11. In the event of the maximum abatement being relevant, can a reasonable reduction of fees be expected and can an indication of the reduction be agreed at this stage?

12. Can an appointment be considered on a no-win, no-fee basis?

CONCLUSIONS

In a recent decision in the Northern Ireland Jurisdiction, a Judge directed that every avenue of negotiation and case settlement should be exhausted before appearing in his court, thereby saving on costly adjournments during the hearing for negotiation purposes. It is the opinion of many construction and legal professionals that the same principal <u>should</u> apply to arbitration.

It would be inappropriate to suggest that every construction dispute resulting in arbitration has a disastrous ending.

It would, however, be appropriate to imply that, in recent times, litigation and arbitration have received criticism from various quarters, with some of the critics openly suggesting that in the event of a dispute arising, all avenues should be explored before deciding on arbitration as a course of action.

This is supported by the wording of a recent information sheet circulated by one of the largest legal contracts consultants in the United Kingdom headed COST EFFECTIVE ARBITRATION and stating: 'The company recognises that high legal costs and the associated adverse cash flow consequence means that arbitration may NO LONGER be recognised as a cost-effective means of pursuing outstanding financial claims by many contractors and sub-contractors'.

The article then goes on to say that the company can undertake arbitration at a greatly reduced cost so as to make arbitration a commercially realistic option for the pursuance of outstanding financial claims. Perhaps the most poignant admission is where the article states 'when combined with other dispute resolution techniques the prospects of settlement can increase dramatically'.

A reputable solicitor when asked to comment on the year's legal developments (1997) wrote 'Is Construction Law Fit For Its Purpose'? The article continues to mention 'The Arbitration Act 1996' stating from now on there will be a lot more arbitration about and the coming of the 'Construction Act' will contain provisions for adjudication in construction contracts. However, with the advent of adjudication and the long-standing process of expert determination, the contractor has a choice.

These comments serve to inform, perhaps warn the industry of the implications emanating from or as result of contractual disputes and their legal pursuit.

A painful example of this point was illustrated in a recent arbitration case, when the contractor claimed £300K, with estimated legal fees of £67K.

The case was settled in favour of the contractor, being awarded a claim of £120K., legal fees of £145K, against actual legal fees of £180K.

The contractor having had to finance the legal fees of £180K. during a 12 month period, a claim reduction of £180K., and a fee abatement of £35K., would have been well advised to suffer the initial loss and ignore the claim.

A wider interpretation of the same advises construction companies of the quagmire of management issues they may face by failing to identify and understand at least the primary mechanisms of dispute resolution.

Spiralling legal costs and their often dire consequences must surely alert the industry of the need to provide education and training as has been relevant for health and safety, quality assurance and the training and employment sectors of industry.

This paper does not suggest that every contractor with projects in excess of several hundred pounds sterling should have a member of staff enrol on a full time law degree course.

However, since market conditions have forced contractors to reduce both prices and profit margins, it would be sensible for contractors in the medium to large bracket to consider participating in a training programme which would include arbitration and other related aspects of contract law. Contractors proficient in contract law will find themselves in a position to manage the process and related costs from the beginning.

Although the main reason for the above prospective training is to prepare for unforeseen legal anomalies, personnel conversant with the legal aspects of construction can be successfully utilised for other duties such as tendering, procurement, resource management and control and as a support team involving construction legalities.

Conversely it would be reasonably safe to assume that contractors finding themselves locked into legal disputes, without the support of trained personnel, could find to their horror ARBITRATION COSTS IN REAL TERMS.

ACKNOWLEDGEMENTS

The author would like to express thanks and appreciation to the following contributors who offered information towards the writing of this paper.

1. KEARNEY, L A Principle, Adrian Kearney Associates, Construction Consultants and Surveyors.

2. FRIZZEL, I A Contracts Manager, Con-Tech Associates.

3. THOMPSON, P G Projects Manager, Con-Tech Associates.

4. BRAINT, A Principle, A B International Marketing and Consultancy Services.

A TECHNIAL RISK ANALYSIS TOOL FOR QUALITY PREVENTION AND CONTROL MEASURES

S Mecca

University of Calabria

Italy

ABSTRACT. Difficult conditions in on site operations can contribute to the risk of non conformance The analysis of many structural failures or non conformities caused by insufficient reliability of construction processes, shows that a great many have been prevented by the implementation of control measures. Quality management according to the requirements of ISO 9000, means a global reduction of quality uncertainty through the planning and control activities. We may intend the quality assurance as grading appropriately the requirements of the quality system in relation with the risk of failure and particularly as grading the quality prevention and control measures, operating roles and acceptance criteria for the design and production of concrete structures. By a systematic analysis we can identify all actions in construction quality management aiming at reducing and preventing risk of failure, intended as specific risk of non conformance to specified characteristics of structural elements, and develop the right organising strategies to increase the operator's reactivity toward failure and non conformity. An experimental Decision Support System for analysing technical risk factors in designing and building concrete structures is presented in its essence. The tool aims to provide a hierarchical knowledge of risk factors toward a systematic approach representing a qualitative non conformance risk analysis. A DSS can help for determining a graduation of levels of performance required to the quality system, which means identifying the specific level of criticality in every project, and every structural element of it, and combining the most appropriate measures of prevention and control, in order to contribute to obtain higher levels of efficiency of performance approach in design and production.

Keywords: Quality management, Technical risk, Performance.

Professor Saverio Mecca is professor of Building Sites Management, Faculty of Engineering, University of Calabria, Italy. Professor Mecca has published widely and participated to international congresses mostly on risk management and innovative site organisation. He is also engaged in consulting and implementing Quality Systems in construction activities and in project management training.

INTRODUCTION

The building process for a specific structure can be divided into two phases:

- the process until a precise, deterministic technical specification;
- the process of building the structure and checking that the specifications are achieved.

In a structural systems failure analysis we need to establish and manage the uncertainties affecting the loads and the strength of material structure. These uncertainties are mostly caused by insufficient reliability in the design or building phases. The analysis of many structural failures shows that many of them were caused by insufficient reliability in the building phase and could be prevented by systematic control .

Difficult conditions in on site operations can contribute to the risk of non conformance. On site, as for the project dynamics, the uncertainty of results is in relation to the prevision and decision capability under the condition of limited knowledge. In design phase some causes of uncertainty may be quantified with statistical data and others have to be assessed subjectively for managing these information in a suitable way for improving design reliability; during the on site building phase the most part of information must be necessary assessed subjectively in qualitative way. According to Thoft-Christensen and Baker [10] the "calculated reliability or failure probability for a particular structure is not a unique property of that structure, but a function of reliability analyst's lack of knowledge of properties of the structure and the uncertain nature of the loading to which it will be subjected in the future". In the on site building phase this condition is magnified, we cannot quantify or calculate reliability or failure probability for a specific structure, but we may only improve the construction system reliability, in a general sense of ability to fulfil all the design specified characteristics of structural elements.

This goal may be obtained by developing methods and tools of quality management specifically oriented and based on integration of different knowledge areas as quality management, performance theory, contingency theory and risk management. Mostly the quality management, technical performance management of construction systems, according to the requirements of ISO 9000 [5], is aimed to a global reduction of quality uncertainty through the planning of prevention and control activities. Safety, serviceability, durability or, better, the performances required to a structure are influenced by the nature of the prevention and control measures which are in operation. In order to improve building system reliability, we need to specify the appropriate (pertinent, efficient and timely) set (plan) of quality prevention and control measures, the operating roles and the acceptance criteria for the production of concrete structures. In this perspective the main functions of quality management are grading prevention, control and nonconformity treatment in relation to the risk level to be considered more or less acceptable according to the effectiveness and efficiency goals of the system and a technical risk analysis tool can be useful for:

- grading appropriately the requirements of the quality system in relation with the risk of failure;
- grading the quality prevention and control plans in design and construction phases according to the risk of failure;
- defining the appropriate organisation for improving technical reliability.

By a systematic analysis of products and processes we can:

- identify and evaluate the risks of failures, intended as specific risks of non conformance to specified characteristics of structural elements,
- define all actions in construction quality management aiming at reducing and preventing risk of failure and develop the right organising strategies to increase the operator's reactivity toward failure and non conformity;
- define all actions in construction quality management aiming at reducing the lack of information and knowledge on failures in building activities.

SCIENTIFIC REFERENCES

The Performance Theory

The performance theory provides an adaptive, contingent and technically non prescriptive framework for building design and construction; its application in buildings allows a translation of human needs to user requirements (safety, security, durability, serviceability, etc.) and a transformation into technical performance requirements and criteria and then into technical features and characteristics of the structural elements (materials, components, etc.). The totality of characteristics and features should bear on its ability to satisfy contractually stated needs. Only a performance concept fully developed in design and construction process enable the evaluation and assessment of innovative building materials, components and systems and the modern approach to quality management and assurance according to ISO 9000 [5]. If applied systematically throughout the building process the performance concept enable the definition of sets of quantitative and qualitative characteristics of structure elements which we can specified in contractual documents and measure and control on site.

The Contingency Theory

In system analysis and contingency theory by Lawrence and Lorsch [6] is demonstrated that reliable and efficient organisations are structured in relation with the environment in which they operate. In organisation studies it is assumed that production systems are efficient if differentiated and graded according to knowledge of the area concerned. Operators in the construction industry work in a turbulent environment, which is made worse by the lack and unreliability of relevant information: a construction project is unrepeatable and nomadic, it is extremely complex from an organisational and technical point of view in its variability and in the intensity of the relationships between the operators and their materials. It is characterised by an insufficient and not-diffused knowledge of the cause and effect relations of construction phenomena and their defects. The extreme complexity of the product and process of construction sites is at the base of the increased vulnerability of such processes, that is to say, the risk of technical failure. This endemic uncertainty tends to conflict with the objectives of quality control of the process and product, unless organisational strategies are developed, which are capable, on one side, of reducing uncertainty through the development of anticipation programmes, and on the other, through the increased ability of the operators to react to unforeseen contingencies, to monitor their own activities and co-ordinate with other participants in the realisation of a "project".

As in Mecca and Torricelli [7] the analysis of environmental, organisational and, mainly, technical causes of failures can be identified as one of the principal actions to be taken in order to reduce the non-conformities. In designing the organisational structure and management strategies most of construction systems don't take account of environmental influences and choose a management and control system fit for a certain, deterministic process. We may observe an ironic phenomenon of the way in which the environmental turbulence and the technical uncertainty that would require a flexible, dynamic and responsive organisational structure, actually cause all the operators of construction process to provide the opposite. Organisational strategies for construction are definable in relation to diverse levels of knowledge, temporarily acquired, of the environment and of the project. Moreover, the organisational strategy of each operator is defined in function of the available information and therefore depends on the collocation, role, competencies and responsibilities of the operator within the construction process.

The Quality Management

The analysis of many structural failures or non conformities shows the limit of traditional formal quality control. Shammas-Toma et al. [9] argued that the "current emphasis of management quality systems (in building design and construction) is on controlling operative performances by trying to detect defects through checking, rather than by managing in such way as to prevent defects." This management style is an application of the principles of Scientific Management and mostly of the rule of separation between supervisors (control) and site operators. The traditional and diffused way of checking quality is based on these principles: all the contractual documents are conceived according to Scientific Management principles, according to a rigid separation between control operators client and contractor operators, the organisational structure of site management and quality control is hierarchical with an emphasis on the formal mechanism of control.

Analysing practical experience and technical literature on structural failures or non conformities we may assume that a great part of them could be prevented by prevention and control measures. As we said before, turbulence conditions in "on site" operations deeply interact with the risk of non conformance to the performances required. As for the project dynamics the uncertainty of results is in relation to the prevision and decision capability under the condition of limited knowledge. The quality management, according to the requirements of ISO 9000 series and Total Quality Management, can aim at a global reduction of quality failures through:

- a global management of the process and of the chain-like relations which exist between the parties to the process, developing the organising strategies able to increase the system's reactivity toward failures;
- an effective self-control of the system providing the people with all the information about the quality or characteristics which are able to satisfy the performances required; establishing feedback between operatives and using finite performances measures, providing operatives with the knowledge and skills necessary to monitor their own work;
- a quality planning based on prevention and control measures and nonconformity treatment graded in relation to the uncertainty of the process and according to the efficacy and efficiency goals of the system.

In this contest a quality plan has to be either a formal contractual document either a planning document substantially:

- appropriate (pertinent, efficient and timely) set (plan) of quality prevention and control measures, operating roles and acceptance criteria for the production of concrete structures
- related to the project realisation and to the construction methods; it must be adapted to the customer's requirements and to the complexity of the activities, graded in relation to failure risks and must represent a complete, clear and not redundant information document for the realisation of activities.

The Risk Management

The risk management is becoming the most important management method and tool to plan a project system reliable, suitable and adequate and subsequently more efficient. Construction project risk management involves a diffuse area of active and continuous research. Numerous techniques are at present available for practitioners for project risk management [2, 3, 4 and 10]. However rarely interest has been focussed on the question of technical risk analysis [2]. The risk management has been developed mostly on cost and time risk, while we have few studies on technical or quality risk, which is the main goal for the client. The quality risks is particularly complex because of the large number of technical characteristics of a building element.

When failures and defects cannot be quantified, the analysis becomes qualitative and tends to describe the plan configuration with its functional relations and the cause and effect relation connected with faults on different levels. Pursuing this objective we can borrow the Failure Method and Effects Analysis (FMEA) [8] specifying it in constructions; a good example in this sense is the pioneer work of SOCOTEC [10]. Other methods we borrow from hazard and waste management [1]. In order to carry out this kind of analysis the research aims to develop a tool able to consider the specificity of construction management and the requirements of technical risks analysis, plays a nodal role in the reliability discipline of project. In particular, the complexity of characteristics associated to the diversity, multiplicity and interdependence of failure causes determine a lack of information on building phenomena; only for few items we may have statistical data to quantify some sources of uncertainty in the on site activities. When we may determine reliable probabilities they have to be thought as an indication useful in a decision-making process, but the most part of the potential failures have to be assessed subjectively.

If subjective judgement must dominate the whole risk analysis process, it is therefore necessary develop a rational method, a systematic analysis of the building elements, of their technical characteristics required, of their criticality for identifying all actions aiming at reducing and preventing risk of failure and develop the right organisational strategies to increase the operator's reactivity toward failure. Becoming qualitative risk analysis tends to describe the plan configuration applied to a specific building and to a contingent organisation with their functional relations and the cause and effect relation connected with faults on different levels. In order to carry out this kind of analysis some tools of general validity are necessary as support of an application which must however consider the specificity of the case study.

A method should starts from experiences of analysis of technical risks in building projects and point out the common matrix to these procedures in the light of the reliability discipline.

The analysis and evaluation of risk must be applied not only to the technical non conformance risk factors, but also to environmental and organising factors, the first ones related to the environmental characteristics where work is carried out, the second ones produced by the specific organisation necessary for the project execution. The analysis of non-quality risk can be the most important tool to plan a quality system suitable and adequate and subsequently more efficient to build in conformity with specifications. Grading the quality system means singling the specific level of criticality in every project and combining the most suitable and efficient measures of control and prevention.

RISK ANALYSIS AND QUALITY MANAGEMENT INTEGRATION

Starting just from a systematic analysis we may identify all the actions in construction quality management to reduce and prevent failure and develop the appropriate organising strategies to increase the operator's reactivity toward failure. Frequently it may be found many interdependencies between different types of failure. According to Shammas-Toma [9] "while the immediate technical causes, such as the misplacement of spacers (in the case of not achieving of the required cover in reinforced concrete) could be reasonably objectively established without recourse to respondent's judgement, the reasons given for the errors having been committed were the result of the complex interactive process referred to above and about which the only data available were bound to be coloured by the perspectives of those who provided them". Project failure, contractual failure, i.e. organisational failure, functional failure i.e. technical failure are deeply connected in risk analysis perspective. Organisational factors weight on the likelihood of technical defects: a subcontractor not satisfied with the contract, time, can disclaim with quality and safety. Inadequate definition of project, lack of clear objectives, lack of communication about problems create the conditions for a high impact of risk in developing an activity. Some characteristics of failure does not affect the finished product, anyway they play an important role in front of technical risk. For instance project error or business failure causes and impacts deeply in the process management on site. The objective for the applied research is how failure in planning can produce specific and general risks, what is the relation between organisational and technical defects.

Quality management strategies integrated with the performance theory permits us to formulate a full list of requirements for project activities. The quality management tools can operate a deep analysis of the performances required from every activity in the project either off site either on site.

We can so identify a performance level of an activity or a grade of satisfaction of a final or intermediate client and connect it with a risk of failure, damage, performance loss. The research aims to apply a technique composed from the following tools:

a) *Functional analysis* to identify hierarchical trees of elements. We can so list the elements of the construction process, order they in relation to their relative importance. Subsequently we need to identify a sub-elements chain necessary to achieve a goal.

b) *Requirements analysis* of inputs and outputs of the activities on design phase through a technical and organisational risk analysis to obtain specific requirements from specific risk of non conformance and in order to define the best strategy to deal the tender phase and to manage the on site activities,
c) *Requirements analysis* of inputs and outputs of the activities on tender phase management in order to obtain the an acceptable quality level trough negotiation, collaborative and concurrent activities in the specification of the quality plans and subcontractors, teamwork, materials qualification,
d) *Requirements analysis* of inputs and outputs of the activities on site management in order to prevent failure through the quality plan or to correct eventual non-conformance through appropriate treatment procedure.
e) *Failure risk identification* following the fundamental principle of FMEA approach articulated on many project levels of details, applied to the organisational and technical domain through cause and effects diagrams analysis, supported with interfaces studies to obtain lists of critical element related to their conditions and effects. We apply Ishikawa diagrams backward to identify the conditions for the failure and forward for represent chains of effects.
f) *Risk evaluation* applied on the list of criticality (critical points) emerging from the failure analysis. The risk analysis provide a risk value for every criticality through the study of both organisational and technical risk factors. The risk analysis involves either the on-site activities criticality either the outputs deriving from design and planning phase and from tender phase. Risk evaluation of the off site phases provide the general and specific risk values weighting on production phase.
g) *Risk reduction* through quality management tools for prevention and control to avoid non conformance. By monitoring risk conditions and the effective cause of events of failure we can define prevention plans and control plans to reduce failure, defects and their consequences on the construction process. The residual risk is evaluated for a risk traceability. For example a residual risk in tender phase in a subcontractor selection is transferred as a general risk in the on site activities risk analysis.
h) *Quality management* appropriate for an effective and efficient management of quality. Type of control and prevention measures in relation to the local and the general risk level are selected. Control plans, prevention measure plans, subcontractor selection, information management are graded pursuing an efficiency resources oriented objective.

A MODEL OF A DECISION SUPPORT SYSTEM IN THE RISK ANALYSIS FOR THE QUALITY PLANNING

Planning quality according to the technical risks evaluation is a challenging task because:

- Risk is not an objective measurement but involves many factors. There is no completed formulated model for combining these factors to achieve detailed quantitative evaluation.
- There is a great deal of uncertainty in each factor. The uncertainty is caused by construction process or by external factors interacting with the construction site.
- Expertise in risk assessment is mostly based upon judgmental knowledge. The judgmental ability results from professional training and is the source of heuristic rules not explicitly formulated.

Hence we can formulate a dual aim for technical risk analysis: it should be capable either of synthesising many factors to reach an overall evaluation or analysing many factors correspondent to adequate actions.

The knowledge based system is comprehensive of an articulated Hierarchical Structure of risk factors of which the elements are:

- nodes for synthetic factors and leafs nodes for subordinate factors;
- criteria assigning values to the factors;
- simple rules to summarise the values and calculate the node factors values;
- connections between risk factors and the procedures of the quality plan levelled in relation to the risk level assigned.

The knowledge tree of the risk factors consists in a hierarchical structure in which the top node representing the total risk at a critical point is comprised of three lower subclasses: likelihood, gravity, visibility. Three lower level nodes are referred to every subclass: environmental factors, organisational factors and technical factors. That means, it is required a mixed evaluation of these three kind of risk to obtain a judgmental measure of likelihood, gravity and visibility of a non-quality event. Synthesising the likelihood, the gravity and visibility, it can be assessed a judgmental numerical value referred both to failure conditions and to failure effects.

The environmental risk factors are related to the environmental characteristics where work is carried out. They are supported by three nodes; geographic factors, regulations-standards factors and social-economic factors. Furthermore each node is based on two nodes. Organisational risk factors are produced by the organisation necessary for the project execution. They support two classes of attributes: off site attributes or contractual and on site attributes. The off site risks class is comprehensive of five factors. The on site attributes class contains two subclasses: the first related to the organisational structure - three factors -, the second related to organisational mechanism of integration - five factors. The technical risk factors support three levels of analysis: whether it concerns a site, a technical subsystem or a technical component. If at the first level the criterion denotes a high risk value, the analysis continues in the second level and, if necessary, in the third. Each level is supported by a WBS of activities or product elements. The analysis of the technical attributes examines possible imperfections in planning and deriving pathologies.

To evaluate a risk factor corresponding to a node, grades of attributes from the sub-nodes must be combined to obtain the grade of the node at the top. The grade of an attribute can be obtained using a specified criteria referred to the main class of risk (environmental, organisational) and in case of technical risk factors using a criteria referred to the level of the analysis.

By ordering the risk factors we obtain a list of project criticality. Hence the quality management process the collection of criticality and connected information, through planning and prevention phases and subsequently feedback from the correction and analysis of failures causes.

AN EXPERIMENTAL QUALITATIVE RISK ANALYSIS TOOL

The research program is directed to provide tools and procedures for quality and risk management in design, tender and construction activities to an engineering company specialised in concrete structures. A tutorial tools for quality and risk management in concrete structure is going to be implemented.

Goals

The scientific goal is obtain a failure risk reduction in managing the realisation of structure elements through a quality planning supported by an experimental Decision Support System for analysing both organisational and technical risk factors. The main goals are split in a hierarchical WBS subgoaling the quality performance required from the structure. An experimental Decision Support System for analysing organisational and technical risk factors in designing and building concrete structures is applied. Through prevention and control we can reduce risk at every phase.

Operational Conditions

Technical failure provides a lot conflicts and economic damages to the engineering company. The client general requirements of a concurrent project management, and short time for both design and realisation posed the company in risk condition. In order to the technical problem there a substantially risk connected with poor technical specification of design. Interesting conditions for an experimental risk analysis test are dued to the limited kinds of contracts and to repetitive construction methods which permits to limit the framework of the application field to the validation of the technique. In the systematic approach representing a qualitative non conformance risk analysis, every organisational company levels are involved. That means build a database input for the DSS risk analysis in quality planning.

Through the DSS application a hierarchical knowledge of risk factors for a systematic approach representing a qualitative non conformance risk analysis is provided. The construction manager is supported to determine a graduation of various levels of performance required to the quality system, by identifying the specific critical level in every structure element and combining the most suitable and efficient measures of prevention and control, in order to obtain higher levels of efficiency of performance approach in design, in bid and in construction.

Procedures

The expected output is a graduation of the requirements of the quality system in relation to the risk of failure and particularly as grading quality plans in on site management. A fundamental goal is not a simple technical diagnosis, but a progressive implementation of tools and procedures to support managers in the quality oriented management development. Managers and technicians are trained to observe risk conditions and to a collaborative communication.

Cause and effects diagrams analysis, fault tree of organisational and technical pathologies are used. A risk analysis procedure, constructing the inference tree aiming a condensation of the knowledge based on the organisational and technical risk factors, is implemented. The inference trees are drawn based on general risk analysis grid applied to the non quality risk evaluation.

Design and planning check, supplier evaluation are dealt to manage risks through interfaces analysis. Design and planning review means check errors to reduce risk which weight strongly on the on site management. In tender phase, after the contractor selection, is managed a collaborative design and planning review to prevent every conflict deriving from a substantially misunderstand of plans.

Particular care is focussed on the criticality specified by the risk analysis. Planning documents substantially result from a collaborative risk analysis adapted to the customer's requirements and to the complexity of the activities, graded in relation to failure risks. The main functions of quality plan prevention, control and nonconformity treatment are concentrated on the critical events and every construction partner can play his own role to avoid failure and conflicts toward a substantially performance goal acquisition an toward a processes quality improvement.

Results

The main result is a quality planning scheme the design phase and to manage requirements and performances of the tender and construction phases. We are obtaining a prototype of a decision support system which can be implemented in software to assist construction managers in quality planning. We are experimenting in pilot studies through analysis forms and check-lists of technical organisational, environmental risk factors. The informations obtained will be used to increase the tool reliability and usability by site people. The prosecution of the research is concerned to support firms modifying organisation as well, achieving action to reduce failures, from the first project phases until the on site management of technical criticality.

CONCLUSIONS

The analysis of non-quality risk could be the most important tool in managing quality plans to obtain a suitable and adequate and subsequently more efficient system to build in conformity with specifications. A common classification of risk factors can help technicians and managers in organisational and technical risk analysis. Through a Decision Support System a systematic analysis of criticality and an automation and simplification of quality plan management are possible.

Combining automation with a decision system we can obtain quality plans which are suitable to specific project condition and efficient and able to reduce structural failures dued to construction activities. The technical risk assessment in building construction emphasises the role of the project quality planning for the client's satisfaction and should be one of the main tools auditing quality systems.

REFERENCES

1. ADELI H. ed. by, Expert System in construction and structural engineering, Chapman & Hall, London, 1988.

2. COURTOT, H. Présentation d'une grille de lecture des méthodologies de gestion des risques d'un projet. IAE de Paris , http://panoramix.univ-paris1.fr/GREGOR/97-14.html, 1996.

3. FLANAGAN, R, AND NORMAN, G. Risk management and construction. Blackwell, Oxford, 1993.

4. GIARD, V. Gestion de projets. Economica, Paris 1991.

5. INTERNATIONAL STANDARD ORGANISATION. Quality Management and Quality Assurance Standards. Guidelines for selection and use, 1994

6. LAWRENCE, P R, AND LORSCH, J. W. Organisation and Environment. Managing Differentiation and Integration, Division of Research, Graduate School of Business Administration, Harward University, Boston, 1967

7. MECCA, S, AND TORRICELLI, M C. Qualità e gestione del progetto nella costruzione. Alinea, Firenze, 1996.

8. NICHEL, J. Technische und methodische Hilfsmittel zur Verbesserung der Fehler-Moeglichkeits und Einfluss-Analyse (FMEA). Shaker, Aachen, 1992.

9. SHAMMAS-TOMA, M, SEYMOUR, D E, AND CLARCK, L. The effectiveness of formal quality management systems in achieving the required cover in reinforced concrete. In Construction Management and Economics, vol. 14, pagg. 353-364

10. SOCOTEC. Réussir la qualité dans la construction. Edition du Moniteur, Paris, 1992.

11. THOFT-CHRISTENSEN, P, AND BAKER, M J. Structural Reliability Theory and Its Applications. Springer-Verlag, New York, 1982

THE RISK AND RESPONSIBILITY OF STRUCTURAL ENGINEERS – A CASE STUDY OF A STRUCTURAL FAILURE

R I Gilbert

University of New South Wales

Australia

ABSTRACT. The Silverton Building was a multistorey reinforced concrete flat slab building constructed in 1983 in Canberra. In 1989, it was evacuated due to doubts arising regarding its structural adequacy. The floor slabs were deflecting excessively and the top surfaces of the slabs were excessively cracked. The curtain wall forming the external cladding of the building was also leaking extensively. The building was demolished in 1994. Litigation between the various parties began in 1989 and continued until 1997. This paper presents a broad overview of the case, including the writer's opinion of the causes and consequences of the failure of the floor slabs. It also highlights the risk and responsibilities of the structural engineers involved in the design and checking of the structure and the retribution imposed on them. The design and construction of the Silverton Building and the events that followed provide a wonderful case study for anyone interested in the causes and consequences of structural failure. The aim here is to examine the case, not with a view to allocate blame or identify negligence, but to use the case as a means of identifying and emphasising the multitude of risks associated with the practice of structural engineering.

Keywords: Case study, Cracking, Deflection, Expert opinion, Flat slab, Litigation, Responsibility, Risk, Structural engineers, Structural failure.

Professor R Ian Gilbert is Head of the School of Civil and Environmental Engineering at the University of New South Wales. Prior to joining the University, he worked as a structural design engineer for two Sydney based consultants. For the past twenty-four years, he has taught structural engineering and undertaken research in the area of concrete structures. His publications include four books and over one hundred refereed papers in the area of reinforced and prestressed concrete structures. He has served on Standards Australia's Concrete Structures Code committee BD/2 since 1981 and also on the Composite Code Committee BD/32 and he was actively involved in the development of AS3600. He was an expert witness in the Silverton Case.

INTRODUCTION

The Silverton Centre office building was designed in 1982 and constructed in 1983 in Canberra, Australia. It had a basement and eight other occupiable levels and was constructed as a conventional reinforced concrete flat slab building. A plan of the typical floors (levels 2 to 7) is shown in Figure 1. The external cladding of the building above the ground level (the curtain wall) consisted of aluminium panels and glass windows supported on an aluminium frame fixed to the reinforced concrete frame. The developer was Silverton Ltd. The building was leased to the Commonwealth of Australia who occupied it soon after completion. Silverton sold the building to the Commonwealth Bank Superannuation Fund at the end of 1983. The Bank commissioned their consulting engineer to undertake a pre-purchase inspection of the building in September 1983.

Soon after the building was occupied, early in 1984, some defects were noticed and reported by the tenants, including apparent deflection of the floor slabs, some cracking of the concrete structure and leaking of the curtain wall. These reports were investigated in 1984 and, with regard to the slab deflection, the conclusion generally was that the deformation of the floors was significant, but was not unusual for this type of construction.

In 1988, some refurbishment of level 4 of the building resulted in the top surface of the slab being exposed, revealing significant and widespread cracking. As a consequence, advice was sought from a prominent engineering expert, who examined the floor slabs in January 1989.

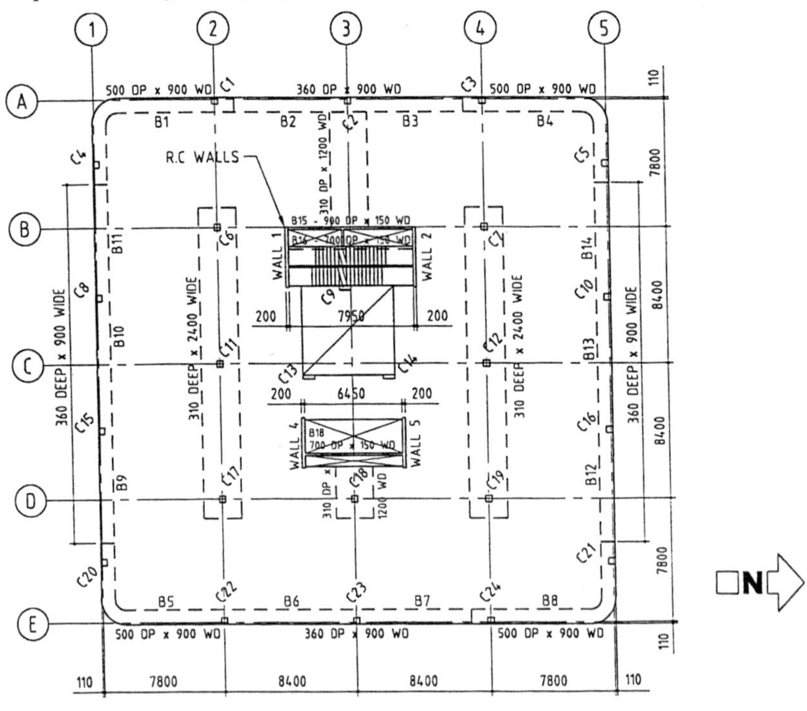

Slab thickness: 210mm throughout
Internal Columns: 520 x 520 typical
External Columns: 400 x 400 typical
Columns C9, C13 & C14: 750 x 250

Figure 1 Plan of typical floor slab - Silverton Building [2]

The expert reported that the severe cracking in the slabs adjacent to a number of internal columns was consistent with incipient flexural failure in the slab-column connections. He went on to say that the presence of this form of cracking could increase the danger of shear failure, which may be sudden, violent and without warning. He recommended that propping be provided immediately to relieve the loads at all interior column heads and to transmit a portion of the floor loads to the basement. Faced with and relying on this advice, the Bank ordered that the building be evacuated. The building remained unoccupied until it was demolished in 1994.

After its evacuation, the building was subjected to detailed investigation by a large number of structural experts. Initially, the investigations concentrated on the severe cracking at the column heads and the likelihood of punching shear failure. It was also observed that the floors had apparently deflected well beyond the code deflection limits and that the curtain wall was leaking severely. Expert opinion was divided on the structural adequacy (ie. strength) of the floors. However, a load test on one of the most extensively cracked column heads was conducted prior to demolition of the building in October 1994. This revealed that the shear capacity conformed to code requirements and that the likelihood of shear failure was very remote. The doubt on the structural adequacy of the slab-column connections, which was the main reason for the evacuation of the building in 1989, was thus removed. The investigations then focussed on the serviceability failure of the slabs and curtain wall, the causes, the allocation of blame and whether or not the structure could have been rehabilitated.

The failure of the Silverton Building, and the subsequent legal dispute over the causes of the failure and the apportioning of blame, was an expensive and tragic episode for the developer, owner, tenant, builder, architect, structural engineer, checking engineer and suppliers, but was a gold mine for the lawyers. Now that the dust has settled, however, the design and construction of the building and the events that followed provide a wonderful case study for anyone interested in the causes and consequences of structural failure.

This paper presents a broad overview of the case, including the writer's opinion of the causes and consequences of the failure. It also highlights the risk and responsibilities of the players involved, particularly the structural engineers, and the retribution imposed on them.

The writer was involved in the case as an expert witness acting for one of the parties. The comments made herein, therefore, are his perception of events, as seen from one side of the dispute, and must be regarded as such. It is not a complete picture, but it is hoped that it is fair summary. The aim here is not to examine the case with a view to allocate blame or identify negligence, but to use the case as a means of identifying and emphasising the multitude of risks associated with the practice of structural engineering.

THE DISPUTES BETWEEN THE PARTIES

Several concurrent legal actions proceeded in the Supreme Court of the Australian Capital Territory. In 1989, the Commonwealth of Australia (the tenant) sued the Commonwealth Bank (the owner) and Silverton Ltd (the developer). The Commonwealth's action against the Bank and Silverton was based on the alleged wrongful termination of their lease in 1989. The Bank had refused to let the Commonwealth occupy the building after its evacuation in 1989 and Silverton was the party with whom the Commonwealth had originally negotiated the lease. The Bank concurrently sued Silverton for alleged breaches of the sales contract, because the building was alleged to be unserviceable and about to collapse.

In both these proceedings, contributions were sought from the builder, the architects, the structural engineers, the Bank's checking engineers, and the designer and installer of the curtain wall.

In April 1995, the Commonwealth, the Bank, the builder and Silverton settled some aspects of the litigation between them. The Bank admitted liability to the Commonwealth and damages were assessed in court. The Bank settled its claim against Silverton and the builder. Silverton then became the moving party seeking relief from the various third parties; the architect, the design engineer, the checking engineer and the designer of the curtain wall.

The Supreme Court appointed a referee to inquire and report on the technical issues in the dispute. The Reference began in May 1995 and the hearing of evidence from both lay and expert witnesses lasted for 39 days. The transcript of the hearing covered some 3300 pages. The referee's report was produced in December 1995 and the conclusions related to technical issues (ie. the causes of the defects in the building) were adopted by the court.

The case was heard before the Supreme Court in October 1997, with many of the experts again appearing to give evidence. An agreement was reached between Silverton and the design engineers, Silverton and the checking engineers and Silverton and the curtain wall supplier in November 1997 before the judgement of the court was handed down. Accordingly, Silverton asked the court to dismiss its claims for contribution.

The Structural failure of the Silverton Building was a 'serviceability failure' and the consequences and costs were great. Far greater than they could or should have been. The building itself was estimated to be worth AUD$20m in 1994, the loss of income and other losses associated with it being unoccupied from 1989 to 1994 was estimated at AUD$25m and the legal costs of all the parties is here estimated in excess of AUD$20m.

The causes of the failure are many. In some respects, from an engineering point of view, what could have gone wrong did go wrong. There were problems associated with the design, the construction and the materials supplied and used. All, in my opinion, contributed to the serviceability failure of the floor slabs. However, the structural problems could have been relatively easily and relatively inexpensively solved, in my view. The problems were aggravated by inappropriate diagnoses of the problems, the order to evacuate the building and the circumstances surrounding that order, and the legal system that tried to resolve the problems.

CAUSES – THE TECHNICAL ISSUES RELATED TO EXCESSIVE CRACKING AND DEFLECTION

Structural Design

The typical floors, shown in plan in Figure 1, were designed in 1982 in accordance with the provisions of the then Australian code AS1480-1974 [1] using the CEANET F-slab computer program. The selection of the thickness of the typical floor slabs and the slab thickenings was initially made based on past experience [2]. The long-term deflections calculated with the aid of the computer program were well within the deflection limit of span/250. In addition, the slabs satisfied the deemed-to-comply span to depth requirement of AS1480.

In short, the floor slabs satisfied the code requirements for both strength and serviceability, provided one accepts that the computer modelling realistically represented the structure. However, this is not to say that the design did not contribute to the failure. The procedure specified in AS1480 for the calculation of the long-term deflection of floor slabs failed to account for the loss of stiffness caused by time-dependent cracking which almost inevitably occurs due to shrinkage induced tension in the concrete. The same comments also apply to the deflection calculation procedures specified in the current Australian code AS3600-1994 [3] and many other international codes. The procedure often grossly underestimates the deflection of slabs and this was the case for the Silverton slabs.

In the case of the Silverton floors, the exterior columns on the northern and southern sides of the building were offset by half the slab panel width (see the column layout in Figure 1). Therefore, the usual methods for deflection calculation involving analysis of equivalent frames must be modified to account for the unorthodox support layout in this structure. In addition, the deemed-to-comply provision (ie. the limiting L/d ratio) was never intended to be applied to a slab with such a layout of supports, but this was not made clear in the code.

It was well known in research circles and in the literature in 1982 that the limiting L/d ratios for slabs in AS1480-1974 were unconservative. Yet Standards Australia reissued the code in 1982 with these provisions unaltered. It was argued quite successfully in court that this was a clear signal to practising engineers that the provisions were adequate. Although the referee concluded that a competent engineer in 1982 should have known that the provisions were unreliable, his conclusion was not accepted by the court.

Recognising that the slabs were slender, and in an attempt to reduce the deflection, the designer included more tensile steel than was required for strength and included compressive steel in the column strips. In addition, 15mm precamber was specified for the slab bands and corner edge beams. Despite these measures, the apparent long-term deflection in many locations was excessive and this was, in part, caused by the slab thickness selected in design (even though the slab thickness complied with the code). It must be emphasised that the slab thickness was not the sole reason for the excessive deflection of the Silverton floor slabs.

Who then is responsible in such a case, the engineer who complied with the code, but did not know of recent developments in the literature or Standards Australia who issued a code with an inadequate provision? The answer is not at all clear and is still to be tested in court.

Concrete Quality

Numerous factors associated with the quality of the concrete also contributed to the excessive cracking and excessive deflection of the floor slabs. The evidence suggested that the concrete had a higher than usual sand content. Such over-sanded mixes generally have a high water demand and this often results in bleeding and higher than usual shrinkage and creep. The measured slump of the concrete often exceeded the maximum specified slump of 90mm [4], and this indicates a high water content. In addition, compressive tests on cores taken from the actual structure in 1989 indicated that the concrete only just complied with the specified 28 day strength of 25 MPa. The strength of the concrete in the actual structure in 1983 is a matter for speculation. It is reasonable, in my view, to conclude that the concrete was not of high quality and that it probably suffered from higher than usual creep and shrinkage deformations. Excessive shrinkage is also evidenced by the wide unserviceable cracks that developed with time.

In my view, excessive shrinkage and not "incipient flexural failure" was the main cause of the unserviceable cracks that led to the evacuation of the building in 1989 and initiated the events that followed. Excessive shrinkage, and the resulting time-dependent cracking, was also in part responsible for the excessive deflections.

Who is responsible? The specification of concrete is the responsibility of the structural engineer. The properties of concrete vary from city to city and depend among other things on the quality of the local aggregates. Anecdotal evidence suggests that Canberra concretes in the early 1980's were generally of poorer quality than the concretes manufactured in many other Australian cities, with higher creep and shrinkage. Structural engineers often specify concrete by 28 days characteristic strength alone. This is a risky practise, since both serviceability and durability depend heavily on the quality of the concrete supplied.

Construction

There was some evidence to suggest that the strength gain in the concrete was slowed by the cold temperatures on site. The floors were constructed through the winter months (May to August 1983) with generally below freezing overnight temperatures. This may well have caused a much slower gain in strength and stiffness in the over-sanded, over-wet concretes than would normally be expected. With a new floor being constructed on average every 12 days, the slabs may have been heavily loaded during construction, by props supporting the concrete floors above, at a time when the concrete had not yet developed its full strength or stiffness. Some experts suggested that a significant portion of the apparent deflection was actually caused by settlement of the supporting props during construction.

It appeared that little or no protection was given to the top surface of each slab after casting and the curing technique, if indeed there was one, is unknown. In addition, excessive concrete cover to the top reinforcement exacerbated the crack widths on the top surface.

In my view, these factors were also in part responsible for the serviceability problems. The construction procedures for any structure, including propping and back propping, curing and finishing, need engineering input. How long should the props be kept in place? Over how many floors should props be included? These questions need engineering calculation and judgement and should not be left to a building foreman to answer. The engineer for the Silverton building had no input into these decisions. This is not an uncommon situation, since often clients will not pay for these 'extra' engineering services.

CAUSES – OTHER ISSUES

The Order to Evacuate and the Role of the Experts

As it turned out, the doubts raised by the Bank's expert concerning the strength of the slab-column connection were eventually dispelled. The slabs had adequate strength. A question arises regarding the risk and responsibility of experts in such a matter.

The expert's report unquestionably led to the evacuation of the building and this precipitated the various legal actions. The expert saw a building with floor slabs that were deflecting significantly and were badly cracked.

Without knowing details of the concrete quality, the construction procedures, the actual steel layout in the slabs (as distinct from what appeared on the drawings), workmanship etc., his conclusions should not be criticised. After all, many others agreed with him even after a far more detailed study of the structure, and the issue of strength was only finally laid to rest after a load test to failure was undertaken in 1994.

Reports written by a significant number of structural experts contained such comments as "… the whole of the concrete structure is severely distressed; so much so that no further load should be placed upon it" and "I have never before seen a concrete structure in such distress" and many other similar comments. However, other experts were less concerned and, while agreeing that the building was suffering excessive deflection and cracking, suggested that other buildings of similar construction in Canberra were exhibiting distress not dissimilar to that of the Silverton building.

Many of the experts' comments, made in good faith and expressing an honest opinion, did inflame the situation. A statement like "this is the worst I have seen" does not in fact say very much at all, unless the speaker has seen a great number of such structures. To be statistically significant, in terms of the probability of failure normally considered acceptable in structural design, the speaker will have needed to inspect literally thousands of flat slab buildings. No single person has that experience.

No significant complaints were lodged by the tenant concerning deflection or cracking of the floor slabs during the period 1985 to 1988. Certainly, the exposure of the level 4 slab that revealed the excessive cracking was not done because of concerns relating to serviceability. It was for the purposes of installing a computer floor. Up until evacuation, the slabs were fulfilling their function as office floors. They were apparently deflecting more than would normally be considered satisfactory and they had cracked excessively, but the same is true for many other existing flat slabs (perhaps not to the same extent as the Silverton slabs).

It was the expert opinion, and the evacuation, that effectively stopped the building acting for the purpose it was intended. The expert opinion, therefore, also contributed to the serviceability failure. In hindsight, a light-weight levelling material applied to the top surface of the slabs and recarpeting would probably have seen the building still in operation today.

That the parties and their experts could not come to this conclusion is largely due to the adversarial nature of the dispute and a lack of cooperation between the parties. The parties and their experts followed the lawyer's approach of "who is responsible" rather than the engineering approach of "how do we fix it". Considering the eventual costs to all the parties, only the lawyers could possibly be happy with that decision.

The Legal System

The legal system in which we operate is geared towards resolving disputes. In the Silverton case, the disputes were eventually resolved, but very slowly and at considerable expense. The case is not unique in this respect.

The total legal costs for all the parties were greater than the value of the building and, in my view, more than 20 times the cost that would have been required to rectify the concrete structure.

The various parties commissioned experts to investigate and report on the causes of the defects in the structure and also to provide critical commentaries of the reports produced by other parties. The volume of material produced was considerable. The parties then identified to the court the particular documents on which they would rely in either proving the claim being made or defending against it. Fairly obviously, in any dispute, the various parties are unlikely to identify to the court any document produced by their experts that may damage their case, so in many disputes, the documentation tendered is incomplete and one-sided. Often the experts are asked to comment only on one aspect of the case and their opinion on other important issues is never sought. It may of course be obtained in cross-examination if the right questions are asked, but in many cases the right questions are not asked.

At the special Reference conducted in 1995, issues over which all experts should have agreed were clouded by questions posed by the lawyers in such a way that consensus was very difficult to achieve. Initially, there was an expectation that the experts would find many areas of common ground and that the Reference would clarify many issues and, hence, greatly assist the court. The lawyers were concerned that their clients' interests may be compromised if the experts were given free rein to express their views and so they carefully drafted the questions and guided their experts. It is unfortunate that the experts and the Referee let the lawyers have as much control over the proceedings as they did, particularly since the lawyers had considerable difficulty in comprehending some of the technical issues. In the end, the experts failed to agree on many issues and the Referee, who himself was an experienced structural engineer, was faced with the difficult tasks of bringing down findings that would almost certainly meet with heavy criticism from some quarters.

At the trial, those aspects of the referee's findings relating to technical issues could not be re-opened and the parties and their experts were not able to challenge any of the referee's conclusions on which they did not agree. Therefore, the legal system did not allow a full and exhaustive airing of those issues of most interest to the engineering community. The lawyers will argue that the legal system is not intended to do this.

THE STRUCTURAL ENGINEER – RISK AND RESPONSIBILITY

For structural engineers, the Silverton Case poses a number of important questions. Three different questions are considered here:

(i) Are the provisions of existing codes still applicable when new research shows them to be inadequate?

(ii) Considering the consequences of structural failure, are the roles and the responsibilities of the structural engineer clearly defined and is current practice good enough?

(iii) Who in their right mind would want to be a checking engineer?

Because the dispute between Silverton and the engineers was settled before the judgement of the court was handed down, these and related questions remain largely unanswered, but they are certainly worthy of consideration.

Design Codes and the Structural Engineer

The objective of the structural engineer is to design a structure that will have adequate strength and durability, be serviceable, have adequate fire resistance and satisfy any special requirements related to its intended use. To achieve this, structural engineers rely on their knowledge of structural engineering and the behaviour of structures. This knowledge is gained by formal education and by experience. They are guided by the applicable codes and make use of appropriate software and other design aids, which facilitate the calculations.

Due to the competitive pressures in modern structural engineering practice, there appears to be a disturbing trend to rely more and more on codes of practice and design software and less and less on knowledge of structural behaviour. Many structural engineers seem to have very little time to read the current literature and to keep abreast of new developments. This is a recipe for disaster. In a paper presented to the Australian profession in 1998, I wrote [5]:

> "The reliance of our profession on Standards and "cookbooks" is also unprofessional, with engineers sometimes misinterpreting code rules and requirements because they are unaware of the underlying principles and the limits on their applicability…"

As has already been mentioned, it was reported in the literature on several occasions through the late 1970s and early 1980s that the serviceability provisions for slabs in AS1480-1974 were inadequate. Yet the then Standards Association of Australia reissued the code with the provisions unaltered. This was a clear message to practising engineers (although an incorrect one) that the provisions were acceptable and could be used. What is a consulting engineer to do in the face of such conflicting advice? The answer depends on whether it is viewed by an engineer or a lawyer. My view is that an engineer should rely on his/her knowledge and engineering judgement to assess the problem. For any structure, the responsibility to ensure adequate strength and serviceability rests with the structural engineer, not with the code writers or the developers of software. The reliance on empirical procedures and computer models, no matter how highly endorsed, involves an unacceptably high risk, unless they have been carefully assessed and their applicability to the structure at hand is assured.

Having said that, the dimensions of the Silverton slabs and the reinforcement used in design were not the only reasons for the excessive deflection and cracking. The slabs were very flexible and significant deflection could be anticipated, but if the concrete had been of higher quality with acceptably low creep coefficients and shrinkage strain, the deflection and the crack widths would not have been as great. In addition, if the concrete had been protected from the prolonged periods of below zero temperatures during construction and adequately cured, and if the slabs had not been heavily loaded during construction while they were still relatively immature, the deflection and extent of cracking would not have been as great.

By not specifying a high quality concrete with low creep and shrinkage characteristics, the designers of any slender floor are taking an unacceptable risk. Structural engineers should not assume that concretes in all cities, regions and countries are the same. In my view, structural engineers are largely responsible for problems arising from inadequate material specification, and ignorance of local conditions and local material characteristics is a rather poor excuse. In my view, it is quite remarkable that more problems of this nature have not occurred considering the profession's general attitude towards the specification of concrete.

The serviceability of the Silverton slabs was adversely affected by the environmental conditions during construction and the construction procedures.

Apart from checking that the reinforcement was adequately placed prior to concreting, the structural engineer had almost no input into the construction process.

This is true for many structures. Considering that the structural engineer is responsible for the strength and serviceability of a structure, control must be exercised by structural engineers during construction. By not insisting on input into the construction procedures and by not supervising all aspects of construction, engineers risk becoming embroiled in litigation and being called on to justify their designs in court.

The Checking Engineer

So far, I have not discussed in any detail the role and responsibility of the checking engineer. In the Silverton Case, the Bank commissioned its engineer to undertake a pre-purchase inspection and to report on the condition of the structure and its services. The engineer was never asked to check the structural design. After an inspection of the architectural and structural drawings, the engineer undertook a one-day inspection of the building and prepared a report for the Bank. In essence, the structure was given a clean bill of health. At the time of their inspection in September 1983, there is no evidence to suggest that the building was suffering any distress, whatsoever. Although not asked (or paid) to check the design, the engineer undertook some preliminary design checks and opened their report with "(We) were commissioned to review and assess the design". This appears to have been a very costly choice of words. As a consequence, both the Bank and Silverton relied on the engineer's report and claimed the engineer was in part responsible for their losses.

In 1983, the engineers invoiced the Bank for a few hours of a senior design engineer's time and the cost of a one day trip from Sydney to Canberra. In 1997, their legal costs were about AUD$3m and they settled with Silverton for an undisclosed amount. Clearly, the engineers were not paid a fee that was commensurate with the responsibility and risk they were taking. Structural engineers who undertake pre-purchase inspections, design checks or structural certifications clearly do so at their own risk, which is considerable. In preparing their reports, legal advice is recommended.

CONCLUSIONS

A review of the Silverton Case has been presented and the causes and consequences of the failure of the floor slabs have been explored. The causes were many, including issues relating to design, construction and concrete quality. Other causes were related to the expert opinion that was provided, the order to evacuate the building and the legal system in which the litigation proceeded. Some of the issues associated with the risk and responsibility of structural engineers have also been canvassed.

The risks taken by structural engineers can be reduced by a more professional attitude towards the acquisition of knowledge, less reliance on empirical design procedures and software of doubtful applicability, and greater attention given to engineering both the material specification and the construction processes.

Considering the risks and the responsibility taken by structural engineers, the fees being charged for their services are staggeringly low. If the profile and credibility of the profession is to be raised, engineers must charge fees commensurate with their risks and responsibility.

ACKNOWLEDGEMENT

The writer has drawn on information contained in a number of documents, in particular reference [2], and the use of this material is gratefully acknowledged.

REFERENCES

1. STANDARDS ASSOCIATION OF AUSTRALIA, "SAA Concrete Structures Code", AS1480-1974, Sydney, 1974.

2. TAYLOR, P J and GIBBONS, K. "The Silverton Case", Structural Engineering Conference on Innovation, IEAust, Canberra, March 1998, p 263.

3. STANDARDS AUSTRALIA, "Concrete Structures". AS3600-1994, Sydney, 1994.

4. RYAN, W G. and Associates, "Investigation into Quality of Concrete: Silverton Centre, Canberra", March, 1989.

5. GILBERT, R I. "How do we get New Knowledge into Practice? The Future", Concrete Institute of Australia seminar on Implementing Concrete Research into Practice, February 1998, pp 83-89.

LATE PAPER

RESIDUAL SERVICE LIFE OF REINFORCED CONCRETE STRUCTURES

S M Skorobogatov

Urals State Academy of Railway Transport

Russia

ABSTRACT. To create the reliability theory for a separate structure the author resorted to the physical peculiarity of strength texture of massive concrete, that is to the crack hierarchy and to the information entropy resulted from the second law of Thermodynamics. The information entropy taking into account the size of a structure, discloses the indeterminacy of a crack pattern in reinforced concrete structures. A curve of the information entropy concides with a curve of serviceability reserve of a structure. It permits to use curves of the information entropy for obtaining the scale coefficient of a massive structure.

Keywords: Information entropy, Indeterminacy, Crack hierarchy, Massive concrete structure, Serviceability, Catastrophe.

Professor S M Skorobogatov works at the Building Structures Section in the Urals State Academy of Railway Transport. While a professor in the Urals State Technical University he took charge of a separate investigation on the fundamentals of the endurance theory for deformed bar reinforcement and on principles of designing high strength deformed bars. At present, he is the author of the catastrophe theory for concrete and reinforced concrete structures. He has published widely and served on many Technical Committees. He is now a corresponding member of the State Academy of Architecture and Building Sciences.

INTRODUCTION

To protect the national economy against sacrifices from possible catastrophes of unique oversize structures it is recommended, while designing, to calculate serviceability as an assessment of all stages of work of a structure with the new lowered design tensile strength taking into account the scale of a structure. Calculation on reserve of serviceability means design on longitudinal crack resistance and provides durability of a structure. The case is that longitudinal compression stresses in reinforced structures under the design load can sometimes cause formation of very dangerous mezo-cracks developing into destructive macro-cracks and long crack.

To prevent structure from destructive longitudinal cracks, the designer must use the new serviceability criterions. This method based on using the well-known two equilibrium equations ($\Sigma X=0$, $\Sigma M=0$) takes into account a cinematic diagram of deformation of a plane section (assumption of plane section hypothesis), tensile diagram of reinforcement steel, and stress-strain curve of concrete.

THE VALIDITY OF USING TERMS OF CATASTROPHE THEORY IN ANALYSIS OF STAGES IN WORK OF A STRUCTURE

In geology there are many classifications of catastrophes. In appearance, catastrophes are classified as natural, technical and natural-tehnical. In the duration of attack, they fall into fast-acting and slow-acting (flood, drought). However, it is known that the objective truth is always specific. The author of this work supposes that use of elements of catastrophe theory in analysis of fracture accumulation and serviceability will be most fruitful.

The case is that the catastrophe is a spasmodic change in the form of a sudden reaction of the structural system to smooth change under external condition. It should be noted, most likely, that such a change does not depend upon human will and consciousness.

The main property of the catastrophe is being considered is the hypothetical character of its manifestation. That is why, the author consider only fast-acting catastrophic phenomenon. This is in contrast to only determinational or stochastical characters.

The fracture accumulation in conrete, corresponds mostly to the increasing hierarchy in the developing cracks pattern. It proves that the main property of concrete in a separate structure is of probability nature of its strength on the background of its size.

The first and the main property of catastrophic phenomenon is absence of a reliable predecessor to the concealed hotbed of the slowly growing event. The gradual process of fracture accumulation and the cracks hierarhy are the basis for searching more reliable predecessor of catastrophe than time and deformation in reliability theories.

Reasons of a breakdown can occur if the building codes and rules are violated at the following stages of construction:

1. Defining the design magnitude of strength, partial safety factors for materials, loads and work condition.
2. Developing the building design.
3. Manufacturing the building elements.
4. Erection of structures and buildings.

By partially unloading and repeat loading a concrete specimen we get a small hystoresis loop on a stress and strain curve of concrete and slipping down to the low repeat curve. Possible bifurcation on deformation curve is illustrated in Figure 1. Further, work on only the low repeat curve make it possible to refer to such an elementary kind of catastrophe known in mathematical theory of catastrophe as "a gather". But taking into consideration the crack hierarchy with various levels in size of crack, it is best to use information entropy with the formula by C E Shannon. Figure 2 shows possible bifurcation on curves of information entropy for disclosing indeterminacy in crack pattern.

PHYSICAL BASIS OF THE CATASTROPHE THEORY NORMALIZED CRACK HIERARCHY DEPENDING ON SIZES OF STRUCTURES

In the fracture mechanics of concrete, reinforced concrete structures with size of 0.3...0.6 m contain specific cracks of three orders: millimeters, centimetres and tens of centimetres and consequently three levels of crack hierarchy. There exists a polimodal shape of the distribution of cracks lengths in concrete during the process of cracking and fracture accumulation.

The availability of three or more modas in distribution corfirms a three or more steps mechanism of forming cracks in massive reinforced concrete structures.As the case stands, we are to consider that a concrete mass converts into hierarchy system of enclosed each into other grains, grain groups and blocks.

Sizes of blocks in between cracks vary within wide range but in the average by 3.5 times. Taking into account the magnitude of the "jump" of 3.5 we shall have the following nominal sizes of concrete blocks and levels of crack hierarchy: 30 cm for the third level, one metre for the fourth level, 3.5 m for the fifth level, 12.0 m for the sixth level, 42.0 m for the seventh level and so on.

Crack patterns of the 4th, 5th and 6th levels can be confirmed by many investigations on reinforced concrete structures in scientific and technical literature. Verification of the proposed crack hierarchy can be found in practice of designing oversize structures (dams, buldings, landing strips).

The distance between through expansion joints in dams is usually equal to 35...40 m that corresponds to the 7th level of the crack hierarchy. The distance between partly through (incision) joints is equal to 10...12 m that corresponds to the 6th level of the crack hierachy.

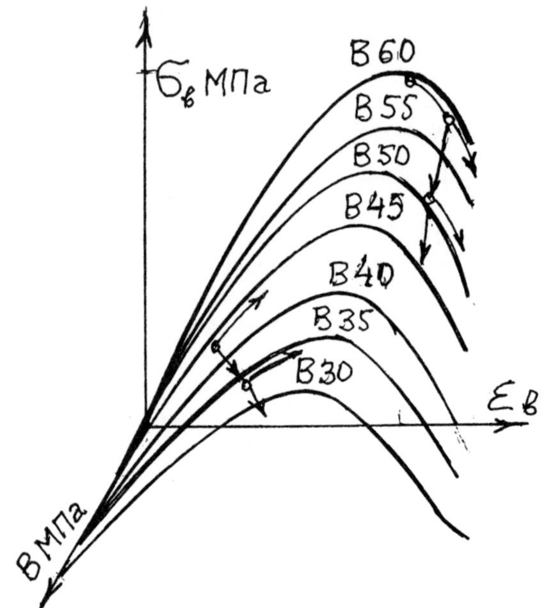

Figure 1 Possible bifurcation on deformed curves of concrete

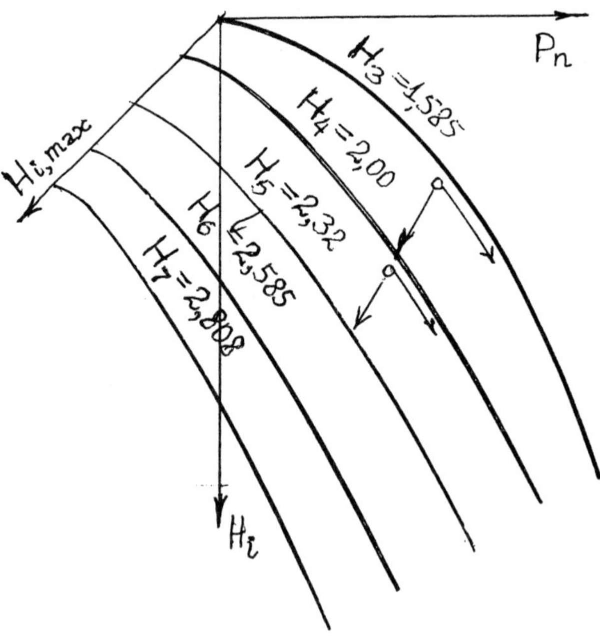

Figure 2 Possible bifurication on cures of information entropy for disclosing indeterminacy in crack pattern

MATHEMATICAL BASIS OF THE CATASTROPHE THEORY USING THE INFORMATION THEORY IN DISCLOSING INDETERMINACY OF CRACK PATTERN

As mentioned above, at the loading a structure, indeterminacy in a cracks pattern is gradually being dissapearing. The process of elimination of indeterminacy by the information and communication theory is considered as a measure of cognition or calculating magnitude of informative entropy or negentropy.If the number of brittle fractured ties between grains or blocks is treated as signals of their fracture, then there appears a possibility to use informative entropy H_i. In this case, the process of gradually increasing cracks may be described with one of informative formulas by C E Shannon.

It will give a measure of indeterminacy for multi step independent communication from many sources (see the mathematical theory of communication and information). Such an abstruction as applied to strength texture of concrete made it acceptable because the quantity of information is measured with a number (bit) not depending on absolute magnitude and kind of information. Just as the volume of a body does not depend on it's shape.

As mentioned above, the local character of fracture of ties between grains is the first concept of the physical basis of the theory. The second concept of the theory is the admittion of the multi step developing cracks pattern and accordingly multi level cracks under loading. These two physical concepts are extremely important for the kind and the length of C E Shannon's formula.

Omitting the bulky deduction of the equation, we give the final formula by C E Shannon:

$$H_i=-p_1\log_2 p_1 - ... - p_2\log_2 p_2 - ... - p_n\log_2 p_n\ , \qquad \text{at } \sum_1^n P_i=1.0 \qquad (1)$$

For widespread beams and structures with depth of cross section of 0.60 m and less we have three step crack hierarchy and consequently the following formula:

$$H_i = - p_i\log_2 p_1 - p_2\log_2 p_2 - p_3\log_2 p_3\ , \qquad \text{at } \sum_1^3 P_i=1.0 \qquad (2)$$

Where P1, P2, and P3 = probabilities of fractured ties according to micro-, mezo- and macro-level of crack hierarcy; $p_1\log_2 p_1$, $p_2\log_2 p_2$ and $p_3\log_2 p_3$ = contributions made accordingly to micro-, mezo and macro- texture information entropy as a measure of disclosing indeterminacy of rupture of a structure. The curves of information entropy with formula by C E Shannon are similar in appearance to curves of serviceability (Figure 3).

At equal probabilities of all the components of P1=P2=P3=0.333 and at $\Sigma P_i = 1.0$, we have the largest magnitude of information entropy $H_i = 1.585$ bits. At the three step information, the largest magnitude $H_i = 1.585$, bits, means the greatest indeterminacy of the fact of rupture. When rupture, $H_i = 0$, the indeterminacy of rupture is equal to zero. Between the process of disclosing indeterminacy in a crack pattern and stress - strain state of a structure there is a definite causul and effect connection.

Magnitude of $R_{bt,\,i}$, $R_{b,\,i}$ and $R_{bt,\,ser,\,i}$ are found by dividing the main design magnitude R_{bt}, R_b, and the characteristic magnitude $R_{b,ser}$ from the Building Code by the new partial safety factor γ_i for the size of a structure:

$$R_{bt,\,i} = R_{bt} / \gamma_i,$$

where $\gamma_i = H_i / H_3$.

As mentioned above, basis magnitudes of the R_{bt}, R_b, $R_{bt,\,ser}$ corresponds to the 3rd level of the crack hierarchy accepted as nominal one.

PRINCIPLES OF CALCULATION OF RESERVE OF SERVICEABILITY AT ALL STAGES OF THE WORK OF A STRUCTURE

Serviceability conception made it possible to check the first limit state (strength of a structure), the second limit state (load accepted for crack and deflection calculation) and to introduce a new calculation on longitudinal crack resistance with criterion of serviceability Hser. At that, the longitudinal crack in compression concrete is dangerous for a highly reinforced structure during its service time.

The primary purpose of the new exact method of calculation of stress-strain state was to establish and to prove the cause and effect relationship between stress-strain state of a structure and the information entropy accounted for the phenomenon of disclosing the indeterminacy in a crack pattern of a structure. On the basis of analysis of flexural elements a stress-strain state of compression and tension zones in a structure can be determined by the new serviceability criterion:

$$\mathbf{H}_{ser} = \frac{\sigma_{bc}\xi_{bt}}{R_{bt}\,\xi_{bc}}, \text{ bin} \qquad (3)$$

where σ_{bc} - stress in the extreme fibre of compression zone of an element;
R_{bt} - actual or mean tensile strength of concrete.
ξ_{bc},ξ_{bt} - relative depths of compression and tensile concrete zones of flexural element.

As for magnitudes and characters of the curve of Hser and the curve of the information entropy Hi coincide then it should be recognized that serviceability implies fracture accumulation and consequently levels of development of cracks (Figure 3). For beams of h=8...30 cm under three step crack hierarchy at the beginning of loading we have Hser=1.585.

At the end of failure we have Hser-> 0 and relatively Hi-> 0. This means the fact of failure has been known and there is no lack of information concerning the failure. As the beam is approaching failure, the difficulty of prognosis, the problem in assessment, or the measure of lack of information concerning the failure are gradually disappearing.

For the moment of serviceability there is a criterion of serviceability Hser=1.376+0.015. Sometimes that moment is less then the design moment calculated according to Building Code. There are two reasons for that, the first reason is the early forming microcracks in concrete of low and middle concrete grades and the second reason leading to necessity of calculation of serviceability additionally to the traditional design is the large exceeding actual stress σ_{bc} - against the value of σ_{bc} in the rectangular stress compression block (or alike to it) accepted in Building Codes.

Thus a serviceability H_{ser} is a criterion for disclosing crack pattern or crack indeterminacy. To stop fracture accumulation means to limit the magnitude of H_{ser}.

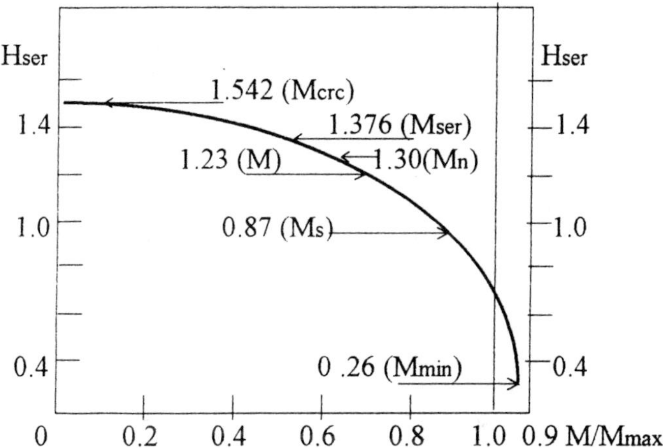

Figure 3 Serviceability reserve diagram H_{ser} under three step crack hierarchy in dependence upon loading M/Mmax

CALCULATION OF RESERVE OF SERVICEABILITY OF A DAMAGED OR WEAKENED STRUCTURE

Only two the main parameters: the depth of the developed "normal" crack x_{crc} and the elongation of reinforcement steel ε_s in cross section obtain, on the whole, the stress-strain state of an element under bending and eventually its reserve of serviceability. The depth of a crack x_{crc} can be measured in an element immediately during inspection. To obtain the value of the relative strain of the longitudinal reinforcement steel ε_s it is necessary to measure the width of a crack a_{crc}.

For transmission from the value of a_{crc} to the stress of reinforcement steel σ_s, and to the strain ε_s we can use the formula from Russian Building Code. After some changes from the formula for a_{crc} we shall finally have

$$\varepsilon_s(\Delta\varepsilon_s) = \frac{a_{crc}}{\delta\, \varphi_1\, \eta 25.68\, (3.5\text{-}100\mu) \cdot \sqrt[3]{d}}, \qquad (4)$$

where the coefficient δ, φ_1, η, μ are taken from the Russian Code on concrete structures.

For prestressed concrete structures we have the increment of strain $\Delta\varepsilon_s=\varepsilon_s-\varepsilon_{sp}$ where ε_{sp} - the strain caused the precompression of concrete. Using a plain section hypothesis it is possible to modify the position of the neutral axis as follows: $x_s=x_{crc}+x_{bt}$ where $x_{bt}=\varepsilon_{bt}x_{crc}/\varepsilon_s=0.00015\ x_{crc}/\varepsilon_s$ or $x_{bt}=0.00015\ x_{crc}/\Delta\varepsilon_s$. Here $x_{crc}=h_{crc}$, the depth of the greatest crack can be measured at the inspection of the beam.

Then using the triangular similarity we can continue our calculation of the strain in the extreme compression fibre $\varepsilon_{bc}=x_{bc}\varepsilon_s/x_s=(ho-x_s)\ \varepsilon s/x_s$ or for prestressed structures $\varepsilon_{bc}=(ho-x_s)\Delta\varepsilon_s/x_s$. Then we obtain the stress σ_{bc} (see above). The arithmetical mean compression strength must be obtained on the basis of experimental control of concrete surface.

To analyse a structure with the criterion of H_{ser} it is necessary to use the values of H_{ser} from the Table. For example, for serviceability moment we have H_{ser} = 1.3764 bit and for the nominal moment H_{ser} = 1.30 at concrete B35. If we have H_{ser} = 0.87...0.70, then the bending moment is near the rupture or at stage of rupture. Above last level, the strongly expressed fracture of concrete is observed. There is confluence of mezo-cracks and their overgrowing into macro-cracks and still further into long cracks. It implies the transmission from pseudo-plasticity to brittleness.

Such an element being under inspection is not subjected to rehabilitation. It can be rehabilitated only by external structural system with another static space diagram, for example, by subdiagonal (strut-framed beam).

Table 1 Assessment of reserve of serviceability Hser in accordance with the level of the crack hierarchy (the size of a structure) for strengthening, rehabilitation and reconstruction

State of work of a structure	Level 3 8 - 30 cm			Level 4 0.3 - 1.0 m			Level 5 1.0 - 3.5 m		
	Class of concrete B, Ma								
	35	50	60	35	50	60	35	50	60
M_{crc}	Crack moment								
	1.542	1.716	1.832	1.946	2.165	2.312	2.256	2.510	2.680
M_{ser}	Serviceability moment								
	1.376	1.531	1.635	1.737	1.933	2.063	2.012	2.239	2.390
M_n	Nominal moment								
	1.30	1.447	1.544	1.640	1.825	1.948	1.902	2.116	2.259
M	Design moment								
	1.230	1.369	1.461	1.552	1.727	1.844	1.800	2.003	2.138
M_{max}	Concrete fracture moment								
	0.700	0.779	0.832	0.883	0.983	1.049	1.020	1.135	1.212

In statically indeterminable structures the rupture process can follow the descending line in stress-strain curve of concrete when H_{ser}=0.70...0.264. In this case a weakened element can not also be rehabilitated. Thus large-size structure when "h" is more than 0.3 m it is recommended to use enhanced magnitudes of H_{ser} from the Table.

CONCLUSIONS

1. State of work of a weakened or damaged structure with cracks in a cross section is determined reliably with the help of serviceability criterion H_{ser}.

2. The proposed method will be useful and convenient for examination, investigation, as well as, for purpose of rehabilitation, reconstruction of various kinds of structures especially of large-size, unique structures of great importance.

3. Calculation of serviceability (longitudinal crack resistance) can stop the danger of fracture accumulation and prevent unique structures from catastrophes.

REFERENCES

1. SKOROBOGATOV S M. Fundamental of Catastrophe Theory for Design of Oversize Structures - " Beton and Zhelezobeton "(Concrete and Reinforced Concrete). Moscow, 1993, No. 10, pp 26-28.

2. SKOROBOGATOV S M. Design of Structures using Crack Indeterminacy and Information Entropy. Proceedings of The International conference " Concrete 2000 ".- University of Dundee, Scotland. - 7 - 8 - September 1993.

3. SKOROBOGATOV S.M. Calculation of Reserve of serviceability of off - shore platform Column // Proceedings of International Congress " Concrete in the service of mankind", Dundee, Scotland, 24 - 28 june 1996.

4. SKOROBOGATOV S.M., KHAYKOV A.A. Assessment of Expedience of Rehabilitation and Restoration of damaged structures on the Base of serviceability (longitudinal crack resistance) // " Proceedings of the International Conference on maintenance and Durability of concrete structures", Hyderabad, march 4 - 6, 1997.

5. SKOROBOGATOV S.M., KHAYKOV A.A. Catastrophe Theory for Durability Design of oversize concrete structures // Proceedings of the International Conference on " Maintenance and Durability of concrete structures ", Hyderalad, march 4 - 6, 1997.

6. SKOROBOGATOV S.M. About the necessity of working out an additional chapter to Building Code on Design of oversize reinforced concrete structures // Transaction of higher educational institutions. Construction. - Novosibirsk: 1998. - № 3. - pp.45-51.

7. SKOROBOGATOV S.M. About an draft of the addition to the Building Code on Design of Reinforced Concrete structures // Transaction of higher educational institutions. Construction. - Novosibirsk: 1998. - № 3. - pp. 52 - 56.

8. SKOROBOGATOV S.M. Recommendation on design of reserve of serviceability of reinforced concrete structures damaged with unknown load // Transaction of higher educational institutions. Construction. - Novosibirsk: 1998. - № 6. - pp.4 - 7.

INDEX OF AUTHORS

SUBJECT INDEX

This index has been compiled from the keywords assigned to the papers, edited and extended as appropriate. The page references are to the first page of the relevant paper